EARTH SCIENCE

EARTH SCIENCE
VOLUME II

THE
EARTH'S SURFACE
AND
HISTORY

EDITOR
JAMES A. WOODHEAD
Occidental College

EDITORIAL BOARD

DAVID K. ELLIOTT, VOLUME II
Northern Arizona University

DENNIS G. BAKER, VOLUME IV
University of Michigan

ANITA BAKER-BLOCKER, VOLUME IV
Applied Meteorological Services

RENÉ DE HON, VOLUME I
University of Louisiana at Monroe

CHARLES W. ROGERS, VOLUMES III & V
Southwestern Oklahoma State University

SALEM PRESS, INC.
Pasadena, California   Hackensack, New Jersey

MANAGING EDITOR: Christina J. Moose
PROJECT DEVELOPMENT: Robert McClenaghan
MANUSCRIPT EDITORS: Doug Long, Amy Allison
ACQUISITIONS EDITOR: Mark Rehn
RESEARCH SUPERVISOR: Jeffry Jensen
PHOTOGRAPH EDITOR: Philip Bader
ASSISTANT EDITOR: Andrea E. Miller
INDEXERS: Melanie Watkins, Lois Smith
RESEARCH ASSISTANT: Jeffrey Stephens
PRODUCTION EDITOR: Cynthia Beres
PAGE DESIGN AND LAYOUT: James Hutson
ADDITIONAL LAYOUT: William Zimmerman
GRAPHICS: Electronic Illustrators Group

Library of Congress Cataloging-Publication Data

Earth science / editor, James A. Woodhead.
    p. cm.
Expands and updates Magill's survey of science: earth science series.
Includes bibliographical references and indexes.
Contents: v. 1. The physics and chemistry of earth — v. 2. The earth's surface and history — v. 3. Earth materials and earth resources — v. 4. Weather, water, and the atmosphere — v. 5. Planetology and earth from space.
ISBN 0-89356-000-6 (set : alk. paper) — ISBN 0-89356-001-4 (v. 1 : alk paper) — ISBN 0-89356-002-2 (v. 2 : alk. paper) — ISBN 0-89356-003-0 (v. 3 : alk. paper) — ISBN 0-89356-004-9 (v. 4 : alk. paper) — ISBN 0-89356-005-7 (v. 5 : alk. paper)
1. Earth sciences. I. Woodhead, James A. II. Magill's survey of science. Earth science series.

QE28 .E12 2001
550—dc 21

00-059567

First Printing

CONTENTS

1
CONTINENTS AND CONTINENTAL PROCESSES

CONTINENTAL CRUST

Continental crust underlies the continents, their margins, and isolated regions of the oceans. Continental crust is distinguished from its counterpart oceanic crust by its physical properties, chemical composition, topography, and age. The creation and eventual modification of continental crust is a direct function of plate tectonics.

PRINCIPAL TERMS

ASTHENOSPHERE: a layer of the Earth's mantle at the base of the lithosphere

CRUST: the outermost shell of the lithosphere

LITHOSPHERE: the outer, rigid shell of the Earth, overlying the asthenosphere

MANTLE: the region of the Earth's interior between the crust and the outer core

MOHOROVIČIĆ (Moho) discontinuity: the seis-mic discontinuity, or physical interface, between the Earth's crust and mantle

OROGENESIS: the process of mountain-range formation

PLUTON: a deep-seated igneous intrusion

TECTONICS: the study of the assembling, deformation, and structure of the Earth's crust

PHYSICAL PROPERTIES AND TOPOGRAPHY

The Earth's crust exists in two distinct forms: continental crust, or sial, and oceanic crust, or sima. Oceanic crust is characterized by the dense, basic, igneous rock, basalt, while continental crust is an assemblage of sedimentary, metamorphic, and less dense, silicon-rich igneous (granitic) rocks. Oceanic crust makes up the floors of the Earth's ocean basins. Continental crust underlies the continents and their margins and also small, isolated regions within the oceans. The total area of all existing continental crust is 150×10^6 square kilometers. In total, continental crust covers about 43 percent of the Earth's surface and makes up about 0.3 percent of its mass.

Continental crust is distinguished from its counterpart, oceanic crust, and from underlying mantle by its physical properties and chemical composition. In addition, the continental crust and oceanic crust contrast in topography. The Earth's major topographic features range from the highest mountain on the continental crust (Mount Everest, at 8,848 meters) to the deepest ocean trench (the Mariana Trench, at 10,912 meters). The difference in average elevation between the two crustal forms is quite pronounced. Continental crust varies in thickness from 10 kilometers along the Atlantic margin to more than 90 kilometers beneath the Himalaya mountain system. On average, continental crust is about 35 kilometers thick. Seismic studies of the Mohorovičić (Moho) discontinuity indicate that oceanic crust is on average 5-8 kilometers thick. Continental crust averages a height of 0.9 kilometer above mean sea level, while oceanic crust averages a depth of 3.8 kilometers below that datum.

This difference in levels is attributed to the fact that despite being thin, oceanic crust comprises the majority of the Earth's crust and has a density (3.0-3.1 grams per cubic centimeter) greater than that of continental crust. While continental crust is thicker than oceanic crust, it is less dense (2.7-2.8 grams per cubic centimeter) and comprises less crustal surface area. For the most part, continental crust lies near sea level or above it. It is thickest where it underlies places of great elevation, such as mountain ranges. It is thinnest where it lies below sea level, such as along continental shelves. There are exceptions to this pattern of thickening and thinning. The relatively flat basins of the oceans are transversed by 2-kilometer-high ridge systems, and areas of continents where intraplate volcanism is active often display thinning where the crust is stretched by rising hot mantle material. Rising hot material also makes for more buoyancy and raises the surface elevation yet maintains a thin crust. The Basin and Range Province of the western United States is a good example; the crust beneath the mountains is of relatively normal thickness, yet the elevation is high.

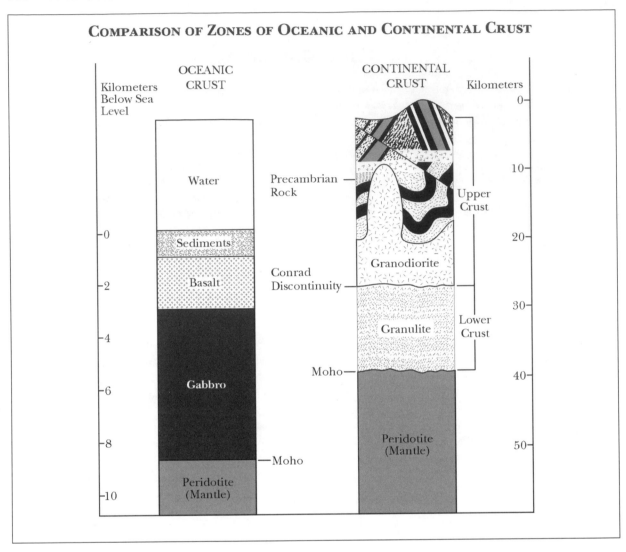

COMPARISON OF ZONES OF OCEANIC AND CONTINENTAL CRUST

COMPOSITION OF OCEANIC AND CONTINENTAL CRUSTS

Differences in the vertical structure and rock composition between oceanic and continental crust are pronounced. The structure of oceanic crust has a prominent layered effect that seismic waves can readily detect. The layers are attributed to petrologic differences between basalt, gabbro, and peridotites that comprise the layers. Continental crust has a more complex layered structure, and the contacts between layers are not well defined. Continental crust is separated into upper and lower zones. The upper zone is usually highly variable in composition, with the top few kilometers of material being any combination of unmeta-

morphosed volcanic or sedimentary rocks, to medium-grade metamorphics such as quartzites and greenschists. Below this immediate layer, the upper zone of continental crust is typically regarded as either granodiorite or quartz diorite. This assumption is based on seismic wave travel times. The upper zone of the crust is separated from the lower zone by a change in seismic velocity similar to that which separates the asthenosphere from the crust itself. This intracrustal boundary is called the Conrad discontinuity. The composition of the lower continental crust is less well known because of the relatively few places where outcrops are available for study. Observations made on rocks in the most deeply eroded regions of Precambrian shields led researchers to believe that the lower zone is composed of granulite. Granulite is a rock of intermediate-to-basic composition, containing mainly pyroxene and calcium feldspars. The velocities of seismic waves through granulite compare favorably with seismic velocities observed passing through rocks of the lower zone. Such circumstantial evidence favors granulite as the composition of the lower continental crust.

When taken on average, the overall chemical composition of continental crust corresponds to that of an intermediate igneous rock with a composition between andesite (tonalite) and dacite (granodiorite). Because igneous rocks of this type are added to continental crust from the mantle at convergent (destructive) plate boundaries, igneous activity of this kind is thought to be responsible for the majority of growth to the continental crust. The accretion rate for new continental crust forming at destructive margins is estimated at 0.5 cubic kilometer per year. Based on calculations of existing continental crust surface areas, ages, and thicknesses, an accretion rate of 0.5 cubic kilometer per year can account for only about half of the existing continental crust. It has been concluded that, while continental crust has formed throughout geological time, the accretion of new continental crust must have occurred at higher rates at different times during the past.

CRUSTAL ROCKS

Rocks of the continental crust formed throughout nearly the complete 4.6-billion-year history of the Earth. These rocks can be grouped into three main components: orogenic belts, Precambrian shields, and continental platforms. Orogenic belts (orogens) are long, broad, linear-to-arcuate (curved) areas of deformed rocks. The deformation occurs to the crust during uplift and usually includes faulting and sometimes the formation of plutons and volcanoes. The result of the deformation is the creation of a mountain system. The deformations affect thick sections through the crust and leave permanent scars that can be recognized long after the uplifted mountains are eroded away.

Precambrian shields consist of deformed crystalline igneous and high-grade metamorphic rocks more than 544 million years old. These shields are the eroded roots of ancient orogens. Continental platforms are regions of relatively underformed,

younger sedimentary or volcanic rocks overlying Precambrian basement. These platforms, while nestled within the continental interior and isolated from internal strain, still typically warp into broad regional structures, usually basins or domes. Shields and platforms can form a stable nucleus to continental masses. These stable regions are called cratons. Examples of a shield and sedimentary platforms forming stable craton regions are the Canadian Shield, the Michigan Basin, and the Ozark Uplift.

CRUSTAL PLATES

The Earth's crust is a solid, rigid layer of mobile plates that comprise the uppermost part of the lithosphere. There are seven major plates, several minor plates, and numerous microplates. These plates appear to float on the plastic upper mantle of the Earth, called the asthenosphere. The vertical boundary between the asthenosphere and the crust is called the Mohorovičić discontinuity, or the Moho. The Moho is a zone less than 1 kilometer thick in some places but several kilometers thick in others, where the velocity of seismic waves changes from about 7 kilometers per second in the crust to about 8 kilometers per second in the mantle. This change in seismic velocity is caused largely by a change in composition between the crust and mantle. Rocks of the mantle are rich in iron and magnesium but poor in silicon, making them denser than the silicon-enriched overlying crust.

The movement of crustal plates upon the denser asthenosphereere is believed to be caused by complex convection currents deep within the mantle. The upper zone of the Earth's crust, in which plate movement takes place, is called the tectonosphere. As the plates move about the tectonosphere, they interact with one another. The plates tear apart (rift), collide, slide under (subduct), or slide against each other (transform fault). The active edges of the plates are called plate boundaries. The interaction of crustal plates at plate boundaries, in addition to the cyclic phenomena of sedimentation, metamorphism, and igneous activity, makes the crust the most complex region of the Earth. These activities process and reprocess crustal material and lead to the diversity of physical and chemical properties observed in crustal rocks. The rocks of the crust indicate that these processes have taken place throughout geological time and further suggest that the crust has grown in bulk at the expense of the upper mantle.

PLATE MARGINS

The oceanic and continental crusts interact along their margins. The margin may be passive, in that stresses are no longer deforming it, or the margin may be active, in which case it is a zone of seismic and tectonic activity. Along passive margins, the transition between continental and oceanic crust is gradual; a good example is the Atlantic shelf along the eastern coast of North America. The best example of an active margin between continental and oceanic crustal plates is the Pacific basin. Around the margin of the Pacific basin, relatively dense oceanic crust is being actively subducted beneath the lighter continental crust. On the continental side, mountain ranges rise (the Andes) and island arcs are formed (the Aleutians), both dominated by active volcanism. Active margins also exist where two continental plates collide. When continents come together, there is little subduction because both plates are of low density. While igneous activity is less prominent than along convergent plate boundaries, the degree of deformation along the margins can be extreme; there is often considerable uplift involved. The contacting continental plates can either slide past each other along a transform fault (such as the San Andreas fault) or act like two cars in a head-on collision. As the plates collide, the crust shortens and the intervening sea floor is uplifted, folded, faulted, and overthrust. The most dramatic example of such an interaction of continental masses is the Himalaya mountain system.

Regions such as the Himalaya of Asia and the Alps of Europe are known as suture zones. They mark the boundaries where two plates of continental crust have collided. At suture zones, oceanic crust is subducted until the ocean basin separating the two continents disappears, and a violent collision takes place. Moving only centimeters per year, the two continents ram into each other. The deformation to the plates during such a collision can be quite dramatic. The two continental plate margins that collide already have thick, mountainous continental margins along their active subduction zones. As the collision takes place, mountain range meets mountain range, and a new, higher, and more complex set of mountains is created. The already

thicker-than-average crusts beneath the two colliding continental edges combine to form an even thicker crust to support the newly uplifted mountains. This process is complicated by secondary magmatic activity that adds buoyancy and uplift. At such suture zones, the continental crust is at its thickest, and mountain peaks reach their most spectacular heights. Continental crust thus forms convergent (subducting) boundaries with oceanic crust and can also collide or slide alongside other continental plates at transform boundaries.

RIFTING

Continental crust can form one other tectonic boundary within its plate margin: It can split and form a spreading zone (rift) similar to the spreading ridges of oceanic crust. Plate interaction at a continental margin may influence the crust hundreds of kilometers inland. If forces within the plate work to stretch the crust, thinning it markedly, crustal faults may develop along the thinning zone. The crustal fault blocks that form will begin to subside as the crust continues to be stretched. Because the upper mantle is also being stretched, material from the lower mantle rises to take its place. This material is hotter and raises the temperature of the surrounding rock. The result is the formation of a magma zone beneath the thin crust of the rift zone. If the magma reaches the surface, volcanic activity similar to that seen at ocean ridge systems develops. Basaltic lava flows to the surface and begins to force the sides of the rift apart. Sometimes the divergence ends as a result of a shift in the overall dynamics of the plate. If that happens, the rift may leave a scar only a few tens of kilometers wide. Some examples are the Midcontinental rift system of North America, the Rhine Valley of Europe, and the East African rift valley. If the rift continues to expand, a new ocean basin/plate is formed. The Atlanic basin is an example of continued rifting of a continental plate to form an active oceanic plate.

In some instances, rifting of two continental bodies occurs near the margin of an older continental margin, and fragments are rifted away from the main continental body. When that happens, small plateaus of continental crust (microplates) become partially submerged in the ocean or become surrounded by oceanic crust. One example is the Lord Howe Rise of the South Pacific. The highest part of the Lord Howe microplate surfaces above the ocean as New Zealand.

ANDESITE MODEL

Earth scientists have used studies of seismic velocity waves to define the boundaries and limits of continental crust and, through exhaustive field investigations and geophysical analysis, have made reliable estimates of the crust's composition. The processes responsible for the formation and dynamic nature of continental crust, however, have remained elusive. To explain their observations, Earth scientists have come to rely on their present understanding of plate tectonics and the related processes of volcanism and orogenesis.

The Andesite Model is a tectonics-based explanation for the formation and growth of continental crust. The model can be stated as follows: The growth of continental crust results from the emplacement or extrusion of largely mantle-derived magmas formed at destructive plate margins. The process begins at the ocean ridges, where melted mantle peridotite rises to the surface as basaltic lava, forming new oceanic crust. The oceanic crust moves away from the ridge by way of seafloor spreading. The spreading is caused by the constant extrusion of more basaltic lava at the ridge. Eventually, the oceanic plate encounters a continental plate and, because of the oceanic crust's greater density, descends below the continental plate. The oceanic crust descends at an angle of 30-60 degrees, forming deep trenches along the continental margin. The descending plate eventually reaches a seismically active region of the mantle known as the Benioff zone. At the Benioff zone, the subducting plate melts, producing a chemically complex, destructive margin magma (andesitic). This lighter, less dense andesitic melt rises through the mantle and into the overriding continental crust. The rising melt creates large plutons within the crust or breaks through to the surface to form andesitic volcanoes. Many large andesitic stratovolcanoes surround the Pacific basin and form the Ring of Fire. Around the Ring of Fire, andesitic lava is erupted and added to the surface of the continents. The volcanoes of the Andes, Cascades, Indonesian Arc, Japan, and Alaska, having such familiar names as Krakatoa, Rainier, and Fujiyama, are the birthplaces of new continental crust.

RESEARCH AND SAFETY APPLICATIONS

The study of continental crust and its related processes is important to Earth scientists because the development of continental crust appears to be a terrestrial phenomenon, that is, one not observed on other planets in the solar system. Furthermore, the continental crust of the Earth provided a platform on which the later stages of the evolution of animal and plant life occurred. Without it, life would have been restricted to ocean basins and isolated volcanic islands, and evolution would have taken a drastically different course.

The memory of early Earth history can only be found in the continental crust. Since oceanic crust records an age no older than 200 million years, the continental crust is scientists' only link with the 4.1-billion-year-old geological record of the Earth. Studies of the continental crust also allow scientists to venture educated speculations as to the beginnings of the solar system some 4.6 billion years ago.

Although the mass of continental crust is small compared to the overall mass of the Earth, it contains substantial amounts of all minerals and elements that are necessary for life to continue on Earth. Additionally, continued investigations into the dynamic processes that form continental crust aid scientists in understanding many of the geological hazards that plague humans. Earthquakes and volcanoes, two of the Earth's most destructive forces, are directly related to the processes that form and shape continental crust. By establishing a more complete understanding of the nature and functions of continental crust, scientists can better prepare and warn citizens of impending geological hazards.

Randall L. Milstein

CROSS-REFERENCES

BIBLIOGRAPHY

Brown, G. C., and A. E. Mussett. *The Inaccessible Earth.* Winchester, Mass.: Allen & Unwin, 1981. An excellent source of general information about continental crust and crustal processes. For the undergraduate student.

Chernikoff, S., and R. Vekatakrishnan. *Geology: An Introduction to Physical Geology.* Boston: Houghton Mifflin, 1999. The authors provide a overview of scientists' understanding of the Earth. Includes the address of a Web site that provides regular updates on geological events around the globe. Contains a chapter on continental crust.

Dolgoff, Anatole. *Physical Geology.* Lexington, Mass.: D.C. Heath, 1996. A comprehensive guide to the study of the Earth; extremely well illustrated with a glossary and an index. Although this is an introductory text for college students, it is also useful for the interested layperson. Contains a chapter on the continental crust.

Dott, Robert H., Jr., and Donald R. Prothero. *Evolution of the Earth.* 5th ed. New York: McGraw-Hill, 1994. This basic textbook on historical geology is aimed at students of geology. However, it is very readable by anyone with a background in science. Presents an up-to-date account of the Earth's history from the viewpoint of plate tectonics. Includes a glossary.

Meissner, R. *The Continental Crust: A Geophysical Approach.* San Diego, Calif.: Academic Press, 1986. Highly technical approach to the study of continental crust. Graduate student reading level.

Taylor, S. R., and S. M. McLennan. *The Continental Crust: Its Composition and Evolution.* Oxford, England: Blackwell Scientific, 1985. Highly technical, more advanced text about the continental crust. For the graduate-level geology student.

CONTINENTAL DRIFT

Continental drift—the horizontal displacement or rotation of continents relative to one another—is the modern paradigm that describes and accounts for the distribution of present-day continents and associated geological formations and phenomena, including mountain ranges, mineral deposits, volcanoes, and earthquakes.

PRINCIPAL TERMS

CONTINENTAL CRUST: the outermost part of the lithosphere, consisting of granite and granodiorite

CONVECTION CELL: a mechanism of heat transfer in a flowing material in which hot material from the bottom rises because of its lesser density, while cool surface material sinks

EARTHQUAKE: the violent motion of the ground caused by the passage of a seismic wave radiating from a fault along which sudden movement has occurred

FAULT: a fracture in the Earth's crust along which there has been relative displacement

GONDWANALAND: a hypothetical supercontinent made up of approximately the present continents of the Southern Hemisphere

LAURASIA: a hypothetical supercontinent made up of approximately the present continents of the Northern Hemisphere

LITHOSPHERE: the outer layer of the Earth, situated above the asthenosphere and containing the crust, continents, and tectonic plates

OCEANIC CRUST: the outer part of the lithosphere, consisting mostly of basalt

PALEOMAGNETISM: the science of reconstructing the Earth's former magnetic fields and the former positions of the continents from the magnetization in rocks

PANGAEA: a supercontinent made up of all presently known continents, which began to break up in the Mesozoic era

PLATE: a large segment of the lithosphere that is internally ridged and moves independently over the interior, meeting in convergence zones and separating at divergence zones

TECTONICS: the study of the movements and deformation of the Earth's crust on a large scale

EARLY CONTINENTAL DRIFT THEORIES

Continental drift is the guiding model for the mechanisms driving the geologic forces near the surface of the Earth. This theory is the simplest explanation for the behavior of the Earth's crust and the distribution of continents and their associated topographic features. The theory is useful not only in decoding the history of the Earth but also in predicting future observations.

The idea that continents may have occupied different geographies in the past was developed by Alfred L. Wegener as early as 1910. Wegener observed that the coastlines of the Americas corresponded with those of Europe and Africa in a jigsaw puzzle fashion. He learned that similar fossils had been discovered on both sides of the Atlantic Ocean, and he proposed that the splitting up of a supercontinent and the drift of its pieces could explain this data. He continued to refine his ideas

and published a book, *Die Entstehung der Kontinente und Ozeane* (1915; *The Origin of Continents and Oceans*, 1924). By his account, about 200 million years ago, at the end of the Permian period, there existed a single supercontinent that Wegener called Pangaea. This supercontinent, he theorized, broke apart, and the various pieces drifted; for example, North America and South America moved westward from Europe and Africa, creating the Atlantic Ocean.

Wegener had an American rival, Frank B. Taylor, who in 1910 published his own theory of mobile continents. Interestingly, Taylor's starting point was not the physical similarity of the Atlantic coastlines but the pattern of mountain belts in Eurasia and Europe. Yet Taylor's hypothesis, like Wegener's, soon faded from scientific memory.

There was, however, much physical evidence to suggest that continental drift was a feasible theory.

First, the physical fit was a good one; in addition, the discovery of similar fossils, mineral deposits, glacier deposits, and mountain ranges seems to show a correlation across the oceans. Wegener described these correlations: "It is just as if we were to refit the torn pieces of a newspaper by matching their edges and then check[ing] whether the lines of print run smoothly across." Efforts to confirm the hypothesis were interrupted by World War I and, soon thereafter, the Depression and World War II. The theory of continental drift thus retained its marginal status until the 1950's.

The major stumbling block to its acceptance was the need to describe a plausible mechanism for driving the continents. Pushing continents around requires a tremendous amount of energy, and no model proposed was acceptable to the geophysics community. The only plausible suggestion came from British geologist Arthur Holmes, who tentatively proposed that thermal convection within the mantle could split the continents and drive them across the surface. Holmes was one of the most respected geologists of his time, and the scientific community paid attention to his idea. Yet he could offer no evidence to support it, and continental drift thus remained merely an interesting possibility in the eyes of most scientists.

Wegener had a more difficult time gathering an audience. To most of the geologic community, he seemed to be an outsider attempting to restructure the science. For example, in the publication of the 1928 American Association of Petroleum Geologists symposium, R. T. Chamberlain quotes a remark made by a colleague:

> If we are to believe Wegener's hypothesis we must forget everything which has been learned in the last 70 years and start all over again.

That, however, is exactly what happened in the 1950's, as the strength of the hypothesis eventually became evident. The hypothesis thus lived on the fringes of the scientific community and was supported by a minority of geologists, most of whom worked in the Southern Hemisphere.

MID-ATLANTIC RIDGE

Worldwide interest in the origins and evolution of the planet's features culminated in the observation of the International Geophysical Year from July, 1957, to December, 1958. The result of this effort was that in almost every area of research, and especially in geology, scientists found the Earth and particularly its oceans to be very different from what they had imagined. One of the most interesting features studied was the Mid-Atlantic Ridge, an investigation that would lead to the understanding of plate tectonics.

The existence of a submarine ridge in the Atlantic had been recognized in the 1850's by Matthew Maury, director of the U.S. Navy's Department of Charts and Instruments. The British expedition aboard the HMS *Challenger* (1872-1876) also recorded a submarine mountain. The

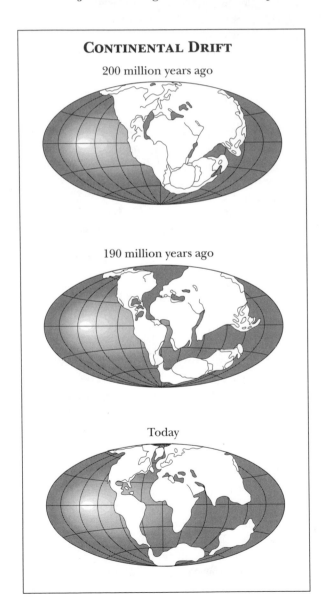

CONTINENTAL DRIFT

200 million years ago

190 million years ago

Today

next advance came in the 1920's with a German expedition led by Nobel laureate Fritz Haber. The expedition utilized an echo sounder to map the ocean floor. In 1933, German oceanographers Theodor Stocks and Georg Wust produced a detailed map of the ridge, and they noted a valley that seemed to be bisecting it. In 1935, geophysicist Nicholas H. Heck found a strong correlation between earthquakes and the Mid-Atlantic Ridge.

Oceanic exploration resumed after World War II as a predominantly American venture. The data collected pointed to an array of seemingly unrelated phenomena. In 1950, Maurice Ewing of the Lamont-Geological Observatory discovered that no continental crust existed beneath the ocean basins. In 1952, Roger Revelle, the director of the Scripps Institute of Oceanography, and his student A. E. Maxwell measured the heat flow from the Earth's interior and discovered that it was hotter over the oceanic ridges. Additional data from Jean P. Rothe, director of the International Bureau of Seismology, revealed a continuous belt of earthquake centers associated with this submarine mountain range, which extends from Iceland through the mid-Atlantic, around South Africa, and into the Indian Ocean to the Red Sea. In 1956, Maurice Ewing and Bruce C. Heezen mapped a

large area of this submarine mountain range and confirmed the existence of a rift valley bisecting the mountain crest. A peculiar faulting style was discovered in association with the range in 1959 by Victor Vacquier. The mountain range was offset by a large transverse fault that ran for hundreds of miles but did not extend into the continents. In 1961, Ewing and Mark Landisman discovered that this ridge system extended throughout the world's oceans, was seismically and volcanically active, and was mostly devoid of sediment cover.

The paleomagnetic researches of University of Manchester scientist Patrick M. S. Blackett and his student Keith Runcom proved central to understanding the relationships among these phenomena. Their studies of fossil magnetism suggested that the position and polarity of the Earth's magnetic field had once been very different from its present orientation. These data could only make sense if one assumed that the continents had shifted relative to the poles and to one another.

By the end of the 1950's, it was clear that then-current geologic theories had failed to predict or explain these seemingly unrelated phenomena, and the new data required a new theory. In 1960, Harry Hammond Hess proposed a simple model to explain the data. He suggested that seafloor

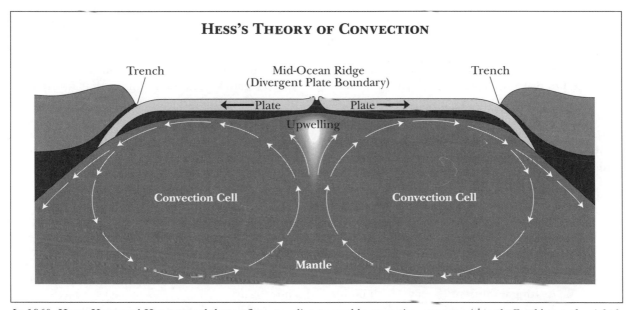

In 1960, Harry Hammond Hess proposed that seafloor spreading powered by convection currents within the Earth's mantle might be the cause of the motion of the continents. Eventually, the theory of continental drift, as expressed in the theory of plate tectonics, emerged as the dominant paradigm for the Earth sciences.

spreading powered by convection currents within the mantle might be the cause of the motion of the continents. Hess's theory, though simple, was radical; it bore out Chamberlain's earlier insight that previous geologic models would have to be discarded and that the geologic community would have to reinterpret and test all of its data in the light of the new model. This did not come easily to the science community; eventually, however, the theory of continental drift did emerge as the dominant paradigm for the Earth sciences.

HISTORY OF DRIFT THEORIES

The lengthy process leading to the acceptance of the theory of continental drift is not unusual in the history of science. Often, a hypothesis has to wait upon the development of technology or upon accidents of timing for its observations to be tested.

The observation that there was a relationship between the coastlines of the Americas and those of Europe and Africa can be traced to Francis Bacon, who in 1620 noted the similarities of shape. The idea was further enhanced by a French monk, François Placet, who in 1666 suggested that the Earth's landmasses had split as a result of the biblical flood, thus separating Europe and Africa from the Americas. The idea was repeated by a German theologian, Theodor Lilenthal, during the eighteenth century. In 1800, Alexander von Humboldt suggested that the oceans had eroded the land to further divide the continents. The drift theory was then supported by Evan Hopkins, who in 1844 proposed the existence of a "magnetic fluid" that circulated to drive the continents. In 1857, American geologist Richard Owen published *Key to the Geology of the Globe*, in which he proposed that the Earth was originally a tetrahedron that had expanded in a great cataclysm, breaking the crust and expelling the Moon from the Mediterranean.

These ideas and hypotheses were interesting but largely unsupported by physical evidence; Wegener's and Taylor's hypotheses, proposed in the early twentieth century, were thus fundamentally different from those of the past. Still, it was the development of new technological tools that illuminated the phenomena and eventually supported the drifting continent theory. Much of this new technology was developed as a result of World Wars I and II and the technological race of the Cold War.

Wegener collected data from paleontology to show a correlation between continents and to illustrate that the continents had been in different latitudes in the geologic past. Further, he suggested an experiment to confirm the theory; the experiment failed not because the idea was inappropriate but because the experimental error resulting from his crude equipment was greater than the phenomena he was trying to measure. His experiment, conducted in 1922 and again in 1927 and 1936, involved the measurement of the time it took radio signals to travel across the Atlantic. The measurements failed to reveal a widening of the Atlantic through progressively longer travel times. Upon the advent of satellite and laser technology, however, widening was detected.

MOUNTING EVIDENCE FOR WEGENER-HESS

Eventually, wartime technology such as sonar was applied to scientific applications. Sonar is the underwater version of radar; an energy pulse is sent out, and its reflection from the seafloor is recorded. These data can be translated via computers to create either a profile or contour map. Literally thousands of soundings were made over thousands of kilometers of ocean in an effort to construct a map of the ocean floor in the greatest possible detail. These data began to produce a map that revealed rather remarkable features, including a continuous 64,000-kilometer mountain range that had a valley running along its crest. Hess realized the significance of the valley on top of the mountain range: It was a tensional, or "pull-apart," feature. The same forces that formed the mountain chain were also pulling it apart.

Other technologies were also contributing to the investigation. After World War II, a worldwide network of seismographs was deployed, not so much for recording earthquakes as to listen for atomic explosions. These new and sensitive instruments mostly recorded earthquakes at plate boundaries and revealed their outlines. (Interestingly, no nuclear powers camouflaged their atomic blasts as earthquakes by detonating them at a plate boundary.) The earthquake pattern was a fingerprint of plate activity and evidence of a dynamic crust.

Other compelling data came from monitoring internal heat flow. The temperature of the Earth increases with depth. At the core, the temperature is more than 4,000 degrees Celsius. This heat flowing from the interior can be measured; the hottest crustal areas were found to be above the junction of plates that are spreading centers.

More traditional geologic sampling of the subsurface was conducted by retrieving core samples from below the ocean depths. These physical samples were analyzed according to the type of sediment and the age of fossils present. The findings were surprising: The oceans are very young compared to the continents, and the sediments on the mid-oceanic ridges are thin to nonexistent, while the sediments next to the continents are kilometers thick. Therefore, not only are the oceans young, but they are also youngest in the middle and oldest next to the continents.

FOSSIL MAGNETISM

The straw that broke the back of opposition to the theory of continental drift came with the study of fossil magnetism. Sedimentary and igneous rocks offer a record of the orientation of the Earth's magnetic field through time, as iron particles within them are incorporated into their structures as they form. The decoding of these fossil magnetic fields suggested that the magnetic poles have reversed themselves and that the continents have wandered through the latitudes. If the poles reversed and the continents wandered, then a mirror image of polar reversal correlating with submarine topography should be present on both sides of a spreading center such as the Mid-Atlantic Ridge. This was exactly what was observed. The model was no longer merely an explanation of previously observed phenomena; it had also been shown to be capable of predicting future observations. Earth scientists thus came to perceive the idea of drifting continents and plate tectonics as the unifying model of geologic phenomenon.

The development of the continental drift theory is a story of how science works in a period of paradigm revolution. This period began with Alfred Wegener's work in 1912 and ended with Harry Hess's discoveries in 1960. The model Hess developed gave a new explanation of virtually all geologic phenomena at or near the surface of the Earth. Within its field, the theory has had an impact comparable to that of Charles Darwin's theory of evolution in the field of biology. In essence, the theory of continental drift accounts for the global distribution of the continents, the birth and death of oceans, and the distribution of earthquakes, volcanoes, and mountain ranges. It also leads explorers to mineral and fossil-fuel deposits.

Richard C. Jones

CROSS-REFERENCES

Abyssal Seafloor, 651; Cenozoic Era, 1095; Continental Crust, 560; Continental Growth, 573; Continental Rift Zones, 579; Continental Shelf and Slope, 584; Continental Structures, 590; Continents and Subcontinents, 595; Earth's Crust, 14; Earth's Magnetic Field, 137; Earth's Mantle, 32; Earthquake Distribution, 277; Faults: Transform, 232; Gondwanaland and Laurasia, 599; Heat Sources and Heat Flow, 49; Hot Spots, Island Chains, and Intraplate Volcanism, 706; Lithospheric Plates, 55; Magnetic Reversals, 161; Magnetic Stratigraphy, 167; Mountain Belts, 841; Ocean Basins, 661; Oceans' Origin, 2145; Plate Margins, 73; Plate Motions, 80; Plate Tectonics, 86; Rock Magnetism, 177; Spreading Centers, 727; Subduction and Orogeny, 92; Supercontinent Cycles, 604.

BIBLIOGRAPHY

Engle, A. E., H. L. James, and B. F. Leonard, eds. *Petrologic Studies: A Volume in Honor of A. F. Buddington.* New York: Geological Society of America, 1962. The primary source for Hess's theory of continental drift.

James, Harold Lloyd. *Harry Hammond Hess.* National Academy of Sciences Biographical Memoirs 43. New York: Columbia University Press, 1973. This biographical sketch yields insight into Hess and his synthesis of the oceanic data to form a new paradigm.

LeGrand, H. E. *Drifting Continents and Shifting Theories.* New York: Cambridge University Press, 1988. Traces the development of the continental drift paradigm and the work of people who contributed to the data collection and debate.

Press, Frank, and Raymond Siever. *Understand-*

ing Earth. 2d ed. New York: W. H. Freeman, 1998. This comprehensive physical geology text covers the formation and development of the Earth. Contains extensive sections on plate tectonics and its effects on paleogeography. Readable by high school students, as well as by general readers. Includes an index and a glossary of terms.

Scientific American editors. *Continents.* San Francisco: W. H. Freeman, 1973. An excellent resource for landmark papers from 1952 to 1970. Illustrates the progression of the geologic "revolution."

Sullivan, W. *Continents in Motion: The New Earth Debate.* New York: McGraw-Hill, 1974. Well written and illustrated with drawings and photographs. Develops like a detective story rather than a textbook.

CONTINENTAL GROWTH

Continents are believed to have increased in size during the Earth's history by accretion of additional crustal material along their margins. This process has played a significant role in the formation of valuable mineral deposits such as gold, silver, copper, gas, and oil.

PRINCIPAL TERMS

FAULT: a fracture in rock strata with relative displacement of the two sides

FOLD: an upward or downward bend in layered rock strata

GEOSYNCLINE: an elongate subsiding trough in which great thicknesses of sedimentary and volcanic rocks accumulate

GRANITE: a light-colored crustal rock produced by the underground cooling of molten rock

LATERAL ACCRETION: the process by which crustal material is welded to a shield by horizontal compression

MANTLE: the Earth's 2,900-kilometer-thick intermediate layer, which is found beneath the crust

PLATE TECTONICS: a theory that describes the Earth's outer shell as consisting of individual moving plates

SHIELD: a continental block of the Earth's crust that has been stable over a long period of time

SUBDUCTION: the process by which one crustal plate slides beneath another as a result of horizontal compression

GEOSYNCLINES

Geologists believe that the continents have increased in size during geologic time by the accretion of additional material along their margins. This additional material usually consists of younger rocks deposited in a deeply subsiding belt known as a geosyncline, which is then welded to the continent by compressive forces. In some cases, the additional material may represent portions of a preexisting continent—or even an entire continent itself—that has been "drifted in" by the mechanism known as plate tectonics.

The idea of geosynclines dates back to the work of two nineteenth century American geologists. In the 1850's, James Hall pointed out that the crumpled strata of mountain ranges along the continental margins were thicker than the equivalent strata in the continental interiors. In 1873, J. D. Dana suggested the term "geosyncline" (literally, "great Earth downfold") for elongated belts of thick sedimentary rocks deposited along the continental margins. Further geologic fieldwork, primarily in the Alps and in the Appalachians, showed that the sedimentary rocks of the geosynclines had been deformed by compressive forces emanating from the ocean basins. By the 1950's, the generally accepted picture of continental growth was that of a stable continental interior, called the shield or craton, surrounded by increasingly younger belts of deformed rock. Each belt was believed to represent a geosynclinal sequence that was deposited in a bordering trough and then welded to the shield by lateral accretion. Yet, no satisfactory mechanism for the source of the compressive forces from the ocean could be discovered.

In North America, the central shield is called the Canadian Shield. Its exposed portion occupies the eastern two-thirds of Canada, the U.S. margins of Lake Superior, and most of Greenland. The Canadian Shield is the largest exposure of Precambrian rocks in the world, consisting predominantly of igneous and metamorphic rocks. There is also a buried portion of the Canadian Shield extending westward to the Rocky Mountains and southward to the Appalachian Mountains, the Arbuckle Mountains in Oklahoma, and into Mexico. This buried portion of the shield has a thin cover of largely Paleozoic sedimentary rocks deposited in shallow transgressing seas. These Paleozoic rocks are still flat-lying except where they have been gently warped into broad domes and basins. Along the margins of the Canadian Shield, four belts of deformed sedimentary rocks represent the former geosynclines. Geologists named the deformed belt on the eastern side of the shield the Appalachian geosyncline, the belt on the south side the Ouachita geosyncline, the belt on the west side the Cordilleran geosyncline, and the belt on the north side the Franklin geosyncline. Because of compressive forces emanating from the ocean basins, overturned folds and thrust faults are present in all four geosynclinal belts.

Similar patterns of continental growth are found elsewhere in the world. Each continent has at least one shield. These include shields in South America, Africa, northern Europe, Siberia, eastern Asia, India, Australia, and Antarctica. The remnants of geosynclinal belts are located adjacent to these shields. The most famous of these geosynclinal belts is the Tethyan geosyncline, found along the southern margin of Europe and Asia. The present-day Alps and Himalaya have risen out of this geosynclinal belt.

PLATE TECTONICS

During the 1960's, the concept of plate tectonics gradually emerged as the result of the work of oceanographers trying to explain the origin of the planet's major seafloor features. These features include mid-ocean ridges rivaling the largest mountain ranges on the Earth and volcanic island chains with associated deep oceanic trenches that rim the Pacific. The plate tectonics theory has revolutionized not only the field of oceanography but also the field of geology. According to the plate tectonics theory, the surface of the Earth is covered by a series of rigid slabs or plates that are capable of moving slowly over the Earth's interior. Geologists recognize six or seven major plates, each one usually containing a continent. The plates are presumed to behave as separate units, and where plates jostle each other, intense geologic activity occurs along their boundaries.

Three different types of activity are believed to take place at plate boundaries. Divergence occurs where two plates are moving apart. The result of the divergence of two continental plates is believed to be the formation of a new ocean basin, a process referred to by geologists as seafloor spreading.

When two plates are moving toward each other, the result is convergence. In this case, three possibilities arise, depending on the nature of the plate boundary. If two continental plates converge, the intervening oceanic sediments are believed to be compressed into a new mountain range, such as the Himalaya. If a continental plate runs into an oceanic plate, the oceanic plate is believed to slide beneath the overriding continental plate, producing a deep oceanic trench. Geologists refer to this process as subduction. Finally, one oceanic plate may override another oceanic plate, producing a trench and adjoining volcanic island arc.

The third type of movement found along plate boundaries occurs when two plates slide past each other horizontally, just as trains pass each other in opposite directions on adjacent tracks. Such movement may proceed continuously or in a series of abrupt jerks, depending upon the amount of friction encountered along the plate boundary. The abrupt jerks result in the type of Earth movement known as earthquakes, and these are particularly associated with the famous San Andreas fault in California.

Exposed portions of the Precambrian Canadian Shield, scraped bare by continental glaciation. (© William E. Ferguson)

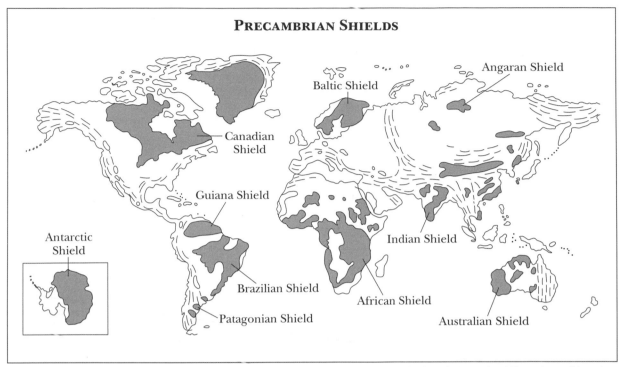

PRECAMBRIAN SHIELDS

Precambrian shields are the oldest exposures of rock in the world, generally accepted as the foundations of stable continental interiors, surrounded by increasingly younger belts of deformed rock. The Canadian Shield is the largest exposure of Precambrian rocks in the world, consisting predominantly of igneous and metamorphic rocks.

The underlying causes for plate movements are not well understood. Geologists speculate that convection cells of rising and sinking material in the mantle (the Earth's intermediate layer) may carry the plates slowly along. Nor is the mechanism by which rigid plates are able to slide across the Earth's interior well known. Presumably there is a "plastic" layer in the upper mantle that provides the necessary lubrication for the crustal plates to move.

The significance of plate tectonics for the concept of continental growth has been recognized by scientists. Because most of the jostling takes place at plate boundaries, that is where they find downwarped geosynclinal belts, earthquakes, volcanic activity, and recently formed mountains. On the other hand, the stable plate interiors are places where little jostling takes place and thus where the quiescent continental shields are located.

CONTINUING CONTINENTAL GROWTH

An example of a present-day geosynclinal belt that is being squeezed between two converging shields is the Tethyan geosyncline. This geosyncline is believed to have originated between the Eurasian and African Shields during the Mesozoic era. It must have resembled a broad tropical seaway extending from the Caribbean eastward through the Mediterranean and the Himalaya to Indonesia on the borders of the Pacific. Thick marine sediments accumulated on the floor of this seaway, and they are preserved today as richly fossiliferous limestone sedimentary rocks.

During Cenozoic time, geologists believe, the convergence of several continental plates initiated the destruction of this seaway. The Indian subcontinent, for example, is believed to have drifted northward until it collided with Asia, producing the Himalaya, the highest mountain chain on the Earth. A second collision occurred as the Arabian plate (a minor subplate) drifted north to collide with Asia Minor, forming the Zagros and other mountains. Finally, Africa is believed to have drifted northward, resulting in the compressive forces that have produced the Alps. The result of these collisions is the near obliteration of the old Tethyan seaway. The only relics of it that survive are the Caspian Sea, the Black Sea, and the Mediterranean Sea. In addition, the sedimentary rocks that accumulated in the geosyncline have been folded and thrust northward against the Eurasian continental platform. Scientists have a clear picture, therefore, of continental growth taking place as a result of crustal plates moving toward each other, with the intervening sediments being welded to the shields by lateral accretion.

As indicated earlier, continental growth can also result when a portion of a preexisting continent, or even an entire continent, collides with another continent. An example of such a collision is believed to be provided by the formation of the Ural Mountains at the end of the Paleozoic era. Geologists now believe that these north-south trending mountains, which have been eroded down to their roots, were formed by the compression of sediments deposited in a seaway lying between Europe and Asia. In other words, the present-day continent of Eurasia, which is twice as large as any other continent, was once two separate continents.

In the cases of the destruction of the Tethyan geosyncline and the formation of the Ural Mountains, the role of plate tectonics seems clear because the plates that did the moving can be identified. Sometimes, however, relationships are not so apparent—for example, in the case of the deformation of North America's Appalachian geosyncline and the thrusting of its Paleozoic sediments against the shield. No continental plate lies along the East Coast of North America that might account for the compression. As a result, students of plate tectonics have postulated an elaborate scenario that involves two stages. They assume, first, that the Atlantic Ocean closed at the end of the Paleozoic era as a result of the collision of Europe with North America and then that Europe moved eastward again, resulting in the reopening of the Atlantic because of seafloor spreading.

The deformation of the Cordilleran geosyncline along the West Coast of North America offers no such problems. It can be explained by the collision of the North American plate with the Pacific Ocean plate. According to plate tectonics, the Pacific Ocean is a separate plate even though it lacks a continent. Thus, the deformation of the Cordilleran geosyncline has resulted from the sliding of the Pacific Ocean floor beneath the North American continent in the process known as subduction.

STUDY OF CONTINENTAL GROWTH

Scientists have studied the subject of continental growth in many ways. Foremost among them has been field investigations in the rock strata found along the margins of the shields. Using the fossils contained within these rocks, as well as radioactive dating, geologists have pieced together a detailed history of geosynclinal accretion. By analyzing the geometry of the folds and faults, scientists have also been able to infer the direction from which the compressive forces came.

An example of such geologic field investigations is seen in the deciphering of the rocks of the Canadian Shield. These rocks constitute the largest outcrop of Precambrian strata exposed anywhere in the world today, and to early workers they appeared to be a hopeless tangle of similar-looking igneous and metamorphic rocks. After years of painstaking research, however, scientists have been able to identify an orderly sequence of mappable rock units within the Canadian Shield, so that at least four distinct cycles of deposition and mountain making during Precambrian time are recognized.

Another way in which scientists have approached the subject of continental growth is by examining the rock types that compose the shields and ocean floors. They have found that the continental rocks are largely granitic and are rich in silica, aluminum, and potassium. These rocks also have a slightly lower average density than do the rocks underlying the ocean basins, and they stand higher, as if both were floating on interior layers of the Earth. By contrast, the rocks of the ocean floors consist of slightly heavier basalt lava and related volcanic rocks.

To everyone's surprise, the granitic rocks of the continents have not proved to be the oldest rocks on the Earth. Radioactive dating indicates that slivers of seafloor rocks incorporated in the granites claim this distinction. Thus, scientists have concluded that the shields do not represent parts of the Earth's original crust but have been built up through time by a process of lateral accretion. Their granites may have come from the reworking of seafloor rocks.

A third way in which scientists are investigating

the subject of continental growth is through detailed study of the seafloor itself. This study began with the Deep Sea Drilling Project in 1968 using the *Glomar Challenger,* a drillship that was retired in 1984. The program continued under the name Ocean Drilling Project, utilizing the *JOIDES Resolution,* a 143-meter drillship capable of recovering samples of rock and sediments from depths of 9,000 meters under the ocean surface. The findings of these ships have been of great significance for plate tectonics.

All evidence seems to point to a very young age for the ocean basins—less than 200 million years old, which is less than one-twentieth the Earth's presumed age of 4.7 billion to 5 billion years. Furthermore, it appears that the Atlantic and Indian Oceans are widening, while the Pacific is shrinking, which, if true, means that the crustal plates will eventually collide, thus providing the mechanism for further continental growth.

MINERAL DEPOSITS

The same processes that have thrust former geosynclinal belts against the shields have also produced rich mineral deposits in the resulting mountain chains. These mineral deposits fall into three categories: metals, such as gold, silver, and copper; nonmetallic deposits, such as certain abrasives, gemstones, and the building stones granite, marble, and slate; and the important energy resources petroleum, natural gas, and anthracite coal.

A good example of a metal deposit found in the deformed rocks of a former geosyncline is California's famous Mother Lode. This zone of gold veins, which is more than 200 kilometers long but barely 1 kilometer wide, can be traced along the western slopes of the Sierra Nevada, a mountain range that has risen out of the former Cordilleran geosyncline. The gold discoveries—which attracted the "forty-niners" to California and led to the rapid growth of San Francisco and neighboring cities—were nuggets and flakes of gold derived from these veins and washed down into the sand and gravel deposits of rivers at the foot of the mountains. The early settlers realized the gold was from a source upstream, and they called this source the Mother Lode (literally, "parent vein") Eventually, the settlers traced the streams up to their headwaters in the Sierra Nevada and discovered the Mother Lode itself.

The fabulously rich oil deposits of the Middle East are another example of an economic resource related to continental growth. The Tethyan seaway, which stretched from the Caribbean to the Pacific during Mesozoic time, was the site of extensive deposits of thick limestone sedimentary rocks. These limestones are now oil-bearing and have been caught in the closing vise between the northward-moving Arabian plate and the portion of the Eurasian continent known as Asia Minor. Because of this compression, the limestones have been shaped into a series of gently undulating folds. Migrating oil has been trapped in the crests of the upfolds (technically known as anticlines), where it is obtained by drilling wells down into the anticlinal structures.

In 1988, more than 40 percent of the world's oil imports came from the Middle Eastern oil fields, a situation that has enabled the Arab nations to wield a political and economic influence far out of proportion to their geographic size or number of inhabitants. A dramatic example of this influence was provided in the 1970's, when these nations paralyzed the free world's economic system with an oil embargo. Even though the major consumers of Middle Eastern oil are Western Europe and Japan, the dislocation in world oil supplies had severe consequences in the United States as well. Americans were asked to turn down their thermostats, the nationwide speed limit was reduced to 55 miles per hour, and automobile companies were told to improve the gas mileage of their cars. As the ripple effects of the oil shortage spread through the United States' economy, a recession was triggered that cost people their jobs and set off a major stock market decline. Despite these experiences, and abortive attempts to market electrically powered automobiles, later generations of U.S. car drivers migrated back to low-mileage cars, trucks, and recreational vehicles. The appetite for Middle Eastern oil has allowed subsequent manipulation of world oil supply, as again occurred early in the year 2000, when oil production was limited and prices again rose dramatically.

Donald W. Lovejoy

CROSS-REFERENCES

BIBLIOGRAPHY

Chernikoff, S., and R. Vekatakrishnan. *Geology: An Introduction to Physical Geology.* Boston: Houghton Mifflin, 1999. The authors provide a overview of scientists' understanding of the Earth. Includes the address of a Web site that provides regular updates on geological events around the globe. Contains a chapter on the development of continents.

Cloud, Preston. *Oasis in Space: Earth History from the Beginning.* New York: W. W. Norton, 1988. A definitive, one-volume synthesis of Earth history by a distinguished geologist and gifted writer. Illustrated with more than three hundred maps, photographs, and diagrams. Continental growth is a major theme, with heavy emphasis on the Precambrian era. Suitable for college-level readers and laypersons with some technical background.

Dolgoff, Anatole. *Physical Geology.* Lexington, Mass.: D. C. Heath, 1996. This is a comprehensive guide to the study of the Earth. Extremely well illustrated and includes a glossary and an index. Although this is an introductory text for college students, it is written in a style that makes it understandable to the interested layperson. Contains a chapter on continental growth.

Dott, Robert H., Jr., and Donald R. Prothero. *Evolution of the Earth.* 5th ed. New York: McGraw-Hill, 1994. This basic textbook on historical geology is aimed at students of geology. However, it is very readable by anyone with a background in science. Presents an up-to-date account of the Earth's history from the viewpoint of plate tectonics. Includes a glossary.

Plummer, Charles C., David McGeary, and Diane H. Carlson. *Physical Geology.* Boston: McGraw-Hill, 1999. This is a straightforward, easy-to-read introduction to geology intended for those with little or no science background. Discusses the development of the continents. Includes many excellent illustrations, as well as a CD-ROM.

Scientific American. The Dynamic Earth. New York: W. H. Freeman, 1983. A collection of eight outstanding articles written by leading scientists who have been involved in the development of the plate tectonics theory as well as the unified view of the Earth. Excellent color photographs, maps, and line drawings. Suitable for college-level readers and the interested layperson.

Windley, Brian F. *The Evolving Continents.* New York: Wiley, 1977. A good source book for the serious student of continental growth. Data are provided for selected shields and fold belts throughout the world, with major emphasis on the Precambrian, Europe, the Alps, and areas outside North America. Photographs, however, are lacking. Suitable for college-level readers with some technical background.

Wyllie, Peter J. *The Way the Earth Works: An Introduction to the New Global Geology and Its Revolutionary Development.* New York: Wiley, 1976. A concise introduction to plate tectonics, suitable for high-school-level readers and the interested layperson. The supporting evidence for plate tectonics is explained in more detail than is customary in most textbooks. Many excellent diagrams have been especially prepared for this text in order to illustrate the basic concepts.

CONTINENTAL RIFT ZONES

Continental rift zones are places where the continental crust is stretched and thinned. Distinctive features include active volcanoes and long, straight valley systems formed by normal faults. Continental rifting in some cases has evolved into the breaking apart of a continent by seafloor spreading to form a new ocean.

PRINCIPAL TERMS

ASTHENOSPHERE: a layer in the upper mantle beneath the lithosphere that behaves as a fluid, permitting the overlying plates to move

CRUST: the outer layer of the Earth, composed of silica-rich, low-density rock, which in continental areas ranges from about 25 to 70 kilometers in thickness

FAULT SLIP: the direction and amount of relative movement between the two blocks of rock separated by a fault

FAULT: a large fracture or system of fractures across which relative movement of rock bodies has occurred

FOOTWALL: the rock body located below a non-vertical fault

GRABEN: a roughly symmetrical crustal depression formed by the lowering of a crustal block between two normal faults that slope toward each other

HALF-GRABEN: an asymmetrical structural depression formed along a single normal fault as the downthrown block tilted toward the fault

HANGING WALL: the rock body located above a nonvertical fault

LITHOSPHERE: the outer shell of the Earth, including both the crust and the upper mantle, which behaves rigidly over time periods of thousands to millions of years

LITHOSPHERIC PLATES: segments of the lithosphere that are similar in size to continents; these plates form a mosaic that covers the Earth's surface

MANTLE: the iron- and magnesium-rich, silica-poor part of the Earth beneath the crust

NORMAL FAULT: a fault across which slip caused the hanging wall to move downward relative to the footwall

CHARACTERISTICS OF RIFT ZONES

Continental rift zones are areas where the continental crust has been stretched and thinned. They are characterized by long valleys bounded by faults (rift valleys), by active volcanoes both within and adjacent to the rift valleys, by earthquakes, and by hot springs and other manifestations of unusually high temperatures near the Earth's surface. Rift zones are sometimes regions of high elevation, so that the margins of the rift valleys are high mountain ranges. Continental rifts are considered to be an expression at the surface of hot, partially molten rock in the mantle buoyantly rising beneath a continent.

Continental rifts are commonly linear valley systems that trend at a high angle to the direction that the crust has been stretched. Examples of linear rifts are the Rio Grande rift in New Mexico and Colorado, the Rhine Valley in northern Europe, and the East African rift system in Ethiopia, Kenya, and Tanzania. Other continental rift zones are broad areas of alternating linear valleys and mountain ranges such as the Basin and Range Province of western North America.

GRABENS AND HALF-GRABENS

The basic architectural unit in the upper crust of continental rift zones is the half-graben. Half-graben valleys collect sediment eroded from the adjacent, relatively uplifted fault block. The resulting sedimentary accumulations are wedge-shaped, thicker at the place of greatest subsidence adjacent to the bounding fault and gradually thinner away from the fault. Some rift valleys are bounded by normal faults on both sides to form grabens rather than half-grabens. In most cases, however, the amount of slip is much larger across one of the two bounding faults, and the smaller fault is generally interpreted to represent minor modification of the basic half-graben form. The size of a half-graben basin is determined by the length and amount of slip across the main normal fault that formed the basin. Sizes vary, but a typical major half-graben basin in a continental rift zone is 50 to 200 kilometers in length and 20 to 50 kilometers across. Slip across the bounding fault of a large half-graben is typically several kilometers and may exceed 10 kilometers.

Linear continental rift zones are chains of half-grabens, linked end to end, with from one to three half-grabens occurring side by side across the rift. The linked half-grabens define a major valley system along which a large river system commonly develops, such as the Rio Grande and the Rhine River. The bounding normal faults of the end-linked half-grabens commonly alternate in dip direction, so that the asymmetry of the half-graben basins reverses from one half-graben to the next down the rift valley. Broad rift zones such as the Basin and Range Province are also composed of half-grabens. In these areas, however, the half-grabens are arrayed side-to-side, such that the rift zone may be ten or more half-graben units wide, as well as being linked at the end. In contrast to the reversing asymmetry of end-linked half-grabens, laterally adjacent half-grabens commonly have the same asymmetry.

Normal slip along faults stretches the crust horizontally. Estimates of the amount of stretching across continental rifts vary from a few kilometers across linear rifts such as the Rhine Valley to hundreds of kilometers across broad rift zones such as the Basin and Range. The higher estimates of extension predict extreme thinning of the crust if the crust does not change volume in the process. Although the crust in rift zones is thinner than normal (usually about 25 to 30 kilometers), it generally is too thick to be consistent with constant-volume stretching of the amount indicated by surface observations.

RIFT-ZONE VOLCANISM

The discrepancy in crustal thickness is probably explained by addition of new rock to the crust during rifting, resulting from intrusion and extrusion of magma derived from the upwelling mantle below. A significant fraction of rift-zone volcanism is basalt, which represents new crustal rock extracted from the mantle by partial melting. The amount of mantle-derived magma trapped within the crust to form intrusions probably greatly exceeds the amount erupted at the surface, so that although the amount of new crust formed during rifting may be quite large, the precise amount is not yet known.

Basaltic volcanism commonly occurs together with eruptions of more silica-rich rocks, particularly rhyolite. The silica-rich rocks are believed to have formed from the melting of the continental crust by the heat carried into it by the mantle-derived basalt. This "bimodal" association of basalt and rhyolite has been considered to be a distinctive characteristic of continental rifts; however, studies have documented important exceptions. For example, much of the volcanism during rifting in the Basin and Range Province formed rocks of intermediate composition (specifically dacite) rather than a bimodal suite. The intermediate volcanic rocks formed mainly by mixing of basaltic and rhyolitic magmas. Therefore, the appearance of bimodal and intermediate volcanism seems to depend on whether the basaltic magmas and rhyolitic magmas remain separate or are mixed together. This mixing and what causes (or prevents) it are a focus of modern research.

Although it has been established that a close relationship exists between crustal stretching and volcanism, the nature of the relationship is controversial. Two main possibilities have been presented, the "active rift" model and the "passive rift" model. It is uncertain if one or the other of these models, or a combination of them, is most correct. In the active rift model, upwelling in the asthenosphere causes rifting of the lithosphere. Hot, partially molten mantle rock rises buoyantly beneath a continental plate, releasing basaltic magma, which in turn rises into and through the plate. Heat from the basaltic magma warms the lithosphere, causing it to expand, resulting in uplift of the Earth's surface in that area. Over geologic time, the heated lithosphere behaves roughly like a fluid, spreading out from the elevated area. In the upper crust, this spreading occurs by normal faulting.

In the passive rift model, stretching of the lithosphere causes the upwelling of the mantle and the basaltic volcanism. The lithosphere thins as it is stretched, allowing the underlying asthenosphere to well up passively beneath. The upwelling reduces the pressure in the asthenosphere, because the weight of the overlying thinned lithosphere is less than that of lithosphere of normal thickness. Because melting temperatures decrease as pressure decreases, the pressure decrease induces melting of the asthenosphere and basaltic volcanism.

CONTINENTAL DRIFT

Continental rifts are commonly thought of as features that lie within a lithospheric plate. In some instances, however, continental rifting evolves into seafloor spreading, breaking the plate in two and forming a new ocean basin at the site of the rift. Therefore, continental rifting might be viewed as the beginning of continental drift. In other cases, however, rifts cease activity without causing continental breakup; these are sometimes called "failed rifts." It appears that there is a critical threshold at which a change occurs from continental rifting to seafloor spreading, but the nature of this threshold is still obscure.

The North Atlantic Ocean and its margins form a classic example of the results of this rift-to-drift process. At the beginning of the Triassic period, about 245 million years ago, there was no North Atlantic Ocean; the continents of North America, Europe, and Africa were joined together to form a part of the supercontinent Pangaea. During the Triassic, a broad continental rift zone formed that was similar to the modern Basin and Range Province of western North America. After 20 or 30 million years of rifting, the continent began to break apart, and seafloor spreading began, first between Africa and North America and later between Europe and North America. The record of continental rifting before the North Atlantic Ocean formed is left in the continental margins that surround the ocean. The continental shelf areas that surround the Atlantic are underlain by continental crust that contains numerous half-graben rift basins of the Triassic period.

STUDY OF RIFT ZONES

Continental rifts and the processes that form them have been studied by virtually every geological, geochemical, and geophysical technique available. Applications of these techniques fall into three main categories: field studies of surface ge-

ology, laboratory analysis of samples collected in the field, and geophysical field studies. Study of any natural phenomenon starts with basic fieldwork, in this case making maps of the rock bodies and their interrelationships as exposed at the surface. These data allow the geologist to draw inferences about the three-dimensional arrangement of rock bodies and the geologic history that led to that arrangement.

Field geologists' ability to formulate detailed and predictive interpretations from surface observations has been expanded greatly by new and improved laboratory measurement techniques. Potentially applicable techniques span all Earth science, because tectonic analysis of a region involves synthesizing all pertinent observations in a single integrated framework. Of particular importance are paleontology and isotopic geochronology, which provide information about timing of geological events and involve petrological and geochemical techniques that are used to infer the depth, rock type, and pressure and temperature at the source regions of rift-related volcanic rocks.

Most geophysical field techniques used in the study of continental rifts utilize seismology, the study of the way sound waves pass through the Earth. Earthquake seismologists analyze the spatial distribution of earthquakes to find which areas are tectonically active at present. Seismic refraction studies, using both earthquakes and artificial explosions as sound sources, are used to determine the thickness of the crust and its large-scale structure. Seismic reflection studies use artificial sound sources such as explosions or specially designed vibrator trucks to provide more detailed information about the internal structure of the crust.

The best approach in tectonic research, in studying continental rifts or any other type of feature, is to blend all these sources of information into an integrated scheme. For example, the three-dimensional structure of the crust can be inferred best by combining seismic reflection and refraction data with data regarding variations in the strength of the Earth's gravity field (which reflects variations in rock density at depth) and with surface geologic mapping.

SIGNIFICANCE

Modern continental rift zones present significant seismic and volcanic hazards. The largest earthquakes along continental rift zones generally measure from 7 to 7.5 on the Richter scale. They are therefore smaller than the magnitude 8 or greater earthquakes that occur along large strike-slip faults, such as the 1906 San Francisco earthquake along the San Andreas fault, or the magnitude 9 or greater earthquakes that occur along subduction zones, such as the 1960 earthquake in Peru. A large rift-zone earthquake is nevertheless capable of producing great destruction in areas near the epicenter, especially if buildings are not built to withstand earthquake stresses. Damaging earthquakes in Greece, for example, reflect continental rifting in the area of the Aegean Sea.

Continental rifts have been the sites of some of the Earth's largest explosive volcanic eruptions. Major eruptions from large rift-related volcanic centers can be literally thousands of times larger than the May, 1980, eruption of Mount St. Helens. Because no historic eruption of this size has occurred anywhere on the Earth, it is hard to estimate how much damage would result from such an eruption. At least three, and perhaps more, such large explosive volcanic centers in the western United States were the sources of huge explosive eruptions within the last 1 million years and are still volcanically active (Yellowstone National Park in Wyoming, Long Valley in California, and

the Valle Grande in northern New Mexico). Similar active centers are present on other continents.

Ancient continental rift zones are important sites of metallic mineral deposits, mainly formed at or near the explosive volcanic centers just mentioned. A large portion of the gold, silver, copper, lead, and zinc deposits in the Basin and Range Province of the western United States formed beneath or adjacent to rift-related volcanic centers. Petroleum accumulations are found in some half-graben basins in continental rift zones, such as the Great Basin, the Rhine Valley in Germany, and the Pannonian Basin in Hungary and Romania.

John M. Bartley

CROSS-REFERENCES
Basin and Range Province, 824; Continental Crust, 560; Continental Drift, 565; Continental Growth, 573; Continental Shelf and Slope, 584; Continental Structures, 590; Continents and Subcontinents, 595; Earth's Crust, 14; Earth's Lithosphere, 26; Earth's Mantle, 32; Earthquake Distribution, 277; Faults: Normal, 213; Forecasting Eruptions, 746; Gondwanaland and Laurasia, 599; Heat Sources and Heat Flow, 49; Lithospheric Plates, 55; Plate Margins, 73; Plate Tectonics, 86; Supercontinent Cycles, 604; Volcanic Hazards, 798; Yellowstone National Park, 731.

BIBLIOGRAPHY

Chernikoff, S., and R. Vekatakrishnan. *Geology: An Introduction to Physical Geology*. Boston: Houghton Mifflin, 1999. Provides an overview of scientists' understanding of the Earth. Includes the address of a Web site that provides regular updates on geological events around the globe. Contains sections on rifting and the development of rift systems.

Courtillot, Vincent, and Gregory Vink. "How Continents Break Up." *Scientific American* 249 (July, 1983): 42. A brief, nontechnical account of the authors' ideas about the transition from continental rifting to seafloor spreading. It focuses on large-scale patterns related to lithospheric plate movement rather than on processes and products of crustal rocks formed in continental rifts.

Dolgoff, Anatole. *Physical Geology*. Lexington,

Mass.: D. C. Heath, 1996. This is a comprehensive guide to the study of the Earth. Extremely well illustrated and includes a glossary and an index. Although this is an introductory text for college students, it is written in a style that makes it understandable to the interested layperson. Contains a section on the development of the African rift valley system and continental rifting in general.

Dott, Robert H., Jr., and Donald R. Prothero. *Evolution of the Earth*. 5th ed. New York: McGraw-Hill, 1994. This basic textbook on historical geology is aimed at students of geology. However, it is very readable by anyone with a background in science. Presents an up-to-date account of the Earth's history from the viewpoint of plate tectonics. Includes a glossary.

Plummer, Charles C., David McGeary, and Diane H. Carlson. *Physical Geology.* Boston: McGraw-Hill, 1999. This is a straightforward, easy-to-read introduction to geology intended for those with little or no science background. Discusses the development of the African rift valley system and rifting in general from the point of view of plate tectonics. Includes many excellent illustrations, as well as a CD-ROM.

Quennell, A. M. *Rift Valleys: Afro-Arabian.* Stroudsburg, Pa.: Hutchinson, Ross, 1982. This volume reprints selected articles from professional journals to provide overviews of the geology and geophysics of the East African-Arabian continental rift and of the development of thought regarding its origins. Although the reproduction quality of the articles is fair to poor and the selection of papers is somewhat uneven, the book makes some articles originally published in obscure places more accessible. Suitable for college students.

Rosendahl, B. R. "Architecture of Continental Rifts with Special Reference to East Africa." *Annual Review of Earth and Planetary Sciences* 15 (1987): 445. This paper discusses the crustal architecture of the East African rift as an illustrative example for continental rifts in general. The main foci are geometries of faults in continental rifts and what insight into such faults is derived from the seismic reflection data that Rosendahl and his students have collected in the East African rift. It is written at an advanced college level, but the illustrations are clear and informative for the general reader.

CONTINENTAL SHELF AND SLOPE

The continental shelf and slope mark the continental margins in the ocean. They are repositories for much of the weathered rock material eroded and transported by rivers and wind. In addition, they serve as major reservoirs for petroleum and a variety of other mineral resources.

PRINCIPAL TERMS

SEDIMENT: solid matter that settles on a surface; sediments may be transported by wind, water, and glaciers

SHELF DAMS: geologic formations that hold back sediments on the continental shelf

SUBMARINE CANYONS: channels cut deep in the sediments by rivers or submarine currents

TRENCH: a long, narrow, and deep depression in the ocean floor, usually with steep sides and often adjacent to island arc systems and continental landmasses

TURBIDITY CURRENTS: fast-moving submarine avalanches of inorganic sediment

ORIGINS OF THE CONTINENTAL SHELF

The continental shelf of the ocean has been called the submerged shoulders of the great landmasses. The shelf, together with the adjacent, seaward continental slope, separates the land from the great depths of the sea. Thus, the shelf is part of a dynamic transition zone that has changed markedly over the millennia; it is also a zone that has been contested by a number of nations eager to maintain what are considered Rights of the Sea. These rights include the free passage of ships, access to valuable submerged minerals, fishing rights, and military intelligence gathering. The shelf is a nearly flat area that marks the submerged edges of continents. It slopes gently toward the ocean basins, but the slope is so slight that it is not discernible. The width may range from 30 meters in some locations to more than 100 kilometers in others, with the average width being 65 kilometers. The nature of the adjacent landmass often dictates the shelf width—broad next to low-lying land and narrow next to rugged, mountainous land.

The massive continental glaciers that covered the Earth some 2 million years ago during the Pleistocene epoch helped shape the continental shelf. Wide, deep shelf areas with rugged topography are found in those parts of the world that were covered by ice sheets during the Pleistocene. As the glacial fronts advanced over the face of the land, they bulldozed vast amounts of the Earth's surface to be deposited many hundreds of kilometers away. The glaciers ground huge boulders into gravel, gravel into sand, and sand into a fine, dust-like "glacial flour." This debris was deposited on the surface of the continental shelf. The shelf itself was dry land during these glacial excursions. The ice sheets, some 3 kilometers high at their maximum, had stored so much of the oceans' waters that sea level was lowered nearly 150 meters below the present level. The immediate impact on the coastal zone was rough, turbulent surf with great waves crashing on the exposed shore. As a result, the beach sediments of these shores were coarse, with an abundance of gravel, cobbles, and boulders.

The broad expanse of the exposed coastal lowlands encouraged the development of wetlands and ponds. Vegetation that accumulated in the wet areas formed peat, which is dredged occasionally from depths of 50 meters, and sometimes from depths of 50 to 100 kilometers, off the present coast. In addition to the wetlands, the exposed shelf zone supported forests of spruce, fir, pine, and oak. Humans lived and hunted a variety of game in these forests some 15,000 to 20,000 years ago. Then the glaciers began to melt, and the sea level reached its present extent about 5,000 years ago. Fish now swim through marine waters where birds once flew among the towering trees. The nearshore deposits on the present-day continental shelf also contain the fossilized remains of giant mastodons, woolly mammoths, and other huge land mammals that once grazed over the coastal plains in vast numbers. In addition to biological deposits, old sand dunes and the rounded pebbles of ancient beaches have been found on the shelf.

SEDIMENTS

The continental shelf is covered with a veneer of sediments that vary greatly in depth. The submerged shelf areas receive most of the weathered rock debris from the erosion of the continents. Indeed, deep exploration has revealed thousands of meters of sediments that have accumulated on the sedimentary rock forming the edges of the continents. Most of the sediments of the shelf are classified as relict; that is, they are not representative of the environment of today but were laid down thousands and perhaps millions of years ago. About 70 percent of continental shelf sediment was deposited during the past 15,000 years. During the million-year extent of the Pleistocene epoch, there were four major lowerings of sea level, followed by flooding, as the glaciers waxed and waned. The flooding from the last major stage led to the accumulation of the sediments found on the continental shelf today.

Marine sediments are classified by a number of criteria. These include origin, particle size, density and shape, mineral composition, and color. Sediments that originate from the erosion of land formations are termed terrigenous. These include sand, gravel, silt, and clay. Sediments that are formed by the accumulation of the shells and skeletons of animals are termed biogenous, or bio-

genic; those that are formed directly from chemicals in seawater are termed hydrogenous, or authigenic. Most of the sediments on the continental shelf are terrigenous; a few are biogenic or mixtures of the two.

At one time, it was believed that sediments on the continental shelf showed size gradations—coarse sands and gravels close to shore, with finer particles farther from shore. The finest particles, the silts and clays, formed a "mud line" along the seaward edge of the shelf. Research conducted during World War II, however, revealed that this progression from coarse to fine is rarely found. Samples of the sediments retrieved from the continental shelf reveal that they consist mostly of coarse sand. The sediment particles are stained red by iron deposits. Frequently, the shelf sediments also contain the empty shells of clams, whelks, and other marine animals that live close to the shallow waters of the coastal region. Occasionally, masses of broken shells accumulate along the outer edge of the continental shelf. Fields of cobble and boulders dot some of the shelf areas. These massive structures represent rocks transported by glaciers and deposited when the glaciers melted.

Sediments are deepest on those shelf areas that received masses of sand and gravel deposited during glacial meltback. Georges Bank, off Massachusetts, and the shallow North Sea between the British Isles and northeastern Europe feature much glacially deposited debris. The varied topography of these glaciated shelf areas includes banks, chan-

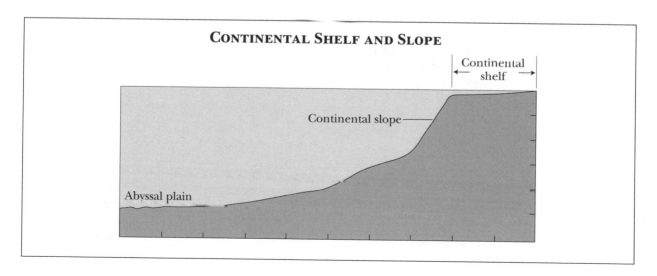

CONTINENTAL SHELF AND SLOPE

Continental shelf

Continental slope

Abyssal plain

nels, and deep basins. An example of a broad but unglaciated shelf, however, is found off the coasts of New York and New Jersey. The deep shelf sediments there are generally smoother, although they may be sculpted by strong currents to form low-relief ridges and, occasionally, deep submarine canyons.

The West Coast of the United States, in contrast, is relatively clear of sediment on the shelf. There is only a thin veneer of materials, and the shelf shows little evidence of glaciation. Strong ocean currents can also sweep the shelf clear of sediments. These currents contribute as well to the formation of narrow, or even absent, continental shelf areas. The East Coast of southern Florida is a good example of this. There, the Gulf Stream, often flowing at 11 kilometers per hour, sweeps close to the mainland. This strong current has prevented normal shelf development. The Gulf Stream is credited with sweeping sediments from the Blake Plateau, located several hundred meters below the sea surface.

CONTINENTAL SHELF DAMS

The sediments that carpet most of the continental shelf are kept in place by a variety of dams along the shelf break. The break marks the boundary between the gently sloping continental shelf and the steeper continental slope. The dams thus trap the sediments and hold them against the continents, preventing them from spilling downslope into the deep-sea basins. Continental shelf dams may be formed by volcanoes, coral reefs, or salt domes. Often, the dams result when massive blocks of basement rock are thrust up by powerful forces in the Earth's interior. The dams and the sediments behind them are often eroded by waves, glacial ice, or powerful local currents. The waves and currents ridge the sediments in a series of terraces that are parallel to the shore. Cutting across the terraces are channels that were eroded by streams during the glacial periods, when the shelves were exposed.

Major modern river systems, including the Hudson River and the Congo River, cut deep channels in the exposed shelf sediments. These channels are easily traced across the continental shelf off the east coast of North America and the west coast of Africa. The Hudson River channel, for example, is so large that it has not yet filled with sediment, despite the active rate of erosion by the riverine system. The great scar of the canyon extends across the broad continental shelf, cuts through the dam at the shelf edge, and plunges down the continental slope. This canyon is a major feature of both the shelf and the adjacent slope.

CONTINENTAL SLOPE

The continental slope is a well-delineated geologic separation between the flat shelf and the moderate grade of the continental rise. Of the three features, the slope is the steepest. It has an average angle of 4 degrees (with a range of between 3 degrees and 20 degrees); in many oceans, this steep boundary extends all the way to the floor of the deep ocean basin. It is one of the largest topographic features of the Earth's surface and may extend nearly 4,000 meters from the depths of the seabed to the shelf edge. Perhaps the most dramatic continental slope area is below the narrow continental shelf off the West Coast of South America. There, the slope wall drops precipitously for 8,000 meters into the Peru-Chile Trench. This slope is similar to others in that it features craggy outcroppings and is relatively bare of sediments because of the steepness.

The continental slope is steepest in the Pacific Ocean basin. There, it averages more than 5 degrees, while in the Atlantic and Indian Ocean basins, it averages about 3 degrees. The Pacific Ocean continental slope is associated with the geologic processes that form the coastal mountain ranges and the deep ocean trenches. The word "slope" does not adequately describe the great and varied topography of these oceanographic regions. Underwater landslides, submarine earthquakes, and the processes of subsurface erosion have produced a variety of topographic features. The continental slope is incised by numerous valleys and canyons, some rivaling the Grand Canyon in size. They are long and short, straight and branched, and may be cut through solid rock as well as through the sediments that may carpet the slope. Many of the canyon walls are cut by side canyons in a variety of sizes.

Most of the canyons that cut across and down the slope are continuations of the canyons of the continental shelf. These are easily traceable to continental landforms and are believed to have been eroded by rivers or glaciers during the Pleistocene epoch, when the shelves were exposed. Many canyons on the slope, however, are nowhere connected to shelf canyons and may, in fact, be far removed. It is believed that these canyons may have been (and continue to be) eroded by fast-moving, powerful turbidity currents. The currents may be triggered by earthquakes or by an excessive buildup of sediment on steep areas of the slope. Moving rapidly downslope, the currents—fast-moving submarine avalanches of mud, sand, fine gravel, and water—erode the walls of the slope and carve out the submarine canyons. In 1929, an earthquake caused a turbidity current on the shelf and slope off Newfoundland that was estimated to have moved 700 kilometers at speeds of 40 to 55 kilometers per hour. The millions of tons of abrasive sediments that periodically sweep through existing canyons effectively erode the walls and floors. As the currents move downslope, the velocity decreases and the sediments spread out to be deposited at the base of the slope as deep-sea fans.

EXPLORATION OF THE SEAFLOOR

Early mariners' curiosity about the seafloor was limited to practical matters—for example, finding the channel for entering a harbor. The earliest instrument that yielded any information about the seafloor was the sounding lead. This club-shaped device, about 20 to 30 centimeters long, was attached to a long hemp or cotton line marked in fathoms. (One fathom is approximately 2 meters long.) The base of the lead, the part that touched the ocean floor, had a shallow, cup-shaped indentation that was packed with lard or tallow. When the lead was dropped over the side of the vessel and touched bottom, markers on the retrieving line indicated the depth of water. When it was hauled back on deck, the fat in the cup usually brought back a small sample of the seafloor sediment. With no other information available to them, mariners (and even the first curious scientists) believed the ocean basin to be a smooth, unrelieved depression whose bottom dropped off to unmeasurable and unimaginable depths.

The early research was limited to mapping the coast and making soundings in water less than 200 meters. The principal concern was safe navigation. Toward the end of the nineteenth century,

however, the laying of transoceanic telegraphic cables, including those from North America to Europe, spurred greater interest in making more accurate and detailed surveys. The pioneering global voyage of the British research vessel *Challenger*, from 1873 to 1876, included many soundings in areas previously not studied, but research methods were tedious and progress was slow. The soundings were made with hemp rope, which frequently broke under the strain of its own weight. The line and its weight—often a cannonball—were laboriously hauled up with a capstan turned by the crew. Later, twisted wire rope and single-strand wire, often graduated in diameter, reduced the time required for lowering and reeling in the line and weight. An innovation that involved releasing the weight after it had touched bottom, lessening the load, further reduced the retrieval time. By the beginning of the twentieth century, there were still fewer than ten thousand soundings in waters deeper than 2,000 meters and only about five hundred in waters deeper than 5,500 meters. Such data provided information about the depths involved but gave no indication of the submarine topography.

TECHNOLOGICAL ADVANCES

In the 1920's, the introduction of sonic devices for making depth soundings revolutionized the procedure. These devices, variously called sound navigation and ranging (sonar), or simply depth sounders, transmitted a sound impulse from the ship to the ocean floor. The impulse "bounced" off the seafloor and was received by an instrument that translated the round-trip time to depth. Later refinements of the device provided paper traces with near-photographic representation of the ocean floor over the entire span of the basins. For the first time, the grandeur of the seafloor topography was revealed. It showed that the land beneath the sea surface is nearly as rugged and sculpted as any part of the dry land. The sonic devices could also penetrate the surface sediments to reveal their depth and complexity. Further refinements made it possible to probe the basement rock beneath the veneering sediments.

As valuable as the sonic data are, nothing is more revealing than samples of the ocean floor over the continental margins. The earliest specimens were collected with rugged iron dredges dragged over the bottom. As with the early hemp sounding lines, the dredges took hours to drop, tow, and retrieve. All too often, the dredge closed up before it reached the bottom. At other times, powerful underwater currents twisted the line into an impossible tangle, and again, the dredge failed to gather bottom samples. Despite these hazards, dredges did collect much valuable material and are still widely used. A variety of grabs lowered from vessels are used to collect bottom samples in specific locations. As with the dredges, the grabs often retrieve biological specimens with the bottom sediments. Corers also collect bottom sediments. These devices are dropped or thrust into the bottom sediment to collect a cylinder of sample sediment.

Perhaps the most dramatic sampling utilizes deep-drilling equipment. Special vessels drill into the bottom sediments, penetrating to depths of more than 2,000 meters. The vessel *Glomar Challenger* successfully drilled into the continental slope in water depths of nearly 3,000 meters. Manned submersibles such as *Alvin* are fitted with maneuverable arms and collecting baskets, still and video cameras, and viewing ports. Thus, the observations of scientists aboard the submersibles are supplemented by photographs and specimens of the ocean floor.

Albert C. Jensen

CROSS-REFERENCES

Abyssal Seafloor, 651; Continental Crust, 560; Continental Drift, 565; Continental Glaciers, 875; Continental Growth, 573; Continental Rift Zones, 579; Continental Structures, 590; Continents and Subcontinents, 595; Deep-Sea Sedimentation, 2308; Earth Resources, 1741; Gondwanaland and Laurasia, 599; Manganese Nodules, 1608; Ocean Basins, 661; Ocean Drilling Program, 359; Ocean-Floor Drilling Programs, 365; Ocean-Floor Exploration, 666; Offshore Wells, 1689; Oil and Gas Exploration, 1699; Petroleum Reservoirs, 1728; Sand, 2363; Sediment Transport and Deposition, 2374; Sedimentary Mineral Deposits, 1637; Supercontinent Cycles, 604; Turbidity Currents and Submarine Fans, 2182.

BIBLIOGRAPHY

Backus, Richard H., and D. W. Bourne, eds. *Georges Bank*. Cambridge, Mass.: MIT Press, 1987. This is a detailed and informative book about one of the most extensively studied continental shelf areas in the world. Well illustrated with photographs, drawings, and maps that supplement fifty-seven articles written by experts. The chapters on marine geology and physical oceanography are particularly well prepared and explain complex ocean basin processes to the nonspecialist.

Borgese, E. M. *The Mines of Neptune: Minerals and Metals from the Sea*. New York: Harry N. Abrams, 1985. Recommended for anyone with an interest in seabed mineral resources, this volume discusses the recovery of mineral resources from the seafloor. Includes color and black-and-white photographs and drawings that show mineral samples and the equipment used to collect them.

Champ, Michael A., William P. Dillon, and David G. Howell. "Non-living EEZ Resources: Minerals, Oil, and Gas." *Oceanus* 27 (Winter, 1984/1985): 28-34. This article describes the seabed's mineral resources in an easy-to-understand style. It also discusses the exclusive economic zone (EEZ) and American possession of sand and gravel, phosphorite, ferromanganese nodules, and oil and gas on the continental shelf and abyssal plains.

Leet, L. D., Sheldon Judson, and M. E. Kaufman. *Physical Geology*. 7th ed. Englewood Cliffs, N.J.: Prentice-Hall, 1987. This textbook covers such marine geology-related subjects as seafloor topography, plate tectonics, seamounts, and the effects of waves and currents on ocean-floor sediments. A useful reference volume for the general reader.

Milliman, J. B., and W. R. Wright, eds. *Environment of the U.S. Atlantic Continental Slope and Rise*. Boston: Jones & Bartlett, 1987. This book discusses the topography and geology of the region. It contains an extensive bibliography and will serve as a useful reference for students of Earth science.

Plummer, Charles C., David McGeary, and Diane H. Carlson. *Physical Geology.* Boston: McGraw-Hill, 1999. This is a straightforward, easy-to-read introduction to geology intended for those with little or no science background. Discusses the development of the continental shelf and slope from the point of view of plate tectonics. Includes many excellent illustrations, as well as a CD-ROM.

Press, Frank, and Raymond Siever. *Understanding Earth.* 2d ed. New York: W. H. Freeman, 1998. This comprehensive physical geology text covers the oceans and their mineral resources, as well as the formation and development of the continents. Readable by high school students, as well as by general readers. Includes an index and a glossary of terms.

Rabinowitz, Phillip, Sylvia Herrig, and Karen Riedel. "Ocean Drilling Program Altering Our Perception of Earth." *Oceanus* 29 (Fall, 1986): 36-40. Useful for those interested in the methodology of research in the ocean basins, this article discusses studies that used ocean-floor drilling and a deep-water camera able to photograph a volcano nearly 3 kilometers below the ocean surface in order to gain a better understanding of the nature of ocean-floor sediment. Suitable for high school readers.

Segar, Douglas. *An Introduction to Ocean Sciences.* New York: Wadsworth, 1997. Comprehensive coverage of all aspects of the oceans and the oceanic crust, including methods of seafloor exploration. Readable and well illustrated. Suitable for high school students and above.

Wertenbecker, W. "Land Below, Sea Above." In *Mysteries of the Deep,* edited by Joseph J. Thorndike. New York: American Heritage, 1980. This discussion of seafloor processes is very understandable and informative. Covers the history of ocean-floor research, the impact of volcanism and deep-sea currents in shaping the benthic zone, seamounts and other submarine topography, and methods for mapping the ocean floor. Excellent photographs and drawings are included.

CONTINENTAL STRUCTURES

The continental crust is layered on a large scale and in most places can be divided into upper granitic and lower gabbroic layers. A variety of rock types and structures are superimposed on this compositional layering. The crust thus is a mosaic of geological terranes that have been assembled to form the present continents.

PRINCIPAL TERMS

BASEMENT: a term that refers to the crystalline, usually Precambrian, igneous and metamorphic rocks that occur beneath the sedimentary rock on the continents

CONTINENTAL SHIELD: the oldest exposed Precambrian rocks that form the nuclei of the continents

CRATON: the part of the continent that is covered with a variable thickness of sedimentary rock but that has not been affected by mountain building

CRUSTAL DISCONTINUITY: a boundary within the crust that is detected by a change in the velocity of seismic waves and that results from the different densities of crustal layers

GEOLOGIC TERRANE: a crustal block with a distinct group of rocks and structures resulting from a particular geologic history; assemblages of terranes form the continents

MOHOROVIČIĆ DISCONTINUITY: the boundary between the crust and the upper mantle that was first defined by a rapid change in seismic velocities; it separates low-density crust from the denser mantle

RIFTING: a process of crustal extension or separation that is accomplished by a series of faults involving down-dropped blocks in the central portion, forming a large valley

CRUSTAL DISCONTINUITIES

The continental crust contains rocks of many different ages that are complexly related to one another. These rocks were formed and emplaced by a wide range of geological processes operating from the earliest Precambrian (approximately 4.0 billion years ago) to the present. As such, the continental crust preserves the most complete history of the development of the Earth. Most of what can be seen of the continents is limited to the surface outcrops and to information from some deep mines and drill holes. These observations have shown that the continents consist of a veneer of sediments and sedimentary rocks overlying a complex basement of igneous and metamorphic rocks.

On a large scale, the continental crust appears to be a horizontally layered mass of variable thickness. The base of the crust is defined by the Mohorovičić discontinuity, the boundary where the rapid increase in seismic wave velocity marks the beginning of the mantle. The depth to the Mohorovičić discontinuity, and therefore the continental crustal thickness, varies between 15 and 80 kilometers. The continents are thickest under the great mountain systems and appear thinnest where the crust is submerged beneath sea level along continental margins or where it has been subjected to rifting. Over much of the interior of the continents, the Mohorovičić discontinuity appears to undulate gently, yielding crustal thicknesses between 25 and 45 kilometers.

In most places, seismic refraction studies have shown that below the sedimentary veneer, the continental crust is divided into two, three, or four layers defined by crustal discontinuities. In general, the lower crust has a higher seismic wave velocity and has a composition that is more mafic than the upper crust. Comparisons of seismic velocities suggest that the lower crust is gabbroic. On the other hand, the upper crust has the composition of granite or granodiorite. While the terms used for the crustal layers are igneous names, scientists think that most of these rocks have been metamorphosed. The existence of this layering is confirmed by two kinds of observations. First, in some mountain ranges, more than 10 kilometers of uplift of early Precambrian rocks has occurred. The rocks exposed in these eroded mountains are granodioritic to dioritic gneisses similar to those deduced from the seismic studies. Second, independent geophysical studies, including seismic refraction, gravity and magnetic anomalies, and heat-flow measurements, confirm the trend to more basic rocks in the deep crust.

ORIGIN OF CRUSTAL LAYERING

The origin of this layering has been speculated upon for many years. The occurrence of denser (gabbroic) material at the base and lighter (granitic) rock at the top of the crust is consistent with what is expected based on segregation of materials by gravity. While the mechanism by which this gravitational segregation developed is not known with certainty, major possibilities that have been considered include remobilization, partial melting, and upward migration of magmas throughout much of geologic time. The low melting fractions form magmas, which migrate upward and crystallize, resulting in rocks that are usually lighter. Thus, igneous processes may lead to the concentration of the less dense granitic fraction to the top of the crust, which is consistent with the distribution of radioactive elements that has been deduced from heat flow and from measures of natural radioactivity. The second idea proposes that the crust has gradually thickened with time as a result of subcrustal deposition of basic material out of the mantle. The last materials to come from the mantle might be denser and more gabbroic than the first, more granitic, upper layers. Whatever the cause, some horizontal layering occurs in all areas of the continental crust. In places, three layers will change gradually into two or into four over distances of a hundred kilometers. In other cases, more sudden lateral changes occur.

Considerable detail on the structure of continental layering has been derived from deep seismic reflection studies. These studies have shown the existence of thick sections of layered rock within the Precambrian crust. These layered rocks have been interpreted as piles of volcanic and sedimentary rock that have been metamorphosed and preserved within the crust. Elsewhere, major reflectors appear to represent thrust faults within the upper or middle sections of the crust, attesting the ability of tectonic events to affect more than the uppermost crustal layers.

GEOLOGIC TERRANES

The continental crust is a mosaic of subcontinental, geologic terranes with different ages, different rock types and structures, and different geologic histories. For many years, this mosaic was recognized in the continental shields, where age determinations on the outcropping rocks allowed the division of the exposed Precambrian rocks into provinces. Within the Canadian Shield of North America, the structures of the younger provinces crosscut earlier structures, and the igneous and metamorphic processes associated with the development of the younger province rework the rocks in older provinces. A classic example occurs in Quebec and in eastern Ontario, where there is a juxtaposition between the early Superior Province, dating from 2.8 to 2.5 billion years before the present, and the linear, northeast-trending Grenville Province, dating from 1.2 to 0.9 billion years before the present. While studies of the shields had demonstrated the existence of crustal provinces, much of the crustal mosaic on the craton was hidden beneath sedimentary rocks of Paleozoic and younger ages. Currently, scientists aided by geophysical surveying recognize many provinces within the continental interiors, based on the character of gravity and magnetic anomalies and on observed crustal layering.

Along the margins of continents or within modern mountain ranges, continental structures have been studied extensively. These regions have been most affected by the most recent cycle of seafloor spreading and plate motions. Thus, geologists have found structures within the continents related to divergent plate motions, involving normal faulting and volcanic activity; to convergent motions, leading to thrust faulting, volcanic activity, metamorphism, and batholith formation; and to transform boundaries such as that found along the San Andreas fault in California.

A large number of "foreign" or "suspect" geologic terranes have been found within the younger mountain belts. These foreign terranes are often geologically and structurally distinct from adjacent terranes. These terranes are similar to the Precambrian provinces of the continental crust in that they have been assembled with other crustal components to form the continents as they are today. The boundaries between the foreign terranes or between crustal provinces are similar in that they often involve thrust faults that may affect a significant thickness of the crust and that may even displace the Mohorovičić discontinuity and involve the mantle.

RIFT SYSTEMS

Continental structures indicate that continents have been assembled over a long period of time and that the process is an ongoing one. The latest cycle of plate motions tells scientists, however, that continents are subject to forces and processes that tend to disassemble them as well. While these forces often lead to continental breakup, in many cases they are only partially successful. In these cases, scars are left on the continental crust that record the event. In general, the disruption of the continent is marked by thinning or rifting of the crust and by injection of material from the mantle.

Where the continental breakup is incomplete or stalled, the rift zone and injected igneous rocks are preserved as bodies of dense, highly magnetic rock that cut across the layering of the continental crust. One such zone, which extends within the midcontinent of the United States from Lake Superior through Minnesota and Iowa on into Kansas, is the largest continuous section of a Precambrian rift system that has been largely disrupted by later tectonic events. Other failed rift systems include faults and intrusions in the lower Mississippi River Valley and in Kentucky. Such failed rift systems usually involve the intrusion into the crust of large volumes of mafic igneous rock. Such dense intrusive masses may have considerable influence on later geologic events, such as the formation of depositional basins and the location of inland seas.

GRAVITY AND MAGNETIC ANOMALIES

The structure of continents has been defined by geologic mapping, by radiometric age determinations, by remote-sensing techniques, and by geophysical methods. By far the most important in characterizing the deep crust or areas covered with sedimentary rocks are the geophysical methods. These include all the methods described in general geophysics textbooks, but the most important are probably gravity and magnetic anomalies and seismic refraction and reflection profiles.

Gravity anomalies allow the scientist to investigate the mass distribution in the subsurface. Observations are made with a gravimeter, which measures values as small as 0.00001 centimeter per second squared (0.01 milligal), which is approximately one part in one hundred million of the Earth's total gravity. Surveys are done where elevation and locations are well known, and the data are evaluated by standard equations. The gravity data allow one to model the densities, depths, sizes, and shapes of rock bodies in the subsurface.

Magnetic surveys may involve measurement of the total magnetic field or simply the strength of the field in the vertical direction. The magnetometer commonly measures values of the magnetic field of one part in fifty thousand of the Earth's field. The magnetic anomaly allows investigation of the magnetic susceptibility of rock in the subsurface. The susceptibility is the property of a material that causes it to reinforce or to move into a magnetic field the way a nail will move toward a magnet. The permanent magnetic effect of the rocks, termed the remanence, can also be evaluated as is done for the rocks created at the crests of the mid-ocean ridges. Data are corrected for the natural variation of the Earth's field as a function of location and are evaluated by a standard set of equations which allow the scientist to characterize bodies at depth. The magnetic method is especially good for the study of the crust because the mantle has very low values of the magnetic properties and does not influence the data greatly.

Gravity and magnetic data are often analyzed together for a region. The combination of the physical properties of density and the magnetic susceptibility restricts the possible rock types rather well. The patterns, or "fabrics," of magnetic and gravity anomalies have been used to define different crustal provinces. The coincidence of large gravity and magnetic anomalies has been one of the best ways to locate and evaluate dense mafic volcanic rocks and intrusions associated with the crustal rift zones.

SEISMIC INVESTIGATIONS

Seismic investigations have provided the most detailed information about the structure of the continental crust. Two methods have been used, and both involve the use of a seismometer with a number of sensors (or phones) and a seismic energy source. The first involves measuring the arrival times of seismic wave energy from a distant source. By this method, termed seismic refraction profiling, a series of phones that will detect ground vibrations is laid out in a straight array. The seismic energy may come from an earthquake or a human-made source, but in either case it travels through the Earth at a velocity characteristic of the rocks through which it passes. The time that it takes the seismic wave to reach the individual phones on the array is analyzed graphically or by a simple set of equations. The velocity structure and layering of the subsurface is then determined. These layer velocities may be interpreted in terms of specific rock types that have been measured in the laboratory.

The second seismic method involves reflection of the seismic waves off of surfaces at depth. Seismic reflection profiling is done with a shorter array of seismic phones and usually uses an artificial energy source. The times for waves to reflect back to the surface are measured and are translated into depths using the rock velocities. In general, more detail can be seen in layered regions by using the reflection method. On the other hand, in areas with few layers, refraction may give sufficient information.

REMOTE SENSING

An additional method for the study of crustal structure uses remote sensing. Remote sensing is the use of electromagnetic radiation from ultraviolet to infrared and of radar wavelengths to survey the surface. Observations are taken from satellites or from aircraft and may be processed by computer. The resulting photographs or computer images are studied to characterize the Earth's surface. One of the primary discoveries of this method has been the existence of long linear features, called lineaments, on the Earth's surface. Some lineaments are only vaguely seen, while others are very obvious. The lineaments often cut across rocks of different ages and types and often truncate major topographic features. Comparison of lineament maps with magnetic and gravity maps shows good correlations in many areas. The lineaments appear to be related to deep basement structures, or zones of weakness. Throughout geologic time, the reactivation of these old zones may have affected the overlying rocks and geologic processes, whatever their types or ages. These basement structures have also served as conduits for ore-forming solutions through time, and thus lineament analysis is used by many in the exploration for natural resources.

ECONOMIC SIGNIFICANCE

Although the upper sections of the continental crust can be studied through standard geologic mapping and borehole analysis, the largest part of the crust is understood through several geophysical techniques. Primary among them are studies based on heat-flow, seismic, gravity, and magnetic observations. These studies have demonstrated that the continental crust is strongly layered, with the lighter and lowest melting components toward the top of the crust. This chemical separation tells scientists that most of the economically valuable elements, such as the precious and base metals as well as the energy-related resources, have already been concentrated upward.

Geophysical studies have also allowed scientists to locate lines of weakness within the mosaic structure of the continents. Basement structures include large vertical fault systems that cut through much of the continental crust. Occasionally, these structures may be reactivated by modern-day geologic conditions, causing seismic activity far from the margins of the major plates. The historic earthquakes in the Mississippi Valley are of this type. In addition, the existence of the deep basement faults may be important in the localization of ore deposits. Economic geologists have found that many large ore districts or metal provinces are related to lineaments apparently caused by basement structures.

Donald F. Palmer

CROSS-REFERENCES

BIBLIOGRAPHY

Barazangi, M., and Larry Brown. *Reflection Seismology: The Continental Crust.* Washington, D.C.: American Geophysical Union, 1986. This book contains a number of advanced articles by specialists in the field of crustal studies. Of great value to the nonspecialist are the many reproductions of seismic reflection profiles and the excellent illustrations showing the nature of the continental crust and crustal structure in selected areas throughout the world.

Bott, M. H. P. *The Interior of the Earth.* 2d ed. London: Edward Arnold, 1982. This textbook gives an excellent nonmathematical introduction to seismic refraction and reflection profiling and to other geophysical methods as they relate to continental structures and composition.

Dawson, J. B., D. A. Carswell, J. Hall, and K. H. Wedepohl. *The Nature of the Lower Continental Crust.* Palo Alto, Calif.: Blackwell Scientific Publications, 1986. This volume is a collection of advanced articles on the character of the lower crust with regard to seismic profiling, electrical surveying, heat-flow data, and geochemical and petrologic studies.

Dennis, J. G. *Structural Geology: An Introduction.* Dubuque, Iowa: Wm. C. Brown, 1987. This basic text in structural geology discusses a wide variety of structures important in crustal studies. Excellent photographs and diagrams show the complexity of structures within the upper crust and in mountain ranges.

Dott, Robert H., Jr., and Donald R. Prothero. *Evolution of the Earth.* 5th ed. New York: McGraw-Hill, 1994. This basic textbook on historical geology is aimed at students of geology. However, it is very readable by anyone with a background in science. Presents an up-to-date account of the Earth's history from the viewpoint of plate tectonics. Includes a glossary.

Hatcher, Robert D., Jr. *Structural Geology: Principles, Concepts, and Problems.* 2d ed. New Jersey: Prentice Hall, 1995. This undergraduate textbook contains a comprehensive overview of stress and strain, as well as a discussion on isostatic equilibrium. Intended for the more advanced reader.

Press, Frank, and Raymond Siever. *Understanding Earth.* 2d edition. New York: W. H. Freeman, 1998. This comprehensive physical geology text covers the formation and development of the Earth. Contains extensive sections on plate tectonics and the development of the Earth's crust. Readable by high school students, as well as by general readers. Includes an index and a glossary of terms.

Robinson, Edwin S., and Cahit Coruh. *Basic Exploration Geophysics.* New York: John Wiley & Sons, 1988. This intermediate-level text describes exploration methods. Good treatments of the geophysical methods by which continental structures have been detailed are provided throughout.

CONTINENTS AND SUBCONTINENTS

Continents are large landmasses with elevations that are considerably higher than that of the surrounding crust. Subcontinents are smaller landmasses that converged over time to form the large continents familiar today. Because of this, continents have a wide variety of terrains and landforms.

PRINCIPAL TERMS

CRATON: a stable, relatively immobile area of the Earth's crust that forms the nucleus of a continental landmass

CRUST: the thin layer of rock covering the surface of the Earth; solid and cool, the crust makes up the continents and floor of the ocean and may be covered with thick layers of sediments

MAFIC ROCKS: rocks that contain large amounts of magnesium and iron, found mainly in the oceanic crust and upper mantle

MANTLE: the region of the Earth between the dense core and the thin crust; the mantle makes up most of the volume of the Earth

OROGENIC BELT: an area where mountain-forming forces have been applied to the crust

PLATE TECTONICS: the process that causes the continents and large unbroken land areas within the oceanic crust, called "plates," to move slowly along with currents of rock in the upper mantle

SEDIMENTS: rocks and soil that have been eroded from their original positions by forces of weather

SUBCONTINENT: an area of land that is less extensive in size and has a smaller variety of terrains than a continent

EARTH'S DIFFERENTIATION

The word "continent" comes from the Latin *continere*, which means "to hold together." Continents are large landmasses composed of lighter rocks that ride on top of the more dense rocks in the mantle, somewhat like a cork in water. This results in areas on the Earth's surface that are higher than sea level, producing dry land. The Earth was not formed with these continents in place; a long, complicated process resulted in the formation of the landmasses familiar today.

When the Earth was first formed approximately 5 billion years ago, it was a molten ball composed of all the elements. The high temperature of the planet was the result of the heat released from several sources: the process that formed the planet, decay of radioactive elements, and intense meteoritic bombardment. This molten state allowed the different elements, and the compounds they form, to differentiate, or separate.

This differentiation process is similar to mixing different kinds of oil with water in a bottle. Shake up the bottle, and the different liquids will be mixed together. Let it sit, and they will begin to form layers. The water is densest and will form a layer on the bottom. Each of the different oils will then form a separate layer above the water, with the layer of the densest oil being on top of the water and the least-dense oil forming the uppermost layer.

Similarly, when the Earth underwent differentiation, the densest material, mainly iron and nickel, sank toward the interior and formed the planetary core. Compounds and elements that were medium in density settled on top of the denser core and formed the layers of the planetary mantle. The least-dense compounds floated to the surface and eventually formed the crust, seawater, and atmosphere. These least-dense compounds consisted principally of the elements silicon, oxygen, aluminum, potassium, sodium, calcium, carbon, nitrogen, hydrogen, and helium, with lesser amounts of other elements.

FORMATION OF THE EARTH'S CRUST

The Earth today has two kinds of crust: the heavier, thinner crust under the oceans and the thicker, lighter continental crust. The oceanic crust was created between 4.2 billion and 4.5 billion years ago, has an average density of 2.9 grams per cubic centimeter, and consists mainly of mafic rocks. Mafic rocks are made of minerals that consist mainly of magnesium and iron. The most common kind of mafic rock in the oceanic crust is basalt, a dark, hard stone. The dark maria on the face of the Moon are the result of basalt that was able to reach the Moon's surface after large meteor impacts. Beneath the crust is the upper mantle, which has a density of approximately 3.3 grams per cubic centimeter and consists of mafic rocks that contain an even larger percentage of magnesium and iron; hence, they are called ultramafic rocks. This layer formed at about the same time as the oceanic crust.

Mixed with these two layers were even lighter materials, mainly compounds of silicon, oxygen, and aluminum, but the high temperature of the planet did not allow these materials to start solidifying until about 4 billion years ago. When this occurred, the first continental rocks began to form, although they were continuously broken up. The oceanic rocks were mainly basalt; the continental rocks were mainly granite with a density of about 2.7 grams per cubic centimeter. The cooling process was slowed by the formation of crystal structures within the oceanic crust and upper mantle that forced out certain rare-earth elements, including the radioactive elements. These elements had to go somewhere and are thus found concentrated in continental rocks. Continental granites contain about ten times as much uranium as the oceanic basalts and about one thousand times as much as the upper mantle rocks. Heat released by the decay of this concentration of radioactive elements helped keep the continental rocks molten longer than the oceanic rocks.

This period in the Earth's history was also characterized by a large amount of volcanic activity; large chains of volcanoes formed archipelagoes of islands. As a result of plate tectonics, these islands moved around on the surface of the Earth and were eventually reabsorbed back into the Earth's interior at subduction zones. Subduction zones occur where an oceanic crustal plate meets continental plate. The denser, heavier oceanic plate is forced under the other plate and into the upper mantle, where it is melted and returned to the surface through volcanic activity. Given enough time, this process will completely recycle the oceanic crust; today none of the original oceanic crust remains.

Sometime about 4 billion years ago, the intense meteoritic bombardment suddenly came to an end; the surface of the planet began to cool more quickly, and more island chains were formed. However, the still-molten, lighter continental rocks sometimes flowed into large cracks called fissures in the volcanic islands and provided them with additional buoyancy. When these islands, riding on the surface of a plate, reached the subduction zones, they were too light to be subducted and were instead scraped off by the other plate. As the other plate continued to move along, it continued to scrape off more of the light islands; over time, a large amount of this lighter material accumulated in front of the plate. Eventually, that plate was subducted by another plate, and the light continental material it had collected was added to any collected by the new plate. In this way, the amount of continental crustal material grew until it was large enough to be a subcontinent.

SUBCONTINENTS AND ACCRETION

A subcontinent is an area of land that is too large to be pushed simply by the movement of oceanic crustal plates. Instead, these large pieces of land ride on top of moving mantle rock deep beneath the surface. Yet while subcontinents are extensive, they are still not large enough to be considered continents. Modern examples of subcontinents include the island of Greenland and the Indian subcontinent. The importance of subcontinents is that they will eventually collide with one another. When they do, they can stick together, forming even larger areas of land and, eventually, continents. This process is called accretion.

When subcontinents or continents collide, the event is something like an automobile crash in very slow motion. The two large bodies are moving and do not stop immediately; they continue to plow into each other, causing the rock to bend, fold, and lift, forming mountain ranges. Areas where this has occurred in the past are called orogenic (mountain) belts; the process of mountain building is known as orogeny. The Himalaya are an example of the result of this process. The Indian subcontinent took millions of years to move from southeast Africa to its current position on the southern side of Asia. When it collided with Asia, the force was enough to raise a giant plateau, with the towering Himalaya on top. Large areas between the orogenic belts are called cratons. The American Midwest between the Appalachian and Rocky Mountains is an example of a craton.

This movement of the landmasses continues even after the formation of continents. The continents continue to ride on top of currents of rock in the mantle, slowly making their way across the Earth's surface. Currently, continents are moving at a rate of about 5 to 10 centimeters per year, but this rate was faster in the past. Over millions of years, the continents have been able to move great distances, and the shape and distribution of continents in the past did not resemble the global features of today. At times, the continents were together to form supercontinents that lasted for millions of years before breaking apart.

VOLCANISM AND EROSION

In addition to accretion, two other major forces at work on continents are volcanism and erosion. Volcanism is a result of plate tectonics, and most volcanism thus occurs along the edges of the landmasses where subduction occurs or where fault lines are found. Volcanism recycles material that has been subducted and adds to the mass of the continents. Erosion, meanwhile, wears down landforms. Rain, ice, heat, wind, and flowing water all work to break apart the rocks and slowly wash the surface material away. Some of these sediments are washed to sea, while others collect in low-lying areas on the land. These sedimentary deposits can be several kilometers thick and eventually turn into sedimentary rock.

STUDY OF CONTINENTS

Continents are vast in size and complexity. Likewise, the study of these landmasses is also vast and complex. Much of the work in learning about continents is hampered by the fact that scientists can easily sample only the thinnest top layer of the crust; moreover, much of the evidence of past activity is destroyed through erosion. As a result, the study of the continents is a slow process involving many scientists using a large variety of techniques and instruments.

The most basic method of studying the continents involves studying the layers of rocks. Sometimes these layers can be seen from the surface, and other times scientists must use drills to remove core samples. Sometimes these layers lie flat, while other times they are at all angles. By studying these layers, it is possible to learn what they are made of, how they were made, and even when they were made. Eventually, it becomes possible to conclude that various layers are related and sometimes even constitute the same layer. For example, it is possible that a layer of rock found in North Dakota is identical to a layer of rock found in Nebraska, hundreds of kilometers away. In this way, geologists are able to build maps showing where these layers of rock can be found.

Maps like this reveal much about the past of the land. If a type of rock found in the desert is made of material found only on the bottom of swamps, geologists can deduce that the desert was once a swamp and that the climate in the area was once different. It might even be possible to track the change from swamp to desert by examining the different layers, although sometimes the layers are destroyed through erosion. If the ages of the different layers are known, it is possible to build a story line showing how the swamp changed to desert over a period of time. Also, the angle of the layers tells about what happened to the land. If the layer is horizontal, then it has probably been undisturbed since it formed. If it is tilted or folded over on itself, then some kind of force was applied to the rocks. Also, if a layer of rock is found on one continent and also found on another continent, then it can be concluded that the two continents were together when the layer was formed.

Rocks themselves also provide clues. What are the rocks made of? If they are sedimentary rocks, then the material in them existed in some other rocks before. Where were those rocks? If the rocks are basalts, it is possible to conclude that there was volcanic activity in the area at one time. The chemical structure can tell much about the temperatures and pressures to which rocks have been exposed over the years.

Instruments on spacecraft can make measurements over very large regions. This not only speeds up the process but also provides new views and evidence not previously possible. This information can be included in computer models in attempts to determine the forces at work in the formation of continents. In this way, clues can be found that were previously unsuspected. Once something is suggested by a computer model, scientists can investigate it to find scientific evidence to support or refute it. Through these and other processes, geologists gradually learn more and more about the Earth and its history.

Christopher Keating

CROSS-REFERENCES

Alps, 805; Appalachians, 819; Archean Eon, 1087; Basaltic Rocks, 1274; Batholiths, 1280; Continental Crust, 560; Continental Drift, 565; Continental Growth, 573; Continental Rift Zones, 579; Continental Shelf and Slope, 584; Continental Structures, 590; Displaced Terranes, 615; Earth's Crust, 14; Earth's Differentiation, 20; Earth's Lithosphere, 26; Earth's Structure, 37; Elemental Distribution, 379; Evolution of Earth's Composition, 386; Geoarchaeology, 1028; Gondwanaland and Laurasia, 599; Granitic Rocks, 1292; Himalaya, 836; Island Arcs, 712; Lithospheric Plates, 55; Plate Tectonics, 86; Proterozoic Eon, 1148; Rocky Mountains, 846; San Andreas Fault, 238; Sierra Nevada, 852; Subduction and Orogeny, 92; Supercontinent Cycles, 604; Transverse Ranges, 858; Ultramafic Rocks, 1360.

BIBLIOGRAPHY

Dott, Robert H., Jr., and Donald R. Prothero. *Evolution of the Earth.* 5th ed. New York: McGraw-Hill, 1994. Chapters 6 through 8 provide a good account of the early history of the planet and the formation of the crust. Chapters 10 and 11 discuss the formation of cratons and orogenic belts. A well-written, well-illustrated text suitable for college and advanced high school readers.

Moores, Eldridge, ed. *Shaping the Earth: Tectonics of Continents and Oceans.* New York: W. H. Freeman, 1990. This collection of readings from *Scientific American* magazine presents a variety of well-written articles by experts in their respective fields. Individual articles are devoted to plate tectonics, mountain forming, and crustal formation, among other topics. Good illustrations and an index covering all articles is provided. Suitable for college and advanced high school readers.

Plummer, Charles C., David McGeary, and Diane H. Carlson. *Physical Geology.* Boston: McGraw-Hill, 1999. This is a straightforward, easy-to-read introduction to geology intended for those with little or no science background. Discusses the development of continents and subcontinents through time. Includes many excellent illustrations, as well as a CD-ROM.

Stanley, Steven M. *Earth and Life Through Time.* New York: W. H. Freeman, 1986. Chapter 7 provides a good discussion of plate tectonics, while chapter 8 covers mountain building. The formation of the Earth, the crust, and the continents is discussed in chapters 9 through 11. Well written, well illustrated, well indexed, with appendices that provide additional details about specific topics. Suitable for college and advanced high school students.

Taylor, S. Ross, and Scott M. McLennan. "The Evolution of Continental Crust." *Scientific American,* January, 1996, p. 274. A comprehensive discussion of the origin of the continental crust and the evolution of the continents. Suitable for high school readers.

Weiner, Jonathan. *Planet Earth.* Toronto: Bantam Books, 1986. This companion volume to the Public Broadcasting Service television series *Planet Earth* covers many aspects of the Earth. Well illustrated; suitable for high school readers.

GONDWANALAND AND LAURASIA

Earth scientists theorize that the present-day continents were produced from the division of a supercontinent, called Pangaea, into two gigantic landmasses, Gondwanaland and Laurasia, which continued to fragment. The continents moved into their present locations in accord with a phenomenon known as continental drift, which scientists believe is still taking place.

PRINCIPAL TERMS

BASALT: a hard, dense volcanic rock

CARBONIFEROUS PERIOD: the fifth of the six periods in the Paleozoic era; it preceded the Permian period

CRATON: the crystalline portion of a continent; it may have a sedimentary rock veneer

FATHOM: a unit of length equal to 1.8 meters, used principally in the measurement of marine depths

GEOMORPHOLOGY: a term meaning the study of the origin of landscapes

ISOBATH: the contour lines of continental slopes

PALEOZOIC ERA: the era immediately before the Mesozoic era; it included the Cambrian, Ordovician, Silurian, Devonian, Carboniferous, and Permian periods

PLACER: a deposit of sand or gravel containing eroded particles of valuable minerals

TROUGH: a long, narrow depression, as between waves or ridges

FORMATION OF THE CONTINENTS

The existence of an ancient supercontinent was first postulated by Austrian geologist Edward Suess in 1885. He called this great continent Gondwanaland, after a region in India inhabited by the Gonds, an aboriginal tribe. Whereas Suess proposed that much of Gondwanaland had sunk, in 1929, German geophysicist Alfred Lothar Wegener suggested that a supercontinent that he called Pangaea had broken up and that the individual continents had drifted to their present positions. In 1937, Alexander Du Toit envisioned two primordial continents: Laurasia in the north and Gondwanaland in the south; subsequently, both parts broke again to form the present continents. Edward Bullard, J. E. Everett, and A. G. Smith's computerized assembly of the ancient continents in 1965 seems to verify Du Toit's continental arrangement.

The standard explanation of the formation of the continents begins 300 million years ago, at the end of the Carboniferous and start of the Permian geological periods, with the formation of Pangaea, a V-shaped supercontinent. The northern extension of Pangaea, called Laurasia, straddled the equator and included North America, Europe, and Asia; the southern arm of Pangaea, called Gondwanaland and comprising South America, Africa, India, Australia, and Antarctica, was positioned so that its northern parts were in the tropical latitudes and its southern parts were beneath the polar ice cap. Although the Tethys Sea lay between Laurasia and Gondwanaland, they were linked from northwest Africa to North America and southern Europe.

The total area of Pangaea, when measured down to the 1,000-fathom isobath, seems to have been 200,000 square kilometers, or roughly 40 percent of the Earth's surface. When the future continents were still part of Pangaea, they were to the south and east of their present locations. If New York had been in existence at the time, it would have been on the equator and at longitude 10 degrees east instead of 74 degrees west. Spain would have been near its present longitude, but it, too, would have been on the equator. Japan would have been in the Arctic, and India and Australia would have bordered the Antarctic.

This loose grouping of partly linked continents remained constant for the next 150 million years, although a few changes were slowly occurring. During this period, Gondwanaland gradually drifted northward. In addition, Gondwanaland, which had formed a huge basinlike area, was

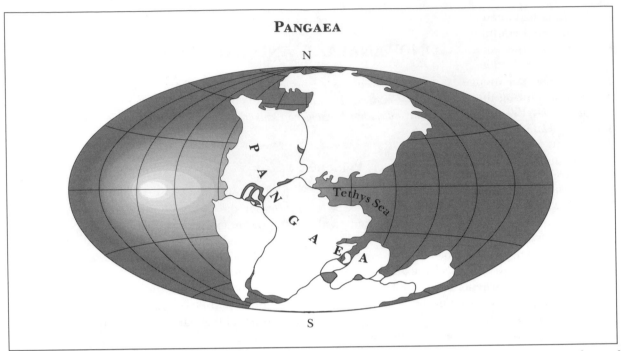

PANGAEA

N

Tethys Sea

S

The standard explanation of the formation of the continents begins 300 million years ago, at the end of the Carboniferous and start of the Permian geological periods, with the formation of Pangaea, a V-shaped supercontinent.

slowly being eroded, and the debris was deposited into the basin itself or into the long troughs that surrounded Pangaea. Toward the end of this period, this enormous basin was transformed into a collection of separate basins.

This scene was violently disrupted approximately 160 million years ago when gigantic floods of basaltic lava spread out on all of the southern continents except for South America, where lava flows did not occur until 40 million years later. As a result of the lava flow, the cracks from which the lava was released sank, forming basins. When the water in these basins evaporated, it left behind salt deposits that now border most of the southern continents.

The predecessors of the present deep oceans were formed between India and Somaliland as the areas sank even lower. Similar seas spread along the coast of India some 10 million years later, separating it from Australia. The seas reached southwest Africa roughly 120 million years ago, the Congo 110 million years ago, and Nigeria 105 million years ago. While the seas were spreading, the Benue Trough, which cuts across the bulge of Africa from Nigeria through Algeria, was slowly be-

ing filled. Instead of driving a great distance down the trough, however, the seas moved westward, where they merged with marginal seas that had been moving eastward between the bulge of Africa and northern Brazil. Thus, when the seas finally met, 92 million years ago, Africa became separated from South America. The Indian Ocean opened approximately 160 million years ago, and the South Atlantic Ocean opened roughly 120 million years ago, but neither ocean really began to widen until about 100 million years ago.

BREAKUP OF GONDWANALAND

Between 100 and 80 million years ago, Gondwanaland began to break up. South America rotated away from Africa before it drifted westward to its present position. At approximately the same time, India rotated away from Africa, moved northward, and collided with Asia. The Himalayan Mountains are a direct result of this collision. Before finally separating into their present positions during the past 50 to 60 million years, Australia and Antarctica drifted away from Africa. After rotating slightly, Africa moved northward to encroach upon Europe.

Laurasia had been lying lower than Gondwanaland at the beginning of this 300-million-year period, with the exception of the Appalachian-Caledonian mountains, the Russian Urals, and the eastern Siberian mountains, which all gradually wore down during the next 50 million years. During the next 150 million years, the seas that covered most of North America slowly receded to the south and west. Shallow seas also covered Europe west of the Urals. These seas, which spread northward from the Tethys, shifted their location periodically during the next 200 million years. A large portion of northeast Asia was covered with marine sedimentary troughs. Some of these troughs cut China off from the rest of Asia at the beginning of this period. Shallow seas spread out from these troughs over parts of China and western Siberia. The first strong geological activity occurred in Siberia 200 million years ago, when flood basalts up to 2.5 kilometers thick spread out over 500,000 square kilometers.

The main framework of fractures that would eventually form the North Atlantic was already in place 300 million years ago. Actual separation, however, began farther south. Even though the northern continents were linked to Gondwanaland at the beginning of this period, North America separated from Africa more than 200 million years ago, after the eruption of some volcanoes in eastern North America and Morocco. The proto-Central Atlantic was formed as a result of this rupture. Although Europe was still connected to North America after this movement occurred, a shallow sea was formed in which a layer of sediments between 3 and 4 kilometers thick was formed; traces of this sediment can still be found along the Atlantic continental shelf of North America. The Atlantic Ocean widened to one-quarter of its present width and connected to the Labrador Sea some 120 million years ago, separating Greenland and Canada. While the Atlantic continued to expand some 70 to 80 million years ago, the opening began between Europe and Greenland after the eruption of flood basalts.

STUDY OF CONTINENT FORMATION

Scientists have employed a wide variety of methods to study the formation of the continents. Glaciation has long been used both to determine the original fracturing of the great landmasses and to estimate the rate of their separation. The close grouping of the glaciated areas provides some of the most convincing evidence for continental drift. Glacial scratches on rocks found in Gondwanaland point to the movement of ice from areas that are now submerged under the ocean. In addition, some materials deposited in the glacial drift, such as Brazilian diamonds, are completely foreign in the lands where they now repose and, therefore, provide proof for continental drift.

The study of terrestrial vertebrates also provides clues to the existence of Pangaea and Gondwanaland. After the land bridge theory fell into disrepute, contiguity of the continents early in the Earth's history seemed to be the only way to explain the fact that fauna is comparatively homogeneous throughout the world. The most convincing evidence for continental drift from terrestrial vertebrates is provided by an early Permian reptile called *Mesosaurus*, whose remains are found only in South America and South Africa. Since this reptile was equipped for swimming only in shallow

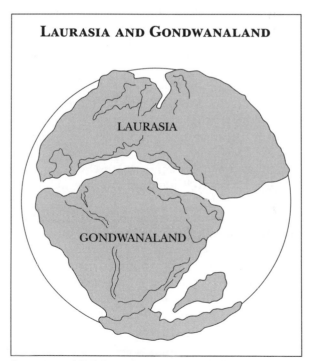

LAURASIA AND GONDWANALAND

LAURASIA

GONDWANALAND

Laurasia and Gondwanaland were formed after the division of the supercontinent Pangaea. North America separated from Africa more than 200 million years ago, after the eruption of some volcanoes in eastern North America and Morocco. The proto-Central Atlantic was formed as a result of this rupture.

fresh water, it probably could not have swum the distance between the two continents.

The evolution of plants also supplies evidence for the existence of a supercontinent. The presence of araucarian pines in South America, the Falklands, Australia, and the South Pacific has led scientists to the inference of former land connections. The distribution of the *Glossopteris* land plants in the Lower Gondwana Formations is larger than that of the glacial deposits. Evidently, the cold climate stimulated the growth of this unique plant. These plants could not have developed independently of one another in widely separated continents but could have thrived in a supercontinent covered by enormous ice sheets.

Modern technology has provided scientists with more advanced methods of exploring the possibility of a supercontinent. Gathering paleomagnetic data is one such method. Magnetic particles in the seafloor rocks recorded the direction of the Earth's magnetic field when the rock hardened. As the ocean floor formed, the direction of the field reversed itself from time to time. By paleomagnetically determining the pole positions for sedimentary and igneous rock in different continents, scientists have been able to determine the relative positions of the continents for any period of time. Some of the data that they have acquired in this fashion indicate the existence of Pangaea.

Examination of Precambrian cratons also provides invaluable information regarding continental drift. Drift can be demonstrated by matching pieces of cratons across facing continents. The cratons between Guyana and West Africa, for example, fit very well together and are well documented.

The complicated process of reassembling the continents has been greatly facilitated by computers. In 1965, following a 500-fathom isobath, Bullard, Everett, and Smith used a computer to fit the continents together into their original configuration. Even though computerized models have brought an air of mathematical precision to the task of reconstructing Pangaea, there is still considerable disagreement about certain features. For example, did Madagascar lie against east or southeast Africa when they were part of Pangaea?

Significance

The study of ancient landscapes is of great value to the economy. The same techniques that geomorphologists devised to comprehend fossil landforms and deposits can also be brought to bear to locate ore deposits. Geomorphologists have also applied their methods to detect placers and to discover petroleum. Through the use of structure contour maps, the search for an "oil pool" almost always involves geomorphology applied to problems of fluid entrapment.

Knowledge of ancient landscapes also aids scientists in predicting the occurrence of natural catastrophes. Scientists have mapped out a worldwide system of ocean trenches and ocean ridges, which were probably made when the oceans were formed and the continents moved to their present locations. They have observed that many deep earthquakes occur beneath oceanic trenches and that volcanic activity is concentrated along the ocean ridges.

Finally, scientists have used the theory of continental drift to predict what the Earth's surface will look like millions of years from now. They predict that the California coast may tear away from the mainland and drift north toward Alaska, that Africa and South America may move even farther apart, and that Australia may collide with Asia.

Alan Brown

Cross-References

Continental Crust, 560; Continental Drift, 565; Continental Growth, 573; Continental Rift Zones, 579; Continental Shelf and Slope, 584; Continental Structures, 590; Continents and Subcontinents, 595; Fold Belts, 620; Ocean Ridge System, 670; Plate Margins, 73; Plate Motions, 80; Plate Tectonics, 86; Supercontinent Cycles, 604.

Bibliography

Dott, Robert H., Jr., and Donald R. Prothero. *Evolution of the Earth.* 5th ed. New York: McGraw-Hill, 1994. This basic textbook on historical geology is aimed at students of geology. However, it is very readable by anyone with a background in science. Presents an up-to-date account of the Earth's history from the viewpoint of plate tectonics. Includes a glossary.

Garner, H. F. *The Origin of Landscapes.* New York: Oxford University Press, 1974. The chapter entitled "Ancient Landforms and Landscapes" contains only a brief explanation of the Gondwanaland theory but provides an excellent chart detailing the geological evidence for the existence of Gondwanaland. The textbook also contains a very fine glossary and index. It is intended for college students.

Sullivan, Walter. *Continents and Motion: The New Earth Debate.* New York: McGraw-Hill, 1974. This book provides an excellent history of the development of the theories regarding the origin of the continents and continental drift. Written for the general reader.

Wegener, Alfred. *The Origins of Continents and Oceans.* Translated from the 4th rev. German ed. by John Biram. Mineola, N.Y.: Dover, 1966. Written by the originator of the continental drift theory, this book is highly technical but invaluable for the Earth scientist who is interested in the basis for most of the current theories regarding Gondwanaland.

White, M. E. *The Greening of Gondwana.* Sydney: Reed Books, 1996. This very well illustrated book has color and black-and-white photographs of Australian fossil plants, with an emphasis on the Gondwana flora.

Windley, Brian F. *The Evolving Continent.* 2d ed. New York: John Wiley & Sons, 1995. The chapter entitled "Pangaea: Late Carboniferous-Early Jurassic" examines the various types of data from which Earth scientists have deduced the existence of Pangaea. Its technical language makes it suitable primarily for college students and Earth scientists.

SUPERCONTINENT CYCLES

Supercontinent cycles, which recur over periods of 400 million to 440 million years, are helpful in understanding the distribution of certain natural resources, fluctuations throughout geologic time of sea level and climates, the process of mountain building, and the evolution of life.

PRINCIPAL TERMS

CONTINENTAL DRIFT: the theory that continental fragmentation and displacement caused the creation of new ocean basins and the formation of mountain ranges

CONVERGENCE: the process that occurs during the second half of a supercontinent cycle, whereby crustal plates collide and intervening oceans disappear as a result of plate subduction

DIVERGENCE: the process of fracturing and dissecting a supercontinent, thereby creating new oceanic rock; divergence represents the initial half of the supercontinent cycle

OPHIOLITE SUITE: a unique vertical sequence of peridotite rock overlain by gabbro, basalt, and oceanic sediments representative of ancient seafloor material

PLATE TECTONICS: a theory describing the Earth's surface as composed of rigid plates continually in motion over the interior, causing earthquakes, mountain building, and volcanism

SEAFLOOR SPREADING: the continual creation of new seafloor bedrock along mid-ocean ridges through the process of ascending thermal currents

SUPERCONTINENT: a single vast continent formed by the collision and amalgamation of crustal plates

WILSON CYCLE: the creation and destruction of an ocean basin through the process of seafloor spreading and subduction of existing ocean basins

CONTINENTAL DRIFT THEORY

One of the most persistent questions relating to the historical development of the Earth concerns the processes whereby extensive mountain chains, such as the Himalaya, Appalachian, and Ural ranges, have been formed. During the nineteenth century, the planet-contraction hypothesis, supported by the doctrine of permanence of continents and ocean basins, suggested that such linear features of the Earth's crust were comparable to the wrinkles within the skin of a dried apple. After the discovery of radioactivity in 1896, studies suggested heat formed by radioactive decay of rock minerals approximately equaled heat lost to the atmosphere by the gradual cooling of the Earth from its supposed original liquid stage. This balance of heat gain and loss did not support a shrinking-Earth concept. With the additional knowledge that linear mountain ranges had formed during different stages of geologic history, rather than simultaneously, the contraction hypothesis gradually lost favor.

Between the publication of his 1915 book *Die Entstehung der Kontinente und Ozeane* (*The Origin of Continents and Oceans*, 1924) and his death during an expedition to the Greenland icecap in 1930, Alfred L. Wegener, a German meteorologist and geologist, gained international repute as the father and chief advocate of the theory of continental drift. Wegener postulated that mountain ranges were created by the collision of large blocks of continental crust moving through oceanic crust, following fragmentation and dispersion of the vast supercontinent he called Pangaea (Greek for "all lands"), surrounded approximately 200 million years ago by the universal ocean Panthalassa (from *thalassa*, Greek for sea). An intriguing set of evidence supports this theory, including the unusual degree of geometric fit of present-day continents (especially South America with Africa) and, through the reassembly of Pangaea, the reconstruction of truncated salt, fossil-reef, glacial-deposit, and mountain-range trends. While many scientists at the time became "pro-drift," many others questioned what possible mechanism could displace solid continental crust through equally solid oceanic crust. Various Earth forces, ranging from centrifugal to tidal to rotational axis wobble, were investigated and rejected as inadequate by physicists and mathematicians. By the 1930's, continental drift as a theory was no longer considered viable and was increasingly mentioned only within the context of the history of science.

EXPLORING THE OCEAN FLOOR

Following World War II, a new era of Earth investigation began, focused primarily on the ocean basins. Employing newly developed military technology, including methods of water-depth sounding (fathometry) and magnetic-body detection, surplus military aircraft and surface-vessel equipment began to collect a wide array of information. By the International Geophysical Year of 1957-1958, ocean-floor depth-profile analyses confirmed the existence of a previously unknown, 65,000-kilometer-long global submarine mountain range system traversing the Atlantic, Pacific, Arctic, and Indian Oceans. This feature defied immediate explanation.

During the same period, ocean-evaluation programs sponsored by both private and U.S. government interests began the task of measuring the magnetism of the ocean floor. By the mid-1950's, sufficient data had been collected off the West Coast of the United States to reveal a repetitive north-south pattern representing alternate zones of above-average and below-average magnetism. Soon similar patterns were shown to exist in the Atlantic and Indian Oceans.

The discovery of these unexplainable phenomena prompted the collection of any form of additional data that would help explain the existence of ocean floors dissected by a universal mountain range and masked by symmetrical magnetic patterns. New oceanographic programs gathered information on bedrock temperature, radiometric age, and ocean-sediment thickness and studied the worldwide distribution of earthquakes and volcanic eruptions.

PLATE TECTONIC THEORY

By the early 1960's, broad-based analyses of these various forms of oceanographic data were being conducted in Canada, the United Kingdom, and the United States. Gradually, consensus began to form that perhaps at least some of the ideas of Alfred Wegener were worthy of reconsideration. Paramount among these resurrected ideas was that of the assembly of the supercontinent Pangaea. Rather than postulating continental crust as floating through oceanic crust, the revised continental drift theory, termed "plate tectonics," envisioned the outer layer of the Earth as divided into a series of major plates, each composed of both continental and oceanic bedrock.

As examples, the North American plate is made up of the continent of North America (including Greenland), the western half of the north Atlantic Ocean, and eastern Siberia, while the Indian-Australian plate is composed of the continent of Australia, the country of India, and portions of the Pacific and Indian Oceans. Major plate displacement was considered possible because of the movement of global thermal-convection cells, which form within the mantle of the Earth and rise toward the surface until they are blocked by the presence of a supercontinent. The blockage of further transmission of heat, caused by the insulating nature of continental rock, divides the convection current into lateral, horizontally directed segments, which gradually dome by thermal expansion and then dissect the supercontinent.

As the new subcontinent begins to diverge, the separating void fills with high-density gabbro and basalt-type rock, which forms the floor of a newly developed ocean. Because the Earth is neither expanding nor contracting in size, continuing divergence cannot proceed indefinitely without experiencing resistance caused by the convergence of antipodal plates. The effect of plate convergence depends on whether such plate margins are oceanic (basaltic) or continental (granitic) in nature. Where margins are oceanic, the more dense margin will subduct, or plunge under, the less dense margin. Where margins are both continental, subduction is unlikely, and the result is massive folding and faulting (earthquaking). Finally, where a continental margin converges with an oceanic margin, the latter, being more dense, will subduct beneath the former. In all three possible convergence cases, crust shortening is accomplished. Rock volume harmony results, as the formation of new oceanic crust through plate divergence is matched by the destruction of older crust through the process of subduction.

Processes of divergence and convergence are believed to have been ongoing throughout a large portion of geologic time and to continue today. Divergence, accompanied by new-ocean development in region A, continues simultaneously with convergence in region B, resulting in the gradual destruction of region A ocean by way of subduction. Eventually, region A ocean will cease to exist, and the cycle of ocean birth and death will be complete. This sequence of events, during which it is estimated 2.6 square kilometers of ocean floor rock is created and destroyed each year, is termed a "supercontinent cycle" (also known as a Wilson cycle, after the Canadian geologist J. Tuzo Wilson, an early advocate of the plate-tectonics theory). Conversely, the dispersion and amalgamation of continental crustal masses, as opposed to oceanic masses, constitutes the principal phases of what has been termed the Pangaean cycle. The operation of plate tectonics continuously creates and recycles ocean basins, while continental regions increase in age geologically even as they are agglomerated and dissected by ongoing seafloor spreading.

FUTURE SUPERCONTINENT CYCLES

A maxim of geology states that the validity of any hypothesis or theory is determined by the degree to which that concept can be examined through the analyses of extant geology. Where, then, might there be modern-day examples of a supercontinent cycle in its various stages of tectonic development?

The Great Basin of the western United States has been portrayed as a model of very early continental rifting that may, in the future, separate the North American plate from a newly constituted Pacific plate. The approximately one hundred block-faulted mountains composing this geographic terrane are caused by the same extensional forces responsible for the early continental-rift stage of dissection of a supercontinent.

Similar forces have formed the more structurally advanced rift systems of East Africa, the most illustrative of which are those broad-basin and steeply dipping escarpment topographies of Tanzania, Kenya, and Ethiopia, which continue to yield a record of mammalian and hominid evolution. The fresh waters that partially cover these rift valleys, such as Lake Tanganyika, are evidence of a late stage of continental rifting. To the north, the central valley or rift of the Red Sea is filled with salt water characteristic of an incipient ocean developing during an early stage of oceanic rifting separating the Arabian from the African plate.

Finally, the Atlantic Ocean is a mature example of oceanic rifting representative of the midpoint of a supercontinent cycle. According to the concepts of plate tectonics, the Atlantic Ocean has been created through some 200 million years of seafloor spreading driven by extensional rifting of Pangaea. This stage, representing the first half of a supercontinent cycle, will in theory be followed by assembly of the world's continents over the next 200 million years into a new supercontinent. This assembly may have already begun, as India, an island subcontinent up to approximately 35 million years ago, has since that time been colliding with the Eurasian plate, resulting in the formation of the Himalaya Mountains.

EVIDENCE OF PAST SUPERCONTINENT CYCLES

Following the general acceptance of the plate tectonic theory in the 1960's, many ideas were advanced regarding the existence and nature of pre-Pangaea supercontinents. Since the supercontinent cycle not only creates but also destroys oceans, pre-Pangaea supercontinents must be reconstructed from continental geologic data that become, with increasing geologic age, more difficult to interpret. Certain criteria have been developed in the attempt to discern a pattern of supercontinent cycling since the earliest periods of geologic time.

The convergence of continents will largely eradicate any intervening ocean. Such loss by subduction is seldom complete, as attested by the presence of remnants of preexisting ocean floor rock contained within the deformed (suture) zone caused by plate collision. Basalt, gabbro, and olivine-rich rocks, termed an "ophiolite suite," are scraped off, or "obducted" from, the subducting ocean floor and thus preserved in the developing mountain belt. Obducted ophiolitic rock from the former Tethys Sea, which lapped onto the eastern shores of Pangaea, is present in the Alps and the Himalayas, while ophiolites within the Ural Mountains of central Asia are evidence of an ocean that was destroyed by the collision of Baltica and Siberia, continental masses that preceded the formation of Pangaea. The presence, location, and age-dating of ophiolite suites are helpful in the identification of former supercontinents.

Paleontological and high-pressure rock evidence collected from the Appalachian Mountains—created by the convergence of proto-Africa (the earliest form of Africa) with proto-North America—suggests the existence approximately 500 million years ago of a proto-Atlantic Ocean. The existence of this body of salt water, the Iapetus Ocean, attests the existence of an associated supercontinent.

The presence of exotic terranes forming the collisional edge of former crustal plates in Japan, New Zealand, and the Apennines of northern Italy is further evidence of former supercontinents and their cycles. These terranes, packages of rock possessing similar mineral and fossil character, are accreted to enlarging supercontinents by the same obduction process as ophiolites.

TWO SCHOOLS

There are two major contemporary schools of thought regarding supercontinent cycles. The oldest Earth specimen found to date is the 3.96-billion-year-old metamorphic continental rock forming a portion of the Northwest Territories of Canada. This and slightly younger rock terranes from Greenland, Antarctica, and Australia may have combined to form the first amalgamated continental masses. John Rogers, a professor of geology at the University of North Carolina, proposed the existence 3 billion years ago of an early subcontinent he calls "Ur" (from the German for "original"). Approximately 500 million years later, a second subcontinent, Arctica (predecessor to Canada, Greenland, and eastern Russia) formed, followed another 500 million years later by Baltica (proto-Western Europe) and Atlantica (eastern South America and western Africa). Then, 1.5 billion years ago, plate tectonic forces formed the sub-supercontinent Nena (from the Russian for "motherland") through the merging of Baltica and Arctica. This lengthy chain of events culminated 1 billion years ago in the formation of the first supercontinent, Rodinia (also known as proto-Pangaea), as a result of the joining of Ur,

Atlantica, and Nena. After a period of stability lasting some 300 million years, Rodinia subdivided, forming numerous proto-continents and Iapetus, the proto-Atlantic. Finally, the reassembly of these proto-continents and subsequent subduction of Iapetus brought about the creation of the second supercontinent, Pangaea, about 250 million to 300 million years ago.

A second school of thought differs principally in the suggestion that plate tectonic processes did not begin until some 2.5 billion years ago, on the occasion of the development of the first distinct oceanic and continental crust. Prior to that time, the very high temperature of the Earth and the relative thinness of primordial crust forestalled the onset of supercontinent-cycle processes such as divergence, convergence, and subduction. While differing in detail, both schools of thought generally recognize Rodinia and Pangaea as supercontinents.

The length of a typical supercontinent cycle is variously estimated at from 400 million to 440 million years. Such a cycle would constitute three phases. Once formed, a supercontinent would exist for 100 million to 120 million years before the accumulation of thermal convection heat initiated crustal dissection. During the second phase, lasting from 150 million to 160 million years, maximum dispersal of subcontinents would take place, with resultant development of new oceans. Finally, subduction would gradually destroy the intervening oceans over a period of 150 million to 160 million years, creating a new supercontinent and terminating the cycle.

Albert B. Dickas

CROSS-REFERENCES

BIBLIOGRAPHY

Dalziel, Ian W. D. "Earth Before Pangaea." *Scientific American* 272 (January, 1995): 58-63. An easy-to-read account of the nomadic wanderings of the North American plate prior to the assembly of Pangaea.

Davidson, J. P., W. E. Reed, and P. M. Davis. *Exploring Earth: An Introduction to Physical Geology.* Upper Saddle River, N.J.: Prentice-Hall, 1997. Chapters 6 through 11 are an excellent undergraduate-level introduction to the natural consequences of plate tectonics and their role in the supercontinent cycle.

Nance, R. Damian, T. R. Worsley, and J. B. Moody. "The Supercontinent Cycle." *Scientific American* 259 (July, 1988): 72-79. Combining specialties of tectonics, oceanography, and geochemistry, the authors discuss in a clear manner several supercontinent cycles and their effects on climate, evolution, and geologic changes.

Nicolas, A. *The Mid-Ocean Ridges.* New York: Springer-Verlag, 1995. A review presenting the European view of the construction and destruction of ocean basins. Written at the knowledgeable adult level.

Sullivan, W. *Continents in Motion.* New York: American Institute of Physics, 1991. The author, a science editor for *The New York Times*, presents a recommended compilation of the history and dynamics of plate tectonics. Suitable for the educated layperson.

2
CRUSTAL AND SURFACE FEATURES

AFRICAN RIFT VALLEY SYSTEM

The African rift valley system is characterized by its elongated basins, which cut across a region dotted with domes that extends from South Africa to the Red Sea. Most lakes of the region are located in the rift basin.

PRINCIPAL TERMS

ASTHENOSPHERE: the layer of the Earth that lies beneath the lithosphere and is partly composed of melt

CARBONATITE: an igneous rock with abundant carbon in its makeup

CINDER CONE: a cone-shaped mound made of volcanic granules

FAULT: a fracture in a rock associated with rock movement or sliding

GRABEN: a linear topographic depression caused by subsidence along faults

HALF-GRABEN: a structural element by which a rift system is formed, consisting of an arcuate ridge that bounds a depression and is formed by normal faulting

HYDROTHERMAL FLUID: a natural hot steam that seeps through the ground

LITHOSPHERE: the top rock layer of the Earth, ranging from 70 kilometers in depth in the African continent to 21 kilometers in the African rift

MAGMA: a melt from which igneous rocks are formed

NORMAL FAULT: a fault in which the rock block on top of an inclined fracture surface, also known as a fault plane, slides downward

PYROCLASTICS: fragmentary igneous rocks that are formed by the forceful ejection of volcanic materials into the atmosphere

RIFT PROPAGATION: the lateral movement of a rifting process that leads to the prying open of a section of the lithosphere, accompanied by the formation of igneous rocks

---

RIFT ZONE CHARACTERISTICS

A continental rift is a linear topographic depression that may develop into an ocean as the bounding regions drift into two separate continental fragments. The African rift valley system is one of several continental rifts; others include the Rio Grande, the Baikal, and the Rhine Graben. It is a long system that extends from South Africa to the Red Sea coast.

The African rifts are marked by depressions that cut across domes, such as the East African and the Afro-Arabian domes. The East African dome encompasses parts of Tanzania, Uganda, and Kenya and is dissected by two rift branches. The eastern branch is discontinuously traceable to the Main Ethiopian Rift, which is one of three rifts of the Afro-Arabian dome. The other two rifts are the Red Sea and the Gulf of Aden. The three rift arms of the Afro-Arabian dome meet in a triangular depression, the Afar. The central structures of the three rifts overlap in the Serdo Block, a roughly square area of about 90 kilometers per side.

The rift basin margins may be distinct and marked by cliffs in some places. At Dalol, about 100 kilometers north of the Serdo Block, the rift floor is about 3,000 meters below the rift rim, and there is a 5,000-meter-thick layer of salt that was deposited in the last 4 million years. Since salt forms at sea level, the Afar floor must have been slowly sinking (subsiding) about 5 kilometers in 4 million years. Clearly, the rift floors have formed by subsidence relative to the rim.

Subsidence occurs as the rock block on top of an inclined fracture surface slides downward, a movement that geologists call normal faulting. Rift basins are formed by a series of normal fault movements that generally produce grabens and half-grabens. A half-graben consists of an arcuate ridge that bounds a depression and is formed by normal faulting; it is the basic building unit of continental rifts. A typical half-graben is about 100 kilometers long and less than 4 kilometers wide. Half-grabens might be arranged facing or opposite or staggered in the rift basins. The thickness of the rock layer (lithosphere) at the rift basins ranges from 21 to 30 kilometers, less than one-half its thickness elsewhere on the continent.

RIFT VOLCANOES

Geologists believe that the lithosphere lies above a partially molten layer, the asthenosphere. The melt, also known as magma, rises from the asthenosphere through fractures and flows to form layered basalt, or it oozes to form volcanic domes and cones, or it is ejected explosively to spray volcanic ash and fragmental rocks called pyroclastics. Some volcanic rocks known as carbonatites are formed from magmas that originate at great depths and that contain abundant carbon dioxide. Also, from deep within the mantle, sodium- and potassium-rich magma forms volcanic rocks termed alkaline igneous rocks. However, the dominant igneous rocks of the rift basins are tholeiites, which are comparatively rich in magnesium and the magma of which originated at comparatively shallow depths. Some of the rift volcanoes issue substantial quantities of volcanic gas. Volcanic gases and steam (hydrothermal fluids) also seep through fractures unrelated to volcanoes.

Igneous and volcanic activity is not restricted to rift basins. Huge volcanic edifices are formed outside the rift basins. A string of such volcanic structures is found on either side of the rift rims. Examples of such volcanoes include Mount Kilimanjaro of Tanzania and Ras Dashen of Ethiopia. The rift-related volcanic activity started some 23 million years ago outside the rift, although most of it has been limited to the rift basin in the last 6 million years.

In addition to igneous rocks, the rift basin is covered by sediments and sedimentary rocks. At cliffed rift basins, boulders, cobbles, gravel, and granules accumulate at the bottoms of the cliffs. Rivers descend into rift basins and wind along the rift until they empty into lakes. Such rivers transport detrital sediments such as sand, silt, and mud, which are deposited within the river channel, in overbank floodplains, and at the lakes. In addition to detrital sediments, evaporation of rift lakes produces chemical sediments such as salt and gypsum. Rift basins, particularly the parts that are close to seas, can be flooded by the sea. Before volcanic rocks isolated it, the northern part of the Afar was covered by a shallow sea. Evaporation of that sea yielded copious amounts of salt deposits and underwater volcanic flows. Thus, rift deposits include chemical and detrital sedimentary rocks that may be interbedded with volcanic flows or pyroclastics.

PLATE TECTONIC THEORY

Geologists have suggested that the origin of the rifts can be explained by plate tectonic theory, according to which the lithosphere is segmented into discrete plates. The plates move, and the boundary type of the plates is identified by the direction of movement of neighboring plates. In divergent plate boundaries, neighboring plates move away from each other; in transform boundaries, neighboring plates slide alongside each other; and in convergent boundaries, the plates collide against each other.

The movement of the plates is considered to be guided by convection within the asthenosphere, which in turn results from movement of molten material to equalize the temperature within the asthenosphere. Hot molten material from the deeper part of the asthenosphere rises, pushes on the lithosphere above, and then diverges beneath the lithosphere. The lithosphere is carried along by the diverging asthenosphere as if it were luggage placed on a conveyor belt. Geologists believe that the African rift system is a divergent plate boundary at which Africa is tearing apart. According to this bulge-rift model, the rift borders are elevated from the surrounding areas because of the initial bulge before the rifting process was established. The discrete domes along the rift would then indicate areas at which randomly rising deeper and hot asthenosphere encountered the lithosphere that was expanded and buoyed because of the heat.

The Afro-Arabian dome, with its higher elevation, shape, and triple rifts, has attracted the attention of many geologists. Much as the crust of a pie placed in an oven would bulge up and form fractures, the rifts of the Afro-Arabian dome are considered to have formed as a result of the heat of the ascending plume. This theory is proposed as an explanation of why triple rifts are present in many places before a continent is split apart and an ocean fills the gap. According to this dome-rift model, the Red Sea and Gulf of Aden, which are now occupied by seas, have begun to open up as

Saudi Arabia is splitting and drifting away from Africa, whereas the third rift, the Main Ethiopian Rift, is not. While the dome-rift model requires that the Main Ethiopian Rift be a failed arm of a rift, there is evidence that shows that it is a divergent boundary with a spreading rate of about 1 centimeter per year. Moreover, the exact triple point, the Serdo Block, is on land. The Serdo Block is linked by a line of volcanoes to the Red Sea and the Gulf of Aden, which mark plate boundaries between the Saudi Arabian plate to the north, the Ethiopian plate to the southwest, and the Somalia plate to the southeast.

The central rift system of the Gulf of Aden is connected to a mid-oceanic rift system, the Carlsberg Ridge, in the Indian Ocean southeast of Saudi Arabia. Scientists have found that the average age of igneous rocks along the central Gulf of Aden Rift is progressively younger toward the Serdo Block. They suggest that the origin of the Gulf of Eden Rift is to be explained by a rift-propagation model. That is to say that a rifting, once begun somewhere, will move laterally. According to geologists, the submarine Carlsberg Ridge might have initiated the "burn" of the African continent on the east side of where the Gulf of Aden is currently located, and that rift has propagated toward the Serdo Block.

In the northern Red Sea, geologists have found that sedimentary rift deposits contain older rock particles in younger strata. It is as though, initially, young rocks from a source area were eroded, and their particles were deposited in the rift. Subsequently, the rim of the rift was uplifted, older rocks were exposed in the source area, and their particles were deposited in the rift basin atop sedimentary layers that contain younger rock particles. Geologists propose that rifting probably arose because the Arabian plate collided with Eurasia to form the Zagros Mountains in the north and that the lithosphere is tearing apart to form the Red Sea as a consequence. The comparatively thin lithosphere at the Red Sea is then heated by the asthenosphere and made buoyant so that the rift borders rise to higher elevations. This is called the rift-bulge model.

STUDY OF THE AFRICAN RIFT SYSTEM

Scientists have used a diverse set of instruments to study the different aspects of the African rift system. Aspects studied include the shape and structure of the landscape, the thickness of the lithosphere, the magnetic properties of the rocks, and the mineral and chemical composition of the rocks.

Ground surveys using various instruments are taken from different ground stations until a whole region is surveyed. The data can then be used to produce contour maps that show the landscape on a two-dimensional piece of paper. Mapping of remote areas awaited the use of aerial photographs. Overlapping photographs of a region are taken by cameras mounted on airplanes that fly along parallel lines. The overlapping photographs are then viewed under stereoscopes that have suitably mounted mirrors and lenses that permit the viewing of the region in its three dimensions. The locations and heights of the ground are measured from the photographs. Topographic maps are made from these measurements after checking some of them by actual field examination, a process also called ground truing.

Satellite images of various sorts help delineate structures and textures. Energy-sensing devices mounted on satellites are used to detect the energy that is emitted from the ground. The type of emitted energy is identified by its wavelength, and the data are converted into numbers, or digitized. The digitized data, along with the coordinates of the source region, are transmitted to receivers on the ground. Maps produced from these data are enhanced by false colors or shading and patterning to highlight particular features. Linear arrangements of volcanic cones and their relationships to other structures, the presence of major fault zones, and other features across a region can be identified from satellite images.

A primary instrument used to study the thickness of the lithosphere is the gravimeter, which is a mass suspended by a spring and encased in a suitable container. The attraction between the mass in the gravimeter and the Earth helps determine the gravity values of a region. Since rocks have lower densities than the asthenosphere, a thick lithosphere has a lower gravity value than a thin lithosphere. Scientists have found that the rift basins have high gravity values, and they have used these values in conjunction with suitably devised models to estimate the thickness of lithospheres. The African continent has a lithosphere that is generally

70 kilometers thick, but in the African rift basins the lithosphere is between 30 and 21 kilometers thick, the thinnest part being in the Afar.

Habte Giorgis Churnet

CROSS-REFERENCES

Basin and Range Province, 824; Carbonatites, 1287; Continental Drift, 565; Continental Rift Zones, 579; Displaced Terranes, 615; Evolution of Humans, 994; Fold Belts, 620; Folds, 624; Geoarchaeology, 1028; Geosynclines, 630; Geysers and Hot Springs, 694; Gravity Anomalies, 122; Joints, 634; Ocean Ridge System, 670; Ophiolites, 639; Plate Margins, 73; Plate Tectonics, 86; Spreading Centers, 727; Thrust Belts, 644.

BIBLIOGRAPHY

Chernikoff, S., and R. Vekatakrishnan. *Geology: An Introduction to Physical Geology.* Boston: Houghton Mifflin, 1999. The authors provide a overview of scientists' understanding of the Earth. Includes the address of a Web site that provides regular updates on geological events around the globe. Contains sections on rifting and the development of rift systems.

Compton, R. C. *Geology in the Field.* New York: John Wiley & Sons, 1985. Provides excellent descriptions of a variety of geologic field methods. Chapter 7 describes the use of aerial photographs and other remotely sensed data.

Dolgoff, Anatole. *Physical Geology.* Lexington, Mass.: D. C. Heath, 1996. This is a comprehensive guide to the study of the Earth. Extremely well illustrated and includes a glossary and an index. Although this is an introductory text for college students, it is written in a style that makes it understandable to the interested layperson. Contains a section on the development of the African rift valley system.

Dott, Robert H., Jr., and Donald R. Prothero. *Evolution of the Earth.* 5th ed. New York: McGraw-Hill, 1994. This basic textbook on historical geology is aimed at students of geology. However, it is very readable by anyone with a background in science. Presents an up-to-date account of the Earth's history from the viewpoint of plate tectonics. Includes a glossary.

Plummer, Charles C., David McGeary, and Diane H. Carlson. *Physical Geology.* Boston: McGraw-Hill, 1999. A straightforward, easy-to-read introduction to geology intended for those with little or no science background. Discusses the development of the African rift valley system. Includes many excellent illustrations, as well as a CD-ROM.

Verbyla, D. L. *Satellite Remote Sensing of Natural Resources.* New York: Lewis, 1995. Chapters 1 and 2 of this well-written and authoritative book give the basics of satellite imaging and processing. Chapter 3 deals with spectral regions; chapters 4, 5, 6, and 7 describe corrections and classifications; and chapter 8 discusses accuracy assessment. In an age of computer application and the expanding use of satellite data, this book provides an excellent source on methodology and information.

DISPLACED TERRANES

The concept of displaced or exotic terranes explains how hitherto inexplicable regions of crust arrived at their present locations. It also provides an approach to a more detailed understanding of how continents grow by accretion of their crust through collision with smaller tectonic bodies.

PRINCIPAL TERMS

ACCRETION: the process of growth of a larger crustal unit, such as a continent, by collision with smaller tectonic terranes, such as volcanic arcs or microcontinents

COLLAGE TECTONICS: a complex patchwork of different types of terranes thought to represent a region in which accretion has joined together suspect terranes

CRATON: a large, geologically old, relatively stable core of a continental lithospheric plate, sometimes termed a continental shield

LITHOSPHERIC PLATE: one of a number of crustal plates of various sizes that comprise the Earth's outer crust; their borders are outlined by major zones of earthquake activity

MICROCONTINENT: an independent lithospheric plate that is smaller than a continent but possesses continental-type crust; examples include Cuba or Japan

OBDUCTION: a tectonic collisional process, opposite in effect to subduction, in which heavier oceanic crust is thrust up over lighter continental crust

PLATE TECTONICS: the branch of geology that describes many crustal phenomena, such as volcanism, in terms of movements and interactions of large and small crustal units called lithospheric plates

SUBDUCTION: the process by which one lithospheric plate, usually a continental one, collides with another, typically oceanic, plate and overrides it, causing it to dive below the continent

SUTURE ZONE: a narrowly definable region of the Earth's crust thought to represent a place where two lithospheric plates have collided and subsequently been joined together

TERRANE: any sizable, discrete region of the Earth's surface crust that is the product of tectonic forces; examples include island or volcanic terranes

CHARACTERISTICS OF DISPLACED TERRANES

The concept of displaced terranes was developed by geologists to explain how anomalous regions of continental crust may have originated. Such areas were discovered to possess indications that their sites of origination differed from their present locations. Most were found to have one or more of a distinctive suite of features, such as a fossil record, mineralogy, stratigraphy, or structural pattern that was basically foreign to the surrounding or adjacent continental rock units. As an additional clue that such regions may have been added to a continental landmass at a date later than the original formation of the landmass, they were usually found to have boundaries that displayed structural deformation, also suggesting subsequent emplacement. Thus, every indication pointed to starting points remote from their present geographic locations.

To complicate matters, geologists differentiate the term "terrain" from "terrane." The former term is used to describe an area of surface topography, while the latter is reserved usually for description of a region's subsurface. Whatever the precise terminology employed, the concept remains the same: Regions of questionable lineage are variously termed terranes of suspect, exotic, or displaced nature. Earth scientists, working within the plate tectonic theoretical system, consider such terranes as products of the collision of a continental lithospheric plate with lesser plate bodies or other entities such as island arcs. From various lines of physical evidence, researchers have concluded that such terranes have undergone a process called accretion.

ACCRETION

In the crust of the Earth, according to plate tectonic theory, all the member units, called lithospheric plates, are in some type of interaction with one another. They are either spreading apart from some common center, as in the case of a mid-ocean ridge system (for example, in the mid-Atlantic Ocean), colliding with another unit with various end products and effects, or sliding alongside of one another along transform faults, such as in the San Andreas fault zone system. Slow convection cells within the Earth are considered the propelling force behind the global tectonic system, driving the plates apart at one point in the heat exchange cycle, together at another, and alongside in still others.

In a scenario involving direct collision and not translational or separating motion, a plate, depending on its particular material composition, and thus its relative density and thickness, will behave in various ways. If it is of about the same density and thickness, it may ram up into a linear, folded mountain range, as in the case of the Himalaya. If of a different relative density and thickness, it may be jammed below the forward end of the oncoming plate, termed the leading edge, in which case it experiences a process called subduction. It also may be jammed above the leading edge, in which case it experiences a process termed obduction. Continental crust, because of the general composition of its basement rocks, is somewhat lighter than oceanic crust. This difference is also described as one in the specific gravities of the two types of rock compositions. Thus, in a typical interaction between the two types, oceanic crust is subducted beneath continental material. That is not always the case, however, and exceptions apparently exist, the mechanics of which are still poorly understood.

Accretion is believed to occur when a leading edge of a continent either obducts heavier oceanic crust, adds other nonoceanic and noncontinental material such as volcanic island arcs, or encounters smaller units of generally similar continental material termed microcontinents. In all these cases, the encountered terrane becomes incorporated into the forward-moving portion of the continental lithospheric plate. In this manner, for example, North America and other continents have added many thousands of square kilometers of area. Evidence indicates that in the case of North America, at least 25 percent of its surface landmass is constituted of exotic terranes. An extreme case is the Alaskan region, which is believed to be composed of about fifty distinct displaced terranes that make up almost one-half of its area.

Orogenic belts, or deformed mountain belts, are sometimes listed as another example of suspect terrane accretion onto the edge of a continental, lithospheric superstructure. The superstructure itself is termed a continental craton or shield and is believed to have a great age in relation to microcontinents and other relatively transitory phenomena. The age disparity is along the order of billions of years for cratons as compared to less than 100 million years for a microcontinent— probably the smaller units, from creation to destruction or accretion, last an average of only tens of millions of years. Various past orogenies, or mountain-building episodes, are interpreted as evidence for microcontinent collision and accretion with regard to different cratons.

SOUTHERN APPALACHIAN OROGENIC BELT

One example is a scenario devised to explain the evolution of the southern Appalachian orogenic belt in the United States. This area is structurally complex and, prior to the acceptance of plate tectonic interpretations, generally defied any hypothesis that satisfactorily explained what its geologic history might have been. According to the scenario involving accretionary tectonics, the Cambrian period (about 544 to 505 million years ago) of the early Paleozoic era witnessed, among other things, a tectonic rifting event that resulted in the proto-Atlantic Ocean and production of various microcontinents, among them one termed the Piedmont. As time passed, the Ordovician period (about 505 to 438 million years ago) saw the onset of crustal subduction and the closure of the marginal sea. The Ordovician period gave way to the Silurian period (about 438 to 408 million years ago), during which the Piedmont microcontinent collided with and accreted to North America, resulting in the Taconian orogeny.

Further subduction resulted in another collision during the Late Devonian period (about 380 to 360 million years ago), which involved North America with another microcontinent termed Avalonia. This accretionary episode resulted in the Acadian orogeny. Further subduction continued on into the Late Paleozoic era (about 360 to 245 million years ago), resulting, at its close, in the Permian period collision of a large continent called Gondwanaland with North America and the generation of extensive overthrust faulting. This event is known as the Appalachian, or Alleghenian, orogeny. Subsequent to these events, the Mesozoic era (about 245 to 66 million years ago) witnessed still more large-scale changes, such as the rifting action that caused the present Atlantic Ocean to appear and widen, a phenomenon still in evidence. Because of the compounding of all these large-scale tectonic events, the geology of Appalachia is an intricate and confusing affair, with a number of exotic terranes of different ages in close proximity to one another.

COLLAGE TECTONICS

This patchwork of accreted terranes has been described as collage tectonics. Such a collage includes not only jammed-together microcontinents and volcanic arcs but also narrow units termed suture zones, which are thought to be actual relic lines of collision. The presence of suture zones is an additional line of evidence used to substantiate the existence of an accretionary event. Suture zones are easily identified because they display a narrow area of intensely metamorphosed and deformed basalts and ultramafic rocks. The rocks, such as ophiolites, and structures of such zones are interpreted as representing the remains of the last vestiges of unsubducted oceanic plate material that once separated a microcontinent and the continent proper. The vestigial, dense, ocean rocks have been jammed up onto the lighter, less dense, continental rock.

Other areas of the world in which collage tectonics is prominently expressed are in the regions of the northern North American Cordillera and eastern Siberia. This area is largely composed of a number of named and as yet unnamed suspect terranes, including, as mentioned, a signif-

icant part of Alaska. One of the better understood terranes in this region is one that has been called Wrangellia, a narrow, sinuous terrane thousands of kilometers in length that stretches from south-central Alaska to about the northern border of the state of Washington.

Evidence indicates that the anomalous terrane of Wrangellia has been displaced from 35 to 65 degrees of latitude northward from an unknown equatorial area. Some geologists, relying on similarities among fossils, believe Wrangellia originated in the South Pacific near the latitude of present-day Indonesia. Wrangellian basement rocks are oceanic basalts overlain by a sequence of limestones of the Middle and Upper Triassic period (about 220 to 208 million years ago). These limestones are stratigraphically inconsistent with the rocks of the other suspect terranes surrounding or adjacent to Wrangellia, not to mention the actual continental rocks to the east. The exotic, displaced terrane of Wrangellia and its neighboring members of the tectonic collage attest the long accretionary history that North America, along with all other continents, has experienced probably throughout most of geologic time.

STUDY OF DISPLACED TERRANES

Geologists researching the problem of suspect or displaced terranes have recourse to a large ar-

The Wrangell Mountains of Alaska, part of the island arc Wrangellia, which collided into and became part of the North American continent. (© William E. Ferguson)

ray of techniques and methods to analyze physical evidence. Firsthand, on-site fieldwork or recording sites are the preferred methods of data acquisition. In the case of the terranes of North America, investigations into the geologic history of former or active continental margin areas where suspect terranes abound were at first economically motivated for mining or oil exploration purposes. Later, academic fieldwork increased the database upon which the theory of displaced terranes was developed. The database was greatly expanded by new techniques involving seismic and other higher-technology approaches that were motivated by such public considerations as earthquake risk and the increasing need for new mineral resources for expanding industries and populations.

Among the more prominent techniques employed in the study of displaced terranes are biostratigraphy and the comparison of fossil records of terranes to their host continents. Fossil floras and faunas, both microscopic and macroscopic, can be used to correlate rock layers and date them in relation to one another. This technique, termed relative dating, is one of the original dating methods used in geology. Fossil correlation and the comparison of the temporal and geographic distribution of certain particularly useful, key fossil species, called index fossils, were, until the development of radiometric or absolute dating, the cornerstone of all geologic dating and still are of great utility.

The study of anomalous, extinct marine faunas of microorganisms such as the fusulinids has been a key factor in the establishment of the concept of suspect terranes. Permian period fusulinids occur anomalously in a particular zone of limestones that can be traced from Alaska to California. The next nearest fossil matches to these faunas occur in Japan and southeastern Asia. Fossil assemblages such as the fusulinids and others have helped emphasize the exotic nature of the suspect terranes.

After World War II, radiometric techniques were used to refine the dating of fossils and the strata or sediments in which they were found. Radiogenic isotopes are different nuclear species of an element that are unstable and that slowly change from one element to another, in the process producing radiation at a fixed, long-term rate. Using the natural decay rate of certain of these isotopes occurring in rocks and fossils as a kind of natural, internal clock, geologists have refined the results of relative dating; they have also been able to cross-check the results of both types of dating. Radiometric dating has proven to be of inestimable worth in the dating of most rocks and fossils and, consequently, in the analysis of accreted terranes.

Evidence from paleomagnetism also grew in importance for the study of suspect terranes and the acceptance of the theory of plate tectonics in general. Paleomagnetism is based on the principle that certain minerals, such as iron oxides, are responsive to the Earth's natural magnetic field. Many igneous rocks, such as basalts, possess these minerals and, once the molten mineral cools down to a certain point (for example, following an eruption where they are extruded or blasted free), the oxides orient themselves with regard to the global field and remain in this orientation indefinitely unless, once again, further heated above the critical temperature. Thus, such minerals are natural indicators of past magnetic orientations and the parent rock's respective, past, relative geographic locations. Such fossil magnetism, referred to as remanent magnetism, is one of the primary proofs of seafloor spreading and continental drift. The phenomenon, like radiometric dating, paleontology, and stratigraphy, has proven useful for tracing the possible paths and former locations of suspect terranes.

Frederick M. Surowiec

CROSS-REFERENCES

BIBLIOGRAPHY

Dolgoff, Anatole. *Physical Geology.* Lexington, Mass.: D. C. Heath, 1996. This is a comprehensive guide to the study of the Earth. Extremely well illustrated and includes a glossary and an index. Although this is an introductory text for college students, it is written in a style that makes it understandable to the interested layperson. Contains a section on the development of accreted terranes using the Appalachians and the Himalaya as examples.

Dott, Robert H., Jr., and Donald R. Prothero. *Evolution of the Earth.* 5th ed. New York: McGraw-Hill, 1994. This basic textbook on historical geology is aimed at students of geology. However, it is very readable by anyone with a background in science. Presents an up-to-date account of the Earth's history from the viewpoint of plate tectonics. Includes a glossary.

McPhee, John. *In Suspect Terrain.* New York: Farrar, Straus & Giroux, 1982. McPhee's treatment of sometimes complex geological subjects in this book is identical to the light and easy-to-understand approach he uses in the other two popular books he has written on the Earth sciences. Useful as well as a basic, good introduction to topics involving displaced terranes and associated concepts. Good for readers at all levels of geologic knowledge.

Miller, Russell, and the editors of Time-Life Books. *Continents in Collision.* Alexandria, Va.: Time-Life Books, 1983. Part of a series of lavishly illustrated books dealing with atmospheric, geologic, and astronomic sciences for the general public. A useful book for those who wish to know more about the general subject of plate tectonics and how and what evidence influenced the majority of contemporary scientists to accept this theoretical system. Offers examples from around the world, with highly informative photographs and artwork. Featured are extensive chapters on the mechanics of tectonics, which include an explanation of suspect terranes and how they fit into current knowledge pertaining to continental growth. An appropriate reading choice for those with only a minimal science preparation, high school and above.

Press, Frank, and Raymond Siever. *Understanding Earth.* 2d ed. New York: W. H. Freeman, 1998. This comprehensive physical geology text covers the formation and development of the Earth. Includes extensive sections on plate tectonics and discusses the development of displaced terranes using the Appalachians and northwestern North America as examples. Readable by high school students, as well as by general readers. Includes an index and a glossary of terms.

Redfern, Ron. *The Making of a Continent.* New York: Times Books, 1983. Treats a broad range of geological subjects concerned with the evolution of the North American continent. Good use of photographs, diagrams, and other illustrations to convey all the Earth science theories and concepts involved. A substantial chapter on suspect terranes is featured. Assumes no prior exposure to geology.

Van Andel, Tjeerd H. *New Views on an Old Planet.* New York: Cambridge University Press, 1985. Approaches the history of the Earth, including the evolution of the oceans, atmosphere, continents, and major groups of organisms, through the agency of the fossil record and various geological lines of evidence such as geochemistry. Plate tectonics is discussed, as well as displaced terranes, which the author refers to as "exotic terranes." Suitable for all readers at a high school or college level who possess some foundation in the physical sciences.

FOLD BELTS

Fold belts are linear regions of the Earth's crust that have been squeezed into folds and cut by faults, similar to the way a loose rug on a floor may be wrinkled when pushed at one end. They form some of the most spectacular and scenic mountain chains, including the Alps, the Himalaya, and the Rocky Mountains. The compressive forces that produce fold belts are generated when large slabs of the Earth's crust converge.

PRINCIPAL TERMS

ANTICLINE: a folded structure created when rocks arch upward; the limbs of the fold dip in opposite directions, and the oldest rocks are exposed in the middle of the fold

DÉCOLLEMENT: the detachment surface beneath a fold belt, usually located in a weak layer of rock such as shale; the maximum depth to which the rocks are folded and faulted

FLAT: that portion of a thrust fault where the fault is subparallel to adjacent layers of rock; also called treads, flats are usually located in weak rocks such as shale or evaporite (salt, anhydrite, gypsum)

RAMP: that portion of a thrust fault where the fault cuts across a layer of relatively stiff rock at a higher angle than does the rest of the fault

SYNCLINE: a folded structure created when rocks are bent downward; the limbs of the fold dip toward one another, and the youngest rocks are exposed in the middle of the fold

THRUST FAULT: a fault that is inclined at a low angle (less than 45 degrees), with rocks above the fault having moved up and over younger rocks below the fault; such faults are generally the result of compressive forces

FORMATION OF FOLD BELTS

Fold belts are one of the most common type of mountain belts. They form scenic and spectacular young mountains such as the Alps, the Himalaya, and the Rocky Mountains, as well as older mountains, such as the Appalachians. By comparison with volcanic mountains, which form by the eruption of molten rock at the Earth's surface, fold belts are produced when the Earth's crust is squeezed by compressive forces into a series of folds and faults.

Fold belts were first recognized in the mid-nineteenth century in the Swiss canton of Glarus. Here, Arnold Escher von der Linth found evidence that the Earth's crust had been heaved up and moved laterally for scores of kilometers, an unbelievable concept for its time. Subsequent Alpine geologists modified the details of Escher's interpretation, but by the late nineteenth century, the concept that some mountains are composed of great folds and laterally displaced rocks was generally accepted. Thus, the discipline of structural geology was born in the discovery of fold belts.

PLATE TECTONIC THEORY

The formulation of the theory of plate tectonics in the 1960's provided an explanation for the location and origin of fold belts. Plate tectonics is based on the concept that the Earth's crust and upper mantle are divided into a mosaic of large slabs, or "plates," that are in constant motion relative to one another. Plates can move away from each other, as along mid-ocean spreading ridges; they can laterally slide past each other, as along the San Andreas fault in California; or they can move toward each other, with one plate sliding beneath the other. This latter process is called subduction, which can occur in three ways: Dense, heavy oceanic basalt can subduct beneath lower-density (lighter) continental crust, as along the West Coast of South America; oceanic crust can subduct beneath more buoyant oceanic crust, typified by the Mariana Islands in the western Pacific Ocean; and continental plates—which, because of their mutual low density, have difficulty subducting beneath each other—can collide and hence push the crust up into large mountains.

Fold belts are associated with all three subduction processes, but the largest are formed by continent-continent collisions. For example, the Appalachian Mountains were created more than 200 million years ago by the continent-continent collision of North America with Western Europe and North Africa, and the geologically young Himalaya are being uplifted today by the collision of India with southern Asia.

When subduction occurs by one of these three processes, fold belts may be formed in two general localities. Assume that there are two plates (plate A and plate B) moving together and that plate A is subducted beneath plate B. The point where plate A bends down and begins to be subducted beneath plate B is called the trench. As plate A slides deeper and deeper beneath plate B, it is heated by the Earth's internal heat and eventually undergoes partial melting. The molten rock, or magma, rises through the overlying plate B and eventually forms a chain of volcanoes called a magmatic arc. Fold belts may be formed on the trench side of the magmatic arc, in which case they are called fore-arc fold belts, or they may form in plate B behind the arc, when they are called back-arc fold belts. The geological character of these two settings is drastically different, resulting in very different fold belts.

TYPES OF FOLD BELTS

Fore-arc fold belts contain an abundance of deep-water oceanic sediment and sometimes the upper part of oceanic crust that was scraped off the subducting plate (plate A) as it descended into the subduction zone beneath plate B. In addition, fore-arc fold belts may contain accreted terranes, or masses of rock that were too large to be subducted. Examples of fore-arc fold belts include the Olympic Mountains of Washington State, the island of Barbados in the lesser Antilles, and the island of Java in Indonesia.

In contrast, back-arc fold belts form farthest from the trench on the back side of a magmatic arc. These fold belts are usually composed of well-stratified, shallow marine and nonmarine rocks; they do not normally contain accreted terranes. Back-arc fold belts are known for their tremendous oil and gas production, such as in southwestern Wyoming and Alberta, Canada. One of the best known back-arc fold belts is the Rocky Mountain system of western North America, which

formed between 150 million and 55 million years ago, when a large subduction zone was located along the West Coast. The Rocky Mountain system forms the topographic backbone of the North American continent, extending more than 8,000 kilometers in length from the Brooks Range in northern Alaska, through western Canada, western Montana, western Wyoming, eastern Idaho, central Utah, southern Arizona, and into Mexico. The Rocky Mountain fold belt is also called the fold and thrust belt, or overthrust belt.

FOLDED ROCKS AND FAULTS

The internal anatomy and style of deformation associated with fold belts has been studied extensively for more than a century, yet there is still much that scientists do not understand. Fold belts are, by their nature, composed of highly deformed rocks that make geological interpretations difficult at best. The most common geological structures encountered in fold belts are folded rocks. Folding is generally the result of compressional forces that squeeze the rock in ridges and valleys, or anticlines and synclines, respectively. Anticlines are formed when rock layers are arched upward, whereas synclines are downfolds, or depressions. The name "anticline" refers to the flanks of a fold that are inclined away from each other, while the flanks of a syncline are inclined toward each other. The process of folding may be envisioned by imagining a flat rug lying on a flat floor; if the rug is pushed at one end, it will be wrinkled into a series of anticlines and synclines; however, the floor beneath the rug will not be wrinkled. The interface between the rug and the floor is called the detachment surface, or décollement, and is located in a weak rock layer such as shale in many fold belts.

In addition to folds, fold belts are commonly cut by faults. Faults are narrow zones of brittle fracture that displace rock layers in lieu of folds. The detachment surface between the rug and floor in the previous example is a type of fault. In addition, faults may cut up from the detachment surface and displace sections of the rug (like tearing the rug). These faults are called thrust faults, and they commonly displace deeply buried, older rocks over younger rocks beneath the fault. The Lewis thrust fault, which forms Glacier National Park in northwest Montana, is one of the world's

An anticline on the banks of the Potomac River. The name "anticline" refers to the flanks of a fold that are inclined away from each other, hence taking the shape of an arch. (U.S. Geological Survey)

logical map through fieldwork or from aerial photographs; sometimes computerized images and photographs from space satellites are very useful in mapping large regions. Next, the geologist must describe the relative motions of rock layers that occurred to produce the structures; this step is called "kinematic analysis." Finally, once the geometry and kinematic sequence are understood, the geologist may ponder the forces and stresses that created the deformation; this step, called dynamic analysis, is often the goal of a structural investigation. Forces and stresses are often, but not always, associated with movement of the Earth's large plates. Fold belts have contributed much to the understanding of how plates move; in turn, plate tectonics has offered an explanation for the origin of great fold belts.

Fold belts are of interest to the layperson for two basic reasons. First, fold belts form some of the Earth's most spectacular scenery; the Alps, the Himalaya, the Rocky Mountains, and the Appalachian Mountains are all examples of fold belts. More practically, fold-belt mountains have nurtured many resources that are valuable, if not essential, to civilization, such as water, timber, animals for domestication and meat, and natural resources from the rocks themselves (for example, gold, copper, and oil). Most of the Earth's great rivers—such as the Indus, Ganges, and Brahmaputra of India; the Rhone and Rhine of Europe; and the Columbia and Missouri-Mississippi of North America—flow from snowfields high in fold-belt mountains. Mountains have thus been a source of renewing resources for humankind.

David R. Lageson

classic thrust faults. Thrust faults and folds are intimately linked in fold belts. As thrust faults move, they tend to cut upward at a steep angle across stiff, hard rock formations such as sandstone. Such areas are called ramps and, as the rock layers move up and over a ramp, they become folded into an anticline called a ramp-anticline or fault-bend-fold. There are often several stiff layers within a stratigraphic section, so the profile of most thrust faults resembles a staircase with many ramps or steps. The flat regions between ramps are simply called flats, or treads, and synclines are commonly formed above these.

STUDY OF FOLD BELTS

Fold belts offer spectacular exposures of contorted rocks on the face of a mountain that provide much information and inspiration to the field geologist. Geologists have traditionally studied fold belts by constructing detailed geological maps in the field. Field mapping has always been the first step in understanding any geological phenomenon. A structural geologist (one who studies deformed rocks) must first understand the geometry of deformed rocks—that is, the shapes and attitudes into which they have been deformed—which is best done by constructing a detailed geo-

CROSS-REFERENCES

BIBLIOGRAPHY

Boyer, S. E., and D. Elliott. "Thrust Systems." *American Association of Petroleum Geologists Bulletin* 66, no. 9 (1982): 1196-1230. An excellent summary of thrust faults in fold belts such as the Alps, the Rocky Mountains, and the Appalachian Mountains.

Hatcher, Robert D., Jr. *Structural Geology: Principles, Concepts, and Problems.* 2d ed. Englewood Cliffs, N.J.: Prentice-Hall, 1995. This undergraduate textbook provides a comprehensive overview of the development of fold belts. Intended for the more advanced reader.

Hsu, K. J., ed. *Mountain-Building Processes.* New York: Academic Press, 1982. A compilation of outstanding papers dealing with the tectonic evolution of the world's great mountain belts, including the Alps, the Appalachians, and the Canadian Rocky Mountains.

McClay, K. R., and N. J. Price, eds. *Thrust and Nappe Tectonics.* Boston: Blackwell Scientific, 1981. An excellent collection of papers that present modern thinking on the origin and structural geology of fold belts throughout the world.

Plummer, Charles C., David McGeary, and Diane H. Carlson. *Physical Geology.* Boston: McGraw-Hill, 1999. A straightforward, easy-to-read introduction to geology intended for those with little or no science background. Has a section on the development of fold belts. Includes many excellent illustrations, as well as a CD-ROM.

Press, Frank, and Raymond Siever. *Understanding Earth.* 2d ed. New York: W. H. Freeman, 1998. This comprehensive physical geology text covers the formation and development of the Earth. Includes extensive sections on plate tectonics and the development of fold belts. Readable by high school students, as well as by general readers. Includes an index and a glossary of terms.

Voight, B., ed. *Mechanics of Thrust Faults and Décollement.* Benchmark Papers in Geology 32. Stroudsburg, Pa.: Dowden, Hutchinson and Ross, 1976. A collection of classic articles pertaining to the mechanics of thrust faulting in fold belts.

FOLDS

Folds are the warping or bending of strata, foliation, or rock cleavage from an original horizontal or undeformed position into high and low areas. A fold is generally considered to be a product of deformation. Much of the folding of the Earth's crust takes place near or along lithospheric plate boundaries and is considered to be a result of compressional stress.

PRINCIPAL TERMS

ANTICLINE: an arched upward fold of stratified rocks from whose central axis the strata slope downward; at the center, it contains stratigraphically older rocks

AXIAL PLANE: a surface connecting all hinges of a fold that may or may not be planar; that is, the axial plane of a fold may vary from a flat plane to a complexly folded plane

AXIS: a line parallel to the hinges of a fold, also called fold axis or hinge line

CRESTAL PLANE: a plane or surface that goes through the highest points of all beds in a fold; it is coincident with the axial plane when the axial plane is vertical

FLANKS: a term describing the sides of a fold, also called limbs, legs, shanks, branches, or slopes; anticlines share syncline flanks, and synclines share anticline flanks

HINGE: the line of maximum curvature or bending of a fold

PLUNGE: the inclination and direction of inclination of the fold axis, measured in degrees from the horizontal

SYNCLINE: a downward bent fold of stratified rock from whose central axis the strata slope upward; at the center, it contains stratigraphically younger rocks

TECTONICS: the study of the form, pattern, and evolution of large-scale units of the Earth's crust, such as basins, geosynclines, and mountain chains

TROUGH: a line occupying the lowest points of a bed in a syncline; the trough plane connects the lowest points on all beds

ROCK DEFORMATION

Folding is a common type of deformation seen in crustal rocks. It is obvious in the dipping (sloping) beds of mountains, where fold axes most often extend parallel to the length of the mountain chains. The folded beds may be observed in cliffs, roadcuts, and quarries, where the observer can see not only the steeply dipping beds but also some of the smaller anticlines (upward-arched folds) and synclines (downward-arched folds). When viewed from the air, mountain chains are often found to be composed of sinuous ridges of sedimentary strata, such as limestone and sandstone, which are more resistant to erosion, while valleys are underlain by rocks such as shale, which are more susceptible to erosion. This phenomenon can easily be seen from the highway west of Denver, Colorado, in the front ranges of the Rocky Mountains, or near Harrisburg, Pennsylvania, where Interstate Highway 81 follows a ridge north for miles toward Scranton.

In the North American midcontinent region, folds are less obvious, and the beds dip with inclinations of as little as 0.5 degree and upward to about 5 degrees; the folds are usually only discernible by surveying techniques. Occasionally there are very large folds in this region, hundreds of miles across and with very low dips. These folds are more circular than elongate in shape and are referred to as basins (downfolds) and arches, or domes (upfolds). Examples are the Michigan Basin, the Cincinnati Arch, and the Nashville Dome.

ROCK MECHANICS

Rock mechanics is the study of the mechanical behavior of rocks; one branch of this study is concerned with the response of rocks to force fields in their geologic environments. Associated with this field is the study of plate tectonics, which relates the large-scale structures of the Earth to these forces. For example, in the crust of the Earth there are differential forces caused by the interaction of the lithospheric plates associated with the processes of plate tectonics along with hydrostatic (fluid) and lithostatic (gravity) pressures. Each may cause folding. The processes of plate tectonics form folded mountain chains at subduction zones. Differential lithostatic pressures associated with differences in specific gravity of different rocks and the effects of gravity form folds as in salt domes; lithostatic pressure, however, may act only as a confining pressure.

Rocks under stress behave in an elastic, plastic, or brittle manner, as do all solid materials. These three properties are related by the sequence of occurrence during deformation. Solid materials, when first subjected to a force, behave elastically; that is, they change shape or volume, but if the force is removed they return to their original shape. If, however, the force, instead of being removed, is increased, permanent deformation occurs: First the material changes shape plastically, and then it becomes brittle, breaking apart. Factors that affect the manner in which a rock behaves are time, temperature, confining pressure, and the pore pressure of water. Higher temperatures, water in pore space, lower confining pressures, or a longer length of time during which the rocks are subjected to a constant force tend to weaken rocks. Conversely, lower temperatures, lack of water, higher confining pressure, or a short period of time during which the force is applied tend to strengthen rocks.

Classification of fold types is a means by which structural geologists group the various kinds of folds in order to understand them and their origins. There are five major divisions in fold classification: descriptive or geometric, morphologic, by mechanics of origin and internal kinematics, by external kinematics and tectonic forces, and by position in the tectonic framework.

DESCRIPTIVE CLASSIFICATIONS

Descriptive or geometric classifications are based primarily on shape—that is, the attitude of the limbs, axial plane, and hinge line of the fold. The major use is for geologic mapping. The classification most commonly used is attitude of the axial plane, or the appearance of a fold in cross section or vertical section normal to its axis. In this classification there are symmetrical folds, in which the axial plane bisects the fold; asymmetrical folds, in which the limbs have different dips and are therefore asymmetrically disposed about the axial plane; overturned folds, in which one limb has been rotated or tilted through the vertical so that the original bed is upside down; and recumbent folds, which have been rotated so that their axial planes are nearly horizontal. Plunging folds are those in which the axis of the fold is not horizontal. Upright folds have vertical axial planes, and inclined folds have inclined axial planes.

Another descriptive classification is based upon fold symmetry: Orthorhombic folds have the axial plane and the plane normal to the axis of the fold as symmetry planes (mirror planes); monoclinic folds have either the axial plane or the plane normal to the axis of the fold as a symmetry plane but not both; and triclinic folds are folds with no symmetry planes.

A third descriptive classification is based on the orientation of the axes of a fold. It includes cylindrical folds, which are described by the rotation of a line parallel to itself at a fixed distance from a cen-

A syncline in weathered shale in Johnson County, Tennessee, with a left limb that dips to the right and a vertical right limb. (U.S. Geological Survey)

TYPES OF FOLDS

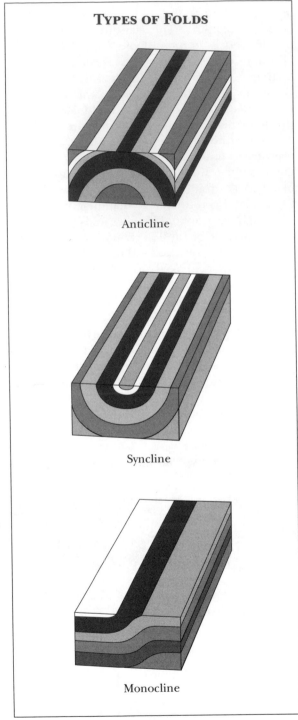

Anticline

Syncline

Monocline

Three primary folds are anticlines, *arched upward folds of stratified rocks from whose central axis the strata slope downward;* synclines, *folds of stratified rock from whose central axis the strata slope upward; and* monoclines, *folds in which the strata dip or flex from the horizontal position in one direction only and are not a part of an anticline or syncline.*

tral point and with parallel hinges (noncylindrical folds are folds with randomly oriented axes), and conical folds, which have axes divergent from an apex. A descriptive classification based upon flanks of folds includes isoclinal folds, with both flanks essentially parallel; open folds, which may be folded more tightly without rock flowage; and closed folds, which may not be folded more tightly without rock flowage. Monoclines are folds in which the strata dip or flex from the horizontal position in one direction only and are not a part of an anticline or syncline.

MORPHOLOGICAL CLASSIFICATIONS

Morphological classifications are based upon the shape of the fold in regard to depth, map view, and spatial relationships with adjacent folds. The changes in shapes and patterns formed by folds are not always apparent when an individual fold is viewed in the field; therefore, the distinction is made between descriptive and morphological classifications. Concentric or parallel folds maintain a constant thickness of beds, which means an anticline will decrease in size downward, whereas a syncline will decrease in size upward. Similar folds are bent into similar curves and do not increase or decrease in size downward but maintain curves by thinning of flanks and thickening of crests and troughs. Disharmonic folds are different beds in a sequence of strata that have different amplitudes and wavelengths. Supratenuous folds die out downward. Nappes are large recumbent anticlines generally isolated by thrust faults.

Folds may also be considered in groups. Structural salients are observed when a sequence of folds is viewed in map view, and the curved fold axes are oriented convex toward the outer edge of the fold belt. Embayments are observed when a sequence of folds is seen in map view, and the curved fold axes are concave toward the outer edge of the fold belt. Homomorphic folding is the condition where folds cover an entire area, whereas idiomorphic folding is the condition where there is an intermittence of fold locality and often the folds are not linear in habit. Anticlinoriums are a series of anticlines and synclines forming a large arch, whereas synclinoriums are a series of anticlines and synclines forming a large trough. Both are generally tens of kilometers across. *En echelon* folds are a series of folds whose lengths are not extreme but that have axes that overlap in an oblique manner.

CLASSIFICATION BY MECHANICS OF ORIGIN AND INTERNAL KINEMATICS

Fold classifications based upon mechanics of origin and internal kinematics describe the processes of folding in rocks. The mechanisms of folding reflect the elastic, plastic, or brittle manner of deformation dominant at different times during folding. Most folds appear, superficially, to be the result of plastic deformation. If, however, one takes a plastic material and pushes it from both sides, only the edges are deformed. In the case of mountain chains, there are often wide fold belts, which suggests that stress was transmitted equally across the width of the fold belt—the result of elastic deformation. The study of folds in thin sections (a thin slice of rock 0.03 millimeter thick prepared for viewing under the microscope), however, may show folding developing along minute shear planes and therefore must be the result of brittle deformation.

Flexure folding or flexural-slip folding, at times called true folding, is a form of elastic deformation. It can be demonstrated by taking a sheet of paper and folding it into 1-inch accordion pleats. When stretched out a little, then compressed slowly, the folds become tighter; when released, the folds will spread out again—elastic rebound. In folding the paper, permanent folds were induced, which is plastic deformation, but the entire "fold belt" behaves elastically. What happens vertically can be demonstrated by taking a book and opening it to the middle, flat on the tabletop. The pages will be bent or folded on each side of the binding into anticlines. Folding of the book has proceeded by each page sliding upward relative to the page below, resulting in a narrow edge of all pages being visible at the side. Here again, there is only elastic deformation, because the pages of the book are not folded permanently. Shear folding—an example of brittle deformation—results from

A railroad cut in eastern Glacier National Park, Montana, reveals a drag fold in the Kootenai Formation. (U.S. Geological Survey)

minute displacements along closely spaced fractures, which can be demonstrated with a stack of playing cards. Draw a vertical line in the middle of one side when the cards are neatly stacked and another line on the opposite side, then hold the deck of cards on end so that the line is horizontal. Next, indent the middle of the deck on the upper side forming a U-shaped trough. The original line now is bent into a synclinal form by each card slipping relative to the adjacent card.

CLASSIFICATION BY EXTERNAL KINEMATICS AND TECTONIC FORCES

Fold classifications based upon external kinematics and tectonic forces focus on external causes of folding. Block folding results from block uplift. Injection folding results from pluton emplacement. Folds caused by general crumpling can be attributed to two basic hypotheses: the tangential-compression hypothesis, or bench vise concept, and the vertical-tectonic concept, with material movement resulting from specific gravity differences. The first hypothesis assumes crustal shortening; the second assumes little or no difference in crustal length. Gravity gliding results when uplift occurs and rock masses move downslope along faults, producing folds and nappes. Rotational folding results from the rotation of blocks of the Earth's crust, which are generally bounded by major faults. Regional coupling results from drag on blocks of the Earth's crust sliding past one another. Differences in specific gravity produce folds in conjunction with the force of gravity. Two examples are salt anticlines and domes. Differential compaction produces folds where beds are draped over areas of less compaction. Ice shove may produce folds when a glacier encounters frozen, loosely consolidated rocks and sediments and shoves them into folds. Geosynclines are mobile downwarpings of the Earth, generally elongate but also basinlike, measured in hundreds of kilometers, which subside as sedimentary and volcanic rocks accumulate to thicknesses of thousands of meters. This part of the tectonic cycle is often followed by orogeny, or mountain-building processes. Geosynclines are related to subduction zones of the plate tectonics theory. Geanticlines are mobile upwarpings of crustal material of regional extent; the term is particularly applied to anticlinal structures developed in a geosyncline.

CLASSIFICATION BY POSITION IN THE TECTONIC FRAMEWORK

Classification of folds based upon position in the tectonic framework is a way of describing components of a mountain chain. There have been few attempts to devise a formal classification, but there are fold types characteristic of particular tectonic regimes. A belt of crustal folds and metamorphism, generally centrally located within a mountain chain, is a region of shear folding vertical uplift. Examples include the Pennine Nappes of the Alps and the Piedmont and New England upland provinces of the Appalachian Mountains. The outer belt of shallow folding and thrusting of mountain chains is characterized by sedimentary rocks folded into anticlines and synclines with associated thrust faults and is often underlain by a plane of décollement. Examples of this portion of a mountain chain include the Valley and Ridge Province of the Appalachian Mountains, the Front Ranges of the Rocky Mountains, and the Jura Mountains of Europe.

Charles I. Frye

CROSS-REFERENCES

African Rift Valley System, 611; Appalachians, 819; Continental Structures, 590; Displaced Terranes, 615; Experimental Rock Deformation, 208; Fold Belts, 620; Geosynclines, 630; Joints, 634; Mountain Belts, 841; Oil and Gas Exploration, 1699; Oil and Gas Origins, 1704; Ophiolites, 639; Petroleum Reservoirs, 1728; Plate Tectonics, 86; Regional Metamorphism, 1421; Rocky Mountains, 846; Stress and Strain, 264; Subduction and Orogeny, 92; Thrust Belts, 644.

BIBLIOGRAPHY

Badgley, Peter C. *Structural and Tectonic Principles.* New York: Harper & Row, 1965. Perhaps the finest text in structural geology at the time it was published. The tectonics section is dated; nevertheless, it has a very fine section on folds, folding, and rock mechanics. Written at the level of an advanced college student.

Billings, Marland P. *Structural Geology.* 3d ed. Englewood Cliffs, N.J.: Prentice-Hall, 1972. Perhaps the easiest-to-read introductory text to the field of structural geology for the college student. Although an older book, it is essentially as modern as any text, except for the final chapter on geophysics. Provides perhaps the most comprehensive introductory descriptions of folds available.

Dennis, John G. *Structural Geology.* Dubuque, Iowa: Wm. C. Brown, 1987. A college-level introductory text to the field of structural geology. Includes a review of modern geotectonics and a good section on folding.

Hamblin, W. K., and J. D. Howard. *Exercises in Physical Geology.* Edina, Minn.: Burgess, 1986. An excellent introductory laboratory manual for beginning college courses in geology. Offers one of the better treatments of folds, with fine colored diagrams; however, it contains no information beyond the very elemental descriptions of folds.

Hatcher, Robert D., Jr. *Structural Geology: Principles, Concepts, and Problems.* 2d ed. Englewood Cliffs, N.J.: Prentice-Hall, 1995. This undergraduate textbook covers folds in three chapters. Intended for the more advanced reader.

Plummer, Charles C., David McGeary, and Diane H. Carlson. *Physical Geology*. Boston: McGraw-Hill, 1999. This is a straightforward, easy-to-read introduction to geology intended for those with little or no science background. Includes many excellent illustrations, as well as a CD-ROM.

Press, Frank, and Raymond Siever. *Understanding Earth*. 2d ed. New York: W. H. Freeman, 1998. This comprehensive physical geology text covers the formation and development of the Earth. Includes extensive sections on structural geology and folds in particular. Readable by high school students, as well as by general readers. Includes an index and a glossary of terms.

Suppe, John. *Principles of Structural Geology*. Englewood Cliffs, N.J.: Prentice-Hall, 1985. Designed as a concise introduction to the deformation of the Earth's crust and written for the advanced college student. Besides the usual material in a structural geology text, this book has a section on regional structural geology of the Appalachian Mountains, with good illustrations of folds, as well as a similar chapter devoted to the Cordilleran ranges of the western United States.

Tarbuck, Edward S., and Frederick K. Lutgens. *The Earth*. Westerville, Ohio: Merrill, 1987. An example of many college freshman books in physical geology, with a minimal treatment of folds and an excellent section devoted to plate tectonics.

GEOSYNCLINES

Geosynclines are linear basins in which vast thicknesses of sedimentary and volcanic rocks accumulate. All geosynclines were once believed to evolve similarly into mountains. With the discovery of plate tectonics, however, geosynclines were found to form in several ways.

PRINCIPAL TERMS

CONTINENTAL CRUST: the upper 50 kilometers of the Earth below continents; it is lighter and older than oceanic crust

CONTINENTAL MARGIN: the edge of a continent next to an ocean basin that is both exposed on land and submerged below water; it is composed of a continental shelf and slope

FAULTING: the process of sliding rocks along a fracture

FOLDING: the process of bending initially horizontal layers of rock so that they dip

ISLAND ARC: a chain of volcanoes next to an oceanic trench in the ocean basins; an oceanic plate descends, or subducts, below another oceanic plate at island arcs

OCEANIC CRUST: the upper 5 to 10 kilometers of the Earth below ocean basins; it is heavier and younger than continental crust

OROGENIC BELT: a linear belt of folded and faulted rocks; a mountain belt

SUBSIDENCE: the sinking of the Earth's surface because of the weight of a load such as unusually thick piles of sediments

DEFINITION

Geosynclines are linear depressions on the Earth's surface characterized by the huge thicknesses of sedimentary and, occasionally, volcanic rocks that accumulate in them. The sedimentary sequences in geosynclines typically are ten times thicker than in neighboring regions. The word "geosyncline" comes from the Greek roots *geo, syn,* and *klinein,* which mean "earth," "together," and "to incline," respectively. In a syncline, sedimentary layers are folded such that they dip toward the central axis at the bottom of the fold, much like the limbs in the letter *u.* An anticline is the opposite; layers dip away from the axis at the top of the fold.

HISTORY OF GEOSYNCLINE THEORY

The concept of the geosyncline was proposed to explain thick linear sequences of faulted and folded sedimentary rocks in mountain belts. Geosynclines were thought to be essential precursors to and causes of orogeny (mountain building). The thickness and linearity of sediments in a given mountain belt reflected the shape and depth of its precursor geosyncline: The greater the depth, the higher the mountains. This causal relationship led to the geosynclinal, or tectonic, cycle, which held that all geosynclines progressed through the same sequence of steps and became mountains. After one mountain belt developed, a new geosyncline formed nearby, and the process repeated. This concept predicted that continents grew at a constant rate by the addition of sequential mountain belts to their peripheries. As a corollary, anywhere huge thicknesses of sediment occurred in the present, mountains would form in the future.

Geosynclines as a phenomenon were introduced in 1859 by American geologist James Hall, who noted that the Appalachian Mountains were an elongate chain of folded shallow-water sedimentary rocks that were much thicker than their temporal equivalents in the Mississippi Valley. From this, Hall deduced that all mountain ranges were preceded by the development of long, narrow depressions that contained unusually thick sediments. These depressions formed preferentially along continental margins and evolved into mountains primarily by subsidence; the weight of the sediments forced the surface of the Earth to sink, or subside. The subsidence, in turn, reduced the width of the surface above, thereby folding the sediments in the depression.

In contrast, James Dana proposed in 1866 that these depressions (geosynclines) and complementary ridges (geanticlines) resulted from the contraction of the surface of a cooling Earth. Early unequal contraction formed the ocean basins and the continents. Lateral forces (as opposed to the vertical forces invoked by Hall) produced during continued contraction caused the surface of the Earth to bend at the boundary between oceanic and continental crusts, a zone of weakness. This created downward (geosynclines) and upward (geanticlines) buckles along continental margins. Eventually, the contraction folded the geosynclines into mountains. These theories assumed that the continents and oceans were permanent features that developed early in Earth history.

The idea of geosynclines as essential to mountain building was enthusiastically embraced by most geologists, but the models of their internal character and their locations were not. For example, most American scientists, who worked primarily on the Appalachian Mountains, believed that geosynclines were composed of shallow-water marine strata and were initiated at the boundary between continents and ocean basins. Most European scientists, however, asserted that geosynclines were the locus of deep-water marine sedimentation and abundant volcanic activity and formed between two continents, citing the Alps as evidence. Others discovered geosynclines on continents and adjacent to island arcs. Still others argued that geosynclines were wedges, not downfolds, and should be renamed geoclines. The most heretical proposed that not all geosynclines became mountains.

Nevertheless, the concept remained the primary model for the surface of the Earth prior to the advent of plate tectonics. With the acceptance of the plate tectonic paradigm in the mid-twentieth century, geosynclines were reinterpreted to form in a variety of tectonic settings and, in most cases, to be the result, not the cause, of orogeny.

TYPES OF GEOSYNCLINES

Two of the most important types of geosynclines are miogeosynclines and eugeosynclines, which according the early theories were envisioned as parallel belts separated by a geanticline. Both underwent subsidence, but eugeosynclines sank deeper and experienced active volcanism. Each had a characteristic sedimentary succession within it. Miogeosynclines contained coarse-grained rocks that were deposited in shallow water, much like those in the Appalachian Mountains. In contrast, eugeosynclines consisted of a sequence of volcanic rocks overlain by fine-grained deep-sea sedimentary strata that were, in turn, overlain by coarse-grained shallow-water sedimentary rocks such as in the Alps. The assemblage of volcanic rocks and deep-sea sediments, called ophiolites, were perceived as remnants of oceanic crust.

Despite the evolution of early theory, four aspects of geosynclines remained unresolved. First, the source of the sediment that filled geosynclines remained a mystery. It was easy to imagine that erosion of mountains would provide debris to a nearby geosyncline, but geosynclines were assumed to predate mountains. To circumvent this problem, elevated areas bordering geosynclines were proposed. The existence and subsequent disappearance of these borderlands, however, could not be established. Second, mountains were discovered to be composed of several geosynclines of different ages and rock types. Individual geosynclines frequently were truncated. Mountains, therefore, were called mobile belts in order to reflect the apparent mobility of the geosynclines within them. The mechanism by which the different geosynclines were juxtaposed and truncated was unknown. Furthermore, these differences violated a corollary to the geosynclinal cycle: that all mountain belts are the same because they develop in the same way. Third, the varied locations of geosynclines—along continental margins, adjacent to island arcs, and in the interiors of continents—could not be explained by either subsidence or by a cooling Earth model. Finally, and perhaps most important, geosynclines did not compare favorably with potential modern analogues. The thick accumulation of sediment along the eastern continental margin of North America was more than 200 million years old and underformed. If geosynclinal theory were correct, this sedimentary sequence should evolve into mountains, something it showed no evidence of doing.

PLATE TECTONICS

The introduction of continental drift, seafloor spreading, and, eventually, plate tectonics resolved many of these problems. Continental drift proposed that 200 million years ago the continents were joined as a supercontinent, Gondwanaland, that subsequently fragmented, causing the continents to drift across the surface of the Earth. To fill the gaps produced by the drifting continents, ocean basins grew by the solidification of molten rock that rose vertically through the Earth and emerged on the surface to form new ocean floor, a process known as seafloor spreading. The Atlantic Ocean thus grew between Africa and North America during the rifting of Gondwanaland.

Plate tectonics stated that new ocean floor would be consumed and returned to depth in the Earth by subduction in oceanic trenches at convergent plate boundaries along either island arcs or certain continental margins. Continental drift, seafloor spreading, and plate tectonics were appealing; as theories, they provided mechanisms for orogeny and for truncation of geosynclines in mountain belts. Rifting of the continents along a mountain belt would truncate geosynclines and produce a tectonically quiet continental margin such as eastern North America. Orogeny would occur where new seafloor descended at island arcs or at tectonically active continental margins, such as western South America. Continental collisions would produce mountain belts (and a precursor geosyncline, according to the geosynclinal cycle) between two continents, thereby explaining the Alps. If collisions occurred repeatedly along the same continental margin, geosynclines of different ages and rock types could be juxtaposed, thus generating mobile belts. In addition, the proximity of Africa to North America approximately 200 million years ago provided a sediment source for the geosyncline of the East Coast of North America. This source did not disappear; it drifted away as the Atlantic Ocean basin opened.

MODERN UNDERSTANDING OF GEOSYNCLINES

The formulation of plate tectonics required reassessment of the geosynclinal concept. Because of the different tectonic and sedimentary environments predicted by plate tectonics, the idea that all geosynclines formed in a similar fashion and became mountains had to be abandoned. Indeed, eugeosynclinal successions most closely resemble rock associations in oceanic trenches, which are the sites of much orogenic activity, whereas miogeosynclinal successions are comparable to sequences on continental shelves and slopes, which are tectonically quiet.

Thus, two types of geosynclines, previously thought to lie adjacent to each other and to evolve together, were discovered to represent two entirely different settings. Miogeosynclines form along Atlantic-type, passive continental margins, such as eastern North America, which are ancient plate boundaries that reflect an earlier episode of continental rifting and plate separation. Their evolution into mountains is incidental. Conversely, eugeosynclines develop along Pacific-type active continental margins, such as western South America, or along island arcs, such as Indonesia, which are modern plate boundaries and the sites of plate convergence.

Furthermore, modern analogues of geosynclines are composed of wedge-shaped, not V-shaped, sedimentary accumulations. The concept of the original geosyncline as a downfold that contained vast thicknesses of sediment and that evolved into mountains is no longer valid. Linear belts of sediment accumulation that are morphologically similar to geosynclines, however, are common, but they do not all occupy the same environments nor do they all become mountains.

Pamela Jansma

CROSS-REFERENCES

BIBLIOGRAPHY

Hatcher, Robert D., Jr. *Structural Geology: Principles, Concepts, and Problems*. 2d ed. New Jersey: Prentice-Hall, 1995. This undergraduate textbook provides a comprehensive overview of the development of geosynclines. Intended for the more advanced reader.

Parker, Sybil P., ed. *McGraw-Hill Encyclopedia of the Geological Sciences*. New York: McGraw-Hill, 1988. The discussion on geosynclines is brief but to the point. Useful primarily for the illustrations and the list of references. Suitable for college-level students.

Plummer, Charles C., David McGeary, and Diane H. Carlson. *Physical Geology*. Boston: McGraw-Hill, 1999. This is a straightforward, easy-to-read introduction to geology intended for those with little or no science background. Discusses plate tectonics and its relation to the development of geosynclines. Also deals with andesitic volcanism. Includes many excellent illustrations, as well as a CD-ROM.

Press, Frank, and Raymond Siever. *Understanding Earth*. 2d ed. New York: W. H. Freeman, 1998. This comprehensive physical geology text covers the formation and development of the Earth. Readable by high school students, as well as by general readers. Includes an index and a glossary of terms.

Schwab, F. L., ed. *Geosynclines: Concept and Place Within Plate Tectonics*. Stroudsburg, Pa.: Hutchinson Ross, 1982. A collection of the significant scientific papers on geosynclines. The original manuscripts of James Hall and James Dana are reprinted. Individual papers are grouped by the years in which they were originally published. Each of these sections has a general introduction and overview by the book's editor that are very useful. The papers themselves require some knowledge of geological jargon. The introductions to each section can be read by the layperson.

Seyfert, Carl K., ed. *The Encyclopedia of Structural Geology and Plate Tectonics*. New York: Van Nostrand Reinhold, 1987. Contains articles of various lengths. The one on geosynclines is particularly long and well illustrated. All the different types of geosynclines, including eugeosynclines and miogeosynclines, are described in detail. Although the terminology can get confusing, this reference is the source for the classification of geosynclines. Suitable text for college-level students.

Uyeda, Seiya. *The New View of the Earth*. San Francisco: W. H. Freeman, 1978. Relates the development of the ideas of continental drift and seafloor spreading to the paradigm of plate tectonics; text is filled with accounts of the personalities involved in the evolution of plate tectonic theory. The place of the geosynclinal, or tectonic, cycle is discussed. Each concept is presented in detail and with clarity. Recommended for the high school or college student with a keen interest in geology.

Wilson, J. Tuzo, ed. *Continents Adrift*. San Francisco: W. H. Freeman, 1972.

_____. *Continents Adrift and Continents Aground*. San Francisco: W. H. Freeman, 1977. These two volumes are collections of articles on plate tectonics, continental drift, and seafloor spreading that were originally printed in *Scientific American*. The amount of overlap in the two volumes is very small, which amply illustrates the rapid advances in plate tectonics in the late 1960's and early 1970's. Geosynclines are referred to throughout the volumes although specific articles about the geosynclinal cycle appear in both volumes. Suitable for the layperson with an interest in continental drift, seafloor spreading, plate tectonics, and geosynclines. Illustrations.

JOINTS

Joints form when rocks undergo brittle failure, usually during expansion. Their orientations, physical features, and patterns of occurrence can be used to help infer the physical conditions present at the time of failure. Because joints are important conduits for fluids, particularly oil and water, that move beneath the surface, understanding their formation and occurrence has economic benefits.

PRINCIPAL TERMS

CHEMICAL WEATHERING: changes in rocks produced by reactions with fluids near the surface of the Earth

COLUMNAR JOINTING: the formation of columns, often with hexagonal cross sections, as joints grow inward from the outer surfaces of cooling igneous rock bodies

CONJUGATE SHEAR SETS: two sets of joints that make angles with each other of close to 60 degrees and 120 degrees

EXFOLIATION: the splitting off of curving sheets from the outside of a body of rock; also called sheeting

EXTENSION: expansion, or stretching apart, of rocks

FRACTOGRAPHY: the study of fracture surfaces to determine the propagation history of the crack

JOINT: a fracture in a rock across which there has been no substantial slip parallel to the fracture

CHARACTERISTICS OF JOINTS

Joints are the ubiquitous cracks found in nearly every outcrop. They are unquestionably the most common structure at the surface of the Earth. They vary in size from microscopic fractures visible only within an individual grain in a rock to fractures kilometers in length, some of which are responsible for the magnificent scenery of Arches National Park in Utah. If one smashes a rock with a hammer, one produces joints. Joints also form as molten rocks solidify and cool, as weathering alters the volume of the outer layers of a rock, and even as erosion removes overlying layers, reducing the weight on the layers below and permitting them to expand and crack.

There is no appreciable slip across a joint. (Failure surfaces accommodating large amounts of slip are faults.) Often, the same mineral grain can be observed on both sides of a joint, neatly cut in two but otherwise not disturbed. This indicates failure in extension; such joints are sometimes called extension joints. The sides of the fracture moved away from each other but did not slide past each other at all. The forces producing such joints were literally pulling the rock apart. An engineer might call these tensile forces. Geologists work with rocks that are nearly always in a state of compression, however, and true tension is uncommon. Therefore, geologists usually call such forces the forces of least compression. The direction of least compression is the direction in which extension occurs. That is, as the crack opens and the sides move away from each other, they will move in the direction of the least compressive force. Consequently, the plane of an extension joint will be perpendicular to the direction of the least compressive force.

FORMATION OF JOINTS

Such extension may be produced in a variety of ways. Most obvious are mechanisms involving large-scale deformation of the rock. The folding of a unit of rock causes extension on the outside of the fold. Often, extension joints are found fanning around the "nose" of a fold. The injection of molten rock into cooler rock, and its subsequent cooling, can fracture the cooler rock, producing joints. Extension is also produced during weathering when, because of chemical reactions at the surface of a rock, the surface layer expands more than the interior does. This expanded surface pulls away from the rest of the rock much as an onion skin pulls off the outside of an onion. This process is called exfoliation or sheeting. One classic example of this is Half Dome in Yosemite National Park, California.

Contraction, too, can cause joints. The familiar cracks that form in dried mud puddles are an example: Moist mud contracts as it dries, and cracks in the surface result when the forces involved in this contraction overcome the cohesive strength of the mud. As rock cools, such as when a molten, igneous rock body solidifies and then cools further, it contracts. The cooling and contraction are greatest at the surface of such a body. The polygonal patterns of cracks that form on such surfaces are very similar to those seen in dried mud. As the hot rock continues to cool, these cracks extend into the interior of the body. This may result in spectacular columns, such as those seen at Devils Postpile National Monument in California. The process is called columnar jointing.

The most common way joints form, however, is probably when rocks that equilibrated at depth are brought to the surface, either by mountain-building forces or when erosion removes the overlying rocks. This means of forming joints was proposed by Neville Price in his 1966 book *Fault and Joint Development in Brittle and Semi-Brittle Rock*. Although the model cannot be allied directly at any particular location because the deformation history, local topography, and other factors vary too much from place to place, it is instructive to consider the process in general terms.

How can vertical uplift produce horizontal extension? Consider large suspension bridges such as the Verrazano Narrows in New York City or the Golden Gate Bridge in San Francisco. The vertical towers supporting these bridges diverge from one another by about one-hundredth of a degree because of the curvature of the Earth. The tops of these towers are farther apart than their bases, so a rope that exactly reached between the bases would have to stretch a bit if it were raised to their tops. If it were unable to stretch, it would break.

Would uplift of 5 kilometers be sufficient to produce joints in a typical rock? If the Earth's circumference is 40,074 kilometers, the circumference of a circle lying 5 kilometers beneath the surface would be 40,043 kilometers. If that circle were brought to the surface, it would have to be stretched by 31 kilometers, or by 0.078 percent. This amount of stretch might not seem like much, as a block of rock 100 meters long would need to extend only 7.8 centimeters. However, if one tried to stretch the block of rock by attaching a gigantic pulling apparatus on it, experimental data show that the block would break before stretching that much.

In addition to this geometric extension, the rock would generally cool as it came up to the surface, contracting and making more joints in the process. Because the state of compression at depth would likely be different from that at the surface, the changes that occur during uplift might encourage or inhibit joint formation, depending on local conditions. Still, if all the joints currently at the surface of the Earth are considered, it appears as if the majority of extension joints

Exfoliation in granite near Shuteye Peak in the Sierra Nevada, Madera County, California. (U.S. Geological Survey)

may form by uplift. The fact that joints form during uplift and erosion does not mean, however, that they are necessarily unrelated to the structural history of the rocks in which they occur. Deformed rocks often contain stored-up energy, much like the energy stored in a spring, which was caused by the deformation. This energy, usually called residual stress, can influence the development of joints. Thus, joints that form hundreds of millions of years after a rock was initially deformed will often occur in patterns and orientations clearly related to that deformation.

SHEAR JOINTS

Some joints are not formed strictly by extension. These joints develop as a series of cracks, called shear joints, which break the rock into diamond-shaped pieces. These pieces slide slightly past one another, accommodating the deformation. Careful examination of these joints may show some slight offset of grains across the joint, but the displacement across any one joint is small. The cumulative effect across hundreds of joints, however, can be considerable. Often, these joints occur in two parallel sets, with angles of about 60 degrees and 120 degrees between them. Such sets are called conjugate shear sets.

Shear joints are not perpendicular to the direction of least compression, but it is possible to determine the direction of compression from the orientation of the conjugate shear sets. The direction of intermediate compression is indicated by the line of intersection of the joints. The direction of least compression bisects the obtuse angle between the sets, and the direction of maximum compression bisects the acute angle. In terms of a diamond-shaped piece, the direction of least compression is the short way across the diamond, and the direction of greatest compression is the long way across the diamond. It is not uncommon for conjugate shear sets to occur in conjunction with extension joints. In this case, the extension joints will be parallel to the long axes of the diamonds.

Consider the joints that might be associated with a fold that forms in a horizontal layer of rock not too far beneath the surface. The fold will form as the layer buckles in response to forces acting along it—in the north-south direction, for example—which is similar to the way a playing card flexes when one squeezes the edges between one's fingers. Early in the deformation process, least compression in the east-west direction results in extension joints running north-south and shear joints running northeast and northwest at a 30-degree angle. Eventually a buckle develops, folding the layer and producing extension fractures in the east-west direction. Much later, erosion and uplift may bring parts of this layer to the surface. The direction of the least compression at that time, which may have no relation at all to the forces that originally produced the fold, will control the vertical extension joints that develop because of this uplift. Finally, weathering and the vagaries of the topography at the time the rock is exposed to weathering control the exfoliation joints that will follow the shape of the exposed surface.

FRACTOGRAPHY

Fracture patterns—often called "decorations"—on the joint surfaces can yield useful information about the speed of fracture growth and the direction in which it grew.

Joints in granite cut by veins of quartz, feldspar, and biotite, Las Animas Canyon, San Juan County, Colorado. (U.S. Geological Survey)

This field of study is called fractography and has been developed by ceramic engineers concerned with reconstructing the brittle failure of glass and ceramic objects in order to improve their design. It can be directly applied to the study of joint surfaces.

When a fracture begins to grow, it starts with a low velocity but accelerates quickly. While it is moving slowly, the front of the fracture is usually a smooth curve, and the decoration it leaves on the fracture surface may be perfectly smooth, called the mirror region, or slightly frosted in appearance, in which case it is called mist hackle. If the crack grows intermittently, arrest lines may result. These curves show where the crack front was at different times when it temporarily stopped growing. As it increases in speed, the fracture front divides into a number of fingerlike projections. These commonly move a bit beyond the initial plane of the fracture as the fracture continues to grow. The result is a pattern on the surface of the fracture that has long been called plumose structure by geologists but is known as twist hackle by fractographers. It looks very much like a feather. The directions in which the fracture grew are shown by the directions of each slightly offset, curving element. Many of these features can be seen on building stones, flagstones, and slate floor tiles.

FLUIDS AND JOINTS

Joints provide conduits for the movement of fluids beneath the surface. Just as cracks in a pot permit water to leak through the pot, joints in the bedrock greatly enhance the rate at which water, oil, natural gas, and other fluids move through it. Near the surface, water is the fluid most likely to move through joints. As it does so, it is likely to attack the rock on both sides of the joint, chemically weathering it. This process enlarges the joint, increasing the flow of water through it, which in turn causes it to be weathered further, and the process continues. When the jointed rocks are limestone, the result may be elaborate systems of caverns, such as Carlsbad Caverns in New Mexico. Maps of such caves clearly demonstrate that joints controlled their development. In areas underlain by less soluble rock, joints may provide access to groundwater resources. By studying joint patterns displayed on geologic maps, aerial photographs, or satellite images, hydrologists are sometimes able to see where the natural underground flow of water may be greatest, and they can exploit this knowledge in their search for water.

Similarly, petroleum geologists seek conditions where joints may facilitate the movement of oil and gas toward potential well sites. Because the rocks of interest to them are often much deeper than those with useful water resources, petroleum geologists may be forced to guess the location of joints at depth. Although the surface traces of joints seen on maps can help, it is often necessary to apply an understanding of how and why joints form in order to predict where they may be at depth. In some cases, artificial joints are produced by pumping fluids under very high pressure into the rocks.

Otto H. Muller

CROSS-REFERENCES

Aerial Photography, 2739; African Rift Valley System, 611; Aquifers, 2005; Building Stone, 1545; Continental Rift Zones, 579; Displaced Terranes, 615; Fold Belts, 620; Folds, 624; Geosynclines, 630; Groundwater Movement, 2030; Igneous Rock Bodies, 1298; Karst Topography, 925; Landsat, 2780; Landslides and Slope Stability, 1501; Mining Processes, 1780; Oil and Gas Exploration, 1699; Ophiolites, 639; Stress and Strain, 264; Thrust Belts, 644; Weathering and Erosion, 2380.

BIBLIOGRAPHY

Hatcher, Robert D., Jr. *Structural Geology: Principles, Concepts, and Problems.* 2d ed. Englewood Cliffs, N.J.: Prentice-Hall, 1995. This undergraduate textbook covers joints and faults in six chapters. Intended for the more advanced reader.

Marshak, Stephen, and Gautam Mitra, eds. *Basic Methods of Structural Geology.* Englewood Cliffs, N.J.: Prentice-Hall, 1988. Chapter 12, "Analysis of Fracture Array Geometry," by Arthur Goldstein and Stephen Marshak (18 pages), gives a good description of joint surface morphology, an overview of the field methods that have been used in studying them, and a summary of graphing and statistical techniques. The treatments are brief, but references are complete. Seems to have been designed for the reader who already knows something about joints but wishes to learn more about how to study them. College level.

Plummer, Charles C., David McGeary, and Diane H. Carlson. *Physical Geology*. Boston: McGraw-Hill, 1999. This is a straightforward, easy-to-read introduction to geology intended for those with little or no science background. Includes many excellent illustrations, as well as a CD-ROM.

Press, Frank, and Raymond Siever. *Understanding Earth*. 2d ed. New York: W. H. Freeman, 1998. This comprehensive physical geology text covers the formation and development of the Earth. Readable by high school students, as well as by general readers. Includes an index and a glossary of terms.

Price, Neville J. *Fault and Joint Development in Brittle and Semi-Brittle Rock*. London: Pergamon Press, 1966. This 176-page book is a classic in its field. Most of the work done on joints since its publication has been done by people who have read this book, so its influ-

ence is great. Suffers, however, from having been written before modern experimental methods for studying rock deformation had been fully developed and from its use of the British system of measurement. Contains only four photographs, of somewhat limited usefulness, and other figures are schematic, though clear. Not overly technical, with a minimum of mathematical derivations, this book gives a fair summary of what was known about joints at the time it was written. Suitable for college-level students.

Suppe, John, ed. *Principles of Structural Geology*. Englewood Cliffs, N.J.: Prentice-Hall, 1985. Chapter 6, "Joints," presents an excellent overview of the classification, appearance, and formation of joints. The sixteen photographs show all manner of joints at a variety of scales, and some of the fourteen line drawings provide good examples of how joint studies have been used in interpreting structures and the state of stress in the crust. At times, the discussion becomes a bit technical, and several references are made to chapter 4, "Fracture and Brittle Behavior," which is more technical yet. A useful digression concerns the state of stress in the upper and lower crust and in the mantle. College-level reading.

OPHIOLITES

Ophiolites are a unique assemblage of rocks found in many mountain belts throughout the world. They were formed in the oceans and subsequently transported to land during mountain-building processes. Ophiolites are useful to geologists as indicators of the location of ancient oceans. They are important to society because they are major sources of asbestos, nickel, and copper minerals.

PRINCIPAL TERMS

IGNEOUS ROCK: a rock formed when magma cools and forms minerals; it can form on the surface of the Earth when volcanoes erupt, or it can form at depth, without reaching the Earth's surface

LITHOSPHERE: the upper, rigid 100 kilometers of the Earth that forms the moving plates; beneath oceans it is called oceanic lithosphere, and beneath continents it is continental lithosphere

MAFIC: a rock rich in iron and magnesium minerals; ultramafic rocks are very rich in these minerals

OLIVINE: a mineral consisting of magnesium, iron, silicon, and oxygen

PERIDOTITE: an igneous rock consisting of minerals rich in magnesium; serpentinized peridotite has been heated such that the mineral olivine is partially converted into serpentine

PILLOW LAVA: the lava that is formed when molten rock erupts into water and cools in the shape of a pillow

PLATE TECTONICS: the theory that the upper part of the Earth consists of a number of "plates," or rigid parts, that move relative to one another across the surface

SERPENTINE: a mineral consisting of magnesium, iron, silicon, oxygen, and water

CHARACTERISTICS OF OPHIOLITES

Ophiolites are unique assemblages of rocks that have fascinated geologists for centuries. The word "ophiolite" is derived from the Greek *ophis*, meaning snake or serpent. The term first appeared in the geological literature in the 1820's, when Alexandre Brongniart of France used it to describe rocks called serpentinite, which are made entirely of the mineral serpentine. The term "ophiolite" is appropriate because serpentinite, like some snakes, has a mottled green appearance.

The usage of the term "ophiolite" changed early in the twentieth century when the close association of serpentinite or serpentinized peridotite with deep-water sediments and with pillow lavas was noted. This assemblage of three rock types became known as the Steinmann Trinity. In the late 1960's, with the advent of plate tectonics as a unifying theory in geology, the definition of ophiolite again changed and was used to describe the Steinmann Trinity plus other rocks arranged in a particular sequence. At the same time, the interpretation that ophiolites formed in the oceans be-

came the cornerstone for reconstructing the history of mountains. As defined for modern usage by a Geological Society of America conference in 1972, "ophiolite" refers to a distinctive assemblage of mafic and ultramafic rocks. A completely developed ophiolite is 10 to 12 kilometers thick and covers an area of hundreds of square kilometers.

Rocks in ophiolites occur in layers in the following order (starting from the bottom and working up): peridotite, gabbro, sheeted dike complex, pillow lavas, and chert. Peridotite is an ultramafic igneous rock formed below the Earth's surface, dark in color, consisting of large grains (greater than 5 millimeters) of the minerals olivine (an olive-green mineral rich in magnesium, iron, silicon, and oxygen) and pyroxene (a black mineral rich in iron, magnesium, silicon, and oxygen). In many ophiolites, peridotite has been subjected to heat and hot water since its formation, resulting in some of the olivine minerals changing to serpentine, some of which is asbestos.

Gabbro is a mafic igneous rock formed below the Earth's surface that has a salt-and-pepper ap-

Gabbro is a mafic igneous rock formed below the Earth's surface that has a salt-and-pepper appearance. This example is mostly labradorite with some dark pyroxene. (©William E. Ferguson)

FORMATION OF OPHIOLITES

Ophiolites can form in two different plate tectonic environments, both of which are in oceans. In one environment, such as in the middle of the Atlantic Ocean, two plates are moving away from each other. As they do so, magma from the interior of the Earth moves up to the bottom of the oceans and cools to form long mountain chains called oceanic ridges (for example, the Mid-Atlantic Ridge). The solidified rock becomes part of an outer layer of the Earth called the oceanic lithosphere. The newly formed oceanic lithosphere then moves away from the ridge at which it formed, resulting in widening of the ocean. (For example, the Atlantic Ocean is growing wider as North America moves westward, away from Europe.) Ophiolites are believed to be remnants of the top part of oceanic lithosphere that was created at ancient oceanic ridges. All the igneous rocks in ophiolites formed by the cooling of magma produced deep in the Earth and spewed forth at an oceanic ridge. Pillow lavas form when the magma erupts into seawater; sheeted dikes, gabbros, and peridotites form when that magma cools at various depths below the surface. Chert forms when fine particles, including mineral grains, animal parts, and plants, fall to the bottom of the newly created ocean floor to form sedimentary deposits.

A second environment where ophiolites may form is the region where two plates are moving toward each other. In this environment, oceanic lithosphere that formed at oceanic ridges disappears down into the interior of the Earth, resulting in melting at depth, the production of highly explosive volcanoes, and the generation of dangerous earthquakes. This general area is known as a subduction zone. When oceanic lithosphere has just begun to subduct (descend into the Earth's interior) beneath another oceanic lithospheric

pearance. It consists of large grains (greater than 3 millimeters) of the minerals pyroxene and feldspar (a white mineral rich in calcium, sodium, aluminum, and oxygen). Sheeted dike complex is made up of mafic igneous rocks formed just below the Earth's surface, similar to gabbro but with smaller mineral grains (less than 3 millimeters). Pillow lavas are mafic igneous rocks formed on the Earth's surface as lava erupts into water, similar to gabbro but with much smaller grains (less than 1 millimeter). Chert is a sedimentary rock formed in deep water consisting of very small grains (some too small to be seen with a microscope) of various minerals and organic remains.

Not all ophiolites exhibit the complete sequence of rocks; some were formed with certain layers missing, and others were dismembered subsequent to their formation. A very important observation is that beneath all ophiolites there is a fault (fracture in the earth) separating them from underlying rocks. Furthermore, there is evidence that ophiolites have probably moved hundreds of kilometers along those faults.

plate, ophiolites may form in the region above the subducting plate. The processes involved in formation of these ophiolites are the same as at an oceanic ridge, though the environments are different.

MOVEMENT OF OPHIOLITES

Another important part of the geological history of ophiolites is how they are moved from their place of formation in the oceans to their current location on continents. The key lies in understanding how mountains are built. Ophiolites have been described from all the world's major mountain belts: the Appalachian Mountains of eastern North America, the Rockies in western North America, the Alps in Europe, the Himalaya in Asia, the Ural Mountains, the Andes in South America, and the mountains of Papua, New Guinea. They range in age from about 600 million to 15 million years old. The movement of ophiolites from the ocean floor to mountains is related to plate tectonics. Subduction zones, where plates are moving toward each other, provide the mechanism for emplacement of ophiolites.

To get a better idea of how ophiolites may have been put into place in the past, scientists consider what might happen to the Pacific Ocean in the future. The Pacific plate (that is, the oceanic lithosphere beneath the Pacific Ocean) is moving westward toward, and subducting beneath, the Asian plate along the Japanese islands, resulting in volcanoes and earthquakes along the zone. The Pacific plate is also subducting eastward beneath western North America. Therefore, the Pacific Ocean will become smaller, the North American continent will move closer to the Asian continent, and they will eventually collide. During this time, probably 99.99 percent of the Pacific oceanic lithosphere (potential ophiolites) will be lost to the interior of the Earth at the subduction zones. At the time of the collision of North America and Asia, however, a small portion (less that 0.01 percent) of the oceanic lithosphere may not slip back into the interior but instead may be pushed up onto the continents, because when the two continents collide, mountains are being built, and some of the Pacific oceanic lithosphere, sandwiched in the zone of collision, may be pushed up with the mountains. These portions of the Pacific oceanic lithosphere will then be called ophiolites.

In a similar fashion, the ophiolites that are now visible in mountain belts are tiny remnants of ancient oceanic lithosphere pushed onto continents during ancient mountain building.

The occurrence of ophiolites in mountains has helped geologists understand how mountains are built. Ophiolites lie in narrow bands that continue for thousands of kilometers along a mountain belt. This narrow band represents the zone where two continents have come together to build the mountains. Thus, by locating ophiolites, geologists can identify and then characterize the two (or more) continents that existed separately before the mountains were built. For example, in the Appalachian Mountains of eastern North America, a narrow band of ophiolites extends from western Newfoundland, Canada, through Quebec, Canada, into Vermont and as far south as Alabama in the United States. This zone may represent the boundary between the ancient North American continent to the west and the ancient African continent to the east; the two ancient continents were plastered together during the building of the Appalachians over a span of time between 500 million and 350 million years ago.

STUDY OF OPHIOLITES

Geologists have used a variety of techniques to study ophiolites, including field mapping, seismic studies, electron microprobe techniques, and X-ray techniques. Field mapping of ophiolites involves identifying rock types, noting their locations on maps, and ascertaining the relationship of the ophiolites with neighboring rocks; it is the basic tool used by the geologist to analyze the origin of ophiolites, and it precedes any laboratory techniques.

Seismic studies have been employed to compare ophiolites to the modern oceanic lithosphere. Seismic waves travel through the Earth after earthquakes; they can also be created by human-made explosions or by hitting the ground hard with a sledgehammer. Seismic waves travel at different speeds through different rock types. Thus, if scientists can identify the velocities of seismic waves as they travel through different types of rock, the type of rock can be identified even though it cannot be seen. In the oceans, seismic velocity measurements through rocks beneath the ocean floor have shown that the modern oceanic lithosphere is made up of layers that correspond exactly in rock type and in thickness to what is observed in ophiolites, which is very good evidence to support the interpretation that ophiolites formed as portions of oceanic lithosphere.

X-ray techniques have been used to determine the chemical composition of rocks in ophiolites. In a technique called X-ray fluorescence, X rays bombard a powdered rock sample and interact with electrons of the chemical elements to produce secondary X rays. The secondary X rays generated by specific elements can be isolated and counted to determine the amounts or percentages of specific elements present in the rock. Results of such studies on ophiolites have shown that pillow lava and sheeted dike rocks are similar to lavas erupted in modern oceans. In particular, low percentages of potassium oxide and of light rare-earth elements are common to both ophiolitic rocks and modern oceanic rocks.

The electron microprobe is also used extensively to help interpret the origin of ophiolites. In the electron microprobe, electrons hit a rock sample at high speed and interact with electrons in chemical elements in the rock to produce X rays. In the same way as in X-ray techniques, amounts of the chemical elements can be determined. In electron microprobe work, however, the rock sample, rather than being a crushed or powdered specimen, is a thin section (0.03 millimeter thick) of an intact rock that can be viewed through a microscope. Thus, minerals in the rock can be seen, selected, and analyzed for their chemical constituents.

Chemical analyses of minerals are most useful in the peridotite and gabbro layers of ophiolites.

For example, chemical analyses of olivine, pyroxene, and feldspar from gabbro layers and upper parts of peridotite layers have shown that these layers precipitated from a magma of the same composition as the pillow lavas and sheeted dikes. Therefore, nearly all the igneous rocks in ophiolites can be related to one magma, part of which cooled at shallow depths (about 5 kilometers below the surface) to produce gabbro and peridotite and some of which was injected into higher levels in the Earth to produce the sheeted dikes and pillow lavas. The lower parts of the peridotite layer, however, must have originated in a different way. The amounts of aluminum in the mineral pyroxene reveal that the lower peridotite layer formed under high pressure at considerable depth, as much as 30 kilometers below the surface of the Earth— good evidence for the suggestion that ophiolites that now exist on the surface of the Earth must have been moved since the time of their formation.

Raymond A. Coish

CROSS-REFERENCES

African Rift Valley System, 611; Displaced Terranes, 615; Fold Belts, 620; Folds, 624; Geosynclines, 630; Hydrothermal Mineralization, 1205; Igneous Rock Classification, 1303; Joints, 634; Lithospheric Plates, 55; Mountain Belts, 841; Oceanic Crust, 675; Plate Margins, 73; Plate Motions, 80; Plate Tectonics, 86; Sedimentary Rock Classification, 1457; Spreading Centers, 727; Sub-Seafloor Metamorphism, 1427; Subduction and Orogeny, 92; Thrust Belts, 644; Ultramafic Rocks, 1360.

BIBLIOGRAPHY

Chernikoff, Stanley. *Geology: An Introduction to Physical Geology.* Boston: Houghton Mifflin, 1999. This is a good overview of the scientific understanding of the geology of the Earth and surface processes. Contains information on the formation of ophiolites. Includes a link to a Web site that provides regular updates on geologic events around the globe.

Coleman, R. C. "The Diversity of Ophiolites." *Geologie en Mijnbow* 63 (1984): 141. A very good review of the state of knowledge on the origin of ophiolites. Suitable for more advanced students of geology.

_____. *Ophiolites: Ancient Oceanic Lithosphere?* New York: Springer-Verlag, 1977. The most detailed treatment of ophiolites available. Includes sections on field descriptions, chemistry of the rocks, and models to explain the origin of ophiolites. Also includes detailed descriptions of four major ophiolites in the world. Suitable for those with previous exposure to geology.

Dolgoff, Anatole. *Physical Geology.* Lexington, Mass.: D. C. Heath, 1996. This is a comprehensive guide to the study of the Earth. Extremely well illustrated and includes a glossary and an index. Although this is an introductory text for college students, it is written in a style that makes it understandable to the interested layperson. Contains a section on the formation of ophiolites.

Dott, Robert H., Jr., and Donald R. Prothero. *Evolution of the Earth.* 5th ed. New York: McGraw-Hill, 1994. This basic textbook on historical geology is aimed at students of geology. However, it is very readable by anyone with a background in science. Presents an up-to-date account of the Earth's history from the viewpoint of plate tectonics. Includes a glossary.

Francheteau, Jean. "The Oceanic Crust." *Scientific American* 249 (September, 1983): 130. Summarizes information on how modern oceanic crust is formed. Relevant to ophiolites because they are thought to have formed as ancient oceanic crust. Suitable for advanced high school and college students.

Gass, Ian G. "Ophiolites." *Scientific American* 247 (August, 1982): 122. A comprehensive treatment of the evolution of ideas on the origin of ophiolites. Well illustrated with colored diagrams and photographs. Suitable for advanced high school students, college students, and interested adults.

Hamblin, W. K. *The Earth's Dynamic Systems.* 4th ed. New York: Macmillan, 1985. A very good general text to introduce readers to the field of geology. Chapters 17 through 20 are particularly informative on plate tectonics, ocean-floor processes, and mountain building. Suitable for anyone interested in geology.

THRUST BELTS

Thrust belts are long, narrow zones composed of many thrust faults that record significant horizontal displacements of the continental crust. Thrust faults are surfaces that emplace older rock above younger rock. Exploration of thrust belts has led to the discovery of vast reserves of oil and natural gas.

PRINCIPAL TERMS

CONTINENTAL CRUST: the upper 50 kilometers of the Earth below continents; it is composed primarily of rock that is lighter than that which makes up oceanic crust

CONTINENTAL MARGIN: the edge of a continent next to an ocean basin; it is both exposed on land and submerged below water

FAULTING: the process of fracturing the Earth such that rocks on opposite sides of the fracture move relative to each other; faults are the structures produced during the process

FOLDING: the process of bending horizontal layers of rock so that they dip; folds include anticlines, which are arches, and synclines, which are shaped like the letter U or V

GEOSYNCLINE: a major depression in the surface of the Earth where sediments accumulate; geosynclines lie parallel to the edges of continents and are long and narrow

OROGENIC BELT: a mountain belt composed of a core of metamorphic and plutonic rocks and an adjacent thrust belt

DISCOVERY OF THRUST BELTS

Thrust belts are arcuate zones in which the surface of the Earth has been shortened in its horizontal dimension, primarily by faulting. The zones are expressed as mountain ranges hundreds of kilometers wide and thousands of kilometers long. They are composed of faulted sedimentary rock and are integral parts of the linear orogenic belts that mark the edge of every continent. Active thrust belts are areas of intense deformation, rugged topography, and frequent earthquakes. Thrust faults are the principal features of thrust belts and are gently inclined surfaces along which older rocks, originally near the bottom of a pile of sediments, are pushed up and over younger rocks at the top. The direction in which the rocks move is perpendicular to the length of the thrust belt. Thrust belts frequently merge with fold belts. Fold belts are similar to thrust belts except that the principal structures formed are folds several kilometers wide. Fold and thrust belts form along continental margins, which are also convergent plate boundaries, and record significant horizontal movements within the continental crust.

Thrust faults were first recognized in the mid- to late nineteenth century in the Scottish Highlands. Geologists working there prior to the middle of the nineteenth century believed that sedimentary strata were interbedded with metamorphic rocks, an interpretation that arose from a lack of knowledge about the process of metamorphism. It is now known that during metamorphism, heat transforms sedimentary rocks into metamorphic rocks. The transformation occurs gradually both in space and in time.

Rocks adjacent to each other should therefore exhibit only slight differences in metamorphism, an implication that led to the conclusion that interbedding of the metamorphic rocks and the sedimentary strata in Scotland was unlikely. At about the same time, another important discovery was made: The metamorphic rocks high in the sediment pile contained older fossils than did the sedimentary strata below. Because the sedimentary sequence was not believed to be upside down, the idea of great gliding of the older, metamorphic rocks up and over the younger, sedimentary strata was proposed. This idea of thrust faulting was quickly accepted and applied to other areas, including the United States, where, in the early twentieth century, Bailey Willis documented that rocks greater than 500 million years old were thrust above strata fewer than 100 million years old in Glacier National Park.

CHARACTERISTICS OF THRUST BELTS

Thrust belts are elongate areas of faulted and relatively unmetamorphosed sedimentary strata that border linear zones of metamorphic and plutonic rock (a pluton is a body of rock that crystallizes from a liquid below the surface). Together, the thrust belt and the zone of metamorphic and plutonic rocks define two parts of an orogenic belt: the core of the orogenic belt, where temperatures were high enough to transform sedimentary strata into metamorphic rocks, and the thrust belt, where temperatures were too low for metamorphism. Deformation in the core of the orogenic belt is complex. In contrast, the deformation in the neighboring thrust belt, although still severe, is fairly simple to unravel. Thrust belts taper from thicknesses on the order of 10,000 to 20,000 meters next to the metamorphic core to less than 5,000 meters at the opposite side, defin-

ing a wedge of faulted sedimentary strata that is inclined gently toward the core. Sediments closest to the core typically have been thrust farther than 100 kilometers.

In most thrust belts, deep-water oceanic rocks are thrust above shallow-water marine strata. The thrusts originate where the deep-water rocks were deposited and move toward the region of shallow marine sedimentation. The area toward which the thrusts move is termed the foreland and is away from the core of the orogenic belt, which is called the hinterland. The thrust faults are not planar surfaces of uniform orientation but are composed of a series of flat faults that parallel layering in the sedimentary rocks and ramp faults that cut obliquely across layering to join flat faults at different depths. The rocks above a ramp fault move upward, whereas rocks above a flat fault move horizontally, relative to those below. The depth of the

In Glacier National Park, rocks more than 500 million years old have been thrust above strata less than 100 million years old. (PhotoDisc)

flat segments of a single thrust fault decreases toward the foreland, imparting a staircase shape to the thrust fault and forcing deep-water rocks to climb to the surface, away from the core of the orogenic belt. The sliding of rocks along a ramp produces folds of a characteristic shape, broad anticlines and narrow synclines, as the rocks bend to accommodate the motion.

After one thrust moves older rocks up and over adjacent younger sediments, a new thrust parallel to the old one forms in the younger strata on the foreland side of the old thrust. Parallelism of the two thrusts allows the thrust slice (the rock between the two thrusts) to be tens of kilometers long. The horizontal distance between the old and new thrusts, usually on the order of hundreds of meters, defines the width of the thrust slice and depends primarily on the thickness of the pile of rocks being faulted. The new thrust then carries the younger strata and the old thrust toward the foreland and above still younger sediments. This process, called piggy-backing, continues such that the thrust slices in the thrust belt are stacked like a series of cards. Because piggy-backing is relatively straightforward, deformation in thrust belts is well understood. The cumulative horizontal displacement resulting from piggy-backing can be as large as 200 kilometers.

DÉCOLLEMENT

The many thrusts in a thrust belt root into one flat fault at a depth of 10 to 20 kilometers, called the detachment surface or the décollement. Each separate thrust begins as a ramp fault that joins the décollement to a flat fault closer to the surface. The style of deformation above the décollement differs from that below. Thrusting occurs only above the décollement; rocks below, however, resist faulting and deform by changing their shape. The depth of the décollement commonly is controlled by a discontinuity that marks a change from brittle material above to ductile (changes shape without breaking when subjected to stress) below.

An example of a material that can be either ductile or brittle, depending on its temperature, is candle wax or paraffin. When warm paraffin is squeezed, it simply changes shape without breaking. Cold paraffin, however, fractures under compression. The warm paraffin is ductile; the cold

paraffin is brittle. Rocks behave similarly. Such a transition from brittle to ductile material exists at depth in the continental crust where sedimentary strata lie above metamorphic or plutonic rocks. The sedimentary strata are brittle and are termed "cover." Conversely, the metamorphic and plutonic rocks are ductile and are called "basement." Because faulting in thrust belts is restricted to the brittle "cover" rocks, the deformation is known as thin-skinned. Thin-skinned thrusting significantly shortens the horizontal and thickens the vertical dimensions of the upper 10 to 20 kilometers of the continental crust. The effects of thin-skinned thrusting on basement rocks is largely unknown.

ACTIVE THRUST BELTS

Active thrust belts are localized along the edges of continents, which are also convergent plate boundaries. The shortening of the upper 10 kilometers of continental crust by thin-skinned thrusting provides a mechanism by which the horizontal compressive forces generated during the convergence of two plates can be accommodated. Essential to the formation of thrust belts are thick sequences of layered rock such as those that accumulate in geosynclines. If the rock being deformed is not thick enough, it will not transmit the compressive forces necessary for the development of thrust faults. Examples of active thrust belts include the foothills to the Himalaya in India and Nepal where the Asian and Indian plates are colliding, the Andes of South America where the Pacific and South American plates are converging, and the Transverse Ranges of California where the Pacific plate and North American plates are converging slightly.

Large basins filled with sediment border all these continental margins. The Appalachian Mountains of Tennessee and North Carolina are an ancient thrust belt. There, slices of sediment that were deposited in geosynclines next to North America were thrust westward 225 million years ago as the continents of Africa and North America collided. The Appalachian Mountains are part of an orogenic belt that is greater than 3,000 kilometers long and almost 500 kilometers wide. In contrast to Tennessee and North Carolina, the northern part of the Appalachian Mountains in Virginia and Pennsylvania is a fold belt. The folds are several kilometers wide and are responsible for the valley and ridge topography characteristic of the northern Appalachian Mountains. Another ancient thrust belt is the Rocky Mountains of Idaho and Wyoming. It was there that piggy-backing was first documented.

Pamela Jansma

CROSS-REFERENCES

African Rift Valley System, 611; Continental Crust, 560; Displaced Terranes, 615; Faults: Thrust, 226; Fold Belts, 620; Folds, 624; Geosynclines, 630; Joints, 634; Mountain Belts, 841; Ophiolites, 639; Plate Margins, 73; Plate Tectonics, 86; Seismic Reflection Profiling, 371; Stress and Strain, 264; Subduction and Orogeny, 92; Well Logging, 1733.

BIBLIOGRAPHY

Finkl, Charles W., ed. *The Encyclopedia of Field and General Geology.* New York: Van Nostrand Reinhold, 1988. This reference contains many useful articles about field and geophysical investigations and oil and natural gas exploration, including petroleum geology, petroleum exploration geochemistry, geological survey and mapping, and geophysical methods. The article on geophysical methods is much less technical than are the other two sources cited here. Well illustrated. Suitable for college-level readers.

Hatcher, Robert D., Jr. *Structural Geology: Principles, Concepts, and Problems.* 2d ed. Englewood Cliffs, N.J.: Prentice-Hall, 1995. This undergraduate textbook covers folds in three chapters. Intended for the more advanced reader.

Plummer, Charles C., David McGeary, and Diane H. Carlson. *Physical Geology.* Boston: McGraw-Hill, 1999. This is a straightforward introduction to geology intended for those with little or no science background. Discusses the development of thrust belts. Includes many excellent illustrations, as well as a CD-ROM.

Short, N. M., and R. W. Blair. *Geomorphology from Space: A Global Overview of Regional Landforms.* NASA SP-486. Washington, D.C.: National Aeronautics and Space Administration, 1986. This book contains beautiful pictures taken by various satellites that orbit the Earth. Text accompanies each picture and explains the tectonic setting. Many thrust belts are shown. Although the text is fairly technical, the photographs are worth examining by anyone. Recommended for college-level students.

Skinner, Brian F., and Stephen C. Porter. *The Dynamic Earth: An Introduction to Physical Geology.* New York: John Wiley & Sons, 1989. This book contains excellent descriptions of thrusts and thrust belts. Beautiful illustrations abound. The text is similar to *The Earth* by Press and Siever but has an approach that is geared more to the practical aspects of geology, such as mineral and petroleum exploration. Suitable for senior high school or college-level students.

3
OCEANIC CRUST AND ITS FEATURES

ABYSSAL SEAFLOOR

The abyssal plains of the oceans lie beyond the continental margins at depths greater than 2,000 meters. They are thought to be the flattest areas on the Earth and are carpeted with thick layers of sediment. Their greatest economic value lies in the metallic minerals that form part of these sediments.

PRINCIPAL TERMS

MANGANESE NODULES: lumps of minerals consisting mostly of iron, manganese, nickel, and copper that form on deeper parts of continental shelves

SEDIMENT: solid matter, either organic or inorganic in origin, that settles on a surface; it may be transported by wind, water, or glaciers

SUBMARINE CANYON: a channel cut deep in the seafloor sediments by rivers or submarine currents

TERRIGENOUS: originating from the weathering and erosion of mountains and other land formations

TRENCH: a long, narrow, and very deep depression in the ocean floor

TURBIDITY CURRENT: a current resulting from a density increase brought about by increased water turbidity; the turbid mass continues under the force of gravity down a submarine slope

DEEP OCEAN FLOOR

The abyssal plains of the deep ocean floor represent the flattest surface areas on the Earth. They are far flatter than any plain on land. Geologists define an abyssal plain as having a slope ratio of less than 1:1,000. Abyssal plains occupy about 40 percent of the ocean basin floor and are widespread in the major ocean basins, the Gulf of Mexico, and the Mediterranean Sea. The peculiar topography of the abyssal plains is the result of deep sediments deposited by turbidity currents. Additional sediments are derived from the rain of biological material from the surface. At one time, mariners and researchers believed that the entire ocean basin beyond the continental margin was a flat, featureless plain. Subsequent studies using sonic devices, deep-sea cameras, submersibles, and other instruments revealed a rugged, varied topography over most of the ocean floor. The research also supported the conclusion that abyssal plains represent less than one-half the area of all the ocean basins.

It was, however, not always thus. As the continents drifted apart and the ocean basins were formed following the breakup of the supercontinent Pangaea about 150 million years ago, the ocean floor was well contoured. The dominant feature of what are now the abyssal plains was broad areas of low hills. Weathering of the continental landmasses produced an abundance of sedimentary material that eroded into the oceans. The coarse material settled on the continental shelves and partly on the continental slopes. The fine material drifted farther offshore and settled on the continental rise and the adjacent abyssal hills. In the course of time, the hills were covered by the sediment to become the abyssal plains. Remnants of the ancient terrain exist beyond the abyssal plains in the form of abyssal hills less than 1,000 meters high, steep-sided seamounts, and flat-topped seamounts. Seamounts frequently jut above the ocean surface as islands. Guyots are seamounts that have been eroded by ocean waves. Also called table mounts, guyots may be 1,000 to 1,500 meters below the sea surface. In some areas of the ocean, the abyssal plains are cut by deep, narrow trenches whose bottoms may lie many kilometers below the surface of the sea. The deepest is the Mariana Trench in the southwest Pacific. At 11,000 meters below the surface, it is the deepest place on the Earth.

Abyssal plains are most abundant in the Atlantic Ocean and Indian Ocean basins. They usually form next to the edges of the continental margins rather than near the centers of the basins. The Pacific Ocean basin features a few abyssal plains, but for the most part the Pacific basin exhibits the relict, rough topography. There, the abyssal hills, thinly covered with sediment, rise 200 to 400 meters above the basin floor. Deep-sea drilling in the northeast Pacific has revealed basalt as the major rock type of the abyssal hills.

Several explanations have been offered for the thin sedimentation of the Pacific abyssal plains. One theory suggests that since there are relatively few large rivers that drain into the Pacific Ocean, transport of sediments is reduced; therefore, the sediments deposited since the Pacific was formed were too sparse to bury the hilly topography. In addition, magmatic arcs formed in some areas, creating marginal seas where sediments discharged by rivers were trapped. Another explanation points to the submarine trenches as possible traps for the sediments where the trenches lie between the continental margins and the abyssal plains. Sediment-laden turbidity currents, flowing down the steep sides of the continental slopes, plunge into the trenches and dump the sediment load. With much of the sediment going into the trenches, little of it flows out over the plains. A third possibility is the powerful "storms" that sweep across the ocean floor in places, scouring it in some regions and reforming the sediments in others. Oceanographic instruments moored on the ocean floor at depths of 4,800 meters have detected massive bottom currents flowing at a rate of more than 0.5 meter per second. Labeled as storms by researchers, the turbulent conditions can rage for about one week, lifting loads of sediment and moving them elsewhere.

SOURCES OF SEDIMENT

Each year, nearly 15 billion tons of weathered rock is eroded from the land and carried by rivers and streams to the sea. Some of this material is deposited in vast deltas, such as those found at the mouths of the Mississippi and Amazon Rivers. A large proportion is trapped on the continental shelves. A few billion tons are transported over the great depths to settle on the abyssal plains. They are joined by a rain of calcareous or siliceous skeletons of microscopic drifting organisms called plankton, which are abundant in the upper, sunlit portion of the water. These biological remains add

about 3 billion tons of sediment annually.

In addition, the sediment includes particles swept up by strong winds blowing over the deserts of the world, such as the Sahara. These particles travel great distances through the atmosphere and settle over the vast expanse of the oceans. They settle slowly to the bottom to form part of the carpet of sediment. Sediments may also be transported from the continental shelf onto the abyssal plain. This process has been detected in the Gulf of Mexico, where underwater landslides break off masses of sediment deposited by the Mississippi River. During times of low sea level, ocean waves breaking at the shelf edge cause the sediments to collapse and slide down the steep continental slope. From there, the sediments fan out in the deep ocean over the abyssal plain.

The thin carpet on the ocean floor includes mineral matter derived from a variety of sources. It includes the ash erupted by volcanoes thousands of kilometers away and extraterrestrial material in the form of meteorites. Some minerals precipitate directly out of the seawater and accumulate as crystals or nodules, including manganese nodules. These valuable, mineral-rich nodules in places obscure the sediments beneath them. Because of the biological skeletons in them, these sediments have a very fine texture and are classified as oozes, which are named after the dominant organism represented: globigerina ooze, radiolarian ooze, and so on. The oozes are calcareous if the dominant organisms were foraminifera and pteropods (animals with chalky skeletons); they are siliceous if the remains were derived from radiolarians or diatoms.

Red clay is a very common sediment on the abyssal plains. It consists of the finest-grained particles eroded into the sea. It is the most durable of the sediment types on the ocean floor. Calcareous skeletons dissolve rapidly as they sink toward the floor and are scarce in the sediments. Siliceous skeletons are limited in their distribution. They are found mainly under the productive surface waters of the polar and equatorial zones.

Sediments accumulate very slowly on the abyssal plains. The amount and time involved ranges from a few centimeters to only a fraction of a millimeter per one thousand years. Clay particles accumulate at a rate of less than 2 millimeters every one thousand years; shells may accumulate at a rate of 20 millimeters every one thousand years.

SEDIMENT CLASSIFICATION

Abyssal plain sediments are classified by geologists as terrigenous, biogenic, hydrogenous (or authigenic), and cosmogenic. The latter include meteorites and tektites, small rounded objects composed almost entirely of glass. Although their origin is unknown, tektites are believed to be meteoric.

Terrigenous sediments are the clays, sands, and gravels derived from the land. Most of them are lacustrine sediments; that is, they were transported by water as material eroded by rivers and streams. Some, however, were transported by glaciers during the Pleistocene epoch. Many sites of such sediments, including Georges Bank, off New England, and the Grand Bank, off Newfoundland, are the terminal moraines of such glaciers. During the thousands of years since they were deposited, these sandy moraines have been worked and reworked by ocean currents and waves and deposited farther out onto the abyssal plains. Some glacial deposits have been moved thousands of kilometers by floating icebergs. As the continental glaciers, such as on Greenland and Antarctica, travel toward the sea, they erode and transport sand, silt, and gravel. These are incorporated in the ice and, as the glacial ice breaks off into the sea as icebergs, the sediments are carried off by ocean currents. Ultimately, when the icebergs melt, they drop their mineral burdens into the sea to become part of the seafloor sediments. Many of the fine components of the terrigenous sediments, including the silts and clays, are aeolian, or wind-blown, sediments.

The biogenic sediments owe their origin entirely to materials derived from organisms in the water column. They include the tests, or shells and external skeletons, of phytoplankton and zooplankton. As the organisms die in the sunlit, or photic, zone, their remains fall to the ocean floor as "snow." Many biogenic sediments are highly fossiliferous. These fossil-rich deposits enable scientists to date the sediments and to interpret ocean temperatures in the geologic past. Carbon 14 dating has also been used to establish the relative age of biogenic sediments where carbonaceous minerals are present.

Hydrogenous, or authigenic, sediments are those that have precipitated out of the seawater solution. These sediments are particularly common on abyssal plains in the Pacific. They include phosphorite nodules and manganese nodules. Some hydrogenous sediments are found in the vicinity of hot springs near mid-ocean ridges. The minerals in the metal-rich water spewing from the springs are carried by deep ocean currents across the seafloor. The metals precipitate out and add to the layering on the plain. The nodules grow by accretion around some hard nucleus, usually a pebble or a fossilized shark's tooth. The temperature of the water from the hot springs generally is about 10 to 15 degrees Celsius (compared to the ambient deepwater temperature of 2 degrees Celsius). Several extremely hot springs, however, have been found with vent waters about 300 degrees Celsius. Hot springs on the ocean floor in the vicinity of the Galápagos Islands are believed to be above a massive magma chamber. Here the molten mantle of the Earth is estimated to be between 1,200 and 1,400 degrees Celsius. Some springs, really more like seeps, have been found oozing water at ambient temperature. These so-called cold seeps also release metal-rich water. Seamounts on the abyssal plain in the southwest Pacific Ocean basin feature cold springs spewing out mineral-laden water.

For decades, scientists considered the abyssal plains to be among the most unchanging environments on the Earth. The water temperature is a near-constant 2 or 3 degrees Celsius year-round. The salinity is unvarying, and the darkness is constant and total. The plains are generally tectonically stable as well. They were therefore frequently looked on as a safe repository for a variety of wastes. Nations dumped quantities of obsolete chemical and biological weapons and some nuclear materials on the abyssal plains, certain that they would remain there forever and do no harm to the human race. As land dumps became unavailable, the abyss was even considered as a place to dump industrial and domestic wastes and high-level nuclear wastes. Research has demonstrated, however, that the abyss is not as unchanging as it was once believed to be.

EARLY OCEAN-FLOOR EXPLORATION

The abyssal plain, of all the ocean's benthic features, is the only one that has borne out the theory that the ocean floor was a flat, featureless expanse. The few scattered soundings made toward the end of the nineteenth century for transoceanic telegraphic cables yielded small amounts of data. Later, the British oceanographic vessel *Challenger* made many soundings in areas not previously probed. During a global voyage, the *Challenger*'s crew laboriously lowered long hemp ropes and, later, single-strand wire to great depths. The ropes and wires were retrieved slowly with capstans turned by the crew. Individual soundings of a few thousand meters often required several hours to complete. Sometimes the hemp rope broke under its own weight, and the entire effort was wasted. Despite the obstacles, however, data about the seafloor accumulated. As the nineteenth century ended, some ten thousand soundings had been made in water deeper than 2,000 meters and about five hundred in water deeper than 5,500 meters.

Information about the creatures of the abyssal plain was being gathered as well, although slowly. Retrieval of damaged submarine cables brought back a host of organisms attached to the cable sheathing. This evidence dismissed forever the belief in an "azooic zone," a depth limit in the ocean below which no life existed. The *Challenger* also had deployed dredges and trawls that dragged over the ocean floor. Bizarre-looking fish, worms, and clamlike animals came up in the collecting gear, as did intriguing samples of the rocks and minerals that carpeted the abyssal plain. These collections, however, were made at great expense of time and equipment. The dredges and trawls, like the sounding lines, took hours to drop and retrieve. Frequently the gear turned upside down and never collected anything. Sometimes the lines and cables were twisted by swift currents and left in a hopeless tangle that, again, collected nothing. Modern oceanographic geologists continue to use dredges, although the hazards are much the same; the gear is still one of the mainstays of research on the ocean floor.

MEASUREMENT OF SEDIMENT

Several different kinds of "grab" are used to scoop large samples of the sediments, but more intensive sampling is done with "corers." These long pipes are dropped or pushed into the sediment to collect a cylinder of the material. Stretched out in the laboratory and cut lengthwise, the core samples expose the millennia-long history of the sediment. The texture, particle size, color, and chemistry of the sample can be measured and correlated with specific dates. Drills operated from surface ships have penetrated the rock under the abyssal plains to depths of more than 2,000 meters.

The depth of the sediments on the floor of the abyssal plains has been measured with a variety of techniques. The oldest technique involved tossing sticks of dynamite into the ocean behind a moving vessel. Special instruments measured the time it took for the vibrations set off by the explosion to reach the ocean floor and bounce back. Since rock and sediment of differing material reflected the vibrations differently, careful analysis of the echoes could reveal the nature of the sediments and their depth. The dangerous practice of using dynamite was replaced by the bouncing of harmless subsonic signals off the ocean floor. This same technique had long been used to measure the depth of water over the ocean floor. Remarkably detailed soundings of the ocean floor, with printouts of the surface features, are made with multiple scanning devices towed by research vessels. These devices include a multibeam system called Sea Beam and a special side-scanning device called Gloria. Satellites are able to survey the oceans in just a few days using a radar altimeter to produce maps of the ocean surface from which the seafloor topography can be deduced. This has made it possible to map features in areas not covered previously by ships.

Cameras—still, motion-picture, and video—loaded with black-and-white or color film have captured the features of the abyssal plains. They show ripple marks from submarine currents and the remains of ancient volcanic eruptions. This pictorial record is a valuable adjunct to the physical specimens collected.

Rock and sediment samples and photographs offer dramatic evidence of the nature of the abyss, but no technique surpasses sending humans to the bottom to make on-the-spot assessments of what is being collected or photographed. Several manned submersibles, such as the U.S. minisubmarine *Alvin* and its French counterpart *Cyana* have carried researchers many kilometers into the depths of the sea. These craft have enabled their human passengers to view the ocean floor first hand and to collect and photograph materials systematically. In addition, remotely operated vehicles (ROVs) are being used extensively and may eventually supersede manned vehicles.

Albert C. Jensen

CROSS-REFERENCES

Biogenic Sedimentary Rocks, 1435; Chemical Precipitates, 1440; Clays and Clay Minerals, 1187; Continental Shelf and Slope, 584; Deep Ocean Currents, 2107; Deep-Sea Sedimentation, 2308; Manganese Nodules, 1608; Mid-Ocean Ridge Basalts, 657; Ocean Basins, 661; Ocean Drilling Program, 359; Ocean Ridge System, 670; Ocean-Floor Drilling Programs, 365; Ocean-Floor Exploration, 666; Oceanic Crust, 675; Oil and Gas Exploration, 1699; Plate Tectonics, 86; Seamounts, 2161; Sediment Transport and Deposition, 2374; Sedimentary Mineral Deposits, 1637; Seismic Reflection Profiling, 371; Turbidity Currents and Submarine Fans, 2182.

BIBLIOGRAPHY

Borgese, E. M. *The Mines of Neptune: Minerals and Metals from the Sea.* New York: Harry N. Abrams, 1985. The recovery of the mineral resources of the seafloor is described in a lively, interesting style. The wealth of excellent illustrations—color and black-and-white photographs and drawings—shows samples of minerals and the equipment used to gather them. Recommended for anyone with an interest in seabed mineral resources.

Charlier, R. H., and B. L. Gordon. *Ocean Resources: An Introduction to Economic Oceanography.* Washington, D.C.: University Press of America, 1978. Examines the metallic and nonmetallic resources of the ocean floor and the economics of recovering them. Covers hard minerals, polymetallic nodules, tin, and sand and gravel. The possible use of the deep sea bed as a dumping ground is considered as well. Readable by upper-level high school students.

Dolgoff, Anatole. *Physical Geology.* Lexington, Mass.: D. C. Heath, 1996. This is a comprehensive guide to the study of the Earth. Extremely well illustrated and includes a glossary and an index. Although this is an introductory text for college students, it is written in a style that makes it understandable to the interested layperson. Contains a section on the development of ocean basins.

Dott, Robert H., and Donald R. Prothero. *Evolution of the Earth.* 5th ed. New York: McGraw-Hill, 1994. This basic textbook on historical geology is aimed at the student of geology; however, it is readable by anyone with some science background. Presents an account of Earth history from the viewpoint of plate tectonics. Includes a glossary.

Gross, M. Grant. "Deep-Sea Hot Springs and Cold Seeps." *Oceanus* 27 (Fall, 1984): 2-6. This issue presents a number of papers that describe the ocean-floor vents that spew metallic minerals in the great depths. Many of the observations of the abyssal zone around the vents were made and reported by scientists aboard submersibles.

Maury, M. F. *The Physical Geography of the Sea, and Its Meteorology.* Cambridge, Mass.: Harvard University Press, 1963. A reprint of the classic oceanographic textbook first published in 1855. Maury is considered the American father of the science of oceanography, and the volume is considered the first real textbook on the subject. It discusses the depths of the ocean, the basin and bed of the Atlantic Ocean, sediments, and how samples and data are collected. The style is sometimes archaic, but the reader will gain a perspective on the history of ocean floor science that is well worth the effort.

Plummer, Charles C., David McGeary, and Diane H. Carlson. *Physical Geology.* Boston: McGraw-Hill, 1999. This is a straightforward, easy-to-read introduction to geology intended for those with little or no science background. Discusses the development of the deep ocean basins from the view of plate tectonics. Includes many excellent illustrations, as well as a CD-ROM.

Rabinowitz, Phillip, Sylvia Herrig, and Karen Riedel. "Ocean Drilling Program Altering Our Perception of Earth." *Oceanus* 29 (Fall, 1986): 36-40. Deep-sea drilling has provided Earth scientists with a better understanding of the nature of the sediments of the ocean basins. The studies described in the article made use of ocean-floor drilling and a deep-water camera able to photograph a volcano almost 3 kilometers below the sea surface. A good article for those interested in the methodology of research in the ocean basins. Suitable for high school students.

Segar, Douglas. *An Introduction to Ocean Sciences.* New York: Wadsworth, 1997. Comprehensive coverage of all aspects of the oceans and the oceanic crust. Readable and well illustrated.

Suitable for high school students and above.

Wertenbecker, W. "Land Below, Sea Above." In *Mysteries of the Deep*, edited by Joseph J. Thorndike. New York: American Heritage, 1980. An easy-to-read, informative discussion of the nature of seafloor processes. Discusses the history of ocean-floor research, the impact of volcanism and deep-sea currents in shaping the benthic zone, seamounts and other submarine topography, and how the ocean floor is mapped. Excellent photographs and drawings. Of value to anyone interested in the nature of the ocean floor.

MID-OCEAN RIDGE BASALTS

Mid-ocean ridge basalts are slowly extruded and build up oceanic ridges so that they or their coarse-grained equivalent make up the bulk of the oceanic crust. The basalts form by the melting of peridotite in the upper mantle. Most of these basalts are characterized by lower potassium than basalts formed in other tectonic environments.

PRINCIPAL TERMS

BASALT: a dark-colored, fine-grained to porphyritic igneous rock

GABBRO: the coarser-grained equivalent of basalt

MAGMA: molten rock material or melt mixed with crystals that occurs below the Earth's surface

PERIDOTITE: a coarse-grained rock that consists of olivine, pyroxene, and garnet; it is likely the main rock in the upper mantle of the Earth

PORPHYRITIC ROCK: igneous rock with large mineral crystals embedded in much finer-grained minerals

THOLEIITIC BASALT: a type of basalt with calcium-rich plagioclase, monoclinic pyroxene, orthorhombic pyroxene, and olivine or quartz; it has low potassium concentrations compared to the more alkali-rich basalts

UPPER MANTLE: the mantle occurs between the core and the crust of the Earth; portions of the upper part of the mantle are believed to have a zone of partial melting that may produce melts that crystallize to basalts

COMPOSITION OF BASALTS

Plate tectonics is the theory that the Earth's crust is divided into about twelve plates that are being created at divergent boundaries on one edge and destroyed by subduction at another edge. Divergent boundaries are ones in which basaltic magma is being produced and added as new oceanic crust to the plate as it moves slowly away from this boundary. Large mountain chains called mid-ocean ridges, which encircle the Earth mostly below the ocean's surface, are formed at the divergent boundaries. The plate gradually moves away from the divergent boundary and is eventually subducted or thrust under another plate.

Basalts are dark, fine-grained to porphyritic rocks that contain calcium-rich plagioclase, pyroxene, olivine, and minor spinel. Basalts or their coarser-grained equivalents, gabbros, are the main rock types produced at oceanic ridges. Basalts and gabbros, then, make up most of the oceanic plate that moves in conveyer-belt fashion out from the oceanic ridges. The basalts at oceanic ridges are reasonably homogeneous; for example, they mostly range from 48 to 52 weight percent silicon dioxide. Most of the basalts at oceanic ridges are in a subgroup of basalts called tholeiitic basalts. They have the same composition as other basalts except they contain orthorhombic pyroxene and generally have low potassium contents.

Tholeiitic basalts are also found in other tectonic environments with other rock types, such as on ocean floors away from plate boundaries (as in the Hawaiian Islands), subduction zones, and continental rifts. They can also occur as flood basalts on continents. The main difference between most oceanic ridge tholeiitic basalts and tholeiites formed in other tectonic environments is that the oceanic ridge tholeiites have lower amounts of potassium and certain trace elements (rubidium, barium, and the light rare-earth elements) than the others.

VARIATION IN COMPOSITION: FAST-MOVING VS. SLOW-MOVING RIDGES

The variation in composition of mid-ocean ridge basalts has not been studied in the same detail as basalts at the surface have been studied, since they usually occur several kilometers below the surface of the ocean. Detailed sampling in some places, however, suggests interesting variations in composition and style of eruption.

Oceanic ridges vary a great deal in the speed at which they separate. Fast-moving ridges, such as the East Pacific Rise, spread at rates of 8 to 16 centimeters per year, and there is no distinct valley produced at the highest portions of the ridge. Also, the width of the volcanic zone is narrow, suggesting a narrow but large magma chamber below the surface. In contrast, slow-moving ridges, such as the Mid-Atlantic Ridge, move at rates of only 1 to 4 centimeters per year, and there is a central valley about 8 to 20 kilometers wide and 1 to 2 kilometers deep with a wider zone of volcanic activity. This suggests that the magma chambers below the surface are wider than those at fast-moving ridges.

In the case of fast-moving oceanic ridges, sheets of lava are extruded fairly rapidly and continuously along a low-relief ridge. In contrast, slow-moving ridges have lavas slowly extruded at only certain points along the ridge, where the ridge is built to fairly high relief. In addition, fast-spreading ridges have frequent eruptions that are fairly homogeneous in composition, while slow-moving ridges have less frequent eruptions that are more heterogeneous in composition. The fast-moving ridges are believed to have narrowly focused magma chambers where the magma is injected along a narrow region all along the ridge crest, while slow-moving ridges are believed to have only a few, less focused magma chambers that erupt lavas only at different points along the ridge over a wider area.

Slow-moving ridges seem to exhibit regular variations in the composition of basalts from the center of the ridge out to the valley walls. Basalts in the center of the rift have a greater abundance of large olivine crystals relative to plagioclase and monoclinic pyroxene crystals than those closer to the valley walls. The volcanic glasses in the center of the ridge also contain less silica and potassium than those near the valley walls. Spreading of ridges at speeds intermediate between these two extremes has characteristics intermediate between those of the fast- and slow-moving ridges.

VARIATION IN COMPOSITION: ERUPTION STYLES

Volcanic eruption rates at undersea ridge crests are not continuous, and they are of low volume compared to eruptions in other tectonic environ-ments. Nevertheless, the continued eruption of the tholeiitic basalts over long periods of time on the ridge crests results in the largest total volume of basaltic eruptions of anywhere in the world.

Basalts formed at mid-ocean ridges likely form by partial melting of rocks in the upper mantle of the Earth. The melts formed may be modified in composition by the crystallization of minerals that settle out of the melt, thus changing their composition.

The main rock in the upper mantle is probably peridotite. Up to 30 percent melting of this rock produces tholeiitic basaltic melts similar in composition to many found at oceanic ridges. The peridotite must rise upward from some depth in the mantle as a blob or plume. The peridotite is close to its melting point, so it can begin to melt as the pressure is reduced on rising to shallow depths of about 80 kilometers below the surface. This melting produces a liquid-crystal mush that melts further as it rises to shallower depths. Eventually, a point is reached where most of the melt can leave the crystals behind as the liquid rises into the oceanic crust.

The large differences in composition among tholeiitic basalts formed in different regions may be explained by differences in the composition of the peridotite source and by differences in the degree of melting of the peridotite. For example, the relatively high potassium, rubidium, barium, and light rare-earth element concentrations along the Mid-Atlantic Ridge near Iceland can be explained by the melting of previously unmelted peridotite with high concentrations of these elements. Most peridotite in the upper mantle below oceanic rises, however, appears to have melted at least once before. Melting of peridotite causes it to be depleted in potassium, rubidium, barium, and light rare-earth elements as these elements move into the melt. Thus, melting of this peridotite a second time produces the abundant tholeiitic basalts depleted in these elements.

The smaller differences in concentrations of elements across the valley at the top of the mid-ocean ridge may be explained by the crystallization and settling out of minerals such as olivine, pyroxene, and plagioclase from the melt. These minerals have different compositions from those of the melt, so that as they settle out of the melt or float, the composition of the melt slowly changes.

For example, as olivine crystallizes, it picks up less silica than does the melt, so the melt will gradually increase in silica as the olivine settles out. The melt and some suspended crystals may periodically squirt upward along fractures as crystallization takes place. Some melt and crystals may reach the surface of the ridge as a lava flow, or some may crystallize completely within the fractures below the surface.

Over long time periods, such processes result in the zonation of the oceanic crust. Depths of about 4 to 8 kilometers below sea level contain mostly coarse-grained gabbros with varied amounts of olivine, pyroxene, and plagioclase. Depths of 2 to 4 kilometers below sea level contain mostly fine-grained basalts produced by periodic extrusions of lava on the seafloor ridge. Between the gabbros and basalts, there are a lot of injections of inter-mediate-grain-size gabbros along fractures. Sometimes these rocks are so abundant that it is difficult to tell the composition of the original rock.

INTERACTION OF BASALT WITH SEAWATER

Some basalts sampled on the seafloor appear to have had sodium added to them and calcium, iron, and silica removed from them. These basalts likely came into contact with hot seawater enriched in sodium. Hot springs have been observed to exist along the East Pacific Rise. The waters emitted at the hot springs have temperatures of up to 350 degrees Celsius. Some hot springs have a black or gray, smoky appearance. The dark color is the result of the precipitation of tiny sulfide and oxide minerals from the hot water as it interacts with cold seawater. As a result of this activity, large columns of sulfide minerals are built up, and a unique ecosystem of bacteria, giant worms up to 2 meters long, clams, and crabs occurs near the vents of hot water. They obtain their energy from the heat of the water.

These gray and black smokers appear to be the surface expression of an elaborate hot-water plumbing system in which seawater percolates down fractures in an area several kilometers wide along the ridge axis. The seawater likely percolates downward to depths of at least 2 to 3 kilometers and is gradually warmed to temperatures of 400 to 450 degrees Celsius. The seawater then runs out of fractures and eventually percolates back up toward the surface, and some is liberated at the ridge axis as smokers. During the course of its circulation through the fractures in the rocks, the hot water likely reacts with the basalts and forms new minerals.

SIGNIFICANCE

Some geologists study the potential ore deposits in association with mid-ocean ridge basalts to learn about their origin. In some places, for example, copper sulfide deposits are formed as a precipitate around the vents of hot springs. The island of Cyprus has abundant copper sulfide deposits that now occur at the surface that were originally formed this way. These deposits have been mined since ancient Greek times.

Manganese is also given off by these hot waters, but not much of it is deposited around the vents. Instead, the manganese slowly precipitates out of seawater in manganese nodules on large portions of the ocean floor. Although not yet mined, these nodules are a potentially abundant source of manganese and other metals. Other ore deposits associated with mid-ocean ridge basalts are related to the precipitation of minerals out of the basaltic melts below the rise. In some cases, the mineral chromite (an ore of chromium) is one of the first minerals to crystallize out of the magma. Since this mineral is dense, it settles out of the melt as pods and layers. Some of the pods are small, but others may contain up to several million tons of chromium. One such deposit is now exposed at the surface in Oman.

Geologists also study the mid-ocean ridge basalts to build theories about how they form and evolve. Much evidence must be accumulated to build such theories. For example, experiments in furnaces that approximate the conditions in the upper mantle suggest that the main rock in the upper mantle, peridotite, is the only rock that can melt to produce basaltic melts. Also, certain isotopic and trace element concentrations in the basaltic melts are consistent with their origin by melting of peridotite. Estimates of temperatures and pressures in the upper mantle suggest that peridotite is close to its melting point in many places. Experiments in furnaces suggest that the release of pressure on peridotites under these conditions as they rise toward the surface will cause them to begin to melt.

Robert L. Cullers

CROSS-REFERENCES

Abyssal Seafloor, 651; Basaltic Rocks, 1274; Eruptions, 739; Igneous Rock Classification, 1303; Lithospheric Plates, 55; Magmas, 1326; Metasomatism, 1409; Ocean Basins, 661; Ocean Ridge System, 670; Ocean-Floor Exploration, 666; Oceanic Crust, 675; Ophiolites, 639; Plate Margins, 73; Plate Tectonics, 86; Plumes and Megaplumes, 66.

BIBLIOGRAPHY

Blatt, H., and R. J. Tracy. *Petrology.* New York: W. H. Freeman, 1996. This introductory text about the formation of rocks is somewhat technical. The section on mid-ocean ridge basalts is very detailed. Contains a glossary and index.

Davidson, J. P., W. E. Reed, and P. M. Davis. *Exploring Earth.* Upper Saddle River, N.J.: Prentice Hall, 1997. This introductory geology text has a chapter on divergent plate margins. Includes a glossary and an index.

Murck, B. W. and B. J. Skinner. *Geology Today.* New York: John Wiley and Sons, 1999. This readable introductory geology text includes a section on mid-ocean ridge basalts, as well as a glossary and index.

Nicolas, A. *The Mid-Oceanic Ridges.* Berlin: Springer-Verlag, 1995. This nicely written account of the formation of oceanic ridges and associated ore deposits is suitable for the introductory reader. Includes a glossary.

OCEAN BASINS

Ocean basins contain basaltic crust produced by seafloor spreading at mid-ocean ridges, which may be covered with a thin layer of oceanic sediments. Seafloor sediments and rocks in the oceans may contain a record of the history of the development of ocean basins. Ocean basin deposits have provided evidence supporting the theories of seafloor spreading and plate tectonics.

PRINCIPAL TERMS

BASALT: a dark-colored, fine-grained rock erupted by volcanoes, which tends to be the basement rock underneath sediments in the ocean basins

BIOGENIC SEDIMENTS: the sediment particles formed from skeletons or shells of microscopic plants and animals living in seawater

DEPOSITION: the process by which loose sediment grains fall out of seawater to accumulate as layers of sediment on the seafloor

LITHOSPHERE: the outermost layers (the crust and outer mantle) of the Earth, which are arranged in distinct rigid plates that may be moved across the Earth's surface by seafloor spreading

MAGNETIC ANOMALIES: linear areas of ocean crust that have unusually high or low magnetic field strength; magnetic anomalies are parallel to the crest of the mid-ocean ridges

MID-OCEAN RIDGE: a continuous mountain range

of underwater volcanoes, located along the center of most ocean basins; volcanic eruptions along these ridges drive seafloor spreading

RIFTING: the splitting of continents into separate blocks, which move away from one another across the Earth's surface

SEAFLOOR SPREADING: a theory that the continents of the Earth move apart from one another by rifting of continental blocks, driven by the eruption of new ocean crust in the rift

SEISMIC REFLECTION: study of the layered sediments in ocean basins by bouncing sound waves sent into the seafloor off the different rock layers

SEISMIC REFRACTION: examination of the deep structure of the ocean crust using powerful sound waves that are bent into the crustal layers rather than being immediately reflected back to the ocean surface

SEAFLOOR SPREADING

Ocean basins make up one-third of the Earth's surface, and the rocks and sediments in these basins may preserve an important record of the past history of the oceans. Earth materials in the ocean basins consist of a layer of volcanic basalts produced by seafloor volcanic eruptions at the mid-ocean ridges, which may be covered by layers of marine sediments and sedimentary rocks. The shape of individual ocean basins may be changed as a result of the interactions of lithospheric plates, such as plate collisions, plate accretion, and plate destruction. Ocean basin rocks and sediments may contain valuable deposits of metals and other economic minerals, which may represent important natural resources that could be extracted by humans at some time in the future.

Eruption of volcanic basalts in ocean basins makes up an important part of seafloor spreading. The creation of new oceanic crust by volcanic eruptions along the mid-ocean ridges provides the driving force to move blocks of continental lithosphere across the Earth's surface. For example, the separation of South America from Africa during the past 200 million years has been driven by the creation of the South Atlantic Ocean by seafloor spreading along the Mid-Atlantic Ridge between these two continents.

Ocean basin shapes may be altered by plate interactions. As lithospheric plates are rifted and move apart from one another, new ocean basins are created between the continental landmasses. In contrast, lithospheric plates may run into each other, and plate collisions cause the ocean basin between continents to be destroyed by subduction, in which crustal slabs are forced downward into the mantle and are remelted. An example is seen in southern Europe, where the collision of the northward-moving African plate with the Eurasian plate has caused the Mediterranean Sea to become shallower and narrower at the same time that crumpling of the edges of the continents has caused mountain building of the Alps in Europe and the Atlas Mountains in northern Africa.

PRESERVATION OF MAGNETIC AND VOLCANIC RECORDS

The volcanic basement rocks in ocean basins may preserve a record of the Earth's magnetic field during the past, through the record of oriented magnetic minerals contained within basalts. When basalt that erupted at mid-ocean ridges cools as a result of exposure to cold seawater, magnetic minerals within the igneous rock are aligned with the Earth's magnetic field and are "locked" into position by the crystallization of adjacent mineral grains. Thus, the alignment of magnetic minerals within seafloor basalts in the ocean basins acts as an enormous magnetic tape recorder, which preserves a record of the alternating reversals of the magnetic field of the Earth. Oceanographers investigating the magnetism of the seafloor during the 1950's discovered the existence of long, straight areas of ocean crust with unusual magnetic properties. These linear magnetic anomalies were parallel to the mid-ocean ridges but were offset from the ridge crests. The anomalies are symmetrical around the mid-ocean ridges: Anomaly records on both sides form identical "mirror images" of each other. This finding supports the theory of seafloor spreading, which predicted that the creation of new oceanic crust at mid-ocean ridges would cause rifting and separation of previously cooled basalts to either side of the ridge.

Seafloor spreading theory further predicted that the older the anomaly, the farther it has been pushed away from the ridge crest. This prediction was proved by deep-ocean drilling, which drilled into and determined the age of sediments immediately atop specific magnetic anomalies in the ocean crust. Estimates of the rate of creation of new oceanic crust along the ridges may be calculated from the distance between the crest of the mid-ocean ridge and specific magnetic anomalies whose age

has been determined. These calculations have proven that creation of new oceanic crust does not occur at a constant rate through time and that there have been episodes of rapid seafloor spreading and of slower spreading during the past.

Examination of the chemistry and mineral composition of the volcanic rocks of an ocean basin may provide a record of the history of volcanic eruptions at the mid-ocean ridges and help geologists to determine the chemistry and type of igneous rocks being erupted at any point in the past. Understanding the mineral content of seafloor crust erupted at specific times in the past allows geochemists to make predictions about the nature of the deeper portions of the Earth's crust. Chemical changes in seafloor basalts may reflect similar changes occurring in the lower crust or upper mantle of the planet.

SEDIMENT DEPOSITION

In addition to preserving historical information in the harder igneous rock basement, ocean basins provide a record of sediment deposition during the past. These regions of the ocean floor are among the flattest areas on the surface of the planet and have minimal relief: Most ocean basins are smooth and nearly flat, with less than 1 meter of vertical altitude change in 1 kilometer of horizontal distance. Their smoothness is the result of the burial of the blocky, irregularly faulted volcanic basement rocks beneath layers of slowly accumulating mixtures of biogenic sediments, turbidites, and other sediment particles derived from continental sources. Newly erupted basement rocks are gradually covered by oceanic sediments, so there is an overall correlation between crustal age and total sediment thickness within an ocean basin. Verification of this relationship by deep-ocean drilling provided further support for seafloor spreading.

Sedimentary rock layers provide information on the history of deposition in the ocean basins by their structure and by the fossils preserved in sediment layers. Marine geologists examine the types, sizes, and sorting of the individual grains that compose seafloor sediments. Geologists attempt to determine the sources of sediment particles deposited in ocean basins and to analyze both the changes in sediment particles as they fall through the water column and changes occurring on the seafloor after the sediments are deposited. Fine-grained particles derived from continental sources may be carried far out to sea by the winds to be deposited in the deep ocean basins. Also, biogenic particles either may dissolve as they sink through the oceans or may be dissolved on the seafloor by deep-ocean water masses.

Paleontologists study the fossils buried within layers of sedimentary rock. Marine sediments deposited in water depths shallower than the carbonate compensation depth have abundant microscopic fossil remains of ancient one-celled plants and animals (plankton) that lived in the shallow water of the oceans during past geologic time. As these organisms died, their remains sank to the seafloor to become an important part of the sedimentary rock layers. Examining the record of fossils preserved in seafloor sediments is like reading the pages of a book containing the history of the ocean basin: Changes in the type and number of ancient fossil organisms and fossil assemblages may be preserved in the microfossils contained in seafloor sediment layers.

Perhaps one of the most fascinating aspects of the study of seafloor sediments in the deep-ocean basins is that these materials contain significant amounts of micrometeorites and extraterrestrial material. Micrometeorites may fall on the continents, but their scarcity and small size make them difficult to identify. In deep-ocean basins far from the continents, however, sediment accumulates at a much slower rate as a result of a combination of the distance from continental sediment sources and the dissolution of biogenic sediment particles by corrosive bottom waters. As a result, deep-ocean sediments tend to be fine-grained red clays, which have few to no fossils and which may be deposited at rates as slow as 1 millimeter per million years. In these red clays, extraterrestrial materials may make up a significant portion of the sediment particles because of the extremely slow sediment deposition rates.

STUDY OF OCEAN BASINS

Because the ocean basins contain a variety of geologic materials, including igneous, sedimentary, and metamorphic rocks, a number of different techniques are used in the study of ocean basins, depending upon the specific feature of interest. Methods that are suitable for the study of one aspect of the ocean basins may be completely useless for obtaining information about other features of the basin. Some of the methods used include acoustic profiling, seismic reflection and refraction studies, dredging, sediment coring, and deep-ocean drilling. In addition, information on ocean basins may be derived from ancient seafloor deposits that have been uplifted and are presently found above sea level on the continents.

The overall shape of the ocean basin and the water depths of individual parts of the basin may be studied by acoustic profiling, or echo sounding. This technique uses an acoustic transponder (a sound source) mounted on the hull of an oceanographic vessel to emit sound waves, which travel down through the water until they are reflected back by the seafloor up to a shipboard recorder, which measures the total time between emission of the sound pulse and its return. The water depth is equal to one-half the total time (sound must go down, then up again), multiplied by the speed of sound in seawater. Profiles of the ocean basin's shape are obtained by continuously running the echo sounder while the ship is sailing across the basin.

While acoustic profiling gives the water depth of the ocean basin, the energies of the sound waves are insufficient to provide information about the buried structure of the ocean floor. The shape and thickness of the basement rocks and the sediment cover on the floor of an ocean basin may be examined by seismic reflection profiling, which is somewhat similar to echo sounding. In seismic reflection studies, a large energy source (such as the explosion of a dynamite charge) is released in seawater to create high-energy sound waves, which move down through the ocean with sufficient energy to penetrate the sediment layers of the seafloor before they are reflected back up to the vessel by the different sub-bottom layers. Reflection profiling is also made by continuously producing these high-energy sound waves while sailing across a basin to obtain a record of the thickness and geometry of the sediment layers and the harder basement rocks of the seafloor.

The deep structure of basement rocks is investigated by seismic refraction and magnetic studies, which may be made by one oceanographic vessel and a stationary floating recorder (sonobuoy) or by two vessels. In seismic refraction studies, extremely large energy sources are released in the ocean to create powerful sound waves, which have the ability to penetrate through seafloor sediment layers into the deeper layers of igneous basement rocks. Sound waves penetrating layers in the seafloor are bent (refracted) into the layer and travel through it for a certain distance before they are refracted back up to the ocean surface. An acoustic recorder at the surface measures the depth of sound penetration below the seafloor and the time elapsed since the explosion. Refraction profiles may be done by exploding charges off the stern of a moving vessel, using a stationary sonobuoy as the recording device, or refraction profiles may be made in a "two-ship" experiment, where one vessel acts as the "shooter" and the second vessel is the stationary recorder. By alternately "leapfrogging" past each other, two ships may make a much longer continuous reflection profile than is possible with only one vessel and a sonobuoy. Magnetic studies measure tiny changes in gravity or magnetic field strength using instruments towed behind a research vessel. This can provide information on the presence of volcanic rock overlain by less dense sediment.

STUDY OF SEDIMENT DEPOSITION

The history of sediment deposition preserved by marine sediments in ocean basins may be studied by obtaining long cores of seafloor sediments, either by sediment coring or by ocean-floor drilling programs. Once a long sediment core is obtained from the ocean floor, the sediment particles and fossils within the sediments are studied layer by layer in order to examine the sedimentation history of the basin. Younger sediments are placed atop older layers, so by beginning with the uppermost layers of the sediment core and continuing into deeper layers, the geologist can examine the record of progressively older deposits in the ocean basin.

Direct examination of the basement rocks of the ocean basins may be made by dredging rocks from the mid-ocean ridges, by drilling through the sediment cover to take cores of the volcanic basalts, or by studying portions of the ocean floor that have been uplifted above sea level by tectonic activity. Dredging uses a wire mesh bag attached to a rigid iron frame, which is towed on a long cable behind a vessel to obtain rock samples from the seafloor. As the ship moves across the surface, it drags the dredge along the bottom, and seafloor rocks are broken off by the frame and caught in the wire mesh bag attached to the rear of the dredge.

Deep-ocean drilling programs have provided long basalt cores that have been drilled from the seafloor in different ocean basins. The deepest seafloor borehole drilled in the oceans by the Deep Sea Drilling Project, site 504B, located near the Galápagos Islands in the eastern equatorial Pacific, has been extended to a depth of 1,350 meters below the surface of the seafloor. Drilling at this location recovered 275 meters of seafloor sediments, and more than 1 kilometer of seafloor basalts have been penetrated.

Information about the ocean basins has also been provided by uplifted sections of seafloor, located in areas of plate collisions. These ancient seafloor deposits, or ophiolite sequences, are found on the island of Newfoundland in eastern Canada, on the island of Cyprus in the eastern Mediterranean Sea, and on the island of Oman in the Persian Gulf, among other locations. Rocks in the ophiolite sequences are an important natural resource, because they contain copper and many other valuable metals interspersed between basalts and igneous rocks. These ancient ocean-floor deposits, which have been uplifted above sea level by the collision of two lithospheric plates, may contain enormous reserves of rare metallic minerals. The metallic deposits in ophiolites were originally deposited as vein minerals between the volcanic basalts in the deeper portions of ocean crust. By understanding the factors controlling the formation of ophiolites, scientists may be able to predict other locations where these rocks may be found, and humans will be better able to utilize these valuable minerals in the future.

Dean A. Dunn

CROSS-REFERENCES

BIBLIOGRAPHY

Anderson, Roger N. *Marine Geology: A Planet Earth Perspective.* New York: John Wiley & Sons, 1986. A textbook discussing various aspects of oceanography, whose content is aimed at readers with minimal scientific background.

Dolgoff, Anatole. *Physical Geology.* Lexington, Mass.: D. C. Heath, 1996. This is a comprehensive guide to the study of the Earth. Extremely well illustrated and includes a glossary and an index. Although this is an introductory text for college students, it is written in a style that makes it understandable to the interested layperson. Contains a section on the development of the ocean basins.

Dott, Robert H., Jr., and Donald R. Prothero. *Evolution of the Earth.* 5th ed. New York: McGraw-Hill, 1994. This basic textbook on historical geology is aimed at students of geology. However, it is very readable by anyone with a background in science. Presents an up-to-date account of the Earth's history from the viewpoint of plate tectonics. Includes a glossary.

LeGrand, H. E. *Drifting Continents and Shifting Theories.* New York: Cambridge University Press, 1988. A review of the history of the "modern revolution in geology," which culminated in the development of the theory of global plate tectonics.

Plummer, Charles C., David McGeary, and Diane H. Carlson. *Physical Geology.* Boston: McGraw-Hill, 1999. This is a straightforward, easy-to-read introduction to geology intended for those with little or no science background. Discusses the development of the deep ocean basins from the point of view of plate tectonics. Includes many excellent illustrations, as well as a CD-ROM.

Segar, Douglas. *An Introduction to Ocean Sciences.* New York: Wadsworth, 1997. Comprehensive coverage of all aspects of the oceans and the oceanic crust. Readable and well illustrated. Suitable for high school students and above.

Seibold, Eugen, and Wolfgang H. Berger. *The Sea Floor: An Introduction to Marine Geology.* New York: Springer-Verlag, 1982. A textbook covering geological oceanography, designed for freshman-level college courses for students with minimal backgrounds in science, which covers all the information attainable by studying ocean basins.

Van Andel, Tjeerd H. *New Views on an Old Planet: Continental Drift and the History of Earth.* New York: Cambridge University Press, 1985. A survey of the theories of continental drift and plate tectonics, written for an educated lay audience.

OCEAN-FLOOR EXPLORATION

The exploration of the ocean floor by probes, submersibles, and remotely operated vehicles is a relatively recent human enterprise. Much of what is known about the floor of the oceans from its geography, the details of its geology, and its unique biology has been discovered since 1960.

PRINCIPAL TERMS

ACOUSTIC ECHO SOUNDING: a method of determining the depth of the ocean floor that measures the time of a reflected sound wave and relates that to distance

CARTOGRAPHY: the science of mapmaking; maps or charts of the ocean floor are developed by linking the individual points of ocean-floor depths

CHEMOSYNTHESIS: the synthesis of organic substances by living organisms through the energy of chemical reactions

CORING DEVICES: devices that drill into ocean-floor sediments to provide scientists with information on the composition of the seabed

FRACTURE ZONES: areas that define the edges between the continents on the seafloor

GEOMORPHIC DOMAIN: major underwater features that define the appearance of a seafloor area

HYDROTHERMAL VENTS: areas where very hot water is expelled from volcanically active vents on the seafloor

REMOTELY OPERATED VEHICLE (ROV): a submersible operated from a remote location; for example, a robot that explores the seafloor while operated by tether from a surface ship

SOUNDING: the measurement of depth; a sounding line is a line used for the measurement of depth

TELEPRESENCE: the ability of a human to explore an area remotely by live television

EARLY EXPLORATION

Scientists know more about the surface of the Moon than about the floor of the Earth's oceans. The reason for this surprising lack of knowledge is twofold. First, more than one-half of the planet lies at depths of more than 3 kilometers, greater than humans can explore at first hand. Second, engineering devices (crewed or uncrewed) that can travel to such great depths under positive control became technologically possible only relatively recently. Yet, from these remotely operated devices, scientists are discovering an often alien environment in a vast and unexplored landscape that covers three-quarters of the Earth.

The first explorers of the ocean floor charted harbor basins by the topography they could actually see or from soundings taken with weighted lines, which represented only a tiny fraction of the Earth's ocean bottom. The first recorded mid-ocean sounding was accomplished by the Spanish explorer Ferdinand Magellan in 1521. He spliced two sounding lines together and lowered them over the side of his ship in the Tuamotu Archipelago until they ran out. Although the 365 meters of line did not reach bottom, Magellan immodestly declared that he had discovered the deepest part of the ocean. The first modern sounding was taken in 1840 in the South Atlantic, measuring a depth of 4,434 meters. The first map of the ocean floor was constructed using approximately seven thousand soundings in 1895. All these soundings were taken with line and weight, which was the only method available at that time to explore the ocean floor.

The first comprehensive ocean-floor soundings were made possible with the advent of acoustic echo sounding in 1920, but the method was not widely used until the 1940's. Considering every recorded sounding prior to World War II, there was, on the average, only one sounding per 2,500 square kilometers of ocean bottom. The use of submarines for warfare became widespread during the 1940's; they required the ability to determine the exact ocean-floor topography. With the subsequent vast improvement of echo sounding and sonar technologies, a new age of exploration was engendered. By the early 1970's, a combination of satellite navigation (which enabled accurate geographical positioning of soundings) and precise echo sounders enabled a rapid charting of the ocean floor using a science called cartography. Sunlight does not penetrate to depths greater than about 100 meters under the very best of circumstances. Hence, for scientists to chart the geography of the seafloor, they must combine individual soundings to form seafloor charts, ultimately covering vast areas.

A picture has developed of the seafloor that has revealed a widely variable seascape, at least as mutable as the continents but quite unlike above-water, continent-born landscapes. The underwater topography is unaffected by the powerful erosional forces found on the surface; the submarine world is shaped by forces that are unique to the oceans. The major submarine features (called the geomorphic domain) are deep-sea trenches, rifts (deep valleys with steep sides), flat-topped undersea volcanic cones, and fracture zones of very long linear cracks and fissures.

Oceanographic research ships have prowled the oceans for the past two centuries. Such ships as *Challenger, Discovery*, and *Endeavor* used lines and weights to determine depths. With a slight modification, they used the same lines and weights to carry sampling devices to the ocean floor to obtain specimens for later examination on the surface. Such remote-sensing devices were called coring devices, and some were complex enough to carry several types of sampling equipment, from water to temperature probes and even sea-life traps of various kinds. Nevertheless, all these sampling devices were not usually a part of a single probe and, more often than not, were sent down independently.

CREWED SUBMARINES AND ROVs

Humans have also studied the ocean floor personally, first as free divers, then using the breathing apparatus invented in the 1940's. Oxygen and nitrogen are toxic and dangerous even at 100 meters, however, thus limiting a diver with a self-contained breathing apparatus to above this level. Thus, it became necessary to invent manned submarines that could dive deeper. There submarines are a form of remote-sensing device in that they shield the people within them from the tremendous ocean pressures by thick hulls.

Such a submarine is the *Alvin*, which is capable of dives as deep as 6,100 meters. Like most submersibles of this type, *Alvin* has a robot hand that extends from the body to enable the operators inside to perform work outside the submarine. This famous submersible has also allowed investigators to observe life-forms around submerged volcanic vents on the ocean floor. One of *Alvin*'s most famous dives was made in 1986 to investigate the sunken hull of the *Titanic* in 4,000 meters of water in the North Atlantic. It was on this dive, however, that the limitations of such a manned submersible became apparent. The trip from the surface to the ocean floor alone required more than two hours and the same time to return. That limited the amount of time *Alvin* had to work on the ocean floor to merely a few hours.

Oceanographers and other

The use of remotely operated submersibles, such as Alvin—*shown here in 1971 with its original support ship* Lulu—*has made it possible to examine phenomena at great ocean depths.* (National Oceanic and Atmospheric Administration)

scientists were thus led to invent devices called re-motely operated vehicles (ROVs). These devices perform work on the ocean floor with as much dexterity as a human inside a submersible. ROVs carry cameras with capabilities from still photo-graphs to live television. Some maneuver inde-pendently on tethers and have remotely operated arms. One inventor of such devices, and the dis-coverer of the *Titanic*'s final resting place, is Rob-ert Ballard of Woods Hole Oceanographic Institu-tion in Massachusetts. Ballard, with funding from the U.S. Navy and the National Geographic Soci-ety, invented three different types of remote-sens-ing ROVs for exploring the ocean floor. Ballard's ROVs include the *Argo*, a camera sled equipped with still and live cameras mounted on a platform towed behind a surface ship. Others are the *Jason* and *Jason Jr. Jason Jr.* rides to the ocean floor on *Ja-son*, then detaches and is capable of executing a wide range of movements. According to Ballard, such devices free humans from the inherent dan-gers of piloting crewed submersibles to great depths, eliminate the excessive travel time from and to the surface, obviate the need for a complex manned vehicle, and give surface operators a "telepresence" on the ocean floor.

OCEAN-FLOOR CHARACTERISTICS

Undersea remote-sensing devices have revealed that the ocean floors are complex, distinctive, ac-tive landscapes with vast mountain ranges, plains, highlands, valleys, and active volcanic vents. Far from being a static, quiescent place, the ocean floor is a dynamic, constantly active geologic area of considerable interest to geologists.

The Earth's continents drift across the planet in a slow, never-ending flux measured in billions of years; vital information describing this process may be found at the fracture zones that mark the boundaries between the Earth's continental plates. On such fracture zones, most of the Earth's earthquakes are centered. In addition, most of the planet's active volcanoes lie on the ocean floor. Only by extensive on-site study of these regions will scientists fully understand what they seek to know about the planet's earthquakes and volca-noes. Because the very character of the Earth's crust is different at these continental boundaries (as opposed to the upper, higher regions of crust on the continents' surface levels), scientists need

to study the nature of the active regions far below sea level.

At the fracture zones, magma from deep inside the Earth wells up to near the surface, heating the cold seawater to very high temperatures (hot enough, in some cases, to melt lead) and spewing forth mineral-rich deposits from hydrothermal vents into the water. At these places, strange life-forms have been discovered that do not rely on sunlight (photosynthesis) for their survival. They are entirely dependent on the minerals from the vents, surviving in a metabolic state called chemo-synthesis. One of these creatures, called a tube worm, was discovered by Ballard in 1977 onboard the *Alvin* at a depth of 6,100 meters near hydro-thermal vents off the Galápagos Islands.

Such astonishing discoveries give rise to ques-tions that challenge fundamental biological as-sumptions. For example, did the life-forms evolve independently from surface, photosynthetically supported plants, and if so, what does that por-tend for the possibility of chemosynthetically evolved life throughout the universe? If such de-velopment is possible, then there may be vast res-ervoirs of life in the universe that have evolved in the absence of a neighbor star, long considered the most basic requirement for the development of life-forms.

Exploration of the ocean floor has applications for the economies and energy needs of modern civilization. On the vast ocean-floor regions lie metal ores (such as manganese nodules) that may one day supply a significant percentage of indus-trial requirements for this resource. Also, a blan-ket of very cold water (near or slightly below freez-ing) lies on much of the ocean floor, which may one day be exploited for its energy transfer capac-ity in ocean thermal energy power plants.

SIGNIFICANCE

As spectacular as is the emergent image of the ocean floor derived from the millions of sound-ings taken thus far, it remains merely a coarse out-line of the total picture of the underwater do-main. As remote-sensing devices improve, five fundamental goals in addition to basic soundings will be accomplished: a more complete under-standing of the extent and details of available re-sources on the ocean floor; an expansion of hu-mankind's telepresence on the deep frontiers of the ocean floor effected from remote locations; continued exploitation of the available resources on the ocean floor, such as petroleum, mineral, and food resources; collection and categorization of a vast amount of basic scientific information on the geology, available energy, and biology of the seafloor; and ongoing exploration to discover and investigate life-forms previously unknown or entirely unsuspected.

These discoveries will both enrich science's base of knowledge about the Earth and make the ocean's resources available to the world's peoples. The technology developed to explore the ocean floor will also be employed in other scientific endeavors. For example, such remote-sensing capabilities will probably be adapted for use in space exploration.

Dennis Chamberland

Cross-References

Abyssal Seafloor, 651; Deep-Sea Sedimentation, 2308; Hot Spots, Island Chains, and Intraplate Volcanism, 706; Life's Origins, 1032; Manganese Nodules, 1608; Mid-Ocean Ridge Basalts, 657; Ocean Basins, 661; Ocean Drilling Program, 359; Ocean Ridge System, 670; Ocean-Floor Drilling Programs, 365; Oceanic Crust, 675; Oceans' Origin, 2145.

Bibliography

Ballard, Robert D. *The Discovery of the Titanic.* New York: Warner Books, 1987. Ballard, as the world's leading expert on seafloor ROVs and telepresence, describes the expedition that discovered the *Titanic.* The remote sensing of the ocean floor is discussed in detail, along with its future potential. A beautiful and important photographic work.

Dolgoff, Anatole. *Physical Geology.* Lexington, Mass.: D. C. Heath, 1996. This is a comprehensive guide to the study of the Earth. Extremely well illustrated and includes a glossary and an index. Although this is an introductory text for college students, it is written in a style that makes it understandable to the interested layperson. Contains a section on the development of the ocean basins and the methods used to explore them.

Dott, Robert H., Jr., and Donald R. Prothero. *Evolution of the Earth.* 5th ed. New York: McGraw-Hill, 1994. This basic textbook on historical geology is aimed at students of geology, but it is readable by anyone with a background in science. Presents an up-to-date account of the Earth's history from the viewpoint of plate tectonics. Includes a glossary.

Koblick, Ian G., and James W. Miller. *Living and Working in the Sea.* New York: Van Nostrand Reinhold, 1984. Although this work is primarily directed to seafloor habitats, it offers a valuable discussion of the rigors of ocean floor exploration. Also discusses a wide range of ocean-floor exploration techniques and equipment. Of textbook quality, well indexed and illustrated.

Segar, Douglas. *An Introduction to Ocean Sciences.* New York: Wadsworth, 1997. Comprehensive coverage of all aspects of the oceans and the oceanic crust, including seafloor exploration. Readable and well illustrated. Suitable for high school students and above.

Weiner, Jonathan Weiner. *Planet Earth.* New York: Bantam Books, 1986. This work gives an excellent accounting of historical seafloor exploration, from the earliest voyages of *Challenger* and the first soundings that revealed the vast seascape below. A beautifully illustrated and well-indexed volume for all readers interested in the human exploration of planet Earth.

OCEAN RIDGE SYSTEM

The ocean ridge system is a complex chain of undersea volcanic mountains that are found in all the oceans. These mountains contain rift valleys along their axes, which are believed to be spreading centers from which continental motion takes place. All existing evidence, such as volcanic activity, the flow of heat from within the Earth, and various types of faulting and rifting, supports modern theories about seafloor spreading, tectonic plates, and continental drift.

PRINCIPAL TERMS

ASTHENOSPHEREHERE: a zone of rock within the mantle that has plastic flow properties attributable to intense heat

BASALT: a heavy, dark-colored volcanic rock

CONVERGING PLATES: a tectonic plate boundary where two plates are pushing toward each other

DIVERGENT PLATES: a tectonic plate boundary where two plates are moving apart

METALLOGENESIS: the process by which metallic ores are formed

SEDIMENTS: the solid fragments of rock that have been eroded from other rocks and then transported by wind or water and deposited

SEAFLOOR SPREADING

The ocean ridge system is a complex chain of mountains about 80,000 kilometers in length that winds through the ocean basins. These mountain ranges vary from a few hundred to a few thousand kilometers in width and have an average relief of 0.6 kilometer. Studies of the seafloor indicate that ocean ridges are found in every major ocean basin. They are composed of basalt and covered by various types of sediments. Many of the ridges have narrow depressions that extend thousands of kilometers along their axes. Heat probes lowered into these rifts indicate much higher temperatures than on the flanks of the ridges. Another very significant finding has been that the rocks that make up the seafloor are much younger than those that make up the continents. This finding countered the pre-1960's belief that the rocks of the ocean basin were more ancient than those of the continents.

In the 1960's, the theory of seafloor spreading suggested that new seafloor is constantly being added by volcanic activity at the ocean ridges. The theory of plate tectonics proposes that the Earth's crust is divided into several major plates. These plates extend down into the Earth's mantle to a hot, semimolten zone known as the asthenosphere. Since the rock that composes the tectonic plates is less dense than the rock that forms the mantle, the plates may be considered to be float-ing on the asthenosphere. Because the interior of the Earth is much hotter than the surface, a flow of heat toward the surface is a constant process. This method of heat transfer, known as convection, is a density type of current. Material such as molten rock or hot air or water is less dense than the same material in a cooler state. As a result, it flows upward, with cooler material filling in below. When the hot material reaches a higher level and releases its heat, it too returns to the depths to be reheated. It is this process, along with the relatively low density of the tectonic plates, that causes the plates to drift apart.

The location and symmetry of the mid-ocean ridges, especially in the Atlantic and Indian Oceans, suggests the configuration of the continents before they began to drift apart. The modern theories of plate tectonics and seafloor spreading are in basic agreement with the theory of continental drift as proposed by Alfred Wegener in 1915; however, modern ideas regarding the mechanics of drift differ. The basic concept of seafloor spreading at the oceanic ridges proposes that tension cracks form in the crust at these spreading centers. Molten rock from the mantle then flows upward through these fissures, both forming the volcanic ridges and creating new seafloor. As the fissures widen, new crustal material moves away on both sides, and additional new seafloor is created. It is in this way that the mid-ocean ridges such as the Mid-Atlantic Ridge, the East Pacific Rise, the Antarctic Rise, and the Carlsberg Ridge of the Indian Ocean were formed. The spreading away from the oceanic ridges takes place at a rate of about 2 centimeters per year in the Atlantic and about 5 centimeters per year in the Pacific.

As new seafloor is being created at the ridges, old seafloor is being destroyed at continental margins. Here, old seafloor is subducted beneath continental plates. At these converging plate boundaries, old seafloor is forced downward into the mantle, where it undergoes remelting. This molten rock may then find its way back to the surface through cracks or fissures in the overlying rock. Volcanic action is the result of this material reaching the Earth's surface. Because old seafloor is destroyed in this manner, it is now understood why rocks that make up the ocean floor are relatively young. Deep-seafloor drilling projects in the Atlantic and Indian Oceans have failed to find rock or fossil samples that are older than the Jurassic period of Earth history, which ended about 140 million years ago.

RIFT ZONES

The present ocean ridges show large offsets in some areas. This phenomenon results from transform faulting. As spreading of the plates took place at divergent boundaries, fracture zones developed at right angles to the axes of the ridges. Displacements of the ridges along these faults produced the observed structure.

The narrow rift valleys are believed to have been caused by downfaulting along divergent plate boundaries. Some rift zones, such as the one associated with the Carlsberg Ridge of the northern Indian Ocean, link up with continental rift zones. This rift has been shown to be connected with the African rift zone. Studies conducted during the 1970's indicate that extensive amounts of volcanism and seismic activity exist along rift zones.

HOT SPRINGS AND SMOKERS

Also found along oceanic ridges are hot-water springs. The existence of these springs, or smokers, had been predicted, but direct evidence for them was not gathered until the early 1960's, when metal-bearing sediments were discovered on the East Pacific Rise. The first actual observation of a smoker took place in 1980 by the crew of the deep-diving research submarine *Alvin*. The *Alvin* was part of an underwater research program being conducted near the Galápagos spreading center in the eastern Pacific. Researchers were surprised to find a rather extensive plant and animal community living near the smoker at a depth too great for photosynthesis to be a factor. Clams as long as 30 centimeters were found, as well as white

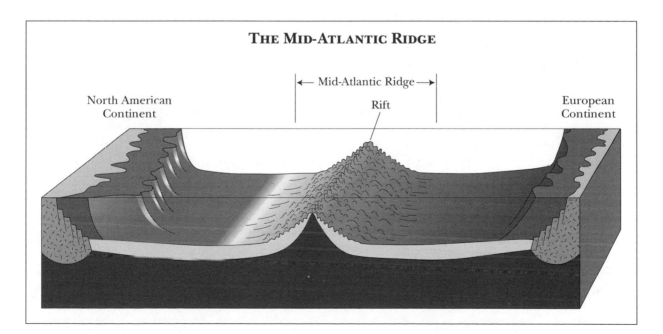

THE MID-ATLANTIC RIDGE

North American Continent

← Mid-Atlantic Ridge →

Rift

European Continent

crabs and tube worms some 3 meters long.

Smokers have been found to emit great quantities of sulfur-enriched waters. Dissolved within the acidic water are various types of metals. These metals are dissolved from the rocks as the super-heated water moves toward the surface. Metals such as copper, zinc, iron, silver, and gold are extracted and concentrated into a supersaturated fluid. When this fluid is discharged from a smoker at the seafloor, it will precipitate to form an ore body if cooling takes place rapidly. These massive sulfide deposits are sometimes made permanent when volcanic eruptions cover them with basalt flows.

MID-ATLANTIC RIDGE

Although most of the knowledge regarding the nature of the sea bottom has been gained since the 1960's, the Mid-Atlantic Ridge system has been known to exist since echo-sounding studies were done after World War I. By 1960, it had been determined that the Mid-Atlantic Ridge was continuous with other oceanic ridge systems around the world. The same cannot be said of the rift system that was discovered in the Mid-Atlantic Ridge. Profiles taken across ridge systems do not always indicate such rift depressions. This anomaly can probably be explained by considering the possibility that these rifts have been filled with volcanic material and therefore go undetected.

A rather extensive study was made along the Mid-Atlantic Ridge about 200 miles south of the Azores in 1974. It was found that the rift valley is bordered by a series of steep slopes that appear to be high-angle faults. Many open vertical fissures were observed, some of them as much as 8 meters across. Although no active volcanism was observed during this study, mounds of pillow lava along the fissures indicated that volcanism had taken place. The positioning of the faults and fissures indicated that the Mid-Atlantic Ridge is spreading outward east and west from its central axis.

EAST PACIFIC RISE

The East Pacific Rise, which is part of the system of mid-ocean ridges, stands some 2 to 3 kilometers above the ocean floor and is thousands of kilometers wide. The slopes in this area are not as great as those of the Mid-Atlantic system, but the mechanism of ridge formation is the same. It has

been suggested that part of the East Pacific Rise extends under the North American continent and that the San Andreas fault may be part of this system. Evidence supporting this possibility is that the rate of displacement along the fault is comparable to the rate of ridge spreading on the East Pacific Rise near the Gulf of California.

STUDY OF THE OCEAN FLOOR

The ocean floor has been investigated by the use of sonar since the early twentieth century. Using this technique, sound waves from a device are sent into the sea in all directions. As these waves strike objects, they are reflected back to the source. By analyzing the reflections, scientists are able to determine the nature of the seafloor. This technique was refined during World War II for the purpose of locating enemy submarines. By the early 1950's, new maps and charts of the seafloor had been made using sonar.

In the 1960's, a plan to drill a deep hole through the crust of the Earth to the mantle was proposed. Since this deep hole was intended to intersect the Mohorovičić discontinuity (named after Andrija Mohorovičić, the Croatian geologist who discovered it), the project was called Mohole. Since the crust of the Earth is much thinner under the ocean basins than it is under the continents, the ocean was the most logical location in which to drill. Although the 6-kilometer-deep hole was never drilled, several preliminary holes were. Samples taken from this drilling revealed the presence of gray, claylike sediments overlying dark, heavy basalts.

Project Mohole was later replaced by a plan to drill many holes at various locations into the rock beneath the ocean depths. This project, under the direction of various American oceanographic institutes, became known as the Deep Sea Drilling Project. Much of the drilling was done from the deck of the now-retired *Glomar Challenger* of the company Global Marine. More than six hundred holes were drilled in the Atlantic, Pacific, and Indian Oceans and the Mediterranean Sea, revealing a wealth of data on the nature and evolution of the ocean basins.

In the past, ocean researchers had to gather data from the decks of ships by lowering various tools to the bottom to gather samples—there was a need for direct observation by the oceanographer. In 1948, the first minisubmarine, or bathyscaphe, made an unmanned dive to 759 meters. In 1954, a manned vessel was taken to a depth of 4,050 meters. During the 1960's, interest in these submersibles grew quickly in the United States. Since that time, submersibles such as the *Aluminaut* and *Alvin* have been used in various deep-sea research projects. Submersibles have been used for on-site studies of the rift zone of the Mid-Atlantic Ridge and of the hot-spring smokers of the East Pacific Rise.

Another technique, known as marine seismology, has been employed to study the ocean ridges. Oil companies have used seismic studies to explore for oil deposits on continents, and it was only a matter of time before this technology was adapted for study of the ocean depths. The process involves making a sound explosion in the sea, which is accomplished by the use of an air gun. The sound waves are produced by the rapid release of compressed air. The waves then travel through the water to the bottom, where they are reflected to seismometers located on the ocean floor. The data are collected, and a seismogram is generated by a computer. This technique has been used to locate active magma chambers in rift zone areas.

ECONOMIC RESOURCES

Studies of submarine rift areas that have revealed the hot-water springs and smokers have shown these areas to offer potential economic riches. Smokers discharge water that contains sulfur in solution as well as various types of metals such as copper, iron, zinc, silver, and gold. These metals precipitate to the sea bottom and, if they

are cooled rapidly enough, form deposits. If the ore bodies are covered with volcanic material shortly after their deposition, they become protected from the erosive processes of seawater. In time, these deposits drift from oceanic spreading centers to become parts of continents. Many mineral deposits formed in just this manner have been mined since ancient times.

In the late 1950's, metal-rich sediments were first found along mid-ocean ridges. By the late 1960's, metalliferous muds were found at the bottom of the Red Sea, the Gulf of California, and the East Pacific Rise near the Galápagos Islands and south of Baja, California. Beginning in 1977, marine geologists were able to observe the formation of metal deposits while observing smokers from the research submarine *Alvin.* Interest in ridge deposition of metals centers on the processes of metallogenesis. Mining of such deposits, however, even with advanced technology, is not profitable. It will no doubt be considered in the future, however, as the copper deposits in the area of the Galápagos spreading center alone are estimated to be worth in excess of $2 billion.

David W. Maguire

CROSS-REFERENCES

Abyssal Seafloor, 651; Basaltic Rocks, 1274; Continental Rift Zones, 579; Hydrothermal Mineralization, 1205; Igneous Rock Bodies, 1298; Lithospheric Plates, 55; Magmas, 1326; Mid-Ocean Ridge Basalts, 657; Ocean Basins, 661; Ocean-Floor Exploration, 666; Oceanic Crust, 675; Plate Margins, 73; Plate Motions, 80; Plate Tectonics, 86; Spreading Centers, 727; Sub-Seafloor Metamorphism, 1427; Subduction and Orogeny, 92.

BIBLIOGRAPHY

Anderson, Roger N. *Marine Geology: An Adventure into the Unknown.* New York: John Wiley & Sons, 1986. A well-written, somewhat technical volume on the geology of the oceans. Contains excellent chapters on metallogenesis and mineral deposits. This volume is suitable for the college student of geology or the informed layperson.

Dolgoff, Anatole. *Physical Geology.* Lexington, Mass.: D. C. Heath, 1996. This is a comprehensive guide to the study of the Earth. Extremely well illustrated and includes a glossary and an index. Although this is an introductory text for college students, it is written in a style that makes it understandable to the interested layperson. Contains a section on the development of the oceanic ridge system.

Dott, Robert H., Jr., and Donald R. Prothero. *Evolution of the Earth.* 5th ed. New York:

McGraw-Hill, 1994. This basic textbook on historical geology is aimed at students of geology. However, it is very readable by anyone with a background in science. Presents an up-to-date account of the Earth's history from the viewpoint of plate tectonics. Includes a glossary.

Segar, Douglas. *An Introduction to Ocean Sciences.* New York: Wadsworth, 1997. Comprehensive coverage of all aspects of the oceans and the oceanic crust. Readable and well illustrated. Suitable for high school students and above.

Skinner, B. J., and S. C. Porter. *The Dynamic Earth.* New York: John Wiley & Sons, 1989. A well-written, well-illustrated volume dealing with general topics in physical geology such as rocks, minerals, erosion, Earth resources, sedimentation, and tectonics. Ideal for the college-level introductory physical geology course.

Wicander, R., and J. S. Monroe. *Historical Geology.* New York: West Publishing, 1989. A very well-illustrated volume dealing with such subjects as geologic time, origin and interpretation of sedimentary rocks, and a detailed account of the historical geology of the Earth through various time periods. Excellent for a first-year college course in historical geology.

OCEANIC CRUST

The oceanic crust is that portion of the outer layer of material forming the Earth that underlies the world oceans. This crust is a dynamic layer, primarily composed of basalt, where new, submarine mountain ranges are continuously formed and old ocean floors are destroyed.

PRINCIPAL TERMS

FRACTURE ZONES: large, linear zones of the seafloor characterized by steep cliffs, irregular topography, and faults; such zones commonly cross and displace oceanic ridges by faulting

HYDROTHERMAL VENTS: seafloor outlets for high-temperature, mineralized springs that are associated with seafloor spreading centers and that are often the site of deep-sea, chemosynthetic biological communities

OCEAN TRENCHES: long, deep (greater than 6,000 meters deep), and narrow depressions in the seafloor with relatively steep sides; these features mark the boundaries between ocean crust and continental crust and are associated with the subduction of oceanic crust

OCEANIC RIDGES: long, narrow elevations of the seafloor, some 2 to 3 kilometers higher than the surrounding ocean basins, that are associated with the creation of new seafloor material

PLATE TECTONICS: the theory of mobility within the Earth's crust that accounts for mountain building at ocean ridges, spreading of the seafloor, and subduction at ocean trenches by dividing the crust into a series of plates that interact by colliding, rifting, or sliding past one another

SEAFLOOR SPREADING: the process whereby crustal plates move away from mid-ocean ridges, creating new crustal material as molten rock moves upward through rifts at the ridge crests

SEAMOUNTS: isolated elevations on the seafloor, usually rising to higher than 1,000 meters, that are commonly the shape of an inverted cone reflecting their volcanic origin; a flat-topped seamount is known as a guyot

SEISMIC ACTIVITY: a disturbance of the crust caused by earthquakes or Earth movements, often associated with zones of seafloor subduction and ocean-ridge formation

SUBDUCTION: the process whereby old seafloor that was produced millions of years earlier at the ocean ridges is forced under continental crust in the vicinity of trenches

EARTH'S STRUCTURE

The Earth is not homogeneous from its center to its surface; rather, it is composed of three concentric layers: the core, the mantle, and the crust. The core of the Earth is composed of a dense mixture of nickel and iron, with a solid inner portion and a liquid outer portion. The core is extremely hot, ranging from above 6,600 degrees Celsius at the center to 4,500 degrees Celsius in the outer core. The core extends from a depth of about 2,900 to 6,378 kilometers (the center of the earth) and accounts for 31.5 percent of the Earth's mass. The next layer, the mantle, consists of less dense rock and holds 68.1 percent of the mass.

The material of the mantle has the properties of iron-magnesium silicate rock rich in olivine and pyroxene minerals. The mantle is about 2,870 kilometers thick and is cooler than the core (1,500 to 3,000 degrees Celsius). The mantle also has two zones. The lower portion is presumed to be essentially rigid, but the upper mantle, or asthenosphere, is more plastic and flows when stressed. The asthenosphere extends to a depth of 700 kilometers and is likely the site of molten magma formation. The outermost layer, the crust, is the less-dense outer shell of the Earth. Also known as the lithosphere, this layer consists of granitic continental crust and basaltic oceanic crust. The crust is underlain and likely fused with a layer of heavier mantle rock. The boundary between the crust and mantle is known as the Mohorovičić discontinuity (or Moho); the boundary occurs under continents at depths of 10 to 70 kilometers but only 5 to 10 kilometers under the oceans.

Above the Moho, both the oceanic and continental crusts have properties that resemble basalt. Basalt, a common volcanic rock found extensively on the Earth's surface, is composed of silicates of calcium, magnesium, and iron. These rocks have an average density of 3 grams per cubic centimeter. Under the continents, but not the oceans, the basalt is overlain by a rock layer with properties similar to those of granite. Granite is a common igneous rock composed of silicates of aluminum and potassium. Such rocks are lighter in color and weight than basalt, with an average density of 2.8 grams per cubic centimeter. Thus, the continental crust "floats" as massive blocks on a layer of basalt. Because the densities of the continental and oceanic crusts are not greatly different, approximately 93 percent of the continental blocks are submerged in the underlying basaltic crust. Continental blocks are analogous to floating icebergs of various heights in that the Moho is pushed deeper under continental mountain ranges than it is under flat coastal plains. The Moho assumes a shape that reflects the surface of the continent but that is exaggerated to nine times greater. The bottoms of the continental blocks must rise as material is eroded to the sea, thus keeping the exposed-to-submerged ratio constant. This flotation phenomenon is known as isostasy, and the rising process is called isostatic adjustment.

CHARACTERISTICS OF THE OCEAN CRUST

One of the most remarkable characteristics of the oceanic crust is its structural uniformity. Essentially, marine sediments overlie igneous rock, which forms three distinct layers. The sediments vary considerably in thickness and in composition. Shell material and debris from marine plants and animals form dominant sediments around the equator and near the polar seas; detritus from the land and glacial deposits are common near the continents; and chemical precipitates (oozes) are found elsewhere. The oceanic ridge crests are generally free of sediments. From the flanks of the ridges to the continents, the sediments generally increase in thickness to more than 3 kilometers at the continental margins. The three igneous layers are each relatively uniform in composition and thickness. The upper layer has been penetrated by deep ocean drilling and is known to be composed of basaltic lavas, 1 to 2.5 kilometers thick. The basal layer, directly overlying the mantle, is thin (0.5 kilometer) and presumably formed of layered peridotite. Peridotite is a dense ultrabasic igneous rock consisting mainly of olivine minerals. Similar rocks are thought to be the principal constituent of the mantle. The main (middle) layer is 5 kilometers thick and has properties consistent with a gabbroic composition. Gabbro is a coarse-grained igneous rock consisting mainly of plagioclase feldspar, pyroxene, and olivine minerals. It is the deep-seated equivalent of the overlying, fine-grained basalt. In certain areas, metamorphism and hydrothermal processes have formed other rock and mineral types including amphibolite, greenschists, zeolites, and serpentine. The surface features of the oceanic crust include such interesting and interrelated topographic features as oceanic ridges, fracture zones, seamounts, abyssal plains, deep-sea trenches, and island arcs. The existence of each of these features can be explained by the concepts of plate tectonics and seafloor spreading.

OCEANIC RIDGE SYSTEM

The oceanic ridge system is the major topographic feature of the ocean basins, extending 80,000 kilometers as a continuous range throughout all the oceans. Ocean ridges generally rise 2 to 3 kilometers higher than do the bordering ocean basins. The ridge system in the Atlantic and Indian oceans lies equidistant between the adjacent continents, whereas in the Pacific Ocean, the system is highly asymmetric with respect to the continents. Passing out of the Indian Ocean between Antarctica and Australia, the ridge system continues eastward across the southern Pacific then arcs northward toward the South American continent. Here known as the East Pacific Rise, it eventually passes into the Gulf of California and presumably under the Basin and Range Province of the western United States to reemerge in the Pacific Ocean as the Juan de Fuca Ridge off British Columbia. Although almost entirely submarine, the ridges do rise above sea level at a few places in the Atlantic and Pacific oceans where recently active volcanos have formed islands (for example, Iceland, Tristan da Cunha, and the Galápagos Islands). The ridges are also seismically active, with frequent tensional earthquakes of intermediate strength. Such earthquakes are generally restricted to the oceanic crust within a few kilometers of the ridge crest. The crest of the ridges in the Atlantic and Indian oceans are characterized by a central rift valley that is commonly 2 to 3 kilometers deep and 20 to 30 kilometers wide. Volcanism occurs along the centerline of the rift valley, which is also the site of most of the seismic activity. In the Pacific Ocean, earthquakes are confined to a similar narrow zone, but volcanism appears to have been much greater. Here, lava flows have filled the central rift valley to a large part so that the ridge crest appears smooth.

The ocean ridge system is clearly a continuous feature on a global scale, but when viewed in detail, the ridge crests are frequently offset by fracture zones. For example, the Mid-Atlantic Ridge has no fewer than forty such zones. These fracture zones are steeply cliffed features that vary in width from a few to 50 kilometers. They are mainly confined to the oceanic crust and only rarely approach the continental margins. Fracture zones are only seismically active along that portion of the fault line between the offset ridge crests; the segments extending toward the continents are seismically quiet. Earthquakes between the crests are associated with transverse motion, indicating that the fracture zones are the result of faults in which each side moves horizontally but in opposite directions. These displacements of oceanic crust are known as transform faults.

SEAFLOOR IRREGULARITIES

Throughout the world's oceans, beyond the flanks of the ridge systems, numerous irregularities rise from the seafloor. Small volcanic extrusions that rise less than 1 kilometer from the ocean floor are known as abyssal hills. Larger volcanic features that reach 1 kilometer or more are called seamounts. Seamounts that have flat tops are known as guyots. The flattening is thought to have resulted when seamounts were near sea level and subjected to wave attack and erosion. The composition of seamounts is closely related to their proximity to oceanic or continental crust. For example, near the center of the Pacific Ocean, the seamounts (including those that broach the sea surface to become islands) are composed of

basaltic-type rock characteristic of oceanic crust, whereas along the margin of the Pacific Ocean the islands are composed of the granitic-type rocks that are found on the continents. The boundary between these two regions has been named the Andesite Line for the type of rocks found in the volcanic mountains of South America. In tropical areas, seamounts often support coralline reefs in the form of atolls. Atolls are generally circular in plan, consisting of a central lagoon surrounded by a narrow carbonate reef dotted with elongated islands. Presumably, atolls form as volcanic islands subside at a rate that is matched by the upward growth of the encircling reef.

ABYSSAL PLAINS AND OCEAN TRENCHES

The relatively flat surfaces of the ocean floor, which extend from the mid-ocean ridges to either the marginal trenches or the continental slopes, are known as abyssal plains. Excluding the trenches, these plains are the deepest portion of the ocean. Abyssal plains account for nearly 30 percent of the Earth's surface, comprising 75 percent of the Pacific Ocean basin and 33 percent of the Atlantic and Indian Ocean basins. Oceanic rises are areas of the ocean floor that are elevated above the abyssal plain, distinctly separated from a continental mass, and that are of greater areal extent than are typical seamounts or abyssal hills. General oceanic rises lie at least 300 meters above the surrounding ocean floor. Rises are not seismically active and are thought to result from the uplifting of oceanic crust associated with volcanic hot spots (source areas for magma in the upper mantle). Examples are the Bermuda Rise in the North Atlantic Ocean and the Chatham Rise in the southwestern Pacific Ocean.

The oceanic trenches are the deepest parts of the oceans. They are elongate, narrow, and commonly arcuate in shape with the convex side facing to sea. With few exceptions, they occur at the margins of ocean basins. By convention, an ocean deep must be at least 6,000 meters below sea level to be considered a trench. Trenches are found in the Atlantic and Indian Oceans but are most common in the Pacific. The deepest are found in the western Pacific, where the Mariana Trench plunges to a depth of 11,033 meters. The largest, however, is the Peru-Chile Trench adjacent to South America; it is 5,900 kilometers long, averages 100 kilometers wide, and extends to depths below 8,000 meters. Most trenches are associated with island arc systems or with volcanic ranges adjacent to continents. Examples include the Guam and Saipan islands west of the Mariana Trench and the Andes east of the Peru-Chile Trench. Island arcs are volcanic belts that parallel the trench on the continental side. The profiles of the trenches are asymmetrical, with steep sides toward the island arcs. Trenches are also areas of high earthquake activity, low gravitational pull, and low heat flow from the Earth.

PLATE TECTONICS AND HYDROTHERMAL VENTS

The concepts of plate tectonics and seafloor spreading provide the mechanisms necessary for creating the features of the ocean floor. Plate tectonics proposes that the Earth's lithosphere is composed of several plates of differing shapes and areas that glide over the plastic asthenosphere. Convection cells caused by radioactive decay of isotopes in the molten rocks of the mantle are the driving mechanism for this motion. These cells circulate the heat upward, causing upwelling in the mantle. The movement of the plates results in areas of separation where magma flows to the surface, creating the volcanic mountains of the mid-ocean ridges. Thus, the ocean floor spreads outward from the ridge crest expanding the dimensions of the ocean. This process occurs in the Atlantic Ocean at the expense of the Pacific Ocean. Where plates collide, such as off the coast of Southeast Asia and South America, deep trenches are formed as oceanic crust is forced under (or subducts) the lighter continental crust. The process is associated with volcanism as the oceanic crust is remelted and island arcs are formed by the accompanying submarine eruptions. The rate of spreading affects the form of the ridge system. Rapid spreading (up to 5 centimeters per year) produces a broad, relatively low ridge without a deep central valley, such as that found in the East Pacific Rise, west of South America. Slow spreading (1 to 3 centimeters per year), on the other hand, results in a high-relief ridge with a deep central rift valley such as that of the Mid-Atlantic Ridge.

Hot waters are discharged by hydrothermal vents at active mid-ocean ridges. The chemical composition of ocean water and deep-ocean sediments is influenced by seawater circulating through hot oceanic crust, formed by volcanic eruptions. Some seawater enters the oceanic crust through faults and eventually reaches the vicinity of the magma chambers below the spreading center where molten rock collects before eruption. Reactions with hot basalt charge the seawater with metallic sulfides and remove magnesium and other elements. The hot water then flows into the ocean through irregular, chimneylike vents up to 10 meters high. The vent mounds are made of silica, native sulfur, and metallic sulfide minerals. The bright-colored chimneys and their surrounding deposits resemble valuable ore deposits of copper, zinc, and other metals that are found on the continents. This phenomenon may also be an important process in regulating the chemical composition of seawater, as well as providing a chemical base for a deep-sea biological community that uses chemical energy rather than sunlight to produce organic compounds (chemosynthesis instead of photosynthesis).

SEISMIC REFRACTION AND REFLECTION

The oceanic crust covers about 70 percent of the Earth's surface, yet it has received relatively little attention. For example, deep-sea drilling and sampling of the crust have been completed at only one site for every 500,000 square kilometers of ocean floor. The great depth of the oceans, the tremendous logistical problems of working on and in the sea, and the high cost of oceanic research have all acted to limit the amount of scientific information that is available. Technological advances in the second half of the twentieth century have permitted researchers to explore the oceanic crust with remote sensing techniques as well as through diving excursions to the ocean floor.

The structure of the Earth and particularly the oceanic crust have been investigated indirectly by seismic refraction and reflection methods. Studies in the early 1950's showed that the crust was composed of several layers on the basis of the velocity of sound within each layer. Ocean sediment transmits sound at 2 kilometers per second, basalt transmits at 5.1, gabbro transmits at 6.7, and peridotite transmits at 8.1. Precise measurements of the length of time required for a seismic shock wave to penetrate these layers have permitted the thickness of each layer to be calculated.

Accurate and detailed maps of the ocean floor (bathymetric charts) have been compiled from enormous collections of sounding data. By the early twentieth century, electronic devices called precision depth recorders (PDRs) became available for oceanographic surveys. These early surveys gave the first realistic view of the ocean's major surface features. Side-scanning sonar, a later development, has provided three-dimensional illustrations of small-scale features of the seafloor. This type of bathymetric data can be recorded continuously as the research ship is under way, and locations can be determined precisely from navigational satellites.

TECHNOLOGICAL ADVANCES

Ocean crustal rock is an average of 7 kilometers thick. This great thickness, plus the hundreds of meters of sediment and thousands of meters of seawater overlying the crust, have made the direct sampling of this rock very difficult. Deep-ocean drilling has, however, permitted samples to be collected from considerable depths below the ocean floor. Beginning in 1968, the Deep Sea Drilling Project (DSDP) has extensively explored the oceanic crust from the ship *Glomar Challenger,* operated by Global Marine. Sponsored by the National Science Foundation and the Office of Naval Research, this project has drilled a total of 160 kilometers of cores. The deepest penetration into the ocean floor was 1.7 kilometers; the deepest water in which drilling took place was 7,000 meters. More than 840 sites were drilled in all parts of the world oceans. Scientists from all over the world participated in this project, which confirmed much of the theory of how the Earth's crust moves. DSDP also provided significant data on the age of the ocean basins and the rates of seafloor spreading.

Other important deep-sea data have been gathered by research submersibles and airborne remote sensors. Starting in the 1960's, submersibles such as Woods Hole Oceanographic Institution's *Alvin* have made some remarkable oceanographic discoveries, especially along the mid-ocean ridges. Much of the knowledge of hydrothermal vents has been obtained through submersible observations. A submersible possesses numerous advantages over surface vessels, including direct observation and sampling. Such vessels are, however, dependent on surface ships for support and transport to the dive sites. Aircraft magnetometer surveys have yielded valuable information on paleomagnetism and the Earth's gravitational field. These data showed that the polarity of the Earth's magnetic field is recorded in the crustal rocks as the ocean floor is formed. Thus, a record was revealed that demonstrated seafloor spreading through mirror-images of polar reversal patterns on the east and west sides of the Mid-Atlantic Ridge.

Satellites have also played a part in the exploration of the oceanic crust. The short-lived Seasat satellite carried a sophisticated altimeter that could measure the precise distances between the satellite and the ocean surface. Slight differences in sea level were observed that correspond to ocean deeps and density anomalies (such as accumulations of dense rock). For example, the sea stands higher over mid-ocean ridges and lower over trenches. Satellite images are now being used to map previously unknown topographic features of the ocean floor.

Charles E. Herdendorf

CROSS-REFERENCES

Abyssal Seafloor, 651; Deep-Sea Sedimentation, 2308; Earthquake Distribution, 277; Earth's Core, 3; Earth's Crust, 14; Earth's Mantle, 32; Heat Sources and Heat Flow, 49; Hydrothermal Mineralization, 1205; Lithospheric Plates, 55; Mid-Ocean Ridge Basalts, 657; Ocean Basins, 661; Ocean Drilling Program, 359; Ocean-Floor Drilling Programs, 365; Ocean-Floor Exploration, 666; Ocean Ridge System, 670; Plate Tectonics, 86; Seamounts, 2161; Spreading Centers, 727.

BIBLIOGRAPHY

Chernikoff, Stanley. *Geology: An Introduction to Physical Geology.* Boston: Houghton Mifflin, 1999. This is a good overview of the scientific understanding of the geology of the Earth and surface processes. Includes a chapter on the oceanic crust. Includes a link to a Web site that provides regular updates on geologic events around the globe.

Dolgoff, Anatole. *Physical Geology.* Lexington, Mass.: D. C. Heath, 1996. This is a comprehensive guide to the study of the Earth. Extremely well illustrated and includes a glossary and an index. Although this is an introductory text for college students, it is written in a style that makes it understandable to the interested layperson. Contains a chapter on the oceanic crust.

Edmond, John M., and Karen Von Damm. "Hot Springs on the Ocean Floor." *Scientific American* 248 (April, 1983): 78-84. A discussion of the role of deep-ocean hydrothermal springs in depositing metallic ores and sustaining life in the absence of sunlight.

Gross, M. G. "Deep-Sea Hot Springs and Cold Seeps." *Oceanus* 27 (Fall, 1984). An introduction to a special issue devoted to the exploration, geochemistry, chemosynthesis, and biology associated with water discharge on the ocean floor.

Kennett, James P. *Marine Geology.* Englewood Cliffs, N.J.: Prentice-Hall, 1982. A comprehensive treatment of the geology of the seafloor, including ocean morphology, geophysics, plate tectonics, oceanic crust, and marine sediments.

Plummer, Charles C., David McGeary, and Diane H. Carlson. *Physical Geology.* Boston: McGraw-Hill, 1999. This is a straightforward, easy-to-read introduction to geology intended for those with little or no science background. Discusses the seafloor and the oceanic crust. Includes many excellent illustrations, as well as a CD-ROM.

4
VOLCANOLOGY AND VOLCANIC PROCESSES

CALDERAS

A caldera is a large depression, more or less circular in form, caused by the collapse of a volcano during or after eruption. With the exception of impacts by asteroid-sized meteorites, the largest caldera-forming eruptions represent the most catastrophic geologic events known. Ancient calderas are sites of many of the Earth's ore deposits, and recently formed calderas are important resources of geothermal energy.

PRINCIPAL TERMS

HOT SPOT: a volcanic center that has persisted for tens of millions of years and that is thought to be the surface expression of a rising plume of mantle material

IGNIMBRITE: an igneous rock deposited from a hot, mobile, ground-hugging cloud of ash and pumice

PLATE TECTONICS: a theory that describes the Earth's outer layer as consisting of large, in-dependently moving fragments

PYROCLASTIC: pertaining to volcanic material formed by explosion

SHIELD VOLCANO: a volcano in the shape of a flattened dome, broad and low, built by flows of very fluid basaltic lava

STRATOVOLCANO: a volcano constructed of layers of lava and pyroclastic rock; also called a composite volcano

CALDERAS AND CRATERS

The term *caldera* (Spanish for "kettle" or "cauldron") was used by inhabitants of the Canary Islands to refer to all natural depressions, including the island's volcanic craters and calderas. The term was introduced into the geologic literature in the nineteenth century to describe volcanic depressions. There remains, however, a debate over the difference between craters and calderas.

In general, craters are caused by the explosive removal of material, while calderas form by the subsidence of the surface during or immediately after explosive volcanism. Because subsidence structures are usually larger than craters, many geologists consider all volcanic depressions larger than 1 mile (or 1 kilometer, to some) to be calderas. Other geologists prefer to emphasize origin; they use "craters" for depressions produced by explosions and "calderas" for all collapse depressions. It is only when explosion structures are extremely large or calderas are small that a problem exists. Geologists, for example, are about evenly divided on whether the large depression produced by the 1980 eruption of Mount St. Helens is, by origin, a crater or, by size, a caldera. One of the small ironies of science is that all volcanologists agree that one of the world's best examples of a caldera is Oregon's Crater Lake.

In general, then, calderas form when the support provided by the underlying molten rock or magma is removed, either by eruption or by withdrawal to a lower level. Three classes of calderas are common: those located at the summits of basaltic shield volcanoes, such as Hawaii's Mauna Loa; those that "behead" andesitic stratovolcanoes, such as Crater Lake; and those that contain the source vents for widespread layers of rhyolitic ash, such as Wyoming's Yellowstone caldera. In addition, planetary geologists have discovered shield-like volcanoes with summit calderas on Mars, Venus, and Jupiter's satellite Io.

BASALTIC SHIELD VOLCANOES

The smallest terrestrial calderas are associated with basaltic shield volcanoes, such as those in the Hawaiian Islands. Shield volcanoes are composed mostly of layers of basalt, a dark-colored volcanic rock rich in magnesium and iron but relatively poor in silica. The Hawaiian Islands are the exposed southeastern end of a largely submarine mountain range of volcanic origin. Volcanic islands and submerged seamounts can be traced for nearly 6,000 kilometers to where they disappear into a deep oceanic trench off Alaska's Aleutian Peninsula. Plate tectonic theory and age determinations performed on volcanic rock suggest that

VOLCANIC ERUPTION AND CALDERA FORMATION

Beginning of eruption
at summit

Lava flow and deposition;
eruption at lower elevations

Subsidence or collapse
of summit

Cooling; cessation of
activity

the entire chain took nearly 80 million years to be produced as the Pacific Ocean floor moved at an average rate of about 8.6 centimeters per year over a subcrustal magma source, or hot spot. Hawaii is now located over the hot spot and contains two active shield volcanoes, Mauna Loa and Kilauea. Each shield volcano has a summit caldera from which emanate radial rift zones marked by recent lava flows, minor vents, and lines of craters. A magma chamber is located at a relatively high level within each volcano.

Although basaltic shield volcanoes erupt frequently, their eruption style is the mildest known. A typical eruptive sequence begins with the magma reservoir within the volcano gradually filling and producing a measurable inflation of the volcano's summit. Swarms of small earthquakes

caused by magma movement occur below the impending vent site. The eruption commonly begins with lava fountaining at the summit. As magma works its way along the rift zones to erupt at lower elevations, activity ceases at the summit. The continued removal of magmatic support causes a part of the summit area to subside along arcuate faults, forming a caldera.

Summit calderas are slightly elliptical in outline, with flat floors and steep walls. Because of the frequency with which basaltic shield volcanoes erupt, summit calderas have a complex history of collapse, uplift, and infilling by later lava eruptions. Kilauea's summit contains several collapse features, including Kilauea caldera, the major structure. Its approximate dimensions are 4 by 3 kilometers, and its average depth is about 100 meters. A second, smaller caldera, Kilauea Iki, is within Kilauea caldera, and both structures are surrounded by arcuate faults along which minor collapse has occurred. Mauna Loa's summit caldera is similar in size and also consists of multiple collapse zones.

Shield volcanoes with summit calderas are by no means restricted to the Hawaiian Islands. They are commonly found where large outflows of fluid basalt occur, and prominent examples are found in Iceland and the Galápagos Islands. Summit shield calderas on what appear to be basaltic volcanoes are also found on Mars. The most impressive example is Olympus Mons, probably the largest volcano in the solar system. It is more than 600 kilometers across and 23 kilometers high, and its summit contains an 80-kilometer-wide caldera complex. Scientists speculate that Mars has hot spots but does not have independently moving plates. They believe Olympus Mons may have been volcanically active for 1.5 billion years as basaltic magma was fed upward from its mantle source.

STRATOVOLCANOES

Calderas are also associated with stratovolcanoes. Stratovolcanoes, with slopes of every grade, most closely resemble the stereotype of the volcano. Lava and pyroclastic material accumulate around a central vent to produce mountains rising as much as 5 kilometers above their bases. The rapid erosion of material from these lofty peaks, sometimes as disastrous mudflows, produces aprons of sediment around the volcanoes' flanks. They are the most abundant type of large volcano on the Earth's surface and the characteristic volcanic landform found on the island arcs and continental margins fringing the Pacific Ocean. Although andesite, a dark- to medium-colored volcanic rock with an intermediate silica content, is the most common type of rock erupted, stratovolcanic eruptions produce a wide range of magma types.

Stratovolcanoes show pro-

longed periods of dormancy broken by eruptive phases that range from mild degassing to catastrophic eruptions that greatly alter or destroy the volcano's shape. Large eruptions from stratovolcanoes are commonly associated with the emplacement of rocks called ignimbrites. During an ignimbrite eruption, parts of the cone may be blasted away, or the volcano may founder into an immense caldera.

Historic caldera-forming eruptions have been impressive. The eruption of Krakatau in modern-day Indonesia in 1883 took place on a deserted volcanic island, yet a giant wave produced by the volcano's collapse killed more than thirty thousand people on neighboring shores. The great Tambora eruption of 1815 caused the deaths of more than ninety thousand people, either directly by eruption or by the ensuing famine. Prehistoric eruptions must have been even more spectacular. The Bronze Age eruption of Santorini in the Mediterranean Sea, for example, has been linked to the decline of the Minoan civilization and, thus, may have changed the course of Western history.

ANDESITIC AND RHYOLITIC STRATOVOLCANOES

Crater Lake, Oregon, has contributed much to the understanding of calderas and serves as a good example of caldera formation on andesitic stratovolcanoes. Crater Lake is a circular caldera

A caldera on Mount Pinatubo, Philippines, which has become a crater lake. (U.S. Geological Survey)

approximately 10 kilometers in diameter. The average depth of the lake is about 600 meters, and the surrounding cliffs rise from 150 to 600 meters. Wizard Island, a small cinder cone, rises 225 meters above the level of the lake. The eruption that formed Crater Lake occurred approximately 6,845 years ago, following thousands of years of intermittent activity that built a large stratovolcano that geologists call Mount Mazama. It is estimated that the cone was approximately 3,500 meters high and was capped by glacial ice.

Detailed field studies around Crater Lake have shown that the initial eruption was from a single vent, which fed ash and pumice into an eruption column that reached into the stratosphere and drifted with the prevailing wind. As the eruption intensified, so much material was emplaced into the cloud that, despite its heat, the cloud's density exceeded that of the surrounding air; it gravitationally collapsed to feed ground-hugging clouds of incandescent ash and pumice. These ash flows had great mobility and moved at hurricane speed to deposit ignimbrite around Mount Mazama. When about 30 cubic kilometers of magma had been expelled, the roof of the magma chamber collapsed to form a caldera. Venting, however, continued to eject another 20 cubic kilometers of magma from multiple vents located along the ring-fracture system bounding the caldera. The caldera continued to subside as venting progressed. Much of the ash fell back into the depression and mixed with rock that was sliding from the oversteepened walls to pile upon the caldera floor. A small cinder cone, Wizard Island, subsequently formed on the caldera floor, and Oregon's abundant rainfall produced the caldera lake.

Stratovolcanoes grade with increasing silica content to volcanoes composed mostly of rhyolite, a silica-rich volcanic rock that is usually light in color. Although eruption frequency of the more silicic

The eruption that formed Crater Lake occurred approximately 6,845 years ago, following thousands of years of intermittent activity that built a large stratovolcano that geologists call Mount Mazama. A small cinder cone, Wizard Island, subsequently formed on the caldera floor, and Oregon's abundant rainfall produced the caldera lake. (© William E. Ferguson)

volcanoes tends to decrease, their eruption volume and caldera size increase. Rhyolitic volcanoes are dominated by ignimbrite eruptions, and they tend to look very unlike volcanoes as most people picture them. A rhyolitic volcanic field consists of a rhyolitic ignimbrite plateau, punctuated here and there by large calderas. In most of these structures, the caldera floor has resurged or been uplifted and arched to form what is known as a resurgent dome. Resurgence, combined with the effects of sedimentation, continued volcanism, and erosion, may even make the caldera difficult to detect. Hundreds of such calderas are known or await discovery, hidden among the rhyolitic ignimbrites of western North America. Many of these ancient calderas are associated with that region's important ore deposits. Most calderas of this type tend to be circular, but the largest examples are elongated; their irregular shape may be caused by their piecemeal collapse into larger, more complex magma chambers or may reflect stresses in the crust. The largest calderas are more properly called volcano-tectonic depressions. The Lake Toba caldera, a volcano-tectonic depression on the island of Sumatra, is 100 kilometers long by 35 kilometers wide, the largest caldera yet recognized. Dormant periods between minor eruptions at rhyolitic volcanoes may be measured in thousands of years. Repeated large eruptions may be separated by 1 million years and can form compound structures such as the Yellowstone caldera, a good example of a recently active, rhyolitic volcano.

YELLOWSTONE NATIONAL PARK

Yellowstone National Park, famous for its hot springs and geysers, takes its name from volcanic rocks altered to bright colors by hot water and steam. The park's volcanic rocks belong to an igneous province composed mostly of basalt and rhyolite that extends southwest into Idaho's Snake River plain. Like Hawaii's, these volcanic rocks appear to be related to a hot spot. The Yellowstone hot spot, however, underlies the thicker and more silica-rich crust of the North American continent, and it is this continental crust that is believed to be the source for the enormous amounts of rhyolite.

Geologists of the U.S. Geological Survey have shown that the Yellowstone area has been the site of three large caldera-forming eruptions and numerous smaller eruptions during the past 2 million years. The ash emplaced within several tens of kilometers of the calderas was hot enough to anneal, or weld, into hard, rhyolite-capped plateaus. Remnants of the more widely dispersed ash have been found as far away as Texas. The last caldera-forming eruption occurred 600,000 years ago, when 1,000 cubic kilometers of magma was expelled, and a caldera 45 kilometers wide and 75 kilometers long was formed over the partly drained magma chamber. Within a few thousand years, magma elevated the caldera floor and arched it into the two resurgent domes contained in this complex structure. Over the past 600,000 years, much of the caldera has been filled with lava and sediment. Part of it is now covered by Yellowstone Lake, but the area remains thermally and seismically active.

Eric R. Swanson

CROSS-REFERENCES

BIBLIOGRAPHY

Cas, R. A. F., and J. V. Wright. *Volcanic Successions, Modern and Ancient: A Geological Approach to Processes, Products, and Successions.* London: Allen & Unwin, 1987. A comprehensive college-level volcanology textbook that covers all aspects of pyroclastic volcanism and is available in university and large public libraries.

Decker, Robert, and Barbara Decker. *Volcanoes.* New York: W. H. Freeman, 1997. This readable and well-illustrated text is geared toward the general public. The authors explain geologic phenomena in a clear and simple manner. Includes a good section on calderas. The appendices list the world's 101 "most notorious" volcanoes, as well as Internet sources for volcano information.

Decker, Robert W., Thomas L. Wright, and Peter H. Stauffer, eds. *Volcanism in Hawaii.* 2 vols. Denver, Colo.: U.S. Geological Survey, 1988. Published to commemorate the seventy-fifth anniversary of the Hawaiian Volcano Observatory, the two volumes that constitute this work are a treasury of information derived from years of research on all aspects of Hawaiian volcanism, and comprise the most comprehensive collection of scientific articles available on the topic, including the islands' calderas. Although written for professionals, many of the sixty-five reports are not above the level of the interested nonspecialist.

Francis, Peter. *Volcanoes: A Planetary Perspective.* Oxford: Clarendon Press, 1993. This is a university-level text, but it is written in a style that makes it accessible to a wide range of readers. The author covers the visible evidence of volcanoes, as well as the theory of why they develop. Illustrations, index, and reference list.

Schubert, Gerald, ed. *Journal of Geophysical Research* 89, no. B10 (1984). This special volume was published by the American Geophysical Union to commemorate the one hundredth anniversary of the eruption of Krakatau. Of general interest are the introduction, a paper on extraterrestrial calderas, and several regional studies. The last paper in the volume is of special interest because it is the most complete description available of large silicic calderas.

Sigurdsson, Haraldur, ed. *Encyclopedia of Volcanoes.* San Diego, Calif.: Academic Press, 2000. Contains a complete summary of the scientific knowledge of volcanoes, including eighty-two well-illustrated overview articles (one devoted to calderas), each of which is accompanied by a glossary of key terms. Although this is a college-level text, it is written in a clear and comprehensive style that makes it generally accessible. Cross-references and index.

Smith, Robert B., and Robert L. Christiansen. "Yellowstone Park as a Window on the Earth's Interior." In *Volcanoes and the Earth's Interior.* San Francisco: W. H. Freeman, 1982. This paper is one of a collection of articles on volcanology originally published in the journal *Scientific American.* The book is widely available and includes sections on volcanoes and plate tectonics, volcanic eruptions, and volcanoes as sources of information about the Earth's interior.

Tilling, Robert I., Christina Heliker, and Thomas L. Wright. *Eruptions of Hawaiian Volcanoes: Past, Present, and Future.* Denver, Colo.: U.S. Geological Survey, 1987. One of a series of general-interest publications prepared by the U.S. Geological Survey to provide information on the Earth sciences. Well illustrated and factual, it is probably the best source on Hawaiian volcanism available to those with no prior geological knowledge.

FLOOD BASALTS

In some of the world's continental regions, layer after layer of basalt lava was erupted at various times in the Earth's history to form extensive accumulations of thick lava flows. These features are called flood basalts (or plateau basalts). In many cases, flood basalts were generated as the continents were torn apart to form ocean basins during the processes of continental drift and seafloor spreading.

PRINCIPAL TERMS

BASALT: dark, fine-grained, silicate igneous rock crystallized from lava flows

CONTINENTAL DRIFT: the horizontal movement of continental masses relative to one another, caused by plate movement involving both the crust and the upper mantle

LITHOSPHERIC PLATE: a segment of the rigid crust and upper mantle that moves horizontally, sliding past, away from, or under other plates, resulting in mountains, volcanoes, rift valleys, and earthquakes

MAGMA: molten silicate liquid plus any crystals, rock fragments, and gases trapped within that liquid

MID-OCEANIC RIDGES: linear rift areas that bisect all the world's ocean basins; oceans are floored by basalt that pours out of these ridges and spreads out to either side

PERIDOTITE: the most common rock type in the upper mantle, composed of dense, iron- and magnesium-rich silicate minerals

RIFT: an area where tensional forces tear the crust and upper mantle apart, producing valleys or basins, generally with abundant volcanic activity

SEAFLOOR SPREADING: the expansion of the ocean floor with the creation of new material by extrusion of basalt lava at mid-oceanic ridges

VOLCANIC ROCKS

Basalt is a dark-colored, relatively iron-rich rock that commonly occurs as lava flows such as those on the islands of Hawaii. It is produced by partial melting of dense, iron- and magnesium-rich rocks in the Earth's mantle called peridotites. Volumetrically, basalt is the most important igneous rock type (rocks crystallized from molten magma) on the Earth in that it constitutes the floors of all the Earth's oceans, makes up most of the world's oceanic islands, and has poured out on the continents as well. The most impressive of these continental outpourings are the flood basalts, also known as plateau basalts. The last of the great flood-basalt areas (the Columbia River Plateau) ceased erupting lava some 10 million years ago, with activity lasting about 11 million years. Since then, no area on the Earth has experienced such tremendous outpourings of basalt lava over such a short period of geologic time.

Volcanic rocks are responsible for several types of surface features. The most familiar are volca-noes, generally cone-shaped edifices that spew out lava from a central crater or vent. Wherever basaltic lavas form volcanoes, they are always low-profile forms that extend laterally over immense areas. Called shield volcanoes because they resemble giant shields lying on the ground, they occur on the Hawaiian Islands and on many other islands and continents. Other kinds of lava form relatively high, pointed mountains that do not spread out so much at their bases. Fujiyama in Japan is a good example of this kind of volcano (called a stratovolcano), as are many of the Cascade volcanoes of the Pacific Northwest in the United States.

CHEMICAL COMPOSITION OF LAVA

To understand why some volcanoes are low and spread out while others are high, graceful cones requires a knowledge of the chemical composition of the lavas that make up the respective volcanoes. Basalt lava has a low abundance of the important component silica, while the lava of which stratovolcanoes (mostly andesite) are composed is relatively enriched in silica. Silica forms molecular chains in lavas called polymers that tend to become entangled with one another in silica-rich magmas, making them sticky and resistant to flow. Basalt, on the other hand, flows quite readily over the land surface because its silica polymer chains are lubricated, so to speak, by other chemical constituents in the magma such as iron, magnesium, calcium, and even water. For this reason, when basalt flows out on the land, it tends to spread over wide areas much as water or thin oil would, which produces the low-profile shield volcanoes typical of basaltic terrains. Stratovolcanoes are composed of materials that do not flow readily, so the lava tends to congregate near the central vent, eventually building up a high, conelike structure.

The tendency for basalt lava to flow readily leads to a second type of volcanic eruption called a fissure flow. A fissure flow is simply a flow of lava that emanates from an elongated fissure (in many cases, a fault or other fracture) and flows away quickly to either side. Instead of a volcanic cone, this process produces horizontal sheets of basaltic lava that eventually harden to form a layer of black basalt rock. Fissure flows seldom occur alone. After the first flow cools and hardens, another flow will pour out to cover it, followed by others over some period of time. A single flow may cover tens of thousands of square kilometers and measure up to 100 meters thick or more. Total accumulations of basalt in fissure-flow areas may reach several kilometers in thickness and cover an area of many thousands of square kilometers. These areas are termed flood or plateau basalts, and nearly every continent has at least one example of these impressive geologic features.

RIFTING

Nearly all the major flood-basalt regions in the world cover vast areas, and most occur on or near the margins of continents rather than in the interiors. The exceptions are the Lake Superior and Siberian regions, which occur far from any ocean basins. The occurrence of most flood basalts near continental margins is no mere coincidence. Geologists who study flood basalts are convinced that they arise during rifting episodes in which continental landmasses are literally pulled apart to make seas and, eventually, ocean basins. Tension

produced during rifting causes the formation of deep fractures (faults) that penetrate through the continental crust to the mantle. Dense rocks in the mantle called peridotite are very hot, in some areas sufficiently hot to make silicate liquid, a liquid that is compositionally basalt.

When faults penetrate into these partially molten rocks, pressure is released, causing the rocks to melt even more and the resulting magma to squeeze up through the faults. Eventually, this magma is joined by similar liquids produced near the deep faults, and much of it makes its way to the surface to be extruded as basalt lava flows. Some basalt magma, however, becomes trapped below the surface to crystallize as coarse-grained rocks called gabbro. Deeply eroded flood-basalt areas display excellent exposures of these "intrusive" rock bodies that, without doubt, accompany all flood basalts at depth.

During rifting events, hardened basalt lava flows produced during earlier eruptions are later split apart by subsequent rifting and covered over by younger lava flows. This process is repeated over and over again as the rift progressively tears the continent in two. Eventually, the continent splits into two parts with an ocean basin in between, an important part of the processes of continental drift and seafloor spreading. In fact, an ocean basin is composed of the basalt lava flows that pour out from the rift area between two continents; the ocean basin expands by "spreading" apart at an underwater rift, new ocean floor issuing forth as basalt lava. Oceans form deep basins that hold water because basalt is a relatively dense material compared to continents (composed on average of low-density granitic rocks) and, thus, the basalt slowly sinks into the mantle below to form a basin depression.

Flood basalts occur on continental margins because they are some of the earliest basalt flows to be erupted during rifting. Once the ocean basins are formed, these flood basalts are left high and dry on opposite sides of the rift on the continents that have slowly drifted apart over millions of years. The rift area itself, complete with deep faults and basalt eruptions, sinks below the watery depths to become what geologists call mid-oceanic ridges.

What of the flood basalts that occur in the interiors of continents? The Lake Superior and Siberian flood basalts erupted over a very short period of geologic time in rift regions that seem to have been aborted at an early stage, before a sea or ocean had time to develop. For example, the Lake Superior basalts were erupted 1.1 billion years ago in a rift that has been traced below the surface from the Lake Superior area southwest to the middle of Kansas. The basalt itself is exposed on the north and south shores of Lake Superior, but the complete exposed system also includes a giant intrusion of gabbroic rocks called the Duluth complex. Inexplicably, this rift system ran out of energy before it succeeded in cutting the North American continent in two, an event that would have had a profound affect on present-day geography.

FLOOD BASALT AREAS

Flood basalt areas that are clearly associated with the formation of present-day ocean basins include the Paraná area in Brazil, the Deccan basalts in India, the Brito-Arctic area (Northern Ireland, Scotland, and the east coast of Greenland), and the Karoo area in South Africa. The Paraná has basalt flows of equivalent age in southwest Africa, extruded when Africa and South America were joined together about 100 million years ago. The Paraná basalts were extruded prior to the separation of Africa from South America, resulting in the formation of the South Atlantic Ocean.

The Deccan Plateau in India was made by outpourings of basalt during the separation of southern India from eastern Africa about 65 million years ago that eventually helped to make the Indian Ocean. At about the same time that the Deccan basalts were erupted, rifting occurred in the present vicinity of northern Ireland and Scotland, which were attached to Greenland at that time. Basalts and gabbroic intrusions proliferated in the British Isles, including the small Inner Hebrides islands of Rhum, Skye, Mull, and others that attracted considerable attention from British geologists in the early twentieth century and later. The Karoo basalts resulted from the separation of Antarctica from southeastern Africa about 150 million years ago to make the southern Indian Ocean.

The best-known flood-basalt area in North America is the Columbia River Plateau of Washington and Oregon in the northwestern United States. Within a period of somewhat less than 2 million years (short in geological terms), between 17 and 15 million years ago, basalt flows

were extruded that covered more than 220,000 square kilometers of land to depths up to 10 kilometers (average thickness is 1 kilometer). The total area covered by basalt flows in eastern Washington, northern Oregon, and west-central Idaho is larger than the entire state of Washington. The name of this region comes from the Columbia River, which carves a rugged, scenic gorge through the basalt flows as it winds its way south across central Washington, then turns west to form the boundary between Oregon and Washington. As it cuts down through the many lava flows, the river reveals the layer-cake aspect of this thick volcanic pile that somewhat resembles the layered sedimentary rocks of the Grand Canyon.

The origin of the Columbia River Plateau is not as straightforward as the other occurrences already described. It is known that until about 400 million years ago, the western margin of the North American continent was located near the present western border of Idaho. Crustal blocks currently to the west of this boundary are "exotic," having been moved into their present positions as microcontinents that formed in earlier times to the south. In some boundary areas between these microcontinents and the ancient continental margin, oceanic crust (basaltic) is known to lie not far below the surface. It is probably no coincidence that the main eruption areas of Columbia River basalt flows are located near one of these thin-crust plate junctures where mantle-derived basaltic magma could be tapped relatively easily.

SUBDUCTION ZONES

What caused basalt lava to pour out in those eruption areas? Sometime around 24 million years ago (during the early Miocene epoch), the North American continent, moving inexorably to the northwest, overrode the mid-oceanic ridge that lay just west of the continent in the present-day Pacific Ocean. Before that happened, the North American lithospheric plate was overriding the Pacific Ocean plate that was diving under the continent to produce a subduction zone. In subduction zones of this type, basaltic oceanic material dives deeper and deeper under the continent until it is heated up to melting. Resulting magmas rise up on the continental margin to make volcanos similar to those in the Cascade Mountains of the northwest United States. Volcanoes similar to the

Cascades were erupting in the area of the present Sierra Nevada range in California and Nevada prior to the mid-oceanic ridge being overridden by the North American plate, but they ceased to erupt sometime after this event because subduction was replaced by horizontal movement manifested by the present San Andreas fault. Shortly after, however, the basalts of the Columbia River Plateau began pouring out, lasting from 17 million to 6 million years ago (95 percent of the basalt erupted in the first 3.5 million years).

An oceanic ridge is an especially hot, active area in which basalt pours out from fractures caused by intense tensional (tearing) forces. Many geologists believe that this very hot material sliding under the continent melted to produce the Columbia River basalt flows. It is also important that this area was being stretched and pulled apart by the rising Cascade Mountains to the west and by the rise to the east of granitic rocks in the Idaho batholith. This stretching effect between two rising crustal masses was probably enough to release pressure on uprising magmas, allowing them to rush to the surface to produce numerous lava flows in the Columbia River Plateau area.

Noncontinental Flood Basalts

Flood basalts can occur in places other than continental areas. If the basalt fissure flows produced at mid-oceanic ridges can be considered flood basalts, then the Earth's ocean basins are covered with flood basalts from shore to shore.

Geologists, however, restrict the term "flood basalt" to those that occur in continental areas only. The dark areas on the Moon (lunar maria) are considered flood basalts, having erupted in huge impact basins more than 3 billion years ago. Similar areas on the planets Mercury, Venus, and Mars are also flood basalts. In fact, it turns out that the grand majority of all basalt in the solar system beyond the Earth occurs as flood-basalt flows. The much rarer shield volcanoes, such as those in the Hawaiian Islands and, on Mars, Olympus Mons and its kin, are caused by the unusual situation in which magma is concentrated along only certain areas of the access fractures. In most cases, however, magma pours out all along the fracture, producing fissure flows, which accumulate over time to make what are termed flood basalts.

John L. Berkley

Cross-References

Basaltic Rocks, 1274; Calderas, 683; Continental Crust, 560; Continental Rift Zones, 579; Eruptions, 739; Geysers and Hot Springs, 694; Hawaiian Islands, 701; Hot Spots, Island Chains, and Intraplate Volcanism, 706; Igneous Rock Bodies, 1298; Island Arcs, 712; Lava Flow, 717; Lithospheric Plates, 55; Lunar Maria, 2550; Magnetic Reversals, 161; Ocean Basins, 661; Ocean Ridge System, 670; Oceanic Crust, 675; Plate Motions, 80; Plate Tectonics, 86; Ring of Fire, 722; Shield Volcanoes, 787; Spreading Centers, 727; Yellowstone National Park, 731.

Bibliography

Ballard, Robert D. *Exploring Our Living Planet.* Washington, D.C.: National Geographic Society, 1983. This book is a comprehensive account of the relationships between volcanic and tectonic (for example, mountain-building, rifting) processes and is illustrated with full-color photographs and diagrams. Well worth perusing for its description of plate tectonic processes and their relationship to volcanic activity in the Earth. Basaltic rocks and volcanism, although not flood basalts per se, are discussed in some detail. Very well written and indexed. Easily enjoyed by professionals and laypersons alike.

Decker, Robert, and Barbara Decker. *Volcanoes.* New York: W. H. Freeman, 1997. This readable and well-illustrated text is geared toward the general public. The authors explain geologic phenomena in a clear and simple manner. Includes a good section on calderas. The appendices list the world's 101 "most notorious" volcanoes, as well as Internet sources for volcano information.

Francis, Peter. *Volcanoes: A Planetary Perspective.* Oxford: Clarendon Press, 1993. This text is for university-level students, but it is written in a style that makes it accessible to a wide range of readers. The author covers the visible evidence of volcanoes, as well as the theory of why they develop. Includes a good section on flood basalts. Illustrations, index, and reference list.

Ragland, Paul C., and John J. W. Rogers, eds. *Basalts.* New York: Van Nostrand Reinhold, 1984. This book is a compendium of mostly journal articles reprinted directly from their source publications. One article, by Peter R. Hooper, entitled "The Columbia River Basalts" (reprinted from *Science*, volume 215, 1982), is worth reading by those who wish to explore the topic of flood basalts in some detail; it is an excellent summary of most of what is known about this well-studied flood basalt area. Location maps and stratigraphic section diagrams show time relationships and other information about the various flow units identified in the area. Includes a reference section that allows for even deeper research into the topic.

Sigurdsson, Haraldur, ed. *Encyclopedia of Volcanoes.* San Diego, Calif.: Academic Press, 2000. This book contains a complete summary of the scientific knowledge of volcanoes and includes eighty-two well-illustrated overview articles (including one on flood basalts), each of which is accompanied by a glossary of key terms. Although this is a college-level text, it is written in a clear and comprehensive style that makes it generally accessible. Cross-references and index.

Smith, David G., ed. *The Cambridge Encyclopedia of Earth Sciences.* New York: Cambridge University Press, 1981. Without question one of the finest compendiums on the Earth sciences for the layperson. The full-color diagrams and photographs are excellent, and the text is authoritatively composed by an all-star cast of experts in their fields. Although it has only a brief section on flood basalts (page 206), this source includes material on related topics, such as basalt volcanism and plate tectonics, that makes it an important resource for understanding the kinds of phenomena that lead to the production of flood basalts. Provides a comprehensive glossary of terms and an index.

Time-Life Books. *Volcano.* Alexandria, Va.: Author, 1982. One volume from the outstanding Planet Earth series, written and illustrated with the nonspecialist in mind. Beautiful color diagrams and photographs and entertaining narrative present a fairly comprehensive treatment of world volcanism. Flood basalts are treated on pages 62-63 and on page 70. Includes a very detailed index.

GEYSERS AND HOT SPRINGS

Geysers are a type of hot spring that periodically erupt steam and hot water. They are the surface expressions of vast underground circulation systems, where constituents from underground rocks are dissolved in the hot fluids, carried to the surface, and deposited. The world's active thermal areas are natural laboratories where ore-forming processes can be observed at first hand.

PRINCIPAL TERMS

CONVECTION: the transfer of heat by mass movement, such as by the flow of hot water and steam

FUMAROLE: a vent that emits only gases

GEOTHERMAL GRADIENT: the rate at which temperature increases with depth in the Earth

HYDROSTATIC PRESSURE: the pressure imposed by the weight of an overlying column of water

HYDROTHERMAL or GEOTHERMAL: general terms that refer to natural systems of hot fluids that circulate underground

RHYOLITE: a type of silica-rich volcanic rock that is uncommon on the Earth but occurs almost universally beneath hydrothermal areas

SINTER: a variety of nonprecious opal that forms around chloride hot springs and geysers

SUBLIMATE: solid, crystalline material that is deposited directly from the vapor state; crystals of native sulfur around fumarole mouths are an example

TRAVERTINE: a variety of hot-spring limestone that is deposited from alkaline waters

WATER TABLE: the level below which all rocks are saturated with water

LOCATIONS OF HOT SPRINGS AND GEYSERS

The term "geyser" derives from an old Icelandic word, *gjose*, which means "to erupt." Great Geyser, a spouting hot spring in southwestern Iceland, is the namesake of similar features around the world. A true geyser is a type of clear, boiling spring that periodically erupts mixtures of steam and hot water. Geysers are among the rarest and most spectacular of natural phenomena and are found in only a few regions of the world. Notable geyser areas occur in Iceland, Chile, the North Island of New Zealand, Japan, and Kamchatka. The most famous area is Yellowstone National Park in Wyoming, which contains more than ten thousand thermal features—more than in all the rest of the world. The three hundred geysers of Yellowstone account for 60 percent of the world's geysers.

HOW THEY WORK

The source of heat that drives hydrothermal systems is magma, solid but still-hot rock that is 5 to 10 kilometers beneath the Earth's surface. Not surprisingly, nearly all the world's major hydrothermal systems are found close to active or potentially active volcanoes. Although some hydrothermal features, such as crater lakes on active volcanoes, are heated by steam and gas that are evolved directly from small, shallow magma bodies, most hydrothermal phenomena are the surface expressions of immense underground convection cells of hot water and are indirectly linked to their magmatic heat source. Heat from magma or hot rock is conducted into the surrounding rocks and from there into groundwater that circulates through the rocks along fractures or through permeable strata. Although surface hot springs occur only within local areas, their underground circulation systems are tens of kilometers across and extend several kilometers deep.

The water in hot springs begins as rain and snowfall, which percolates several kilometers down into the Earth's crust through permeable volcanic rocks and sediments. The normal geothermal gradient (the rate at which temperature increases with depth in the earth) of the continental crust is about 20 degrees Celsius per kilometer, but in hydrothermal regions such as Yellowstone National Park, the geothermal gradient is ten to thirty times that value. The water becomes heated, and because hot water is more buoyant than the cold water that continually displaces it underground, a huge subterranean convection system is established. The hot water rises to the surface, usually along faults or through connected pores in the rock. Depending on the permeability of the rocks, the volume of water, and the amount of heat, the complete cycle from snowflake to hot spring may take centuries or millennia.

CHEMICAL COMPOSITION

The chemical composition of hot-spring waters is quite variable. Although some dissolved constituents may come directly from magma, laboratory experiments and chemical analyses of spring waters have shown that most of the dissolved matter comes from the underlying rocks. Thus, the compositions of hot-spring waters are strongly controlled by the compositions of the rocks through which they circulate, and such rocks are usually volcanic in origin. Rhyolite (a silica-rich type of volcanic rock) is almost universally found beneath the world's major hydrothermal regions, including Yellowstone. The rhyolite is the source of silica that forms the mineral deposits that line underground channels and that are deposited on the surface.

HOT-SPRING CLASSIFICATION

There are three major classes of hot springs, based on their fluid types (liquid- or vapor-dominated), fluid compositions (acidic, neutral, or alkaline), and surficial deposits (sinter, mud, or travertine).

The first type is chloride springs, including geysers, which discharge clear water that is at or near its boiling point and is neutral or slightly

HOW A GEYSER WORKS

Geysers are caused when (A) a mass of hot rock causes water trapped below the Earth's surface to approach, but not reach, the boiling point, which is higher than at the surface, due to the pressue at depth. (B) As water above is heated, it reaches its boiling point and exits the surface vent as steam, which in turn reduces the pressure below and allows the deeper water to reach the boiling point, resulting in an eruption of water as a geyser and the reduction of pressure. The cycle then begins again.

alkaline. The most important dissolved constituents in the water are several parts per thousand of sodium chloride (common salt) and up to a few hundred parts per million of silica (hydrous silicon dioxide). The chloride is very soluble and remains in solution, but the silica readily comes out of solution during cooling of the water, and siliceous sinter is deposited on the surface. A variety of common opal, sinter takes many forms including nodular masses, terraces, cones, and round "geyser eggs" that form only in the turbulent pools adjacent to geyser vents. Sinter is white or pale pink when first deposited but turns gray with age. Hot-spring formations are quite delicate, and when deposition ceases, they commonly break down into tiny white chips. Also carried in neutral-chloride waters are smaller amounts of potassium and lithium salts, boric acid, fluorides, and sulfates, together with traces (parts per million or billion) of heavy metals such as copper, lead, zinc, silver, gold, manganese, and thallium. The associated steam is mostly water vapor, with a small amount of carbon dioxide, and traces of other gases such as hydrogen sulfide, hydrogen, nitrogen, methane, ammonia, and hydrogen fluoride.

Acid-sulfate systems constitute the second major class of hot springs, including mud pools, mud pots, and fumaroles (vents that emit only gases). Acid-sulfate hydrothermal systems are so hot that the water table boils underground. As a result, these systems are said to be vapor-dominated, as opposed to the other two types, which are liquid-dominated. Steam and carbon dioxide are the two major fumarole gases, but a trace of hydrogen sulfide gas plays a very important role. The hydrogen sulfide is readily oxidized to form sulfur dioxide gas (with its acrid smell), native sulfur, and sulfuric acid. The latter product accounts for the extreme acidity of acid-sulfate fluids, and the sulfuric acid vigorously leaches the surrounding rocks and soil. The main product of this chemical attack is kaolinite, a white clay mineral that accounts for features ranging from turbid pools to viscous mud volcanoes, depending on the proportions of mud and water. Tiny black grains of pyrite (iron sulfide) or graphite (elemental carbon) commonly form in the pools, giving the mud a gray color and forming a black surface scum. Mud pools can

Steam rises from an old geyser near Mammoth Lakes, California, after a swarm of more than one thousand earthquakes struck in November, 1997. (AP/Wide World Photos)

be further colored by yellow, orange, and red iron oxides. Yellow, needlelike crystals of native sulfur are a common sublimate around the mouths of fumaroles.

The third type of hot spring is less common than the others. Liquid-dominated alkaline springs occur only where hydrothermal fluids have passed through underground limestone formations and are rare in comparison to the other two types. Calcium carbonate, the major constituent of limestone, becomes dissolved in the water, is carried to the surface, and is ultimately deposited as mounds or terraces of travertine, a banded variety of hot-spring limestone. The water temperatures of alkaline springs are usually well below the boiling point of water, and any bubbling is caused by carbon dioxide gas, not steam.

HYDROTHERMAL ENVIRONMENTS

The physical and chemical differences between liquid-dominated chloride systems and vapor-dominated acid-sulfate systems are largely caused by their positions with respect to the water table. In hydrothermal areas, the water table stands higher than elsewhere because of the relative buoyancy of upwelling hot water. Chloride springs occur where the water table emerges at the surface and the water boils off into the atmosphere, leaving deposits of sinter. On higher ground where the water table is deeper, only steam and gases may reach the surface because the water table is boiling underground. Acidic fumaroles (dry gas vents) are therefore typical features.

Within and around thermal basins, the high soil temperatures and acid vapors often destroy plant roots, and pits filled with carbon dioxide are death traps to birds and small animals. Geysers and hot springs are also sustainers of life, however, as shown by the birds and mammals that congregate near the year-round open water and green vegetation of hydrothermal features. Certain species of plants, bacteria, and even entire food chains of colorful algae, flies, and rare spiders are wholly dependent on the supply of warm water.

The brilliant coloration of hot springs results from several factors. Brightly colored iron oxides and other metal compounds may form locally around hot springs, but most hot-spring deposits are rather dull. Hot springs sometimes show zones of colors that vary with water depth, a result of

Unlike most geysers, Echinus Geyser, located in the Norris Geyser Basin at Yellowstone National Park, spouts extremely acidic water. (U.S. Geological Survey)

blue skylight blending with reflections from yellow sulfur crystals or algae that line the pool. Although the vivid surface coloration of hot springs partly results from mineral deposits, microorganisms are largely responsible. Certain species of algae and cyanobacteria (blue-green algae) thrive at temperatures up to about 75 degrees Celsius, and they live in the relatively cool outflow channels of hot springs. Different species have different colors and prefer different temperatures; thus, green, brown, red, and yellow color patterns are formed in the variable temperatures of outflow channels.

NECESSARY CONDITIONS

Geysers are short-lived and fragile features. Many have become inactive during historic time as a result of natural events and human activity. Of the ten major geyser areas that are truly outstanding, only three—in Yellowstone, Iceland, and the former Soviet Union—remain essentially undisturbed. Three have been destroyed by the nearby construction of dams, and four have been altered by nearby geothermal development. After the 1959 Hebgen Lake earthquake (magnitude 7.3) occurred near Yellowstone, hundreds of geysers and hot springs changed their patterns of activity. Any disruption of the delicate plumbing systems by humankind or nature involving the extraction of fluids, injection of fluids, or perturbation of the water table will almost always yield unpredictable and possibly detrimental consequences to nearby thermal features.

A bubbling mud pot. (© William E. Ferguson)

Geysers require very specific physical and thermal conditions. There must be a potent source of heat, usually magma or hot but solidified rocks. There must be abundant underground water to form a deep convection system, and there must be a focused pathway (usually a major fault or fracture in the Earth's crust) for the hot fluids to rise toward the surface at temperatures close to boiling. Near the surface, there must be a shallow, cavernous storage system for the underground water and steam. The storage system must have a constricted surface opening in order to propel its fluids into the air. Finally, the storage system must be capable of recharging and reheating between eruptions.

When these conditions are not met, features other than true geysers develop. If there is no focusing pathway to the surface, fluids may remain underground. If the heat source is not hot enough, or if there is too much cool groundwater, then warm springs may result. If underground temperatures are very high but there is too little groundwater present, then dry steam vents (fumaroles) will form. If the surface opening is too large or if the shallow reservoir allows free circulation, then instabilities may not develop, and the hot spring may simply boil but not erupt; such vents are called perpetual spouters.

GEYSER ERUPTIONS

The causes of geyser eruptions are complex, and no single theory can explain the detailed activity of all geysers. The phenomenon, however, can be described generally in terms of the boiling behavior of water. Geysers are unstable, neutral-chloride hot springs that contain boiling or near-boiling water. The underground plumbing systems of geysers can be visualized as cavernous, vertical pipes that extend for tens to several hundred meters beneath the surface. On the surface, the water column is boiling at atmospheric pressure, and its temperature is thus about 100 degrees Celsius. At a depth of several hundred meters, the water is under considerable hydrostatic pressure and may have a temperature of several hundred degrees. When the water in any part of the system is heated close to its boiling point, then a very small drop in pressure will cause the water to "flash" explosively into steam.

One way to achieve the necessary pressure drop is to decrease the weight of the overlying water column, causing a slight drop in the hydrostatic pressure, which can happen as a consequence of boiling in the upper levels of the water column: Because steam bubbles are much lighter than water, any replacement of water by steam will result in a decrease of the column density, which will be immediately felt as a pressure drop at deeper levels. Many geyser pools begin to bubble and their water levels rise within a few moments of eruption, as the water is displaced by rising steam bubbles, some of which become trapped in the roofs of cavernous chambers. At some point in the plumbing system, flashing eventually occurs, and a mixture of steam and water is propelled out of the orifice. The eruption of fluids lowers the pressures in successively deeper levels of the system, and flashing occurs as a downward-propagating chain reaction. When all the eruptible fluids have been ejected from the reservoir, the eruption ceases. Among geysers that occur in groups, eruptions are often highly irregular as a result of interconnected plumbing systems that allow the shunting of hot fluids into one or another plumbing system.

The periodicity of geysers (the length of time between eruptions) depends on how long it takes the plumbing system to be recharged with hot water (which is largely controlled by the permeability of the surrounding rocks and the volume of the geyser reservoir), the time required for this water to reheat back to its boiling point, and the amount of fluid that was discharged during the previous event. Eruptions of long duration represent thorough evacuation of the reservoir; the time required

for subsequent recharge will thus be lengthened, and the succeeding eruption will usually occur after a longer-than-normal period. Some irregularity is therefore typical of most geysers, including Old Faithful, but the timing and vigor of an impending eruption can often be predicted quite accurately if the nature of the previous eruption is known.

Geyser eruptions can sometimes be induced by the addition of soap to the vent because the surface tension of the water is lowered and frothing occurs more readily. This practice is illegal in virtually all protected natural geyser areas, as it can do irreparable harm to the orifice and plumbing system. Similarly, throwing coins or other objects into geyser pools is potentially harmful to the orifice or the plumbing.

William R. Hackett

CROSS-REFERENCES
Calderas, 683; Earth Resources, 1741; Flood Basalts, 689; Hawaiian Islands, 701; Heat Sources and Heat Flow, 49; Hot Spots, Island Chains, and Intraplate Volcanism, 706; Island Arcs, 712; Lava Flow, 717; Ring of Fire, 722; Spreading Centers, 727; Yellowstone National Park, 731.

BIBLIOGRAPHY

Bryan, T. S. *The Geysers of Yellowstone.* Boulder: Colorado Associated University Press, 1986. This 300-page book is written by a former park employee who is thoroughly familiar with the thermal areas of Yellowstone. Geared to the nonspecialist, it has several very good introductory chapters on geysers and hot springs and an index, glossary, and annotated bibliography. Much of the book is devoted to descriptions of individual thermal features in the park; thus, it is useful to tourists as well as to researchers of geothermal phenomena.

Decker, Robert, and Barbara Decker. *Volcanoes.* New York: W. H. Freeman, 1997. This readable and well-illustrated text is geared toward the general public. The authors explain geologic phenomena in a clear and simple manner. Includes a good section on calderas. The appendices list the world's 101 "most notorious" volcanoes, as well as Inter-

net sources for volcano information.

Elder, John. *Geothermal Systems.* New York: Academic Press, 1981. This 508-page book gives a quantitative treatment of the subject. It is written for scientists and engineers who would be involved in projects designed to exploit the internal heat of the Earth. Emphasizes active geothermal systems within the context of global geologic processes.

Francis, Peter. *Volcanoes: A Planetary Perspective.* Oxford: Clarendon Press, 1993. This is a university-level text, but it is written in a style that makes it accessible to a wide range of readers. The author covers the visible evidence of volcanoes, as well as the theory of why they develop. Includes a section on geysers. Illustrations, index, and reference list.

Rinehart, J. S. *Geysers and Geothermal Energy.* New York: Springer-Verlag, 1980. This 223-page book should be the first stop for detailed information about most geothermal phenomena. Coverage is very complete and includes the locations, physics and chemistry, mineral deposits, and historical behavior of the world's major hydrothermal areas. Other chapters cover human influence and the use of geothermal fluids. Has an index, data tables, an extensive bibliography, and abundant black-and-white illustrations. A well-written, thorough reference work for people with good general-science backgrounds.

Sigurdsson, Haraldur, ed. *Encyclopedia of Volcanoes.* San Diego, Calif.: Academic Press, 2000. This book contains a complete summary of the scientific knowledge of volcanoes. It contains eighty-two well-illustrated overview articles (including several on geysers and geothermal systems), each of which is accompanied by a glossary of key terms. Although this is a college-level text, it is written in a clear and comprehensive style that makes it generally accessible. Cross-references and index.

White, D. E. "Some Principles of Geyser Activity Mainly from Steamboat Springs, Nevada." *American Journal of Science* 265 (1967): 641-684. Although this journal article is written for geologists, no bibliography of geysers and hot springs would be complete without at least one citation of the work of Donald E. White. A classic reference on geyser activity. The journal is available at most university libraries.

HAWAIIAN ISLANDS

The Hawaiian Islands were formed almost entirely from active volcanoes on the ocean floor. Hawaiian lava is of a basaltic composition, forming two distinct flow types. The lava flows are younger in age southward in the island chain, with current volcanic activity taking place on the southern part of the island of Hawaii. The age progression and alignment of volcanoes are evidence for a hot spot located beneath the islands.

PRINCIPAL TERMS

AA FLOW: a basaltic lava flow with a surface characterized by large angular blocks

BASALT: a dark-colored, fine-textured igneous rock rich in iron and magnesium with a low percentage of silicon

CALDERA: a large crater resulting from subsidence or collapse, with a diameter many times greater than its depth

HARMONIC TREMOR: a movement or shaking of the ground accompanying volcanic eruptions

LAVA TUBE: a cavern structure formed by the draining out of liquid lava in a pahoehoe flow

PAHOEHOE FLOW: a lava flow having a ropy or billowy surface of basaltic composition

PILLOW BASALT: spheroidal masses of igneous rock formed by extrusion of lava under water

PYROCLASTIC ROCKS: rocks formed in the process of volcanic ejection and composed of fragments of ash, rock, and glass

SHIELD VOLCANO: a large, gently sloping flat lava cone with the shape of a shield and composed of numerous flows of basaltic lava

TEPHRA: a general term that describes all volcanic ejecta

HAWAII AND VOLCANISM

As a group, the Hawaiian Islands are remarkable in terms of their volcanism and geological development. Stretching from Kure Island in the northwest to Hawaii Island in the southeast, the islands, also known as the Hawaiian Archipelago, span 2,400 kilometers in length. In the northwestern part of the chain, the islands rise some 5,000 kilometers above their ocean floor bases; those located in the southeastern portion are higher with Mauna Kea and Mauna Loa, Hawaii's highest mountains, reaching up to 9,000 meters. The major islands, which include Hawaii, Maui, Oahu, Kauai, Molokai, and Lanai, make up only the last 650 kilometers of the chain. The island of Hawaii is among the largest volcanic islands in the world, ranking second only to Iceland. The geological history of the islands spans many millions of years; the islands are youthful, however, when compared to the history of the Earth. Fractures were formed across a narrow northwest region of the ocean floor from which molten rocks or lava issued forth, solidified, and gradually built the volcanic mountains layer by layer. The lava that builds such massive structures consists of thousands of thin flows.

The lavas that compose the Hawaiian Islands are of a dark, iron-rich material called tholeiitic basalt, and, to a lesser extent, of alkalic basalt. Tholeiitic basalt has an abundance of silica. In the molecular arrangement of silica, silicon and oxygen atoms form a four-sided crystal called a tetrahedron; the silicon atoms, in the center, are surrounded by four oxygen atoms. The tholeiitic basalt has a low percentage of the alkalines, sodium and potassium. Alkalic basalts, in contrast, have a lower percentage of silica. Both tholeiitic and alkalic basalts usually contain olivine, a dark greenish-black glassy mineral composed of iron-magnesium silicates.

Hawaiian eruptions originate at great depths in the ocean and tend to flow rather than to explode, given the restraining effects of the large weight of ocean water upon gas expansion. Lava flows tend to be dense and to form a broad shield structure. As the shield structure approaches sea level, the confining water pressure decreases, and explosive eruptions become more common. As lava rises above sea level to a subaerial environment, material called tephra is ejected along with large fragmental material called pyroclastics. Lava flows still accrete most of the magma to the volcano, as tephra comprises less than 1 percent of eruptions. Under the enormous weight of many lava flows, the summit of the shield eventually collapses along fractures or rift zones, forming a caldera.

At this point, the number of eruptions decreases, and the erupting magma shows a marked change in composition. The magma, although still of basaltic composition, has more explosive power and is richer in alkalines with a higher percentage of gases than is tholeiitic basalt. As eruptions become enriched with alkalines and gases, the lavas become more viscous, forming small, steep-sided volcanic structures called spatter cones. Volcanic activity may pause for as long as 2 million years before resuming. Erosion, subsidence, and reef building are the main processes during this stage.

TYPES OF LAVA

When volcanism resumes, the basalts are depleted in silica and form a very fluid lava. The

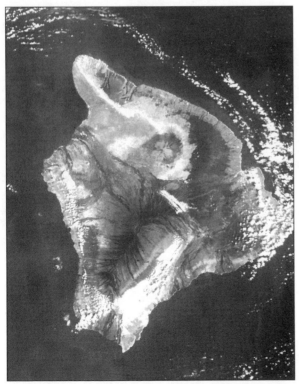

Landsat image of the big island, Hawaii, which rests on a plate that is moving across a "hot spot." (U.S. Geological Survey)

fluid basaltic lava tends to pour out from structures called lava fountains and is of two distinct types. Pahoehoe lava has billowy, smooth, or ropy surfaces; aa lava has a rough, spiny, or blocky surface. Occasionally, the two types intergrade, and classification becomes difficult. Some flows actually change from pahoehoe to aa as they move downslope. Although the reverse does not occur—aa lava does not change to pahoehoe lava—the flow of pahoehoe near its margins will sometimes burrow beneath an aa flow, resembling such a change. Pahoehoe lava has a greater tendency to flow, and as it continues downslope—stirring, mixing, and losing volatile gases—the lava becomes more viscous and tends to change to aa. Pahoehoe and aa lava are also differentiated by small gas-shaped bubbles within the flows called vesicles. Vesicles tend to be spheroidal in shape but are twisted or irregular in the aa flows. The high fluid content of the pahoehoe lava exerts less distortion on the gas bubbles than does the aa lava. Some flows have moved at speeds up to 55 kilometers per hour, but the entire flow as a whole advances more slowly.

The ropy nature of pahoehoe flows results from a dragging out and wrinkling of the still plastic and not yet solidified lava by the liquid lava moving beneath the flow. As the center of the flow moves faster, the ropy texture curves outward in the direction of the flow, creating small, lobe-shaped protuberances that are either smooth or hummocky in appearance. The more spectacular of the pahoehoe flows are tunnels known as lava tubes. The flowing lava crusts over on the surface, forming a roof while continuing to flow beneath. Smaller tributary tubes branch off from the main tubes, allowing lava to reach the margins of the flow. The flowing lava freezes inward from the outer margins while the center remains flowing. Active movement of the liquid is confined to a cylindrical pipelike region near the center of the flow. As the supply of lava diminishes, the liquid no longer completely fills the interior cavity. While the supply of lava continues to diminish, the level of the liquid surface within the cavity lowers to form a flat floor in the tube.

The blocky surface of an aa flow covers a massive dense interior. The central portion of the flow does the actual flowing, while the blocky top is merely carried along. As the flow advances, irregu-

Mauna Ulu, on the island of Hawaii. A geologist stands in front of a large, solidified flow of aa lava. (U.S. Geological Survey)

lar fragments called clinker drop down from the top and are buried by the advancing lava, resulting in a layer of clinker on the bottom of the flow. The clinker fragments covering the flows tend to be very sharp and spiny and have been known to cause painful cuts and even to slash leather boots.

Pillow lava can be created when the water is of sufficient depth to allow the confining pressure to prevent an explosive eruption or if the surface of the lava is cooled by contact with ocean water to form an outer crust over a molten core. This lava is made up of ellipsoidal masses the size and shape of pillows. Close examination reveals a radial internal structure.

HAWAIIAN HOT SPOT

The volcanic series of the major volcanoes of the Hawaiian Islands have been dated by radiometric methods. Results show that the lava flows of volcanoes are progressively older toward the northwest part of the chain: On the island of Hawaii, for example, the age of flows is 500,000 years, while on the island of Midway, the age of flows is 11 million years. This age progression, along with the chain's linear trend, indicates that the newest volcanic activity should occur toward the southeast. The Hawaiian Island chain is only one of several linear chains in the Pacific, including the Marshall-Gilbert and Tuamotu island chains, that dis-

play the northwest alignment.

It has been suggested that the Hawaiian Islands developed as the Pacific plate moved northwestward over a deep-seated melting region, or hot spot, within the mantle. A hot spot, or mantle plume, would supply magma to the overriding lithospheric plate, creating volcanoes. Volcanoes that had already developed would gradually move away from the hot source and eventually become dormant. If the hot spot location within the mantle were fixed for a relatively long time, the lithosphere would transport the volcanoes beyond the source. As this process continued, a succession of volcanic mountains would develop in the direction of plate motion. Although they are responsible for less than 1 percent of total volcanic activity, hot spots have a number of distinctive features that correlate well with the Hawaiian chain: They tend to occur in isolated regions distant from the major lithospheric plate boundaries; they are located in regions of crustal uplift and produce linear island chains that show an age progression; and they produce volcanoes that are basaltic in composition.

The exact origin and nature of the Hawaiian hot spot is not known, but it is possible to speculate on its size and location when compared with the positions of the Hawaiian Islands. Hawaiian volcanoes show alignment with respect to two different loci. A locus of volcanic centers includes the islands of Kauai, West Molokai, Lanai, East Hawaii, and the new seamount, Loihi. Another locus of points falls across East Molokai and West Maui, and includes Mauna Kea and Kilauea. Volcanoes falling on the two loci that are within the given region may experience simultaneous eruptions; it may then be said that the size of the Hawaiian melting spot approximates the distance between volcanic pairs or loci.

A serious problem with the hot spot hypothesis is the so-called rejuvenated volcanism, or volcanic activity that resumes after a period of dormancy. The Honolulu volcanic series on the island of Oahu

Lava from Mauna Ulu enters the ocean west of Apua Point, island of Hawaii. (U.S. Geological Survey)

hot spot has drifted at rates of 1 to 2 centimeters per year. Plotting the locations of hot spots with respect to the Hawaiian volcanoes gives significant positional errors. If hot spots are not stationary features, determining the nature of plate motion beyond 30 million years ago becomes difficult.

Geologists tend to agree, though, that whatever the nature and size of the Hawaiian hot spot, its present position is southeast of Kilauea and Mauna Loa, near the Loihi seamount. Located 40 kilometers east of Hawaii's South Point and 5,000 meters under the ocean, Loihi's long base and oval summit seem appropriate for its name, which is the Hawaiian word for "long." Although its rate of growth is unknown, it is estimated that its summit will break through the surface in the next 100 to 10,000 years. Loihi's caldera generally lacks large marine life, perhaps because of the high carbon dioxide emitted as well as the instability of the terrain as the result of earthquakes, eruptions, or rock slides on an active volcano. Temperatures of as much as 48 degrees Celsius have been measured on the summit as compared with 22 degrees Celsius for the surrounding seas at that depth.

Michael Broyles

generated new magma after a quiet period of more that a million years, with the probable location of the hot spot some 500 kilometers away. In addition, hot spots may not be entirely stationary; for example, evidence indicates that the Hawaiian

CROSS-REFERENCES

Basaltic Rocks, 1274; Calderas, 683; Eruptions, 739; Flood Basalts, 689; Forecasting Eruptions, 746; Geysers and Hot Springs, 694; Hot Spots, Island Chains, and Intraplate Volcanism, 706; Island Arcs, 712; Lava Flow, 717; Lithospheric Plates, 55; Pyroclastic Rocks, 1343; Recent Eruptions, 780; Ring of Fire, 722; Seamounts, 2161; Shield Volcanoes, 787; Spreading Centers, 727; Volcanic Hazards, 798; Yellowstone National Park, 731.

BIBLIOGRAPHY

Borg, James C. "Birth of an Island." *Oceans,* August, 1987, p. 27. Exciting documentation of the exploration of the caldera of Loihi, an active undersea volcano that may well become the next Hawaiian island. The dive was accomplished with the minisubmarine *Alvin,* with a full narrative description of the exploration of a hydrothermal vent. Suitable for the layperson interested in ocean research.

Decker, Robert, and Barbara Decker. *Volcanoes.*

New York: W. H. Freeman, 1997. This readable and well-illustrated text is geared toward the general public. The authors explain geologic phenomena in a clear and simple manner. Includes a good section on the Hawaiian Islands. The appendices list the world's 101 "most notorious" volcanoes, as well as Internet sources for volcano information.

Francis, Peter. *Volcanoes: A Planetary Perspective.*

Oxford: Clarendon Press, 1993. This is a university-level text, but it is written in a style that makes it accessible to a wide range of readers. The author covers the visible evidence of volcanoes, as well as the theory of why they develop. Illustrations, index, and reference list.

Holcomb, Robin T., James G. Moore, Peter W. Lipman, and R. H. Belderson. "Voluminous Submarine Lava Flows from Hawaiian Volcanoes." *Geology,* May, 1988: 400. This article discusses the nature of submarine lava flows as revealed by sonar imaging in the Hawaiian trough east and south of the island of Hawaii. Charts, photographs, graphs, and tables summarize research investigations. An excellent reference for geologists that is suitable for college-level students.

Sigurdsson, Haraldur, ed. *Encyclopedia of Volcanoes.* San Diego, Calif.: Academic Press, 2000. This book contains a complete summary of the scientific knowledge of volcanoes. It includes eighty-two well-illustrated overview articles (including one on calderas), each of which is accompanied by a glossary of key terms. Although this is a college-level text, it is written in a clear and comprehensive style that makes it generally accessible. Cross-references and index.

Stearns, H. T. *Geology of the State of Hawaii.* Palo Alto, Calif.: Pacific Books, 1985. A revised, expanded, and comprehensive book on the geology of the Hawaiian Islands. A separate chapter is devoted to each of the major islands with accompanying geologic map. Photographs, charts, drawings, and maps are abundantly used throughout this edition. Written for both the professional geologist and the layperson with an interest in volcanoes and Hawaii.

Tilling, Robert I., Christina Heliker, and Thomas L. Wright. *Eruptions of Hawaiian Volcanoes: Past, Present, and Future.* Denver, Colo.: U.S. Geological Survey, 1987. One of a series of general-interest publications prepared by the USGS to provide information on the Earth sciences. Well-illustrated and factual, it is probably the best source on Hawaiian volcanism available to those with no prior geological knowledge.

Time-Life Books. *Volcano.* Alexandria, Va.: Author, 1982. A pictorial atlas of the major volcanoes and their destructive effects, including Mount Pelée, Mount Vesuvius, Mount St. Helens, and the volcanoes of Hawaii and Iceland. A full chapter is included on the major eruptions as well as the history of volcanology in Hawaii. The layperson and beginning science student should enjoy the edition, with its many full-page color photographs.

HOT SPOTS, ISLAND CHAINS, AND INTRAPLATE VOLCANISM

Hot spots are small, isolated regions of volcanism that are stationary compared to the drifting lithospheric plates that comprise the surface of the solid Earth. The most common volcanic site for a hot spot is within the interior regions of oceanic plates away from all convergent plate boundaries. Hot spots can exist at a given location for tens of millions of years. After an oceanic plate moves over a hot spot, a line of extinct volcanic islands, seamounts, and undersea guyots trail behind the hot spot for a distance of several thousand kilometers. Age dating of the volcanic materials at various distances along the island chain allows the calculation of the absolute velocity of the plate's motion over the hot spot.

PRINCIPAL TERMS

ASEISMIC: lacking earthquake activity

ASTHENOSPHERE: the layer within the Earth's mantle that displays convective motion while in the solid state; the convective motion, which results from the cooling of the Earth's interior, causes the overlying lithosphere to move on the Earth's surface

LITHOSPHERE: the brittle, outer layer of the Earth that is broken into individual pieces called "plates"; each plate is composed of the crust and the rigid portion of the upper mantle

MID-OCEANIC RIDGE: a volcanic ridge on the floor of the ocean basins where a divergent boundary between two oceanic plates occurs; often called a "rise" when it is not centered in the ocean basin

SUBDUCTION ZONE: a lithospheric plate margin where a converging oceanic plate is forced to slide down into the asthenosphere under the opposing converging plate; a subduction zone forms a deep trench on the ocean floor

CHARACTERISTICS OF HOT SPOTS

A hot spot is a circular region on the Earth's surface that has a radius of about 1,000 kilometers. The region, especially if it occurs in an ocean basin, is usually elevated above its surroundings by about 1,000 meters. An active volcano or a tight cluster of volcanoes caps the uplifted area. Measurements of the amount of heat being radiated from the Earth show that hot spots are regions of high heat flow from the interior of the Earth. Most hot spots display a stronger gravitational attraction than their surroundings, indicating that dense mantle material may have risen into the lithosphere under the area. The region appears to be relatively stationary in reference to the Earth's rotational poles, and, thus, hot spots do not move with the lithospheric plates that move across the Earth's surface. The plate is said to move "over" the hot spot.

Volcanic chains associated with oceanic hot spots may have as much as 80 percent of their volcanoes submerged below sea level. Submerged volcanoes are called seamounts. Many seamounts are characterized by steep sides, having a 22 degree slope, and a flat top, which can be more than 20 kilometers across. This flat-topped, submerged volcanic structure is called a "guyot." It is believed that guyots form from normal conical-shaped volcanoes. Oceanic lithosphere that moves over a hot spot becomes elevated, allowing the hot spot to form an island. Continued movement of the lithosphere would bring the now-extinct volcanic island off the apex of the hot spot, where the summit could be truncated by wave action and marine erosion. When the lithosphere has moved completely off the elevated region of the hot spot, the eroded volcano becomes fully submerged below the level of wave action to be preserved as a guyot in the volcanic chain.

Hot spots are known to have a long volcanic life, with the products of their volcanism forming a chain of igneous landforms that can extend hundreds of kilometers down the plate in the same direction that the plate is moving. Most of the hot spots in the Pacific Ocean basin have produced deposits of volcanic rocks for 25 million years or longer. The hot spot occurring at Hawaii has existed for more than 70 million years and has produced a 4,000-kilometer chain of extinct volcanoes on the Pacific Ocean floor that includes the Emperor and Necker string of seamounts, as well as the series of islands found at Midway and Hawaii. The ages of the extinct volcanoes within the chain are older in the direction that the plate moves over the hot spot.

NUMBER AND LOCATION OF HOT SPOTS

There are nearly fifty well-documented hot spots currently producing active volcanoes, and nearly three times that number of volcanoes have at one time or another been suggested as hot spots. Of these fifty, about twenty are located in the Pacific Ocean basin, fifteen are in the Atlantic Ocean, about five exist in the Indian Ocean, and fewer than ten occur on all of the continents. The Canadian geophysicist J. Tuzo Wilson developed the theory of hot spots in 1963, and by 1967 he

had compiled a list of 122 hot spots that had been active within the last 10 million years. When the historic record is included, about 50 percent of all hot spots developed in continental regions, and 50 percent have been in ocean basins.

Hot spots provide considerable information about plate motions, making it important to locate hot spots in a plate tectonic framework. The Earth's surface is covered with more than thirty lithospheric plates. The majority of the world's earthquakes occur along rather narrow belts, and these belts mark the boundaries between adjacent plates. Plates have three types of boundaries: divergent boundaries (where the two plates move directly away from each other); convergent boundaries (where the plates move toward each other); and transform fault boundaries (where plates slide past each other, with each moving parallel to the boundary between them). About one-fourth

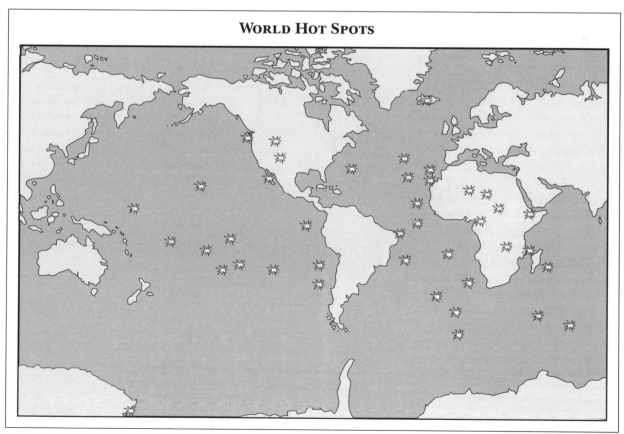

WORLD HOT SPOTS

Hot spots—small, isolated regions of volcanism that are stationary compared to Earth's drifting lithospheric plates—are most common within the interior regions of oceanic plates, away from convergent plate boundaries. They can exist at a given location for tens of millions of years.

of all hot spots occur on or near divergent boundaries at either mid-oceanic ridges (Iceland, for example) or in continental rift valleys (the Afar hot spot in the Red Sea region of northern Africa). About one-tenth of the currently active hot spots occur on transform fault boundaries (the Azores are a good example). The remaining two-thirds of the hot spots occur in the interior regions of a plate away from all boundaries (for example, the island of Hawaii in the Pacific Ocean basin). Hot spots do not occur at convergent boundaries.

DISTINGUISHING A HOT SPOT

Most of the world's volcanoes do not develop at hot spots. A few key characteristics allow hot spot volcanism to be distinguished from other types of volcanoes. A long, linear belt of active volcanism commonly develops at both convergent and divergent plate boundaries. Within these bands of active volcanoes occurring at plate margins are intermixed extinct volcanoes, whereas the volcanic chain extending from a hot spot has active volcanoes only at the head of the chain, with extinct volcanoes making up the chain itself. Volcanism at plate margins develops in a line that parallels the plate boundary, whereas the volcanic chain of a hot spot trends at an angle of nearly 90 degrees to both convergent and divergent boundaries.

Volcanism at plate boundaries is associated with active earthquakes, but there are no earthquakes along the hot spot chain. This lack of earthquakes in the volcanic chain of a hot spot is such a distinctive feature that a hot spot trace is often called an "aseismic" ridge in geological literature.

Subtle differences in terminology are often used to distinguish the tectonic setting for volcanism. The more abundant, active belt of volcanoes associated with an oceanic convergent boundary is called a volcanic "arc" (such as Japan). A volcanic chain occurring at a divergent boundary in the ocean basins is called either a "mid-oceanic ridge" or a "rise" (such as the East Pacific Rise). The term "volcanic island chain" usually refers to the volcanic trace of an oceanic hot spot.

MANTLE PLUMES

The causes for the formation of hot spots remained unknown until Jason Morgan hypothesized the existence of mantle plumes in 1971. A mantle plume is a thin, cylindrical column of mantle material that originates near the boundary of the core and mantle. The lower mantle is thought to be fairly stagnant. This nonconvecting, lower mantle is called the mesosphere. Thermal irregularities in the liquid outer core cause stalklike columns of hot rock material to rise and penetrate vertically through the mesosphere, the asthenosphere, and possibly into the lower portions of the lithosphere before the plumes begin to melt. The more buoyant magma continues to rise through fractures in the upper lithosphere to emerge as volcanoes. A hot spot is the direct surface manifestation of a mantle plume.

Researchers have been able to map mantle plumes passing through the upper mantle under various hot spots using waves that are generated by earthquakes. Mantle plumes that would be about 100 kilometers wide after 150 million years have been simulated in laboratory experiments. Studies have also shown that there is a very strong correlation between the distribution of hot spots and the occurrence of topographic highs on the top of the outer core.

ABSOLUTE PLATE MOTIONS

The stationary nature of mantle plumes and the long life of the associated hot spots allow several aspects of plate motion to be determined. The velocity that a plate has as it passes over a hot spot can be calculated by dividing the distance the plate has traveled by the length of time that has passed while traveling that distance. Plate velocities are usually expressed as centimeters per year. The concentration of various radioactive elements in the lava of a volcano allows for the dating of how long ago the volcanic rock formed. In a volcanic chain, the lava found on an extinct volcano was formed when the volcano was at the hot spot, and the rock's age will indicate how much time has elapsed since the volcano was over the hot spot. The distance the volcanic rock has traveled since it was formed is merely the distance from the hot spot to the current position of the volcano within the chain. Velocity analyses for the Hawaiian Islands yield a consistent velocity for the Pacific plate of 10 centimeters per year for the past 30 million years. Similar values have been calculated for the Cook-Austral island chain, which leads away from the Macdonald hot spot located in the South Pacific.

The volcanic island chain generated by a hot spot is a record of the direction in which the plate has moved. The line formed by the volcanic trace (pointing away from the hot spot) provides the compass bearing of the direction that the plate has moved. When multiple hot spots occur on the same plate, their volcanic traces typically form a series of mutually parallel lines. All five island chains caused by hot spots in the South Pacific show the same west-northwest bearing, which corresponds to the direction of movement of the Pacific plate.

When a hot spot occurs on a divergent boundary, it can form two simultaneous volcanic chains, one on each of the plates that are forming. The Iceland and Tristan da Cunha hot spots, which occur on the Mid-Atlantic Ridge, each have two aseismic ridges leading away from the active volcanic islands. One of the two ridges trends in an easterly direction across the eastern Atlantic basin in the direction of movement of Europe and Africa, while the other ridge has a westerly bearing, showing the movement direction of North America and South America.

The volcanic trace can record changes in the direction a plate has moved. The volcanic trace associated with the hot spot located under Yellowstone National Park in Wyoming has recorded a change in direction of motion for the North American plate. During the past 15 million years, this hot spot has caused the development of the Snake River volcanic plain, which trends to the southwest from Yellowstone. Magnetic studies have shown that the North American plate, which drifted westward for more than 200 million years, took a sharp turn to the southwest about 14 million years ago (verifying the evidence from the hot spot trace of Yellowstone).

The classic example of a hot spot recording a change in the direction of plate motion is Hawaii. The Hawaiian-Emperor chain has a distinct bend at Midway Island. The younger Hawaiian chain has trended west-northwest for nearly 40 million years. The Emperor chain was formed from 70 million to about 40 million years ago and has a trend of due north.

Jason Morgan used the twenty-five hot spots (and their volcanic traces) that are associated with the Atlantic Ocean and its neighboring continents to reconstruct the plate movements of the last 200

million years. His hot spot reconstruction of how North America and South America split apart from Europe and Africa to form the Atlantic Ocean agrees well with magnetic studies of plate motions for these continents.

TECTONIC EVOLUTION

As an oceanic plate is formed at a divergent boundary of a mid-oceanic ridge, it moves across an ocean basin and is destroyed as it is subducted into a trench at a convergent plate boundary. When a hot spot forms under this moving plate, the volcanic island chain that forms will eventually be carried into the trench where the plate is being consumed. Often the volume of volcanic material in the chain is so large that it disrupts the normal subduction process.

There are three effects that hot spot traces can have when they reach a convergent boundary. The most common effect is to cause a kink, cusp, or sharp bend in the normally smooth shape of a volcanic arc. This is well displayed by the Emperor seamounts, which are the trace of Hawaiian hot spot. The seamounts are being subducted along the Kurile and Aleutian trenches southwest of the Alaska Peninsula. At their point of entry into the trench, the volcanic arc changes its orientation from northwest to northeast, a bend of almost 90 degrees.

As the volcanic island chain is subducted, it can disrupt the mechanism that generates arc volcanism, causing a gap in the active volcanoes associated with an arc. In the southeastern Pacific basin, the Nazca plate is being subducted beneath the South American plate along the Peru-Chile Trench. The "arc" of active volcanoes corresponds to the Andes Mountains, which run the entire length of western South America. The volcanic chain called the Nazca Ridge trends in an easterly direction from the Easter Island hot spot. The Andes Mountains have an 800-kilometer gap without an active volcano where the Nazca Ridge enters the trench.

In numerous cases the lavas associated with the volcanic chain are so abundant that they cannot be subducted beneath the oncoming plate. The hot spot material can stop the subduction process, in which case the volcanoes of the island chain will "dock" or weld themselves to the leading edge of the opposing plate. These pieces of old hot spot traces that have been added to the leading edge of a continental plate are given a variety of terms: accreted terranes, displaced terranes, or exotic terranes. The western edge of the North American continent has picked up more than twenty displaced terranes during its 200-million-year western march, and about one-half of these represent old hot spot materials from ancient ocean basins.

HOT SPOTS WITHIN CONTINENTAL PLATES

Plates move over time, while hot spots are stationary. Thus, a hot spot that forms under a continent may have the continent drift away from the plume, while the hot spot becomes an oceanic spot with an aseismic ridge leading back to the continent. An example is the Trinidad hot spot in the southwestern Atlantic Ocean, which has an easterly trending aseismic ridge (Columbia seamounts) that goes under Brazil. The hot spot trace over the continent corresponds through the locations of most of the explosive, deep-seated, kimberlite diamond pipes that represent outgassing from the mantle. From 120 million years ago to about 60 million years ago, the hot spot was under Brazil, whereas for the last 60 million years, the hot spot has created the Columbia aseismic ridge in the Atlantic basin. North America moved northwest over the Great Meteor hot spot 100 million years ago, causing kimberlites and volcanoes in New England. Today the hot spot is located well out in the Atlantic Ocean, and the New England seamount chain shows the trend it has followed since the continent drifted west off the hot spot.

Hot spots appear to be important in the breakup of continents. The rising mantle plumes weaken the continental plate and provide breakage points. When a continent is rifted apart, the break begins as a zigzag rift valley that switches direction at each continental hot spot. A modern example is the Ethiopian hot spot, which occurs at the angle between the Red Sea and the Gulf of Aden, where the Arabian platform is being rifted away from the African continent.

As two continents continue to split apart and the pieces diverge from each other, the rift valley evolves into an ocean basin, and the hot spots change from a continental environment to being mid-oceanic ridge hot spots. When the supercontinent Pangaea was rifted apart 200 million years ago, the hot spot now located at Tristan da Cunha went through this evolution. The hot spot began under Pangaea and caused a large lava plateau. As the rift valley formed, the plateau was split in half: One half is the Paraná plateau on the east coast of South America, and the other half is the Etendeka plateau on the west coast of Africa. As the continents drifted away from each other, the hot spot became situated atop a mid-oceanic ridge with two traces back to both of the diverging continents. Tristan da Cunha is on the Mid-Atlantic Ridge, and it has produced both the aseismic Rio Grande Ridge (trending off to the west and connecting to the Paraná on the South American plate) and the Walvis Ridge (trending off to the east and connecting to the Etendeka).

INTRAPLATE VOLCANISM

The lavas produced at hot spots are the most common form of intraplate volcanism. Although the actual region of the hot spots makes up an insignificant percentage of the Earth's surface, the island chains, submarine aseismic ridges, lava plateaus, and rift lavas generated by hot spots are estimated to cover nearly 20 percent of the planet's surface.

Over millions of years, the movements of plates can take old volcanic rocks from an oceanic hot spot and add them to the leading edge of a continental plate, where later they may be caught in a collision between two converging continents. Thus, volcanic material generated at a hot spot in the interior of an oceanic plate can, over time, find itself situated in a convergent zone between two continental plates. Examples of this occur in the Himalaya Mountains. Researchers have attempted to develop techniques that will identify the tectonic setting where the volcanic rock was originally formed and that will apply no matter where the rock currently is located.

The most promising of these techniques is called the trace-element discrimination diagram. A chemical analysis is made of the rock, and the relative amounts of three rare elements—titanium, yttrium, and zirconium—are compared. The chemistry of the rock is plotted on a triangular diagram in which three regions are identified: mid-oceanic ridges, volcanic arcs, and within-plate lavas. The magmatic processes that occur in mantle plumes are such that the yttrium content is very low compared to titanium and zirconium in the lavas that are generated at hot spots. These elements were selected to identify rocks formed by intraplate volcanism because their relative abundance does not shift when the rocks are subjected to the many tectonic and chemical processes that occur on the surface of the Earth.

Discrimination-type diagrams have been used to explain the origin of intraplate volcanism when current plate configurations do not allow tectonic origins to be deduced. For example, discrimination diagrams indicate that the 30-million-year-old San Juan Volcanic Field, located in the state of Colorado, is the result of magmatic processes associated with a convergent plate boundary. The nearest convergent boundary at the time that these Colorado volcanoes were forming was more than 1,000 kilometers to the west, off the coast of California.

Another cause of intraplate volcanism is the failure of early-formed rifts related to the breakup of a continental plate to fully develop. Such "failed" rifts are commonly associated with old hot spots that occurred within a continental plate. The rift valleys produce an abundance of fissure eruptions, leaving extensive lava plateaus in the interior of the continent. Some examples of major failed rifts with associated intraplate volcanism are the Keweenaw Peninsula of upper Michigan, the Rio Grande Valley in New Mexico and Colorado, the Connecticut and Hudson River Valleys of New England, the Rhine River Valley in Europe, the Benue Trough of western Africa, and the East African Rift Valley in Kenya and Tanzania.

Dion C. Stewart and Toby R. Stewart

CROSS-REFERENCES

Calderas, 683; Continental Rift Zones, 579; Flood Basalts, 689; Geysers and Hot Springs, 694; Hawaiian Islands, 701; Island Arcs, 712; Lava Flow, 717; Mantle Dynamics and Convection, 1608; Plate Margins, 73; Plate Motions, 80; Plate Tectonics, 86; Plumes and Megaplumes, 66; Red Sea, 2254; Ring of Fire, 722; Seismic Tomography, 252; Spreading Centers, 727; Yellowstone National Park, 731.

BIBLIOGRAPHY

Kearey, Philip, and Frederick J. Vine. *Global Tectonics*. 2d ed. Oxford, England: Blackwell Science, 1996. A concise and in-depth analysis of tectonic features, including hot spots and associated mantle plumes. This book requires considerable geological background, and it is written as a textbook for undergraduate geology students. It achieves a good balance between geophysics and geology.

Plummer, Charles C., David McGeary, and Diane H. Carlson. *Physical Geology*. 8th ed. New York: McGraw-Hill/William C. Brown, 1999. This is a superb introductory textbook. The first chapter gives an overview of plate tectonics. Chapters 18 on the seafloor explain aseismic ridges, guyots, and mid-oceanic ridges. Chapter 19 on plate tectonics contains excellent illustration of past hot spot locations and shows how hot spots have rifted continents apart.

Strahler, Arthur N. *Plate Tectonics*. Cambridge, Mass.: Geo-Books, 1998. This very readable book is generally descriptive, with abundant quantitative data, but otherwise is nonmathematical. The introductory chapter is suitable for readers with no geology background. Each chapter begins with an introductory section requiring no background but quickly builds to a detailed documentation of advanced concepts presented at the level of the better science news magazines, such as *Scientific American* or *Nature*.

Wilson, J. Tuzo. "Evidence from Islands on the Spreading of the Ocean Floor." *Nature* 197 (1963): 536-538. This article, the first to put forth the modern concept of hot spots, is of considerable historical importance.

ISLAND ARCS

Island arcs are arc-shaped chains of volcanic islands formed by the collision of two oceanic plates. They are the sites of most of the world's explosive volcanic eruptions and large earthquakes. Tsunami and ash clouds generated by these events can affect people around the globe.

PRINCIPAL TERMS

ANDESITE: a light-colored volcanic rock rich in sodium and calcium feldspar, with some darker minerals

BASALT: a dark-colored igneous rock containing minerals such as feldspar and pyroxene, high in iron and magnesium

BENIOFF ZONE: the dipping zone of earthquake foci found below island arcs, named after Hugo Benioff, the seismologist who first defined it

EARTHQUAKE FOCUS: the region in the Earth that marks the starting site of an earthquake

GRANITE: a light-colored igneous rock containing feldspar, quartz, and small amounts of darker minerals

GRAVITY ANOMALIES: differences between observed gravity readings and expected values after accounting for known irregularities

LITHOSPHERE: the rigid outer shell of the Earth, composed of a number of plates

SUBDUCTION: the process by which a lithospheric plate containing oceanic crust is pushed under another plate

TSUNAMI: a seismic sea wave generated by vertical movement of the ocean floor, caused by an earthquake or volcanic eruption

FEATURES OF ISLAND ARCS

An island arc is a long, arcuate chain of volcanic islands with an ocean on the convex (or outer) side of the arc. Paralleling the arc lies a long, narrow trench with steeply sloping sides that descend far below the normal ocean floor. A map of the world shows that many island arcs occur in and around the Pacific Ocean, such as the Aleutians, Japan, Tonga, Indonesia, New Zealand, and the Marianas. The West Indies are an island arc bordering the Atlantic Ocean. The associated trenches contain the deepest places on the Earth. The Mariana Trench near the island of Guam reaches a maximum depth of 10,924 meters. This is farther below sea level than Mount Everest (at 8,848 meters) is above sea level. These island arc features, though merely topographic ones, demonstrate that island arcs and deep-sea trenches are parts of the same Earth structure.

There are six basic features common to island arc-trench systems: chains of volcanoes, deep ocean trenches, earthquake belts, a shallow sea behind the island arc, large negative gravity anomalies, and rock deformation in later geologic time. Some features are better displayed in one arc than another, but all are present.

VOLCANISM

All island arcs consist of an arc-shaped chain of volcanoes and volcanic islands. Many volcanoes are currently active or have been active in the recent geologic past. Some scientists group island arcs into two types: island arcs composed of volcanic islands located on oceanic crust (such as the Aleutians, Kurils, Marianas, and West Indies) and chains of volcanoes on small pieces of continental crust (such as Japan, Indonesia, New Zealand, and the Philippines). The main difference is that the continental-type arc is older, has had a more complex geologic history, and thus represents a later stage in the evolution of island arcs. The volcanoes of both types produce andesitic magma. Andesite is a light-colored, fine-grained igneous rock composed primarily of sodium- and calcium-rich feldspar. In composition, andesite lies midway between quartz-rich granites and iron- and magnesium-rich basalts. The andesitic magma also contains large amounts of gases that cause extremely explosive and destructive eruptions. Examples of island arc volcanic eruptions are Krakatau, Indonesia, in 1883; Mount Pelée, Martinique, in 1902; and La Soufrière, St. Vincent, in 1902.

DEEP-SEA TRENCHES

On the ocean side of all island arcs lie deep-sea trenches, long, narrow features that parallel the island chains. They have steep, sloping sides extending to great depths. Minor differences occur. Some trenches, such as the Mariana and the Kuril, have V-shaped cross sections and are rock-floored to their bottom. Others, such as the Puerto Rico and southwest Japan trenches, have a flat bottom. Detailed studies have shown that these flat-bottomed trenches are sediment-filled and that the underlying rock floor is also V-shaped.

SEISMIC ACTIVITY

Island arcs are active seismic regions and the sites of many of the world's largest and deepest earthquakes. The region in which an earthquake begins is called its focus. The foci of earthquakes in arc regions lie along a narrow, well-defined zone that dips from near the trench below the island arc. The number of earthquakes generally decreases with depth, with some foci reaching 600 to 700 kilometers below sea level. This dipping seismic zone is called the Benioff zone, after the seismologist Hugo Benioff, who first defined it.

SHALLOW SEAS

Behind the island arc lies a shallow marginal sea; examples are the Sea of Japan, the Philippine Sea, and the Caribbean Sea. Below some marginal seas, the crust is partly continental but becomes oceanic toward the arc. Below other seas, the crust is entirely oceanic. The composition of the oceanic crust beneath the marginal seas is more like andesite than the basalt of the normal ocean floor.

NEGATIVE GRAVITY ANOMALIES

Also common to arc-trench regions are large negative gravity anomalies. Geophysicists have found that the value of gravity over the Earth's surface varies by slight amounts. Most of the variations can be accounted for and result from irregularities in altitude and topography. After observed gravity readings are corrected, however, variations called gravity anomalies still remain. These are caused by differences in rock density from place to place below the Earth's surface. A negative anomaly shows that a greater volume of lighter (less dense) rocks is present in one area than in surrounding ones. Large negative anomalies are associated with the deep-sea trench and imply the presence of a great volume of low-density rocks at depth.

ROCK DEFORMATIONS

Deformation of rocks in the recent geologic history of an island arc is common. Some rocks have been folded, others metamorphosed. Areas of local uplift and subsidence are found that may also be related to shallow earthquakes and to faulting. These features are more easily seen and studied on the island arcs located on continental crust, such as Japan or New Zealand. Deformation, however, is present in all island arcs.

ORIGIN OF ISLAND ARCS

Earth scientists have long sought to explain the origin of island arcs. The remains of ancient marine volcanic islands and volcano-derived sediments are found in the core of the present-day Appalachian Mountains. These rocks and large amounts of continental sediments were compressed, faulted, and folded to form the ancient Appalachians. Yet the relation of the volcanic islands to the mountain-building process was unclear. The development of the concept of plate tectonics has provided an explanation.

The outer portion of the Earth is composed of a number of rigid lithospheric plates. Driven by forces in the mantle and by the sinking of cold, dense crust, the plates pulled over the face of the Earth. A lithospheric plate may contain continental crust, oceanic crust, or (more commonly) both. Island arcs form when the ocean portion of one of two colliding plates is forced under the other. The continuing collision of the plates may eventually result in the creation of a new mountain range, such as the ancient Appalachians. The process in which oceanic lithosphere is pushed under another plate is called subduction. Subduction of the plate causes the geological and geophysical features observed in the island arc system. (It should be noted that oceanic plates can also be subducted beneath continental plate margins. The resulting features are similar to those found in island arcs except that the andesite volcanoes form along the edge of the overriding continental margin. The Andes and Cascade Mountains are the island arc equivalent for subduction beneath a continent. Mount St. Helens is an andesite volcano.)

As the two plates converge, one bends and is pushed under the other. The line of initial subduction is marked by a deep ocean trench. Subduction is not a smooth process. Friction between the subducting plate and the overriding plate and between the downgoing plate and the mantle tries to prevent movement. When frictional forces are overcome, an earthquake occurs. The location of earthquake foci outlines the subducting lithospheric plate. As subduction continues, earthquakes occur at greater depths. The lack of earthquakes below 700 kilometers suggests that this is the maximum depth that the plate can reach before it becomes part of the mantle. The downgoing oceanic plate drags along any deep-sea sediments that have been deposited on it or in the trench area. Both the plate and the sediments are heated, primarily by friction and by the surrounding hotter mantle. At about a depth of 100 kilometers, partial melting occurs, giving a magma rich in sodium, calcium, and silica. This magma mixes with the iron- and magnesium-rich mantle, creating a magma less dense than is the surrounding mantle. Forcing its way upward through zones of weakness in the overlying plate, the magma generates andesite volcanoes. The gravity anomaly associated with the trench is caused by the light crustal rocks of the subducting plate being held (or pushed) down by the overriding plate. The increased volume of less dense rocks produces a large negative anomaly.

Also common to island arc systems are recently deformed rocks. The overriding plate does not slip smoothly over the subducting plate. Rocks in the leading edge of the plate are compressed (pushed together), causing faulting, folding, and uplift. Pieces of the subducting plate can be broken off and folded into the island arc. Heat from the mantle causes metamorphism in the overlying rocks. Behind the island arc, shallow faulting caused by tension (pulling apart) creates earthquakes.

The marginal sea between the island arc and the continent is called the back-arc basin. The presence of these basins is not totally understood. Some, such as the Aleutian and the Philippine basins, were formed from pieces of preexisting ocean. Others have features suggesting that they were once continental crust that has been turned into oceanic crust. Still other basins appear to have been created by interarc spreading, like that seen along mid-ocean ridges. The origin of back-arc basins is a topic of active geologic research.

STUDY OF ISLAND ARCS

In studying island arcs, scientists use a wide range of geological and geophysical techniques. The geological methods generally study the accessible portions of island arcs and include mapping and sample collection and analysis. The geophysical methods study the deep features of island arcs using earthquake and explosion seismology, gravity surveys, and heat-flow measurements. Computers aid in the analysis of data and in the generation of island arc models. Earth scientists studying certain aspects or features of island arcs select a combination of tools most appropriate to their region of interest.

Geologic mapping requires direct access to the rocks forming island arcs. The geologist surveys a region, recording the type of rock found and its extent and the orientation of observed faults and folds. Rock samples are collected for later study. These may be supplemented by drilling to sample rocks below the surface. The field data are transferred to a topographic map and, with the aid of aerial photographs, a geologic map is drawn. Aerial and satellite photographs have become increasingly helpful in the mapping of regions covered by vegetation. The traces of faults and the effects of changing bedrock can be reflected in surface features visible from high altitudes.

The rock samples collected are subjected to chemical and mineralogical analyses. The presence of trace elements or certain minerals can provide clues to the source region of a rock's components or to the thermal history of the rock since it was formed. Minerals containing radioactive elements, such as potassium 40 and rubidium 87, can be used to obtain the age of the rock units. Microscopic analysis of the rocks yields information on their thermal and deformational history. Direct collection of rock samples is limited to the exposed portions of island arcs. Dredging is used for sample collection in shallow ocean regions such as the back-arc basins. The use of submersibles, such as the *Alvin*, operated by Woods Hole Oceanographic Institution, has allowed scientists to photograph and collect rocks and other data from the ocean floor far below sea level, enabling them to extend the study of island arcs. These methods, however, do not reach the regions far below the Earth's surface.

Indirect methods are used by geophysicists to study the deeper island arc regions. The two most widely used methods are earthquake and explosion seismology. Earthquake seismology studies the seismic waves generated by earthquakes and provides an average velocity structure of the crust and mantle. The distribution of earthquakes in arc regions, particularly Japan, led to the discovery of the dipping seismic zone beneath the arc. Scientists also study plate movements to learn the mechanism causing earthquakes.

Explosion seismology uses seismic waves generated by controlled explosions to study the detailed crustal and lithospheric structure of island arcs. Within the ocean regions, this technique has revealed the steep topography of the deep-sea trenches and the deformed rock layers near the base of the overriding plate.

Measurements of the Earth's gravity field can be obtained on land and at sea. The data are corrected for irregularities in altitude and terrain. Remaining differences relate to density variations deep in the crust. The negative anomalies in the trench areas reflect the great thickness of crustal rocks at the plate boundary. In other areas, such as Japan, gravity data suggest a thicker crust or a less dense mantle below the back-arc basin (Sea of Japan) than below the Pacific Ocean.

Heat-flow measurements also reflect regional features. On the average, heat flow is the same over both continental and oceanic regions as a result of a deep, common source. In Japan, one of the most thoroughly studied island arc areas, a region of high heat flow coincides with the distribution of volcanoes and hot springs. A second high below the Sea of Japan suggests that the mantle is hotter than average, perhaps as a result of an interarc spreading center. A zone of low heat flow occurs on the Pacific side of the arc.

Earth scientists seek to unravel the history of island arcs to understand their formation. Computers are used to form models of island arcs so that theories of arc formation can be tested. By modifying the model to fit the observed geological and geophysical data, scientists can increase their overall understanding of the island arc system.

Pamela R. Justice

CROSS-REFERENCES

Andesitic Rocks, 1263; Calderas, 683; Earthquake Distribution, 277; Earthquake Hazards, 290; Earthquake Prediction, 309; Flood Basalts, 689; Geysers and Hot Springs, 694; Gravity Anomalies, 122; Hawaiian Islands, 701; Heat Sources and Heat Flow, 49; Hot Spots, Island Chains, and Intraplate Volcanism, 706; Krakatau, 752; Lava Flow, 717; Mount Pelée, 756; Plate Margins, 73; Plate Tectonics, 86; Ring of Fire, 722; Spreading Centers, 727; Subduction and Orogeny, 92; Tsunamis, 2176; Tsunamis and Earthquakes, 340; Volcanic Hazards, 798; Yellowstone National Park, 731.

BIBLIOGRAPHY

Bolt, Bruce A. *Earthquakes*. New York: W. H. Freeman, 1988. A popular book on the many aspects of earthquakes. Chapters cover distribution of earthquakes, tsunamis, earthquake prediction, and hazard protection planning. Illustrated, with bibliography and index. Suitable for the general reader.

Decker, Robert, and Barbara Decker. *Volcanoes*. New York: W. H. Freeman, 1997. This readable and well-illustrated text is geared toward the general public. The authors explain geologic phenomena in a clear and simple manner. Includes a good section on the development of island arcs. The appendices list the world's 101 "most notorious" volcanoes, as well as Internet sources for volcano information.

Dott, Robert H., Jr., and Donald R. Prothero. *Evolution of the Earth*. 5th ed. New York: McGraw-Hill, 1994. This basic textbook on historical geology is aimed at students of geology. However, it is very readable by anyone with a background in science. Presents an up-to-date account of the Earth's history from the viewpoint of plate tectonics. Includes a glossary.

Francis, Peter. *Volcanoes: A Planetary Perspective.* Oxford: Clarendon Press, 1993. This is a university-level text, but it is written in a style that makes it accessible to a wide range of readers. The author covers the visible evidence of volcanoes, as well as the theory of why they develop. Illustrations, index, and reference list.

King, Philip B. *The Evolution of North America.* Rev. ed. Princeton, N.J.: Princeton University Press, 1977. A revised edition of the classic book on the geology of North America. An excellent discussion of features of an island arc-trench system (the West Indies) appears on pages 84-90. Requires some knowledge of geology.

Lambert, David. *The Field Guide to Geology.* New York: Facts On File, 1988. A concise introduction to basic geologic terms and concepts. Chapter 2 explores the structure of the lithosphere. Chapter 3 discusses volcanoes and igneous rocks. A well-illustrated book suitable for high school students or beginning students at any level. Index and bibliography.

National Research Council. *Explosive Volcanism: Inception, Evolution, and Hazards.* Washington, D.C.: National Academy Press, 1984. A study on all types explosive volcanism, prepared for the National Academy of Sciences. Includes a series of background reports on various aspects of explosive volcanism, including Mount St. Helens and Kilauea. Recommends plans for future research and emergency planning in the United States. Reports are technical but suitable for the knowledgeable reader.

Walker, Bryce S. *Earthquake.* Alexandria, Va.: Time-Life Books, 1982. Extensively illustrated book on earthquakes, their causes, and their effects on humans. Chapter 1 covers the cause and effects of the 1964 Alaskan earthquake. Traces the historical development of scientific attempts to understand earthquakes and to predict them. Index and bibliography. Suitable for the general reader.

LAVA FLOW

The Hawaiian words "pahoehoe" and "aa" are used all over the world to designate two very common types of lava flow. The outward appearance of lava flows tells much about their eruption temperatures, chemical compositions, and viscosities. The surfaces of lava flows vary from smooth and glassy to rough and clinkery, depending on the viscosity of the lava and whether it is emplaced on land or under water.

PRINCIPAL TERMS

AA: a Hawaiian term (pronounced "ah-ah") that has been adopted for lava flows with rough, clinkery surfaces

AUTOCLASTIC BRECCIA: the clinkery or blocky rubble that forms on some lava flows

BLOCK LAVA: lava flows whose surfaces are composed of large, angular blocks; these blocks are generally larger than those of aa flows and have smooth, not jagged, faces

BRECCIA: a general term for any deposit composed mainly of coarse volcanic rock fragments

PAHOEHOE: a Hawaiian term (pronounced "pahoy-hoy") that is used in reference to lava flows with smooth, ropy surfaces

PILLOW LAVA: a type of bulbous, glassy-skinned lava that forms only when basaltic lava flows erupt under water

LAVA FLOWS

Aa and *pahoehoe* are Hawaiian words that describe the surface textures of lava flows. Aa is a rough, clinkery variety of lava, and pahoehoe has a smooth, ropy surface. Although aa and pahoehoe are the most common types of lava flow on land, other forms occur if lava is erupted under water or if it is extremely viscous. An explanation of why lava flows assume different forms requires a basic understanding of the physical properties of magma.

MAGMAS

Magma is molten rock that originates by partial melting of either the crust or the mantle of the Earth. "Lava" is a general term for magma that has erupted onto the Earth's surface. Lava flows are bodies of magma that have been emplaced onto the surface as coherent, flowing masses of lava. The viscosity of lava—its resistance to flowage— largely controls the external appearance of the lava. In turn, the viscosity of lava is controlled mainly by its temperature and chemical composition. Most magmas are dominantly composed of the elements silicon and oxygen, which form strong chemical bonds with each other in the magma; together, these elements are referred to as silica. The silica content of magma exerts a ma-

jor influence on magma viscosity. Similar to adding flour to batter, increased silica content means higher magma viscosity.

Although geologists recognize hundreds of kinds of volcanic rocks, there are three fundamental types of magma. Basalt magma is produced in the Earth's upper mantle and is poorer in silica (about 50 percent by weight) and dissolved gases than other types of magma. It erupts at the highest temperatures of all lava, usually about 1,100 to 1,200 degrees Celsius. Because of its low silica content and high temperature, basalt lava has relatively low viscosity, or resistance to flowage. With the consistency of honey or peanut butter (although considerably denser, at around 2.7 grams per cubic centimeter), basalt lava is relatively fluid and usually erupts as thin, sheetlike lava flows.

Andesite magma erupts at slightly lower temperatures than does basalt (around 1,000 degrees Celsius) and is richer in dissolved gases and silica (about 60 percent by weight). Andesite magma has a viscosity much greater than that of basalt. Andesite lava viscosity is similar to that of cold putty or frozen caulking compound, although its density of about 2.6 grams per cubic centimeter is considerably higher. Because of their relatively high viscosity, andesite lava flows are thick and tonguelike in shape and do not travel as far from the vent as do fluid basalt lava flows.

Rhyolite, the third fundamental type of magma, is formed by melting of the continental crust. It is very rich in silica (greater than 70 percent by weight) and often contains high amounts of dissolved gases. Thus, it tends to erupt explosively and to form voluminous deposits of frothy pumice. The viscosity of molten rhyolite lava is extremely high: Even at its white-hot eruption temperature of around 900 degrees Celsius, rhyolite behaves more like solid material than a liquid. It fractures when struck with a hammer, and its viscosity is sometimes similar to that of glacial ice. Thus, rhyolite magma moves very slowly and oozes onto the Earth's surface as thick, pasty lava flows or steep-sided volcanic domes. The latter are steep piles of lava that grow directly on top of volcanic vents because the lava is too viscous to flow away from its source. Because the crystallization of rhyolite is very sluggish, its lava flows and domes are often composed of obsidian, a type of silica-rich volcanic glass.

PAHOEHOE

Pahoehoe seldom forms on lava flows other than basalt, which is the hottest and most fluid type of magma. The smooth surface texture on pahoehoe lava flows is formed during quenching of the flow surface against the Earth's atmosphere. A several-centimeter-thick crust of brittle lava is formed, commonly with a thin skin of glass. The smooth, brittle crust forms a barrier to rising gas bubbles (vesicles), and there is often a frothy zone of round vesicles directly underneath the crust, which rides atop the hot, fluid interior of the lava flow. The lava crust is an insulating barrier that prevents the flow of heat from the interior of the lava flow, which can remain hot for weeks, months, or even years, depending on its thickness. During flowage of the molten interior, the smooth crust is commonly deformed into folds and ridges, as when a carpet is shoved against a wall. The folds and ridges assume many forms, as they are repeatedly stretched and refolded onto themselves. Several varieties of pahoehoe have been recognized. Entrail pahoehoe is commonly formed when lava tubes are breached, spilling their contents as entwined, elongate bulbs of glassy-skinned lava. Shelly pahoehoe forms only near volcanic vents, where centimeter-thick lava crusts are stacked atop one another and are separated by cavernous gas pockets.

Pahoehoe lava flows are able to travel tens of kilometers from their vents because rocks are poor conductors of heat. As the top and sides of the lava flow solidify, an insulating barrier is formed around the interior of the flow, which remains hot and fluid. Like blood within a system of blood vessels, fluid lava moves within circular channels known as lava tubes, and the advancing nose of the lava flow is fed by this network. When eruption of lava ceases, some of the tubes drain out and their roofs may collapse in places, leaving accessible caves with flow marks on the walls and lava dripstones hanging from the ceilings.

PILLOW LAVA

Pillow lava is the subaqueous counterpart of pahoehoe and forms only when fluid basalt lava is erupted under water. Pillow lava has been observed forming around the Hawaiian Islands, and beneath its blanket of sediment, most of the seafloor is composed of this type of basalt lava. When

A lava tube. (© William E. Ferguson)

As eruptions become enriched with alkalines and gases, the lavas become more viscous, forming small, steep-sided volcanic structures called spatter cones, such as these on the Galápagos Islands. (© William E. Ferguson)

basalt lava erupts under water, large quantities of steam are formed, sapping heat from the lava surface. As a result, a glassy skin quickly forms. Back-pressure from the still-fluid lava inside the lava flow eventually bursts the brittle skin, which is shattered into plates, chunks, and small shards of glass. New bulbs of lava ooze out of the cracks in a sort of budding process. Continuation of the process builds entwined masses of bulbous, glassy-skinned lava that look like toothpaste extrusions. The lava is mingled with layers of glassy fragments from the brittle skin of the advancing lava flow. Because it looks like a stack of pillows, the bulbous, glassy lava is appropriately called pillow lava.

Aa

Aa is a type of lava with a rough, clinkery surface. It is typical of basalt and andesite lava flows that are somewhat fluid but are too viscous to have pahoehoe surfaces. Jagged, spiney blocks of lava are formed during crumbling of the viscous mass. The spiney fragments continually rub against one another, eventually forming a thick outer envelope of debris that grades into the hotter, more fluid interior of the lava flow. The clinkery rubble rides on the flow until it eventually tumbles down the nose of the moving flow front: Similar to an advancing tractor tread, the fluid interior of the

lava flow overrides its own debris. Eventually, the interiors of aa flows may become so viscous that they can no longer flow as liquids and instead begin to shear along horizontal fractures (platy joints) that form near their bases. The shear planes are similar to those that develop near the bases of glaciers, and the process is much like spreading out a deck of cards across a tabletop.

Cooling of Lava

Single lava flows can change from pahoehoe into aa, but the reverse of this process has never been observed. When lava is first emitted, it is hottest and most fluid. As the lava cools, loses its gases, and slowly crystallizes, its viscosity is irreversibly increased. The originally smooth lava crust becomes thicker, and slabs of it begin to grind against one another. Eventually, smooth crust can no longer develop because the outer portion of the lava flow has become too brittle.

In addition to cooling and the resulting increase in viscosity, the pahoehoe-to-aa transition has also been observed to occur when lava flows undergo high rates of internal shear. This can occur, for example, when a pahoehoe flow travels over steep terrain such as a cliff face, sometimes continuing as an aa flow at the base of the cliff. Under such circumstances, the lava must flow faster, thus increasing internal shearing. During the pahoehoe-to-aa transition, the brittle lava crust becomes thicker, and the interior of the flow becomes more sluggish, as shown by the generally slower rates of movement of active aa flows (meters to hundreds of meters per hour), as compared to pahoehoe flows (hundreds of meters to tens of kilometers per hour). Pahoehoe flows have been clocked at speeds up to about 60 kilometers per hour in open channels and in lava tubes.

Block Lavas

A third type of surface is formed on lava flows with extremely high viscosities. In some andesite and most rhyolite lava flows, fragmentation of the lava is very thorough because of the high viscosity of the mass. Great volumes of large, angular blocks are formed, each block having relatively smooth (not sharp, clinkery) faces and sharp edges between the faces. Called block lavas, the flows are thick, crumbling masses of fine-grained lava or obsidian that can barely move away from the vent. The noses of block-lava flows are steep (30- to 35-degree) embankments of coarse rubble, and the tops of the flows are pocked with irregular depressions, lava spines, and automobile-sized lava chunks. The jagged rubble of block and aa lava flows is called friction breccia, crumble breccia, or autoclastic ("self-fragmented") breccia. "Breccia" is a general size term that refers to any deposit that is composed of coarse volcanic rock fragments (greater than 64 millimeters).

When the output of block lava is relatively low and magma viscosity is high, steep-sided volcanic domes may form over the vent. Most domes grow from within, by internal expansion, rather than by the addition of surface flows of lava. As a result, the thick carapace (shield or shell) of blocky rubble and lava spines is continually shoved aside, and the debris tumbles down the margins of the growing dome. Volcanic domes are very unstable and tend to collapse unpredictably into piles of blocky rubble. During collapse, ground-hugging avalanches of hot debris are sometimes jetted outward from the dome. Hence, volcanic domes are among the most dangerous of volcanic phenomena. Lava domes have repeatedly formed in the crater of Mount St. Helens since its catastrophic eruption of 1980, and the explosion of a Mount Pelée lava dome destroyed the city of St. Pierre on the Caribbean island of Martinique in 1904.

STUDY OF LAVA FLOWS

Lava flows are a major component of most volcanoes. Together with chemical analyses of lava, the physical features of ancient lava flows have revealed much about how lava flows are emplaced. From laboratory experiments and theoretical calculations based on the chemistry of lava, the viscosity of virtually any type of magma can be derived.

Observations on active lava flows have revealed much about how lava moves. The temperatures of active lava flows can be estimated in several ways. The incandescent color of the lava gives a general indication of temperature, in much the same way that blacksmiths judge the temperature of forged steel: dull red is about 600 degrees Celsius, orange is about 900 degrees Celsius, and golden yellow is about 1,100 degrees Celsius.

A more precise method uses an optical pyrometer to measure the wavelength of visible and near-infrared radiation, which varies uniformly as a function of temperature. The instrument is essentially a telescope in which a wire filament is mounted. The filament is heated by increasing an electric current until its color temperature matches that of the object being viewed; the temperature of the filament (and hence, the lava) can then be calculated.

Both the visual method and the optical pyrometer can give inaccurate results because of atmospheric effects and because only the surfaces of objects are measured (the interiors of lava flows are much hotter than their surfaces). Another method that is less subject to error but considerably more hazardous in use is the thermocouple, a pair of wires of different composition welded together at both ends. When one end of the circuit is immersed in hot material (a difficult procedure in viscous lava), an electrical current is generated, its strength depending on the temperature difference between both ends of the circuit. An ammeter near the cold end of the circuit reads the electrical current, from which the temperature at the hot end of the circuit can be calculated.

Viscosities of active lava flows can be estimated by the use of penetrometers: When a known amount of force is applied to a steel rod of known diameter, the penetration of the rod into the lava will depend on the viscosity of the lava. Again, this is a superficial measurement that may not give an accurate estimate of the viscosity of the fluid interior of the lava. The chemical compositions of lavas are now routinely analyzed, and theoretical viscosities can then be calculated from the chemical data. Similarly, if the chemical composition of a lava is known, its crystallization temperature can be accurately estimated, and such a value is usually close to the eruption temperature of the lava. Field measurements are often used together with theoretical calculations in order to arrive at the best understanding of the flow behavior and physical properties of lava.

William R. Hackett

BIBLIOGRAPHY

Cas, Ray A. F., and J. V. Wright. *Volcanic Successions: Modern and Ancient.* Winchester, Mass.: Allen & Unwin, 1987. This 530-page text is written for geologists at the college level. Presents very thorough coverage of volcanic materials at an advanced level; much of the information on lava flows is readily digested by persons with some college science background. Contains many black-and-white photographs of volcanic features.

Decker, Robert, and Barbara Decker. *Volcanoes.* New York: W. H. Freeman, 1997. This readable and well-illustrated text is geared toward the general public. The authors explain geologic phenomena in a clear and simple manner. Includes a good section on lava flows. The appendices list the world's 101 "most notorious" volcanoes, as well as Internet sources for volcano information.

Francis, Peter. *Volcanoes: A Planetary Perspective.* Oxford: Clarendon Press, 1993. This is a university-level text written in a style that makes it accessible to a wide range of readers. The author covers the visible evidence of volcanoes, as well as the theory of why they develop. Illustrations, index, and reference list.

Sigurdsson, Haraldur, ed. *Encyclopedia of Volcanoes.* San Diego, Calif.: Academic Press, 2000. This book contains a complete summary of the scientific knowledge of volcanoes. It includes eighty-two well-illustrated overview articles (including one on lava flow), each of which is accompanied by a glossary of key terms. Although this is a college-level text, it is written in a clear and comprehensive style that makes it generally accessible. Cross-references and index.

Williams, Howell, and A. R. McBirney. *Volcanology.* San Francisco: Freeman, Cooper, 1979. This 400-page college textbook has the complete descriptive coverage of the book by Macdonald but is more quantitative. Geared to geologists, it nevertheless contains much descriptive information that can be understood by nonscientists. Chapters on the physical nature of magma and on lava flows are excellent, and the text is liberally illustrated with graphs, drawings, and black-and-white photographs. A subject index and reference list are given.

RING OF FIRE

The Ring of Fire is a relatively narrow band of land composed of parts of the continents and island chains that surround the Pacific Ocean; the region is characterized by extensive volcanic and earthquake activity. The ring represents the surface expression of a tectonic boundary in which oceanic crust is forced into the upper mantle, producing molten rock that rises to the surface and forms volcanoes. The stresses imparted to the crustal rocks as they are forced into the mantle provide the energy for numerous earthquakes. Geologic investigations of the Ring of Fire have helped scientists refine the theory of plate tectonics.

PRINCIPAL TERMS

CONTINENTAL CRUST: near-surface rocks primarily composed of low-density, light-colored minerals and constituting the bulk of continental land masses

MANTLE: the zone of the Earth immediately underlying the crust; the uppermost region of the mantle is composed of partially melted and melting rocks

OCEANIC CRUST: near-surface rocks primarily composed of dense, dark-colored minerals that underlie the ocean floors

PLATE: a relatively thin slab of crustal rock, either continental or oceanic, that moves over the face of the globe, driven by currents of circulating molten rock in the underlying mantle

PLATE TECTONICS: a theory that explains the distribution of geologic features and phenomena across the face of the Earth based on the interactions of crustal plates

SUBDUCTION: the process occurring at colliding plate boundaries when denser oceanic crust is forced down into the upper mantle by overriding continental crust

TRENCH: a very deep portion of the ocean where oceanic crust is being forced beneath continental crust

PLATE TECTONICS

The "Ring of Fire" is the dramatic name of a narrow band of the Earth's surface surrounding the Pacific Ocean. This band extends from the islands of the South Pacific, through Japan, to the Aleutian Islands between Asia and North America, down the West Coast of North America, and through Central America and the western portions of South America. The name was applied to the roughly circular region because of its many active volcanoes. In addition to volcanic activity, this region of the Earth's surface exhibits a combination of geologic features that have intrigued Earth scientists for years. The region is characterized by mountain chains with many active and dormant volcanoes that parallel coastlines. Another characteristic of the area is its numerous strong earthquakes. In addition, some of the deepest portions of the world's oceans lie just offshore the volcanically active mountain chains of the region's land masses. Geologists had noticed the correlation between the land masses surrounding the Pacific, the volcanic and earthquake activity, and the oceanic depths long before they had an explanation relating these phenomena. The advent of the theory of plate tectonics provided the mechanism to explain the relationships among these geologic features.

The theory of plate tectonics, developed during the 1960's, views the Earth's surface as divided into a number of crustal plates. Like the pieces of a jigsaw puzzle, the plates fit together at their edges and cover the surface of the Earth; however, these pieces are in motion. Each plate moves slowly over the surface of the Earth, riding on currents of molten rock in the upper mantle, the layer within the Earth immediately underlying the crustal plates.

Geologists have long known that rock temperatures increase with depth in the Earth; for example, temperatures at the bottom of the deepest mines are sometimes uncomfortably hot. Deeper into the Earth—in the upper mantle and beyond—temperatures are so high that most rocks begin to melt. Sources of heat within the mantle are not evenly distributed, however, and circulation systems exist; in areas of high heat, the molten rocks are less dense because they are hotter, and they rise. In cooler areas of the mantle, the molten rocks become denser and sink. This circulation system acts as a conveyor belt, propelling the crustal plates across the surface.

The two major types of crustal plates are composed of either continental or oceanic crust. Because the types of rocks making up these crustal types are chemically different, they also differ in density, or the mass per unit volume of the rock. Continental crust is less dense than oceanic crust. Tectonic plates may be composed of oceanic crust (for example, the Pacific plate), continental crust (for example, the Arabian plate), or some combination of the two types.

PLATE BOUNDARIES

There are three basic types of plate boundaries. Boundaries where two plates are moving away from each other are called divergent boundaries and are probably associated with the areas of rising molten rock within the mantle. A good example of a divergent plate boundary is found in the middle of the Atlantic Ocean, where an underwater mountain chain, the Mid-Atlantic Ridge, runs the length of the seafloor. Plate boundaries where two plates are moving past each other are called transverse boundaries and are characteristically marked by a major fault. An example of a transverse boundary is found in the San Andreas fault zone of Southern California.

The third type of plate boundary occurs where two plates are colliding. These are known as "convergent plate boundaries"; this type of plate boundary is characteristic of the Ring of Fire. At a convergent plate boundary, two plates being moved along on mantle currents collide. At the point of collision, something must give. In many cases where the plates are composed of different crustal types (such as continental and oceanic), the denser crustal type (oceanic) will tend to be pushed beneath the lighter, less-dense continental crust and into the upper mantle. An example of this type of convergent boundary exists along the West Coast of South America. There, the denser oceanic plate is subducted beneath the lighter continental plate. The subduction zone, where the oceanic plate is pushed beneath the continental plate, runs parallel to the coastline. Immediately offshore, the subduction zone is characterized by extremely deep ocean water. As the oceanic plate is subducted, it is bent down into the mantle, and the bottom of the ocean in the subduction zone becomes very deep. These deep linear features in the ocean, called "submarine trenches," are some of the deepest parts of the ocean.

As the oceanic crustal plate sinks deeper into the upper mantle, temperatures rise, and the rocks begin to melt. The molten rock, which is much hotter than the solid oceanic crust it came from, is less dense than the surrounding mantle material, and it tends to rise, like a beach ball held under water. As the molten rock from the oceanic plate rises, it hits the bottom of the overlying continental crust. Working its way through cracks and other openings in the continental crust, the molten rock often forces its way to the surface, where it pours out from a volcano. The volcanic activity tends to occur in a narrow band within the continental plate some distance inland from where the oceanic plate is being subducted. Mountain chains on the continent, composed of many volcanic peaks, are characteristic of this zone.

Another type of convergent plate boundary occurs where two similar crustal types collide. In the case of the Ring of Fire, this type of convergence is found in the South Pacific, where plates of oceanic crust are colliding. The result is that one of the plates again sinks beneath the other. The subducted plate is heated and melts within the upper mantle, and rising melt finds its way through the overlying oceanic crust to the surface of the ocean floor. The molten rock that builds these volcanoes continues to spill onto the ocean floor until the volcano rises above sea level to form a volcanic island. Volcanic island chains, called "island arcs," form with a deep submarine trench offshore.

HOT SPRINGS AND GEYSERS

Another feature of volcanic areas in the Ring of Fire is the presence of hot springs and geysers at the Earth's surface. The abundance of hot and molten rock at relatively shallow depths above a subduction zone provides a perfect setting for hot-spring formation. Water from surface precipitation enters the Earth's surface to become groundwater. Some of this groundwater will penetrate deeply into the Earth's crust and be heated by the rocks and melt associated with the Ring of Fire. When the water is heated, like the heated, molten rock rising from the subducted plate, it becomes less dense and rises toward the surface through cracks and fractures in the Earth's crust to emerge as a hot spring or geyser.

In addition to absorbing heat from the surrounding rocks and melt, the circulating water may dissolve many chemical elements from the rocks. These dissolved metals are carried along with the water through the cracks in the Earth's crust toward the surface. However, as the water rises, it cools, and its ability to keep the elements dissolved decreases. As a result, minerals containing specific elements, especially metals such as lead, zinc, and copper, are deposited along the path of the water. Mineral deposits being mined today are thought to have been deposited from just such hot (hydrothermal) waters.

EARTHQUAKES

Earthquakes are another common feature of the Ring of Fire. Rocks at plate boundaries are subjected to the stresses at that boundary. Rocks are being forced apart, are sliding past each other, or are colliding. Under these pressures, all rocks tend to break along planes called faults. Sometimes the movement along a fault is not smooth. The rocks along the fault get stuck, and the pressures driving the movements build up. Eventually, the built-up pressures become so great that the sticking point along the fault breaks free, and the stored energy is released. The energy release travels through the surrounding rocks as vibrations—similar to the vibrations from a bell after it is struck—producing an earthquake. The intensity of the vibrations are related to the amount of energy released along the fault. The vibrations from a strong earthquake can do severe damage to buildings and roads.

Earthquakes are common in the Ring of Fire because areas of converging plates are ideal locations for rocks to break and form faults. In addition, the pressures of collision are constant and can build up if movement along the faults ceases. Deep earthquakes are characteristic of the Ring of Fire. Faults only occur in solid rocks. In a subduc-

tion zone, solid rocks are forced deep into the upper mantle, an area usually composed of molten or partially melted rock. The faults in the relatively cold subducted plate are the only places in the upper mantle where pressures can build up and be released as an earthquake.

OCEAN SURVEYS

The explanation for the Ring of Fire has depended, in large part, on investigations into the processes of plate tectonics. Geologists have used a broad range of technologies to understand how plate tectonics operates, and that understanding has helped explain the association of geologic phenomena composing the Ring of Fire.

Geologists have mapped the ocean floor with tools as crude as a weighted rope and with more sophisticated tools such as sonar. Sonar uses sound waves generated within the water to measure the distance to any object that can reflect some of the sound waves back to a receiver. The time that elapses between when the sound is made and when the reflected sound is received can be used to calculate the distance to the reflecting object. A map of the ocean floor can thus be created by towing a sound source and sound receiver behind an oceanic research vessel as it crisscrosses the ocean. Data from many such surveys have been compiled into maps of the ocean floor, which clearly show the mid-ocean ridges and the submarine trenches characteristic of divergent and convergent plate boundaries.

Oceanic surveys also often involve the taking of heat measurements. Using a sensitive heat sensor towed near the ocean floor behind a research vessel, scientists are able to map the amount of heat entering the ocean through the oceanic crust. The results of these surveys show that the warmest oceanic crust and the areas where the most heat enters the oceans are associated with the mid-ocean ridges. Conversely, the coolest portions of the ocean floor are those associated with submarine trenches. However, some of the highest heat-flow measurements taken on land are associated with the areas of volcanic activity in island arcs and continental mountain chains within the Ring of Fire.

SEISMOLOGY

Seismologists are geologists who specialize in the study of earthquakes. Seismologists use seismographs, sensitive instruments that measure vibrations in the Earth's crust, to identify where an earthquake has occurred. Seismographs can be used to identify the time earthquake vibrations arrive at a seismic measurement station. Knowing the velocity of earthquake vibrations through rocks, seismologists can calculate the distance from their measuring point to the earthquake source. With only a single measurement point, a seismologist would know only that an earthquake could have been centered anywhere on a circle of that radius. However, with many seismographs around the world, seismologists at each station combine measurements from other stations to locate the point on the Earth's surface directly over the source of the earthquake. This point is called the epicenter of the earthquake. Measurements showing the locations of many epicenters over the years were used to identify the Ring of Fire as an area of high earthquake activity.

Another characteristic of the Ring of Fire is the depth of the earthquake source. Earthquakes occur where two bodies of solid rock are moving past each other, separated by a fault plane, when that movement is temporarily halted. The rocks must be solid; within the Earth's depths, however, temperatures increase rapidly, and solid rock is not thought to exist below a certain depth. The implication is that earthquakes cannot occur below the depths predicted for melting of continental or oceanic crustal rocks. However, seismologists can use seismograph data to determine the depth of an earthquake source. Examination of seismograph data from many earthquakes in the Ring of

Fire clearly indicates that, in addition to relatively shallow earthquakes, there are many very deep earthquakes. The epicenters of the very deep earthquakes are centered inland away from coastlines and submarine trenches. In the context of plate tectonics theory, the data suggest that deep earthquakes are centered within the relatively cold, and solid, rocks of the subducting plate; researchers have theorized that the plate remains solid and able to generate earthquakes because it is being driven into the Earth's depths rapidly enough that the rock does not heat up and melt, as geologists might otherwise predict. Additional analysis of the seismic data has allowed seismologists to differentiate between subduction zones where the subducting slab is entering the upper mantle at a sharp angle, therefore producing a relatively narrow zone of earthquake activity, with deep earthquakes closer to a submarine trench, and those entering the upper mantle at a shallow angle, producing a broader earthquake zone and deep earthquakes farther inland.

Richard W. Arnseth

CROSS-REFERENCES

Calderas, 683; Earth's Crust, 14; Earth's Mantle, 32; Earth's Structure, 37; Earthquake Distribution, 277; Earthquakes, 316; Flood Basalts, 689; Geysers and Hot Springs, 694; Hawaiian Islands, 701; Heat Sources and Heat Flow, 49; Hot Spots, Island Chains, and Intraplate Volcanism, 706; Island Arcs, 712; Lava Flow, 717; Plate Margins, 73; Plate Tectonics, 86; Spreading Centers, 727; Volcanic Hazards, 798; Yellowstone National Park, 731.

BIBLIOGRAPHY

Decker, Robert, and Barbara Decker. *Volcanoes.* New York: W. H. Freeman, 1997. This readable and well-illustrated text is geared toward the general public. The authors explain geologic phenomena in a clear and simple manner. Includes a chapter on the Ring of Fire. The appendices list the world's 101 "most notorious" volcanoes, as well as Internet sources for volcano information.

Erickson, Jon. *Plate Tectonics: Unraveling the Mysteries of the Earth.* New York: Facts On File,

1992. A comprehensive, technical review of plate-tectonics theory, from the initial development of the theory to more recent areas of investigation. The material is presented in the manner of a classroom textbook. Illustrated; includes an extensive bibliography for each chapter subject and a comprehensive index.

Francis, Peter. *Volcanoes: A Planetary Perspective.* Oxford: Clarendon Press, 1993. This is a university-level text, but it is written in a style that makes it accessible to a wide range of readers. The author covers the visible evidence of volcanoes, as well as the theory of why they develop. Illustrations, index, and reference list.

Levy, Matthys, and Mario Salvadori. *Why the Earth Quakes.* New York: W. W. Norton, 1995. A general, readable discussion of the forces creating earthquakes and volcanoes. The text is supplemented throughout with helpful illustrations. The last few chapters discuss case histories of earthquakes, some of the engineering aspects of building in an earthquake-prone area, and the social consequences of earthquake predictions. Includes a glossary of terms and an index.

Ritchie, David. *The Ring of Fire.* New York: Atheneum, 1981. A mixture of science and history devoted to the Ring of Fire. The author combines the history of the development of explanations for the Ring of Fire with descriptions of some of the dramatic, and catastrophic, events associated with the area. Contains a glossary of terms, a bibliography, and an index.

Vogel, Shawna. *Naked Earth: The New Geophysics.* New York: Dutton, 1995. A readable discussion of the role of modern geophysics in exploring the mysteries of the Earth. The author focuses on the roles of individual scientists and the array of technologies used in solving geologic mysteries. The text reads like a mystery, with just enough details to satisfy the more knowledgeable reader. Includes an index.

SPREADING CENTERS

Volcanism occurs along spreading centers, the places where the Earth's crust is rifting apart. This volcanism has formed the mid-ocean ridge system, the largest mountain chain on the Earth. It also contributes to the dense oceanic crust, allowing ocean basins to form. Associated hydrothermal activity produces valuable deposits of copper, zinc, and other metals.

PRINCIPAL TERMS

BASALT: a fine-grained, dark, heavy rock of volcanic origin, primarily composed of calcic feldspars and pyroxene

CONVECTION CURRENTS: the transfer of material caused by differences in density, usually brought about by heating

CRUST: the thin, rigid outer layer of the Earth, extending generally 35 kilometers under the continents and 10 kilometers under the oceans

FISSURE: a fracture or crack in rock along which there is a distinct separation

HYDROTHERMAL: a term meaning related to hot water, particularly involving the production or dissolution of minerals

LAVA: the fluid rock issued from a volcano or fissure and the solidified rock it forms when it cools

MAGMA: molten rock below the Earth's surface that has not erupted and therefore retains its gaseous components

MANTLE: the layer of the Earth's interior between the crust and core, either semisolid (lower mantle) or plastic and nonbrittle (upper mantle)

SEAMOUNT: a submarine mountain 1,000 meters or higher, often a volcanic cone

MID-OCEAN RIDGES

Plate tectonics explains that the Earth's crust is omposed of relatively thin, rigid plates that float on top of the denser, less rigid mantle. Where the individual plates of crust meet, three types of motion are possible. Plates may collide with each other, forming a convergent boundary; they may slide past each other, forming a lateral, offset boundary; or they may move away from each other, forming a divergent boundary. Along divergent boundaries, cracks or rifts in the crust are formed between the plates. Lavas made of basalt seep up and fill these cracks from below, forming new crust. Where the new basalt erupts, great ridges form.

As spreading continues, the newly formed volcanic crust is slowly forced away from the ridge, as if carried on a great conveyor belt. This motion seems to be driven by large-scale convection currents within the mantle, fueled by the release of heat from the Earth's interior. As basaltic crust moves away from the spreading center, it gradually subsides. The basaltic crust produced by volcanism at spreading centers along divergent bound-aries is heavier than the granitic continental crust. Because it is denser, it sits lower in the Earth's surface and forms the great topographic depressions of the ocean basins.

The ridges that form along the spreading centers are interconnected and make up the mid-ocean ridge system. This is a vast mountain chain that extends for more than 60,000 kilometers and spans the globe. The mid-ocean ridge system covers approximately 33 percent of the ocean floor, a surface equal to that of all the Earth's continents. The segment of the mid-ocean ridge system that occurs in the Pacific basin is referred to as the East Pacific Rise. The Atlantic segment of the ridge is known as the Mid-Atlantic Ridge and bisects almost the entire Atlantic Ocean from the Arctic to the Antarctic Circle. Most of the ridges are totally submerged, though a few of the highest peaks (notably Iceland and the Azores) extend above sea level as islands.

The typical mid-ocean ridge stands about 2 to 3 kilometers above the surrounding seafloor. The ridges are broader than most mountain ranges. Individual ridges are typically 1,000 to 4,000 kilometers wide at their base. Unlike most continental mountains, the ridges typically have a central rift that is a few kilometers wide and about 1 kilometer deep. Volcanic activity associated with spreading occurs within this central rift.

UNDERSEA ERUPTIONS

The basaltic lava that erupts along the spreading centers is exceptionally hot (1,000 to 12,000 degrees Celsius) and therefore very fluid compared with other types of volcanic flows. These basaltic eruptions are therefore less violent and less explosive than are other types of volcanic events, and the lava is spread over large areas, similar to eruptions recorded on land in Iceland and, to a lesser extent, those of Hawaii.

At depths of more than 2,000 meters below the sea, volcanic eruptions are considerably less explosive than they are on land. At these depths, the water pressure is so high that the gases released during volcanic eruptions do not explode but immediately go into solution with the seawater. Consequently, below 2,000 meters, there would be no explosions, only effusive outpourings of lavas. This concept is supported by the low percentage of vesicles, which are the spherical voids formed around gas bubbles found in deep mid-ocean ridge basalts.

At depths of less than 2,000 meters, steam, ash, and sulfuric gases may be released. Unless the volcanism occurs at very shallow depths, however, the steam, ash, and gases will be absorbed by the seawater before reaching the surface. Many of these eruptions go completely unnoticed by humans, unless an ocean traveler happens to note a yellow tinge to the sea, the characteristic evidence of volcanism at depth. If the eruption occurs at shallow water depths (less than a few hundred meters), the eruption may be more explosive than if it had occurred on land, because of the secondary explosions caused by the violent vaporization of the water. The eruptions of steam propel the volcanic projectiles higher and faster through the air than they would normally travel. Spectacular ash plumes known as "cock's tail" explosions may arise.

PILLOW LAVA AND PAHOEHOE

The eruptions along the mid-ocean ridge system are episodic and frequently emanate from conical vents of basalt along the central rift. These vents have been measured up to 20 meters high. In some areas, there is evidence of multiple lava flows erupted in quick succession. Many of these flows form thick layers of basaltic "pillows," which are structures that form only under water. Pillow lavas are cemented masses of bulbous, slightly squashed ovoids of rock. They form as the molten lava contacts the cold water (which, at the depths of the mid-ocean ridges, is barely above freezing, at 2 degrees Celsius). As the lava flows, its surface is chilled by the cold water and quickly cools into a bulge of dark, glassy rock. The still-molten interior of the mass may burst under pressure and stretch or may open the pillows, allowing small buds or toothpastelike bulges to protrude.

Along the axis of the central rift, flows of pillow basalts have been recorded up to 200 meters high and 500 to 1,000 meters wide, particularly along the Mid-Atlantic Ridge. These flows form steep-sided, flat-topped ridges similar to the table mountains or möberg mountains formed by eruptions under glaciers.

Smooth or ropy flows of lava known as pahoehoe also occur within the central rift. These flows form when the slopes are too gentle for pillows to take shape. Under water, a flow's outer 10 to 30 centimeters quickly solidify, allowing the lava beneath to advance under this hardened shell. These flows form sheets or lava plains. Single flows emanating from spreading centers have been measured up to 20 kilometers from their source.

Sometimes, a single eruption may create such a rapid outpouring of basalt that a lava pond or lake may form. Some of these lakes have been measured up to hundreds of meters long and up to 5 meters deep. Lava lakes may display a number of interesting features. In places, the surface may have caved in, forming a collapse pit. Often, there are residual hollow basalt pillars, which upheld the roof of the lava lake as it drained and collapsed. Horizontal ribs, like the rings around a bathtub, occur repeatedly on the pillars and on the edges of the lake. These ribs apparently record changes in the lava level.

In addition to the flows, large, circular volcanic seamounts sometimes form near the ridge crests, particularly along the Mid-Atlantic Ridge. These may have been formed from lava with a slightly lower temperature (and therefore lower fluidity), which allowed the basalt to pile up. Some seamounts occur as pairs equally spaced from the central rift, indicating that they were probably once part of a single volcano that split and was separated.

ERUPTION FREQUENCY AND VOLUME

Along spreading centers, the amount of basalt generated and the time between eruptions may vary considerably. The East Pacific Rise is spreading at a relatively fast rate (averaging 6 centimeters per year); the frequency of eruptions at a given location is probably about once every few thousand years, and the volume of material per eruption may reach up to 200 cubic kilometers. The spreading rate along the Mid-Atlantic Ridge is approximately three times slower. Along parts of this ridge, eruptions are expected to occur with an average frequency of once every 14,000 years. Estimates of the amount of material erupted vary considerably. At one submarine location, the rate of lava being generated has been speculated to be 8,700 cubic meters per kilometer per year. An-

other study near Iceland suggested that ten times that amount of fresh material was being released. To put this amount into perspective, that would be enough material to cover the entire landmass of Great Britain with a meter of new basalt every year.

Although volcanism has not been directly observed on the ocean floor, scientists have seen heated fluids (over 350 degrees Celsius) spewing from chimneylike vents along the rifts in the mid-ocean ridges. These chimneys rise from layers of basaltic lava and stand up to 30 meters tall. In some areas, they coincide with the source of the most recent lava flows. Sometimes the fluids emanating from the chimneys are black because of large quantities of sulfide precipitates. The chimneys are called "black smokers" or "white smokers" depending on the clarity of their fluids. They are apparently produced as seawater circulates down through cracks in the basalt to the molten magma beneath. As the water heats up, it dissolves elements such as zinc, iron, copper, and lead as well as silver and cadmium. The hot, mineral-rich wa-

Pillow lava rocks, such as these off Hawaii's shore, are cemented masses of bulbous, slightly squashed ovoids that form when molten lava contacts ocean water. As the magma oozes out of undersea vents, its surface is chilled by the cold water and quickly forms a bulge of dark, glassy rock. (National Oceanic and Atmospheric Administration)

ter rises and is recycled back into the ocean. Upon contact with the cold water, sulfide minerals precipitate, forming chimneys around the rising jets.

This hydrothermal activity supports an incredible concentration of previously unknown animal life, including foot-long clams, fish, crabs, and giant tube worms up to a meter and a half long. These creatures thrive on bacteria that digest the sulfide provided by the vents. This is the only known ecosystem that is not directly or indirectly based on solar energy.

Mary D. Albanese

CROSS-REFERENCES

Basaltic Rocks, 1274; Calderas, 683; Eruptions, 739; Flood Basalts, 689; Geothermal Power, 1651; Geysers and Hot Springs, 694; Hawaiian Islands, 701; Heat Sources and Heat Flow, 49; Hot Spots, Island Chains, and Intraplate Volcanism, 706; Hydrothermal Mineralization, 1205; Island Arcs, 712; Lava Flow, 717; Magmas, 1326; Ocean Basins, 661; Ocean Drilling Program, 359; Ocean Ridge System, 670; Ocean-Floor Drilling Programs, 365; Ocean-Floor Exploration, 666; Oceanic Crust, 675; Plate Tectonics, 86; Ring of Fire, 722; Seamounts, 2161; Yellowstone National Park, 731.

BIBLIOGRAPHY

Ballard, Robert B. *Exploring Our Living Planet.* Washington, D.C.: National Geographic Society, 1983. The chapter "Mountains of the Sea" gives an excellent firsthand account of some of the most illuminating journeys to the mid-ocean ridge system. The chapter "Discovery of Planet Earth" provides a highly readable, informative treatment of plate tectonics. The text is well illustrated with large color photographs and graphics. Contains a glossary and is completely indexed. Excellent reading for anyone interested in volcanism.

Chernikoff, Stanley. *Geology: An Introduction to Physical Geology.* Boston: Houghton Mifflin, 1999. This is a good overview of the scientific understanding of the geology of the Earth and surface processes. Includes sections on the ocean ridge system. Includes a link to a Web site that provides regular updates on geologic events around the globe.

Decker, Robert, and Barbara Decker. *Volcanoes.* New York: W. H. Freeman, 1997. This readable and well-illustrated text is geared toward the general public. The authors explain geologic phenomena in a clear and simple manner. The first chapter discusses the development of the ocean ridge system. The appendices list the world's 101 "most notorious" volcanoes, as well as Internet sources for volcano information.

Dolgoff, Anatole. *Physical Geology.* Lexington, Mass.: D. C. Heath, 1996. This is a comprehensive guide to the study of the Earth. Extremely well illustrated and includes a glossary and an index. Although this is an introductory text for college students, it is written in a style that makes it understandable to the interested layperson. Contains a section on the development of the ocean ridge system.

Dott, Robert H., Jr., and Donald R. Prothero. *Evolution of the Earth.* 5th ed. New York: McGraw-Hill, 1994. This basic textbook on historical geology is aimed at students of geology. However, it is very readable by anyone with a background in science. Presents an up-to-date account of the Earth's history from the viewpoint of plate tectonics. Includes a glossary.

Rona, Peter A., K. Boström, L. Laubier, and K. L. Smith, Jr., eds. *Hydrothermal Processes at Seafloor Spreading Centers.* New York: Plenum Press, 1983. For those readers interested in a technical treatment of the subject delineated in the book's title, this volume contains a compilation of thirty papers ranging from hydrothermal mineralization to the biology of hydrothermal vents. The book is well indexed, and each paper contains its own bibliography.

Rosbacher, L. A. *Recent Revolutions in Geology.* New York: Franklin Watts, 1986. Chapters 3 and 4, "Exploring the Ocean Floor" and "Riches from the Sea," provide a good treatment of the subject for the younger reader interested in the topic of undersea exploration.

YELLOWSTONE NATIONAL PARK

Yellowstone National Park is known throughout the world for its volcanic and thermal features. In the past 2 million years, Yellowstone has been the site of three major volcanic eruptions, reflecting its position over a "hot spot" in the Earth's crust and upper mantle.

PRINCIPAL TERMS

ASH-FLOW TUFF: a sheetlike pyroclastic deposit that is laid down as a hot mixture of gas, crystals, pumice, and volcanic ash

CALDERA: a large, flat-floored volcanic depression that is formed on top of a large, shallow magma chamber during the eruption or withdrawal of magma

HOT SPOT or MANTLE PLUME: a zone of hot, upwelling rock that is rooted in the Earth's upper mantle; as plates of the Earth's crust and lithosphere glide over a mantle plume, a trail of hot spot volcanoes is formed and the Earth's surface bulges upward

OBSIDIAN: a dense variety of silica-rich volcanic glass; the same material as pumice but does not contain gas bubbles

PLINIAN ERUPTION: a violent, explosive type of volcanic eruption named either for Pliny the Elder, a Roman naturalist who died while observing the eruption of Mount Vesuvius in 79 C.E., or for Pliny the Younger, his nephew, who chronicled the eruption

PUMICE: frothy, silica-rich volcanic glass that is commonly produced during explosive eruptions of rhyolite magma

RESURGENT DOME: a broad, oval area of uplift within a volcanic caldera that is marked by the upwarping and fracturing of caldera-filling deposits

RHYOLITE: a viscous, gas-rich type of magma that contains greater than about 70 percent silica; rhyolite erupts nonexplosively as thick, slow-moving lava flows and explosively as widespread sheets of frothy pumice

YELLOWSTONE AND ITS ENVIRONS

The first recorded observations of the Yellowstone country were made by fur trappers in the early nineteenth century, and later government-sponsored surveys confirmed the geological wonders of the area. As a result, much of what is now northwestern Wyoming and adjacent portions of Idaho and Montana were set aside in 1872 as the Yellowstone National Park. The largest thermal area in the world, Yellowstone is best known for its outstanding assemblage of hot springs, geysers, and mud pots.

Yellowstone National Park is an 899,152-hectare, wooded mountain area in the northern Rocky Mountains. The rugged and geologically youthful terrain is mostly of volcanic and glacial origin, with many waterfalls and lakes. Its eastern border is within the Absaroka Range, and the Teton Range and Jackson Hole lie to the south. The northeastern sector of the park is cut by the Grand Canyon of the Yellowstone River. The park occupies the central part of the Yellowstone Plateau, a 6,500-square-kilometer, high plateau that is three times larger than the park and averages 2,500 meters in elevation. Immediately west of Yellowstone, Island Park, Idaho, is a shallow basin that is partly filled with basalt lava flows and sediments. To the southwest, the Yellowstone Plateau drops gradually to the eastern Snake River plain of Idaho, a down-warped region (1,000 to 1,500 meters elevation) of low topographic relief that is largely covered by basalt lava flows.

A strong geological relationship exists between Yellowstone and the Snake River plain, an 800-kilometer-long, low-lying volcanic belt that crosses southern Idaho. Near the Idaho-Oregon border, the oldest volcanism on the western Snake River plain began about 17 million years ago. Since that time, volcanic activity has slowly progressed northeastward toward Yellowstone at an average rate of about 4 centimeters per year. The youngest basalt eruptions on the eastern Snake River plain occurred about 2,100 years ago, and the youngest rhyolite eruptions at Yellowstone occurred around 50,000 years ago.

The most popular explanation for the progression of volcanism toward Yellowstone is that a stationary plume of upwelling, perhaps partially molten rock, is rooted in the Earth's upper mantle beneath the northwestern United States. Although the plume seems to be fixed in place, the tectonic plate of North America has been moving westward for at least 17 million years. Like a candle flame held beneath a moving piece of paper, the volcanoes form an eastward-bound trail, first with explosive rhyolite eruptions and then with basalt lava outpourings. Thus, the old rhyolites of the Snake River plain in southern Idaho are covered almost completely by basalt lava flows, but Yellowstone is the youngest rhyolite volcanic center, and very little basalt has erupted there. The mantle plume, or hot spot, is now considered to lie beneath Yellowstone and is the ultimate source of magma and geothermal heat under that area.

VOLCANIC ACTIVITY

For the past 2 million years, Yellowstone has been the site of huge but infrequent eruptions of rhyolite lava and pumice; the youngest eruptions of rhyolite lava occurred about 50,000 years ago. The products of those eruptions comprise the Yellowstone volcanic field, which formed during three cycles of volcanism. Each cycle began and ended with the eruption of lava flows of rhyolite obsidian, a dense, black, silica-rich volcanic glass. The lava flows issued mostly from curved fractures of the Yellowstone caldera and now cover much of the caldera floor. All three volcanic cycles culminated with explosive eruptions of voluminous rhyolite ash and pumice. The ash and pumice deposits from the three climactic eruptions of the Yellowstone-Island Park area are known as the 2.1-million-year-old Huckleberry Ridge tuff, with an estimated volume of 2,500 cubic kilometers; the 1.3-million-year-old Mesa Falls tuff, with an estimated volume of 280 cubic kilometers; and the 0.6-million-year-old Lava Creek tuff, with an estimated volume of 1,000 cubic kilometers. For comparison, the total volume of material emitted from Mount St. Helens on May 18, 1980, was only 1 to 2 cubic kilometers.

Frothy rhyolite pumice and ash were explo-sively discharged from curved fractures around the rim of the Yellowstone caldera. Plinian eruptions ejected large quantities of rhyolite to heights of tens of kilometers into the atmosphere. Some of the ejecta then cascaded back to the Earth like a water fountain or "boiled over" as hot, ground-hugging flows of ash and pumice, forming three major ash-flow deposits that fill the Yellowstone caldera and blanket the surrounding terrain. The ash flows were sufficiently hot when they came to rest (about 550 degrees Celsius or more) that the particles of pumice and ash were compacted into a dense mass, forming welded ash-flow tuffs. Each of the catastrophic eruptions probably lasted only hours or days. The rapid eruption of voluminous rhyolite pumice from shallow (perhaps 5 to 10 kilometers deep) magma chambers led to the collapse of the roof rocks overlying the magma chambers, forming large rhyolite calderas ringed by 50-kilometer-wide sheets of welded ash-flow tuff. Finer particles of ash and pumice were ejected to great heights and carried away by high-altitude winds. The finest particles probably took months or years to sift out of the atmosphere. Ash-fall deposits from the three major Yellowstone eruptions are found as far away as Louisiana, California, and southern Canada, where they are important marker beds for the dating of Ice Age fossils and glacial deposits.

The Yellowstone caldera—a broad, oval-shaped depression about 40 by 60 kilometers in size—was formed during the climax of the last cycle of volcanism 600,000 years ago, when the Lava Creek tuff was emplaced. Since that time, voluminous lava flows of rhyolite obsidian—the Plateau Rhyolites of Yellowstone—have obscured the caldera rim and now largely cover the caldera floor. The most voluminous Plateau Rhyolite lava flows comprise the Madison and Pitchstone Plateaus in the western and southwestern sectors of the park.

During the Ice Age, large masses of glacial ice formed in northern Yellowstone National Park. Two main periods of ice formation have been identified. An older ice cap formed during the Bull Lake glaciation, about 150,000 years ago. Most of the Bull Lake glaciers flowed to the west, and the West Yellowstone Basin to the west of Yellowstone Park is filled with Bull Lake outwash gravels and other deposits. A later ice cap was formed during the Pinedale glaciation, between about 30,000 and 15,000 years ago. Most of the Pinedale ice flowed to the north because rhyolite lava flows of the Madison Plateau had blocked westward passage of the ice. Both ice caps were probably 1 kilometer or more thick, and much of Yellowstone Park is covered by a thin mantle of glacial deposits. Glacial deposits play an important role in the thermal activity of Yellowstone, since permeable sands and gravels become capped with mineral deposits but remain permeable to hot fluids at depth. The formation of impermeable caprock over permeable strata is important in the development of the shallow plumbing systems beneath hot springs and geysers.

SEISMIC ACTIVITY

Yellowstone is the most seismically active region of North America. During the fifteen-year period between 1973 and 1988, about fifteen thousand earthquakes greater than Richter magnitude 2 were recorded. In 1959, the Hebgen Lake earthquake, centered about 10 kilometers west of Yellowstone and estimated at between 7.3 and 7.5 on the Richter scale, was the largest tremor ever recorded in the Rockies; it caused changes in the behavior of hundreds of geysers and hot springs. The frequent earthquakes of Yellowstone are probably important in maintaining the thermal activity of the region, since permeable fault zones that channel hot fluids to the surface would quickly be sealed with mineral matter if not for regular disturbances from earthquakes. Beneath the Yellowstone caldera, the velocities of earthquake waves traveling through the Earth's crust are measurably decreased, suggesting the presence of a plumelike body of partly molten rock extending downward from a depth of about 8 kilometers. The amount of heat flowing from the ground beneath Yellowstone is estimated as thirty to forty times the heat flow elsewhere in North America and accounts for about 5 percent of the total heat flow from all North America.

RESURGENT DOMES

Since the first precise topographic surveys were done during the 1920's, resurveying has shown that the central caldera floor is slowly warping up at an average rate of about 14 millimeters per year. Two broad areas of uplift, known as resurgent domes, have been identified in the southwestern

and northeastern Yellowstone caldera. Both resurgent domes are about 15 kilometers across and 200 meters high. Resurgence of the caldera floor, together with the high heat flow and seismicity, suggests the presence of rhyolite magma in the upper crust, either left over from the last cycle of volcanism or a new magma batch that may appear at some time in the future as a fourth volcanic cycle. The hot rock, or magma, beneath Yellowstone is the source of heat for nearly ten thousand hot springs and several hundred active geysers. Rain and snowmelt soak into the ground and percolate through fractured volcanic rocks to depths of several kilometers, where the fluids become heated. The hot water then buoyantly rises to the surface along north-trending faults of the northern Rocky Mountains, the ringlike fractures of the Yellowstone caldera, and the fractured rocks around the edges of the two resurgent domes. Hence, the

Old Faithful, perhaps the world's most famous geyser, draws thousands of visitors to Yellowstone. (© William E. Ferguson)

thermal features of Yellowstone are not randomly distributed throughout the park but tend to occur where fluid pathways are available, especially along the rim of the Yellowstone caldera and near the two resurgent domes.

HYDROTHERMAL ACTIVITY

The Norris Geyser Basin is on the northern edge of the Yellowstone caldera, where circular fractures of the caldera rim are intersected by a major, north-trending fault that extends northward from the caldera. The intersection of two major fault zones enhances the upward migration of deep, hot fluids, and Norris is considered to be the hottest and most active geyser basin in Yellowstone. The basin contains a great variety of thermal features, including geysers, clear boiling springs, mud pots, and fumaroles. Porcelain Basin is the world's largest sinter-covered area (sinter is a variety of hydrated silica that forms around certain hot springs). The geothermal waters at Norris are neutral-chloride to slightly acid-sulfate in composition. The content of dissolved sulfates may be a result of the hot waters traveling through underground sedimentary rock formations that are intersected by the faults at depth.

Old Faithful is a famous geyser in the Upper Geyser Basin, one of several major thermal areas along the Firehole River in the western part of Yellowstone Park. The Upper Geyser Basin is situated on the edge of a resurgent dome within the Yellowstone caldera, where fractured volcanic rocks allow the upward passage of hot fluids. The basin occurs in a topographically low-lying area between three lobes of rhyolite lava, and the valley is underlain by permeable sand and gravel deposits from glaciers and rivers; those deposits allow the hot fluids to disperse over a wide area. As a result of such ideal conditions, the Upper Geyser Basin contains the largest concentration of geysers in the world.

The Mud Volcano area is to the north of Yellowstone Lake, on the southwestern edge of a resurgent dome within the Yellowstone caldera. The water table is boiling underground, and fumaroles are a major feature of this vapor-dominated thermal area. The muddy pools and mud volcanoes are composed of clay, formed by the chemical alteration of near-surface rocks and glacial deposits by acid vapors. The source of the surface water is rainfall and steam condensation. Mammoth Hot Springs is one of the few major thermal areas that occur outside the Yellowstone caldera. Hot, underground water flows northward along the same fault system as Norris, cooling to subboiling temperatures and becoming alkaline in composition during its passage through sedimentary rocks. Calcium carbonate is dissolved from the sedimentary rocks and is later deposited on the surface as mounds and rimmed terraces of travertine, a variety of hot-spring limestone. Travertine grows at much more rapid rates than sinter, and substantial changes in the configuration of Mammoth Hot Springs are observed from year to year.

William R. Hackett

Yellowstone is known for its geysers. The Africa Geyser, its spray frozen during winter, melts from its base, where the ice meets the warm water. (U.S. Geological Survey)

CROSS-REFERENCES

49; Hot Spots, Island Chains, and Intraplate Volcanism, 706; Island Arcs, 712; Lava Flow, 717; Pyroclastic Rocks, 1343; Ring of Fire, 722; Spreading Centers, 727; Volcanic Hazards, 798.

BIBLIOGRAPHY

Decker, Robert, and Barbara Decker. *Volcanoes.* New York: W. H. Freeman, 1997. This readable and well-illustrated text is geared toward the general public. The authors explain geologic phenomena in a clear and simple manner. Includes sections on Yellowstone and its hydrothermal activity. The appendices list the world's 101 "most notorious" volcanoes, as well as Internet sources for volcano information.

Fournier, Robert O. "Geochemistry and Dynamics of the Yellowstone National Park Hydrothermal System." In *Annual Review of Earth and Planetary Sciences,* edited by G. W. Wetherill, A. L. Albee, and F. G. Stehli. Vol. 17. Palo Alto, Calif.: Annual Reviews, 1989. This technical and comprehensive review article emphasizes the geochemistry of hot spring waters. It is readable by people with a knowledge of college chemistry. A number of graphs and a summary table of fluid compositions are given. The reference list is extensive and current, but the article has no index or glossary.

Francis, Peter. *Volcanoes: A Planetary Perspective.* Oxford, England: Clarendon Press, 1993. This is a university-level text, but it is written in a style that makes it accessible to a wide range of readers. The author covers the visible evidence of volcanoes, as well as the theory of why they develop. Illustrations, index, and reference list.

Fritz, William J. *Roadside Geology of the Yellowstone Country.* Missoula, Mont.: Mountain Press, 1985. This 144-page paperback is written for the general public. It contains numerous maps and black-and-white photographs of the geological and thermal features of Yellowstone and is meant to be used in self-guided tours of the roadside geology. The introductory chapters are concise, and the book contains a glossary and extensive reference list.

Pierce, Kenneth L. *History and Dynamics of Glaciation in the Northern Yellowstone National Park Area.* U.S. Geological Survey Professional Paper 729-F. Washington, D.C.: Government Printing Office, 1979. This 90-page paperbound publication with four plates is geared to professional geologists who are interested in glacial features; it should, nevertheless, be useful to people with a good background in general science. Outlines the glacial history of northern Yellowstone. Well illustrated with maps and diagrams. Contains an extensive list of technical references plus an index.

Sigurdsson, Haraldur, ed. *Encyclopedia of Volcanoes.* San Diego, Calif.: Academic Press, 2000. This book contains a complete summary of the scientific knowledge of volcanoes. It contains eighty-two well-illustrated overview articles (including several relating to goethermal systems in Yellowstone), each of which is accompanied by a glossary of key terms. Although this is a college-level text, it is written in a clear and comprehensive style that makes it generally accessible. Cross-references and index.

Smith, Robert B., and Robert L. Christiansen. "Yellowstone Park as a Window on the Earth's Interior." *Scientific American* 242 (February, 1980): 104-117. An excellent overview of the geological and geophysical significance of Yellowstone, this article is readable by persons with a sound high school science background. Well illustrated but contains no glossary or reference list.

U.S. Geological Survey. "Geologic Map of Yellowstone National Park." Miscellaneous Investigations Map I-711. Denver, Colo.: Author, 1972. A 1:125,000-scale, color geological map that shows the areal distribution of rock units exposed at the Earth's surface. Understandable by anyone who has taken an introductory geology course. Would be useful for persons wanting to know the general distribution of rock units within the park.

Vink, Gregory E., W. J. Morgan, and Peter R. Vogt. "The Earth's Hot Spots." *Scientific American* 242 (April, 1985): 50-57. This article reviews the locations and criteria for recognition of the world's hot spots, including Yellowstone. Written from a global perspective, it emphasizes the significance of hot spots in the breakup of continents and in reconstruc-

tions of the Earth's plate motions. Readable by persons with a sound high school science background, it is well illustrated but contains no glossary or reference list.

5
ERUPTIONS

ERUPTIONS

No two volcanic eruptions are identical. There are, however, sufficient similarities between some eruptions that they serve as models to describe the activity of the remainder. Quantification of these descriptive models has permitted a scientific basis for more accurate eruption prediction, for interpreting past unwitnessed eruptions, and for assessing the nature of volcanic activity on extraterrestrial bodies.

PRINCIPAL TERMS

CRYSTAL: a solid with a regular atomic arrangement

LITHIC: having to do with rock

PUMICE: pyroclastic rock full of vesicles (spherical voids originally occupied by gas)

PYROCLASTIC ROCK: fragmented rock produced during a volcanic eruption (the term includes essentially all volcanic products except lava flows)

CLASSIFICATION OF ERUPTIONS: PAST ERUPTION STYLES

There are many ways to classify volcanic eruptions, whether by the shape of the vent (linear versus point source), the location (submarine or subaerial), the composition of the erupted material, the form of the erupted material (lava, gas, or pyroclastics), or the style of the eruption. All these categories are interrelated to a certain degree. Volcanologists classify eruptions according to style in order to have tools for readily describing the particular type of activity of any given volcano.

By the 1920's, Alfred Lacroix had defined four distinct eruption styles, which he termed Hawaiian, Strombolian, Vulcan, and Peléan; these terms remain in common usage. The names make reference to volcanoes that typified the style of eruption described. Style terms that have been adopted since that time follow this system. (An exception is the term "Plinian," which makes reference to Pliny the Elder, who died in 79 C.E. during the eruption of Vesuvius that destroyed the town of Pompeii, or to Pliny the Younger, who described the eruption.) For example, the eruption on May 18, 1980, of Mount St. Helens led to the introduction of the term "directed blast eruption" to the volcanological literature (although a similar style had been previously exhibited by the volcano Bezymianny in Kamchatka). The eruption in the 1960's of Surtsey, off the coast of Iceland, led to the definition of the Surtseyan eruption style.

QUANTIFICATION OF ERUPTION STYLES

No attempts to quantify eruption style were made until the pioneering work of George Walker in the early 1970's, with subsequent refinements by Walker and by graduate students who worked with him. In reality, no volcano has a single eruption style. For example, at each of its three vents, the volcano Stromboli might be erupting in a different style at the same moment. Thus, it is more accurate to group together those volcanoes where a single eruption style predominates, those that exhibit a repeated pattern or sequence of eruption styles, and those that have several eruptive styles but no pattern of their occurrence.

Walker's attempt to quantify eruption style is based on five parameters: magnitude, intensity, dispersive power, violence, and destructive potential. Magnitude has to do with either the quantity of material or the amount of energy released by an eruption. Because the purpose of quantifying style is to be able to apply measurable parameters to unwitnessed eruptions, the total volume emitted during a single event is a more useful parameter than is energy release. To facilitate comparison between volcanoes, the actual volume is converted to a dense rock equivalent (DRE), which makes allowance for the voids in the rock. Measured volumes range from less than 0.001 cubic kilometer to 1,000 cubic kilometers. Intensity is a measure of the discharge rate—that is, the volume emitted during a specified period of time. Measured intensities range from less than 0.1 to more than 10,000 cubic meters per second DRE. Intensity usually varies throughout the course of an eruption, so a more applicable measurement might be peak intensity. Although there is a strong correlation between intensity and magnitude, a wide range of magnitudes can exist for a given intensity. Dispersive power has to do with the area covered by the deposits of a single event. This parameter can be strongly affected by winds. Areas covered can range from less than 1 square kilometer to about 10^6 square kilometers. Violence is a measure of the momentum applied to particles in an eruption. Destructive potential refers to the areal extent of the damage to property and vegetation. Neither violence nor destructive potential is particularly applicable to past unwitnessed eruptions, but both are significant in volcanic hazard assessment.

Christopher Newhall and Steve Self developed a volcanic explosivity index (VEI), which attempts to integrate quantitative data with more subjective style descriptions. As its name implies, the index is most applicable to pyroclastic eruptions and is unsuitable for large-scale lava eruptions. Furthermore, the VEI is based partly on eruption column height, which is a function of intensity, and intensity cannot be correlated directly with magnitude. In spite of the fact that calling an eruption a VEI 5 provides little in the way of a mental image of the volcanic activity, it does provide an extremely useful means of computer-based comparisons of the thousands of dated eruptions.

Geoffrey Wadge proposed that lava eruptions could be subdivided into groups defined by magnitude and intensity: high effusion rate with low volume, low effusion rate with high volume, and, perhaps, high effusion rate with high volume and low effusion rate with low volume.

Attempts to quantify eruption style by direct measurement are restricted to subaerial (land) volcanoes on the Earth. It has been estimated, however, that about 75 percent of the material erupted on the Earth is generated in the deep sea, where pressure and temperatures are vastly different from those on land. Only with the development of manned submersibles did the variety of volcanic products on the ocean floor become apparent. "Smokers"—small, chimneylike structures issuing hot, mineral-laden gases—have been filmed. On other solar system bodies that exhibit volcanic activity, such as Io, there may be essentially zero pressure and much lower gravity. Even if eruption magnitude and intensity in such an environment were the same as those of a subaerial eruption, the style of eruption would be drastically different.

STUDY OF ERUPTIONS

With temperatures on the order of a thousand degrees Celsius and velocities of some erupting materials of 100 meters per second, direct observation of active eruptions can be a hazardous undertaking. Fortunately, direct measurements to determine the magnitude and intensity of an eruption can be performed at a distance, using photographic or remote-sensing techniques. The height of the volcanic column can be measured directly, using ground-based or aerial photography, to provide an estimate of the intensity of the eruption. The volume of solid particles in the eruption column can be measured by determining the amount of light that can be passed through it. Velocities of entrained particles can be measured through viewing high-speed films of the eruption column.

Techniques pioneered by Japanese geologists and extended by George Walker involve measurement of the ejected material after an eruption has ceased. These techniques are little different from those applied to sedimentary rocks; they include the measurement of thickness, the maximum grain size, the grain-size distribution, the proportion of different components, the crystal content of pumice, and the density and porosity. Measurements of thickness of a deposit are used to construct an isopach map—that is, a map showing where the deposit thickness falls between a series of upper and lower limits. Thickness measurements are taken at numerous locations, the number being determined by the size of the deposit and the level of accuracy required. The resultant map provides an indication of the deposit volume, the location of the vent, and the way the deposit was dispersed.

Maximum grain-size measurements are useful for determining the energetics of the eruption column and the effects of wind velocities. If numerous grains are measured over the area of the deposit, an isopleth map can be constructed, indi-

(continued on page 743)

A Time Line of Notable Volcanic Eruptions

Date	Location
5,000 B.C.E.	MAZAMA, OREGON: The volcano that became Crater Lake erupts, sending pyroclastic flows as far as 37 miles (60 kilometers) from the vent; 65,395 to 91,553 (50 to 70 cubic kilometers) of material are erupted as a caldera forms from the collapse of the mountaintop.
c. 1470 B.C.E.	AEGEAN SEA: A volcanic eruption and caldera collapse leave a town buried and preserved intact, possibly causing the disappearance of the Minoan civilization on the island of Crete, the alleged location of the "lost continent of Atlantis."
Aug. 24, 79 C.E.	ITALY: Mount Vesuvius erupts, burying Pompeii and Herculaneum. More than 13,000 dead, 4 cities completely buried, 270 square miles (700 square kilometers) devastated.
c. 186 C.E.	NEW ZEALAND: On North Island, a huge eruption with a volcanic explosivity index greater than 6 creates a crater that eventually becomes Lake Taupo.
1169	SICILY: Mount Etna erupts with a volcanic explosivity index of 2 and leaves more than 15,000 dead.
1362	ICELAND: Öroefajökull erupts with a volcanic explosivity index of 5; 200 die.
1586	JAVA, INDONESIA: Kelut erupts with a volcanic explosivity index of 5, leaving 10,000 dead.
1591	PHILIPPINES: Taal erupts with a volcanic explosivity index of 3, leaving thousands dead.
Dec. 16, 1631	ITALY: Mount Vesuvius erupts with a volcanic explosivity index of 4; 4,000 die.
Mar. 11, 1669	SICILY: Mount Etna erupts, leaving more than 20,000 dead, 14 villages destroyed, 27,000 homeless.
1741	ECUADOR: Cotopaxi erupts with a volcanic explosivity index of 2, leaving 1,000 dead.
Sept. 29, 1759	MEXICO: Jorullo erupts with a volcanic explosivity index of 3, leaving hundreds dead.
Oct. 23-28, 1766	LUZON, PHILIPPINES: Mayon erupts with a volcanic explosivity index of 3, leaving 2,000 dead.
Aug. 11-12, 1772	JAVA, INDONESIA: Papandayan erupts with a volcanic explosivity index of 3, leaving 3,000 dead.
Dec., 1779-Jan., 1780	JAPAN: Sakurajima erupts with volcanic explosivity index of 4, leaving 300 dead.
1783	JAPAN: Asama erupts with a volcanic explosivity index of 4, leaving 1,377 dead.
June, 1783-Feb., 1784	SOUTHERN ICELAND: The Laki fissure eruption in Iceland produces the largest lava flow in historic time. Benjamin Franklin speculates on its connection to a cold winter in Paris the following year.
1790	HAWAII: Kilauea erupts with a volcanic explosivity index of 4, leaving 100 dead.
Feb. 10, 1792	JAPAN: Unzen erupts with a volcanic explosivity index of 2, leaving 14,500 dead.
1794	RIOBAMBA, ECUADOR: Tunquraohua erupts with a volcanic explosivity index of 2, leaving 40,000 dead.
Apr. 27, 1812	ST. VINCENT, WEST INDIES: La Soufrière erupts with a volcanic explosivity index of 4, leaving more than 1,000 dead.

(continued)

Date	Location
Feb. 1, 1814	LUZON, PHILIPPINES: Mayon erupts with a volcanic explosivity index of 4, leaving more than 2,200 dead.
Apr. 5, 1815	SUMBAWA, INDONESIA: The dramatic explosion of Tambora, 248.6 miles (400 kilometers) east of Java, the largest volcanic event (VEI = 7) in modern history, produces atmospheric and climatic effects for the next two years; 92,000 die. Frosts occur every month in New England during 1816, the Year Without a Summer.
Oct. 8 and 12, 1822	JAVA, INDONESIA: Galung Gung erupts with a volcanic explosivity index of 5, leaving 4,000 dead.
Jan. 22, 1835	NICARAGUA: Cosigüina erupts with a volcanic explosivity index of 5; hundreds die.
1845	COLOMBIA: Nevado del Ruiz erupts with a volcanic explosivity index of 3, leaving 700 dead.
June 24, 1853	TONGA ISLANDS: Niuafo'ou explodes, leaving 70 dead, village mostly destroyed.
Apr. 24-26, 1872	ITALY: Mount Vesuvius erupts again, leaving 22 dead.
June 26, 1877	ECUADOR: 1,000 humans, thousands of animals dead, buildings and bridges destroyed after Cotopaxi erupts.
Aug. 26, 1883	INDONESIA: A cataclysmic eruption of Krakatau is heard 2,983 miles (4,800 kilometers) away; 36,417 die and two-thirds of the island is destroyed. Pyroclastic flows race over pumice rafts floating on the surface of the sea; many die from a tsunami.
July 15, 1888	HONSHŪ, Japan: Bandai erupts, leaving 461 dead, 70 burned and scarred, several villages buried.
June 23, 1897	LUZON, PHILIPPINES: Mayon erupts, leaving 400 dead, villages and animals destroyed.
May 7, 1902	ST. VINCENT ISLAND, LESSER ANTILLES: La Soufrière erupts, leaving 1,500-1,700 dead, great losses in livestock and crops.
May 8, 1902	MARTINIQUE, CARIBBEAN: Mount Pelée, on the northern end of the island, sends violent pyroclastic flows into the city of St. Pierre, killing all but 4 of the 30,000 inhabitants.
Oct. 24, 1902	GUATEMALA: Santa María erupts, leaving 6,000 dead, animals and crops destroyed, buildings collapsed.
1905-1906	ITALY: Vesuvius erupts, leaving dozens dead, buildings destroyed.
Jan. 4, 1906	NICARAGUA: Masaya erupts; thousands die.
Jan. 30, 1911	PHILIPPINES: Taal erupts; 1,335 dead, 200 injured.
June 6, 1912	ALASKA: Katmai erupts with a volcanic explosivity index of 6. Ash covers the Valley of Ten Thousand Smokes.
June 6, 1917	EL SALVADOR: Boquerón erupts, leaving 450 dead, 100,000 homeless.
May 20, 1919	JAVA, INDONESIA: Kelut's eruption kills 5,500 and destroys many villages.
Apr. 17, 1926	HAWAII: Mauna Loa erupts, killing dozens and destroying the town of Hoopuloa.
Aug. 4-5, 1928	PALUWEH, INDONESIA: Rokatenda erupts, killing 226 and destroying villages and boats.
Dec. 13-28, 1931	JAVA, INDONESIA: Merapi erupts, killing more than 1,300.

Date	Location
Feb. 20, 1943	MEXICO: Paricutín Volcano comes into existence in a cultivated field; eruption continues for nine years.
Jan. 17-21, 1951	GUINEA: Lamington erupts, killing 3,000.
Dec. 2-8, 1951	PHILIPPINES: Hibok-Hibok erupts, killing 500.
Mar. 30, 1956	KAMCHATKA PENINSULA: The volcano Bezymianny erupts with a violent lateral blast, stripping trees of their bark 18.6 miles (30 kilometers) away.
Mar. 20, 1963	BALI, INDONESIA: Mount Agung erupts, killing more than 1,200 and leaving 200,000 homeless.
Nov. 8, 1963-June 5, 1967	ICELAND: Surtsey Island is born from an eruption with a volcanic explosivity index of 3.
Sept. 28, 1965	PHILIPPINES: Taal erupts, killing 200.
July, 1968	COSTA RICA: Arenal erupts with a volcanic explosivity index of 3, killing 80.
Jan.-May, 1973	ICELAND: Hundreds of homes are destroyed after an eruption on Heimaey Island.
Jan. 10, 1977	ZAIRE: Nyiragongo erupts with a volcanic explosivity index of 1, killing more than 1,000.
May 18, 1980	WASHINGTON STATE: Mount St. Helens erupts, killing 57 humans, an estimated 7,000 big-game animals, and destroying nearly 200 homes, more than 185 miles of road, and 4 billion board feet of timber. Ashfall is detected for 22,000 square miles.
Mar. 28-Apr. 4, 1982	MEXICO: El Chichón erupts, killing about 2,000 and injuring hundreds. Hundreds are left homeless, thousands evacuated, 9 villages are destroyed, and more than 116 square miles of farm land are ruined.
Nov. 13, 1985	COLOMBIA: Mudflows from the eruption of the Nevado del Ruiz volcano kill at least 23,000 people.
Aug. 21, 1986	CAMEROON: After building up from volcanic emanations, carbon dioxide escapes from Lake Nyos, killing more than 1,700 people.
June 12-15, 1991	LUZON, PHILIPPINES: Mount Pinatubo erupts, killing approximately 350 (mostly from collapsed roofs); extensive damage to homes, bridges, irrigation canal dikes and cropland; 20 million tons of sulfur dioxide spew into the stratosphere to an elevation of 15.5 miles.
Nov. 22, 1994	JAVA, INDONESIA: Merapi eruption kills at least 31.
Sept.-Nov., 1996	ICELAND: Eruption of lava beneath a glacier in the Grimsvötn Caldera melts huge quantities of ice, producing major flooding.
June 25, 1997	MONTSERRAT, THE CARIBBEAN: Soufrière Hills eruption kills 19, and 8,000 are evacuated.

cating areas in which maximum particle sizes are equal. An isopleth map can sometimes be more useful than an isopach map, for erosion tends to remove fine material from a deposit, lessening its thickness, while larger particles remain. The grain-size distribution is obtained by simply sieving numerous samples. A certain amount of care has to be exercised, because pumice and glass are extremely fragile. The result of this analysis is a determination of the degree of sorting of the deposit. This information can be employed to distinguish between certain types of eruptions.

A deposit contains three components: pumice, lithics, and crystals. The varying proportions of

Eruptions

each of these components throughout the deposit can be used to infer the conditions in the magma chamber prior to eruption, and the relative proportions at different sites can be employed in determining the eruption dynamics. The physical separation of these components can be an extremely time-consuming task, involving sorting by hand under a microscope. The crystal content of pumice is a further indicator of the preeruption conditions in the magma chamber. To determine the crystal content, pumice lumps can be crushed and the more durable crystals removed. Measurements of density and porosity are important for calculating the dense rock equivalent. In the case of pumice, which often has unconnected pore spaces, these measurements can be rather complicated. Simply, samples are variously measured dry in air, after soaking for several days in water, and under water.

ERUPTION PREDICTION

Determination of the type of a volcanic eruption is important for two major reasons: to permit theoretical reconstructions of unwitnessed eruptions and so that scientists can predict the consequences of future eruptions at an active volcano. After the magnitudes and intensities have been established for the unwitnessed eruptions at a specific volcano, such as Mount St. Helens, by combining these data with age data for each deposit, it becomes possible to determine the eruptive history. The volcanologist is then armed with information regarding the largest magnitude eruption that has taken place in the past and the frequency at which volcanic events occur. It is then possible to forecast that a specific volcano is capable of producing an eruption of a given magnitude in the future and to ascertain the probability of such an

event taking place within a given period of time.

If a volcano goes through a regular sequence of eruptive events prior to a paroxysmal eruption, recognition of this precursor activity may assist authorities in their decision to evacuate the local populace. The effects of failure to undertake such an analysis and to heed warning signs is well illustrated by the 1951 eruption of Mount Lamington, which had been regarded as an extinct volcano, in Papua New Guinea. A cloud of smoke was seen above the vent on January 15 of that year; five days later came the paroxysmal eruption, which caused six thousand deaths. In those areas with a high population density or rich agricultural land, insurance brokers are particularly interested in knowing details of possible eruptions in order to minimize their economic losses.

From the scientific viewpoint, types of volcanic activity provide windows into the interior of the planet and allow assessments of what is happening beneath plate margins, continental interiors, and the ocean floors—and beneath the surfaces of alien planets and satellites. Climatic changes and mass extinctions have been attributed to large-scale volcanic eruptions. Long-term changes in the magnitude and styles of eruption reflect important changes taking place in planetary interiors.

James L. Whitford-Stark

CROSS-REFERENCES

Forecasting Eruptions, 746; Io, 2521; Krakatau, 752; Magmas, 1326; Mars's Volcanoes, 1608; Mount Pelée, 756; Mount Pinatubo, 761; Mount St. Helens, 767; Mount Vesuvius, 772; Plate Tectonics, 86; Popocatépetl, 776; Recent Eruptions, 780; Shield Volcanoes, 787; Stratovolcanoes, 792; Volcanic Hazards, 798.

BIBLIOGRAPHY

Cas, Ray A. F., and J. V. Wright. *Volcanic Successions: Modern and Ancient.* Winchester, Mass.: Allen & Unwin, 1987. George Walker's influence is readily evident in this book; one of its authors was his student. An excellent overview of volcanism from a largely sedimentological viewpoint. Aimed at college-level and research students.

Decker, Robert, and Barbara Decker. *Volcanoes.* New York: W. H. Freeman, 1997. This readable and well-illustrated text is geared toward the general public. The authors explain geologic phenomena in a clear and simple manner. The appendices list the world's 101 "most notorious" volcanoes, as well as Internet sources for volcano information.

Francis, Peter. *Volcanoes: A Planetary Perspective.* Oxford: Clarendon Press, 1993. This is a university-level text, but it is written in a style that makes it accessible to a wide range of readers. The author covers the visible evidence of volcanoes, as well as the theory of why they develop. Illustrations, index, and reference list.

Lipman, Peter W., and D. R. Mullineaux, eds. *The 1980 Eruptions of Mount St. Helens, Washington.* U.S. Geological Survey Professional Paper 1250. Reston, Va.: Department of the Interior, 1981. Detailed accounts of the most violent eruption witnessed in the contiguous United States at this writing. Aimed at a research-level audience, it can, however, be read by high school students. Of particular interest to anyone living in the western United States.

Sigurdsson, Haraldur, ed. *Encyclopedia of Volcanoes.* San Diego, Calif.: Academic Press, 2000. This book contains a complete summary of the scientific knowledge of volcanoes. It includes eighty-two well-illustrated overview articles, each of which is accompanied by a glossary of key terms. Although this is a college-level text, it is written in a clear and comprehensive style that makes it generally accessible. Cross-references and index.

Wood, C. A., and J. Kienle. *Volcanoes of North America: U.S.A. and Canada.* New York: Cambridge University Press, 1989. Descriptive accounts of recent and past activities of volcanoes in North America.

FORECASTING ERUPTIONS

Forecasting volcanic eruptions is one of the most important goals of volcanology, as people and property are at risk in volcanically active areas. Although the phenomena are very different, volcanic forecasting is similar in principal to forecasting of the weather: It is based on the statistics of previous events as well as on information about current activity.

PRINCIPAL TERMS

CORRELATION SPECTROMETER (COSPEC): an instrument that measures the output of sulfur dioxide from volcanic gas plumes

EARTHQUAKE SWARM: a number of earthquakes that occur close together and closely spaced in time

TILTMETER: an instrument that precisely measures tilting of the ground surface

TUMESCENCE: a local swelling of the ground that commonly occurs when magma rises toward the surface

VOLCANIC EARTHQUAKES: small-magnitude earthquakes that occur at relatively shallow depths beneath active or potentially active volcanoes

VOLCANIC TREMOR: a continuous vibration of long duration, detected only at active volcanoes

GENERAL AND SPECIFIC FORECASTS

All forecasting of volcanic eruptions is probabilistic and is based on long-term patterns together with present-day observations. Forecasts can be subdivided into two categories, general and specific, depending on how constrained the forecast is. General forecasts are based mainly on statistical averages. For example, if it rains an average of 182 days per year in a given city, on any one day the chance of rain is 50 percent. A general forecast for tomorrow or any other day would report a 50 percent chance of rain, which is hardly a useful prediction but is nevertheless statistically valid. Specific forecasts are based not only on historical statistics but also on current information. Thus, the 50 percent general forecast could be greatly improved by using satellite photographs and information on today's movements of atmospheric pressure systems. In that case, rain might be forecast with a 90 percent chance, and the specific forecast would be much more useful.

A similar rationale can be applied to volcanoes. To make general forecasts of eruptions, it is necessary to know something about the past activity of volcanoes. The problem is that volcanoes have widely differing eruptive patterns. Some erupt frequently, and historical records are therefore reliable indicators of eruption patterns. Other volcanoes erupt so infrequently that the historical record is not a statistically valid sampling of their activity. Some potentially dangerous volcanoes have never erupted during historic time. Although prehistoric eruptive patterns give important long-term information that is needed in general forecasting, it is also necessary to have current information about volcanoes that show signs of awakening. Three types of present-day observations are useful in forecasting volcanic eruptions: geophysical, geochemical, and geological.

GEOPHYSICAL OBSERVATIONS

Indirect physical changes usually accompany the underground movement of magma, such as changes in the magnetic, thermal, or gravitational properties of the volcano; changes in the configuration of the ground surface; seismic activity; and fluctuations in the electrical properties of rocks. All these geophysical phenomena are related to the fact that magma is introduced into the volcano or its plumbing system prior to an eruption.

It might seem that the thermal effects of magma would be obvious, by heating of the volcano's surface. Magma commonly produces only minor temperature changes in the overlying near-surface rocks and soil, however, because rocks are generally poor conductors of heat. Furthermore, any temperature effects are usually masked by heating and cooling during the daily solar cycle. Most rocks, and especially iron-rich volcanic rocks, have a natural magnetization that is acquired during cooling within the Earth's magnetic field. Just as an iron magnet can be destroyed by heating, few rocks can remain magnetized at temperatures greater than about 550 degrees Celsius. The temperatures of magmas range from about 800 to 1,200 degrees Celsius, and molten rock is therefore nonmagnetic. Underground rocks are slowly heated by magma, further decreasing the local magnetic field strength.

As magma approaches within a few kilometers of the Earth's surface, it makes room for itself by shouldering aside the surrounding rocks. As a result, the overlying rocks and soil may bulge upward in a phenomenon known as tumescence. Similarly, deflation of the ground surface commonly occurs when magma withdraws, sometimes signaling the end of an eruption. The shape and extent of the uplifted area are controlled by the configuration of the underground magma body, and the magnitude of tumescence can be substantial: For several months prior to the catastrophic eruption of May 18, 1980, the northern summit region of Mount St. Helens bulged several meters per day, achieving a total of about 100 meters of northward bulging. Cracks and faults also opened on the summit of St. Helens, some a few meters wide and extending for more than 1 kilometer. At other volcanoes where the volume of magma is small or the magma is relatively deep, tumescence may be imperceptible to the eye, amounting to only a few millimeters of uplift over a distance of several kilometers.

The movement of underground magma is also marked by the occurrence of earthquake swarms. Thousands of small earthquakes may occur daily as the magma forces itself upward, fracturing the surrounding rocks at successively shallower depths and sending out seismic waves. Volcanic earthquakes differ in several ways from earthquakes associated with large faults in the Earth's crust. First, volcanic earthquakes are localized beneath active or potentially active volcanoes and occur as a result of local magma injection rather than of large-scale shifting along major faults. Second, volcanic earthquakes are relatively shallow events, usually originating at depths less than 5 or 10 kilometers. Third, volcanic earthquakes have low magnitudes (usually less than 4 on the Richter scale) because

the brittle rocks around local magma bodies are easily fractured.

Volcanic tremor, in contrast to single-shock volcanic earthquakes, is continuous vibration that can last for minutes, hours, or days and has been detected only on active volcanoes. Also called harmonic tremor, the vibrations are thought to be produced by the subsurface flow of magma or by the formation of gas bubbles within shallow magma bodies. At basalt volcanoes such as those of Hawaii, harmonic tremor commonly begins a few minutes prior to the eruption of fluid lava and continues during the surface emission of lava. At volcanoes with more viscous types of magma, such as Mount St. Helens, harmonic tremor does not always occur, and its origins are not well understood. A 1996 survey of earthquake swarms that occurred between 1979 and 1989 showed that 58 percent were followed by eruptions. Those not followed by eruptions had a shorter average duration than those that were.

Other geophysical studies include monitoring the electrical properties of rocks and soil near volcanic vents and measuring any changes in the strength of the Earth's local gravitational field. Small but measurable changes in gravity may accompany the underground injection of magma if the magma is lighter or heavier than the rocks it displaces, or if the ground is uplifted.

GEOCHEMICAL OBSERVATIONS

Geochemical measurements of gas and lava compositions are also useful during volcano monitoring. The types and volumes of gases emitted from lava or fumaroles (vents that emit only gases) commonly change when fresh magma is injected underground. For example, at Kilauea Volcano in Hawaii, the ratio of carbon dioxide to sulfur dioxide gases is greater in fresh magma that arrives from the Earth's mantle than is the ratio of those gases emitted from older magma that has been stored beneath the volcano for some time. Thus, changes in the gas chemistry and temperature of lava lakes and fumaroles can signal the injection of fresh, gas-rich magma that might be more likely to erupt than would older, gas-poor magma.

Studies of the gas plume from the Mount St. Helens lava dome have shown that increased emissions of sulfur dioxide and carbon dioxide occurred when fresh batches of magma were injected beneath the summit crater. Increased rates of sulfur dioxide emission have been measured before several nonexplosive eruptions at Mount St. Helens. The emissions have been interpreted as resulting from rapid degassing of small bodies of magma as they moved toward the surface from a larger, deeper magma chamber. Changes in the percentages of hydrogen, helium, and radon in volcanic gases may also prove useful in signalling eruptions.

Measurements of lava compositions can also be useful for recognizing fresh batches of magma that have moved into the volcano. At Mount St. Helens, each nonexplosive eruption since October of 1980 has added a new flow of viscous lava to the summit crater. The chemistry and mineralogy of lava samples from Mount St. Helens have remained nearly constant through many successive eruptions. This consistency suggests that the magma feeding the eruptions has come from a single, shallow magma body that was injected in 1980 and has not undergone any significant changes in composition since that time. A different scheme of magma supply seems to operate beneath Kilauea volcano in Hawaii, where repeated injections of fresh magma into shallow magma chambers have occurred many times during historical eruptions of that volcano. The arrival of fresh magma causes variations in the temperatures of lava flows and lava lakes as well as variations in the chemical composition and mineralogy of the lavas.

GEOLOGICAL OBSERVATIONS

Events that are too complex to be numerically measured in any simple way can nevertheless be observed and recorded, such as the opening of new fractures or vents, the advent of new fumaroles, lava flows or pyroclastic fountains, and the fluctuation or cessation of lava output. Together with geophysical and geochemical data, such geological observations are an important aspect of eruption forecasting.

Geologists also study prehistoric volcanic deposits, which can give important information about long-term eruption patterns on a time scale of thousands or millions of years. Prehistoric volcanic deposits are sometimes difficult to study because of poor outcroppings of the deposits, the destruction of ancient deposits by erosion, and

difficulties with accurately dating the ancient materials. In spite of these limitations, geological field studies are an important aspect of volcano forecasting because volcanism can be examined on a much longer time frame than is possible with historical records alone. The periods of repose at different volcanoes vary from months to hundreds of thousands of years; some volcanoes erupt so infrequently that the historical record does not accurately represent their long-term behavior. This is true for many of the Cascade volcanoes, including Mount St. Helens, which was active between 1831 and 1857, then lay dormant for more than a century before explosively awakening in May, 1980.

PAST ERUPTIONS AND PROBABILITIES

The statistics of past eruptions, based on historical records and the study of prehistoric deposits, are very important both in finding the average probability of future eruptions and in recognizing any patterns of eruption characteristics. At least three patterns have been recognized from a few well-studied, active volcanoes. The first pattern is characterized by completely random behavior: No matter how long it has been since the last eruption, the average chance of a future eruption remains the same. This is analogous to cutting a deck of cards to get the ace of spades; no matter how many cuts are made, the chance of getting the ace in the next cut remains 1 in 52. Mauna Loa volcano in Hawaii seems to behave this way, and the average chance for a new eruption is about 2 percent each month, regardless of how long it has been since the previous eruption.

The second pattern is like cutting a deck of cards for the ace of spades and throwing away each unwanted card. The chance of getting the ace thereby increases with each new cut because the deck grows smaller. At Hekla volcano in Iceland, the probability of an eruption increases with the passage of time since its last eruption. A third pattern has no analogy in card cutting. At volcanoes such as Kilauea in Hawaii, eruptions tend to cluster together in time, and the probability of a new eruption decreases the longer the time that has passed since the last eruption. This is the opposite of the second pattern. Recognition of one of these time patterns is important in forecasting future eruptive activity, but such recognition depends on having observed a statistically significant number of eruptions at a given volcano—about twenty or so. Unfortunately, only a few of the world's volcanoes are sufficiently active or have been studied long enough for scientists to recognize patterns in their activity. Ironically, eruptions must occur for reliable forecasts to be made.

VOLCANO MONITORING

No single approach to volcano forecasting can yield trustworthy results, but the combined data of geophysics, geology, and geochemistry, interpreted within the context of long-term eruptive behavior of a given volcano, can often be compiled into reliable forecasts. As with any forecasting, the reliability of volcanic forecasts increases with the quantity and diversity of information that is available to volcanologists. The 1980 awakening of Mount St. Helens after more than a century of dormancy was predicted by U.S. Geological Survey scientists in a very general way, leading to an evacuation several weeks before the eruption. After the major eruption, continuous monitoring of smaller, more numerous eruptions led to more specific forecasts, generally reliable to within a few hours or days. Specific, reliable forecasts are also routinely issued by the U.S. Geological Survey at Kilauea volcano, which has erupted dozens of times since the Hawaiian Volcano Observatory was established there in 1912.

The geophysical equipment used in volcano monitoring is diverse, and some instruments are also used for industrial applications. The temperatures of lava and gases can be measured with a thermocouple—a pair of wires of different metals that are welded together at both ends. When one end of the circuit is immersed in hot material, an electrical current is generated, its strength depending on the temperature difference between both ends of the circuit. An ammeter near the cold end of the circuit reads the electrical current, from which the temperature at the hot end of the circuit can be calculated. Another temperature-measuring instrument is the optical pyrometer, which measures the wavelengths of visible and near-infrared radiation that is emitted from the surfaces of incandescent objects.

Magnetometers are used to measure the strength and sometimes the direction of the Earth's magnetic field. Over large regions, the instruments can be used from airplanes, but volcano monitoring usually involves ground measurements at fixed stations that are reoccupied in order to get comparative readings. Tumescence and faulting of the ground are measured with surveying instruments such as steel tapes, transits, levels, or theodolites. A network of survey stations and reflective targets is established on a volcano, and measurements are repeated in order to recognize any deformation of the ground that might be associated with the movement of underground magma. Tiltmeters are used to measure changes in the slope or inclination of the ground surface. Using the same principle as a carpenter's level, electronic tiltmeters employ sensitive bubbles to measure tilt in two directions at right angles; thus, the direction and magnitude of tilting can be monitored. Anchored on the ground, the instruments are extremely sensitive—the amount of tilt is expressed in microradians, the equivalent of lifting one end of a 1-kilometer-long rod by 1 millimeter.

Volcanic earthquakes are monitored with a network of seismometers encircling the volcanic area. The instruments can be tuned to detect very small, unfelt earthquakes with Richter magnitudes less than 1. With each shock, several types of seismic waves are emitted from the site of an earthquake. Because different waves travel at different speeds, their arrival times can be used to calculate the distance between the seismometer and the earthquake center in much the same way that the time lag between lightning and thunder can be used to estimate the distance to the lightning.

Gravimeter readings can be repeatedly taken at fixed stations to measure any changes in the Earth's gravitational force. In one form, the instrument consists of a weight suspended from a sensitive spring; variations in gravity cause variations in the extension of the spring. Several mechanical and optical schemes have been developed to measure this very small deflection, and the most sensitive instruments can detect gravitational changes of 1 part in 10 million. Measurable changes sometimes occur when magma is either lighter or heavier than the underground rocks that it displaces or if the ground surface has been uplifted by the magma. In the latter case, the gravitational force is diminished because the gravimeter is farther away from the Earth's center of mass.

Collections of volcanic gases are made directly at vents with evacuated and flowthrough sampling tubes or indirectly by using remote devices. One such device is the correlation spectrometer (COSPEC), an instrument that measures the output of sulfur dioxide gas from volcanic plumes at a safe distance. The COSPEC can be used from the ground or from an airplane. It measures the amount of solar ultraviolet light absorbed by sulfur dioxide gas in the plume, compares it to a standard, and thus measures the amount of sulfur dioxide. In 1983, sulfur dioxide emissions at Kilauea volcano averaged about 250 tons per day, while Mount St. Helens emitted about 100 tons per day, down from its 1980 peak of about 1,500 tons per day.

William R. Hackett

CROSS-REFERENCES

Earthquake Prediction, 309; Eruptions, 739; Gravity Anomalies, 122; Hawaiian Islands, 701; Heat Sources and Heat Flow, 49; Krakatau, 752; Lava Flow, 717; Mount Pelée, 756; Mount Pinatubo, 761; Mount St. Helens, 767; Mount Vesuvius, 772; Popocatépetl, 776; Pyroclastic Rocks, 1343; Recent Eruptions, 780; Rock Magnetism, 177; Shield Volcanoes, 787; Stratovolcanoes, 792; Volcanic Hazards, 798.

BIBLIOGRAPHY

Bailey, Roy A., P. R. Beauchemin, F. P. Kapinos, and D. W. Klick. *The Volcano Hazards Program: Objectives and Long-Range Plans.* U.S. Geological Survey Open-File Report 83-400. Washington, D.C.: Government Printing Office, 1983. This technical report is available at Federal Repository libraries. It discusses monitoring of volcanoes in the western United States and classifies them by group according to frequency of eruption and within groups according to specific geological hazards. Readable by those with a good high school science background.

Brantley, Steve, and Lyn Topinka, eds. *Earthquake Information Bulletin* 16 (March/April, 1984). This 120-page special issue, *Volcanic Studies at the U.S. Geological Survey's David A. Johnston Cascades Volcano Observatory, Vancouver, Washington,* summarizes the eruptive activity of Mount St. Helens from 1980 to 1984 and describes the successful monitoring program of the U.S. Geological Survey. All monitoring methods and results are discussed in straightforward language, and the book is readable by the general public.

Decker, Robert, and Barbara Decker. *Volcanoes.* New York: W. H. Freeman, 1997. This 320-page paperback is geared to the general public. Easy to use, it includes a glossary, selected references for each chapter, an index, and appendices on the world's individual volcanoes and volcano information centers. The discussion of volcano forecasting is excellent.

Fiske, R. S. "Volcanologists, Journalists, and the Concerned Local Public: A Tale of Two Crises in the Eastern Caribbean." In *Explosive Volcanism: Inception, Evolution, and Hazards.* Washington, D.C.: National Academy Press, 1984. Essentially a case study, this article is an informed narrative and discussion of the interactions among scientists, the press, civil authorities, and the public during two volcano crises. No reference list. Readable by all.

Francis, Peter. *Volcanoes: A Planetary Perspective.* Oxford: Clarendon Press, 1993. This is a university-level text, but it is written in a style that makes it accessible to a wide range of readers. The author covers the visible evidence of volcanoes, as well as the theory of why they develop. Contains a good section on eruption forecasting. Illustrations, index, and reference list.

Heliker, C., J. D. Griggs, T. J. Takahashi, and T. L. Wright. *Earthquakes and Volcanoes* 18 (1986). This 71-page special issue summarizes the history and work of the Hawaiian Volcano Observatory. Written in nontechnical language, it is an excellent companion to Brantley and Topinka's 1984 special issue on Mount St. Helens. Thoroughly interesting, it contains excellent photographs and graphics throughout.

Sigurdsson, Haraldur, ed. *Encyclopedia of Volcanoes.* San Diego, Calif.: Academic Press, 2000. This book contains a complete summary of the scientific knowledge of volcanoes. It contains eighty-two well-illustrated overview articles (including one on calderas), each of which is accompanied by a glossary of key terms. Includes an extensive section on eruption forecasting. Although this is a college-level text, it is written in a clear and comprehensive style that makes it generally accessible. Cross-references and index.

Tazieff, H., and J. C. Sabroux, eds. *Forecasting Volcanic Events*. Developments in Volcanology 1. New York: Elsevier, 1983. Although this 635-page volume is rather technical and is geared to geologists, many of the articles are of a general nature and should be readable by people with a solid high school science background. Contains an extensive reference list but no index.

KRAKATAU

Residents of Sumatra and Java thought that the quiet volcano Krakatau was extinct until 1883, when it violently erupted, destroying three hundred villages and generating sea waves that killed more than thirty-six thousand people. Perhaps the loudest noise in recorded history, the paroxysmal eruption was heard more than 4,600 kilometers away, and dust from the explosion reddened skies around the globe.

PRINCIPAL TERMS

BASALT: a type of volcanic rock that cooled from magma relatively low in silica, which enabled it to flow freely

CALDERA: a large circular or oval depression formed by the explosion and collapse of a volcano

EJECTA: material, such as rock fragments and ash, thrown out by a volcano

IGNIMBRITE: volcanic rock, high in the element silica, formed by the cooling of layers of volcanic ash mixed with rock fragments

PLATE TECTONICS: the theory of geology that re-lates volcanism, mountain building, earthquakes, and other phenomena to the movements of "floating" plates of the Earth's crust

PUMICE: a type of volcanic rock that is actually glass filled with air bubbles

PYROCLASTIC: formed by the accumulation of rock fragments and ash

TSUNAMI: a large, destructive wave caused by earthquakes or other short-duration disturbances of the seafloor; tsunamis are characterized by great speed and increased height when piled up in shallow water

EARLY OBSERVATIONS

Before 1883, the island Krakatau was noteworthy only as a landmark to the hundreds of ships sailing from the Indian Ocean through the high-traffic sea lane of Sunda Strait to the ports of Singapore and Jakarta in the China Sea. The 9-kilometer-long island, formed by the volcanoes Perboewatan, Danan, and Rakata, was the largest of four that marked the rim of a prehistoric volcanic crater. Java, to the north of Krakatau, contains forty-nine volcanoes, more than one-half of which have erupted in the past three hundred years or so. Sumatra and the smaller islands of Bali, Lombok, Timor, and Flores also are packed with volcanoes and accompanying geysers and hot springs. There is little argument that Indonesia has the highest concentration of volcanic activity in the world.

Yet the possibility that Krakatau was capable of eruption was ignored by the Dutch and other European colonists of the late nineteenth century. Unlike the neighboring island of Sebesi, Krakatau, a crater rim segment, was not an easily recognized cone-shaped volcano. Forgotten was the historical account of an eruption from two hundred years earlier: a roar like thunder, earthquakes, and

a burned forest. Ignored were the legends of fifteen centuries earlier, the Javanese chronicles, that describe an enormous eruption and accompanying fires, shaking, and fearful roars from the mountain called Kapi. This same account explains the division of a single preeruption island into Sumatra to the north and Java to the south by the rising ocean that now ran between them as the Sunda Strait. The ancient chronicles are supported by contemporary geologic evidence. Volcanic deposits from the nearby Java Trench and the huge crater formed by Krakatau and neighboring islands imply the existence of an earlier gigantic volcano about one-half the size of Rhode Island.

Accounts of volcanic activity on Krakatau before the 1880's are sparse, but eyewitness accounts of the eruptions of August 26 and 27, 1883, and studies of the aftermath are extensive. Passing ships and local residents detail the deafening sounds, the darkness, falling pumice, choking ash, bolts of lightning, the eruption itself, and the subsequent destruction by 12- to 40-meter-tall tsunami waves. The tremendous noise, startling red sunsets, and other unusual phenomena were heard and seen around the world.

KRAKATAU'S REAWAKENING

Krakatau began its return to life on May 20, 1883, when it ejected pumice, steam, and ash. During the next three months, residents became accustomed to the continual artillery-like booms that rattled doors and windows, the shaking from earthquakes, and the views of molten lava and columns of dust. Observers from an exploration party to the island witnessed new vents on the crater floor of Perboewatan, the smallest of the three volcanoes.

At 1:00 P.M. on Sunday, August 26, the first of a series of violent explosions shook the volcano. Within one hour, a 27-kilometer-high black cloud hid Krakatau from view. For the next twenty-three hours, eruptions burst forth every ten minutes and grew more ferocious through the night. Passing ships reported hellish conditions. From 65 kilometers away, the schooner *Sir Robert Sale* witnessed a great vapor cloud, lit by bursts of forked lightning, "like a large serpent rushing through the air." On the nightmare journey past Krakatau, Captain Watson of the *Charles Bal* kept a log that details falling hot pumice stones and the buildup of muddy ash that kept his crew busy shoveling for their lives. If a sailor paused in his duty, he became more aware of the hot, choking sulfurous fumes and the almost supernatural pink glow, called St. Elmo's fire, on the rigging above. On the horizon, sailors observed flashes of lightning jumping back and forth from Krakatau to the sky.

The volcano's eruption was not yet over. At 5:30 A.M. on Monday morning, Krakatau radically changed styles of eruption in a grand finale of four explosions. The third explosion, at 10:02 A.M., brought down the island. Where there had been several peaks, the highest at 4,267 meters, a gaping crater, 300 meters below sea level, appeared. That was not known, however, until twenty-four hours later, when the pitch-black darkness, produced by about 9 cubic kilometers of ash and rocks ejected an estimated 80 kilometers into the atmosphere, lifted.

Two-thirds of the island's disappearance was not, however, the main subject of concern. The explosions, equivalent to 1.5 million tons of dynamite, generated sea waves (tsunamis) that drowned more than thirty-six thousand residents and air waves that produced what is considered the loudest sound in human history. The largest sea wave, cresting at 40 meters, or the height of a twelve-story building, swept away the coastal town of Merka in Sumatra one hour after the explosion. This first giant wave—and nine others—drowned about three hundred towns and villages on Java and Sumatra. Ships close to shore, such as the gunboat *Berouw*, were lifted out of the water and carried 3 kilometers inland to lay stranded 9 meters above sea level. Blocks of coral weighing 600 tons were carried 11 kilometers inland. The waves circled the globe twice, rocking ships in South African harbors 8,000 kilometers away as they passed.

THE LOUDEST NOISE EVER HEARD

Before and immediately after the colossal 10:02 A.M. explosion, continual thunder deafened Java and Sumatra residents. During the later eruptions, the nearby islands, thickly blanketed in ash, heard nothing, even though windows and walls were blown out. Yet on Rodriguez Island 4,800 kilometers to the west, in Australia 3,200 kilometers to the south, and in places in between, such as Celebes, Ceylon, and the Philippines, loud explosions were heard. People on more than one-thirteenth of the Earth's surface, about 40 million square kilometers, heard the noise. At no other time in recorded history has a sound been heard at such a distance without technological amplification. If Krakatau had exploded in Denver, Colorado, residents from Alaska to Florida would have heard the noise. Pressure differences on barographs at a number of widespread locations recorded seven distinct airwaves bouncing back and forth up to five days later.

During nearly twenty-four hours of the eruptions on Sunday and Monday, ash completely darkened the area within 200 kilometers of Krakatau. Ships as distant as 6,000 kilometers to the northwest swept ash from their decks. Clouds of ash, an estimated 8.5 cubic kilometers in volume, spewed 50 kilometers into the stratosphere, where they were transported globally in the jet stream. This volcanic veil of dust caused spectacular atmospheric effects around the Earth lasting for three years. Exceptionally colored and prolonged sunsets visible to three-quarters of the world population became common; the Sun, and sometimes the Moon or Venus, often appeared blue or green; and a white or bluish white corona, called Bishop's Ring, circled the Sun. These finely ground ash particles created more than the gaudy optical displays. The ash became a filter to solar radiation and reduced heat reaching the Earth by 1 percent. Global temperatures, averaging 0.05 degree Celsius lower in 1884, decreased the British growing season by one week. For the next four years, reduced solar radiation and resulting lower-than-average global temperature were recorded.

AFTERMATH

News of Krakatau's disastrous eruption spread quickly. Telegraph-cable carried the news to Singapore and then to the world. Only after the dark cloud of ash over Sumatra and Java had dispersed was the magnitude of the death toll known. Ships returned to Great Britain and the United States with blocks of pumice and ash samples; tidal gauges and barographs recorded the passage of sea and airwaves. Almost around the globe, every "man on the street" witnessed garish sunsets and unusual optical effects. Within three weeks after Krakatau's final blast, a group of scientists commenced detailed studies on the eruption, subsequent tsunamis, and global atmospheric effects.

LOUDEST, NOT BIGGEST

Unequivocally, Krakatau made the loudest noise of any volcano in history. Certainly it was one of the most violent. An eruption of Krakatau's energy and duration occurs about once or twice every one hundred years; yet in the context of global volcanology, other volcanoes have superseded Krakatau in the amount of ash and magma ejected and the size of caldera created. The 1815 eruption of Tambora, also in Indonesia, released about fifty times more energy and produced five times as much pyroclastics (mixed pulverized rock or ash and lava). In Indonesia, blasted by 17 percent of the world's volcanoes, Krakatau's eruption was not unusual.

Elementary plate tectonics, the theory that the Earth's outer crust is composed of moving plates or blocks, provides an explanation for why Indonesia is the most volcanically active region in the world. Indonesia rides on the Indian Ocean plate, which plunges northeastward beneath the buoyant continental Eurasian plate at geologically rapid speeds of up to 6 centimeters per year. A 3,000-kilometer stretch of volcanoes in Sumatra,

Java, the Lesser Sunda Islands, and northward toward the Philippines marks the collision zone, the Java Trench, between these two plates. Volcanoes that lie along this type of contact, called a subduction boundary, are more explosive and produce lavas distinct from volcanoes exploding within a plate or under water where two plates pull apart.

ALL BOTTLED UP

Krakatau resembles Mount Vesuvius in the Mediterranean and Mount St. Helens in the western United States in the properties of the lava emitted, the explosiveness and collapse into a empty magma chamber, and the enormous amount of ash and other ejecta shot into the atmosphere. Krakatau blew out pyroclastics that welded into a type of rock called ignimbrite, rich in silica. An estimated 18 to 21 cubic kilometers of ignimbrite and 8.5 cubic kilometers of ash poured out during the eruptions. Unlike the more easily flowing basaltic lavas of the Hawaiian volcanoes, high-silica lava is thick enough to plug up the vent behind itself and build up pressure in volcanic gases unable to escape.

At the time of the eruption, mining engineer R. D. M. Verbeek, sent to research the volcano, conjectured that the exceptional explosiveness of Krakatau was caused by seawater seeping into a collapsed magma chamber, where it was converted to steam. Later twentieth century studies argue that significant amounts of seawater did not enter the vent, which had not collapsed during the early stage of the eruptions. Scientists suggest that a partially solidified plug developed to block the vent. Pressure built up and eventually exploded into the Monday series of four eruptions, each producing a pyroclastic flow into the surrounding sea.

One significant remnant of such an explosion is the giant caldera formed by the collapse of the volcano into the drained magma chamber within. Krakatau collapsed under water into the Sunda Strait, where a 6-kilometer-diameter caldera edged by three small islands marked its resting place. The flooded caldera was not immediately identified. At first, scientists assumed that the island blew itself apart during the series of explosions. If this were so, then newly formed volcanic rocks should be of the same composition as the preeruptive Krakatau. Yet analysis of the ash and pumice rock indicated that 95 percent of this material formed from cooled lava rather than from pulverized rock. The missing island was, therefore, underwater in the caldera, not deposited as new strata or carried around the globe as ash.

Cory Samia

CROSS-REFERENCES

Atmosphere's Global Circulation, 1823; Calderas, 683; Eruptions, 739; Forecasting Eruptions, 746; Hot Spots, Island Chains, and Intraplate Volcanism, 706; Island Arcs, 712; Lava Flow, 717; Mount Pelée, 756; Mount Pinatubo, 761; Mount St. Helens, 767; Mount Vesuvius, 772; Ocean Ridge System, 670; Plate Motions, 80; Plate Tectonics, 86; Popocatépetl, 776; Pyroclastic Rocks, 1343; Recent Eruptions, 780; Shield Volcanoes, 787; Stratovolcanoes, 792; Tsunamis, 2176; Tsunamis and Earthquakes, 340; Volcanic Hazards, 798.

BIBLIOGRAPHY

Decker, Robert, and Barbara Decker. *Volcanoes.* New York: W. H. Freeman, 1997. This readable and well-illustrated text is geared toward the general public. The authors explain geologic phenomena in a clear and simple manner. Includes a good section on the explosion of Krakatau. The appendices list the world's 101 "most notorious" volcanoes, as well as Internet sources for volcano information.

Francis, Peter. *Volcanoes: A Planetary Perspective.* Oxford: Clarendon Press, 1993. This is a university-level text, but it is written in a style that makes it accessible to a wide range of readers. The author covers the visible evidence of volcanoes, as well as the theory of why they develop. Illustrations, index, and reference list.

Ritchie, David. *The Ring of Fire.* New York: Atheneum, 1981. Author relays many eyewitness accounts of destructive forces, such as volcanoes and earthquakes, that rim the Pacific Ocean. The section on Krakatau summarizes for the lay reader scientific studies of the eruption and aftermath.

Sigurdsson, Haraldur, ed. *Encyclopedia of Volcanoes.* San Diego, Calif.: Academic Press, 2000. This book contains a complete summary of the scientific knowledge of volcanoes. Includes eighty-two well-illustrated overview articles, each of which is accompanied by a glossary of key terms. Although this is a college-level text, it is written in a clear and comprehensive style that makes it generally accessible. Cross-references and index.

Simkin, Tom, Richard S. Fiske, and Sarah Melcher, eds. *Krakatau 1883: The Volcanic Eruption and Its Effects.* Washington, D.C.: Smithsonian Institution Press, 1983. The definitive work, between 1883 and 1983, on the volcano. The Smithsonian Institution and the U.S. Geological Survey collected and reinterpreted information on Krakatau for the one hundredth anniversary of the eruption. Most articles are condensed versions of the original studies.

Time-Life Books. *Volcano.* Alexandria, Va.: Author, 1982. Like other volumes in the Planet Earth series, geared to the general Earth science reader, this issue is extremely readable and enhanced by excellent graphics. The eruption of Krakatau is viewed within the framework of contemporary plate tectonics theory.

Waters, Tom. "Return to Krakatau." *Discover* 9 (October, 1988): 64-67. A short article that discusses the contemporary work of ecologists studying the recolonization of Krakatau by plants and animals.

MOUNT PELÉE

Mount Pelée, located on the island of Martinique in the West Indies, gained its notoriety on May 8, 1902, when all but two of about thirty thousand persons in the town of St. Pierre were suddenly killed by the eruption of a hot cloud of ash, rock fragments, and gases called a nuée ardente—the first time such an eruption was recognized.

PRINCIPAL TERMS

ANDESITE: a volcanic rock of intermediate silica content that contains plagioclase (a calcium, sodium, and aluminum-silicate mineral) and often hornblende (a calcium, iron, magnesium, and aluminum-silicate mineral) and biotite (a potassium, iron, magnesium, and aluminum-silicate mineral)

DACITE: a volcanic rock of fairly silica-rich composition with plagioclase, alkali feldspar (a potassium, sodium, and aluminum-silicate mineral), and quartz (a silica mineral)

MAGMA: a naturally occurring liquid that is usually composed of silicate material and contains suspended minerals or rocks

NUÉE ARDENTE: a hot cloud of rock fragments, ash, and gases that suddenly and explosively erupts from some volcanoes and flows rapidly down the volcano's slope

SUBDUCTION: the theory of plate tectonics that assumes that the Earth's surface is divided into a number of large plates that either are colliding or are being pulled apart

VISCOSITY: how readily liquids flow; silica-rich magmas flow less readily than do silica-poor magmas

VOLCANO: a vent at the Earth's surface in which gases, rocks, and magma erupt at the surface and build a more or less cone-shaped mountain

NUÉES ARDENTES

Mount Pelée, on the island of Martinique, is one of many volcanoes located along the sinuous arc of islands of the Lesser Antilles in the West Indies. One oceanic plate is being thrust or subducted underneath another oceanic plate along the Lesser Antilles, resulting in the generation of silica-rich magma such as andesite and dacite. The silica-rich magma is very "stiff," or viscous. Therefore, it will not readily flow up the conduit of the volcano, which may result in a plug of magma in the volcano crater that prevents the escape of magma and gas. The gas pressure inside the volcano may thus build until it explodes at its weakest point.

The explosions of this type often blast horizontally out below the volcanic plug, resulting in an exceedingly hot cloud of glowing ash (fine dust), rock fragments, and gases called a nuée ardente that cascades rapidly downslope. Initial temperatures may be approximately 1,200 degrees Celsius, but the temperature may rapidly drop off as the nuée ardente moves downslope. The rapid expansion of gases after eruption apparently produces much of the cooling, but the nuée ardente of May 8, 1902, at Mount Pelée was still hot enough to soften glass (about 700 degrees Celsius) 8 kilometers from the volcano. The speeds of nuées ardentes average 100 kilometers per hour. Lava does not extrude during a nuée ardente eruption.

This type of eruption was first recognized on May 8, 1902, when Mount Pelée erupted, instantly killing all but two of nearly thirty thousand people in the town of St. Pierre. The nuée ardente eruption is therefore often called a Peléan type of eruption. The journalists and scientists who descended on the area shortly after the eruption left many fascinating eyewitness accounts.

WARNING SIGNS

The many signs of unusual volcanic activity that occurred prior to May 8 are now recognized as warnings of an imminent eruption. The emission of steaming gases was first noticed on April 2 in the upper portions of a stream draining off of Mount Pelée. There were minor earthquakes and an odor of sulfur, and ash was noticed in St. Pierre on April 23. Explosions in the cone of the volcano threw ash and rocks into the air on April 25. People who climbed up to the crater on April 27 said that it contained a small cinder cone, a lake, and columns of steam that were not previously noticed. The fine particles of dust, called ash, thrown out of the volcano during loud explosions gradually became heavier and eventually blocked some roads and forced residents indoors. Many residents left St. Pierre in fear; many from the countryside, however, migrated to the city. As a result, the population of the city was much larger than normal by early May. Authorities issued assurances that residents were in no danger and that they should not panic.

Volcanic Weather

Often, the hot updraft around active volcanoes produces thunderstorms. This often results in a large amount of water suddenly mixing with the abundant volcanic ash. Heavy rains occurred on the southwest slope of Mount Pelée on May 5, which resulted in water that was ponded in one of the craters bursting through the crater walls and forming a scalding torrent that rushed down the river valley at speeds of up to 90 kilometers per hour. As the mudflow, carrying huge trees and

boulders, hit the ocean, it produced a large wave that overturned boats and washed up on the low-lying areas of St. Pierre. Residents of the lowlands panicked and fled to higher areas, which caused more of an exodus from St. Pierre. Explosions on May 6 were quite loud, and the authorities had to block the roads to keep people from leaving the city. The newspaper and governor issued assurances that there was no cause for alarm. By May 7, there were continuous downpours and clouds of ash that produced muddy floods in the swollen streams.

Violent Blasts, Sudden Death

Then, on May 8, the volcano erupted with four loud blasts. One black cloud full of lightning rose upward from the crater. A nuée ardente moved rapidly down the volcano, over the town of St. Pierre, and out over the ocean, killing nearly everyone in its path within two minutes. People died suddenly either from burns or from breathing the heated gas and ash. Houses lost their roofs, and walls 1 meter thick were ripped apart. Most of the ships in the harbor were destroyed. The two ships that were not capsized had a few survivors who were in inner cabins with the doors closed. There were two survivors in St. Pierre, including a prisoner in a windowless cell. A small grate in the door had allowed enough heat into the cell to burn parts of his body. The second survivor was located in an inner room of a house in which all other occupants died. Much of the city had burned after the initial blast; the city itself resembled an ancient ruin. Foundations of houses were left standing mixed with mangled bits of metal and furnishings. Everything was covered with a mantle of ash nearly 1 meter thick. A few tree trunks remained, devoid of their bark and scorched by the heat.

A view from Morne d'Orange of St. Pierre, on Martinique, destroyed after the 1902 eruption of Mount Pelée. Pelée can be see in the distance. (U.S. Geological Survey)

Continuing Eruptions

Mount Pelée continued to erupt more or less continuously for several hundred days.

Nuées ardentes were observed on May 20, June 6, July 9, August 30, and December 6. The nuée ardente of August 30 moved down a different slope and killed several thousand people. The amount of ash given off in these continuous ash eruptions was prodigious. One conservative estimate of the quantity of ash given off in twenty-four hours was that it was about equal to one and one-half times that of the sediment carried by the Mississippi River in one year. Brilliant afterglows and modified sunsets were observed for several years after the eruptions. Deflections of the Earth's magnetic field related to a volcanic eruption were observed for the first time.

A giant obelisk, or spine, of stiff lava gradually projected upward from the crater of Mount Pelée. This spine first appeared in photos taken in June, 1902, and it pushed upward out of the summit for more than one year before it broke up and shattered. The obelisk reached its maximum height of about 330 meters above the crater floor and about 160 meters across at the base in June and July of 1903. It continued to crack and break at the top as it continued to grow in irregular spurts at rates of up to 17 meters per day. The obelisk was reduced to a height of approximately 160 meters by August, 1903. The continued eruptions from August to November reduced it to a pile of rubble.

Mount Pelée was quiet until September, 1929.

Debris on Morne d'Orange, Martinique, after the 1902 eruption of Mount Pelée. (U.S. Geological Survey)

Then it erupted in a pattern similar to that of the earlier eruptions. Fortunately, the one thousand people living in St. Pierre were evacuated at the first signs of volcanic activity, so there was no loss of life. During this period of activity, there were a number of nuées ardentes during the early stages of eruption; none of them, however, was as violent as those that occurred in 1902. As the volcanic activity waned, a new spine grew in the crater near the old spine and gradually crumbled. The new spine never reached the height of the 1903 spine.

SIMILAR ERUPTIONS

The volcano La Soufrière exists on St. Vincent Island, 150 kilometers from Martinique. It also erupts a silica-rich magma and has a pattern of eruption similar to that of Mount Pelée. La Soufrière erupted in 1718, in 1812, only one day before Mount Pelée in 1902, and in 1971. The eruption of La Soufrière on May 7, 1902, had a much smaller and more vertical initial blast than that at Pelée because the blast from La Soufrière occurred through an open rather than a blocked vent. The resultant cloud, consisting of sand-sized particles, descended down the slope induced only by gravity; the speed of the flow (about 40 kilometers per hour) was not nearly as great as at Pelée. The initial destruction of buildings in this area was much less than that caused by Pelée. Fewer buildings caught on fire than at Pelée because the cloud was much cooler from its longer contact with the atmosphere. The area covered by the cloud was much larger at La Soufrière than at Pelée because of the vertical nature of the eruption. The people on one side of the mountain in the town of Soufrière noticed that the volcano was erupting, and most evacuated. Clouds covered the other side of the volcano; the people there did not evacuate in time. Approximately two thousand people lost their lives. This variation of a nuée ardente is sometimes called the Soufrière type of eruption.

Another nuée ardente occurred at the Merapi volcano in

Java. Here, a dome or spine built up much like that at Pelée, except that the dome extended out beyond the rim of the crater. The spine would periodically break off, sending a nuée ardente down the side of the volcano. There nuées ardentes moved quickly, like those at Pelée. This variation of a nuée ardente is also known as the Merapi type of eruption.

STUDY OF MOUNT PELÉE

Information about the nature of volcanic eruptions such as Mount Pelée's comes from two types of observation. One is the visual observation of the eruption while it is occurring. The second is the examination of the deposits of erosional features that result from the eruption. These two types of observation were made during and after the eruption of Mount Pelée.

The stiff spine that protruded from the volcanic vent at Mount Pelée has been seen at other Hlocations that also have silica-rich and viscous magma. It is believed that the viscous magma plugs the vent of the volcano in these types of eruption. The gas pressure builds until gas-rich magma is suddenly released at a low angle. During the sudden decompression, the magma froth—broken up into ash, fine dust, or large rock fragments—discharges like a cannon. Observers report deafening explosions during these eruptions.

In a 1930 eruption of Pelée, one trained observer, standing only 35 meters away from the eruption, confirmed that a lateral explosion did occur. In lateral explosions, the solid particles and magma are believed to release water vapor rapidly and continuously. In this way, the magma freezes, so little or no magma is left after the eruption's initial release of pressure. Much of the flow's initial speed is a result of this lateral blast. The lower portion of the gas-particle mixture takes on speed and moves rapidly and fluidly down the low portions of valleys on the cooler cushion of air beneath it.

Thus, nuées ardentes behave much like running water or avalanches, except that they move faster. The continual release of hot water vapor from the particles may cause rapid expansion and compression of the gas-particle fluid and add to its downslope speed. Some of the dust-sized particles rapidly escape upward from this moving gas-particle fluid and rise in a glowing, black cloud. These glowing avalanches erode gullies along the steep flanks of the volcano and scratch rocks parallel to their flow. Little material is deposited until the avalanches reach the lower slopes of the volcano and slow down. The deposits usually consist of a random mixture of various-sized particles—from fine ash to blocks several meters wide. Apparently, there is enough turbulence to keep the particles stirred up. Some deposits, however, have coarser material concentrated at the bottom, with the finer material near the top. A thin layer of dust sits atop the main deposit as this material slowly settles from the overlying ash cloud.

Geologists who study volcanoes observe the eruption processes and magma compositions as they are ejected at the surface. One volcanologist, Frank A. Perret, set up an observation shed on the slopes of Mount Pelée during the 1929 eruptions near the area where a number of nuées ardentes descended down the slopes. The edge of one of these flows hit his shed and nearly killed him. He survived to publish his observations and conclusions on the flow mechanisms of nuées ardentes. His theories provided the basis for the first real understanding of these processes.

Robert L. Cullers

CROSS-REFERENCES

Andesitic Rocks, 1263; Eruptions, 739; Forecasting Eruptions, 746; Krakatau, 752; Magmas, 1326; Mount Pinatubo, 761; Mount St. Helens, 767; Mount Vesuvius, 772; Phase Changes, 436; Plate Margins, 73; Popocatépetl, 776; Pyroclastic Rocks, 1343; Recent Eruptions, 780; Shield Volcanoes, 787; Stratovolcanoes, 792; Volcanic Hazards, 798.

BIBLIOGRAPHY

Bullard, Fred M. *Volcanoes of the Earth.* Austin: University of Texas Press, 1976. A very readable and nontechnical discussion of the location and the origin of volcanoes of the world, suitable for the layperson. Includes a long section on nuées ardentes. Contains many pictures and illustrations. A glossary defines unfamiliar terms.

Decker, Robert, and Barbara Decker. *Volcanoes.* New York: W. H. Freeman, 1997. This readable and well-illustrated text is geared toward the general public. The authors explain geologic phenomena in a clear and simple manner. Includes a good section on Mount Pelée. The appendices list the world's 101 "most notorious" volcanoes, as well as Internet sources for volcano information.

Francis, Peter. *Volcanoes: A Planetary Perspective.* Oxford: Clarendon Press, 1993. This is a university-level text, but it is written in a style that makes it accessible to a wide range of readers. The author covers the visible evidence of volcanoes, as well as the theory of why they develop. Illustrations, index, and reference list.

Heilprin, Angelo. *Mount Pelée and the Tragedy of Martinique.* Philadelphia: J. B. Lippincott, 1903. A readable account of the eruption of Mount Pelée by a trained observer who arrived shortly after its eruption of May 8, 1902. Many photographs of the destruction, the spine, and one of the subsequent nuées ardentes. Can be easily read by the layperson.

Hill, Robert C. "On the Volcanic Disturbance in the West Indies." *National Geographic* 13 (July, 1902): 223-266. One of the best-illustrated summaries of the eruption and destruction at Pelée, written in lay terms.

Kennan, George. "The Tragedy of Pelée." *Outlook* 71 (June 28-August 16, 1902): 769-777. Another eyewitness account of the eruption and destruction at Pelée. Suitable for the layperson.

Perret, Frank A. *Eruption of Mount Pelée, 1929-1932.* Washington, D.C.: Carnegie Institution of Washington, 1935. An eyewitness account of the smaller eruptions of Pelée during 1929-1932 by an electrical engineer who was a volcanologist for many years. He explained the origin of nuées ardentes by observing them as close as 30 meters. Suitable for the layperson. No glossary.

Russell, Israel C. "The Recent Volcanic Eruptions in the West Indies." *National Geographic* 13 (July, 1902): 267-284. An illustrated description of the eruptions on Martinique and the St. Vincent Islands. Suitable for the lay reader.

Sigurdsson, Haraldur, ed. *Encyclopedia of Volcanoes.* San Diego, Calif.: Academic Press, 2000. This book contains a complete summary of the scientific knowledge of volcanoes. It contains eighty-two well-illustrated overview articles, each of which is accompanied by a glossary of key terms. Although this is a college-level text, it is written in a clear and comprehensive style that makes it generally accessible. Cross-references and index.

MOUNT PINATUBO

The eruption of Mount Pinatubo in June, 1991, was one of the most violent eruptions of the twentieth century. The use of modern techniques of eruption prediction and danger assessment allowed the safe evacuation of nearly 150,000 people from the area adversely affected by the eruption.

PRINCIPAL TERMS

ASH: small fragments of volcanic material less than two millimeters in diameter formed by explosive ejection from a vent

CALDERA: a large circular depression around a summit vent that typically forms by collapse when large volumes of magma are rapidly ejected

DOME: a small, steep-sided mass of volcanic rock formed from nonexplosive, viscous lava that solidifies in or above a vent

FUMAROLE: a volcanic vent from which only gases are emitted

HARMONIC TREMOR: a type of earthquake activity in which the ground undergoes continuous shaking in response to subsurface movement of magma

LAHAR: a volcanic mudflow resulting from the mixing of erupted lava and ash with surface water, rain, or melted snow

PYROCLASTIC FLOW: a turbulent, dense mixture of volcanic gases, magma, ash, and rock that is ejected from a vent and flows rapidly down the flanks of a volcano

STRATOVOLCANO: a large, steep-sided volcano consisting of alternating layers of coherent lava and explosively ejected fragmental material; also called a "composite volcano"

DORMANT SINCE 1541

Mount Pinatubo is one of the nearly three hundred active volcanoes that rim the Pacific Ocean Basin in a narrow belt called the Ring of Fire. This stratovolcano had been dormant since the Philippines were first settled by the Spanish in 1541. The awakening of Mount Pinatubo was of considerable concern to the United States because Clark Air Base was located only 20 kilometers to the east of the summit, and the Subic Bay Naval Station was 40 kilometers to the southwest.

Pinatubo rumbled into activity on April 2, 1991, when a series of small explosions formed a row of craters 1.5 kilometers long just northwest of the volcano's summit. The explosions lasted several hours and deforested an area of several square kilometers. Villagers living on the volcano's flank reported hearing the explosions and awoke the morning of April 3 to find a row of fumaroles extending along the western side of Pinatubo. Civil officials evacuated a circular area in a 10-kilometer radius from the volcano's summit.

SEISMIC WAKE-UP

During all of April and the early part of May, Mount Pinatubo experienced dozens of small earthquakes each day as magma began to move beneath the volcano. Magma fractured the brittle rocks as it cleared a pathway upward to form a shallow magma body. The center of the earthquake activity was about 5 kilometers northwest of the summit and 4 kilometers below the surface.

During May, the earthquake activity increased, with more than 1,800 low-intensity quakes being recorded between May 7 and June 1. During the last two weeks of May, the fumaroles showed a tenfold increase in the emission of sulfur dioxide.

At the beginning of June, a second area began to experience a series of shallow earthquakes several kilometers closer to the summit. The magma was rising beneath the volcano, and it was forcing a conduit to form above the magma chamber. The rate of emission of gas reversed its May trend, decreasing from a production of 5,000 tons of sulfur dioxide on May 28 to a production of 1,800 tons on May 30; by June 5, the total had dropped to only 260 tons. Apparently, the rising magma had sealed the fractures that allowed the escaping gases to reach the fumaroles.

HARMONIC TREMORS

On June 3, a series of small explosions blew ash from the summit area, and the earthquakes moved into a type of activity called "harmonic tremor": a prolonged rhythmic shaking of the Earth, in contrast to the single sharp jolts of earlier earthquakes. Harmonic tremor occurs when underground magma flows in a more continuous manner through established subsurface channels. On June 6, the summit region of the volcano began to bulge outward, with the upper flank showing a measurable increase in the slope of the land surface. As the tilting increased, there was an increase in earthquake activity, leading to a strong shock and explosion on June 7. This explosion sent a column of ash and steam to a height of between 7 and 8 kilometers. The explosion allowed the summit area to stop its inflation temporarily, and there was a brief period of reduced earthquake activity. Based upon these warning signs, the area of evacuation was extended to a radius of 15 kilometers.

Magma first appeared on the surface of the volcano the next morning, June 8. Observers reported that a small lava dome 150 meters in diameter was growing near one of the fumaroles just northwest of the summit. The vents associated with this dome emitted a series of weak ash clouds that rose only to the level of Pinatubo's summit over a three-day period ending on June 12. During this interval, harmonic tremors occurred almost continuously. An examination of the composition of ejected ash revealed that a new, very fluid magma had invaded an old magma chamber that contained residual magma from the last eruption nearly five hundred years ago. This intruding magma signaled the potential of a major eruption, and authorities extended the evacuation area to a radius of 20 kilometers from the volcano. On June 10, fourteen thousand U.S. servicemen left Clark Air Base along with their aircraft, never to return. They left behind three helicopters and a contingent of fifteen hundred security and maintenance personnel.

MULTIPLE ERUPTIONS

The first major explosion of Pinatubo occurred at 8:51 A.M. on June 12, signaling the final and most powerful phase of the 1991 eruption. This very violent phase lasted about ten days, with the most intense activity occurring on June 15 and 16. The June 12 eruption lasted thirty-five minutes and spewed a column of steam and ash to a height of 19 kilometers. A small pyroclastic flow (a turbulent, dense mixture of volcanic gases, magma, ash, and rock that is ejected from a vent and flows rapidly down the flanks of a volcano) traveled a short distance from the vent down the northern flank of the volcano into already evacuated villages along the Maraunot River. Clark Air Base evacuated six hundred maintenance personnel, and Filipino authorities extended the evacuation zone to a 30-kilometer radius. Ash from the eruption was so dense and spread so far that the airport in Manila, more than 50 kilometers away, was forced to close.

Aerial view toward the southern end of Mount Pinatubo, in the Philippines. White plumes rise after phreatic eruptions from fumaroles on the crater floor. (U.S. Geological Survey)

Similar explosions continued over the next few days, producing ash clouds that prevented visibility of the volcano. The arrival of Typhoon Yunya on June 13 meant that most of these explosions were completely unseen from the ground. The associated earthquake activity was recorded by seismic stations, and the presence of an eruption cloud was verified by military weather radar.

Beginning on June 14, the main, violent phase peaked in a three-day period with more than fifty short, violent explosions in which numerous vents erupted simultaneously. The intensity of the earthquakes generated by these eruptions was more than a hundred times greater than those recorded during May. These eruptions grew progressively more violent during this time, sending ash to a height of more than 40 kilometers. At 5:55 A.M. on June 15, the ejected ash switched from a vertical orientation to a more horizontal one. The onset of laterally directed explosions indicated that the summit region was beginning to collapse. Several pyroclastic flows erupted and traveled up to 13 kilometers from the summit. The remaining personnel at Clark Air Base were evacuated that morning.

At about 3:30 P.M. on June 15, the climax occurred. A series of strong earthquakes started, lasting all afternoon and throughout the night. The seismographic equipment on the volcano was destroyed during the night by a series of large pyroclastic flows. Pyroclastic deposits that exceeded 200 meters in thickness formed in many valleys around the volcano. A 2-kilometer caldera formed when the summit area collapsed, causing the volcano to drop by more than 300 meters in elevation.

ASHFALL, MUDFLOW, AND POISONOUS GAS

About 0.8 cubic kilometer of ejected material covered the west-central portion of the island of Luzon. The falling ash blanketed villages and buried crops around the volcano. Many rooftops collapsed under the added weight of the ash, causing the majority of deaths attributed to the volcanic eruption. The total volume of ash ejected was close to 5 cubic kilometers, most of which fell into the South China Sea.

The water runoff from the heavy rains associated with the typhoon mixed with the recently fallen ash and caused disastrous volcanic mudflows known as "lahars." The lahars swept down twelve river valleys, including the Abacan and Sacobia River Valleys, where the town of Angeles City was completely destroyed. Seven other towns were damaged by these mudflows. The July monsoons caused secondary lahars, one of which inundated the city of Pabanlog along the Gumain River. All these lahars destroyed bridges and farmland along the floodplains and caused widespread economic and social disruption. Special early-alert systems detected the lahars and sharply reduced the loss of life. However, the region continued to experience secondary lahars caused by heavy rains, and this threat was expected to last well into the twenty-first century.

The cloud associated with the June 15 eruption released four billion pounds of chlorine gas and forty billion pounds of sulfur dioxide into the stratosphere. By June 25, satellite images showed that a 7,750-kilometer-long cloud of sulfur dioxide had spread across the tropical Northern Hemisphere.

DAMAGES AND DEATH TOLL

After June 16, the eruption slowly grew less intense. There were occasional ash explosions into 1992 that sent ash columns upward of 10 kilometers in height. During July and August of 1991, a dome grew in the fuming caldera that was 300 meters across and nearly 100 meters high. With an average of more than twenty centimeters of ash covering the region around Clark Air Base, base personnel were unable to stop ash infiltration into the jet engines, and the base had to be abandoned.

More than 108,000 homes were partially or totally destroyed. The final death toll for the region was 722; of these, 281 died as a result of ejected material either as ashfalls or pyroclastic flows, 83 died from primary and secondary lahars, and 358 died from disease related to the social turmoil associated with the interruption of the country's infrastructure. The loss of life was undoubtedly compounded by the typhoon that hit the island during the climax of the eruption.

GLOBAL IMPACT

The world's largest volcanic eruptions have all produced global climate and atmospheric changes. Fine volcanic ash and gases are blasted into the high atmosphere and dispersed around the world.

Upper-level winds can keep volcanic ash suspended in the atmosphere for many years. The suspended ash from Mount Pinatubo's eruption produced colorful sunsets worldwide in 1991 and 1992.

The gases and ash combined to produce an aerosol that blocked both incoming sunlight and infrared radiation emitted by the Earth. The loss of solar radiation causes cooling, whereas the absorption of the Earth's transmitted infrared radiation leads to global warming. The two effects did not balance equally, and by 1993, measurements from the National Aeronautics and Space Administration's Earth Radiation Budget Satellite provided the first conclusive evidence of a significant change in global energy as the result of a volcanic eruption. The net effect of a loss of solar radiation and a greater retention of infrared radiation resulted in a period of global cooling in which the average global temperature dropped by one-half degree Celsius (about one degree Fahrenheit). Approximately 2 to 3 percent of the Sun's energy was blocked out, counteracting prevailing global-warming trends and temporarily setting the Earth's climatological clock back to the 1950's.

The gases also altered the chemistry of the upper atmosphere. Three months after the eruption, there was 50 percent less ozone in the tropical stratosphere over an area roughly coincident with Mount Pinatubo's volcanic plume. The ozone layer over the United States was 10 percent thinner than normal, translating to a 20 to 30 percent increase in the amount of cancer-causing ultraviolet radiation reaching the Earth's surface. There was fear that an ozone hole might open over the populated areas of the Northern Hemisphere; by 1996, however, measurements of the levels of ozone-depleting chemicals had dropped to lower-than-average values.

Assessing the Reponse

The successful evacuation of nearly 150,000 people during the Mount Pinatubo eruption required both successful short-term forecasting of eruption events and a well-organized program of danger assessment. The activities that transpired during the ten weeks from the first steam explosion on April 2 to the culminating eruption of June 15 were a model of cooperation between scientific personnel, Filipino authorities, and the local citizenry. Immediately following the first gas

explosion, volcanologists from the U.S. Geological Survey joined forces with other scientists from the Philippine Institute of Volcanology and Seismology to coordinate the scientific work of eruption prediction. They jointly established the Pinatubo Volcano Observatory (PVO).

Modern eruption prediction involves determining the nature of the volcano's past eruptions and monitoring the active volcano for any changes in its physical and chemical behavior. An array of scientific equipment must be deployed around the volcano and the active vents to measure earthquake activity, volumes and composition of emitted gases, changes in the slope of the land surface, changes in the horizontal distance across vents and fissures, and fluctuations in the temperature of the volcano.

Active volcanoes have sporadic eruptions that are separated by time intervals called "periods of repose." Most active volcanoes display a repetitive history in terms of their repose interval, their eruptive violence, and the kinds of materials that they expel. The nature of a future eruption can often be established by an examination of the geological record of previous eruptions.

The PVO team realized the value of knowing the previous eruptive history at Mount Pinatubo and did a rapid geological reconnaissance of the volcano during the month following the initial steam explosion. They found that three previous eruptions had occurred at about 500 years ago, 2,500 years ago, and 4,800 years ago. Each eruption had been dominated by highly explosive activity. Pyroclastic flows had swept down the volcano's flanks, leaving deposits more than 20 kilometers from the summit. Debris from lahars from these prehistoric eruptions were found in six river valleys leading away from the volcano for a distance of more than 40 kilometers.

On the basis of their geological study of past eruptions, the PVO team compiled a hazard map showing the regions that were likely to be affected by pyroclastic flows, ashfall, and lahars. The map was used by both civil defense officials and military commanders during the various stages of the eruption. A map of the areas affected by the actual 1991 eruption shows a remarkable correlation with the pre-eruption hazard map. A videotape depicting the various volcanic hazards was produced by filmmaker and geologist Maurice

Krafft, and the film aired repeatedly on local television.

Danger assessment is a complex task that goes beyond the accurate forecasting of eruptions. For example, in 1985, the Nevado del Ruiz volcano erupted in a region of Colombia that was much less populated than the area around Mount Pinatubo. The eruption generated a mudflow that killed twenty-five thousand people. Although volcanologists were successful in predicting the eruption and recommended evacuation of the region, the civil authorities did not view the impending eruption to be particularly dangerous, and no evacuation occurred.

The PVO team worked closely with civil and military personnel to assess the level of danger posed by the various stages of the Mount Pinatubo eruption. Using the hazard map and the video, the Filipino scientists developed a five-level alert system. Alert level 1 was to be declared when low-level seismic activity was coupled with fumarolic emission. The level 1 alert occurred on April 3, when fumaroles developed in a row along the northwest flank of Mount Pinatubo. Alert level 2 was to be invoked when moderate levels of seismic activity occurred and there was positive evidence for the existence of subsurface magma. Level 2 was reached in late May, when 1,800 earthquakes were recorded within a three-week interval.

Authorities were instructed to interpret alert level 3 to mean that the magma was intruding into the volcano and could eventually lead to a major eruption. Level 3 was to be issued when high levels of gas emission occurred with simultaneous ground deformation. This level of danger was attained on June 5. Level 3 meant that a major pyroclastic eruption could occur within two weeks; it actually occurred ten days later. Alert level 4 was to be issued when extensive harmonic tremors indicated the magma was moving more freely beneath the volcano. Level 4 was announced immediately following the June 7 explosion at Pinatubo that generated the 7-kilometer-high ash column. Civil authorities were told that level 4 meant that a major pyroclastic eruption was possible within twenty-four hours. The highest level of danger assessment was alert level 5, which meant that an eruption was in progress. This level was issued on June 9, after a pyroclastic flow engulfed several evacuated villages on the northern flank of Pina-

tubo. The day after the level 5 alert was issued, the U.S. military evacuated fourteen thousand personnel from Clark Air Base.

LESSONS LEARNED

The violent eruption of Mount Pinatubo was classified as a magnitude 6. The largest recorded eruption has been classified as magnitude 7, and the largest geologically known eruption has been ranked as a magnitude 8. Such violent eruptions as these have historically caused high death tolls. However, as shown by Mount Pinatubo experience, the most dangerous volcanoes can cause far fewer deaths when proper surveillance is employed.

There are nearly three hundred active volcanoes that should be monitored for the eruption of pyroclastic flows and lahars, and many of these volcanoes are located in underdeveloped countries. Surveillance and risk-reduction programs are very expensive, requiring equipment, trained scientists, education programs for local populations, and development of cooperative military and civil planning. The expenses associated with such programs are beyond the capabilities of many less-developed countries. Consequently, the United States, Russia, Iceland, Italy, and Japan lead the world in this field of volcanology.

The United States has two facilities dedicated to monitoring active volcanoes, both staffed by geologists from the U.S. Geological Survey. The Hawaiian Volcano Observatory has been successfully predicting eruptions of Mount Kilauea for more than forty years. After the eruption of Mount St. Helens in 1980, the Cascades Volcano Observatory was established to monitor magma amounts and movements beneath the active volcanoes in Washington, Oregon, and northern California. Japan has twenty-one volcano observatories.

It is very likely that Mount Pinatubo will erupt again, although the volcano has entered a period of repose that may last as long as a few hundred years. The volcano is located on the margin of two converging plates that have produced a very active chain of volcanoes. For example, Mount Mayon lies 300 kilometers to the south of Pinatubo and has had more than forty historical eruptions, with the last as recent as 1993. A key factor indicating continued life for Mount Pinatubo is that a 1996 study found that between 40 and 100 cubic kilometers of magma remained in a reservoir below the volcano's summit. This volume is the largest quantity of magma (by almost a factor of three) ever detected beneath any volcano.

Dion C. Stewart

CROSS-REFERENCES

Eruptions, 739; Forecasting Eruptions, 746; Krakatau, 752; Mount Pelée, 756; Mount St. Helens, 767; Mount Vesuvius, 772; Popocatépetl, 776; Recent Eruptions, 780; Shield Volcanoes, 787; Stratovolcanoes, 792; Volcanic Hazards, 798.

BIBLIOGRAPHY

Bullard, F. M. *Volcanoes of the Earth*. Austin: University of Texas Press, 1976. A good scientific treatment of volcanoes and volcanic products. Well written; does not require a geological background. The author, an experienced volcanologist, relies heavily on his own experiences, photographs, and drawings.

Francis, Peter. *Volcanoes: A Planetary Perspective*. New York: Oxford University Press, 1993. A thorough coverage written for the general reader. The book includes forty pages on pyroclastic flows and twenty-five pages on lahars. Well illustrated; gives numerous historic examples.

Robinson, Andrew. *Earth Shock, Hurricanes, Volcanoes, Earthquakes, Tornadoes, and Other Forces*. New York: Thames & Hudson, 1993. Chapter 4 is devoted to volcanoes and gives a humanistic accounting of many historic eruptions, interwoven with excellent scientific commentary.

Wolfe, Edward W. "The 1991 Eruptions of Mount Pinatubo, Philippines." *Earthquakes and Volcanoes* 23, no. 1 (1992): 5-35. The public report of the successful emergency response of the United States to the geologic danger presented by the 1991 eruption of Mount Pinatubo.

MOUNT ST. HELENS

The eruption of Mount St. Helens in 1980 and the series of geologic and volcanic events that followed provided a unique opportunity for geologists to study a classic volcanic disturbance in detail using modern technological tools.

PRINCIPAL TERMS

CALDERA: the sunken area at the summit of a volcano caused by the internal collapse of magma

FISSURE: an extensive crack or fracture; a linear volcanic vent

MAGMA: molten rock material formed deep in the Earth's interior; when thrown out of a volcano, it is known as lava

PYROCLASTIC MATERIAL: rocks formed from the debris of explosive volcanic eruptions and fragments from the walls of the vents

SUBDUCTION: a process that occurs when one tectonic plate dives beneath another tectonic

plate; it may occur over thousands or millions of years

TECTONICS: the study of the processes that formed the structural features of the Earth's crust, especially the creation and movement of immense crustal plates

TUNNEL VENT: the central tube in a volcanic structure through which material from the Earth's interior travels

VENT: a break or tear on the side of a mountain through which magma and pressure can escape

WARNING SIGNS

In the weeks leading up to May 18, 1980, Mount St. Helens exhibited warnings of an impending eruption, including internal rumbles and small bursts. A large bulge had developed over 98 meters on the north face of the mountain, indicating a highly pressurized buildup of old, trapped magma just below the mountain's surface. As days passed, the bulge grew larger while the tiny eruptions at the summit continued.

What followed, however, could not have been predicted. First, deep internal disturbances forced magma (molten rock) up to the surface of the Earth through the volcano's central, or tunnel, vent. As the magma spread upward, it exerted a large amount of pressure on the sides of the central vent and, as the pressure increased, caused the bulge of old magma to push even farther outward. Then, when the pressure became stronger than the mountain's ability to contain it, the magma blasted through two enormous holes: one at the summit and one on the north flank. The release of energy resulted in a huge explosion.

A MOUNTAIN BEHEADS ITSELF

At 8:32 on the morning of May 18, 1980, an earthquake in the Mount St. Helens region of the

Washington State Cascade Range started a scenario of startling events, beginning with the sudden and spectacular eruption of Mount St. Helens and culminating in the destruction of hundreds of square kilometers of valuable timberland, vacation resorts, and foraging area for wildlife. Considered by many scientists to be a classic series of volcanic activities, the eruption blasted away the top 396 meters (one-third) of the mountain in a cataclysmic explosion that exuded 360 billion kilograms of rock and ash. Forests more than 27 kilometers away were completely leveled; pyroclastic debris (steam, gas-filled rock, and wet ash), moving at the amazing speed of 321 kilometers per hour, flew horizontally across the expanse of the Cascade Range together with suffocating volcanic gases, eventually taking the lives of sixty-one people and an estimated 2 million birds, fish, and animals; twenty-six lakes and 241 kilometers of freshwater streams were destroyed. In the end, the scene in the Cascades was one of utter desolation for some 300 square kilometers.

The loosed fury of the initial blast, comparable to a 10-megaton bomb, immediately tore away one-third of the entire mountain and was followed by a huge landslide. Hot gases, steam, and molten rock shot laterally from the mountain's flank at nearly the velocity of a rifle bullet. Explosion after explosion shook the volcano as debris poured down the mountainside, carrying hot lava, timber, and wet ash. Volcanic ash in heavy, dense clouds spewed out of the two gigantic fissures. Driven by mighty pressures from the Earth's depths, the clouds of ash funneled outward, reaching beyond 33 kilometers in less than two minutes and beginning to spread over the northwestern region of the United States. Mount St. Helens's snow-capped summit quickly turned to water and steam under the heat and pressure of the eruption, releasing 174 billion liters of water down the mountainside. The flood swept mud, solidified lava, tree splinters, pulverized rock, and other material in its path, creating an avalanche that clogged streams and bulldozed forestation as it continued its outward spread. Attacked by volcanic gases, fumes, ash, and intense heat from the bowels of Mount St. Helens, the atmosphere immediately surrounding the erupting volcano became a cauldron of earsplitting thunder and crackling lightning.

There was another eruption on May 25, 1980, spreading an additional ash deposit, up to 70 millimeters thick, for hundreds of square kilometers, mostly westward. Then, in October of the same year, Mount St. Helens erupted once more, this time sending yet more clouds of ash southward. The floor of the caldera grew to more than 260 meters in diameter, with a dome 49 meters high.

ASHES, ASHES

The picturesque valleys around the mountain were smothered with debris. In areas farther from the mountain, gray ash fell like a huge rainstorm, covering the ground with a blanket 1.5 to 2 meters thick. Even four days later, the ash was extremely hot, registering temperatures as high as 323 degrees Celsius. Rescuers had to move swiftly through the deep ash to keep their legs from burning. People in the town of Yakima reported removing 540 million kilograms of ash from streets, rooftops, and other structures. Elsewhere, the weight of the volcanic ash caused flattening of hayfields and similar agricultural problems, resulting in an estimated $300 million worth of damage to crops. Once fallen, the ash dried to a powder that was blown in all directions by new winds. Where the ash was thickest, people had to wear protective face masks to keep from choking. Not all the ice from the peak melted during the first several blasts; huge chunks—some larger than trucks—lay scattered along the valley floors, where they then melted.

The ecological damage from the eruption was widespread. Streams were choked with debris, killing fish and destroying watering holes; dense clouds of ash, composed of tiny glass particles driven like shrapnel by the foaming winds, sanded down the wings of birds and insects and clogged their breathing; the rolling landslides inundated

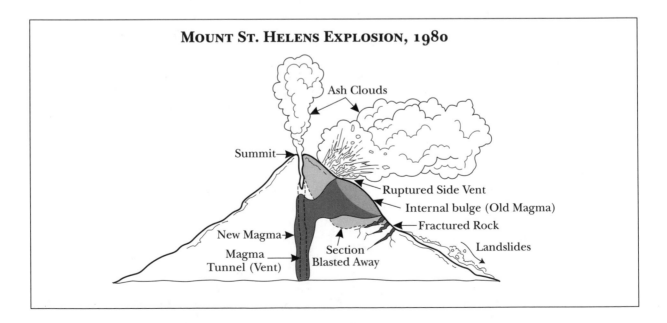

MOUNT ST. HELENS EXPLOSION, 1980

timberland and covered wildlife with a shroud of ooze, ash, gases, and debris. Decades will be required for the forests and lakes to rebuild themselves.

ORIGINS OF THE BLAST

The Mount St. Helens eruption is among the most closely studied volcanic events in history. Scientists were able to record in detail the physical and chemical activities that occur during a cataclysmic volcanic eruption. Eyewitness accounts were available from people who had been caught in the midst of the eruption and narrowly escaped death. These accounts furnish important data about many different aspects of volcanism.

The Cascade Range in western Washington State was created by a subduction process beginning about 80 million years ago. Subduction is one of the main methods by which earthquakes are produced; it occurs where the edges of two tectonic plates meet. The stronger, or perhaps more rigid, plate "dives," or subducts, under the other plate. In the process of subducting, the plates scrape along each other and cause great jolts, which are experienced as earthquakes. In the Mount St. Helens scenario, the tiny Juan de Fuca plate subducted under the larger North American plate (the North American continent), a process that is still occurring. The Pacific plate has been moving closer to the North American continent as the Juan de Fuca plate disappears, and it is now abutted next to the San Andreas fault, which runs across western California. The Juan de Fuca plate is almost entirely subducted.

STUDY OF MOUNT ST. HELENS

More scientific experience and technology were applied to the investigation of the Mount St. Helens eruption than to any other volcanic event in North America. One of the most valuable of all the data sources, however, was eyewitness accounts provided by both trained and untrained observers. Dozens of geologists and volcanologists worked in the immediate vicinity of St. Helens, some paying with their lives. In addition, hundreds of local citizens who witnessed the week-long scenario furnished information. The narratives, offered by citizens who had lived for many years in the region, enabled researchers to construct an almost minute-by-minute catalog of events, complete with time

Mount St. Helens during its 1980 eruption, showing pyroclastic flow. (U.S. Geological Survey)

estimates, sizes, volumes, and intensities—much of which replaced the data that went unrecorded when scientific instruments were destroyed in the eruption.

The most frequently used instrument for studying volcanic activity is the seismometer, which measures the frequency, location, and intensity of tremors. Seismometers were placed in as many as twelve locations near Mount St. Helens, furnishing a constant monitoring system. In this instance, the instruments recorded dozens of fair-sized tremors up to the moment of the 5.0-magnitude earthquake that triggered the eruption.

Another valuable instrument is the geodimeter, which uses a laser beam to measure minute changes in ground swelling, such as the huge subsurface bulge on the mountain's north flank. By monitoring changes in the height of such a dome, geologists can determine its rate of growth or subsidence. A fast-growing bulge would signal an impending crisis, perhaps a vent-creating explosion.

Tiltmeters measure a mountain's angle of slope. They tell geologists about the rate of change of the mountain's growth and the altering angle of its sides. These measurements can help to predict an eruption because internal volcanic pressures and upwelling magma tend to push outward against the crust of the volcano, increasing the slope angle.

Regular cameras are used to record an eruption's progress. Cameras on tripods are wired to receive radio signals that automatically advance the film for the next picture, making it unnecessary for the cameras to be operated by humans. Cameras can therefore be placed in areas that would be life-threatening for researchers, perhaps because of intense heat or because they are in the path of a shock wave. Cameras in various positions can be loaded with color or black-and-white film. Color is extremely important in assessing various activities; variations in grayness in an ash cloud tell geologists about the density of different portions of the cloud. Cameras are also vital in recording the aftermath of an eruption; they help to determine the seriousness of such problems as stream choking, ash fallout, and forest destruction.

Geologists use a stream gauge to record water temperature, amounts of material suspended in the water, and the water's chemical content. By means of such measurements, scientists can deter-

mine the seriousness of the stream choking, its cause, and the rate at which the stream is being filled with alien material. It is important to know the temperature of the stream, too, because of the possible negative impact that sudden temperature changes might have on fish, fish eggs, and other aquatic life.

Finally, a vital instrument for the geologist is the gas sensor, which is used both at ground level and in airplanes. It is important to be able to assess the presence or absence of various gases in the region of an eruption or impending eruption, especially hydrogen, carbon dioxide, and sulfur dioxide. Although these gases are potentially life-threatening, they can reveal subsurface movement of magma.

Thomas W. Becker

CROSS-REFERENCES

Calderas, 683; Earthquake Distribution, 277; Earthquake Locating, 296; Eruptions, 739; Forecasting Eruptions, 746; Global Warming, 1862; Island Arcs, 712; Krakatau, 752; Lava Flow, 717; Mount Pelée, 756; Mount Pinatubo, 761; Mount Vesuvius, 772; Notable Earthquakes, 321; Ozone Depletion and Ozone Holes, 1879; Plate Margins, 73; Plate Motions, 80; Plate Tectonics, 86; Popocatépetl, 776; Pyroclastic Rocks, 1343; Recent Eruptions, 780; Seismometers, 258; Shield Volcanoes, 787; Stratovolcanoes, 792; Stress and Strain, 264; Subduction and Orogeny, 92; Volcanic Hazards, 798.

BIBLIOGRAPHY

Attenborough, David. *The Living Planet: A Portrait of the Earth.* Boston: Little, Brown, 1984. Prompted by the television series of the same name, this book explores the planet as an ecological and environmental entity. An easy-to-read text is one delightful aspect of the book; the excellent accompanying photography is another. The value of this volume lies in its attempt to integrate the Earth's basic attributes. A good place for the interested layperson to begin.

Decker, Robert, and Barbara Decker. *Volcanoes.* New York: W. H. Freeman, 1997. This readable and well-illustrated text is geared to-

ward the general public. The authors explain geologic phenomena in a clear and simple manner. Includes a section on Mount St. Helens. The appendices list the world's 101 "most notorious" volcanoes, as well as Internet sources for volcano information.

Findley, Rowe. "St. Helens: Mountain with a Death Wish." *National Geographic* 159 (January, 1981): 2-65. This series of well-written and spectacularly illustrated articles is perhaps the best narrative account of the Mount St. Helens eruption. Eyewitness accounts, scientific explanations, and interviews with survivors are woven together into an astonishing story. Some of the on-scene photographs carry great emotional impact. The best starting point for the reader interested in Mount St. Helens.

Francis, Peter. *Volcanoes: A Planetary Perspective.* Oxford: Clarendon Press, 1993. This is a university-level text, but it is written in a style that makes it accessible to a wide range of readers. The author covers the visible evidence of volcanoes, as well as the theory of why they develop. Illustrations, index, and reference list.

Redfern, Ron. *The Making of a Continent.* New York: Times Books, 1986. Possibly no book on the subject of Earth science can compare with this volume of excellent photographs, clear and concise diagrams, and compelling text. The author has created a record of the North American continent that lays bare all nature's secrets. All the photography was created by special camera techniques. A basic textbook for anyone interested in the Earth sciences.

Sigurdsson, Haraldur, ed. *Encyclopedia of Volcanoes.* San Diego, Calif.: Academic Press, 2000. This book contains a complete summary of the scientific knowledge of volcanoes. It contains eighty-two well-illustrated overview articles, each of which is accompanied by a glossary of key terms. Although this is a college-level text, it is written in a clear and comprehensive style that makes it generally accessible. Cross-references and index.

MOUNT VESUVIUS

Mount Vesuvius is one of the most famous, most studied, and most influential volcanoes in the world. The science of volcanology began with reports of its activity; several key terms and concepts originated from its example. The science of archaeology also began with Vesuvius because of the preservation of the ruins of Pompeii and Herculaneum as a result of its most famous eruption, in 79 C.E.

PRINCIPAL TERMS

CONE: the hill or mountain, more or less conical, surrounding a volcanic vent and created by its ejecta; it is normally surmounted by a crater

CRATER: the circular depression atop a volcanic cone

ERUPTION: volcanic activity of such force as to propel significant amounts of magmatic products over the rim of the crater

LAVA: molten rock (as opposed to ash) erupted by a volcano; the term originated with reference to Vesuvius

MAGMA: molten rock, normally subterranean

STRATOVOLCANO, or COMPOSITE VOLCANO: a noticeably conical volcano composed of interbedded lava flows, ash, and cinders; also, the remnant of such a volcano

VOLCANOLOGY: the study of volcanoes

MEDITERRANEAN VOLCANOES

When most people think of a volcano, they think of the kind of volcano that Vesuvius is—a distinctly conical mountain that erupts spectacularly at intervals, blowing rocks and ash high into the air while sending fiery lava down its slopes. In actuality, however, there are several different kinds of volcanoes and several different kinds of eruptions. Vesuvius, the best-known example of what is called a stratovolcano, or composite volcano, features a prominent cone made up of lava, craters, and ash. Like most stratovolcanoes, it has erupted in more than one way. Two of the variously classified eruption types, the Plinian and the Vesuvian (or Vulcanian), were either first or best observed here.

Vesuvius is one of a series of volcanoes irregularly aligned throughout the Mediterranean region, from southern Greece down to central and southern Italy and then to Sicily and its adjacent islands. The best-known Greek volcano, now extinct, is the crater-remnant island of Santorini (or Thera), the eruption of which is sometimes credited with the destruction of Minoan Crete. Aside from Vesuvius, the best-known Italian volcanoes, all of them recently active, are Mount Etna in Sicily and the two Aeolian island volcanoes, Vulcano (for which all other volcanoes are named) and

Stromboli, which has been in constant eruption for more than 2,500 years. Geologists explain this Mediterranean lineage of volcanoes through plate tectonics. According to this theory, the African plate is slowly moving northward, colliding with and sliding under the European plate. Its leading edge plunges downward beneath the European plate and eventually melts, creating large accumulations of liquid rock (magma) that later ascend to express themselves as volcanoes. This level of understanding, however, has only been achieved since about 1970, when plate tectonic theory was accepted.

Vesuvius, located 11 kilometers south of Naples and easily visible from there, is part of a larger volcanic complex surrounding that city. Other parts of the complex include the Phlegraean Fields, a group of nineteen small, closely grouped craters (regarded as the entrance to Hell in classical times); the so-called Temple of Jupiter Serapis at Puzzuoli, a Roman ruin that shows unmistakable evidence of abrupt vertical displacement within historical times (it was utilized by Charles Lyell as the frontispiece to all twelve editions of his *Principles of Geology*, 1830-1875); the Solfatara, a large, steaming crater whose last eruption (in 1198) probably displaced the Temple of Jupiter Serapis; and the buried cities of Pompeii and Herculaneum. All these phenomena are closely associated with Vesuvius and have been immensely influential in forming ideas about volcanoes and how they work.

Vesuvius is the most recently active (1944) member of a group of volcanoes—many of them now deeply eroded and scarcely recognizable—lying roughly parallel to the Apennine Mountains and the Tyrrhenian (west) coast of the Italian peninsula. In mid-Tertiary times (about 30 million years ago), through the agency of plate tectonics, the Apennine chain began to rise, and the Tyrrhenian Basin to subside. The Bay of Naples is also essentially a downwarp, with the Triassic and Cretaceous limestone rocks of the Sorrento peninsula forming one of its uptilted edges; Vesuvius, the Solfatara, and the Phlegraean Fields are located roughly atop its center. Vesuvius probably began as a submarine volcano in the Bay of Naples, then emerged as an island and finally became part of the mainland through the constructive effects of its own eruptions.

BIRTH OF VESUVIUS

Vesuvius originated about 10,000 years ago; there were several cones before the present one. Archaeological evidence attesting human occupation of its slopes goes back to the seventh century B.C.E. The name "Vesuvius" (its origin is unknown) was used around 45 B.C.E. by Diodorus Siculus, a Sicilian writer who recognized the mountain as a former volcano. By around 30 B.C.E., however, the geographer Strabo regarded Vesuvius as a scene of local fires only. He described a flat-topped, barren summit and fertile slopes teeming with agriculture. Thus, no actual eruptions of Vesuvius had been recorded in historic times, and its identification as a volcano had been forgotten.

A destructive earthquake struck the Naples region on February 5, 63 C.E., causing extensive havoc at Pompeii and Herculaneum also. Subsequent tremors continued off and on for the next sixteen years, with one of the strongest occurring on the night of August 24, 79 C.E. It is apparent to modern scientists that such earthquakes are often caused by the upward movement of magma in the vent of an active volcano, but no such understanding prevailed at the time. The most famous volcanic eruption in history therefore took the population it affected almost completely by surprise.

PLINIAN ERUPTIONS

The first reason that the Vesuvian eruption of 79 C.E. is so well known is that there is an eyewitness description of it written some years after the event by Pliny the Younger, in two letters to the historian Tacitus regarding the death of Pliny the Elder, the writer's uncle, who was a distinguished naturalist and public figure. This two-part epistolary account, often reprinted in translation, is the first volcanological field report and more exciting than geological papers tend to be.

According to the younger Pliny, about one o'clock in the afternoon on August 24, a strange pine-tree-shaped cloud, unusually large and rapid, began to ascend over Vesuvius. The Italian pine tree, unlike North American pines, is narrow at the base and becomes increasingly broad as its grows upward. Pliny's description being so apt, an eruption cloud of the same shape today is still called a *pino* (Italian for pine tree); eruptions of this type are called Plinian. Realizing that a natural disaster was at hand, Pliny the Elder went by water to the home of a friend, where he slept for a time, until tremors and falling ejecta from the volcano awakened everyone. A corpulent man, he then died suddenly the next morning, from asphyxiation or possibly a heart attack—the first "scientist" ever to be killed by a volcano. Other portions of his nephew's account describe, for the first time, phenomena now frequently associated with volcanic eruptions: preternatural darkness, the retreat of the sea from its established shores, and repeated bursts of lightning within the eruption cloud.

DESTRUCTION OF POMPEII AND HERCULANEUM

The other reason that the Vesuvian eruption of 79 C.E. is so famous is that the nearby cities of Pompeii and Herculaneum were destroyed (and their remains preserved) by it. Pliny the Younger does not mention either city, though Tacitus may have in a portion of his history no longer extant. There is not another written account before about 230 C.E., when Dio Cassius erroneously claimed that the Pompeiians died while sitting in their theater, a scene later dramatized in Edward Bulwer-Lytton's novel *The Last Days of Pompeii* (1834) and numerous films. Archaeological evidence, however, has proved that Dio Cassius's account is false.

Except for some immediately posteruption attempts at salvage, Pompeii and Herculaneum lay essentially untouched beneath their respective volcanic coverings until well into the eighteenth century, when organized attempts at treasure hunting began. Skeletons of the deceased were then discovered rather frequently, especially at Pompeii, where the modern science of archaeology evolved. Destruction there had taken place so rapidly that the city's life ended with food still on tables and bread baking in ovens. Pompeii is one of the most often visited monuments in Europe because the evidence of its everyday life and the drama of its abrupt termination are apparent to all.

Pompeii and Herculaneum were destroyed in different ways. The greater number of bodies— some of them skeletons, others casts in plaster— are from Pompeii, and they often appear to indicate death by asphyxiation. No lava reached the city, though it was bombarded with ash to a depth of about 1 meter. It is believed, however, that even such an ashfall could not have killed the more than two thousand persons known to have perished there. Current theorizing holds that at least one phase of Vesuvius's eruption (which lasted two days in all) included a nuée ardente, a sudden basal surge of incandescent, heavier-than-air gas that poured out of the crater and ran down its slope like an avalanche, instantly searing and suffocating whatever animate life lay in its path. Pliny the Elder may have died because of it.

Herculaneum, probably a slightly later victim, was inundated not by lava or ash but by mud. The torrential rains that sometimes follow a volcanic eruption (which ejects huge volumes of water vapor) can gather newly deposited loose particles of ash upon the volcano's slopes into a precipitous mudslide (lehar) that will then flow rapidly and in great volume for kilometers. Such a mudslide buried the city of Herculaneum, which remains less excavated than Pompeii because its eruption cover is much harder to dig through and because the modern city of Resina sits atop most of it. Both Pompeii and Herculaneum have fundamentally influenced archaeology, history, and the fine arts.

Plinian eruptions, such as that of 79 C.E., are noted for their explosive violence. The cones of stratovolcanoes, moreover, are surprisingly fragile and will often change significantly from one eruption to the next. In 79 C.E., this fragility was more than usually apparent, as the entire upper cone of Vesuvius was completely destroyed, leaving behind only an incompletely circular ring now known as Mount Somma. From it, the more general term "somma," which applies to similar structures everywhere, is derived. Although some disagreement exists, for most volcanologists the mountain that erupted in 79 C.E. is the present-day Mount Somma and the cone within it now called Vesuvius did not arise until the quieter eruption of 172 C.E.

LATER ERUPTIONS

Since then, there have been dozens of further Vesuvian eruptions, most of them from the central vent but a few from the flanks as well. Various literary sources tell of eruptions in 203, 472, 512, 685, 993, 1036, 1049, 1138, and 1139; there was then nothing of importance for almost five hundred years, though the overnight eruption in the Phlegraean Fields that produced Monte Nuovo in 1538 is attested by no fewer than four eyewitness accounts. This extended period of quiescence ended on December 16, 1631, when, after an explosive beginning, seven streams of lava poured down Vesuvius's slopes, destroying a dozen villages and some eighteen thousand people. Several accounts of this disastrous eruption were written, some of them book-length; it is the first volcanic eruption in history for which there is a published literature.

Vesuvius has been under almost constant observation ever since, with many subsequent outbursts. From 1631 until 1944, however, activity remained unusually frequent. During most of this time (including major eruptions in 1676, 1694, 1707, 1737, 1751, and 1760), Vesuvius was unquestionably the premier geological attraction in the world; it soon became a necessary destination for anyone who aspired to authority regarding the natural history of the Earth. This prominence intensified as a result of the work of Sir William Hamilton, the British representative at Naples, who published highly influential accounts of the Vesuvian eruptions of 1766, 1767, 1770, and 1779. He is arguably the first person who might properly be described as a modern volcanologist.

Other major eruptions followed in 1793-1794, 1822, 1834, 1855, 1867-1868, 1872, 1906, and 1944. During this 150-year period, the nature of observations changed gradually from those of the intrepid amateur (such as Hamilton) to fully sophisticated, rigorous science. One may cite in particular the establishment in 1844 of the Royal

Vesuvian Observatory, a then unique (but since copied) institution located on the very slopes of the volcano. Under Macedonio Melloni initially, the observatory made ongoing monitoring of day-to-day volcanic activity possible and became the training ground for a distinguished series of internationally known volcanologists, who naturally derived many of their concepts from Vesuvius itself. Among others, Luigi Palmieri, R. V. Matteucci, Giuseppe Mercalli, Frank Perret, and Giuseppe Imbo have all made important contributions.

Dennis R. Dean

CROSS-REFERENCES

Basaltic Rocks, 1274; Eruptions, 739; Folds, 624; Forecasting Eruptions, 746; Igneous Rock Classification, 1303; Krakatau, 752; Magmas, 1326; Mount Pelée, 756; Mount Pinatubo, 761; Mount St. Helens, 767; Plate Margins, 73; Plate Motions, 80; Plate Tectonics, 86; Popocatépetl, 776; Recent Eruptions, 780; Seismometers, 258; Shield Volcanoes, 787; Stratovolcanoes, 792; Subduction and Orogeny, 92; Volcanic Hazards, 798; Weathering and Erosion, 2380.

BIBLIOGRAPHY

Decker, Robert, and Barbara Decker. *Volcanoes.* New York: W. H. Freeman, 1997. This readable and well-illustrated text is geared toward the general public. The authors explain geologic phenomena in a clear and simple manner. Includes a section on Mount Vesuvius. The appendices list the world's 101 "most notorious" volcanoes, as well as Internet sources for volcano information.

Francis, Peter. *Volcanoes: A Planetary Perspective.* Oxford: Clarendon Press, 1993. This is a university-level text, but it is written in a style that makes it accessible to a wide range of readers. The author covers the visible evidence of volcanoes, as well as the theory of why they develop. Illustrations, index, and reference list.

Grant, Michael. *Cities of Vesuvius: Pompeii and Herculaneum.* New York: Spring Books, 1974. A reliable text about the cities destroyed by Vesuvius. Includes fine illustrations.

Hoffer, William. *Volcano: The Search for Vesuvius.* New York: Summit, 1982. A travel book centering on Vesuvius; short, accessible, and entertainingly (but responsibly) written. Recommended for high school students.

Leppmann, Wolfgang. *Pompeii in Fact and Fiction.* London: Elek, 1966. This source emphasizes cultural history, that is, the impact of Pompeii on later writers and artists.

Lobley, J. Logan. *Mount Vesuvius.* London: E. Stanford, 1868. Of the several nineteenth century books devoted exclusively to Vesuvius, this one is the best. Useful primarily for its extensive history of Vesuvian eruptions; however, the science is dated.

Sigurdsson, Haraldur, ed. *Encyclopedia of Volcanoes.* San Diego, Calif.: Academic Press, 2000. This book contains a complete summary of the scientific knowledge of volcanoes. It contains eighty-two well-illustrated overview articles, each of which is accompanied by a glossary of key terms. Although this is a college-level text, it is written in a clear and comprehensive style that makes it generally accessible. Cross-references and index.

Sigurdsson, Haraldur, et al. "The Eruption of Vesuvius in A.D. 79." *National Geographic Research* 1 (1985): 332-387. This article is an authoritative but technical account of the most widely known eruption of Vesuvius.

POPOCATÉPETL

The Popocatépetl volcano is located in central Mexico, 40 kilometers east of the outskirts of Mexico City. It is in sight of 30 million people, most of whom are living on the debris from previous eruptions. The volcano has erupted explosively in the past, and it became active again during the 1990's, raising major concerns about the safety of people in this heavily populated area.

PRINCIPAL TERMS

ANDESITE: a volcanic rock with intermediate silica content that contains the minerals plagioclase, biotite, and hornblende; because the molten equivalent of andesite is often viscous, gases can build up in it and produce explosive eruptions

ASH FLOW: a moving mass of hot ash and gas that flows from a volcano; the hot material is deposited as it spreads out on flatter areas

ASHFALL: fine-grained volcanic material that settles out of the atmosphere

LAHAR: a volcanic mudflow formed when volcanic ash and minor coarse volcanic particles mix with the abundant rainfall that often accompanies eruptions

PLINIAN ERUPTION: a rapid ejection of large volumes of volcanic ash that is often accompanied by the collapse of the upper part of the volcano

PUMICE: volcanic glass, usually light in color, that contains so many bubbles that it may float in water

SUBDUCTION: a tectonic process in which one lithospheric plate is underthrust below another plate into the mantle

VOLCANISM IN MEXICO

The Popocatépetl volcano, located in central Mexico, rises to a height of 5,465 kilometers above sea level. Although there is usually snow on its summit, oranges, mangoes, bananas, and palms grow at its base.

The volcano is the result of tectonic activity in the region, where the Cocos oceanic plate is underthrust or subducted below the Caribbean or Americas continental plates. The subduction process produces melting along or above the subducted plate, resulting in the creation of fairly viscous melted rocks such as andesites. These stiff or viscous melts do not readily flow up through the volcano, often resulting in a plug of partially solidified lava that causes a buildup of gases. The gas pressure may eventually become so great that it explodes out of the volcano at its weakest point.

Such an explosion may take the form of a Plinian eruption, in which copious amounts of volcanic ash, gas, and larger particles are blown upward out of the central cone. This type of eruption has occurred at Popocatépetl several times throughout history. As the particles settle out of the atmosphere, they produce ashfalls. Sometimes the ash and gas may be blown out laterally and flow down the sides of the volcano as ash flows. Heavy rains produced by hot updrafts over the hot, central eruptions may mix with the abundant volcanic ash, producing lahars that move rapidly and for many kilometers down stream and river channels.

ERUPTION HISTORY

Scientists believe that Popocatépetl has experienced major Plinian eruptions every one thousand to three thousand years since 4000 B.C.E. Some of the most recent of these have destroyed human settlements. Large eruptions occurred sometime between 3195 and 2830 B.C.E., between 800 and 215 B.C.E., and between 675 and 1095 C.E. Each of these eruptive cycles began with ashfalls and ash flows in small volumes, producing a number of 2- to 10-centimeter-thick volcanic layers. The subsequent phase of Plinian eruptions produced a thick sequence of coarser-grained pumice fragments. It is estimated that the eruption column of fragments and gas would have reached heights of more than 25 kilometers. This would have had a major impact on world climate as the particles spread high into the atmosphere. Some of these major eruptions also produced lahars and hot ash flows that moved rapidly down valleys and were deposited on the plains.

The area around Popocatépetl has long supported fairly large populations because of the fertile soils, the presence of water, and the climate. For example, agricultural and ceremonial towns were established in the area between 1000 and 100 B.C.E. Pottery shards, building foundations, and agricultural furrows dating to this era have been found in pumice deposits, suggesting that the settlements suffered widespread destruction as the result of volcanic activity. Subsequently, areas near the eruptive center declined in population between 100 B.C.E. to 100 C.E., while areas to the north became more populated.

By 750 to 800 C.E., several large cities containing as many as 150,000 people each had been established in the region surrounding Popocatépetl. Two of these, Cholula and Teotihuacán, were large religious centers. Archaeologists have determined that Cholula was abandoned around 800 C.E., a phenomenon that was accompanied by a significant decrease in population in the general area. A variety of reasons were initially proposed for this decline, including drought, invasion, food shortages, and the exhaustion of soil nutrients. However, this time period also correlates with a major Plinian eruption in which more than 3,000 square kilometers of the area were covered by deposits from the eruption. The remains of lahars at the base of the pyramid in Cholula were found to contain pieces of pottery and volcanic glass artifacts. Several researchers who have studied these eruptions have suggested a scenario in which the area was covered by a muddy plain of lahars, with only the great pyramids projecting above the mud.

MODERN ACTIVITY

After a long period of dormancy, Popocatépetl became active again during the early 1920's. The first stages of activity began with steam and ash emissions during the summer of 1920. A lava dome formed on the floor of the crater in the latter part of the same year, then stopped growing.

The dome eventually exploded during a minor eruption in January of 1922.

Relatively little activity occurred at Popocatépetl between 1922 and December of 1994. During that time, researchers noted a gradual increase in earthquake activity and gas emissions enriched in sulfur dioxide. On December 21, 1994, ash emissions also began, which continued into the middle of 1995. The activity then quieted until emissions resumed on March 5, 1996. Ash was observed covering snow and glaciers on the summit, and ash and gas were thrown as high as 800 meters above the crater. A viscous dome of lava about 400 square meters in area was observed on the central crater floor on March 29, 1996, indicating that the lava phase of the eruption had begun. Lava continued to flow until most of the inner crater was filled.

On April 30, a large explosion blew a hole in the surface of the lava dome. This resulted in the deaths of five climbers who had defied warnings to stay away from Popocatépetl. A rain of warm volcanic debris was deposited in nearby towns. Particles 0.5 centimeters in diameter were observed 12 kilometers away from the volcano, while sand-sized material was observed at a distance of 60 kilometers.

Volcanic activity continued sporadically through late 1997 as lava domes periodically built up on the crater floor, only to be destroyed by often spectacular explosions. An explosion on October 28, 1996, at 9:05 A.M., emitted large amounts of ash and gas. This was followed by a lesser explosion the following day at 10:11 P.M. Minor explosions also occurred on December 21, 1996, and between February 17 and June of 1997. An eruption on April 24, 1997, produced an ash and gas plume 13 kilometers high and 8 kilometers wide. It eventually spread out to a plume that was 65 kilometers long. Meanwhile, the lava body first observed within the crater in March of 1996 remained unchanged.

The largest ash emission since the beginning of the eruption cycle in 1994 occurred on June 30, 1997. It began with several earthquakes related to the volcanic activity at 4:56 P.M. The first of the ash eruptions began at 5:11 P.M. and lasted 135 minutes. The second ash eruption began at 7:26 P.M. and lasted for 90 minutes. The ash plume rose to more than 13 kilometers within a few minutes, and ash began falling in towns around the volcano two to three hours later. The ashfall shut down the airport in Mexico City for twelve hours, leading to numerous press reports. A red alert was posted, but no evacuations were ordered. No casualties were reported, but minor lahars were reported in an area with heavy rain, causing the flooding of one house. Several lava tongues up to 2 kilometers long flowed south to southeast down the volcano on July 3 and 4.

More small, sporadic eruptions were reported through the end of 1997, with the most significant eruptions occurring on August 12, in late September, in early October, and in early to late December. The biggest eruption during this period occurred on December 24, during which incandescent material was thrown from the volcano and grass fires were started to the east. Intermittent activity continued through 1998 and 1999, with several significant eruptions. A thirty-second explosion sent ash 5 kilometers into the atmosphere on January 1, 1998. Larger explosions began in late November of 1998, producing ash plumes 3 kilometers high. Ashfalls were reported on the southwest side of Mexico City. Explosions on November 25 broke windows and walls in several villages 60 kilometers away, and large volcanic blocks landed 5 kilometers away. People were warned to stay at least 7 kilometers away from the volcano. Large explosions in December, 1998, shattered windows, fractured walls, opened doors, and knocked tiles off roofs. Popocatépetl continued to be active through 1999. Significant eruptions occurred on March 17, March 23, June 16, and June 22. The eruption on March 17 was large enough to result in evacuation centers being put on alert.

SIGNIFICANCE

Much has been learned about the long-term eruptive cycles of volcanoes through the study of ancient volcanic deposits. If authorities can understand the potential danger of Popocatépetl in terms of its historic eruption cycle, they can set up better evacuation procedures in case of a catastrophic eruption. The authorities have such evacuation and emergency procedures in place should a large eruption cycle begin.

Scientists are also interested in the effects of volcanic eruptions on the global climate. Several of Popocatépetl's major eruptions appear to have been powerful enough to send volcanic ash into the stratosphere, allowing it to spread worldwide. For example, analysis of ice cores from the Greenland ice sheet has suggested that a major eruption occurred somewhere in the world around 822 or 823 C.E., which corresponds to the approximate date of a major eruption of Popocatépetl. Such large eruptions appear to affect the climate of the world by lowering the penetration of sunlight, thus causing a cooling of the surface of the Earth.

Robert L. Cullers

CROSS-REFERENCES
Andesitic Rocks, 1263; Eruptions, 739; Forecasting Eruptions, 746; Igneous Rock Bodies, 1298; Krakatau, 752; Magmas, 1326; Mount Pelée, 756; Mount Pinatubo, 761; Mount St. Helens, 767; Mount Vesuvius, 772; Recent Eruptions, 780; Shield Volcanoes, 787; Stratovolcanoes, 792; Subduction and Orogeny, 92; Volcanic Hazards, 798; Volcanoes: Climatic Effects, 1976.

BIBLIOGRAPHY

Bullard, F. M. *Volcanoes of the Earth.* Austin: University of Texas Press, 1976. A nontechnical and readable discussion of the location and origin of many volcanoes. It contains sections on Plinian eruptions, ash flows, and lahars. A glossary is included.

Macdonald, G. A. *Volcanoes.* Englewood Cliffs, N.J.: Plinian Press, 1972. Macdonald gives detailed accounts of volcanoes and volcanic eruptions. There is a lot of terminology, but

it is generally defined. The book contains sections on Plinian eruptions, ash flows, and lahars, and includes an appendix listing active volcanoes and their locations.

Siebe, C., M. Abrams, J. L. Macias, and J. Obenholzer. "Repeated Volcanic Disasters in Prehispanic Time at Popocatépetl, Central Mexico: Past Key to the Future?" *Geology* 24 (1996): 399-422. This article is in a technical geology journal, but it has a minimum of jargon. It summarizes the main eruptive cycles of Popocatépetl in terms of age and archeological studies.

Stuart, G. E. "Teotihuacan." *National Geographic* 188 (1995): 2-35. This article, which contains numerous photographs, discusses the excavation of Teotihuacan, as well as evidence for its decline.

White, S. E. "Popocatépetl: The Ever-Burning Torch." *Volcano News* 17 (1984): 1-3. An easily read overview of Popocatépetl.

RECENT ERUPTIONS

Volcanoes are, with earthquakes, a most dramatic expression of the fact that the Earth is a living, dynamic planet. Unfortunately, because of previously erroneous concepts regarding them as well as inadequate descriptive techniques, most written records of specific eruptions undertaken before 1800 are now of limited value. Since then, however, a number of particularly instructive examples have been witnessed.

PRINCIPAL TERMS

ASH: rocky, unconsolidated ejecta of sand-grain size

ERUPTION: volcanic activity of such force as to propel significant amounts of magmatic products over the rim of the crater

LAVA: molten rock (as opposed to ash) erupted by a volcano

NUÉE ARDENTE: a sudden basal surge of incandescent, heavier-than-air gas

Although volcanic eruptions are a normal part of the Earth's workings, they have always been regarded with awe because of their unpredictability and power. Only within the last two hundred years has scientific understanding of them been achieved. During that time, certain specific eruptions (and recurring activity of certain volcanoes) have been especially influential.

TAMBORA

Located on the island of Sumbawa, about 160 kilometers east of Java, Tambora volcano erupted in April, 1815—the exact date and preliminary activity are uncertain—creating the greatest explosion and the most powerful volcanic eruption of modern times. There were no surviving eyewitnesses, as virtually the entire population of Sumbawa died. The noise of the explosion was heard 1,500 kilometers away, and its atmospheric effects were worldwide. As clouds of volcanic ash circulated throughout the stratosphere, 1816 became in America and Europe "the year without a summer," during which normal harvests were greatly curtailed.

GRAHAM ISLAND

Another otherwise routine Vesuvian eruption of October 22-26, 1822 (together with its predecessor of 1818), prompted George Julius Poulett Scrope to study the volcanoes of Europe. His *Considerations on Volcanoes* (1825) inaugurated the modern science of volcanology. Graham Island was a submarine volcano that appeared off the southern coast of Sicily for a few months in 1831, only to disappear by year's end. The first volcano of any kind to be observed almost daily throughout its existence, it was considered by such European geological theorists as Scrope and Sir Charles Lyell to be an example of fundamental importance in their debates regarding the formation of volcanoes. The forceful but otherwise routine eruption of Vesuvius on April 24-30, 1872, stimulated Luigi Palmieri, director of the Volcano Observatory on the slopes of Mount Vesuvius, to formulate his influential theory that volcanic eruptions are often characterized by predictable phases. This eruption was also the first major eruption to be photographed.

KRAKATAU

Krakatau was a three-part volcanic island located between Java and Sumatra. After preliminary activity beginning on May 20, 1883, it erupted explosively on August 26 and 27, the reports being heard nearly 5,000 kilometers away. Tsunamis generated by the eruption killed more than thirty-six thousand people. As with Tambora, subsequent atmospheric effects were worldwide, generating much scientific and popular interest. Dutch and British investigators published important studies.

TARAWERA

Following some preliminary earthquakes, the top of domelike Mount Tarawera, on the North Island of New Zealand, split apart on June 10, 1886, exposing the internal structure of the dome and creating a series of steam-blast craters. When fissuring extended beneath Lake Rotomahana, further steam explosions followed, destroying two world-famous hot-spring formations, the White and Pink terraces. Ash and mud from the eruption buried several Maori villages and more than 140 people.

BANDAI-SAN

After more than one thousand years of relative inactivity, Bandai-san volcano (on the island of Honshu) erupted suddenly on July 15, 1888, for less than two minutes initially. The north slope of the mountain was demolished, burying the villages and people of an entire valley with avalanches and mudflows. Although only a steam explosion, it was the worst volcanic disaster in the history of Japan.

LA SOUFRIÈRE

La Soufrière at the northern end of the island of St. Vincent in the West Indies became active on May 6, 1902, expelling its crater lake in a series of mudslides. Its major eruption, on May 7, was the first modern one in which glowing avalanches (nuées ardentes) were recognized. They sped down all sides of the mountain, causing little damage. The thrust of the eruption had been vertical, with the glowing avalanches on its slopes a secondary phenomenon caused only by fallout and gravity. Activity continued into March, 1903.

MOUNT PELÉE

Nearby Mount Pelée, on the island of Martinique, was active from April 24, 1902, to October 5, 1905 (and again from September 16, 1929, to December 1, 1932). After giving ample warning, it erupted violently on May 8, 1902—one day after Soufrière on St. Vincent—but the nuées ardentes this time were ejected straight down the western slope of the volcano. Thus, searing, suffocating clouds of gas and ash rushed like an avalanche to engulf the city of St. Pierre, destroying it and all but two (some say four) of its citizens—about thirty thousand people in all. Second only to the destruction of Pompeii by Vesuvius in 79 C.E., this volcanic disaster was the most famous of modern times. It greatly stimulated interest in volcanoes,

attracted two major figures (Thomas A. Jagger and Frank A. Perret) to their study, established a new type of eruption (the Peléan), and initiated a dispute regarding nuées ardentes that evidence from its own later eruptions resolved.

ALASKAN ERUPTIONS

Accounts of a major Alaskan Peninsula eruption from June 6 to June 9, 1912—the most powerful modern one on the North American continent—differ with age. Current understanding is that the primary effusion radiated from a newly formed volcano named Novarupta on the flank of Mount Katmai rather than from Katmai itself. (There were no nearby eyewitnesses, and no scientific investigations before 1916.) In any case, explosions were heard 1,000 kilometers away, ash fell thickly at a distance of 160 kilometers, and an adjacent valley floor was transformed into a steaming fumarolic basin, the so-called Valley of Ten Thousand Smokes (with rapidly decreasing activity). The old crater of Mount Katmai later collapsed, perhaps because the liquid rock (magma) underlying it had been withdrawn. Aside from its forcefulness, Katmai is significant for being the only example of an ignimbrite (fused rhyolite tuff) eruption in modern times.

LASSEN

Lassen Peak in Northern California is, like Tarawera, a volcanic dome, and one of the largest known. Long dormant, it resumed activity on May 30, 1914, becoming particularly violent one year later, from May 19 to May 22, 1915, when directional nuées ardentes coursed down its shattered slopes to devastate whole forests and create powerful mudslides. Some less important lava flows then followed, with activity continuing into 1917. Prior to the eruption of Mount St. Helens in 1980, the 1915 eruptions of Mount Lassen were the most famous recent ones to have taken place within the United States.

PARÍCUTIN

Parícutin emerged overnight from a cornfield in central Mexico on February 20, 1943. This fragile but rapidly growing cinder cone reached a height of 50 meters the next day; lava began flowing from its base rather than its crater, destroying a nearby farm. In June, 1944, it buried an entire village. After years of spectacular but relatively harmless activity, Parícutin ended as abruptly as it had begun, on February 25, 1952, its height being 410 meters. Parícutin was the first wholly new volcano to arise in North America since nearby Jorullo (1759; poorly observed) and attracted unprecedented scientific attention.

HEKLA

Hekla, the best-known Icelandic volcano, erupted memorably on March 29, 1947, after more than one century of dormancy. Large amounts of ash and other debris were ejected from a fissure, and lava flows poured down the volcano's slopes from a series of fissure-related vents. After nearly thirteen months of continuous activity (ending April 21, 1948), approximately 1 billion cubic meters of lava had been extruded, in addition to immense volumes of ash. Eruptions of this type do not usually endanger human life, but Iceland's livestock and farming industries were severely affected, both from the destruction of arable land and from the widespread emissions of toxic gases.

LAMINGTON

Prior to 1951, Mount Lamington, on the northern side of Papua New Guinea, was regarded as extinct. It consisted of a deeply eroded complex of volcanic domes, with heavy forestation and no record of previous activity. On January 15, however, earthquakes and emissions began, gradually intensifying. The major eruption took place on May 21, featuring a huge eruption cloud and glowing avalanches of the St. Vincent type. These avalanches devastated an area of more than 230 square kilometers and killed about six thousand people. Later eruptions were of the Vulcanian type and did not feature glowing avalanches but generated mudflows. Activity, lasting into 1956, then culminated in dome-building.

BEZYMIANNY

Bezymianny ("no name"), in Kamchatka, was also thought to be extinct. After one month of seismic activity, eruptions began on October 22, 1955. The most violent blast, on March 30, 1956, destroyed the top 200 meters of the volcano and most of one side. Nuées ardentes, followed by huge ash and pumice flows, rushed down adjacent valleys. Dome formation then concluded the vol-

cano's activity, all of which was meticulously recorded by Soviet volcanologists at a nearby observatory.

ILHA NOVA, I AND II

The Azores are a group of volcanic islands in the north Atlantic Ocean. A new vent adjacent to the island of Fayal opened on September 27, 1957, creating the island volcano of Capelinhos. Called Ilha Nova ("new island") I in this manifestation, it lasted only one-half month before disappearing (as Graham Island did in 1831). On November 7, Ilha Nova II appeared, soon joining the island of Fayal as a new peninsula. It then fell into dormancy after little more than one year of activity. The resulting cone resembles others, such as Diamond Head in Hawaii, formed by underwater explosive eruptions.

SURTSEY

The best-studied submarine eruption of all is that of Surtsey, a newly created island in the Westman Islands off the southern coast of Iceland. Named for a fire giant in Norse mythology, Surtsey announced its presence with undersea turmoil on November 14, 1963. An island appeared the next day and grew rapidly, with every step in its progress being closely observed. The main vent became inactive for a time (beginning May 7, 1965), but two associated ones then spewed up smaller independent islands (Syrtlingur and Jolnir) of their own; both subsequently disappeared. All activity ended on June 5, 1967, yet Surtsey continues to be of scientific interest as ecologists note its increasingly diverse complement of plant and animal life.

HEIMAY

Heimay ("home island") is the largest and most permanent of the Westman Islands, the same group to which Surtsey belongs. It is approximately five thousand years old but had never been known to erupt. On January 23, 1977, however, a 2-kilometer fissure opened at the edge of the major town, whose five thousand inhabitants were then evacuated. The new vent, Eldfell, quickly developed a cone and poured lavas down into the deserted village, eventually destroying more than three hundred homes. Accumulations of heavier-than-air noxious gases made salvage operations hazardous and endangered farm animals, many of which had to be destroyed. While various attempts to impede the encroaching lava achieved only limited success, the volcano stopped of its own accord on June 28.

MOUNT ST. HELENS

After erupting nine times between 1831 and 1857, Mount St. Helens, in the state of Washington, was dormant until March 27, 1980, when significant activity began. The major eruption of May 18 blew off the top of the mountain and was said to be the most forceful in the contiguous forty-eight states during historic times.

Mount St. Helens with a mantle of ash, following the 1980 eruption. (© William E. Ferguson)

Mudslides triggered by the eruption then destroyed bridges, logging camps, homes, and thousands of trees. About sixty people perished in one way or another. Further eruptions followed on May 25, June 12, July 22, and August 7, with occasional activity thereafter. These volcanic eruptions were the first in the contiguous United States since those of Mount Lassen. They attracted immense popular interest and were closely monitored by scientists.

EL CHICHÓN

In a now-familiar pattern, El Chichón volcano, in northern Chiapas, Mexico, awoke from a prolonged dormancy to erupt without warning on March 28, 1982. At least ten people who fled into a local church died when earthquakes demolished their haven; others were killed by falling debris and molten rock. A long ash cloud rained cinders across southern Mexico. Later eruptions in April left more people dead and homeless; survivors were then evacuated. Little had previously been known about this hitherto inactive killer.

MOUNT PINATUBO

Until 1991, Mount Pinatubo in the northern Philippines had no recorded eruptions. This was suddenly changed by the second-largest eruption of the twentieth century. More than three hun-

dred people were killed by the eruption, but loss of life would have been much greater if the region had not been evacuated. The Soufrière Hills of the island of Montserrat in the Caribbean Sea consist of a stratovolcano, the first eruption of which began in 1995. Pyroclastic flows were produced, and more than six thousand people were evacuated to the north of the island.

A SMALL SAMPLING

Although these eruption episodes are among those most likely to appear in discussions of modern volcanism— either because of their historical significance or because of the thoroughness with which they were observed—it would be extremely misleading to suggest that they include most, or even the most typical, examples. Thousands of eruptions take place every year, most of which scientists know nothing about because the volcanoes remain hidden within the depths of the oceans. There are still remote areas on the Earth, moreover, in which even a moderate land-based eruption might go unnoticed and therefore unrecorded.

Among the volcano records that do exist, all of the following have erupted twenty times or more since 1800: the three classic Italian volcanoes, Vesuvius, Stromboli, and Etna; Nyamuragia in Africa; Karthala (Grand Comoro) and Piton de la Fournaise (Reunion) in the Indian Ocean; White Island, Tarawera, Tongariro, Ngauruhoe, and Ruapehu in New Zealand; Bam and Manam, New Guinea; Ambrym and Lopevi, New Hebrides; Marapi and Dembo, Sumatra; Anak ("son of") Krakatau, between Sumatra and Java; Gedeh, Gunter, Slamet, Merapi, Semeru, Tengger, Lamogan, and Ruang, Java; Batur, Bali; Soputan and Lokon-Empung, Sulawesi; Apu Siau, Sangihe Islands; Gamalama, Halmahera; Mayon and Taal, the Philippines; Sakurajima, Kirishima, and especially Aso, Kyushu; Yake-Dake, Asama, Zao, and Iwake, Honshu; Oshima, Izu Islands; Komaga-Take and Taramai, Hokkaido; Karymsky and Kliuchevskoi, Kamchatka; Akatan and Shishadlin, Aleutian Islands; Pavlof, Alaskan peninsula; Kilauea and Mauna Loa, Hawaii; Colima, Mexico; Fuego, Guatemala; Izalco and San Miguel, El Salvador; Cerro Negro and Masaya, Nicaragua; Poas and Irazu, Costa Rica; Purace, Colombia; Reventador, Cotopaxi, and Sangay, Ecuador; Fernandina, Galápagos; and Rupungatito, Llaima, and Villarrica, Chile. It will be noted that a number of the most famous volcanoes do not appear; they have erupted either less frequently or not at all since 1800.

STUDY OF VOLCANIC ERUPTIONS

Broadly speaking, data relevant to any given volcanic eruption fall into three separate but related categories, depending upon whether they were collected prior to, during, or after the eruption. Data collected prior to an eruption include the previous eruptive history of the volcano: the types, the frequency, and the magnitude of its known eruptions. Illuminating as this information can be, however, it has not been shown to have reliable predictive value. Yet a volcano on the verge of eruption will often signal its upcoming behavior in a variety of ways. These include increased fumarolic activity, changes in the chemistry of associated waters, seismic disturbances, and tumescence (swelling). Certain changes in electrical and magnetic phenomena have been noted also. Volcanologists, therefore, monitor all of these indicators carefully.

To ensure objectivity, precision, and thoroughness, much of this monitoring is accomplished through instruments. Pyrometers and thermometers, for example, measure the heat of volcanic rocks (the temperature of the cone rises before an eruption) or of associated waters, as in a crater lake. As rocks adjacent to the vent heat up, they become less powerfully magnetic, as surface or aerial surveillance with a magnetometer will establish. In some cases, rising temperature can also be detected through a series of aerial infrared photographs. Unfortunately, none of these methods predicts eruptions definitively.

The two most commonly used instruments (and, on the whole, the most reliable) are the seismometer and the tiltmeter. The seismometer measures and records vibrations within the Earth. When strong enough to be detected by humans, these vibrations constitute earthquakes. It has been known for centuries that seismic activity often precedes volcanic eruptions. Since the invention of the seismometer in the latter nineteenth century, however, it has also been possible to detect fainter, more deep-seated vibrations; the subterranean rise of molten lava within the volcanic vent. Typically, such vibrations become stronger and more frequent as the time of eruption nears. Seismometry, therefore, is one of the most useful observational techniques available to the volcanologist.

As molten lava rises within the vent of an already existing volcano, the cone of that volcano subtly changes shape. Although the changes are generally too small to be obvious, they can be ascertained and verified through the use of a tiltmeter. This instrument can be in any of several forms (a pendulum, for example) but most commonly depends upon the displacement of a liquid, as in a carpenter's level. Laser beams have been used to measure tilt with great precision. On the other hand, some tumescence takes place so rapidly and conspicuously (generally along a seacoast) that detection by the naked eye is possible. Subsidence usually follows in any case, regardless of whether there has actually been an eruption.

Actual volcanic eruptions can last either seconds or days; a period of activity, characterized by frequent eruptions, can continue for years. During all or part of the times involved, it may well not be possible to inspect the volcano closely. First at Vesuvius, then at Hawaii and elsewhere, a number of the most active (or the most dangerous) volcanoes are monitored constantly, by instruments and by humans, at volcano observatories. During a very violent eruption, instruments, observers, and even the observatory itself may be lost. In any case, volcanologists have always relied upon whatever eyewitness data they could gather to supplement their understanding of a volcanic event. Any eruption is a contribution to history. In this regard, the most significant new technique has been the use of photography.

Although photography was invented in 1839, early techniques did not permit the recording of subjects in motion. Thus, no photographs of any eruption exist before 1872, when on April 26, memorable views of Vesuvius were obtained. The first eruption to be photographed in the Western Hemisphere was that of Izalco, El Salvador, in 1894. There were then striking images of nuées ardentes at Mount Pelée by Tempest Anderson and Alfred Lacroix in 1902. Anderson's *Volcanic Studies in Many Lands* (1903) became the first picture book of volcanic phenomena. Eruptions of Mount Lassen in 1914-1915 were the first within the continental United States to be photographed. B. F. Loomis's *Pictorial History of the Lassen Volcano* (1926) is perhaps the earliest example of its kind.

Even in more modern times, it is possible for a major eruption to take place without being observed by humans; Katmai, 1912, is a well-known example. The only recourse then available to volcanologists is that of hypothetical reconstruction based upon the physical evidence. This evidence generally consists of changes in the volcano's cone, adjacent geological effects, and the type and pattern of its discharges, insofar as they can be ascertained.

SIGNIFICANCE

In the early days of geology, it was believed by some that volcanoes were only local phenomena, deriving their power from the ignition of coal beds. If so, then they were of little consequence in the history of the Earth and hardly more than a curiosity within it. Since the latter eighteenth century, when this view was promulgated by Abraham G. Werner, scientists have come to realize how mistaken it is.

In his *Theory of the Earth* (1788), James Hutton first emphasized that the Earth's internal heat was the fundamental driving force underlying its surface permutations. Volcanoes for him were safety valves intended to prevent his heat-engine Earth from blowing up. Unlike certain predecessors, however, he did not emphasize their major constructive role in creating the surface of the Earth. The first theorist to hold essentially modern views regarding volcanoes was George Julius Poulett Scrope in *Considerations on Volcanoes* (1825). As more volcanoes were studied, three facts strikingly emerged. First, volcanoes are not randomly placed on the Earth but are instead grouped into narrow belts—one surrounding the Pacific, in particular. Second, volcanoes are of different types, differing in origin, appearance, and behavior. Third, eruptions are also of different types, though individual volcanoes often produce more than one type, even during the same eruptive episode. By the beginning of the twentieth century, therefore, the understanding of volcanism and its role in the history of the Earth had become much more sophisticated.

Two of the most important twentieth century discoveries about volcanoes were that they are even more numerous in ocean basins than they are on land and that those on land are almost always associated with established or incipient plate boundaries. With very few exceptions, continental volcanoes occur as a by-product of subduction, the process through which one tectonic plate slides underneath another and is melted. Many theorists believe that the Earth is less thermal now than it was in the past—

Houses crushed by an ash-slide following the 1912 eruption of Katmai in Alaska. (U.S. Geological Survey)

that its geological activity in this respect is slowing down. Others insist that the most powerful episodes of volcanicity with which humans are familiar have taken place in relatively recent geological times. However this debate is eventually to be resolved, scientists regard volcanoes as the most visible indicators of the Earth's still-active interior processes.

Dennis R. Dean

CROSS-REFERENCES

Eruptions, 739; Forecasting Eruptions, 746; Hawaiian Islands, 701; Krakatau, 752; Lava Flow, 717; Mount Pelée, 756; Mount Pinatubo, 761; Mount St. Helens, 767; Mount Vesuvius, 772; Popocatépetl, 776; Shield Volcanoes, 787; Stratovolcanoes, 792; Tsunamis, 2176; Tsunamis and Earthquakes, 340; Volcanic Hazards, 798.

BIBLIOGRAPHY

Decker, Robert, and Barbara Decker. *Volcanoes.* New York: W. H. Freeman, 1997. This readable and well-illustrated text is geared toward the general public. The authors explain geologic phenomena in a clear and simple manner. The appendices list the world's 101 "most notorious" volcanoes, as well as Internet sources for volcano information.

Francis, Peter. *Volcanoes: A Planetary Perspective.* Oxford: Clarendon Press, 1993. This is a university-level text, but it is written in a style that makes it accessible to a wide range of readers. The author covers the visible evidence of volcanoes, as well as the theory of why they develop. Illustrations, index, and reference list.

Macdonald, Gordon A. *Volcanoes.* Englewood Cliffs, N.J.: Prentice-Hall, 1972. Similar to Bullard, but more recent and with a larger number of eruption narratives, this remains a very useful book, lucidly written and richly informative.

Sigurdsson, Haraldur, ed. *Encyclopedia of Volcanoes.* San Diego, Calif.: Academic Press, 2000. This book contains a complete summary of the scientific knowledge of volcanoes. It contains eighty-two well-illustrated overview articles, each of which is accompanied by a glossary of key terms. Although this is a college-level text, it is written in a clear and comprehensive style that makes it generally accessible. Cross-references and index.

Sigurdsson, Haraldur, and Steven Carey. "The Far Reach of Tambora." *Natural History* 97 (June, 1988): 67-73. A nontechnical, well-illustrated account of the 1815 eruption and its worldwide effects.

Simkin, Tom, et al. *Volcanoes of the World: A Regional Directory, Gazetteer, and Chronology of Volcanism During the Last Ten Thousand Years.* Stroudsburg, Pa.: Hutchinson Ross, 1981. A book to be consulted rather than read, this one lists all known eruptions for all known volcanoes.

SHIELD VOLCANOES

Shield volcanoes are the products of eruptions of low-viscosity materials. They have been identified on five solar system bodies. The most active volcanoes on the Earth are shields.

PRINCIPAL TERMS

CORONAE: ring structures on Venus consisting of alternating concentric ridges and valleys, higher than the external terrain; possibly volcanic in origin

HOT SPOTS: areas of anomalously hot mantle

IGNIMBRITE: rock formed by widespread deposition and consolidation of block, pumice, and ash flows

MONOGENETIC: pertaining to an eruption in which a single vent is used only once

PATERAE: inverted, saucer-shaped features considered to be of volcanic origin

POLYGENETIC: pertaining to volcanism from several physically distinct vents or repeated eruptions from a single vent punctuated by long periods of quiescence

PYROCLASTIC MATERIALS: broken rock formed by volcanic explosion or aerial expulsion from a volcanic vent

THOLUS: an inverted, bowl-shaped feature considered to be of volcanic origin

VOLCANO CLASSIFICATION

The shield volcano is so called because of its similarity to the ornate shields of the ancient Nordic warriors. More specifically, the name is derived from the Icelandic word *dyngja* (shield). The typical shield volcano is circular to ellipsoidal when viewed from above; its "ornamentation" can take the form of superposed cinder cones, radiating lava flows, a summit or flanking crater or craters, and rift valleys. Viewed from the side, the shield can take forms ranging from inverted saucers to inverted mixing bowls. In this respect, shields differ from the familiar volcano shape represented by the symmetrical, conical Mount Fuji in Japan.

Most shields are basaltic in composition and formed primarily of lavas (more than 98 percent). Exceptions are found in shields that have lavas that are richer in silica and of andesitic composition, such as Hayli Gub in Ethiopia, and those that are poorer in silica, such as the carbonatitic volcanoes of the East African Rift zone. There are also shields composed primarily of pyroclastic materials, although frequently they are separately classified.

One of the earlier attempts to produce a systematic classification of volcanoes was made by Alfred Rittmann in 1936. This scheme did not achieve wide recognition until the second German edition of *Vulkane und ihre Tätigkeit* (1960; *Volcanoes and Their Activity*, 1962) was translated into English. In this text, Rittmann recognized a Hawaiian type and an Icelandic type of shield volcano. A similar subdivision was recognized by Sir Charles Cotton in his text *Volcanoes as Landscape Forms* (1944). In the most influential of the English geomorphology textbooks of the twentieth century, *Principles of Physical Geology* (1944), Arthur Holmes classified the shield volcanoes as "domes of external growth" to distinguish them from the domes of viscous lava built by the addition of material to the interior.

With the acquisition of images of volcanoes on other bodies in the solar system, there began a new phase in volcano classification. These new images provided views of volcanoes produced on bodies with different gravities, atmospheres, compositions, and other important parameters. The volcanologist now needed a database with which to make objective comparisons of the morphologies of the volcanoes on the different bodies. One such attempt was made by James Whitford-Stark in the book *Volcanoes of the Earth, Moon, and Mars* (1975). This classification was made on the basis of the basal diameter versus the height of the shield volcanoes and resulted in the identification of four classes of terrestrial shield volcanoes: Hawaiian, Icelandic, Galápagos, and scutulum. (The last is from the

Latin *scutum*, or "shield.") Lunar domes were shown to be similar to Icelandic shields, whereas Martian volcanoes had dimensions exceeding those of most terrestrial volcanoes.

A more detailed analysis was undertaken by Richard Pike and Gary Clow, who employed a statistical technique called principal component analysis. They first distinguished between polygenetic and monogenetic shields and then erected subgroups on the basis of size, shape, composition, and location. The monogenetic shields, shields produced by only one eruption phase, included the large Icelandic shield, a smaller steep shield, and a low shield. The polygenetic shields included a Hawaiian type, a Galápagos type, and a locally varied type, all of which are characterized by tholeiitic basalts; an oceanic and continental type dominated by alkali-rich basalts; and a group represented by seamounts, or submarine volcanoes. The more silicic of the shields were grouped into alkalic and calcalkalic ashflow plains categories. Pike and Clow also recognized three distinct groups of Martian volcanoes and a group of lunar domes. Volcanic landforms have since been recognized on the Jovian satellite Io and on Venus. Unfortunately, there is a paucity of topographic data for these features, so objective comparisons cannot yet be made.

Fernandina Island, in the Galápagos Archipelago, is topped by a large, low-lying shield volcano. (© William E. Ferguson)

TYPES OF SHIELD VOLCANOES

Small low and small intermediate-slope shields are monogenetic and variously described as scutulum-type shields, lava cones, or low shields. These shields are abundant in the Snake River plain volcanic field of the western United States and are represented by the volcano Mauna Iki, which erupted between 1919 and 1920 on the flanks of Kilauea. Typically these shields have lower slopes of about 0.5 degree and steepen to about 5 degrees near the summit. This slope change has been attributed to a change from extensive, thin, fluid, pahoehoe flows early in the eruption to more viscous, aa flows during the waning eruption. Many have summit craters formed by collapse, and a few have spatter ramparts around the vent. In the Snake River plain, the shields are aligned along fractures and probably represent initial fissure-type eruptions that subsequently contracted to point-source eruptions.

Small steep shields are typified by Mauna Ulu, which was produced in an eruption lasting from 1969 to 1974 and punctuated by a three-and-one-half-month period of quiescence in late 1971 and early 1972. The early phase was marked by lava fountains reaching heights of up to 540 meters, whereas in the second phase, the fountains rarely exceeded 80 meters. Nearly 350×10^6 cubic meters of lava contributed to the final structure. The eruption was characterized by a summit lava lake. Sustained overflows of this lake led to the production of tube-fed lava flows that traveled up to 12 kilometers from the vent. Short-duration overflows resulted in growth of the shield structure. The final structure rose 121 meters above the pre-1969 base and had a basal diameter of slightly more than 1 kilometer. The smaller Icelandic and seamount shields also fall within this category.

Medium low shields are polygenetic constructs represented by calcalkalic ash-flow plains and the smaller of the alkalic ash-flow plains. The calcalkalic ash-flow plains are composed primarily of rhyolitic or dacitic ignimbrites, rather than lavas, surrounding a caldera ranging from 7 to about 60 kilometers in diameter. Toward the caldera, lavas and pyroclastic fall deposits tend to dominate, and the flank slopes increase from about 1 degree to about 8 degrees. These structures are the products of extremely powerful eruptions that, fortunately, have not occurred within historic times. The Venusian coronae, if of volcanic origin, would probably fall within this group. Medium intermediate-slope shields consist of a wide variety of the groupings established by Pike and Clow. Included are the steeper of the ash-flow sheet shields, the majority of the continental alkalic basaltic shields, and the smaller of the oceanic and Martian shields. Medium steep shields are represented primarily by the Icelandic shields, the larger of the seamounts, the steeper continental shields, and the small steep oceanic shields. The small-volume members can be monogenetic, whereas the large-volume members are invariably polygenetic.

Large low shields are represented by the terrestrial Toba volcano, the Martian paterae, and the larger of the Venusian volcanoes, such as Colette and Sacajewea. Large intermediate-slope shields include the majority of the Hawaiian volcanoes, other terrestrial oceanic shields, and the Martian tholii. The mature Hawaiian shields typically have radiating rifts, a summit collapse caldera, and superposed small shields and cinder cones of a more evolved composition than the mass of the volcano. The large steep shield group includes the Galápagos volcanoes and other oceanic shields. The Galápagos shields differ from the Hawaiian shields in being much steeper in the summit region, having concentric rifts, and having a greater portion of more alkaline basalts.

The very large low shield category is currently occupied by one Martian volcano, Alba Patera. It is probable that a number of the Io volcanoes will be found to fall into this category. Very large intermediate shields are represented by the giant Mons volcanoes on Mars (Olympus Mons is the largest volcano in the solar system). Data suggest that these very large shields are dominated by lava flows.

STUDY OF SHIELD VOLCANOES

The morphologies of terrestrial shield volcanoes are readily obtainable from topographic maps, their compositions are obtained via standard geochemical techniques, their ages can be ascertained by radiometric dating, and their interior structures can be determined by observation of eroded structures. On other bodies in the solar system, these quantities are less easily obtained. Topography can be determined by measuring shadow lengths or determining the time delay between the sending and receiving of electromagnetic radiation by such instruments as laser altimeters. Since there are rock samples only from the Moon (excluding an exotic group of igneous meteorites found on the earth), the compositions of other extraterrestrial materials have to be determined by analysis of their spectral characteristics. Relative ages of the outermost layers of extraterrestrial volcanoes can sometimes be determined from the density of superposed impact craters—the more craters, the older the surface. The internal structure of these volcanoes cannot be determined, since none of their parent bodies has as powerful agents of

A lava fountain, or fire fountain, on Hawaii, 1983. (U.S. Geological Survey)

erosion as those found on the Earth.

One of the reasons for assembling morphometric data for terrestrial volcanoes is to ascertain any relationships between the size and shape of the volcano and the forces that led to its construction. Then, by making corrections for the different gravities, atmospheres, and other parameters on the nonterrestrial bodies, one can infer the conditions that led to the construction of their shield volcanoes.

In spite of the wide range in the volumes of shield volcanoes, all appear to be constructed of low-viscosity material, whether it be basaltic lava or silicic ignimbrite. A major difference exists in the volume eruption rates between the lava and ignimbrite shields. The lava shields are formed primarily by an eruption style called Hawaiian, which is characterized by volumetric eruption rates of about 100 cubic meters per second for short durations but less than 1 cubic meter over time spans of a few years for the smaller structures, to about 1 million years for the larger Hawaiian shields. The ignimbrite shields are products of a plinian or ultraplinian eruption style and are thought to have had volumetric eruption rates of the order of 1 million cubic meters per second for eruptions lasting less than one day.

A feature that serves to distinguish the majority of terrestrial shield volcanoes from cone volcanoes is that the shields are commonly not found at plate margins; that is, they are predominantly intraplate volcanoes. It has been suggested that many shields are associated with hot spots, or areas of anomalously hot mantle. The paucity of terrestrial continental shields with heights in excess of 2 kilometers implies that the Earth's oceans are acting as buttresses to prevent the collapse of the oceanic shields. The great heights of the Martian Mons volcanoes is probably, in part, a function of the lower gravity on that planet.

SIGNIFICANCE

Shield volcanoes include the most active volcano on the Earth (Kilauea), the tallest volcano in the solar system (Olympus Mons on Mars), and perhaps some of the highest eruption columns (more than 200 kilometers high, on Io). They have undergone some of the most violent eruptions ever to take place on Earth, such as the eruption at Toba Caldera, Sumatra. On Earth, shields are present on the continents and on the ocean floor, and they occur as oceanic islands. From the seafloor to its summit, Mauna Loa is the largest mountain on Earth, rising some 10 kilometers above its base. Small shields resembling those of Iceland are found on the Moon. Gigantic shields have been identified on Mars. Active shield volcanoes are present on Io, and numerous shieldlike structures occur on Venus.

Chains of terrestrial shield volcanoes extending across a plate and formed by a single, fixed hot spot have been employed to determine the direction of motion of the plate, and the ages of the various volcanoes along the chain have been used to infer the plate's rate of motion. At the large and very large end of the shield volcano scale, inferences can be drawn as to the nature of the planet's interior, such as the internal heat transfer as a function of time and the strength of the outer layers of the planet required to support such huge masses. The heights of shield volcanoes have been employed

Aerial view of snow-covered Mauna Loa, a shield volcano on the big island of Hawaii, with several craters visible. (U.S. Geological Survey)

to determine lithosphere thicknesses, and the horizontal separations between volcanoes have been used to infer the depths in the interior at which magma originated.

The high eruption frequency of terrestrial shield volcanoes, such as those of Hawaii, is of importance to those living in the paths of lava flows, which may destroy their homes. The enormous scale of eruptions associated with the formation of the ash-flow shields has fortunately not been witnessed within historic times, but such eruptions have been relatively frequent throughout geologic history and may recur in the future. On a more positive note, shield volcanoes can provide a source of tourist revenue; furnish new, rich arable land; and endow geothermal energy.

James L. Whitford-Stark

CROSS-REFERENCES

BIBLIOGRAPHY

Decker, Robert, and Barbara Decker. *Volcanoes.* New York: W. H. Freeman, 1997. This readable and well-illustrated text is geared toward the general public. The authors explain geologic phenomena in a clear and simple manner. Includes a good section on Hawaiian shield volcanoes. The appendices list the world's 101 "most notorious" volcanoes, as well as Internet sources for volcano information.

Francis, Peter. *Volcanoes: A Planetary Perspective.* Oxford: Clarendon Press, 1993. This is a university-level text, but it is written in a style that makes it accessible to a wide range of readers. The author covers the visible evidence of volcanoes, as well as the theory of why they develop. Illustrations, index, and reference list.

Lunar and Planetary Institute, Houston, Texas. *Basaltic Volcanism on Terrestrial Planets.* Elmsford, N.Y.: Pergamon Press, 1982. A compendium, by several dozen expert authors, of the majority of information published in English relating to basaltic volcanism. Chapter 5 is particularly pertinent to shield volcanoes. Aimed at a college-level audience.

Sigurdsson, Haraldur, ed. *Encyclopedia of Volcanoes.* San Diego, Calif.: Academic Press, 2000. This book contains a complete summary of the scientific knowledge of volcanoes. It contains eighty-two well-illustrated overview articles, each of which is accompanied by a glossary of key terms. Although this is a college-level text, it is written in a clear and comprehensive style that makes it generally accessible. Cross-references and index.

Simkin, Tom, et al. *Volcanoes of the World.* Stroudsburg, Pa.: Hutchinson Ross, 1981. A summary of locations, dimensions, and eruption histories of all terrestrial historic eruptions. For the reader desiring a detailed source.

STRATOVOLCANOES

A volcano that erupts both cinder and lava is a composite volcano, or stratovolcano. These volcanoes are found at sub-ducting tectonic plate margins and are the most abundant of the large volcanoes. The tallest and most famous stratovolcanoes of the world, such as Vesuvius, Fuji, and Mount St. Helens, have produced the most devastating and violent eruptions.

PRINCIPAL TERMS

ANDESITE: a gray volcanic rock with a silica content of about 60 percent

ASH: fine volcanic ejecta less than 2 millimeters in diameter

CALDERA: a large circular basin with steep walls that resulted from the collapse of a summit or of an earlier volcanic cone

CINDER CONE: a small volcano composed of cinder or lumps of lava containing many gas bubbles, or vesicles; often the early stage of a stratovolcano

PUMICE: a solidified volcanic froth of a glassy texture light enough to float on water

PYROCLASTICS: fragmented ejecta released from a volcanic vent

RHYOLITE: a light-colored volcanic rock composed of a viscous lava containing about 70 percent silica

TEPHRA: all pyroclastic materials blown out of a volcanic vent, from dust to large chunks

VISCOSITY: the resistance of lava to flow; it depends upon the chemical composition, temperature, and crystalline nature of the magma

CHARACTERISTICS OF STRATOVOLCANOES

The most abundant type of large volcano on the Earth is the stratovolcano, also referred to as a composite volcano. These classic "poster volcanoes" constitute 80 percent of all active volcanoes, including Mount Fuji, Mount Vesuvius, Mount St. Helens, and Krakatau. Stratovolcanoes vary in elevation from a few hundred to 4,000 meters or more above base levels, with diameters approaching 40 kilometers. Their slopes vary from 10 to 35 degrees, with an increase toward the summit because of larger amounts of volcanic material deposition there. In humid regions, much of this volcanic material, in the form of tephra, is washed downslope in mudflows, but when transported aerially, very little tephra actually reaches the base.

The shapes of most of these volcanoes are symmetrical, with eruptions from a single pipe vent at the center, although some build up around long fissure vents. Mount Hekla in Iceland is an example of the latter, with a 5-by-10-kilometer oval shape. In the former case, a central crater, usually a funnel-shaped hole, is found on the summit. The central vent tends to remain in the same position despite the explosive events. Often, the crater may be enlarged by wall collapse because of magma withdrawal. A flat central floor in the crater may be more prevalent in humid areas, where heavy rains wash fine sediments downward.

Stratovolcanoes may reach great heights, but these large, symmetrical cones require a riblike structure of lava for support. Pure cinder cones, in contrast, rarely exceed 500 meters in elevation because their symmetry tends to be destroyed by slumping. One eruptive cycle may be an effusive flow of lava, followed in some later period by an explosive event that produces cinder and ash layers. The relative proportions of tephra and lava in stratovolcanoes' structures vary considerably, from pure ash or cinder to pure lava flows. Rarely do volcanoes exhibit classic "layer-cake" geology—that is, alternating layers of lava and ash. Some mountains (for example, Taal in the Philippines) consist of 70 to 80 percent pyroclastics, but others are dominated by lava eruptions.

ERUPTION PATTERNS

Eruption patterns vary from one stratovolcano to another; three basic types are described here. The Vulcanian type of eruption, characteristic of many active volcanoes, produces a solid crust that lasts until the next eruption. Gas pressure builds up in the magma column, eventually blowing out the solidified obstruction. A great explosion may follow, accompanied by a large, dark, cauliflower-shaped eruption cloud. With the reduction of pressure, gas-charged magma is replaced by pumice and ash. After the vent has cleared, lava flows may issue forth from the crater. Vesuvius, the stratovolcano east of Naples, falls within this group of eruptions.

The Strombolian eruption is distinguished by glowing fragments of lava accompanied by a white eruption cloud. In contrast to the Vulcanian eruption, which contains much ash, the Strombolian cloud contains very little ash. The crust that forms over the lava column is very thin, allowing for frequent, mild eruptions. Stromboli, a stratovolcano located on an island west of Italy, is the classic example, but some volcanoes may exhibit Strombolian activity during some portion of their history.

The Peléan eruption gives rise to the expulsion of nuées ardentes. Pumice and ash are characteristic; no liquid lava develops. The magma is under such intense pressure that it is shattered into a fine dust and mixed with superheated steam. This mixture rolls over the top of the crater as an avalanche of red-hot dust. In the final stages of eruption, the gas content of the magma is greatly reduced, no longer breaking the surface but instead pushing upward to form a dome with spinelike projections, as in the case of Mount Pelée on the island of Martinique.

Stratovolcanoes tend to be positioned at the margins of descending or subducting masses of ocean floor or tectonic plates beneath the continents. The western United States has thirty-five volcanoes that have erupted and may erupt again in the future. One region of volcanic activity is found in the Cascade Range, extending from Washington and Oregon to Northern California, with a continuation into Nevada and through Idaho to Yellowstone National Park. Another volcanic zone winds through southeastern Utah into Arizona and Mexico.

In terms of frequency of eruption, volcanoes are divided into two groups. The first comprises those that have experienced eruptions on average every two hundred years and have last erupted within that period of time. Examples in this group are Mount St. Helens, Lassen Peak, and Mounts Shasta, Rainier, Baker, and Hood, all in the Cascade Range. The other group includes those whose last eruption was more than one thousand years ago; their frequency of eruption is greater than one thousand years. Examples are Crater Lake Volcano and Mounts Adams and Jefferson, also located in the northern Cascade Range.

The stratovolcano gains its heat from subducting-plate friction. The magma tends to have a lower temperature and a higher viscosity than does the basaltic magma of shield volcanoes (found at rift zones and located over hot spots). The eruptions are explosive because of the high silica and gas content and high viscosity. Rhyolite and andesite are often present in the flows, along with volcanic products called pyroclastics. Not all stratovolcanoes, however, erupt lava containing such high amounts of silica; Mount Etna in Sicily and Mount Fuji in Japan tend to have a silica composition of 50 to 60 percent in a basaltic lava.

STRATOVOLCANIC ERUPTIONS

Several stratovolcanic eruptions have had significant consequences for people living nearby. Located in the southeastern state of Chiapas, Mexico, El Chichón experienced a series of explosive eruptions on March 28, 1982, killing 187 people and leaving another 60,000 homeless. A large dust cloud, which was monitored by satellite, rose to an altitude of 25 kilometers and lasted for one month after the eruption. Scientists recorded 50 centimeters of ashfall 16 kilometers away and 20 centimeters of ashfall 80 kilometers away. In Palenque, 120 kilometers east of the volcano, ashfall was measured at 40 centimeters.

On the eastern shore of Sicily, Mount Etna, the largest and highest of the European volcanoes, rises 3,320 meters above sea level. Small subsidiary cones around its flanks, up to 1,000 meters high, erupt rather frequently, but only rarely does sufficient energy accumulate to cause an eruption from the summit crater. Early eruptions, several centuries B.C.E., were recorded by the Greeks. Thousands of lives have been lost and several towns destroyed during Etna's history. In 1669, twenty thousand died in an eruption. The town of Mascati was destroyed in 1853, and in 1928 the village of Nunziata was nearly leveled. Explosive eruptions near the summit in 1979 claimed nine lives. Mount Etna is classified as a transitional volcano, transforming from shield to stratovolcano, and has had more than 150 eruptions since the first recorded one in 1500 B.C.E.

On June 6, 1912, one of the greatest eruptions of the twentieth century occurred on the Alaska Peninsula in the Valley of Ten Thousand Smokes. Mount Katmai is a complex stratovolcano, with both a caldera and a lake on its summit. The explosion was heard in Juneau, 1,200 kilometers away. The valley was flooded with 25 cubic kilometers of pumice, ash, and gas and was filled to a depth of nearly 200 meters. In Kodiak, some 160 kilometers away, pumice and ash nearly blocked out the sunlight, reducing visibility to about 2 meters. The upper 1,900 meters of the mountain collapsed, forming a giant caldera more than 2 kilometers wide, which filled with water and became a crater lake. A new vent, Mount Trident, on the west flank of Katmai, became active in 1949 and has since experienced several thick lava flows, the last erupting in 1974.

Stromboli, one of the most active volcanoes, lies on a Liparian island 60 kilometers north of Sicily. It has been in a continuous state of eruption its entire history, although the intensity varies considerably. It was known for centuries to sailors as the Lighthouse of the Mediterranean because eruptive flashes from its summit were visible far out to sea. Even during World War II, it was used by Allied bombers as a navigational aid. The more explosive eruptions occur at about fifteen-minute intervals, with moderate activity in between. The principal crater, located 600 meters up the 900-meter mountain, is always full of semifluid lava and thus does not greatly resist magma pressure from beneath.

Mount Tambora, on the island of Sumbawa, Indonesia, began a series of eruptions on April 5, 1815, that did not end until July of that year. The early explosions sounded like cannon fire and could be heard 720 kilometers away. The greatest eruptive activity occurred on April 11 and 12, when explosions were heard in Sumatra, 1,600 kilometers to the west. There were only a few survivors of a population of twelve thousand on the island, and forty-five thousand other lives were lost on surrounding islands. Volcanic ash was so heavy that there was darkness at noon in Java, 480 kilo-meters distant. The dust sent into the atmosphere obscured sunlight more than had any other volcanic dust within the previous four hundred years. The year 1816 was known as the "year without summer."

Vesuvius is the only active volcano on mainland Europe. The famous eruption of August 24, 79 C.E., was described by Pliny the Younger from a distance of 30 kilometers. His account describes a strange cloud that rose up and away from the mountain as well as roars and explosions coming from deep within the mountain. The residents of Pompeii were overcome with searing ash and toxic gases. Some sixteen thousand lives were lost, mainly by suffocation; the residents of nearby Herculaneum, for example, were overcome by boiling mudflows. Another violent eruption took place in 1631, causing the loss of eighteen thousand lives. During World War II, on March 18, 1944, an eruption forced the evacuation of the city of Naples, and lava flows caused extensive damage at an air base close to the mountain. Mount Vesuvius is a complex stratovolcano, with lava flows generally following explosive eruptions.

VOLCANOLOGY

A major goal in volcanology is the prediction of the time, place, and nature of an eruption to minimize loss of lives and property. A variety of methods are employed, for no single factor is sufficient to warn of impending eruption.

Geophysicists monitor the earthquakes and tremors that accompany volcanic eruptions with seismographs that measure the intensity of earthquake activity. Almost every eruption or increase in volcanic activity is preceded by earthquakes—often very small earthquakes, or microearthquakes, that occur in swarms of hundreds or thousands. A volcanic tremor is almost always present during an eruption and often begins before the surface outbreak. This motion has a natural frequency from 0.5 to 1.0 hertz and produces a very low hum. The source of this noise is not clearly understood, but it may be related to the formation of gas bubbles in the lava.

Small changes in the slopes, shown by distances between markers on the flanks and summits of active volcanoes, are another precursor of volcanic activity. Techniques for detecting such changes include conventional leveling, reflected light beams, and instruments called tiltmeters capable of measuring changes in slope of less than one part in one million. One type of tiltmeter uses brass pots filled with water and fastened to support posts connected by hoses. The posts are placed a few meters apart. Micrometers that measure fractional water-level change are placed inside the pots.

With the rise of magma toward the surface, the Earth's magnetic field is disturbed by molten lava, which loses its magnetic properties at high temperatures. Such changes may be detected by a magnetometer carried aerially over the volcano. Moving magma

Mount Ngauruhoe is a stratovolcano on New Zealand's North Island. (© William E. Ferguson)

within the volcanic chambers may cause changes in the electrical currents within the ground. Sensitive resistivity meters placed under the surface are used to measure electrical conductivity. An increase in temperature of hot springs, fumaroles, and groundwater may signal an eruption; several months before the 1965 eruption of Taal in the Philippines, the temperature of the water inside the crater increased by 12 degrees Celsius. The chemical composition of waters and volcanic emissions may also change during eruptions. An increase in sulfur dioxide or hydrochloric acid in groundwater or surface emissions can precede an eruption.

Direct visual observations are helpful in monitoring volcanoes. Ice and snow on a volcano may melt from the intense heat. A volcano may bulge from movement of magma along one of its flanks, as was observed prior to the 1980 eruption of Mount St. Helens. Active volcanoes are observed from nearby observatories or sentry outposts. Valuable contributions are made by volcanologists in these facilities, but close observation can be very dangerous.

SIGNIFICANCE

The majority of the world's six hundred active volcanoes lie within the Pacific Ocean Basin, with about one-half in the western Pacific region. They form an almost continuous pattern known as the Ring of Fire along the edges of the Pacific. Stratovolcanoes that form along the ocean margins are termed marine stratovolcanoes. Their initial structure resembles basaltic seamounts, with pillow basalts building up on the seafloor, followed by pyroclastics and less effusive eruptions.

Deep marine volcanoes are less explosive than volcanoes on land, because the pressure of the seawater retards the expansion of steam within the magma. Subduction-zone and island-arc volcanoes, on the other hand, contain a high concentration of gases in the upper parts of their magma chambers, making them more explosive than are volcanoes in other locations. The lava tends to have a greater amount of silica, so it is more viscous and resistant to flow. The lava that solidifies within the vent, or throat, forms a hardened plug that is blasted into pieces, or pyroclastics, by the pressure of the gas trapped below. Lava flows that are rough and blocky are termed aa, while those with a smooth, ropy texture are called pahoehoe.

Island-arc stratovolcanoes have had an explosive and dangerous history. Volcanoes such as Krakatau and Tambora in Indonesia as well as Rabaul in Papua New Guinea are examples. It is believed that these volcanoes are economically important for the deposition of metallic sulfides. The magma chambers beneath may be regions where copper, molybdenum, and gold are concentrated during the final consolidation.

It is important to study volcanoes to minimize volcano damage, including loss of lives. Various methods have been devised to reduce volcano damage; these include the digging of channels and construction of levees to divert lava flows. Lava streams have been doused with water in attempts to slow down and solidify the lava. In Java, dams have been constructed to direct volcanic mudflows away from cities and agricultural lands. In Hawaii, the U.S. Air Force has bombed lava flows emerging from Mauna Loa, with limited success.

Some volcanoes eject considerable amounts of ash and dust into the atmosphere. Certain massive volcanic eruptions had a significant effect on global temperatures for months afterward. The 1963 eruption of Mount Agung, a stratovolcano in Bali, Indonesia, ejected clouds of gas and dust 10 kilometers into the stratosphere. The average worldwide temperature dropped 0.5 degree Celsius for three years after the eruption. Whether volcanic eruptions do effect significant changes in the weather and climate cannot be satisfactorily answered until scientists can obtain accurate measurements of the volcanic dust and debris entering the stratosphere and the heating and cooling of the Earth's atmosphere that result from such dust.

Michael Broyles

CROSS-REFERENCES

Andesitic Rocks, 1263; Calderas, 683; Eruptions, 739; Forecasting Eruptions, 746; Island Arcs, 712; Krakatau, 752; Mount Pelée, 756; Mount Pinatubo, 761; Mount St. Helens, 767; Mount Vesuvius, 772; Popocatépetl, 776; Pyroclastic Rocks, 1343; Recent Eruptions, 780; Shield Volcanoes, 787; Spreading Centers, 727; Subduction and Orogeny, 92; Volcanic Hazards, 798.

BIBLIOGRAPHY

Bullard, F. M. *Volcanoes of Earth.* Austin: University of Texas Press, 1984. A definitive study of volcanoes that includes sections on mythology and early speculation, how volcanoes are classified, the types of volcanic eruptions, and the distribution, current activity, theory, and environmental effects of volcanoes. Many diagrams, photographs, and tables make this a valuable resource for the geology student.

Cas, R. A. F., and J. V. Wright. *Volcanic Successions, Modern and Ancient.* Winchester, Mass.: Allen & Unwin, 1987. A very detailed work on the classification of modern and ancient volcanic rocks and volcano source types. Detailed geologic columns and cross sections show types and thickness of volcanic rock successions. A chapter on stratovolcanoes includes tables of dimensions and average lifetime rates. Suitable for the student of geology.

Chester, D. K., A. M. Duncan, J. E. Guest, and C. R. J. Kilburn. *Mount Etna: The Anatomy of a Volcano.* Stanford, Calif.: Stanford University Press, 1985. A comprehensive study of the major volcano on the island of Sicily. An interesting historical chronology of both flank and summit eruptions is included. Another table presents the effects of eruptions and earthquakes on settlements in the region. Includes many black-and-white photographs. Some chapters are technical and are best suited to the geology student, while other sections, especially the historical activity, would be of interest to the general reader.

Decker, Robert, and Barbara Decker. *Volcanoes.* New York: W. H. Freeman, 1997. This readable and well-illustrated text is geared toward the general public. The authors explain geologic phenomena in a clear and simple manner. Includes a section on stratovolcanoes. The appendices list the world's 101 "most notorious" volcanoes, as well as Internet sources for volcano information.

Erickson, Jon. *Volcanoes and Earthquakes.* Blue Ridge Summit, Pa.: TAB Books, 1987. An excellent resource for the general reader. A chronology is given for the major eruptions and their toll in human lives; more detailed descriptions of fourteen different volcanic eruptions follow this table. Chapters have been added on the formation of the solar system and the evidence for plate tectonics and continental drift. Volcanoes that have been detected on the other planets and moons are also described.

Francis, Peter. *Volcanoes: A Planetary Perspective.* Oxford: Clarendon Press, 1993. This is a university-level text written in a style accessible to a wide range of readers. The author covers the visible evidence of volcanoes, as well as the theory of why they develop. Illustrations, index, and reference list.

Sigurdsson, Haraldur, ed. *Encyclopedia of Volcanoes.* San Diego, Calif.: Academic Press, 2000. This book contains a complete summary of the scientific knowledge of volcanoes. It includes eighty-two well-illustrated overview articles, each of which is accompanied by a glossary of key terms. Although this is a college-level text, it is written in a clear and comprehensive style that makes it generally accessible. Cross-references and index.

VOLCANIC HAZARDS

Hazards associated with volcanic activity include explosions that eject liquid magma and rock fragments ranging from dust-sized particles to blocks of solid rock, lava flows, glowing avalanches, ash flows, volcanic mudflows, and gases. Active volcanoes typically lie within relatively narrow zones in close proximity to tectonic plate margins, coinciding with principal earthquake zones. The type of volcanism, with associated hazards, differs from one kind of plate boundary to another.

PRINCIPAL TERMS

ASH FLOW: a density current composed of a highly heated mixture of volcanic gases and ash, which travels down the flanks of a volcano or along the ground surface

COMPOSITE VOLCANO: a volcano built of alternating layers of lava and pyroclastic deposits, along with abundant dikes and sills

LAHAR: a mudflow composed chiefly of volcanic debris on the flanks of a volcano

MAGMA: a body of molten rock, including any dissolved gases and suspended crystals

MUDFLOW: a general term for a flowing mass of predominantly fine-grained earth material

that possesses a high degree of fluidity during movement

PYROCLASTIC: composed of clastic (rock-derived) material formed by volcanic explosion or aerial expulsion from a volcanic vent

SHIELD VOLCANO: a volcano in the shape of a flattened, broad, and low dome built by flows of very fluid lava or ash flows

TEPHRA: a general term for all pyroclastic material and shreds of liquid magma formed by volcanic explosion or aerial expulsion from a volcanic vent

MAGMAS AND GASES

Volcanoes are essentially vents in the Earth's surface through which magma—a combination of molten rock, dissolved gases, and suspended crystals—and associated gases and ash erupt or are ejected from the interior onto the surface of the Earth. The nature of the eruption is largely dependent on the proportion of gas to molten rock and how readily the gas escapes. Eruptions can occur only once or frequently, with the proportion of gas to molten rock varying widely. Hazards associated with volcanic activity commonly occur in the form of lava flows, glowing avalanches, ash flows, volcanic mudflows, tephra falls, glacier bursts, and volcanic gases.

GASES

The latter originate in the magmas the rise out of the Earth to activate the volcano. Gases in magma are soluble at high temperatures, deep in the Earth, but as magma rises, the pressure exerted at depth decreases, and some of the gas is no longer in solution. Gas, which first appears as

small and widely spaced bubbles, increases in proportion and begins to coalesce as the magma continues to rise. As the gas pressures then increase, an explosion is prevented by the confining pressure of the magma and, partially, by the viscosity of the magma.

When the gas pressures are high enough, however, the enclosing liquid bursts, gas escapes, and an explosion occurs. If the confining pressures are low and the enclosing magma is relatively fluid, gas bubbles escape readily, with rather mild eruption of magma at the surface. If the confining pressures are high, however, and continue to increase, with gas bubbles escaping with much difficulty, eruptions are very violent and of an explosive nature. If the magma comes in contact with water, in the form of seawater, surface water, or groundwater, violent steam explosions are possible, as documented at Kilauea Volcano on the island of Hawaii. Magma evolution, however, eventually causes an increase in viscosity and gas content, which leads to progressively shorter and thicker lava flows, formation of domes composed of many individual flows formed by the extrusion of highly fluid lava, and increasingly explosive eruptions.

Volcanoes can be broadly divided into two groups based on magma composition: basaltic volcanism and acidic volcanism. Basaltic volcanism is associated with the oceanic crust, commonly producing large quantities of lava with minimal explosive activity. Those active volcanoes situated on the Hawaiian Islands and Iceland's Kirkefell are examples of this type of volcanism. Acidic volcanism is associated with destructive plate margins and tends to be violent, with an unpredictable pattern of behavior and long periods between eruptions. The relatively high viscosity of the magma can induce a buildup of gas pressures, resulting in massive explosions such as those of Krakatau and, more recently, Mount St. Helens. This type of volcanism can produce flowing gas clouds or surges of pyroclastic material without any warning during periods of prolonged and less turbulent activity. Hot gases charged with varying proportions of red-hot ash can descend a mountainside at speeds greater than 100 kilometers per hour, bringing instant destruction as in the case of Pompeii by the eruption of Vesuvius in 79 C.E.

LOCATION OF VOLCANOES

The most explosive volcanoes exist in zones where subduction of oceanic plates occurs. The volcano develops on the overlying plate within a few tens of kilometers from its edge. The eruptions of this type of volcano are characterized by tephra constituting from 45 to 99 percent of the volcanic product. Mount St. Helens and the other volcanoes that form the Cascade chain are examples of this type of volcanism. Associated hazards include heavy ashfalls, glowing avalanches, ash flows, mudflows, and lava flows.

EJECTA AND LAVA FLOWS

Explosive-type volcanoes result in the ejection of liquid magma or fragments, or a combination of both, into the air. These fragments, referred to as pyroclastic ejecta, or simply tephra, fall back to the ground or are carried away by the wind hundreds to thousands of kilometers distant. Sand-sized and smaller fragments are referred to as ash or dust; fragments measuring 2 to 60 millimeters in diameter are called lapilli, and fragments having diameters greater than 60 millimeters are called bombs or blocks, depending on whether the material was ejected in a liquid or in angular chunks of solid rock, respectively.

Glowing avalanches, nuées ardentes, are rapidly expanding clouds of dust that are typically black in daylight but a glowing dull red at night. The principal part of this phenomenon, however, is not the dust cloud itself but rather the avalanche of incandescent lava blocks, sand, and dust beneath it. Nuées ardentes can travel at great speeds, with documented cases exceeding 150 kilometers per hour. As a result of such great speed, when opposing hill slopes or the bend of a valley wall are encountered, the avalanche is capable of climbing vertically as much as several hundred meters. Ash flows closely resemble glowing avalanches in mobility and cause; however, the flow is composed primarily of an emulsion of bits of very hot glass, referred to as ash, and occasional lumps of pumice. Debris avalanches and landslides result from the sudden downslope movement of loose materials on the flank of a volcano.

MUDFLOWS

Mudflows, which are also known as lahars, are slurries of solid fragments in cool to hot water, associated with explosive-type volcanoes where loose fragmented material is abundant. Flowing downhill under the influence of gravity, the flows are largely controlled by topography and can attain speeds of 90 kilometers per hour, with some documented at 160 kilometers per hour. Mudflows have the notoriety of having destroyed more property than any other single volcanic process. Also attributed to them is the loss of thousands of lives.

GLACIER BURSTS

Glacier bursts occur by volcanic eruptions generated beneath glaciers, as in Iceland. For periods of a few hours, these bursts may achieve volumes greater than that of the world's largest river, with documented cases exceeding velocities of 92,000 cubic meters per second and volumes of liquid exceeding 6 cubic kilometers.

TSUNAMIS

As an effect of volcanic activity, huge waves known as tsunamis may be generated by abrupt displacement of the seabed. Great volumes of water are involved, and the travel speeds of the waves can reach 800 kilometers per hour; a tsunami may arrive at a shore, after being slowed by the shallower water and increasing in amplitude, as a breaker up to 20 meters high. In 1896, a tsunami was responsible for the loss of 27,000 lives in Japan. The tsunami resulting from the explosion of Krakatau in 1883 caused nearly all the deaths from that disaster. Fortunately, tsunamis are relatively infrequent, occurring less than once every ten years; nearly all of them originate in the Pacific Ocean.

EVALUATING VOLCANIC RISKS

Evaluation of volcanic risks and hazards on or near any volcano is heavily dependent on an understanding of its historic behavior; the assumption is that a volcano is most likely to behave like it has behaved in the past. This evaluation is based on the recorded history of eruptions and geological studies of the composition and structure of the cone.

Attention to volcanic eruptions has been increasingly directed toward the assessment of risks to life and property. A problem arises, however, in defining when a volcano is actually dead or extinct and thus presents no further risk. For example, volcanoes can remain quiet or dormant for thousands of years and then, abruptly, erupt. In fact, in some of the most destructive volcanic events, no documentation of previous eruptions exists and the volcanoes were thus considered to be inactive; the eruption of Vesuvius in 79 C.E. is an example.

As far back as the beginning of the twentieth century, maps have been prepared identifying areas around known dangerous or active volcanoes, based on the historical evidence. Noted have been the eruptive characteristics of a particular volcano, especially the effect of topography on pathways of flows. Certain areas are typically coded to determine the types of risks to be expected given differing types of activity. For example, the primary risks associated with the Hawaiian volcanoes (Kilauea and Mauna Loa) are almost wholly from lava flows, whereas with the Cascade and Alaskan volcanoes, the primary risks are from ashfalls and mudflows. Thus, with the former, topography is considered principally, whereas with the latter, meteorological conditions are monitored.

FORECASTING ERUPTIONS

Earth scientists contribute to the reduction of volcanic hazards through an understanding of the nature and geographic distribution of these hazards. Attention presently is directed toward the assessment of risks to life and property. The solution is to minimize these risks for those people who live in close proximity to volcanoes. Thus, the practical goals of volcanology are to be in a position to provide ample warning of the timing, type, and location of future eruptions; to minimize the effects of eruptions and associated phenomena; and to rehabilitate devastated areas in an efficient and effective manner. Great loss of life was averted during the 1991 eruption of Mount Pinatubo in the Philippines by the action of volcanologists, whose warnings led to the evacuation of sixty thousand people prior to the main eruption.

With sufficient warning, people and property can be evacuated. Prediction of the timing of an eruption uses a combination of indicators. Unfortunately, no volcanic district as a whole is moni-

An automobile engulfed by lava flow from Kilauea on the island of Hawaii. (© William E. Ferguson)

tored sufficiently in order to avoid the loss of life and property completely. Monitoring the temperature of fumaroles, hot springs, and crater lakes may serve to warn of an eruption because, in some cases, a rise in temperature has precluded an eruption. Changes in gas discharge and composition also may forewarn of an imminent eruption. Additionally, an alteration of the strength or orientation of the Earth's magnetic field may precede an eruption; this indicator is based on the hypothesis that increased temperature of a volcano should reflect a decrease in the overall strength of its local magnetic attraction.

Innovative methods of attempting to predict volcanic eruptions, which appear to show much promise, have included monitoring of deformation of the ground surface and earthquake behavior. Eruptions are commonly preceded by local earthquakes, which result from the opening of fissures at depth through which the magma rises toward the surface. Earthquakes are also produced by the movement of blocks in and around the volcano as it swells or, in the case of Kilauea, as an indication that the underlying magma reservoir is being filled and an eruption is possible. Volcanic tremors (rhythmic vibrations of the ground surface, reflecting the movement of magma) often occur just before or during eruptions, although tremors may also occur without an eruption. Changes in the Earth's electrical currents within the volcano—where rapid changes in Earth currents have occasionally been observed prior to eruptions—are monitored and inferred to precede a potential eruption. Finally, the disturbed and excited behavior of animals on or near the volcano previous to its eruption has been observed, suggesting that certain animals may sense earthquakes that are too small in magnitude to be felt by humans.

BENEFITS TOO

Volcanoes are probably the most impressive as well as the most devastating of phenomena. Overall, however, volcanoes have done more good than harm by creating thousands of square kilometers of very fertile land surface. In addition, the use of volcanic heat in the form of natural steam and hot water continues to be developed to produce electricity and heat at a low cost and with minimum pollution to the environment. In fact, most episodes of volcanic unrest typically end without an eruption; eruptions are thus the exception rather than the rule.

Nevertheless, millions of people continue to live in close proximity to active volcanoes, risking the potential destruction all too often unpredictably associated with them. As developable land becomes scarcer, urban development progressively encroaches upon these areas. During the last five hundred years, more than 200,000 people have lost their lives as a result of, at minimum, five hundred active volcanoes—and that figure does not include the casualties resulting from tsunamis. Lives were lost not only directly, by the volcanic activity itself, but also by the destruction of food crops and livestock, which led to deaths from starvation.

Stephen M. Testa

CROSS-REFERENCES

Calderas, 683; Eruptions, 739; Forecasting Eruptions, 746; Hawaiian Islands, 701; Krakatau, 752; Magmas, 1326; Mount Pelée, 756; Mount Pinatubo, 761; Mount St. Helens, 767; Mount Vesuvius, 772; Plate Margins, 73; Plate Tectonics, 86; Popocatépetl, 776; Pyroclastic Rocks, 1343; Recent Eruptions, 780; Shield Volcanoes, 787; Spreading Centers, 727; Stratovolcanoes, 792; Tsunamis, 2176; Tsunamis and Earthquakes, 340.

BIBLIOGRAPHY

Bolt, B. A., W. L. Horn, G. A. Macdonald, and R. F. Scott. *Geologic Hazards.* Rev. ed. New York: Springer-Verlag, 1977. Chapter 2, "Hazards from Volcanoes," presents a comprehensive summary account of geological hazards in general, including an excellent discussion of volcanic hazards. The text is elementary and descriptive in nature and aimed at the layperson and the college-level reader.

Decker, Robert, and Barbara Decker. *Volcanoes.* New York: W. H. Freeman, 1997. This readable and well-illustrated text is geared toward the general public. The authors explain geologic phenomena in a clear and simple manner. Includes chapters on forecasting eruptions and reducing volcanic risk. The appendices list the world's 101 "most notorious" volcanoes, as well as Internet sources for volcano information.

Francis, Peter. *Volcanoes: A Planetary Perspective.* Oxford: Clarendon Press, 1993. This is a university-level text, but it is written in a style that makes it accessible to a wide range of readers. The author covers the visible evidence of volcanoes, as well as the theory of why they develop. Illustrations, index, and reference list.

Gribbin, John R. *This Shaking Earth: Earthquakes, Volcanoes, and Their Impact on Our World.* New York: G. P. Putnam's Sons, 1978. This book provides a popularized, well-illustrated discussion of current scientific theory, volcano types and mechanisms, and their impact on human society. Suitable for high school readers.

Lipman, Peter W., and D. R. Mullineaux. *The 1980 Eruptions of Mount St. Helens, Washington.* U.S. Geological Survey Professional Paper 1250. Washington, D.C.: Government Printing Office, 1981. This source presents the early but extensive and comprehensive results of studies of the eruption and associated volcanic events that occurred in 1980, including the geophysical monitoring and studies of volcanic deposits, effects, and potential hazards. Suitable for college-level readers and those more technically inclined.

Sigurdsson, Haraldur, ed. *Encyclopedia of Volcanoes.* San Diego, Calif.: Academic Press, 2000. This book contains a complete summary of the scientific knowledge of volcanoes. It contains eighty-two well-illustrated overview articles (including a sequence of twelve articles on volcanic hazards), each of which is accompanied by a glossary of key terms. Although this is a college-level text, it is written in a clear and comprehensive style that makes it generally accessible. Cross-references and index.

Smith, David G., ed. *The Cambridge Encyclopedia of Earth Sciences.* New York: Crown Publishers, 1981. Chapter 26, "Geologic Hazards," presents a general discussion of volcanic hazards in relation to the two major magma compositional groups. A well-illustrated and carefully indexed reference volume suitable for the general and college-level reader.

6
Mountains and Mountain Ranges

ALPS

The Alps are a mountain range that forms an arc across south-central Europe. Originally, this area was under a sea that lay between Europe and Africa. When Africa began moving in a northerly direction more than 65 million years ago, the rocks were highly folded and finally uplifted to form these mountains, which continue to rise at a rate of 0.5 millimeter per year. Glaciers, streams, and slope processes have eroded them to their present striking appearance.

PRINCIPAL TERMS

CONGLOMERATE: sedimentary rock composed of gravel in a sandy matrix

DEBRIS AVALANCHE: a large mass of soil and rock that falls and then slides on a cushion of air downhill very rapidly as a unit

DEBRIS FLOW: a mass movement of high fluidity in which more than half the solid material is greater than sand size

GLACIER: a mass of ice showing motion and originating from the compaction of snow

GNEISS: metamorphic rock formed under high pressure and temperature

GRANITE: igneous rock originating from the cooling of magma slowly under the ground

LANDSLIDE: a relatively rapid movement of soil and rock downslope

LIMESTONE: sedimentary rock made of calcium carbonate

MORAINE: deposit of glacial till

NAPPE: a complex, large-scale rock fold on its side where some beds are overturned

ROCKFALL: a relatively free-falling movement of rock material from a cliff or steep slope

SANDSTONE: sedimentary rock composed primarily of sand grains

SCHIST: metamorphic rock with subparallel orientation of micaceous minerals that dominate its composition

SHALE: sedimentary rock composed of silt and clay particles

SNOW AVALANCHE: a relatively rapid movement of snow downslope

THRUST FAULT: a break in a rock body at a low angle of inclination where the hanging wall has moved up in relation to the footwall

TILL: unsorted, unconsolidated material deposited directly by glacier ice

LOCATION

The Alps are the famous mountains of south-central Europe. They make a huge arc extending from southern France through Switzerland into Austria, Germany, Yugoslavia, and Italy. This range is approximately 800 kilometers long and covers more than 207,000 square kilometers. It lies about halfway between the North Pole and the equator (from 44 to 48 degrees north latitude). The Alps are essentially part of a large mountain chain that extends from Europe through Greece to the higher ranges of Iran and Central Asia. All these mountains were formed at about the same time.

The Alps are subdivided into the western, central, and eastern sections, and each section contains a number of prominent subranges. The western Alps lie primarily in France and rise from the shores of the Riviera and the plains of It-aly to form subranges of the Maritime, Cottian, Dauphiné, Savoy, and Graian Alps. The central Alps lie mainly in Switzerland with the primary subranges being the Pennine, Bernese Oberland, Lepontine, Glarus, and Rhaetian Alps. The Pennine Alps are the highest and most spectacular of the subranges and lie along the French-Swiss-Italian borders. They extend 95 kilometers from Mont Blanc (4,807 meters), the highest mountain in the Alps, to Monte Rosa (4,634 meters). Also located in this range is the famous glacially carved Matterhorn (4,478 meters), which lies on the border between Switzerland and Italy. Just north of the Pennine Alps, separated by the Rhone Valley, lies the Bernese Oberland, the second-highest subrange of the Alps. The Jungfrau (4,158 meters) is one of the highest and most beautiful mountains in this range. The eastern

Alps have less lofty peaks but are still noted for their spectacular scenery. The subranges are found in several countries: the Bavarian Alps in West Germany, the Dolomite and Carnic Alps in northern Italy, the Julian Alps in Yugoslavia, and the Hohe Tauern, Noric, and Stubai Alps in Austria.

FORMATION

For many, it is hard to conceive of the idea that the rocks that make up the Alps were once under the sea. During the Mesozoic era (about 225-65 million years ago), the continental shorelines were different. A large sea, called the Tethys Sea, lay between the continents of Africa, which formed the south shore, and Europe, which formed the north shore. Spain, Italy, Greece, and Turkey were actually small microplates at the western end of this sea and were not attached to Europe.

A large trough of sediment, or geosyncline, developed during the Mesozoic era in the Tethys Sea between the continents and extended all the way to Indonesia. Sediments eroded from the uplands of the continents collected in this geosyncline and formed sedimentary rocks on top of the basement rock of basalt and intrusions of granite. During the Triassic period (about 225-190 million years ago), some limestones were formed in the geosyncline, but most of the rocks were formed in the Jurassic (about 190-135 million years ago) and Cretaceous (about 135-65 million years ago) periods. Other limestones, marls, shales, and localized deposits of sandstones and conglomerates were formed during this time. Some metamorphic, green, lustrous schists, called Bundner Schiefer, were also created during the Cretaceous period as mafic magma intruded into the sediments.

The Cenozoic era (the last 65 million years) marked the major episode of mountain building of the Alps. The African plate began moving in a northerly direction at the beginning of the Cenozoic. By this time, the Italian microplate, called the Carnic plate, had moved to the center of the Tethys Sea. The intense compressional forces of Africa pushed the Carnic plate northward into the geosyncline sediments. The European continent acted as an immovable object. First, the compression produced complicated folding and faulting of the sedimentary rocks in the geosyncline. The Alps are noted for the highly complicated anticlines (upfolds), synclines (downfolds), recumbent folds (folds on sides), and thrust faults (rock breaks at low angles) created during this time. Most of the intense compression occurred from the Eocene through the Miocene epochs (about 55-5 million years ago). Second, these rocks of the geosyncline between the Carnic microplate and the European continent were uplifted by the compression to form the Alps as the crust was reduced in width by as much as 250-400 kilometers. The rate of closing was approximately 5 centimeters per year. It is believed that this uplifting process is continuing at a rate of 0.5 millimeter per year. The mountain building was not a single episode but was episodic and interrupted by prolonged periods of relative calm. Most of the uplift has occurred in the last 30 million years.

NAPPES

A nappe is a large recumbent fold that may be kilometers across and that is generally bounded by a thrust fault. The largest and most studied nappes in the world are in the Alps. The Alps were the first mountain range in which nappe structures were found to play a predominant role in development. The nappe concept was developed in the late nineteenth century in the Swiss Alps when geologists recognized thick sequences of older rocks thrust faulted over younger rocks, a sequence typical of nappes. To explain these structures, Hans Konrad Escher von der Linth and his student Albert Heim described the Glarus overthrust in terms of a nappe, and the concept was born.

The incredible crustal shortening and the uplift formed these large overlapping nappe structures. For example, the Pennine Alps are made of seven major nappes that came from the central portion of the geosyncline. They form the metamorphic core of the Alps and were formed from very old sedimentary rocks with the help of the intense temperatures and pressures of regional metamorphism during the mountain-building process. The most common metamorphic rocks are gneisses, mica schists, the green Bundner Schiefer schists, phyllites, and slates. The Bernese Oberland Alps are also called the High Calcareous Alps because they are composed mainly of six nappes that originated close to the European side of the geosyncline and are made mainly of limestone. Also in this range are the ultrahelvetic nappes made of flysch, a sandstone formed during the mountain-building process from sediment formed from the erosion of other nappes.

The Pre-Alps are just north of the Bernese Oberland and also are mainly limestone. According to some theories, they originated far to the east of the nappes of the High Calcareous Alps in the geosyncline but slid over them via gravity sliding during the uplift. The nappes that make up the eastern Alps and the western Alps came mainly from the part of the geosyncline closest to Africa. The Jura Mountains, which lie to the northwest of the Alps on the Swiss-French border, also came from the same geosyncline, and they came from the portion closest to the European continent.

The Alps were continually being eroded during their uplift. Based on the sediment deposits at the edges of the Alps, it has been estimated that the mass of the Alps has been reduced during the last 30 million years into about one-quarter of its original volume. Erosion of the Alps created fine-grained sediment that became flysch and a coarse-grained sediment that became a rock called molasse. The Swiss Plateau, just north of the Alps, is composed of abundant flysch and molasse formed from 60 to 10 million years ago. Overall, the Alps are made primarily of sedimentary rocks formed in the Tethys Sea between 225 and 30 million years ago. Limestone is the most abundant of the sedimentary rocks. The central core of the Alps is mainly metamorphic rock. Large intrusions of granite are found near Mont Blanc and the Aar and St. Gotthard massifs and are older than the sedimentary rocks formed in the geosyncline. Other small outcrops of older Paleozoic rocks (rocks older than 225 million years) from the base of

the geosyncline have been found in the Alps. Fossils of ammonites and brachiopods formed during the Mesozoic and Cenozoic are common in the rocks.

ALPINE EROSION: GLACIERS

Since being uplifted, the Alps have been carved to their present shapes and forms through the processes of erosion. Ice erosion during the ice age of the last 2.5 million years has been the major action of change. Water erosion in the form of streams has also provided extensive transforma-

The French Alps. (PhotoDisc)

tion. In this steep terrain, gravity also has played a major role in erosion through landslide and avalanche production.

The Great Ice Age began about 2.5 million years ago as the climate of the world cooled. Glaciers grew and filled the stream-carved valleys of the Alps, some to a thickness of more then 2,000 meters. Only the peaks of the mountains protruded from this large ice field. These large moving masses of ice advanced and retreated numerous times as the climate cooled and warmed, each time eroding the mountains, transporting the rock debris, and eventually depositing it as till. Glaciers carved out the landforms seen everywhere in the Alps. Circular basins on the sides of mountains called cirques, knife-edged ridges between valleys named arêtes, and pointed mountain peaks called horns were carved by the ice.

Many of the peaks listed on a map of the Alps end in "horn" because they have been carved by glaciers. Most of the V-shaped valleys that had been formed by streams before the Ice Age were broadened to a more U-shaped form by the ice. The glacial debris that was carried by the ice was eventually deposited when the ice melted as moraine, a deposition of glacial till. Ridges of moraine on the edges of valleys and arcuate ridges at the ends of valleys are lateral and terminal moraines produced by the glaciers at their maximum extent. Lakes also formed behind some of these terminal moraines in the valleys. The city of Geneva in Switzerland is located on one of these terminal moraine dams formed by the Rhone glacier. Massive deposits of glacial till lie at the edges of the Alps where the glaciers ended.

More than 1,200 glaciers still exist in the Alps today, most of them in the Pennine and Bernese Oberland Alps. They continue to wear down the sides of mountains and transport glacial debris. Most are found at high elevations, but some, such as the Grindelwald Glacier in Switzerland, can descend to an elevation of 1,000 meters. The Aletsch glacier of the Bernese Oberland is considered to be the largest glacier of the Alps; it is 25 kilometers long and has a surface area of 170 square kilometers. Many of the streams in the Alps have a milky green color that is produced by abundant silt from glacial abrasion and shows the glacial origin of the water. Most of the glaciers are observed to be in recession, suggesting a warming of the climate. This trend is attributed to a steady temperature increase caused by human activity, particularly the emission of greenhouse gases and the destruction of forests.

ALPINE EROSION: WATER

Water for great rivers and lakes is abundant in the Alps and comes from snowmelt and rainfall. This water feeds the many powerful streams that erode the valleys, forming V-shapes. Some deep, stream-carved gorges have been formed in the past 10,000 years since the last glaciers melted, with some, such as the Aar and Trient Gorges in Switzerland, more than 150 meters deep. Water has also carved smaller landforms, such as the Swiss Pyramids, near Sion, which are pedestals of glacial moraine with boulders on the top. The four most important rivers in Europe, the Rhone, the Rhine, the Po, and the Danube, all contain abundant water from the melting of snow in the Alps. The St. Gotthard massif at the end of the Pennine Alps has been called the water tower of Europe because it is the headwaters of so many of these great rivers. Abundant large lakes are relics of the ice age and add to the scenic beauty of the Alps. Many of the lakes are deep, as they fill valleys deepened and dammed by glacial action. Lake Geneva is the largest, with a surface area of 580 square kilometers and a depth of 310 meters.

ALPINE EROSION: GRAVITY

Gravity also helps erode the Alps through landslides and snow avalanches. Small landslides and rotational slumps are common on these steep slopes, especially after heavy rainfalls. The world's second-largest dam disaster occurred in 1963, when a landslide fell into the lake in back of the Vaiont Dam in the Italian Alps. A wall of water more than 100 meters high proceeded downvalley and killed 2,600 people in the village of Longarone and the surrounding towns.

Debris flows are formed in steep mountain stream valleys when they flood and erode the valley bottom. It is common to find large human-made levees, or embankments, at the bottoms of these steep valleys to contain the debris flows from hitting villages at the valley bottoms. Debris avalanches are rare but devastating when they occur. In 1881, a total of 115 people were killed when a debris avalanche moved at a velocity of more than

A TIME LINE OF HISTORIC AVALANCHES

Jan. 17, 1718	LEUKERBAD, SWITZERLAND: 55 dead after an avalanche strikes this town at 4,629 feet in the Swiss Alps.
July 12, 1892	ST. GERVAIS, SWITZERLAND: 140 dead when an avalanche strikes the towns of St. Gervais and La Fayet, Switzerland, in an unusual summer occurrence when a massive chunk of La Tête Rousse glacier breaks free and hundreds of tons of ice and debris slide down the 14,318-foot Mont Blanc.
Mar. 23, 1915	BRITANNIA MINE, NEAR VANCOUVER, BRITISH COLUMBIA: 50 dead when hundreds of tons of snow break loose above the Britannia Mine at Howe Sound and fall on mining bunkhouses.
Jan. 28, 1931	BARDONECCHIA, ITALY: 21 dead from a regiment of men who have climbed Mount Galambra to the northeast.
Feb. 11-13, 1952	MELKOEDE, AUSTRIA: 78 dead after a snowstorm that raged for ten days wreaks havoc in many countries, with Austria bearing the brunt.
Dec. 23, 1952	LAGEN, AUSTRIA: 23 dead when a blast of air preceding an avalanche blows a bus of tourists off a bridge on the Flexenstrasse mountain road into the Aflenz River 18 feet below.
Jan. 11-14, 1954	AUSTRIA, GERMANY, ITALY, SWITZERLAND: 145 dead after a succession of avalanches during a fierce winter blizzard buries families, farms, and entire villages throughout the area.
Jan. 10, 1962	ANDES MOUNTAINS, PERU: Estimated 4,000 dead, $1.2 million in crop damage when melting ice causes the edge of a huge glacier on the peak of Mount Huascarán to break away.
Feb. 18, 1965	LEDUC CAMP, NEAR STEWART, BRITISH COLUMBIA, CANADA: 26 dead, 17 seriously injured, and a mining camp mostly destroyed by snow and ice when part of the Leduc glacier slides into an 11-mile mining tunnel, trapping 40 people.
Feb. 10, 1970	VAL D'ISÈRE, FRANCE: 42 dead, 60 injured in what was called the worst avalanche in French history, when more than 100,000 cubic yards of dry-powder snow slides down the Val d'Isère ski resort and smashes through the picture windows of a hotel.
Mar. 19, 1971	CHUNGAR, PERU: 600 dead when an earthquake in the Andes sets off a landslide that pours into a lake, causing water to wash down a lower face, creating an avalanche of snow, mud, and rock.
Jan. 31, 1982	SALZBURG, AUSTRIA: 13 dead when, after two days of unseasonably warm weather, an avalanche rumbles down from a steep, craggy ridge to engulf 19 cross-country skiers.
Mar. 14, 1982	FRENCH ALPS: 16 dead when an avalanche triggered by a sunny warm spell following two days of heavy snowfall engulfs skiers.
Oct. 27, 1995	FLATEYRI, ICELAND: 20 dead as a result of two avalanches: one that descends on a herd of horses in Langidalur, and one that crashes down on the harbor in Sugandafjor, killing two men.
Mar. 16-18, 1996	KASHMIR, INDIA: 72 dead when tons of snow overcome frictional resistance on the deeply cut slopes above the village of Kel, Azad "Free" Kashmir, India; the avalanche uproots a pine forest on its way down.
Feb. 9, 1999	CHAMONIX, FRANCE: 12 dead when a dry-powder avalanche rumbles down Mount Pléceret in the Chamonix Valley, ripping up forests and demolishing chalets in the villages of Le Tour and Montroc-le-Planet.
Feb. 14, 1999	NEAR MOUNT BAKER, WASHINGTON STATE: 2 dead when an avalanche sweeps down a chute just outside the boundary of the Mount Baker Ski Area, carrying away many and burying 2.
Feb. 21, 1999	EVÈLONE, SWITZERLAND: 10 dead after the heaviest snowstorms in the Alps in fifty years end in fatal snowslides in Austria, France, Italy, and Switzerland.
Feb. 23-24, 1999	GALTÜR AND VALZUR, AUSTRIA: 38 dead, 10 houses destroyed, 2,000 trapped after several winters of heavy snowfalls in the Alpine countries of Europe.

300 kilometers per hour through the village of Elm. Rockfall is common at the edges of these steep, U-shaped valleys, and no one ever knows when a boulder may break away and head for the valley bottom. Soil creep, where the soil is moving at 1 to 2 centimeters per year down the slope, is also a problem on most slopes.

Each year, snow avalanches kill between twenty and thirty people in the Alps. The worst avalanches happen when there is a large snowfall on steep slopes that cannot hold the snow. The year of 1951 was called the winter of terror because there were 265 deaths in the Alps and abundant damage. It was a year of plentiful snowfall, with falls of 3 to 5 meters in a three-hour period occurring on numerous occasions. In 1916 during World War I, more than ten thousand soldiers were killed in the eastern Alps by snow avalanches.

With the abundant water seeping into the ground of the numerous areas with limestone bedrock, many caves have been formed in the Alps. The world's deepest cave, the Gouffre Jean Bernard, is found in the French Alps and is 1,535 meters deep. The 140-kilometer-long cave at Holloch, Switzerland, is the second-longest in the world. The Elizabeth Casteret cave in the French Alps is one of the largest single caverns in the world. The largest underground lake in Europe, a flooded cave formed in gypsum between layers of marble and schist, is located at St. Leonard in the Bernese Oberland of Switzerland.

STUDY OF THE ALPS

Compared to most other high mountain ranges of the world, the Alps are highly populated and have been for many years. Scientists have been studying the Alps since the early nineteenth century, making them the most studied mountain range in the world. Through sheer determination, skill, and thoroughness, several generations of geologists have spent hours putting together an incredible story of the origin of these mountains. Many books have been written on the general geography, geology, structure, glaciers, past glaciations, landforms, and snow avalanches of the Alps. The Swiss have always been world leaders in the production of maps. Guillaume-Henri Dufour took thirty-two years to collect data and produce the first topographic map of Switzerland in 1864, and this twenty-five-page document was a master-

piece. At a scale of 1:100,000, it was extremely accurate. By 1920, excellent topographic and geologic maps were available for all parts of the Alps.

Because of the excellent exposures of rocks in the valley walls, the Alps became a favorite field area for geologists from around the world. The Alps provide some of the most exciting places in the world for the study of rocks (petrology) and the study of the folding and breaking of rocks (structural geology). All parts of the mountains have been intensively studied as to the rock types and the folding and faulting. Thousands of geologic maps and cross sections have been produced. These local maps have then been interpreted in relation to the rest of the Alps and their relationship to the original geosyncline deduced. Classic works of the early twentieth century by Albert Heim and Leon Collet are still treated with respect. When most of the rocks had been studied, geologists began to reinterpret sites, and this comprehensive reevaluation analysis continues.

The early work of the structural geologists explained how each part of the Alps came from a particular part of the geosyncline, but it could not provide a mechanism for the movements of these vast zones of rocks. With the acceptance and development of the plate tectonics theory in the 1960's and 1970's, the geologists could then supply a process for the development of the parts of the Alps. The 1970's and 1980's were fruitful for alpine geologists, who combined the very accurate mapping and cross sections of the classical view of the geology of the Alps with the modern ideas of plate tectonics.

Many railroad and automobile tunnels were cored through the Alps during the twentieth century to help travelers avoid high mountain passes. The Simplon, Mont Blanc, Great St. Bernard, Lötschberg, and St. Gotthard Tunnels have given geologists a very important chance to see inside the mountains to confirm their ideas of the origins of different sections of the Alps.

The study of glaciers and the development of the idea of an ice age both had their roots in the Alps. After many hikes in the mountains and studies on glaciers, Swiss naturalist Louis Agassiz wrote *Études sur les glaciers* in 1840. He introduced two radical ideas: Glaciers are actually moving and are eroding as they proceed downvalley; and the Alps had been covered by a vast ice sheet in the past similar to Greenland, and this ice had produced much of the glacial till found in and around the Alps. Agassiz's work laid the foundation for the whole study of glaciology.

Scott F. Burns

CROSS-REFERENCES

BIBLIOGRAPHY

Ager, D. V., and M. Brooks, eds. *Europe from Core to Crust.* London: John Wiley and Sons, 1977. The section on alpine tectonics and their relationships to plate tectonics by Jean Aubouin is very good.

Anderson, J. G. C. *The Structure of Western Europe.* Oxford, England: Pergamon Press, 1978. The chapter on alpine fold belts is superb and puts the geology of the Alps in perspective of plate tectonics. Excellent coverage of geological structures.

Maeder, Herbert. *The Mountains of Switzerland.* London: Allen & Unwin, 1968. This excellent book includes many black-and-white photographs of the Alps. A fine section on geology, with additional sections on plants, animals, and mountain climbing.

Plummer, Charles C., David McGeary, and Diane H. Carlson. *Physical Geology.* Boston: McGraw-Hill, 1999. This is a straightforward, easy-to-read introduction to geology intended for those with little or no science background. There is a section on plate tectonics and its relationship to the development of the Alps. Includes good illustrations and a CD-ROM.

Press, Frank, and Raymond Siever. *Understanding Earth.* 2d ed. New York: W. H. Freeman, 1998. This comprehensive physical geology text covers the formation and development of the Earth. Includes extensive sections on plate tectonics and the development of the Alps. Readable by high school students as well as general readers. Index and glossary.

Rutten, M. G. *Geology of Western Europe.* New York: Elsevier, 1969. The five chapters on alpine Europe constitute one of the best summaries in English of the geology of the Alps. The view is mainly classical.

Trumpy, Rudolph. *Geology of Switzerland.* Basel: Wepf, 1980. This guidebook was put out by the Swiss Geological Commission and is one of the best English summaries of the geology of the Swiss Alps, which is the most complex portion of the Alps. The author is the leading authority on the subject. The classical view of the geology is considered in relation to modern plate tectonics.

Windley, Brian F. *The Evolving Continents.* New York: John Wiley and Sons, 1995. This general text aimed at college students has a chapter that deals with the development of the Alps. Bibliography and index.

ANDES

The Andes are the classic example of mountains formed by subduction beneath a continent and are the modern analogue for many ancient mountain belts. Earthquakes and volcanic eruptions indicate that the Andes are still rising. Because of their geologic setting, the Andes contain abundant minerals, oil, and natural gas.

LOCATION AND CHARACTERISTICS

The Andes are some of the most impressive mountains on Earth, not only for their grandeur but also for their geologic importance. Their evolution over the past 200 million years can be explained by the subduction, or descent, of an oceanic plate below a continent. Subduction is basic to the theory of plate tectonics, which is the fundamental paradigm of the Earth sciences. The ease with which plate tectonics explained the Andes convinced numerous scientists of the validity of the theory and established the mountains as the classic example of the active, or Pacific, type of continental margin. Furthermore, many researchers consider the Andes a modern analogue for the Sierra Nevada Range of California and for the Appalachian Mountains of the eastern United States; these areas are said to have formed along Andean-type margins more than 100 million and 450 million years ago, respectively.

The Andes rise 7,000 meters and stretch 7,500 kilometers along the western edge of South Amer-

ica from Venezuela to southern Chile. The rugged terrain and the topographic relief attest to the youth of the mountains. Elevation drops rapidly from the high peaks of the Andes to the floor of the Pacific Ocean, a few hundred kilometers west, where a trough 7,000 meters deep, the Peru-Chile Trench, lies parallel to the continental margin. The mountains and trench together are known as the Andean arc and are active today. Indeed, the periodic volcanic eruptions are consequences of subduction at the Peru-Chile Trench and are reminders that the Andes are still forming.

The Andes can be divided into three morphological provinces that parallel the South American continental margin. From west to east these are the western cordillera ("cordillera" means "mountain system"), the altiplano, and the eastern cordillera. Each is distinct physiographically and geologically. The provinces are easily delineated in the Andes of southern Peru, Bolivia, and northern Chile, where the features are related exclusively to subduction. North and south of this region, the Andes are more complex, and their geologic evolution is not as straightforward. The descriptions of the provinces, therefore, are for the central Andes.

The western cordillera begins 100 kilometers inland from the coast. It rises 6,000 meters in 50 kilometers and contains Mesozoic and Tertiary volcanic rocks, whose compositions are so distinct that the rocks are called andesites after the mountains in which they lie. The eastern flank of the western cordillera descends 1,000 meters in 15 kilometers to the altiplano, a high plateau (3,800 to 4,500 meters) greater than 150 kilometers wide. Beneath the altiplano is a basin filled with 10 kilometers of Tertiary sedimentary and volcanic rocks.

Superimposed on the Mesozoic and Tertiary rocks of the western cordillera and the altiplano is a chain of huge volcanoes (7,000 meters) that began forming less than 15 million years ago and that presently erupt. The eastern boundary of the altiplano rises abruptly to altitudes near 6,000 meters and marks the transition to the thrust belt of the eastern cordillera. Rocks of the eastern cordillera were deposited along the South American continental margin between 450 million and 250 million years ago prior to the onset of subduction. During that time, the west coast of South America looked like the present East Coast of North America. These ancient continental margin rocks now lie more than 250 kilometers inland from the present continental margin. East of the eastern cordillera is the Brazilian Shield, a piece of old continental crust, or craton, next to which the Andes have grown.

EVOLUTION

Before the evolution of the Andes can be described, the theory of plate tectonics must be summarized. The theory states that about twelve rigid plates define the surface of the Earth. The plates, which are either continental or oceanic, ride on a partially molten layer and move relative to each other at speeds between 2 and 10 centimeters per year. Plates form at mid-ocean ridges where magma (molten rock) from deep in the Earth ascends to the surface and solidifies into new seafloor.

View of the Andes near Quito, Ecuador. (© William E. Ferguson)

To make room for the new material, old pieces of seafloor on opposite sides of the ridge move away from each other. Thus, the mid-ocean ridge is a divergent plate boundary. Across the oceanic plate from the mid-ocean ridge is a trench that marks the convergent boundary where the oceanic plate moves toward another plate, either oceanic or continental, and subducts, or descends, below it. (Because oceanic crust is denser than continental crust, it always subducts.) If the overriding plate is oceanic, a trench develops in the ocean basin, and a chain of volcanic islands (island arc) grows on the overriding plate. If the overriding plate is continental, the trench sits along the continental margin, and the string of volcanoes forms on the continental plate (continental arc).

The Andes are above a subduction zone, the Peru-Chile Trench, which is adjacent to the continental margin of South America between the latitudes of 4 degrees north and 40 degrees south. There, the Nazca plate, one of several plates in the Pacific Ocean, subducts below the South American continental plate at a rate of 6 centimeters per year. Subduction along the South American margin has been active for 200 million years and, during this interval, has produced the three Andean provinces. Three parameters govern the physiography and geology of the provinces: the amount of sediments that descend with the subducting plate in the trench, the angle at which the subducting plate dives below the continent, and the quantity of magma generated by melting of the subducting and overriding plates that is added to the continental crust at shallow depths.

As an oceanic plate subducts, it bends below the overriding plate. This bend forms a huge depression, or trench, in the ocean floor. Sediments on the descending plate either remain attached to it and subduct or scrape against and accrete to the overriding plate. When the descending plate reaches a depth of 100 kilometers, the temperature of the Earth is hot enough to melt small areas of either the subducting or the overriding plate. Because the magma generated during melting is less dense than the surrounding solid rock, it rises through the Earth, where it either crystallizes at shallow depths or erupts on the surface.

In the Andes, the overriding plate is continental. Sediments remain attached to the subducting plate and travel to great depths in the Earth.

These two factors explain the voluminous andesite in the western cordillera. If magma produced at a depth of 100 kilometers is contaminated either by sediments that were subducted with the oceanic plate or by continental crust, it yields volcanic rocks of andesitic composition.

The locus of volcanism on the surface is controlled by the angle at which the subducting plate descends below the overriding plate. If the angle is steep, the plate reaches a depth of 100 kilometers closer to the trench than if the angle is shallow. Changes in the angle, therefore, cause the volcanoes to migrate. In the modern Andes, the angle of subduction varies between 10 and 30 degrees. In the past, the angle was steeper. Thus the recent volcanoes are at the boundary between the western cordillera and the altiplano, whereas the older ones were in the western cordillera closer to the trench.

The addition of magma at shallow depths in the crust pushes adjacent rocks out of the way. This process causes the surface to extend and rift (fracture) immediately adjacent to the volcanic arc and to shorten farther inland. The shortening is accommodated in a thrust belt that develops approximately 200 kilometers continentward of the volcanic arc. In the thrust belt, rocks that accumulated prior to subduction are deformed by faults and folds into mountains hundreds of kilometers long, tens of kilometers wide, and thousands of meters high. The thrust belt in the Andes is the eastern cordillera. The rifting, in turn, forms a trough, or basin, between the volcanic arc and the thrust belt. The basin fills quickly with debris eroded from the adjacent highlands. Lava flows and ash eruptions from the volcanic arc occasionally are large enough to cover the basin, resulting in the interleaving of volcanic and sedimentary rocks. This is precisely the setting of the altiplano, which is a trough between two highlands: the volcanic chain of the western cordillera and the thrust belt of the eastern cordillera.

Prior to subduction, the eastern cordillera was the west coast of South America. The rise of the Andes implies that the width of South America has increased 250 kilometers in 200 million years by the formation of new continental crust. The new crust below the western cordillera has a thickness greater than 70 kilometers, compared with a thickness of continental crust of 30 kilometers below the Brazilian Shield. The large thickness of continental crust, called a root, is required by the principle of isostasy, which states that a mass excess on the surface must be compensated for by a mass deficiency at depth. Because continental crust is lighter than the mantle (the layer of the Earth below the crust), the space taken up by the continental root (which should be taken up by the mantle) is less dense than it should be. Geophysical data indicate that the crust beneath the Andes is homogeneous and is composed of andesite. Subduction below continental margins, therefore, may be a fundamental mechanism by which continents grow.

STUDY OF THE ANDES

Techniques from a variety of disciplines in the Earth sciences, including geophysics, geology, and geochemistry, are used to study the Andes. Those from geophysics document subduction and a thick continental root below South America presently, whereas those from geology reveal subduction in the past. Geochemical methods provide evidence for relationships between subduction, andesitic volcanism, and formation of continental crust.

The important geophysical tools are seismic reflection profiling and earthquake analysis. Seismic reflection profiling uses sound waves to allow scientists to determine the physical properties of the Earth, the most critical of which is density, several tens of kilometers below the surface. Conceptually, seismic profiling is very simple. Sound waves travel downward from a source and reflect back to sensors on the surface when they encounter a change in the physical properties at depth. Because the velocity

Machu Picchu, Peru, high in the Andes, was a center of the Inca civilization. (PhotoDisc)

of the waves depends on the density of the material in which they are traveling, the time the reflected waves take to reach the sensor contains information about the depth of the change and the composition of the rock above it. From this technique, geophysicists learned that the thickness of the crust below the western cordillera and the altiplano in the central Andes was larger than that below the Pacific Ocean and the Brazilian Shield.

Fitzroy National Park in Patagonia, near the southern end of the Andes range. (PhotoDisc)

They also discovered that the compositions of the crust at the surface and of the crust tens of kilometers below were similar.

The analysis of earthquakes involves finding their location, or epicenter. An earthquake occurs when the Earth's crust breaks along a fault and the two pieces on either side of the rupture move past each other. The energy released by the faulting travels as waves through the Earth. The geophysicist finds the epicenter of the earthquake by measuring the different times at which the waves arrive at various places around the world. Epicenters of earthquakes in the Andes define a zone that extends at an angle from the Peru-Chile Trench to depths hundreds of kilometers below South America. This zone outlines the subducting Nazca plate and provides strong evidence that the west coast of South America is a convergent plate boundary. (These zones, called Benioff zones after the scientist who discovered them, occur along all convergent plate boundaries.)

To determine the evolution of the Andes in the past, geologic fieldwork is necessary. In the field, the geologist establishes the different types of rocks that exist and their relationships to each other. The geologist carefully describes the rocks and documents the faults that may indicate that two adjacent sequences of rocks have had greatly different histories. In the Andes, such faults are absent, suggesting that all the rocks formed near one another. Volcanic rocks, most of which were andesites, were the most abundant and were discovered to have erupted throughout the evolution of the Andes. Thus, scientists believe that the tectonic setting of the Andes has remained the same for the last 200 million years.

In addition to geophysics and geology, geochemistry has contributed to an understanding of the Andes. Geochemists in the laboratory melted pieces of basalt and granite, the primary constituents of oceanic and continental crust, respectively. From this experiment, the scientists determined that andesites solidify from magmas that contain significant amounts of continental crust. This led to the conclusion that large areas of the overriding continent melt where the temperature is high, at great depths in the Earth above subduction zones. The andesitic composition of the root below the central Andes indicated that substantial melting of the continental crust has occurred beneath the volcanic chain.

NATURAL RESOURCES

Adventurers have long sought the Andes not only for their beauty but also for their riches, both of which are direct consequences of subduction. The mountains tower above the landscape and host spectacularly large ore, oil, and natural gas deposits. The earliest geological expeditions, in the first half of the nineteenth century, were hampered by the rugged and remote terrain. The technological advances of the twentieth century, however, solved this problem and allowed the exploitation of the natural resources. Chile is known for copper and Bolivia for tin.

The ore deposits of the Andes are hydrothermal: They form where hot, mineral-rich fluids associated with magmas derived from deep in the Earth interact with cold rock at the surface. These fluids can contain dissolved iron, silver, tin, lead, manganese, molybdenum, zinc, tungsten, and copper. The drop in temperature at the surface forces the minerals to precipitate in the surrounding rock, where, given enough fluid, economically significant ore deposits will form. Large amounts of fluid require abundant magmatism. Because most magmatism on the Earth occurs in subduction zones, and the Andes formed by subduction, the mineral wealth of the Andes is no surprise. Predictably, most mines are in the western cordillera and in the altiplano, where magmatism was and is most active.

The vast oil and natural gas fields of the Andes lie east of the mineral deposits in the thrust belt of the eastern cordillera where the rocks have not been heated by hot fluids or magmas. The location of the fields reflects the low temperatures and thick sedimentary sequences necessary for the transformation of organic matter into oil and natural gas. Faulting in the thrust belt created traps that collected the oil and gas in economically viable deposits.

VOLCANIC AND SEISMIC HAZARDS

The plethora of natural resources makes the Andes one of the world's greatest mountain systems. The Andes, however, are important for still another reason: They yield insight into the potential for destruction that exists above subduction zones.

Subduction zones are the locus of most of the large earthquakes and violent volcanic eruptions that presently occur. Investigations of the Andes may prevent loss of life and property that accompanies these natural events. First, geophysical instruments can monitor a volcano to determine when an eruption is imminent. (The number of small earthquakes increases when magma is moving toward the surface.) This technique successfully predicted an eruption at Mount St. Helens, an andesitic volcano in the western United States. Second, detailed examinations of volcanoes and the surrounding area may reveal the preferred slope for the descent of mudslides that are often triggered by eruptions and earthquakes. Through the centuries, mudslides have killed many hundreds of people and destroyed countless villages. Towns in the paths of slides can be moved or evacuated. Finally, analysis of the ground motion during a major earthquake may help civil engineers design buildings that can withstand the shaking.

Clearly, the ultimate goal is to predict earthquakes and volcanic eruptions far enough in advance that adequate precautions can be taken to minimize the loss of life and property. If prediction is successful in the Andes, disasters along other convergent boundaries may be avoided.

Pamela Jansma

CROSS-REFERENCES

BIBLIOGRAPHY

Francis, Peter. *Volcanoes: A Planetary Perspective.* Oxford: Clarendon Press, 1993. This is a university-level text, but it is written in a style that is accessible to a wide range of readers. The author covers the visible evidence of volcanoes, as well as the theory of why volcanoes develop. Includes a section on andesitic volcanism and the development of the Andes. Illustrations, index, and reference list.

James, D. E. "The Evolution of the Andes." *Scientific American* 229 (August, 1973): 61-69. The article discusses the geology of the Andes and the geophysical tools that were used to document subduction below the western margin of South America. The illustrations are good, particularly a set of diagrams that show the continental margin of South America from 400 million years ago to the present. Not technical but does contain a lot of jargon. Recommended for the senior high school or college student.

Parker, S. P., ed. *McGraw-Hill Encyclopedia of Geological Sciences.* 2d ed. New York: McGraw-Hill, 1988. The encyclopedia contains short articles on a variety of topics. The Andes are not included as a separate entry but are discussed briefly in the sections on mountains and geosynclines. Although not particularly useful for specific information on the Andes, it is suitable for learning more about general processes such as subduction or volcanism. Illustrations are good. Recommended for college-level students.

Plummer, Charles C., David McGeary, and Diane H. Carlson. *Physical Geology.* Boston: McGraw-Hill, 1999. This is a straightforward, easy-to-read introduction to geology intended for those with little or no science background. Discusses plate tectonics and its relation to the development of the Andes. Also deals with andesitic volcanism. Includes many excellent illustrations, as well as a CD-ROM.

Russo, R. M., and P. G. Silver. "The Andes' Deep Origins." *Natural History* (February, 1995). This short article discusses the relationship between plate tectonics and the origins of the Andean mountain chain. Intended for the general reader.

Seyfert, Carl K., ed. *The Encyclopedia of Structural Geology and Plate Tectonics.* New York: Van Nostrand Reinhold, 1987. As its title suggests, this reference is a good source for learning more about plate tectonics. Not technical. Recommended for anyone interested in the subject.

Uyeda, Seiya. *The New View of the Earth: Moving Continents and Moving Oceans.* Translated by Masako Ohnuki. San Francisco: W. H. Freeman, 1978. This book is a good reference for those who would like to learn about the "revolution" in Earth sciences, as some have called the discovery of plate tectonics. Anecdotes of scientists who pioneered the theory are presented. The theory also is explained very well. Nontechnical. For the nonspecialist.

Windley, Brian F. *The Evolving Continents.* New York: John Wiley and Sons, 1995. This general text aimed at college students has a chapter that deals with the development of the Alps. Bibliography and index.

APPALACHIANS

The Appalachian Mountains form one of the most prominent features of the North American continent, dominating the topography of the Atlantic Coast. They mark the eastern continental divide, influence the climate of the region, and provide a record of the transformation of the Earth through a vast range of geological epochs.

PRINCIPAL TERMS

CRATON: a part of a continent that has been free of significant structural rupture for a long time; often a region with a thin covering of newer rocks

EROSION: the process by which the surface of the Earth's crust is gradually broken down and worn away

FAULT: a fracture in the Earth's crust along which there has been some displacement or deformation

FOLD: the deformation of rocks caused by external pressure

GLACIATION: the effect of a glacier on the terrain it transverses as it advances and recedes

LITHOSPHERE: the outer, rigid shell of the Earth's crust; it includes the continental plates and the oceans

METAMORPHIC ROCK: rock altered by means of intense heat

OROGENY: a major event in which tectonic processes combine to radically alter the Earth's crust; mountain building

TECTONICS: the process by which the broken shell of the Earth's crust (the lithospheric plates) moves in varying directions to create geological features

LOCATION AND CHARACTERISTICS

The Appalachian Mountains in their present form are a chain of relatively low but steeply inclined ridges that follow the East Coast of North America for roughly 3,000 kilometers; ranging from Newfoundland in easternmost Canada to Alabama in the southeastern United States, they extend across nearly 20 degrees of latitude and more than 30 degrees of longitude. They were named for an Indian tribe known as the Apalachees (or Apalachis) by the Spaniards who followed Hernando de Soto in 1540 across the southern part of the chain, as de Soto had stayed with the tribe in southern Georgia before his expedition into the mountains. The term "Appalachian Mountains" gradually was applied to the entire range, superseding local designations such as "Allegheny," even though the mountains are not a single continuous range but are rather a complex of separate mountain groups whose boundaries tend to overlap. They range from 25 to 125 kilometers in width and are described by geologist and author John McPhee as a "long continuous welt . . . long ropy ridges of the eastern sinuous welt."

Running roughly parallel to the mountain ridges, valleys of varying width divide the highland ranges. Some of them are narrow and deeply wooded, some broad and gently contoured. They all are essentially a part of the Great Valley of the Appalachians, known locally as the Shenandoah or Tennessee Valley, or some even more specific local designation. In most places, the valley is several hundred meters above sea level, while the highest points of the range rise to 2,037 meters (Mount Mitchell) in the Black Mountains in northwest North Carolina, to 2,024 meters (Clingman's Dome) in the Great Smoky Mountains near the Tennessee border, and to 1,917 meters (Mount Washington) in the White Mountains of New Hampshire. There is no continuous crest to the chain; the highest peaks are found either at the center of the mountain range or on its northwest side. This geologically interconnected series of valleys and ridges has been subdivided into geologic provinces with common general features but distinct local characteristics.

In the southern segment, four belts (or provinces) have been described that run the length of the chain. The farthest northwest is the Appalachian Plateau Province, including the Poconos in

Pennsylvania, the Cumberlands in Kentucky, and the Alleghenies in West Virginia. In spite of the term "plateau," the land frequently is cut by deep valleys and rugged hills. The area supports some small-scale farming but is primarily valuable as a source of coal, oil, and gas, which has led to the growth of rust-belt industry around its rim.

The second belt is designated the Valley and Ridge Province because it is most characteristic of the pattern that typifies the entire chain. The most prominent mountain groups in this province are the Clinch Mountains in Virginia, the Shawangunks in New York, and the Kittatinny Mountains in Pennsylvania. The mountains are arranged along stretches of fertile valleys that are among the most productive farming regions in the eastern United States, valleys fed by the rivers that have formed them in the rivers' courses from the higher peaks. The Delaware Water Gap is typical, with fields and woodlots sculpted by the rush of water over land and, above the flatland, with huge rock deposits such as the Martinsburg formation, a very large aggregate of slate used in blackboards and pool tables.

The third province is called the Blue Ridge, running from the low Highlands of New Jersey to a single, massive ridge through the central Appalachians. There, it rises abruptly from the west and drops off just as sharply in the east. From 10 to 20 kilometers across, it diverges into two ridges south of Roanoke, Virginia, where it runs in two roughly parallel chains through the southern Highlands. Between these two ridges, additional connecting ridges (including the Black Mountains) occur before the two border ridges end fairly abruptly in northern Georgia. The well-known Blue Ridge Parkway, built as a Works Progress Administration project in the 1930's, follows the crest of the eastern ridge from Virginia to the Great Smoky Mountain National Park. This section is the least populated of the Appalachians; not especially suitable for agriculture or industry but scenically exceptional, the Blue Ridge Province has many summer resorts scattered amid the previously isolated communities of mountain pioneer families.

The fourth belt is the Piedmont Plateau, which slopes gradually from the Blue Ridge toward the Atlantic coastal plain. The availability of timber and water resources in this belt has led to the development of an extensive furniture industry as well as

textile mills in many of its small cities, especially along the fringes toward the coastal flatlands.

The Northern Appalachians correspond primarily to the Blue Ridge and Piedmont Plateau Provinces but are not as clearly demarcated. The Blue Ridge, while not actually connected, is replicated by the Hudson River highlands through the Berkshire Hills in Massachusetts and the Green Mountains of Vermont. The remainder of New England and the Canadian Maritime Provinces resemble the Piedmont Plateau. Toward the ocean, the land is generally regular, with minor topographic features, while linear ridges and hilly belts predominate closer to the line of the mountains, where the effect of glaciation is more pronounced. During the early days of the United States, this was a thriving agricultural region, but the immense growth of urban centers has transformed it into a manufacturing and communications complex, and extensive efforts are under way to preserve the unique features of the Appalachian wilderness.

FORMATION

The present form of the Appalachians is very different from that maintained during earlier geological eras. The mountains have risen and then been worn down several times within the scope of historical observation—what McPhee calls "the result of a series of pulses of mountain-building, the last three of which have been spaced across two hundred and fifty million years." Before that, the North American continent formed part of a larger, supercontinental entity. During the latter part of the Proterozoic eon (approximately 600 million years ago), the cratonic cores of North America, Europe, and an immense entity known as Gondwanaland separated and began to drift apart. During that time, the evidence of marine deposits in sedimentary rock indicates that the North American continent was a low-lying, relatively flat landmass. The effects of weathering and erosion over a long period produced this level surface, and glaciation contributed to a worn, scarred appearance. Because of what is known as epeirogenic movement (the rise and fall of landmasses), the craton was sometimes covered to a large extent by water; during the Cambrian period (from 544 to 505 million years ago), much of the North American landmass was submerged, and a tropic or subtropic climate existed.

At that time, the first stirrings of the Taconic orogeny began. This surge of energy, an upheaval in which sheets of oceanic lithosphere were thrust toward the continental interior, produced the first "modern" Appalachian chain. In the Ordovician period (from 505 to 440 million years ago), the sedimentary and volcanic rock that had been built up was fractured and crumpled. The deformation (or deconstruction) of the original mountains was so great that slices of the range were transported considerable distances, folding the mountains upon themselves. The general direction of plate motion was westward, and as the high-density material of the upward-thrusting oceanic plate depressed the low-density material of the craton, forcing it downward into what is called a subduction zone, the low-density material eventually rebounded to produce orogenic waves, which further crumpled and folded the mountain chain.

In the Early Paleozoic era, the continents converged, producing the Acadian orogeny (about 350 million years ago). As the North American and African landmasses moved toward a collision, the violence of the Acadian orogeny folded and faulted the sedimentary rock that remained from the erosion of the Taconian orogeny. The intense temperatures and pressures generated by the mountain-forming process also metamorphosed some of the rock, changing the shales into slates, the sandstones into quartzites, and the limestones and dolomites into marble. The final episode of mountain building took place between 280 and 240 million years ago and is known as the Alleghenian orogeny. Occurring quite some time

Alum Cave Bluff, in Great Smoky Mountains National Park near Gatlinburg, Tennessee, where three new minerals were discovered in 1996. (AP/Wide World Photos)

after the initial contact between North America and Africa, it may have been the consequence of an additional impact, possibly with a microplate (wandering island formation) such as New Guinea, Madagascar, Fiji, or the Solomon Island system. These smaller masses, known also as exotic terrains because of their separation from the main continental blocks, may have collided with the North American continent to trigger the most recent orogenic episode. The proto-Atlantic Ocean was probably in the process of closing when the first collision between North America and Africa occurred, but by the Alleghenian orogeny, the continents were closer to relative fixture (or stability), thereby suggesting that another island arc may have been responsible for the orogeny.

In any event, this last orogenic episode was a time of tremendous upheaval, and one of its products was the creation of the extensive anthracite deposits of eastern Pennsylvania. These coal fields were formed when severe pressure folded carbonaceous rock at least one kilometer beneath the surface. Erosion and an eventual balancing of forces that ended the upheavals brought coal seams closer to the surface, which led to the mining industry of the Appalachian region. Similarly, petroleum deposits—the transmuted fossils of ocean algae—were formed when rocks were heated to about 50 degrees Celsius and remained between 50 and 150 degrees Celsius for at least 1 million years. The rocks that were once part of the western Atlantic Ocean were driven by orogenic waves onto the North American continent, and, in the process of mountain building, in some areas and under the proper conditions of temperature and pressure, they formed oil deposits. The richest deposits in the Appalachians were in Pennsylvania and Ohio, where oil seeped out of the ground in such purity that it could be used prior to refining and was once sold as a health tonic. Much of this oil has been extracted, but geologists speculate that there may still be deposits of oil and natural gas off the south Atlantic Coast—the product of the process that formed the Appalachian Mountains.

Although the Appalachians in their present form are not as spectacular as they once must have appeared when their highest peaks stood from 8 to 10 kilometers above the Earth's surface, the terrain is continuously varied and visually engaging. In the deformed, sedimentary Appalachians—the present topographical form of the chain—the rock "not only has been compressed like a carpet shoved across a floor," as John McPhee observes, "but in places had been squeezed and shoved until the folds tumbled forward into recumbent positions." An overview reveals a consistent sinuosity: The mountains bend right into Georgia, left into Tennessee, right into North Carolina, and so on toward Newfoundland. Some geologists speculate that this pattern represents the coastline of North America in Precambrian times.

Leon Lewis

CROSS-REFERENCES

Alps, 805; Andes, 812; Basin and Range Province, 824; Cascades, 830; Coal, 1555; Continental Glaciers, 875; Continental Rift Zones, 579; Fold Belts, 620; Folds, 624; Himalaya, 836; Mountain Belts, 841; Oil and Gas Distribution, 1694; Plate Tectonics, 86; Rocky Mountains, 846; Sierra Nevada, 852; Subduction and Orogeny, 92; Transverse Ranges, 858.

BIBLIOGRAPHY

Dietrich, Richard V. *Geology and Virginia.* Charlottesville: University Press of Virginia, 1970. A good introduction to the specific issues of geological conditions in one region of the Appalachians, with some features for the intermediate-level student of geology. Includes an extensive glossary and a good list of supplementary readings. The black-and-white drawings and photographs are straightforward and historically interesting.

Fisher, George W., ed. *Studies of Appalachian Geology: Central and Southern.* New York: John Wiley & Sons, 1970. Primarily designed for the serious student of geology or the professional in the field, some specific chapters would be of use to the reader interested in examining a particular facet of Appalachian geology. The epilogue by Philip King is a good historical study of geological exploration in the region, and there is a detailed but comprehensible map of the rock distribution and tectonics of the region.

Jannsen, Raymond E. *Earth Science: A Handbook of the Geology of West Virginia.* Clarksburg, W.Va.: Educational Marketers, 1973. Essentially a textbook for an introductory course in the geology of the state of West Virginia, this book is well organized and written with the interested nonprofessional in mind. Comprehensive and thorough, it includes many maps and charts, an excellent geologic timetable, and a selective glossary. Somewhat dated.

Lessing, Peter, ed. *Appalachian Structures: Origin, Evolution, and Possible Potential for New Exploration Frontiers*. Morgantown: West Virginia University, 1972. A series of papers by professionals who have studied the Appalachian region in great detail. Primarily for the serious professional, but the concluding remarks offer insights for the layperson as well.

Lowry, W. D., ed. *Tectonics of the Southern Appalachians*. Roanoke: Virginia Polytechnic Department of Geological Sciences, 1964. Another series of essays by professionals, the introduction and some of the abstracts illustrate the approaches that were being explored in the 1960's.

McPhee, John. *In Suspect Terrain*. New York: Farrar, Straus & Giroux, 1983. The author's work is well known; his *Basin and Range* (1981) interested many people in geology. Like that earlier book, this one explores ancient terrains in juxtaposition to travels in the modern world. While based on McPhee's observations around the Delaware Water Gap, it presents much information about the structure and geological history of the Appalachian chain. An indispensable book for anyone interested in the Appalachian range.

Murphy, J. B., and R. D. Nance. "Mountain Belts and the Supercontinent Cycle." *Scientific American* (April, 1992). This short article discusses the relationship between the breakup and amalgamation of continents through time and the development of mountain belts such as the Appalachians. Intended for the general reader with some scientific background.

Plummer, Charles C., David McGeary, and Diane H. Carlson. *Physical Geology*. Boston: McGraw-Hill, 1999. This is a straightforward, easy-to-read introduction to geology intended for those with little or no science background. Discusses plate tectonics and its relationship to the development of mountain belts. Includes many excellent illustrations, as well as a CD-ROM.

Press, Frank, and Raymond Siever. *Understanding Earth*. 2d ed. New York: W. H. Freeman, 1998. This comprehensive physical geology text covers the formation and development of the Earth. Contains extensive sections on the development of mountain belts through time. Readable by high school students, as well as by general readers. Includes a CD-ROM, an index, and a glossary of terms.

Rogers, John. *The Tectonics of the Appalachians*. New York: John Wiley & Sons, 1970. A specific, detailed presentation of the rock structures of all of the provinces of the Appalachian region. The introduction is informative and accessible to the nonprofessional, and the bibliography is extensive.

BASIN AND RANGE PROVINCE

Alternating linear ranges and valleys characterize the Basin and Range Province of western North America. This topography resulted from crustal stretching that caused large normal faults. The crustal stretching, which occurred over the last 40 million years, is superimposed on a complex record of older geologic events.

PRINCIPAL TERMS

DETACHMENT FAULT: a horizontal or gently dipping, regionally extensive fault; the hanging wall usually contains numerous smaller, steeper normal faults that end at the detachment fault

DIP: the angle between a sloping surface, such as a fault, and a horizontal plane

DUCTILE SHEAR ZONE: a planar zone of rock that accommodates relative movement like a fault, but the movement has occurred by processes of solid-state flow rather than by fracture

FAULT: a fracture or system of fractures across which relative movement of rock bodies has occurred

FAULT SLIP: the direction and amount of relative movement between the two blocks of rock separated by a fault

HALF-GRABEN: an asymmetrical structural depression formed along a normal fault where the downthrown block is tilted toward the fault

HANGING WALL and FOOTWALL: the rock bodies located respectively above and below a fault

INTERNAL DRAINAGE: the condition in which a river system has no outlet; instead it drains into a saline lake or playa (a lake basin that contains water only shortly after a rainstorm)

LITHOSPHERIC PLATES: rigid blocks that make up the outer shell of the Earth, 100 to 200 kilometers thick and generally similar in size to the continents, forming a mosaic that covers the Earth's surface

NORMAL FAULT: a fault across which slip caused the hanging wall to move downward relative to the footwall

PLATE TECTONICS: the theory that Earth movements reflect the relative movements between a small number of rigid lithospheric plates, along narrow zones of deformation called plate boundaries

STRIKE-SLIP FAULT: a fault across which the relative movement is mainly lateral

THRUST FAULT: a fault, usually dipping less than 30 degrees, across which the hanging wall moved upward relative to the footwall

LOCATION AND CHARACTERISTICS

The Basin and Range Province comprises three different but related entities: a geologic province of western North America, the alternating linear mountain ranges and valleys that characterize that province, and the fault-block bedrock structure responsible for the distinctive topography. The Basin and Range Province includes the area between the Rocky Mountains on the east and the Sierra Nevada and Cascade Mountains on the west, from southern Montana, Idaho, and Oregon south to northern Mexico. Crustal stretching in a roughly east-west direction across this area in the last 40 million years is responsible for the distinctive topography and crustal structure. The stretching is the most recent chapter of a long, complex geologic history.

Alternating elongate mountain ranges and valleys (basins) define Basin and Range topography, which occurs not only in the Basin and Range Province but also in other areas of young continental stretching, such as southern Greece and the Yunan Province in China. Ranges and basins are usually similar in extent, commonly 50 to 200 kilometers in length and 20 to 30 kilometers in width. Climate throughout the Basin and Range Province is arid or semiarid. As a result, the ranges are generally rugged, sparsely vegetated, and drained by ephemeral and occasional perennial streams. Topographic relief from range tops to basin bottoms is usually 1 to 3 kilometers. Internal drainage is common, either in an individual basin or with several basins linked in a single internal drainage system. An example of the latter is the Great Salt Lake in Utah, which is the terminus of a drainage system that includes most of the basins and ranges of northwestern Utah.

Ranges and basins are defined on one or both sides by young, and in some instances seismically active, normal faults. The range side of the fault is the upthrown block and the basin side the down-thrown block, so the fault movements are expressed directly in the topography; however, erosion of the ranges and sedimentation in the basins cause topographic relief to be much less than the total displacement across the faults, which can exceed 10 kilometers. Typical Basin and Range faults dip from 30 to 70 degrees. Internal drainage results because subsidence along active normal faults ponds the water running off in streams. The common fault pattern is one in which basins and ranges are faulted on one side only to form a series of half-grabens, or asymmetrical structural depressions. Between the half-grabens are asymmetrical tilted-block ranges that are straight and steep on the faulted side and more gently sloping on the other side which, downslope, forms the bedrock floor of the next basin. Series of tilted-block ranges and half-graben basins are sometimes called "domino-style" fault systems, because the fault-bounded blocks resemble a series of dominos that have been stood on end next to each other and then toppled together. Strike-slip faults also are common in the Basin and Range Province, commonly acting to link together the ends of normal faults. Some of these strike-slip faults are very large. For example, the Furnace Creek fault zone that runs along northern Death Valley in California, linking normal faults in Death Valley to others farther north, may have accommodated 70 kilometers of strike slip.

METAMORPHIC CORE COMPLEXES

Much study of the Basin and Range has focused on ranges called metamorphic core complexes (MCCs) that are not domino-style fault blocks. There are more than twenty MCCs in the Basin and Range, and more may exist but are yet unrecognized. Both the structure and topography of MCCs are controlled by a detachment fault that separates an upper level, usually composed of sedimentary and volcanic rocks, from a lower level made up of metamorphic and plutonic rocks formed deeper in the crust. Rocks of the upper level generally have been intensely stretched along numerous normal faults, such that they have increased in width by several hundred percent. Rocks beneath the detachment fault commonly have flowed at high temperatures and great depth.

Analysis of the flow structures indicates that much or all of this flow also represents a large amount of horizontal stretching. Upper-level faulting and lower-level flow are therefore believed to be related. The detachment fault in an MCC is usually dome-shaped, so that the fault dips gently off the flanks of the range in every direction. Differences in erosion resistance between rock types above and below the detachment fault resulted in the domelike shape of the ranges, which is quite different from that of the more typical fault-block ranges. It is widely accepted that MCCs record extreme horizontal stretching, but since their recognition in the 1970's, their origin and significance have been intensely debated and remain controversial. There are two main issues: the relationship between MCCs and "normal" Basin and Range domino-style faults, and the significance of detachment faults. Regarding the former, one view is that MCCs and domino-style Basin and Range faults represent distinct extensional processes and that an MCC forms instead of large domino-style faults when the crust in an area is especially hot and susceptible to flow. Others consider MCCs to be exposures of what underlies the domino-style fault systems at depth; that is, the domino-style faults end downward at a detachment fault like those exposed in MCCs. In this view, greater than normal stretching in some areas caused the upper crust to be stretched so thin that deeper rocks and structures were exposed as MCCs.

Researchers are similarly split into two camps regarding the significance of the detachment faults. One group considers the detachment faults to represent a horizontal zone along which the cooler upper crust that deforms by faulting is mechanically separated (detached) from the hotter lower crust that deforms by flow. In this case, the detachment fault would act to accommodate differences in the movement patterns of rocks above and below it; it need not accommodate large displacement, but merely local adjustments. The other view is that the detachment fault represents a large, gently dipping normal fault that cuts down through the crust. In this view, as much as 50 to 60 kilometers of horizontal displacement has occurred along detachment faults in MCCs, thus bringing together originally widely separated upper- and deeper-crustal rocks. Consensus tends to favor the latter view of each issue—that is, that MCCs represent the deeper parts of domino-fault systems and that detachment faults are major dislocations that cut through the crust and accommodate tens of kilometers of slip. Much future work will be needed before either of these issues will be considered resolved.

FORMATION

The structure summarized in the foregoing has been superimposed in the Basin and Range upon a long and complex previous geologic history. In-

deed, much of the knowledge of that history can be attributed to the Basin and Range faulting that has uplifted and exposed crustal sections many kilometers thick to allow the geologic record to be read. The oldest rocks in the Basin and Range are scattered occurrences of ancient, deep-seated metamorphic and plutonic rocks as much as 2.5 billion years old. These rocks are exposures of the ancient crust of the North American continent. In the Late Precambrian eon, part of the west side of the North American continent was faulted away and carried by plate tectonics to some other part of the globe. This event formed a new western continental margin facing a new ocean basin. This ancient continental margin presently runs southwestward across southern Idaho and central Nevada, so that the eastern and southern parts of the Basin and Range formed from crust of the ancient continent, but the western Basin and Range was part of a deep ocean basin. From this time until the middle of the Paleozoic era, this continental

Fault-block mountains in the eastern California region of the Basin and Range Province. (© William E. Ferguson)

margin resembled the modern East Coast of North America: a broad, gently sloping shelf, generally tectonically stable and largely covered by shallow seawater.

During the Late Paleozoic and Early Mesozoic eras, lithospheric plate movements caused two major lithospheric blocks to collide successively with this continental margin. The collisions resulted in periods of mountain building along the North American continental margin, known as the Antler and Sonoman orogenies. Other results of these plate collisions included the building outward of the North American continent, such that by 250 million years ago, all of what now lies in the Basin and Range was part of continental North America, and strike-slip faulting that shifted some crustal blocks hundreds of kilometers along the continental margin, substantially changing the shape of the coastline in the process.

Following the Sonoman orogeny, the western continental margin of North America changed completely. By 200 million years ago, the continental margin was the site of underthrusting of oceanic lithosphere (subduction), causing the construction of a high mountain chain along western North America that was similar to the modern Andes in South America. What is now the Basin and Range Province covered an area analogous to modern Peru, Bolivia, and northern Argentina, including the eastern parts of the Andes, its foothills, and some of the adjacent plains. The mountain range was an area of volcanism and magmatic intrusion, now represented by large masses of granite in the Sierra Nevada and much of the Basin and Range. Thrust faulting occurred along the east side of the mountain chain, as it does today on the east side of the Andes. Such thrust faults are especially well known in the areas around Las Vegas, Nevada, and Salt Lake City, Utah.

Formation of the Basin and Range is part of a comparatively recent change in the tectonic pattern of western North America. From about 30 million years ago to the present, the Andes-like plate boundary (still active in the Pacific Northwest) has been changing into a strike-slip plate boundary (transform fault) that includes the San Andreas fault of Southern California. Somewhat earlier, large-scale stretching of the lithosphere began in the Basin and Range. Most of the MCCs formed between 40 and 20 million years ago. At the same time, numerous large volcanic centers appeared in and around the Basin and Range. These volcanic centers were the sources of enormous explosive eruptions that blanketed much of the Western United States with hot ash. Volume estimates of these eruptions reach and even exceed 3,000 cubic kilometers, greatly exceeding the largest known historic eruptions on the Earth. Both extension and explosive volcanism continue to the present, as evidenced by modern earthquakes along Basin and Range faults and by a few major volcanic centers that are still active. Most geologists, however, believe that both the rate of extension and the intensity of volcanism have decreased.

STUDY OF THE BASIN AND RANGE PROVINCE

The Basin and Range Province is one of the foremost areas worldwide for study of the processes by which the continental crust stretches. It is a natural laboratory in which the Earth has done an experiment, and the geologists' job is to analyze the results and to determine just what the experiment was. Three main types of techniques are brought to bear in such studies: field studies of surface geology, laboratory analysis of samples collected in the field, and geophysical field studies. The first step is basic fieldwork, which involves mapping the rock bodies and structures exposed at the surface. The geologist uses this information to make inferences about the spatial relationships of rock bodies and the geologic history that led to those arrangements. In spite of being one of the best-studied continental rifts in the world, at least one-half of the Basin and Range Province's surface geology has yet to be studied in detail. Fundamental questions remain about continental extension processes. Therefore, the surface geology of the Basin and Range will continue to be an important source of new insights for years to come.

Laboratory analytical techniques provide more precise and detailed information about rocks than can be gathered in the field. A complete summary cannot be attempted here because relevant techniques span all of Earth science; however, isotopic geochronology and thermobarometry are particularly important. Isotopic geochronology is based upon measuring the progressive decay of radioactive elements, and the consequent buildup of the products of that decay. This is the principal method of determining numerical ages of ancient rocks, and therefore it is vital in determining when ancient events occurred, for comparing ages from one area to another in order to look for spatial patterns, and for determining the rates at which processes have operated. Thermobarometry uses chemical compositions of minerals in rocks to estimate the temperature and pressure conditions under which the minerals grew. Mineral compositions are generally determined using the electron microprobe on a polished chip of the rock. Thermobarometry is used to find where in the crust deep-seated igneous and metamorphic rocks formed.

Important geophysical field techniques include seismology and studies of potential fields such as gravity and magnetism. Seismologists study the transmission of sound waves through the Earth, including sound waves from earthquakes, and from artificial sources such as explosions. The variations in the time it takes for sound waves to travel from their sources to various receivers (seismometers) and variations in the characteristics of the waves recorded at different seismometers allow seismologists to map out the internal structure of the Earth in terms of the rocks' sound-transmitting properties. For example, it is mainly through seismological studies that researchers know the thickness of the Earth's crust (about 30 kilometers in most of the Basin and Range). Variations in the Earth's gravity field from one place to another can be used to investigate variations in the densities of rocks at depth, and similar variations in the magnetic field indicate changes in the magnetic properties of rocks at depth. It is by combining such geophysical studies with observations of the surface geology that the present understanding of the structure and evolution of the Basin and Range has been achieved.

John M. Bartley

CROSS-REFERENCES

Alps, 805; Andes, 812; Appalachians, 819; Cascades, 830; Continental Rift Zones, 579; Earth's Structure, 37; Geomorphology of Dry Climate Areas, 903; Himalaya, 836; Mountain Belts, 841; Oil and Gas Distribution, 1694; Plate Margins, 73; Plate Tectonics, 86; Rocky Mountains, 846; Seismic Reflection Profiling, 371; Sierra Nevada, 852; Subduction and Orogeny, 92; Thrust Belts, 644; Transverse Ranges, 858; Yellowstone National Park, 731.

BIBLIOGRAPHY

Davis, George H., and Evelyn M. VandenDolder, eds. *Geologic Diversity of Arizona and Its Margins: Excursions to Choice Areas.* Tucson: Arizona Bureau of Geology and Mineral Technology, 1987. This volume contains geological descriptions and road logs for thirty-three field trips in Arizona, about nineteen of which concern the Basin and Range Province. The trips are intended for professional geologists, but some of the content is accessible to others. The observations and interpretations presented are up-to-date, describing a number of key areas for understanding the southern Basin and Range.

Kearey, Philip, and Frederick J. Vine. *Global Tectonics.* 2d ed. Oxford: Blackwell, 1996. This well-illustrated and clearly presented undergraduate text discusses the principles of geology in relation to plate tectonics. The authors present the background necessary for an understanding of the Basin and Range Province. Index and bibliography.

McPhee, John. *Basin and Range.* New York: Farrar, Straus & Giroux, 1981. This is a popular account of a journalist's introduction to geology and geologists, mainly in the context of the Basin and Range Province. The book is readable and entertaining, and the geological descriptions are reasonably accurate, if superficial.

Plummer, Charles C., David McGeary, and Diane H. Carlson. *Physical Geology.* Boston: McGraw-Hill, 1999. This is a straightforward, easy-to-read introduction to geology intended for those with little or no science background. One section is devoted to the Basin and Range Province. Includes many excellent illustrations, as well as a CD-ROM.

Press, Frank, and Raymond Siever. *Understanding Earth.* 2d ed. New York: W. H. Freeman,

1998. This comprehensive physical geology text covers the formation and development of the Earth. Includes extensive sections on the deformation of the continental crust and the development of the Basin and Range Province. Readable by high school students, as well as by general readers. Includes an index and a glossary of terms.

Stewart, John H. *Geology of Nevada.* Reno: Nevada Bureau of Mines and Geology, 1980. This book was written to accompany the 1:500,000-scale geologic map of Nevada (available from the U.S. Geological Survey and the Nevada Bureau of Mines and Geology), by a geologist who has spent much of his career study-ing the Basin and Range. It is organized based on the complete geological history recorded in Nevada, so only the latter part deals with Basin and Range topography, structure, and volcanism per se. For college-level students.

Weide, David L., and Marianne L. Faber, eds. *This Extended Land: Geological Journeys in the Southern Basin and Range.* Las Vegas: University of Nevada Press, 1988. This volume contains descriptions and road logs for eighteen field trips in southern Nevada and adjacent areas. As with the field-trip guidebook for Arizona, the trips are intended for professionals, but some of the book's contents are accessible to the general reader.

CASCADES

The Cascade mountain range extends from Northern California through western Oregon and Washington into British Columbia in Canada. The loftiest peaks of this range are stratovolcanoes, the highest being 4,392-meter Mount Rainier in Washington. The history of the Cascades goes back about 25 million years, but the periodic eruption of once-dormant volcanoes such as Mount St. Helens shows that the range is still evolving.

PRINCIPAL TERMS

ANDESITE: a type of volcanic, igneous rock, typically gray in color; the predominant lava expelled by stratovolcanoes such as those in the Cascades

BASALT: a dark, iron-rich, silica-poor volcanic rock

DACITE: a volcanic rock similar to andesite but containing more silica, typically gray in color

LITHOSPHERIC PLATES: massive horizontal slabs of the Earth consisting of crust and the uppermost upper mantle, capable of movement

by sliding over a semifluid layer in the mantle

RHYOLITE: a light-colored, silica-rich volcanic rock, commonly violently erupted as ash deposits

STRATOVOLCANOES (COMPOSITE VOLCANOES): lofty volcanoes composed of alternating layers of volcanic ash and lava or mudflows

SUBDUCTION ZONE: a linear area representing the line of collision between two lithospheric plates

LOCATION AND CHARACTERISTICS

The Cascade mountain range, named for the great cascades (waterfalls) of the Columbia River, spans 1,139 kilometers through the northwestern United States and southwestern Canada. Its highest peaks are a string of lofty volcanoes that extend from Northern California through Oregon and Washington and terminate in British Columbia. The Cascades lie about 160 to 240 kilometers inland from the Pacific Coast and are separated from the northwestern coast ranges by a broad valley called the Puget Trough in Washington and the Willamette Valley in Oregon. The range is divided into western and eastern sections; the western side consists of older, eroded volcanic rocks of relatively low topographic relief, but the east side, popularly known as the "high Cascades," sports more spectacular—and much younger—volcanic peaks. Because the Cascade Range creates a major climatic barrier, the western slopes are heavily forested, nourished by abundant rainfall. In contrast, the eastern slopes are dry scrublands covered mostly by grass and low bushes. The rugged scenery of the Cascades, including numerous deep valleys, lakes, and ridges, has been sculpted both by stream erosion and by glaciation, which began in

the Pleistocene ice age but which has continued even in recent times.

The high Cascades give the range its highest peaks and grandest mountain scenery. They consist of a string of stratovolcanoes that extend from Northern California (Mount Shasta, 4,317 meters; Lassen Peak, 3,187 meters) to British Columbia (Meager Mountain, 2,679 meters; Mount Garibaldi, 2,678 meters). Other important volcanoes in the Cascades include Mount Hood (3,426 meters; near Portland) and Mount Jefferson (3,199 meters) in Oregon, and Mount Adams (3,751 meters), Mount Rainier (4,392 meters; near Seattle), and Mount Baker (3,284 meters) in Washington. Mount St. Helens (2,500 meters) in southwestern Washington is one of the youngest and the most active of the Cascade volcanoes, having erupted violently in May, 1980, and more than sixty times during the last fifty thousand years. Oregon's famous Crater Lake was created by a volcano that literally blew itself out of existence nearly seven thousand years ago. The crater, the remains of once-mighty Mount Mazama (estimated elevation of more than 3,500 meters), is now an 8-kilometer-wide bowl-shaped collapse depression (caldera) filled with water.

Stratovolcanoes (also known as composite volcanoes) occur worldwide, mostly along continental margins and in near-continent island arcs such as the Aleutian Islands, Japan, and the Philippines. They tend to form lofty, cone-shaped peaks, commonly snow-packed because of their high elevations above sea level. The Andes in western South America, like the Cascades, also exist as a coast-hugging, linear chain of stratovolcanoes.

Stratovolcanoes owe most of their impressive height to the composition of the volcanic magma they emit. This magma is rich in silica (silicon dioxide), the most common chemical component of rocks and minerals in the Earth's crust and mantle. High-silica magmas have high viscosities and thus are generally thick and sticky upon eruption. As a result, erupted concentrations of high-silica volcanic materials tend to pile up in a single area, producing tall mountain peaks. Less viscous basaltic magmas tend to spread out, creating broad, low shield volcanoes such as those that make up the islands of Hawaii.

The volcanic rocks produced by the high-silica magmas of stratovolcanoes are called andesite. Dacite is an especially siliceous (silica-rich) variety of andesite; stratovolcanoes commonly eject one or both of these magma types. stratovolcanoes are noted for their explosive eruptions, and this proclivity, too, can be attributed to the nature of their magmas. Andesite and dacite magmas generally contain significant quantities of water, which, when trapped as steam within this sticky magma, may suddenly explode with the power of several atomic bombs. For example, the 1980 eruption of Mount St. Helens released the estimated explosive energy of about five hundred Hiroshima-type atomic bombs. For this reason, stratovolcanoes, including those in the Cascades, are the most dangerous volcanoes in the world, at times causing catastrophic death tolls and property damage.

GEOLOGIC HISTORY

The geologic history of the present Cascade range is complicated, especially as it relates to other major geologic terrains in the Pacific Northwest. For example, the Cascades can be considered a younger northern extension of the Sierra Nevada range in eastern California. The thick, layered accumulations of basalt (a dark, low-silica, high-iron rock) lava flows of the Columbia River Basalt Plateau also have a relationship to the Cascades, as do the Willamette and Puget Valleys and the Pacific Coast ranges. The Cascades, especially the high Cascade volcanoes, represent one of the most recent additions to the geology of the Northwest; most of these volcanoes have initial eruption ages of less than 1 million years, making them very young by geologic standards. The eruptions of Mount Lassen between 1914 and 1917 and Mount St. Helens in 1980 show that these volcanoes are still actively adding to the mass of the range. On the other hand, the north-south-trending volcanic field on which the current volcanoes are built was first established about 25 million years ago, during the Miocene epoch. Much of the western Cascades consists of the eroded remnants of volcanoes that once dominated the region's skyline but that are now extinct.

Mount Rainier, in the Cascade range. (Digital Stock)

To appreciate the origin of the Cascade range fully, it is necessary to go back in geologic time about 150 million years, a time when the dinosaurs dominated life on the Earth. At that time, much of northwestern Oregon and southwestern Washington were covered by an embayment of the Pacific Ocean that extended nearly to the present state of Idaho. Paralleling the coastline of this embayment was a subduction zone, a linear terrain representing the collision line of two lithospheric plates. The Earth is broken up into "plates," averaging about 100 kilometers thick, that consist of the crust and a part of the upper mantle. These plates move horizontally over the Earth's surface, carrying the continents with them. Where they collide, one plate will slide under the other (a process known as "subduction"). When this process involves a thinner but denser oceanic plate colliding with a thick, buoyant continental plate, the oceanic plate always dives under the continental plate. This movement creates a narrow trench in the ocean off the coast of the continent that fills with sediments scraped off the ocean floor as the plate descends. These crumpled sediments eventually are pushed upward to become coastal ranges. Inland from the coastal ranges about 100 kilometers will be volcanoes produced by the melting of former surface materials sinking to ever deeper and hotter depths. Magma produced by the partial melting of subducted rocks powers the volcanoes of the present Cascades and most other stratovolcanoes.

Approximately 150 million years ago, subduction was occurring along the Oregon coast, parallel to the northeast-trending ocean embayment. This movement produced a northeast-trending mountain range consisting of coastal ranges (including the current California, Oregon, and Washington coast ranges) and a string of volcanoes, now mostly eroded away. These old mountains are now represented by the Blue and Wallowa ranges in eastern Oregon and southeastern Washington and the Klamath range in southwestern Oregon and Northern California. Also at this time, a subduction zone paralleled much of the California coast, producing the present California coast ranges and active andesite stratovolcanoes in the area of the Sierra Nevada in eastern California. As in Oregon, these old volcanoes in the Sierra are now eroded away, but the under-lying magma chambers, now crystallized to granitic rocks, provide evidence of their former existence. Yosemite National Park is an excellent place to view the massive granitic plutonic bodies that once fed volcanoes similar to those in the present-day Cascades.

About 35 million years ago, a strange event occurred that altered the pattern of mountain building and volcanic activity in the Pacific Northwest. The subduction zone that had been trending northeast across Oregon straightened out into a new north-south trend, an orientation it has maintained. About 25 million years ago, a new north-south trending line of volcanoes sprang up parallel to the subduction zone. These volcanoes were similar to the present Cascade volcanoes, but they spewed out an even more siliceous and explosive variety of magma called rhyolite, the volcanic equivalent of granite. This began a very impressive era of explosive volcanism, with the widespread deposition of volcanic ash, that continued for about 10 million years. The remnants of these ash deposits can still be observed in the John Day formation of northern Oregon. All during this time, the volcanic range was gradually being uplifted by the buoyancy of the hot crustal rocks below. To the east, the areas of the present Willamette and Puget Valleys were down-warped as they were squeezed between the uplifting coast ranges to the west and the volcanic arc to the east. These valleys originally were shallow seas that eventually filled with sedimentary rocks and later were covered by basalt lava flows.

During much of the time that volcanoes were erupting in Oregon and Washington, they were also erupting in a nearly continuous band down the length of California in the present area of the Sierra Nevada. About 30 million years ago, however, volcanism suddenly ceased in the Sierra as the former subduction zone transformed into the current San Andreas fault system. The San Andreas, famous for the earthquakes it has wrought in California, is another form of plate boundary called a transform fault. In a transform fault, the lithospheric plates slide past each other but do not subduct. Thus, volcanic activity is limited at such boundaries. The Sierra volcanoes, which had once formed a continuous chain with the early Cascade volcanoes, died with the formation of the San Andreas system. They have since

Boston Glacier, the largest single glacier in the North Cascades, occupying a cirque northwest of Buckner Mountain. (U.S. Geological Survey)

eroded away, leaving behind the granitic rocks that crystallized below them.

About 20 million years ago, volcanism in the present western Cascade area suddenly stopped, replaced some time later by the extrusion of voluminous quantities of very fluid basalt lava that flooded wide areas in Oregon and Washington. This was the event that created the present Columbia River Basalt Plateau. About 12 million years ago, volcanism resumed in the western Cascades but never achieved the level of activity of their explosive early history. The current eastern Cascade volcanoes have resulted from a renewed burst of volcanism over only the last 1 million years; however, the infrequent rate of volcanism in the Cascades probably is signaling a decline in subduction activity, which may eventually lead to the extinction of the Cascade volcanoes in a manner similar to what befell the Sierra Nevada volcanoes.

STUDY OF THE CASCADES

The Cascades are perhaps the best-studied volcanoes in the world. From a scientific standpoint, they are of interest for what they reveal about how continental subduction-zone volcanoes originate and evolve over time. They also have important implications for how continents in general grow

and evolve. From a human-interest standpoint, the study of Cascade-type volcanoes may lead to methods of predicting their potentially devastating eruptions. Scientists—principally volcanologists, geophysicists, and other geologists—study volcanoes by first performing field studies of their lava flows or pyroclastic (explosively ejected) ash deposits. These rock units are investigated by field parties who chart their surface distribution on geologic maps. This information may be used to correlate volcanic materials with tectonic features such as faults, folds, uplifted or down-warped areas, and plate boundaries, or it may be used to estimate the energy of eruptions and the amount of material involved. The history of a volcanic area can be determined by noting the number and order of interlayered deposits of volcanic ash (indicating an explosive event) and lava flows (indicating a relatively quiet event). Samples are also analyzed by radiometric means (potassium-argon, carbon 14, and other techniques) to produce the ages of the rocks. Rock ages allow the precise determination of just when volcanoes were most active and for how long. Other chemical data collected from volcanic rocks can be used to suggest what kinds of parent rocks were melted to produce the magmas and under what conditions.

The prediction of volcanic eruptions is difficult because volcanoes seldom produce warning signals in a reproducible pattern that can be universally applied. Previous experience with volcanic eruptions shows that earthquakes generally precede eruptions, but the strength and frequency of quakes varies from one eruption to the next. Many volcanoes show an actual rise in elevation (caused by rising magma) shortly before an eruption, and such changes can be measured with precision laser transits. Also, the Earth's magnetic field around volcanoes may show a change in intensity shortly before an eruption. These measurements rely on having sophisticated instruments and well-trained operators observing the volcanoes on a nearly constant basis. Highly technical prediction strategies may not be financially or practically feasible in all circumstances in which the threat of volcanic eruptions is a constant concern.

The May 18, 1980, eruption of Mount St. Helens provided an excellent opportunity for intense, close-up study of an active volcano. Not only did this volcano provide a spectacular show of volcanic fury when it finally erupted (with the loss of fifty-seven lives), but it also was located in a uniquely accessible area, close to established government and university facilities with expertise in volcanology. The U.S. Geological Survey made thorough scientific studies of the volcano before, during, and after the catastrophic eruption. One encouraging result of these studies was the discovery that a certain frequency (vibration rate) of earthquake wave called a "harmonic tremor" provided a fairly reliable warning that an eruption was imminent. Harmonic tremors are believed to be generated by magma rising within the central feeder conduit of the volcano. As the magma moves upward, it rubs against the rocks lining the conduit, producing the characteristic vibration of the harmonic tremor. Detection of this special earthquake wave allowed the prediction of five subsequent minor eruptions of Mount St. Helens between 1980 and 1982. It remains to be seen whether this technique has applications beyond this one volcano.

John L. Berkley

CROSS-REFERENCES

Alps, 805; Andes, 812; Andesitic Rocks, 1263; Appalachians, 819; Basin and Range Province, 824; Continental Growth, 573; Flood Basalts, 689; Himalaya, 836; Igneous Rock Bodies, 1298; Island Arcs, 712; Lava Flow, 717; Lithospheric Plates, 55; Magmas, 1326; Mount St. Helens, 767; Mountain Belts, 841; Pyroclastic Rocks, 1343; Recent Eruptions, 780; Rocky Mountains, 846; Sierra Nevada, 852; Stratovolcanoes, 792; Subduction and Orogeny, 92; Transverse Ranges, 858; Volcanic Hazards, 798.

BIBLIOGRAPHY

Alt, David D., and Donald W. Hyndman. *Roadside Geology of Oregon.* Missoula, Mont.: Mountain Press, 1994. One of a series describing informal geology tours that can be conducted by automobile. Other recommended books in the series pertaining to the Cascades are the volumes for California and Washington. The Oregon volume has the best description of the nature and geological history of the Cascades and their volcanoes. All volumes are illustrated with clearly labeled diagrams, maps, and monochrome photographs and include a glossary and recommended reading list.

Decker, Robert W., and Barbara B. Decker. *Mountains of Fire.* New York: Cambridge University Press, 1991. The writers of this book, professional volcanologists, are well qualified to explore the nature of world volcanic terrains, the latest scientific findings, the anatomy of volcanoes, and the reasons why volcanoes erupt where they do. Cascade volcanoes are covered, particularly Mount St. Helens. A nontechnical, highly readable, indepth treatment. Illustrated with more than one hundred color and monochrome photographs and drawings. Features eyewitness accounts of major volcanic eruptions.

King, Philip B. *The Evolution of North America.* Princeton, N.J.: Princeton University Press, 1977. By a renowned veteran geologist with the U.S. Geological Survey. This is the best book to consult for the geologic history of any major geological terrain in North America. Chapter 9 (part 3) deals with the origin and evolution of the Cascades; other sections deal with related areas such as the coast ranges, the Sierra Nevada, and the Columbia Plateau. Illustrated with monochrome maps, diagrams, and sketches. Aimed at readers with some knowledge of geology, but not overloaded with jargon; should be intelligible to college students and college-educated readers.

Plummer, Charles C., David McGeary, and Diane H. Carlson. *Physical Geology.* Boston: McGraw-Hill, 1999. This is a straightforward, easy-to-read introduction to geology intended for those with little or no science background. Chapter 4 deals with the Cascades and the Mount St. Helens eruption. Includes many excellent illustrations, as well as a CD-ROM.

Wood, Charles A., and Jurgen Kienle, eds. *Volcanoes of North America: United States and Canada.* New York: Cambridge University Press, 1993. A major compendium of North American volcanoes, including those of Alaska and Hawaii. Every entry contains information on the type of volcano and its location, dimensions, eruptive history, and composition. The introduction has sections that show the methods used to study volcanoes. The Cascade volcanoes are well represented in this book; the introductory section to the chapter "Volcano Tectonics of the Western U.S.A." has a subsection with a discussion of the Cascades. Illustrated with monochrome photographs, maps, and diagrams. Easily comprehensible by the general reader.

HIMALAYA

The Himalaya constitute one of the greatest physical features on the Earth, containing some of the world's youngest and highest mountains, largest glaciers, and deepest gorges. The complex geology, loftiness, and length of the Himalayan mountain chain affect the climate and life patterns of much of continental Asia.

PRINCIPAL TERMS

CONTINENTAL DRIFT: a hypothesis that attributes the present arrangement of continental shields and ocean floors to the breakup of the original supercontinent, Pangaea

DRAINAGE SYSTEM: a network of stream branches bounded by topographical divides with a common outlet

NAPPE: an underlying rock sheet overturned

forward along a low-angle fault

OROGENY: tectonic activity that results in folds and faults of strata; this process is associated with mountain building

TRANSVERSE VALLEY: a river-cut valley or gorge that runs perpendicular to the main strike direction of a mountain chain

FORMATION AND CHARACTERISTICS

Until approximately 3 or 4 million years ago, most of the Himalayan mountain region, including the highlands of the Tibetan plateau, was covered by the broad, shallow Tethys Sea that lay between present-day India and Tibet. The Himalaya formed as the South Asian plate drifted northward, pushing the sedimentary-rock sea bottom against the inner Asian continent. This process of crustal plate movement, attributed to the theory of continental drift, began during the Jurassic period, about 150 million years ago, and continues today. Such continent-continent collision results in a greatly thickened crust as neither continent can sink. In the Himalaya, the crust is approximately 60 kilometers thick, almost double the average for continents as a whole, and consists of the underthrust Indian plate along with crust that has been shortened because of compression. The actual collision of the northern Indian shield with Eurasia dates to the Tertiary period, some 20 million years ago, and represents the initial period of tectonic activity associated with mountain building, or orogeny.

Orogenic activity of the Himalaya is conventionally divided into three phases. The first of these occurred at the end of the Eocene epoch and into the Oligocene epoch, between approximately 50 and 35 million years ago, when upheaved crystalline rock and sedimentary rock formed the central axis of the Himalaya. The second phase of folded sediments took place during the Miocene epoch, which lasted until approximately 5 million years ago. The raised central part of the range and the outer foothills formed during the third phase, which coincides with the post-Pliocene, between 5 and 2 million years ago. Activity during this phase gave the range its contemporary morphology. Since the last glacial epoch, about 20,000 years ago, the Himalaya have grown more than 1,500 meters (between 7.5 and 10 centimeters per year).

Geographers and geologists differ in their measurements of the length of the Himalaya. Some extend the chain to include the Koh-i-Baba and Safed Koh Ranges of the Hindu Kush in the west and the Assamese highlands in the east. This makes the Himalayan range more than 4,000 kilometers in length—an arc spanning the entire Indian subcontinent, including the Hindu Kush, Karakoram, and Himalayan physiographic divisions. Other scientists place the dividing line at the Oxus River in Afghanistan and include the Karakoram in northern Pakistan as part of the Himalayan mountain system but omit the Hindu Kush in Afghanistan, thus delimiting the chain to 3,000 kilometers. Finally, some experts separate the Himalaya proper from the Karakoram and ranges farther west by using the Indus River catchment as the boundary. This latter division defines the Himalaya as the mountain chain extending between the Indus and Brahmaputra Rivers, a distance of some 2,700 kilometers.

There are no compelling geological arguments that clearly show where to divide the Himalaya from east to west in genetic or structural terms. Geomorphological divisions use rivers, such as the Oxus and Indus Rivers, to make the divisions. Elevation character is also used to distinguish the main Himalayan range from its western extensions. For example, west of longitude 68 degrees west, the mountains rarely exceed 4,000 meters, while most of the Himalaya average between 6,000 and 7,000 meters elevation. Using elevation as a criterion, some experts extend the Himalaya for 3,000 kilometers in a northwest-southeast strike between longitude 68 degrees east and longitude 96 degrees east.

When the Karakoram is included as part of the Himalayan chain, the system includes all fourteen of the world's peaks over 8,000 meters and hundreds over 7,000 meters. Nine of the fourteen tallest peaks are in Nepal, including the world's highest mountain, Mount Everest, at 8,872 meters. The second-highest peak is K2 at 8,611 meters, located in the Karakoram range. The Nepal Himalaya contain several of the next tallest mountains: Kangchenjunga (8,585 meters), Lhotse (8,501 meters), Makalu (8,470 meters), Dhaulagiri (8,172 meters), Cho Oyu (8,153 meters), Manaslu (8,125 meters), Annapurna (8,091 meters), and Kao-seng-tsan Feng (8,013 meters). The remaining peaks over 8,000 meters are in the Karakoram and include Nanga Parbat (8,125 meters), Hidden Peak (8,068 meters), Gasherbrum (8,060 meters), and Broad Peak (8,047 meters).

GEOLOGIC ZONES

The relief range from high to low spots in the Himalayan chain is unsurpassed by other world mountain systems. Using the most liberal defini-

Mount Everest, in the Himalaya, the highest peak in the world at 8,872 meters, as seen from Gokyo Ri peak in Nepal. (AP/Wide World Photos)

tion of the lateral extent of the chain, it is also one of the world's longest systems of mountains. The interplay of geological forces that contributed to the initial formation of this impressive mountain range is still not fully understood. Early geological research emphasized the study of rock layers, or strata, based on fossil findings. More recent scientific investigations have considered the underlying structure of the Himalayan thrust sheets. These studies geologically divide the Himalaya into three zones.

The first zone is the Outer-Himalayan foothills, which rise out of the lowlands of the Indo-Gangetic plains in the south. These hills form a series of ridges of up to 1,300 meters in elevation that strike northwest-southeast, separated by longitudinal depressions called dun valleys. In India, the Outer Himalayan ranges are called the Siwaliks and in Nepal they are referred to as the Churia Hills. Composed of detrital rocks, such as clays, sandstones, limestones, and conglomerates, these hills are a series of broad anticlines (upfolds) and synclines (downfolds) formed from the weathered granitic core of the central Himalaya.

North of the foothills of the Outer Himalaya are the middle ranges of the Lesser Himalaya, which reach elevations surpassing 5,000 meters. This band of intermediate hills, averaging 65 kilometers in width, lies between the Great Himalaya central upthrust region and the outer ranges of the Outer Himalaya. This geological zone is composed of compressed or metamorphosed rock of various ages. Combined with the Great Himalaya, it forms the second geological zone. The lofty peaks and snow summits of the Great Himalaya, in numerous places exceeding 8,000 meters, cap the complicated geology of the main thrust zone. The convoluted geology of this region is partially attributed to great rock sheets, called thrust sheets, or nappes, that have moved many kilometers. In this Great Himalayan region is found the oldest crystalline core material of Tethyan sediment origin in thick layers thrust approximately 160 kilometers south over the Lower Himalaya.

The third geological zone is called the Tibetan zone. Unusually thick crust (up to 120 kilometers) and intensive folds and upthrusts characterize this extensive northward-dipping plateau region that lies north of the Great Himalaya. Extensive basins as well as outcroppings of crystalline rock punctuate the expanse of the Tibetan highland plateau zone.

The mountains of the Great Himalaya and the outer foothills are transversed by valleys that cut across the strike of the range, producing the deepest gorges in the world. These erosional valleys are river-cut and differ from the tectonic dun valleys situated in the Outer Himalaya, which are actually synclinal troughs. The rivers that cut these great transverse valleys originate in the Tibetan zone and predate the Himalaya's orogeny; thus they have been continually eroding while the mountains were uplifting. The main rivers breaching the Himalaya include the Hunza River, at about 1,800 meters in elevation, only 4 kilometers from Mount Rakaposhi (7,788 meters); the Indus River, about 1,200 meters and 22 kilometers from Nanga Parbat (8,125 meters); the Kali Gandaki River, about 1,500 meters and less than 7 kilometers from Dhaulagiri (8,172 meters); and the Trishuli River, about 1,800 meters in elevation and 13 kilometers from Langtang-Lirung Himal (7,245 meters).

The location of the Himalayan mountain range in southern Asia predicts its role in the continental climate and the consequential impact of local weather on surface geomorphology. It presents a barrier to north-south airflow associated with the Asian monsoon, thus producing heavy orographic rainfall along the southern flanks in the summer, when moisture-bearing winds from the southern oceans are forced to rise over the mountains. The north side in Tibet and Central Asia lies in the Himalayan rainshadow. The precipitation in the form of snow at high elevations feeds some of the world's largest glaciers, including the Siachen Glacier (72 kilometers), the Hispar Glacier (61 kilometers), and the Baltoro Glacier (58 kilometers), all located in the Karakoram. Meltwaters from these glaciers and rainfall from lower elevations enter into extensive drainage systems that in turn erode the land to contribute to the greatly dissected topography of the mountains. Precipitation receipts, elevation, and slope aspect vary greatly in this highly complex and dynamic natural system and contribute to a multitude of habitats for more diverse assemblages of flora and fauna as well as for human settlement.

STUDY OF THE HIMALAYA

Scientists use a variety of techniques to study the Himalayan mountain system, including geological and climatological surveys, erosional studies, geoecological mapping, and land-cover analysis. Modern geotechniques have enabled scientists to reconstruct evolutionary stages in mountain building. For example, paleobiology, the study of plant and animal fossil remains, and radiocarbon methods have contributed to more precise dating of rocks and rock units. The geochronological methods are supplemented with geochemical analyses that enable scientists to age rocks and rock strata, ascertain their origins, and monitor their movements. Geological structures are analyzed through geoseismology readings, field interpretation, and mapping of geophysical transverses. By relating structures to rock ages, experts can reconstruct the evolutionary history of such structures. Field survey instruments such as plane tables, compasses, spotting scopes, and altimeters are commonly employed during geological fieldwork.

Climatologists rely on weather data obtained from climate stations that are distributed throughout the mountains, or they conduct on-site climate studies using an assortment of instruments, including rain gauges, thermometers, barometers, wind gauges, and altimeters. Systematic climatological data are not available for many regions within the Himalaya, however, and the understanding of Himalayan weather is incomplete. During the last few decades of the twentieth century, aerial surveys from low-flying aircraft and electronic images from meteorological satellites began providing a wealth of new information.

The extremely difficult terrain and inhospitable climate at high altitudes make fieldwork difficult in the Himalaya. Energy-sensing satellites such as Landsat and Spot provide previously unobtainable information about surface conditions in the region. By computer processing this electronically obtained information, scientists have learned much about areas not frequently visited or studied.

The conventional Earth science study approach has been augmented by multidisciplinary surveys that seek to understand the geoecological integration of geomorphology, vegetation, climate, and land surface processes. Because much of the Himalayan natural environment is inhabited by human populations, these studies tend to be executed with a cultural focus as well. Land-use mapping, agroecological research, natural hazards assessments, and resource management studies complement the field research in the natural and Earth sciences. Scientists map and interpret surface processes such as soil erosion, landsliding, and mudflows on a local level. These slope problems are often attributed to human activity in the inhabited regions, but experts remain unconvinced about the degree of the cause-and-effect relationship between human activity and geoecological processes.

ENVIRONMENTAL CONCERNS

The Himalaya comprise the Earth's most magnificent mountain system. Their importance is reflected in the desire many people have to visit, climb, and conquer the summits; in the lives of the millions of people who inhabit the Himalaya; and in the powerful role that these mountains play in the geology and climate of one of the world's largest continental landmasses. The study of the Himalayan environment not only stimulates further scientific revelations but also kindles a profound sentiment that people throughout the world have acquired for these mountains.

On a practical level, it is clear that the Himalayan region experiences distressing levels of environmental degradation, with adverse consequences for the land and for the people who live within and near it. The problems that the Himalaya face in respect to deforestation, water flow, soil erosion, and landslides directly affect the lives of residents and visitors of these mountain regions. The waters contained in the highland drainage system help to ensure agricultural productivity in the densely populated lowlands in Pakistan, India, and Bangladesh. Scientists' understanding of these highland-lowland interactions is still inadequate to plan properly for the optimal resource usage in the mountains.

While many natural disturbances relate to the fact that these are young and dynamic energy environments, human alterations of the land also contribute in places to surface processes such as soil erosion and landslides. The geoecological linkages between human resource needs and environmental disturbances are important to the people who inhabit these mountains and quite possibly critical to the millions of lowlanders who reside in the lower reaches of the mountain catchments. Only through concise and specific fieldwork by both natural and social scientists can problems be properly identified. The eventual resolution of such problems necessitates scientific understanding and political action on the part of Himalayan countries.

David N. Zurick

CROSS-REFERENCES
Alps, 805; Andes, 812; Appalachians, 819; Basin and Range Province, 824; Cascades, 830; Climate, 1902; Drainage Basins, 2325; Faults: Thrust, 226; Folds, 624; Landslides and Slope Stability, 1501; Mountain Belts, 841; Plate Motions, 80; Plate Tectonics, 86; Rocky Mountains, 846; Sierra Nevada, 852; Soil Erosion, 1513; Thrust Belts, 644; Transverse Ranges, 858.

BIBLIOGRAPHY

Lall, J. S. *The Himalaya: Aspects of Change.* New Delhi: Oxford University Press, 1981. This book is a collection of articles written by social and natural scientists that covers a range of Himalayan topics, from meteorology and geology to cultural systems. The articles are well organized by subject matter. Useful reference for college students and others with a working interest in the region.

MacFarlane, Allison, Rosoul B. Sorkhabi, and Jay Quade, eds. *Himalaya and Tibet: Mountain Roots to Mountain Tops.* Boulder, Colo.: Geological Society of America, 1999. This book, intended for the specialist, is divided into three areas: the Trans-Himalaya, the High Himalaya, and the Himalayan Foreland. Extensively illustrated with clear diagrams and photographs.

Mountain Research and Development 7, no. 3 (1987). This journal issue contains several papers, written by Himalayan scientists, that deal with environmental degradation in the Himalayan region. They are most suitable for college students who want to understand the specific nature of Himalayan environmental problems.

Nicolson, Nigel. *The Himalayas.* Amsterdam: Time-Life Books, 1975. This is a lavishly illustrated general introduction to the entire Himalayan region, from Afghanistan to Bhutan. The discussion of natural history is expanded with anecdotes from the author's travels in the region. Photographs and maps visually orient the reader to the study area. Suitable for a lay reader with no background in Earth science or the Himalayan region. Includes discussions of early explorations of the region by scientists and mountaineers.

Olschak, B. C., A. Gansser, and E. M. Buhrer. *Himalayas.* New York: Facts On File, 1987. This volume of texts, maps, and photographs is one of the best general books available on the Himalaya. Written by several foremost Himalaya scholars working in the fields of geology, mythology, and anthropology. Covers all main aspects of the Himalaya, with excellent discussions of its geology and cultural history. Well written, accurate, superbly designed, and suitable for most nonscientists.

Press, Frank, and Raymond Siever. *Understanding Earth.* 2d ed. New York: W. H. Freeman, 1998. This comprehensive physical geology text covers the formation and development of the Earth. Readable by high school students, as well as by general readers. Includes an index and a glossary of terms.

Windley, Brian F. *The Evolving Continents.* New York: John Wiley and Sons, 1995. This general text aimed at college students has a chapter that deals with the Himalaya and the Cenozoic deformation of Asia. Bibliography and index.

MOUNTAIN BELTS

Mountain belts are products of plate tectonics, produced by the convergence of crustal plates. Topographic mountains are only the surficial expression of processes that profoundly deform and modify the crust. Long after the mountains themselves have been worn away, their former existence is recognizable from the structures that mountain building forms within the rocks of the crust.

PRINCIPAL TERMS

CONTINENTAL SHELF: the submerged offshore portion of a continent, ending where water depths increase rapidly from a few hundred to thousands of meters

EPEIROGENY: uplift or subsidence of the crust within a region, without the internal disturbances characteristic of orogeny

GABBRO: a silica-poor intrusive igneous rock consisting mostly of calcium-rich feldspar and iron and magnesium silicates; its volcanic equivalent is basalt

GRANITE: a silica-rich intrusive igneous rock consisting mostly of quartz, potassium- and sodium-bearing feldspar, and biotite or hornblende

IGNEOUS: from the Latin *ignis* (fire), a term referring to rocks formed from the molten state or to processes that form such rocks

METAMORPHISM: the change in the mineral composition or texture of a rock because of heat, pressure, or the chemical action of fluids in the Earth

OROGENY: the profound disturbance of the Earth's crust, characterized by crustal compression, metamorphism, volcanism, intrusions, and mountain formation

PLATE TECTONICS: the theory that the crust of the Earth consists of large moving plates; orogeny occurs where plates converge and one plate overrides the other

SEDIMENTARY ROCKS: rocks that form by surface transport and deposition of mineral grains or chemicals

SUBDUCTION: the sinking of a crustal plate into the interior of the Earth; subduction occurs at subduction zones, where plates converge

OROGENY

Mountains have many origins. They can be volcanic, like Mount Vesuvius, or they can be formed by vertical movements along faults, as the Sierra Nevada or the Ruwenzori of central Africa were. Some mountains are the result of relatively gentle epeirogenic (deformational) uplift of the crust—for example, the Black Hills or the Adirondack Mountains. The causes of epeirogeny, however, are poorly understood. The great mountain chains of the Earth, such as the Andes, the Rocky Mountains, or the Himalaya, however, formed not only from uplift but also from internal deformation of the crust, volcanic activity, metamorphism, and the intrusion of vast quantities of molten rock into the crust, especially granite and related rocks. These processes are collectively called orogeny, and mountain chains that form from such processes are called orogenic belts. Orogeny is one of the most important consequences of plate tectonics. It occurs when plates collide and one plate overrides the other—a process known as subduction—in response to compressional forces and heating generated by the plate collision.

The Earth's crust consists of two types of plates, called continental and oceanic by geologists. The continents and the adjacent continental shelves are underlain by granitic crust, averaging about 40 kilometers thick. The ocean floors are made of gabbro and basalt, averaging about 5 kilometers thick. The true edge of a continent is not the shoreline, which is constantly changing, but the boundary between continental and oceanic crust. The edge of the continental shelf coincides closely with this boundary.

Many mountain belts form at subduction zones with a continental overriding plate and an oceanic descending plate. The downward bending of the plate creates a deep, narrow trench on the ocean floor, sometimes more than 10 kilometers below sea level. The descending oceanic plate sinks into the Earth's interior, eventually to be reabsorbed. The overriding plate experiences orogeny. All orogenic belts differ in detail, but most have certain major features in common. A typical orogenic belt consists of parallel zones, which may be defined by distinctive rock type, type of metamorphism, level of igneous activity, or type of deformation of the rocks. The zones, generally parallel to the boundary where the two plates collide, are the result of different crustal conditions and processes at different distances from the plate boundary. It is useful to regard orogenic belts as having an "outer" side, adjacent to the plate boundary, and an "inner" side within the overriding plate.

The first zones recognized in orogenic belts were those defined by environment of deposition: an outer zone of thick, deep-water sedimentary rocks and volcanic rocks and an inner zone of thinner, shallow-water sedimentary rocks without abundant volcanic rocks. The nineteenth century American geologist James Hall first described these zones, which he envisioned as parallel troughs formed by downward folding of the crust. Because these troughs were viewed as immense versions of ordinary downward folds, or synclines, James Dana later called the troughs geosynclines. The outer trough was called eugeosyncline, and the inner trough was called miogeosyncline.

The original concept of the geosyncline disturbed many geologists, because it did not quite match the structure of active mountain belts. In 1964, Robert Dietz reexamined the geosyncline concept and showed that the rocks need not have accumulated in troughs. With this insight, it became clear that the rocks of miogeosynclines corresponded closely to those of the continental shelves, while eugeosynclines were a good match to the rocks of many volcanic island chains. Later, it became clear that the rocks of the eugeosyncline often formed separately from the miogeosyncline and were later juxtaposed by plate motions.

Because of these revisions of the original geosyncline concept, many geologists have abandoned the original terms and prefer the terms "geocline," "eugeocline," and "miogeocline" instead. The eugeocline is the outer belt of deep-water sedimentary rocks and volcanic rocks. The miogeocline is the inner belt of shallow-water sedimentary rocks. Beyond the miogeocline is the platform, where thin shallow-water or terrestrial rocks were deposited on the stable interior of the overriding plate. In addition, most orogenic belts have an inner belt of coarse sedimentary rocks deposited late in the history of the orogenic belt. This belt, called the molasse basin, consists of debris eroded from the mountains and deposited at their base. Molasse basins consist mostly of rocks deposited in shallow-water or land environments.

MOUNTAIN BELT STRUCTURE

Much of the structure of a mountain belt is related to processes in the descending plate. As the descending plate reaches a depth of about 100 kilometers, it begins to melt, and molten rock, or magma, invades the overriding plate. In general, volcanic rocks in orogenic belts become progressively richer in silica with increasing distance from the plate boundary, because the rising magma has more time to react with silica-rich continental crust. Also, volcanic and intrusive rocks in mountain belts tend to become more silica-rich over time. Most orogenic belts have a main axis of igneous activity, the igneous arc, where volcanic and intrusive activity are concentrated. The igneous arc is generally on the inner side of the eugeocline. In deeply eroded orogenic belts, intrusive rocks of the igneous arc, usually granitic in composition, are exposed as great masses known as batholiths.

Different thermal conditions in different parts of the overriding plate give rise to two distinct zones of metamorphism. Adjacent to the descending plate, rocks are carried downward to great depths, often 20 kilometers or more, but because the rocks are in contact with the still-cool descending plate, they remain unusually cool. Temperatures in this zone generally average 200 to 300 degrees Celsius, instead of the 500 to 600 degrees Celsius that might be expected at 20 kilometers depth. This low-temperature, high-pressure metamorphism is known as blueschist metamorphism because many of the minerals that form often impart a bluish color to the rocks.

Generally coinciding with the igneous arc is an inner belt of metamorphism where temperatures are high but pressures are moderate. Peak temperatures commonly exceed 600 degrees Celsius, with pressures typically reflecting depths of 5 to 10 kilometers. Such conditions are called amphibolite metamorphism. Adjacent to the region of highest temperature is a region of lower-temperature metamorphism, where temperatures of 400 to 500 degrees Celsius prevail. This type of metamorphism is called greenschist metamorphism, from the greenish color of many of the minerals formed. The outer zone of blueschist metamorphism generally occupies the outer part of the geocline. Amphibolite metamorphism generally coincides with the igneous arc and the inner part of the eugeocline. Greenschist metamorphism commonly extends into the miogeocline.

The deformation of rocks in the overriding plate depends on the nature of the rocks, stress, temperature, and confining pressure. Orogenic belts display several zones of distinctive structures. The most important of these belts are the accretionary prism, zone of basement mobilization, and the foreland fold-and-thrust belt. The eugeocline in general is a region of intense deformation; deformation in the miogeocline is less intense. The outermost edge of the orogenic belt is occupied by the accretionary prism, and it often forms much of the eugeocline. Where the colliding plates meet, sediment is scraped off the descending plate. Other sediment is eroded from the continent and pours into the trench. The sediment from the continent is deposited rapidly, with little weathering or sorting, to form impure sandstone called graywacke.

Submarine landslides and slumps are common in the unstable setting of the trench; the resulting complex of chaotically deposited graywacke is called flysch. A wedge of intensely deformed sediment accumulates on the edge of the continent, much the way a wedge of snow accumulates ahead of a snowplow. This wedge of sediment is the accretionary prism. Frequently, fragments of oceanic crust break off the descending plate and are incorporated into the accretionary prism. These slices of oceanic crust, called ophiolites, are of enormous geologic value. Not only do they mark the location of former subduction zones, but they also provide otherwise unobtainable cross sections of oceanic crust exposed on dry land. The actual contact between the two plates is marked by mélange, a chaotic mixture of broken rock with fragments ranging from microscopic to kilometers in size.

The high temperatures in the igneous arc and amphibolite zone of metamorphism can make the rocks of the crust plastic. That is, the rocks flow like stiff fluids, even though they do not melt. Because the hot rocks are less dense than the cooler crust around them, they rise upward. A mass of rock that flows upward in this manner is called a diapir. The process of heating deep crust (or "basement") so that it rises is called basement mobilization, and evidence of it occurs in many mountain belts. The mobilized crust appears as intensely deformed and highly metamorphosed rock called gneiss. Often the rising mass of gneiss appears to have shouldered the overlying rocks aside, so that the rocks are arched upward with a central core of gneiss. Such a structure is called a gneiss dome.

Compressional forces arising from plate convergence thicken the crust of the overriding plate in a number of ways. Within the accretionary prism, sheets of sedimentary rock are thrust downward beneath the overriding plate, resulting in a stack of faulted slices of rock. These slices may thicken internally by fracturing along small faults and stacking the resulting small slices one above the other, a process called duplexing. Within the igneous arc, crustal thickening occurs when magma invades the crust, increasing its volume. Magma can also be added to the base of the crust, a process called underplating. Heating of the crust within the igneous arc makes much of the lower crust plastic, permitting the plastic crust to be squeezed upward by compression. This process probably assists the upward movement of gneiss domes.

The thickening of the crust results in the uplift of the surface and the formation of topographic mountains. In the foreland, which basically coincides with the miogeocline, deformation is largely a response to events in the active core of the mountain belt. Some of the deformation in the foreland seems to be driven by rising masses of mobilized basement. Rocks of the miogeocline and some of the underlying crust fracture into sheets that are shoved over the rocks beneath. Fractures or faults where one mass of rock overrides another are called thrust faults. These rocks may also be buckled into folds by compressive forces. Rocks nearer the surface often slide off the rising mountain belt. The rocks may break into thin sheets called nappes that stack one atop the other or may crumple into folds. Often the folded rocks have detached from the rocks beneath, much like a carpet slides and folds when a piece of furniture is pushed over it. This process of detachment is called décollement. Because this deformation involves only the surface layers of rock and not the underlying basement rocks, it is called thin-skinned deformation.

OTHER TYPES OF PLATE COLLISION

Other kinds of plate collision result in different combinations of structures. Oceanic-oceanic subduction zones have somewhat simpler orogenic belts. When the descending plate begins to melt, magma rises and breaks through the surface to create a volcanic island arc such as is found in the Aleutian Islands or the Lesser Antilles. Since both plates are oceanic crust, made largely of basalt, the magma also is basalt. Erosion strips sediment off the volcanic islands and dumps much of it in the trench to form an accretionary prism. Over a very long time, the island arc may be built up into a continuous belt of intensely deformed volcanic rocks and sedimentary rocks derived from them. The Greater Antilles and the Isthmus of Panama probably formed this way. Such orogenic belts consist essentially of a eugeocline and igneous arc, with the associated metamorphic zones and deformation structures.

Continent-continent collisions start out as continent-ocean subduction zones, but eventually the convergence of plates brings two continents together, one of which is pushed beneath the other. Continent-continent collisions include the Himalaya, where India is being pushed beneath Tibet, and the Persian Gulf, where the Arabian Peninsula is being pushed beneath Iran. Because continental crust is thick and relatively light, the descending plate cannot be subducted. Instead, one continent rides onto the other, creating a double thickness of crust. Eventually, resistance to further movement may cause plate motions to change on a regional or even global scale.

Usually the overriding continent has had a long history of orogeny before the collision, whereas the other continent may have had none. Orogeny results in such a wide range of structures that it is usually immediately obvious which of the continents is or was the overriding plate. The collision boundary between the two continents, called a suture, may display relics of the former accretionary prism, including melange, fragments of ophiolites, or evidence of blueschist metamorphism.

Often a small block of crust, called a terrane, collides with a larger plate. The terrane may be a volcanic island chain or a small fragment of continental crust. The northern coast of New Guinea is an area where terranes (in this case volcanic island chains) are colliding with a continent. The addition of terranes to a larger plate is called accretion. Terranes are recognizable as distinct blocks of crust separated from adjacent rocks by major faults. In many cases, eugeoclines did not form near their corresponding miogeoclines but as separate terranes. Repeated accretion of terranes can add large areas to a continent. Roughly 1,000 kilometers of the western United States was accreted to North America in the last 500 million years.

After orogeny ceases, it is common for mountain belts to experience a period of crustal extension and faulting. Once the compressional forces that uplifted the mountains subside, many mountain ranges simply cannot support their own mass and begin to spread under their own weight. It requires about 20 million years for erosion to level a mountain range. Nevertheless, long after the topographic mountains are gone, the structures created by orogeny remain. The most conspicuous markers of ancient orogenies are usually the eugeocline and miogeocline, igneous arc, and molasse deposits.

If mountains are worn away within a few tens of millions of years, it follows that the present Urals and Appalachians, the products of continental collisions more than 200 million years ago, cannot be remains of the original mountains. In the case of the Appalachians, this point is clear because rivers such as the Potomac flow across the structures in the mountain belt. Clearly, the Appalachians were once level enough (or buried) that rivers could flow across them. The Appalachians, the Urals, and even the Alps are the result of recent epeirogenic uplift after erosion had largely, or entirely, leveled the original mountains. Why mountain belts sometimes experience renewed periods of uplift long after orogeny ceases is unknown.

Steven I. Dutch

BIBLIOGRAPHY

Cook, Frederick A., L. D. Brown, and J. E. Oliver. "The Southern Appalachians and the Growth of Continents." *Scientific American* 243 (October, 1980): 156-168. Seismic probing has revealed the fault where ancient Africa rode onto North America to create the southern Appalachians. *Scientific American*, which is aimed at the scientifically informed but nonspecialist reader, roughly at college level, is probably the best source of information on recent advances in science for nonspecialists. Its coverage of advances in the Earth sciences since 1970 has been especially thorough.

Howell, David G. "Terranes." *Scientific American* 253 (November, 1985): 116-125. Terranes are small blocks of crust added to a mountain belt by plate collisions. This article describes types of terranes and what happens when they collide.

Jones, David L., A. Cox, P. Coney, and Myrl Beck. "The Growth of Western North America." *Scientific American* 247 (November, 1982): 70-84. The westernmost 1,000 kilometers of North America are a mosaic of at least 200 small blocks added by plate collision in the last 500 million years. This article was written by some of the scientists who were most influential in discovering this process.

Molnar, Peter. "The Structure of Mountain Ranges." *Scientific American* 255 (July, 1986): 70-79. A summary of discoveries about the internal structure of mountain ranges. The apparently solid crust of a mountain range is actually brittle and unable to support its own weight without external compressive forces.

Plummer, Charles C., David McGeary, and Diane H. Carlson. *Physical Geology.* Boston: McGraw-Hill, 1999. This is a straightforward, easy-to-read introduction to geology intended for those with little or no science background. Discusses plate tectonics and its relation to the development of mountain belts. Includes many excellent illustrations, as well as a CD-ROM.

Press, Frank, and Raymond Siever. *Understanding Earth.* 2d ed. New York: W. H. Freeman, 1998. This comprehensive physical geology text covers the formation and development of the Earth, including extensive sections on the development of mountain belts through time. Readable by high school students, as well as by general readers. Includes an index and a glossary of terms.

Spencer, Edgar W. *Introduction to the Structure of the Earth.* 3d ed. New York: McGraw-Hill, 1988. A college-level textbook on structural geology. Chapters 1-3 survey plate tectonics and its role in deforming the crust. Chapter 19 describes mountain ranges that form from continent-ocean plate collision, particularly the structure of western North America. Chapter 20 describes mountains that form by continent-continent collision, with emphasis on the Appalachians and Alps.

Stanley, Stephen M. *Earth System History.* New York: W. H. Freeman, 1999. A comprehensive overview of the development of the Earth. Chapter 9 deals with continental tectonics and the development of mountain belts. Illustrations, glossary, bibliography, and CD-ROM. Readable by high school students.

ROCKY MOUNTAINS

The various mountain ranges and basins that comprise the Rocky Mountains represent a region of tremendous mineral wealth that has played, and continues to play, an essential part in the economic development and prosperity of the United States and Canada. In addition, the ranges constitute a recreational resource of inestimable value. Finally, the Rockies have proven to be an invaluable key to understanding the Earth's past as well as the processes that continue to transform it.

PRINCIPAL TERMS

ANCESTRAL ROCKIES: the mountain ranges that preceded the Rockies by many millions of years and no longer have surface topographic expressions of their former existence

CORDILLERA: a major mountain chain or system of such chains, especially one that is a dominant feature of a continent-sized landmass

CRATON: a large, tectonically stable core area of a continental-type tectonic plate, usually of great age; also termed a continental shield

EPEIRIC SEA: a shallow sea that temporarily (in geologic terms) covers a portion of a craton; also termed an epicontinental sea

LARAMIDE REVOLUTION: an orogeny that occurred between the Late Cretaceous and the Early Tertiary periods of geologic time, instrumental in producing the ranges termed the Rocky Mountains in the United States

OROGENY: a mountain-building episode or event that extends over a period usually measured in tens of millions of years; also termed a revolution

PLATE TECTONICS: a widely accepted theory in geology that contends that the Earth's crust is composed of about twelve moving, rigid plates

SUBDUCTION: a process, according to the theory of plate tectonics, whereby one crustal plate is thrust beneath another at a convergent boundary

TRANSGRESSION: the flooding of a large land area by the sea either by a regional downwarping of continental surface or by a general global rise in sea level

MOUNTAIN FORMATION

Numerous mountain ranges are sometimes grouped under the general definition of the Rocky Mountains by various geologists and geographers. Canadian scientists often define the Rockies differently from the way their counterparts in the United States define the Rockies. During the past few centuries, the defined extent of the Rockies has expanded or contracted to include first one range or feature, then yet another. To define the Rockies in a geologically meaningful way, it is necessary to include those features that are surface expressions of particular systems of formative processes. The Rockies and their geologic history are products of long-term and very large scale processes that shaped not only the North American continent but also the configuration of the Earth's continents and oceans as a whole. Thus, the term "Rockies" will mean the more easterly of those mountain ranges and closely associated features that comprise the North American cordilleran region.

The cordillera is a long, basically continuous highland structure extending from eastern Alaska southward to northern New Mexico. For the majority of their length, the Rockies front the Great Plains to the east and, in the United States, border a geophysical feature called the Basin and Range Province to the west. The Rockies owe much of their present conformation to the Cordilleran orogeny, of which the Laramide orogeny was one of the last great pulses. Thus, a geologic history of the Rocky Mountains is, for the most part, a few localized acts in the greater drama of the Cordilleran orogeny and the formation of North America.

Orogenies are mountain-building episodes or events, occurring over periods of time measured typically in tens of millions of years. In modern geologic theory, they are considered as consequences of large-scale forces operating in the realm of plate tectonics. Plate tectonic theory, aspects of which were once called continental drift, holds that the crust of the Earth is composed of about twelve major rigid units called lithospheric plates. Some of these plates are oceanic while others are continental in nature. All are involved in some type of motion relative to one another.

Plate motion can take various forms: Plates can move away from a common center, slide alongside each other, or collide with each other with various effects. One collision type involves a process called subduction, in which a heavier plate is forced to dive beneath the edge of a lighter plate. This effect is believed to be taking place along the western coast of South America as the edge of a heavier oceanic plate is subducted beneath the lighter continental plate, on which South America rides. The Andes mountain chain is considered a direct result of this subduction. Geologists point to the Andes and the associated subduction as a model for what probably occurred in relation to the formation of the Rockies. In both cases, a long, high belt of mountains possessing similar rocks, minerals, and geologic structures was formed as a product of plate collision.

Much of the evidence of the way the Rockies formed exists in the findings of two subbranches of geology: stratigraphy and sedimentology. Both sciences deal with the layers of rock termed strata and the story they tell about past environments in which deposition and erosion took place. Findings indicate that the ranges known as the Rockies did not exist in their present incarnation before the close of the Mesozoic era, 65 to 70 million years ago. Across great durations of time previous to that period, however, evidence exists for other incarnations of the Rockies occupying somewhat similar geographic orientations. The incarnation just prior to the present range has been termed the Ancestral Rockies. Before the appearance and elimination through erosion of this range, it appears that other Rocky Mountain-like ranges existed at least two other times in Earth history.

These very ancient pre-Ancestral Rockies are evidenced by recrystallized, tightly folded rocks, which represent orogenies estimated to have occurred during the Late Precambrian eon, also termed the Proterozoic eon, which ended about 600 million years ago. Very general ages for two orogenies of these Precambrian Rockies have been estimated at about 1.5 billion and 2.5 billion years before the present. After about 1 billion years before the present, orogenic activity seems to have terminated, and a very lengthy period of crustal stability and erosion appears to have dominated. During this period, the last of the Precambrian Rockies seems to have been totally beveled down to the roots by erosion during a time referred to as the Lipalian interval, which lasted minimally for hundreds of millions of years.

PALEOZOIC AND MESOZOIC ERAS

With the advent of the Paleozoic era about 600 million years before the present, stratigraphic evidence indicates marine conditions dominated in the region formerly occupied by mountains. Sandstones, limestones, and shales indicative of quiet, marine depositional environments prevailed. This situation continued during much of the Early Paleozoic over much of what later would comprise the North American cordilleran region. Beginning with the Late Devonian period in the Late Paleozoic, about 380 million years before the present, and extending into the following Mississippian period, tectonic disturbances began to have effect again. An orogeny called the Antler took place that deposited at least 1,000 meters of sandstone and conglomerate sediments found in such places as present Nevada. These sediments accumulated as the mountains began to erode. Other evidence of this orogeny exists in the form of the clastic (broken rock) layers lying unconformably above older marine strata.

Unconformable stratigraphic rock sections indicate that a break in sedimentary deposition occurred, often as a result of erosional processes outpacing depositional ones, which is exactly what one would expect if a mountain range were rising in an area. Although the Antler orogeny is not directly related to the formation of the present Rockies, it is an event representative of the collage of many smaller orogenic events that contributed, over time, to the Cordilleran orogeny and, thus, to the formation of the Rockies. Like most, if not all, orogenesis, plate tectonic activity is believed to be the driving force behind the Antler event. It is theorized that either a relatively small lithospheric plate called a microcontinent or another tectonic feature called a volcanic arc was being subducted somewhere to the west.

Following the Antler orogeny by tens of millions of years, another orogenic event occurred in the area now part of the western United States termed the Colorado orogeny. This mountain-building event was responsible for uplifting the Ancestral Rockies. During the Pennsylvanian period, about 300 million years ago, two large, mountainous islands or island chains arched up in the Colorado area. They rose above a shallow sea that covered, with the exception of the Appalachia cordilleran area to the east, most of the North American continental landmass. A sea of this epeiric, or epicontinental, type represents a transgression, or advance, of marine waters over a core area of a continental tectonic plate, known as a craton.

The Colorado orogeny produced what are believed to have been two distinct subranges: Front Range and Uncompahgre. A narrow, linear, shallow seaway is thought to have separated them. As the two island ranges eroded, they deposited extensive aprons of sediment into the surrounding seas. These sediments accumulated particularly in the narrow intervening region and are represented by the Fountain and Maroon geologic formations of Colorado. Evaporite basins also formed in the intermontane region as some arms of the surrounding seas became trapped or isolated. Sedimentary environments typical of such basins are represented by gypsum (calcium sulfate) and salt deposits in the west-central Colorado area. Erosion of the Ancestral Rockies continued for tens of millions of years, resulting eventually in their complete leveling. Remnants of these ranges are thought to have persisted in the form of minor hilly topographic relief until the middle of the following geologic era—the Mesozoic, famous for its dominant reptilian fauna such as the dinosaurs.

Tectonic forces had configured the arrangement of the continents into one great landmass by the Late Paleozoic era. Epeiric seas virtually disappeared for a time from North America by the close of the Paleozoic—the Permian period, 280 to 240 million years ago. Arid conditions prevailed for millions of years in western North America and in other areas that made up the supercontinent, which geologists call Pangaea. By the Triassic period, at the onset of the mid-Mesozoic, Pangaea slowly began to break up into smaller units. The following Jurassic period witnessed not only the formation of a large seaway between North America and Eurasia and the southern continents, termed Gondwanaland, but also an intensification of tectonic activity along the western margin of North America. This intensification culminated in the long process of mountain building in the west termed the Cordilleran orogeny.

As the North American continental plate progressively overrode various microcontinents, also known as suspect terranes, as well as volcanic arcs and seafloor to the west, an Andean-type plate margin was initiated. This convergent-type tectonic margin generated the various cordilleran ranges. Widespread crustal deformation and orogenic activity began during the Jurassic near the Pacific Coast with such events as the Nevadan orogeny. Orogenic activity spread eastward, initiating more uplifting activity such as the Sevier event in the Late Cretaceous period, near the close of the Mesozoic era. Finally, the Laramide event was generated as one of the last, most eastward pulses of the Cordilleran orogeny, starting in the Late Cretaceous period. This orogeny lasted for about 20 million years, from roughly 65 million to 45 million years ago, up into the Eocene epoch of the early Cenozoic era. Its end product was the easternmost ranges of the North American Cordillera—the ranges known as the Rocky Mountains.

CENOZOIC AND PLIOCENE ERAS

By the onset of the Laramide orogeny, an epicontinental sea had formed by the transgression of marine waters from the Arctic to the north and the Gulf of Mexico to the south. These seas joined in midcontinent to form a long, wide but shallow, epeiric sea, which left marine sediments behind known generally as the Sundance formation. To the west of these marine strata, a famous, nonmarine sequence of strata formed, termed the Morrison formation, famous for numerous dinosaur fossils. As the land still farther to the west began to uplift, erosion accelerated, depositing continental sediments to the east, prograding, or extending out, into the epeiric, marine sediments. These sediments are termed the Morrison clastic wedge and are the first evidence, in Late Jurassic times (about 180 million years ago), of the eastward spreading effects of the Cordilleran orogeny. Lower Cretaceous period deposits in the central Rocky Mountain area exhibit a much thicker clastic wedge of up to 10,000 meters. This wedge's accumulation coincides with the draining of the epeiric sea from the continental interior and immediately presages the Laramide event.

By the Paleocene and Eocene epochs of the Early Cenozoic, the Laramide orogeny had created the long, high ranges of mountains of the Rockies. Between the ranges, intermontane basins had formed as lowlands between the north-south trending, linear uplifts. These basins began to fill with sediments eroded off the adjacent mountains as well as with river and lake sediments. Numerous basins became large lakes, which, over millions of years, gradually filled completely. During their existence as active lake environments, the larger lakes were hosts to thriving ecosystems of plants and animals. The economically important deposits of oil shale found widely in Colorado, Wyoming, and Utah formed from the petrochemical mineral kerogen (which yields oil when heated) trapped within the shale. The kerogen is believed to have formed primarily from the decay of lake-dwelling planktonic organisms during the lifetime of the larger lake bodies. Indications suggest the lakes were in existence for at least 5 to 8 million years.

The filling of the intermontane basins and lakes is believed to have been completed by the end of the Eocene or perhaps the Early Oligocene epoch, about 35 million years ago. At roughly the same time, regional uplift had come to a halt and the effects of the Laramide orogeny had definitely dwindled to a standstill. Erosion by Oligocene times had made great inroads toward leveling the

Looking south to the Sidenius thrust fault in the Northern Rocky Mountains, Halfway River area. A broad hanging-wall syncline is visible to the left. (Geological Survey of Canada)

landscape of a generally stable tectonic region. For something like 5 to 10 million years, the eastern Rocky Mountain region saw no new uplift, and erosion proceeded to wear the mountains away. A huge, sloping apron of eroded sediments began to develop out onto the Great Plains to the east, forming a deep clastic wedge during the Oligocene and most of the following Miocene epochs of the Cenozoic. The remnants of the once-lofty Laramide summits stood only as minor prominences, sticking out of huge, surrounding aprons composed of their own eroded debris. As erosion proceeded still further, these aprons began to coalesce into a solid sheet of sediment termed the Tertiary pediment, which was once erroneously called the Rocky Mountain peneplain.

This trend was thwarted, however, during the Late Miocene by a great regional upwarping of the crust over most of the Rocky Mountain region as well as adjacent areas. To the west, the huge area of the present Colorado Plateau began to lurch upward toward its present altitude during this period. As the regional uplift proceeded into the Pliocene, rivers and streams were rejuvenated by the increased gradients they now flowed down, and stream-cut erosion accelerated. A lengthy period of canyon cutting ensued, which has continued. This phenomenon of stream rejuvenation by regional uplift explains much of the unusual entrenched meander and superimposed drainage found in the Rocky Mountain region and in the western states at large. The Miocene-Pliocene regional uplift added 1,000 to 2,000 meters of new height to the peaks of the Rockies and also lifted surrounding terrain such as the Tertiary pediment so that these former lowlands have been considerably elevated and seem like anomalous terrain.

Finally, as the cooling of the Pliocene gave way to the intermittently frigid period of continental glaciation dominating the Pleistocene epoch of the past 2 million years, large alpine glaciers formed at higher altitudes in the ranges of the Rockies. At least three episodes of alpine glaciation are believed represented in places such as the Colorado Rockies of the United States. Glaciation helped to scour sedimentary layers from mountaintops, exposing the harder and older crystalline rocks lying beneath. Glaciation also widened and deepened downslope valleys, giving some a U-shaped cross section, characteristic of glacially influenced topography. Glaciers also left their mark on the Rocky Mountain landscape in the form of myriad examples of glacially sculpted terrain: landforms such as cirques (steep-walled basins), moraines (accumulations of rock debris), and razor-sharp ridges called arêtes. The end of the Pleistocene epoch's cycle of glacial advance and retreat is not certain, however, and the few thousand years since the waning of widespread glacial effects has not been enough time for significant geologic processes to leave a new page in the history of the Rocky Mountains.

Frederick M. Surowiec

CROSS-REFERENCES

Alps, 805; Andes, 812; Appalachians, 819; Basin and Range Province, 824; Cascades, 830; Earth Resources, 1741; Himalaya, 836; Mountain Belts, 841; Plate Tectonics, 86; Sierra Nevada, 852; Subduction and Orogeny, 92; Transgression and Regression, 1157; Transverse Ranges, 858.

BIBLIOGRAPHY

Batten, Roger L., and R. H. Dott, Jr. *Evolution of the Earth.* New York: McGraw-Hill, 1988. An in-depth survey of North America from the viewpoint of historical geology. Each geological era and period is treated chronologically with respect to the processes and effects that influenced various regions of the continent. Significant formations and deposits are explained in terms of geologic structure and stratigraphy. Suitable for college readers or others with a good, basic knowledge of geologic principles.

Caputo, Mario V., James A. Peterson, and Karen J. Franczyk, eds. *Mesozoic Systems of the Rocky Mountain Region, USA.* Denver: SEPM, 1994. This multiauthored volume deals with a variety of aspects of the Rocky Mountains, including tectonics, sedimentation, and paleontology. Written for the specialist.

Chronic, Halka. *Roadside Geology of Colorado.* Missoula, Mont.: Mountain Press, 1980. One of the more comprehensive treatments of the landforms found in Colorado that is designed for the general reading public. Like other titles in this series, it is written for the geologically aware or curious traveler or reader. The author assumes no prior geology education or training. Profusely illustrated throughout with numerous maps, diagrams, and photographs. These visual aids in combination with the text effectively explain the geologic history of the middle Rockies of the United States.

_____. *Time, Rocks, and the Rockies: A Geologic Guide to Rocks and Trails of Rocky Mountain National Park.* Missoula, Mont.: Mountain Press, 1984. This source provides numerous illustrations and descriptions pertaining to Rocky Mountain National Park, but with the general intent geared toward explaining the general story of the Rockies through time to both park visitor and nonvisitor alike. Suitable for readers at all levels of geologic knowledge, high school and above.

Kearey, Philip, and Frederick J. Vine. *Global Tectonics.* 2d ed. Oxford: Blackwell, 1996. The well-illustrated and clearly organized undergraduate text discusses the principles of geology in relation to plate tectonics. Index and bibliography.

Lageson, David R., and D. R. Spearing. *Roadside Geology of Wyoming.* Missoula, Mont.: Mountain Press, 1988. A fascinating and informative introduction to the geologic history and processes still at work in this state, which possesses a number of outstanding examples of Rocky Mountain ranges and related geomorphologist features. Well illustrated throughout, like other titles in this series, with highly explanatory geologic and geographic maps, diagrams, and photographs. A useful sourcebook for the general reader on the Rockies and their geologic history.

McPhee, John. *Rising from the Plains.* New York: Farrar, Straus & Giroux, 1986. One of about three books on the subject of American geology by the well-known, "new" or "literary" journalist, McPhee. The author uses a novel approach to discussing geologic concepts by involving the reader in the lives of working field geologists. Excellent introduction to the geologic history and processes of the entire western United States, focusing on the middle and northern Rockies. Suitable for all readers high school and above.

Plummer, Charles C., David McGeary, and Diane H. Carlson. *Physical Geology.* Boston: McGraw-Hill, 1999. This is a straightforward, easy-to-read introduction to geology intended for those with little or no science background. Chapter 20 discusses the development of mountain belts such as the Rockies. Includes many excellent illustrations, as well as a CD-ROM.

SIERRA NEVADA

The Sierra Nevada preserve rocks that formed several kilometers below the Earth's surface during subduction of the Pacific Ocean below North America more than 100 million years ago. Erosion and deposition by streams 50 million years ago concentrated gold in the western foothills discovered in the 1849 California gold rush.

PRINCIPAL TERMS

BATHOLITH: body of rock crystallized from magma, or molten rock; compositions are typically granitic

CRUST: the outermost layer of the Earth; continental crust is 30 to 35 kilometers thick; oceanic crust is 5 to 10 kilometers thick; the greater density of oceanic crust relative to continental crust forces it to subduct

FAULTING: the process of fracturing the Earth such that rocks on opposite sides of the fracture move relative to each other; faults are the structures produced during the process

GRANITE: an igneous rock that is composed of mostly pink and white (quartz) minerals, which frequently has a speckled appearance

METAMORPHIC ROCKS: rocks deposited initially as sediments and subsequently exposed to extensive heat, resulting in their recrystallization

NORMAL FAULT: a steep fault that forms during extension, or stretching, of the Earth's surface; the rock above the fault moves down relative to the rock below

QUARTZ: one of the most common minerals on the Earth's surface; it occurs in many different forms, including agate, jasper, and chert

VEIN: a mineral deposit that fills a crack; veins form by precipitation of minerals from fluids

OVERVIEW

Much of what is admired in the Sierra Nevada—the face of the eastern slope, the deep canyons of Yosemite National Park, the peaks where the interaction of light and shadow inspired the name the "range of light," and the undulating topography of the western foothills—is what makes the range geologically significant. The uplift and erosion that created the Sierran landscape over the last 20 million years exposed the roots of a giant volcanic chain active between 225 million and 80 million years ago. During that interval, the Sierra Nevada resembled the modern volcanic belts of the Andes in South America and the Cascade Range of the Pacific Northwest; however, only the bowels of the Sierran volcanoes remain. The origins of the ancient and modern volcanic chains are similar; all three formed during the important process of subduction, or descent, of an oceanic plate below a continent. Subduction is fundamental to plate tectonics, the basic theory of the Earth sciences. Exposure in the Sierra Nevada of rocks originally formed several kilometers below the surface of the Earth allowed scientists to investigate processes at depth in continental crust above subduction zones and to assess the validity of plate tectonic theory to explain events in the geologic past. Furthermore, the processes inferred for the Sierra Nevada probably occur beneath the Andes and the Cascade Range.

The description of the Sierra Nevada includes the morphology, or physical characteristics that define it, and the geology, or the rocks that compose it. The Sierra Nevada are the longest continuous mountain range in the conterminous United States, extending 640 kilometers north-northwesterly along the eastern border of California. The mountains rise gently in the west from the agricultural belt of central California's Great Valley to the glaciated crest. From the crest, the range drops precipitously to the deserts of the Basin and Range Province in Nevada to the east. To the northwest, the Sierra merge with the Klamath Mountains of Northern California and Oregon. The rocks and faults of the Sierra Nevada can be identified in the Klamath Mountains, suggesting that the two ranges were once one. To the north, the Sierra Nevada terminate against the Cascade

Range of northern California, Oregon, Washington, and British Columbia. The southern Sierra Nevada bend westward into the Transverse Ranges. The Sierra Nevada form a barrier to storms that enter California from the Pacific Ocean and create a rain shadow east of the range crest. Thus, the shallow western slope has pine forests and alpine lakes, whereas the steep eastern slope is barren and dry.

Different processes created the landscapes of the northern, central, and southern Sierra Nevada. In the north, much of the morphology was formed by volcanoes that spewed lava and ash. Glaciers dominated in the central portion, carving great U-shaped valleys. The highest peaks, including the tallest mountain in the lower forty-eight states, Mount Whitney, are in the southern Sierra, where elevations approach 4,250 meters. Here, glaciers and streams were not very active. What is remarkable about the entire Sierra Nevada, however, is their flat top. Looking from a distance toward the crest, the Sierra Nevada appear to be a tableland cut by numerous deep canyons. This flatness led scientists to conclude that the mountains represent uplift of an area that was previously eroded to a uniform level. The Sierra Nevada consist of a series of long belts of metamorphic rocks that were intruded by a large batholith. The belts parallel the axis of the mountain range and grow younger from east to west. Their boundaries are faults across which rocks change age and composition. The metamorphic belts of the eastern Sierra Nevada contain sediments that accumulated on the continental shelf of North America between 500 million and 225 million years ago. In contrast, the sediments in the metamorphic belts of the western Sierra (which are the same age as those of the eastern Sierra Nevada) collected on the floor of the old ocean basin that bordered the ancient continental shelf. Slivers of the old seafloor, called ophiolites, occur with the sediments in the western metamorphic belts.

Yosemite Valley, perhaps the most famous section of the Sierra Nevada, with Half Dome to the right. (© William E. Ferguson)

The batholith that intruded the mosaic of fault-bound belts appears to be a monolith of pink and gray speckled rock. Closer inspection, however, reveals that it consists of numerous small bodies of rock, each of which is a few kilometers across. The bodies are between 135 and 85 million years old and were derived from tens of kilometers depth. The pink and gray rock is granite, a chief constituent of continental crust, and is readily distinguished from the dark-colored metamorphic rocks. The large size of the mineral crystals in the granite suggests that the magma cooled slowly below the surface of the Earth. Granite is estimated to comprise more than 60 percent of the Sierra Nevada. Other rocks in the Sierra Nevada include glacial deposits, volcanic rocks, and stream gravels that are younger than the metamorphic belts and the batholith.

FORMATION

To understand the evolution of the Sierra Nevada, one must have a grasp of the theory of plate tectonics, which states that the surface of the Earth is composed of rigid plates between 5 and 100 kilometers thick. The plates, either oceanic or continental, float on a partially molten layer that allows them to move relative to one another at velocities of a few centimeters per year. Oceanic plates are much younger than continental plates and, in fact, are created continuously along the ridges that line the world's ocean basins. Magma (molten rock) from deep in the Earth erupts at the mid-oceanic ridge and cools, forming new seafloor. To make room for the new material, old seafloor is removed from the ocean basins at trenches. Trenches are on the opposite edges of oceanic plates from ridges and are the sites where seafloor subducts, or descends, below another plate, either continental or oceanic. Ridges and subduction zones, therefore, define two types of plate boundaries: divergent (where plates move apart) and convergent (where plates come together), respectively.

The geology and morphology of the Sierra Nevada tell a story that spans 0.5 billion years—of

Half Dome, in the most famous section of the Sierra Nevada, Yosemite National Park. A glacier that moved over the Sierra Nevada batholith formed the Yosemite Valley and other landmarks during the last ice age. (Corbis)

which subduction dominated the last 200 million years. For the first few hundred million years, however, the area of the Sierra Nevada was quiet. Sediments were deposited in shallow water on the continental shelf of Nevada to the east and in deep water on the Pacific Ocean floor to the west. (The edge of the North American continent during this time was several hundred kilometers farther east.) Approximately 225 million years ago, the scene changed. The Pacific Ocean began to subduct below the North American continent. Sediments on the seafloor scraped against and accreted to the overriding continent in long, linear belts. Pieces of seafloor occasionally accreted with the sediments. Addition of material to its edge forced the continent to grow progressively westward. As subduction continued, the oceanic plate reached depths in the Earth below the continent where the temperature was high enough to cause melting of the interface between the overriding and descending plates. The melting produced small bodies of magma that ascended to the surface. Significant quantities of subducted sediments and continental crust also melted and mixed with the magmas to produce liquids of granitic composition. Most of the magmas solidified at shallow depths between 135 million and 85 million years ago, leading to the large batholith now exposed in the Sierra Nevada. Some of the magmas, however, erupted on the surface as volcanoes. Emplacement of the magmas near the surface raised the temperature sufficiently to recrystallize the sediments into metamorphic rocks.

About 80 million years ago, the magmas cooled, the volcanoes stopped erupting, and the seafloor ceased accreting to the North American continent. For the next 50 million years, erosion dominated. Streams stripped the volcanoes from the surface, exposing the batholith and metamorphic rocks that lay below. The removed material was deposited as soil and gravel in the rolling hills between the Sierra Nevada and the Pacific Ocean. Eventually, the landscape was eroded to a uniform level. Subduction below California stopped completely approximately 25 million years ago. The rise of the modern Sierra Nevada began 20 million years ago. The continental crust between the Sierra Nevada and Utah stretched and broke apart, producing normal faults oriented in an almost northerly direction. This stretched province

of narrow mountains and wide valleys is the Basin and Range Province. The westernmost of the normal faults in the Basin and Range Province defines the eastern edge of the Sierra Nevada. Rocks west of the Sierra fault moved up to create the modern mountains, while those to the east moved down to form a valley. Eruption of lava and ash accompanied the stretching of the continental crust as spaces formed to allow the passage of molten rock from deep in the Earth. Lavas frequently flowed down the old stream channels. New streams circumvented the lavas, leaving them as high grounds in the landscape. Finally, glaciers of the Wisconsin glacial epoch (60,000 to 90,000 years ago) carved the spectacular valleys of the crest, including Yosemite Valley.

The combination of subduction followed by stretching built and is building the modern Sierra Nevada. The Sierra Nevada are still active; large earthquakes on the eastern edge of the range reveal that the mountains are rising and the fault is moving. Magma lies within a few kilometers of the surface below Mammoth Lakes, ready to feed the volcanoes at Mono Craters, which last erupted 6,500 years ago. Streams continuously erode the high peaks trying to bring the land to sea level, thereby restoring the landscape of 0.5 billion years ago.

STUDY OF THE SIERRA NEVADA

A variety of techniques from all disciplines in the Earth sciences are used to study the Sierra Nevada. Two, however, are particularly important: fieldwork and isotope geochronology. By doing fieldwork, investigators map the distribution of rock types, document structures such as faults, and analyze the relative ages of the rocks, primarily by painstakingly describing each rock type and walking throughout the range. In contrast, isotope geochronology allows scientists to ascertain the absolute ages of the different rock types from field samples taken back to the laboratory.

To determine the evolution of ancient rocks, geologic fieldwork is necessary. In the field, the geologist systematically notes the different types of rocks that exist and the contacts among them. Features in the rock that may be chosen for evaluation include the kinds of minerals, the sizes of the different types of particles, and the nature of the boundaries between grains. Contacts are evaluated to determine positions of rock types to establish their relative ages. Generally, rocks at the top of a sequence are assumed to be the youngest; those at the bottom, the oldest. The geologist also watches carefully for faults that may indicate that two adjacent sequences of rocks have had dissimilar histories. Such faults define the boundaries of the western metamorphic belts in the Sierra Nevada. Although the rocks on either side of these faults were deposited in deep water, they have different compositions and ages, leading scientists to conclude that the rocks now juxtaposed in the field were not deposited next to one another but were brought together during faulting. They originally may have been separated from one another and North America by great distances. The term "exotic terrane" frequently is applied to such rock sequences. Accretion during subduction explained this phenomenon. Sediments could arrive at North America from far-off sites, carried by the moving oceanic plate.

Isotope geochronology, or radiometric dating, allows scientists to determine the "absolute" age of a rock by taking advantage of the property of radioactivity. The dates derived from this method differ from those calculated by fossils and the superposition of strata. The latter are "relative" ages based on the geologic time scale. Prior to radiometric dating, the number of years represented by divisions on the time scale were unknown. Radioactivity is the process by which the nuclei of certain chemical elements such as uranium, radium, thorium, rubidium, and potassium break apart and emit radiation. The process yields an atom of a different element. Uranium, for example, changes to lead. Isotopes are species of the same chemical element that differ only in the number of neutrons in their nucleus. Some isotopes are stable; some are unstable and break apart by radioactive decay. Each isotope has a characteristic rate of decay measured as a half-life (the time in years for half the nucleus to decay). Examples of

half-lives include the 47 million years for rubidium to change to strontium and the 1.35 million years for potassium to become argon. Because half-lives can vary by several orders of magnitude, isotopes used for young rocks are different from those used for old. By accurately determining in a mineral the amounts of the radioactive isotope and the atom into which it decays, geologists can calculate the age at which the mineral crystallized. The technique, therefore, is very useful for dating igneous and metamorphic rocks. Minerals crystallize in these rocks when the rocks form. In contrast, sedimentary rocks contain fragments of minerals that were eroded from older areas. Thus, radiometric dating of minerals in sediments yields only the ages of the region from which the minerals were derived.

The ages of and the relationships between the batholith and the metamorphic rocks in the Sierra Nevada were greatly elucidated by isotope geochronology. Intrusion of the batholith was discovered to have spanned tens of millions of years as individual bodies rose one by one. Despite the similar appearance of two granites, their age can differ by as much as 50 million years. Field observations suggested that heat from the batholith metamorphosed the accreted sediments; dating of the metamorphic belts corroborated the field evidence. In addition, isotopic studies established the age of the lavas that postdated magmatism.

Pamela Jansma

CROSS-REFERENCES

Alps, 805; Andes, 812; Appalachians, 819; Basin and Range Province, 824; Cascades, 830; Continental Growth, 573; Earthquakes, 316; Faults: Normal, 213; Faults: Thrust, 226; Himalaya, 836; Isostasy, 128; Mountain Belts, 841; Plate Motions, 80; Plate Tectonics, 86; Rocky Mountains, 846; Seismic Reflection Profiling, 371; Subduction and Orogeny, 92; Transverse Ranges, 858.

BIBLIOGRAPHY

Adams, Ansel. *Yosemite and the Range of Light.* New York: New York Graphic Society, 1979. This collection of photographs was taken by the noted photographer Ansel Adams, after whom a wilderness has been named in the high Sierra Nevada near Mammoth Lakes. Adams spent many years hiking the Sierra Nevada early in the twentieth century prior to the advent of cars in the alpine country. His images are beautiful, and this book is suitable for everyone.

Alt, David D., and Donald W. Hyndman. *Roadside Geology of Northern California.* Missoula, Mont.: Mountain Press, 1975. This book is an excellent guide to the geology of Northern California. Discussion of the Sierra Nevada and the Klamath Mountains occupies more than one-third of the book. Other sections discuss the Coast Ranges, the Cascade Range, and the Great Valley. The section on the Sierra Nevada includes an introduction followed by summaries of the geology along the major federal and state highways that pass through the region. The illustrations are very good, and the text is easy to read. The book is recommended for anyone interested in the geology of Northern California and in the Sierra Nevada, specifically. If the reader enjoys this book, several more are available for the western United States.

Hatcher, Robert D., Jr. *Structural Geology: Principles, Concepts, and Problems.* 2d ed. Englewood Cliffs, N.J.: Prentice-Hall, 1995. This undergraduate textbook provides an overview of the structural development of the Sierra Nevada. Intended for the more advanced reader.

Norris, R. M., and R. W. Webb. *The Geology of California.* New York: John Wiley & Sons, 1976. This book systematically describes the geology of the state. Geological concepts are only briefly introduced because the authors assume considerable prior knowledge of the discipline. Illustrations are good. Recommended for college-level students who want to learn a considerable amount about the geology of California.

Oakeshott, G. B. *California's Changing Landscapes: A Guide to the Geology of the State.* New York: McGraw-Hill, 1978. This book has ex-

cellent illustrations and a fine glossary. Geological concepts are introduced prior to the discussion of the geology of the state. Short sections on geochronology and the geologic time scale are particularly good. Recommended for the college-level student.

Press, Frank, and Raymond Siever. *Understanding Earth.* 2d ed. New York: W. H. Freeman, 1998. This comprehensive physical geology text covers the formation and development of the Earth. Contains extensive sections on the deformation of the continental crust and the development of the Sierra Nevada. Readable by high school students, as well as by general readers. Includes an index and a glossary of terms.

Seyfert, Carl K., ed. *The Encyclopedia of Structural Geology and Plate Tectonics.* New York: Van Nostrand Reinhold, 1987. As its title suggests, this reference is a good source for learning more about plate tectonics. Faulting and folding also are discussed. The text is not technical and is recommended for anyone interested in the subject.

Stanley, Stephen M. *Earth System History.* New York: W. H. Freeman, 1999. A comprehensive overview of the development of the Earth. Chapter 19 deals with the development of the Sierra Nevada. Illustrations, glossary, bibliography, and CD-ROM. Readable by high school students.

Webster, Paul. *The Mighty Sierra: Portrait of a Mountain World.* Palo Alto, Calif.: American West, 1972. This book is a pictorial voyage through the entire Sierra Nevada. The morphology is discussed extensively; the geology is mentioned briefly. Quite a large amount of space is devoted to plant and animal life. Recommended for anyone who wants to get a feel for the high alpine country but does not want to be limited to geological information.

TRANSVERSE RANGES

The Transverse Ranges are a prominent east-west-trending mountain chain that forms a spectacular backdrop behind the most populated region in California. The range is unique in several ways: The east-west orientation departs markedly from the usual north-south trend of other geologic structural blocks in California; there is a greater variety of unusual rocks and structures in the range than elsewhere in the state; and the range has important mineral and energy resources.

PRINCIPAL TERMS

BASEMENT: an underlying complex generally of igneous and metamorphic rocks

REVERSE FAULT: a steeply to moderately inclined fault in which the overlying block of rock has moved upward over the underlying block

RIGHT-SLIP: sideways motion along a steep fault in which the block across the fault appears displaced to the right; left-slip faults are the opposite

TECTONICS: a branch of geology that deals with the study of regional large-scale structural or deformational features and their origins, mutual relations, and evolution

THRUST FAULT: a variety of a reverse fault in which the inclination of the fault plane is at a low angle to horizontal

---

LOCATION AND CHARACTERISTICS

The Transverse Ranges are a narrow mountainous belt in California that extends from Point Arguello, 80 kilometers west of Santa Barbara, eastward for more than 450 kilometers to the Eagle Mountains in the Mojave Desert; they are from 16 to 80 kilometers wide. The Transverse Ranges are a distinctive, if not unique, geologic province of California because of an east-west orientation (which is rare in North America and unique in California), a greater range of rock types and geologic structures than any other region in California, the oldest plutonic and metamorphic basement rocks in California, and the largest oil fields in the state outside the San Joaquin Valley and Los Angeles County.

Initial observations of the overall alignment of the Transverse Ranges might suggest that the province constitutes a single, apparently homogeneous east-west-trending mountainous belt. That is not the case, however, as closer examination reveals that the Transverse Ranges comprise three unrelated segments, each with widely differing geologic histories and origins, that have been brought into alignment coincidentally by the large-scale tectonic regime that is peculiar to Southern California. On the basis of geologic dif-

ferences, the three segments of the Transverse Ranges may be subdivided into the eastern segment, which includes the San Bernardino, Little San Bernardino, Hexie, Pinto, and Eagle Mountains; the central segment, the major portion of which is the San Gabriel Mountains but which also includes the lesser Verdugo, Sierra Pelona, Liebre, and Sawmill Mountains; and the western segment, which consists of the Santa Susana, Santa Monica, Topatopa, and Santa Ynez Mountains and the northern four Channel Islands.

The mountains of the eastern segment of the Transverse Ranges are bounded largely by faults, of which the San Andreas fault forms the southern margin. The eastern segment is about 250 kilometers long and includes the highest elevation in Southern California at San Gorgonio Mountain, 3,505 meters above sea level, in the San Bernardino Mountains. Rocks underlying the eastern segment are predominantly Mesozoic granitic rocks, gneiss and schist of probable Precambrian age, and lesser amounts of fossil-bearing Late Paleozoic metamorphosed marine sedimentary rocks. Most of the summit region of the San Bernardino Mountains is a plateaulike old erosion surface of moderate relief that developed prior to uplift. This surface now stands at elevations of about 1,950 to

2,290 meters. The preservation of the old erosion surface is explained by the comparatively recent and rapid uplift of the range, which is thought to have occurred within only the last 2 million years. A second elevated plateaulike old surface exists in the Little San Bernardino Mountains. This surface has been eroded into the picturesque, bold rock formations for which Joshua Tree National Monument is famous.

The central segment of the Transverse Ranges is dominated by the San Gabriel Mountains, a high, rugged upland about 100 kilometers long and as much as 38 kilometers wide. The range is bordered on the north and east by the San Andreas fault zone and by reverse faults on the south and southwest. The range consists mainly of a crystalline basement complex of intrusive igneous rocks and a variety of metamorphic rocks. Within the rock complex are two generations of early Precambrian gneisses (1,045 million and 1,700 million years old, respectively), which are the oldest rocks in California. Summit elevations of the peaks in the range vary from about 1,950 to 2,450 meters, the highest being Mount San Antonio (Mount Baldy) at 3,072 meters.

There are many faults within the central segment of the Transverse Ranges, some of which have right-slip movement of several tens of kilometers. The range has been uplifted essentially as a unit by broad arching across its northern margins and central part and by reverse faulting along its southern margin. Most of the uplift has occurred within the last 15 to 20 million years. Because uplifting of the ranges of the central segment began long before that of the eastern segment, erosion has had greater time to wear away any uplifted old erosion surfaces. Consequently, no erosion surfaces exist in the San Gabriel Mountains, in contrast to those present in the more recently elevated ranges of the eastern segment. A thick sequence of Cenozoic fossiliferous nonmarine sedimentary rocks and some volcanic rocks occur around the base of the central segment. These rocks have been eroded into the scenic landforms at Devil's Punchbowl County Park near Pearblossom, Vasquez Rocks County Park on State Highway 14, and at Cajon Pass.

Unlike the segments of the Transverse Ranges to the east, the mountains of the western segment are composed almost entirely of marine sedimentary rocks of late Mesozoic and Cenozoic age. Locally within the sedimentary rocks are extensive volcanic rocks, many of which were produced by submarine eruptions. The ranges of the western segment have a combined length of about 190 kilometers. Summit elevations of the mountains are modest, reaching their highest elevation of 2,050 meters at Hines Peak in the Santa Ynez Range.

Part of the western Transverse Ranges is underlain by a large syncline about 190 kilometers long, named the Soledad Basin in the east and the Ventura Basin in the west, where it is contiguous with the Santa Barbara Channel farther to the west. The Ventura Basin, filled with primarily marine sedimentary rocks, merges eastward with the nonmarine sediments of Soledad Basin. The Santa Susana Mountains, South Mountain, Oak Ridge, Ventura Anticline, and Sulphur Mountain are prominent anticlinal hills that have formed in the last 200,000 to 300,000 years within the Ventura Basin. The Ventura Basin is geologically famous for its immense combined thickness of late Mesozoic and Cenozoic marine sedimentary rocks totaling more than 17,600 meters, for the recency of its deformation, and for its great petroleum production.

STUDY OF TRANSVERSE RANGES

Geologists use their full armamentarium of devices and techniques to study the geology of the Transverse Ranges. The most important technique has been field mapping, the process of on-site research on foot, by car, or in an airplane in order to locate and plot the various rock units, faults, and folds on aerial photographs and topographic base maps. Rock and fossil samples are collected and labeled in the field. These samples are taken into the laboratory for detailed examination and identification and compared with samples from other areas of the same or similar geology that have been studied previously. Paleontologists, experts in the study of fossils, may be consulted to assist in fossil identification. Correlation (the determination of age relationships between rock units or geologic events in separate areas) of rocks and fossils in exposures on opposite sides of faults allows the determination of the amount and direction of movement along the faults, assuming that rocks with identical properties and fossils were once a single body.

Added to the work of field geologists is subsurface information obtained from oil wells: drilling logs, electrical resistivity and self-potential logs, gamma ray-neutron logs, cores, and side-wall samples. This subsurface information is immensely useful to geoscientists, enabling them to "see" the geology underground. Other techniques to discover what is underground are seismic reflection and seismic refraction profiling. Seismic reflection profiling is based on the travel time of sound waves excited by precisely timed "thumping" of the ground that are reflected off buried rock layers or fault surfaces. Seismic refraction profiling is based on the travel time of sound waves that are generated by a small controlled explosion. Data generated by these techniques are plotted on graphs, from which the depths and orientations of the various layers and structures in the subsurface geology may be interpreted.

Supplementary subsurface information important in the study of the Transverse Ranges is obtained by a geophysical technique called gravimetry. By using gravimeters, sensitive instruments that can precisely measure very slight differences in the Earth's gravitational field, it is possible to deduce much about the physical characteristics of the rocks underlying the surface. For example, gravimetry reveals that the San Gabriel Mountains lack "roots," lower-density rocks such as granite, which usually exist beneath mountain ranges, where they extend into the mantle for tens of kilometers. Such underlying roots provide the buoyant lifting force that supports mountain ranges as elevated masses above the average elevation of the Earth's surface. Thus, most of the world's mountains literally float on the mantle buoyed by their roots much as the hidden keel of an iceberg supports its visible portion. An explanation of how the high-standing San Gabriel Mountains can exist without such roots comes from another geoscientific discipline, geodetic surveying.

Geodetic surveying is a branch of civil engineering that requires consideration of the Earth's curvature. Instrumental techniques used in geodetic surveying incorporate laser optical devices, pulsed infrared lightwave and microwave electronic surveying equipment. These instrumental techniques produce extremely accurate measurements of distances and angles to a high degree of precision. Precision surveys conducted since the 1970's reveal strains in the Earth's crust across the San Gabriel Mountains that cause the range to be compressed in a north-south direction. Thus, the San Gabriels are supported essentially by being squeezed upward by crustal compression, a process that is much different from that supporting most of the world's mountain ranges.

Radioactive age dating, which determines absolute ages of rocks, is another technique used in the study of the Transverse Ranges. Knowing the rate of decay of radioactive elements allows geochronologists (geoscientists who specialize in radioisotopic age dating of rocks) to determine the ages of rocks in years. The technique is an involved laboratory procedure requiring the use of a mass spectrometer and similar specialized equipment. Radioactive age dating is especially useful in determining the ages of igneous and metamorphic rocks.

ENVIRONMENTAL IMPACTS AND RESOURCES

It is important that scientists and others understand the relation between rock types, slope angles, and watershed management in the Transverse Ranges in order to protect the lowland communities from flooding during the episodic winters of high precipitation. A second concern with the Transverse Ranges is their role as a watershed that provides runoff water used by the residents of the foothill communities that border those ranges.

The mineral wealth of the eastern and central segments of the Transverse Ranges is of economic importance, with numerous mines and mining prospects scattered throughout the province. Mineral deposits in the central segment of the Transverse Ranges have been exploited for almost 150 years. The first gold discovery in California was made in the San Gabriel Mountains, at Placerita Canyon, in 1842. Exploration during the fifty years following the discovery produced important gold discoveries in several areas of both the San Gabriel and San Bernardino Mountains and in some of the lesser ranges of the eastern segment. Iron ore was mined for many years in the Eagle Mountains. Silver, copper, lead, tungsten, and numerous nonmetallic minerals also have been sought and mined from time to time in the central and eastern segments. Limestone deposits in Lower Cushenbury Canyon on the north side of the San Gabriel Mountains are the basis for the current mining activity in the Lucerne Valley limestone district and the existence of the largest cement plant in the world.

Probably even more significant is the great petroleum production from the western segment of the Transverse Ranges. The marine sedimentary rocks comprising the western segment in concert with the faulting, folding, and rapid uplift of the rocks during the last 1 million years not only have created the impressive modern landscape but also have formed the geologic structures that are responsible for the numerous oil fields of the region. Of historic importance is that the first petroleum production in California, in 1857, came from oil seeps near Ventura and that the first oil well drilled in California and put into serious commercial production was completed near Ojai in 1866. This early well is located in what is now known as the Silverthread area of the Ojai field. Even as late as the 1980's, the production from the Silverthread area was still economic, yielding upwards of 1 million barrels of oil per year.

D. D. Trent

CROSS-REFERENCES

Alps, 805; Andes, 812; Appalachians, 819; Basin and Range Province, 824; Cascades, 830; Faults: Normal, 213; Faults: Thrust, 226; Gold and Silver, 1578; Himalaya, 836; Igneous Rock Classification, 1303; Iron Deposits, 1602; Metamorphic Rock Classification, 1395; Mountain Belts, 841; Oil and Gas Exploration, 1699; Petroleum Reservoirs, 1728; Rocky Mountains, 846; Sedimentary Rock Classification, 1457; Seismic Reflection Profiling, 371; Sierra Nevada, 852; Well Logging, 1733.

BIBLIOGRAPHY

Ehlig, P. L. "Origin and Tectonic History of the Basement Terrane of the San Gabriel Mountains, Central Transverse Ranges." In *Rubey*. Vol. 1, *The Geotectonic Development of California*, edited by W. G. Ernst. Englewood Cliffs, N.J.: Prentice-Hall, 1981. This article is probably the most authoritative source of information on the basement rocks comprising the San Gabriel Mountains. The text is suitable for college-level readers who are not intimidated by technical language.

Fife, D. L., and J. A. Minch, eds. *Geology and Mineral Wealth of the Transverse Ranges*. Santa Ana, Calif.: South Coast Geological Society, 1982. Probably the most all-inclusive work on the geology, mineral, and energy resources of the Transverse Ranges. Especially relevant are eight articles by Thomas W. Dibblee, Jr., on the regional geology of virtually the entire Transverse Ranges province. Although written for professional geologists, the maps and diagrams in Dibblee's articles are essential to an understanding of the geology of the province.

Hatcher, Robert D., Jr. *Structural Geology: Principles, Concepts, and Problems*. 2d ed. Englewood Cliffs, N.J.: Prentice-Hall, 1995. This undergraduate textbook provides an overview of the structural development of the San Andreas fault system and the Transverse Range. Intended for the more advanced reader.

Plummer, C. C., and David McGeary. *Physical Geology*. 4th ed. Dubuque, Iowa: Wm. C. Brown, 1988. A first-semester college-level introductory geology textbook that is clearly written and wonderfully illustrated. An excellent sourcebook of basic information on geologic terminology and the fundamentals of geologic processes. The chapters on sedimentary rocks, geologic time, structural geology, and plate tectonics are especially relevant to an understanding of geologic processes involved in the formation of the Transverse Ranges.

Press, Frank, and Raymond Siever. *Understanding Earth*. 2d ed. New York: W. H. Freeman, 1998. This comprehensive physical geology text covers the formation and development of the Earth. Contains extensive sections on the deformation of the continental crust and the development of the Transverse Range in relation to the San Andreas fault system. Readable by high school students, as well as by general readers. Includes an index and a glossary of terms.

Stanley, Stephen M. *Earth System History*. New York: W. H. Freeman, 1999. A comprehensive overview of the development of the Earth. Chapter 19 deals with the development of the Transverse Range. Illustrations, glossary, bibliography, and CD-ROM. Readable by high school students.

Trent, D. D. "Geology of Joshua Tree National Monument, Riverside and San Bernardino Counties." *California Geology* 37 (April, 1984): 75-86. An article intended for the interested layperson. One of the few "popular" sources of information on the origin of the unusual scenery for which the monument is famous.

_____. "Mount Baldy Mining Area, Los Angeles and San Bernardino Counties, California." *California Geology* 39 (August, 1986): 179-184. A review of the mining history and the geology of the higher-elevation regions on Mount San Antonio, the highest peak in the San Gabriel Mountains. Discusses the role of the Vincent thrust fault in localizing deposits of gold in the central part of the range. Intended for the general reader.

7
GLACIAL GEOLOGY

ALPINE GLACIERS

Alpine glaciers are masses of ice and snow that move slowly down from the peaks to produce the spectacular landforms that are associated with high mountain scenery. Steep horn-shaped or pyramidal peaks, rushing meltwater rivers, cliff-sided valleys with waterfalls—one associates these features with alpine mountains that have been eroded by glaciers. Where such glaciers are still active, they may threaten lives and property through catastrophic forward surges and flood-waters, or they may be essential sources of meltwater in dry areas.

PRINCIPAL TERMS

ABLATION: the result of processes, mainly melting (evaporation is also involved), that waste ice and snow from a glacier

CIRQUE: a steep-sided, gentle-floored, semicircular hollow produced by glacial erosion of bedrock high on mountain peaks

EQUILIBRIUM LINE: the boundary between areas of mass balance gain and loss on a glacier's surface for any one year

LITTLE ICE AGE: a short-term cooling trend that lasted from about 1450 to 1850, during which mountain glaciers all over the world advanced considerably beyond their present limits

MASS BALANCE: the summation of the net gain and loss of ice and snow mass on a glacier in a year

MORAINE: a ridge of glacial-ice-deposited till

TILL: an ill-sorted mixture of fine and coarse rock debris deposited directly by glacial ice

ZONES OF ACCUMULATION AND ABLATION

Alpine, or valley, glaciers are long, narrow streams of ice that originate in the snowfields and cirque basins of high mountain ranges and flow down preexisting stream valleys. They range from a few hundred meters to more than 100 kilometers in length. In many ways, they resemble river systems. They receive an input of water in the form of snow in the high parts of the mountains. They have a system of tributaries leading to a main trunk system. The flow direction is controlled by the valley that the glacier occupies, and, as the ice moves, it erodes and modifies the landscape over which it flows.

The essential parts of the mass balance of the alpine glacier system are the zone of accumulation, where there is a net gain of ice, and the zone of ablation, where the ice leaves the system by melting and evaporating. In the zone of accumulation, snow is transformed into glacial ice through a process of metamorphism, or change of form. Freshly fallen snow consists of delicate hexagonal (six-sided) ice crystals, or needles, with as much as 90 percent of the total volume as empty air space. As snow accumulates, the ice at the points of the snowflakes melts from the pressure

of the snow buildup and migrates toward the center of the flake, where it refreezes. Eventually, many small, elliptical grains about the size of BB shot (about 0.45 centimeter in diameter) are formed of recrystallized ice. The accumulation of masses of these ice pellets is called firn, or névé. With repeated deposits, each year the loosely packed firn granules are compressed by the weight of the overlying snow. Meltwater, which results from daily temperature fluctuations and the pressure exerted by the overlying snow, seeps through the open pore spaces between the granules; when it refreezes, it adds to the recrystallization process. Air in the pore spaces is forced out. When the ice reaches a thickness of about 30 to 40 meters, it can no longer support its own weight and yields to slow plastic flow. The upper part of a glacier is thus rigid and tends to fracture, but the ice beneath moves by plastic deformation and flow.

On the surface of the alpine glacier, the boundary between the zone of accumulation and the zone of ablation is approximated by the equilibrium line. Above this line, the surface of the glacier tends to be smooth and white because more new snow accumulates than is lost by melting and

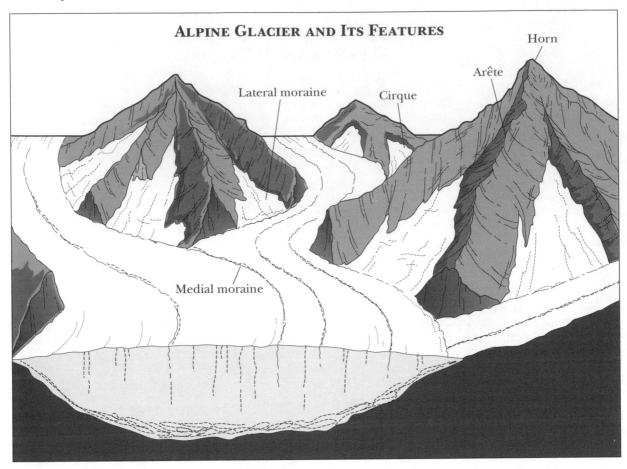

ALPINE GLACIER AND ITS FEATURES

Lateral moraine

Cirque

Arête

Horn

Medial moraine

all the irregularities are soon covered and smoothed with snow. These areas are dangerous to mountain climbers, who can fall into buried fractures or crevasses there. Below the equilibrium line, melting and evaporation exceed snowfall. There, the surface of the ice is rough and pitted and is commonly broken by open crevasses or streaked with rock debris.

The glacier ice flows and slides from the positive accumulation area to the negative wastage zone; the amount of gain or loss in each area constitutes the mass balance. Should a glacier have an excess of one over the other for a time interval of tens to hundreds of years, it will advance or retreat in consequence. As glacial ice flows over the land beneath, it erodes, transports, and deposits vast amounts of rock and soil material. Glaciers erode in two ways: by glacial plucking, in which meltwater beneath the ice freezes blocks onto the passing ice, and by abrasion of the substrate by the rock

blocks held fast in the overlying ice. Eroded material can be carried throughout the glacier, though especially at the bottom and the top, and is later mainly deposited as moraines composed of till near the margins of the glacier, where melting dominates.

LANDFORMS

Alpine glacial landforms of the ice itself include a variety of features attributable to the characteristic fracture and flow of the movement. Fracturing, or breaking, of the brittle ice occurs close the surface of the glacier, where the ice is under lower pressure. Flow of the ice occurs through recrystallization of ice crystals at depth in the glacier, where the greater pressure causes slow change and consequent movement.

At the head of an alpine glacier, where the ice pulls away from the wall of the mountain, a bergschrund (from the German word for "mountain crack") crevasse develops in summer; during winter, it is filled with avalanche snow. Wherever a glacier moves over an irregular rock bed below, the ice on the surface fractures into a variety of other crevasses. Above the equilibrium line, these features are usually covered with snow and thus are dangerous to traverse, but below they are uncovered and of far less hazard to mountain travelers. When crevasses, such as those described, create tumbled cliffs and icefalls in causing flow over a submerged cliff, the (originally French) term "serac" is used to describe them. Below such icefalls, where compressive flow occurs as the ice piles up, the glacier surface is commonly formed into a series of semicircular waves, or bands, called ogives. Ogives are formed when the broken crevasses collect dust and dirt during the summer melt season. In winter, only snow accumulates in the crevasses. Thus, when the dirty seracs close up and move away from the icefall, they form a dark band, whereas the snow-filled crevasses form a light band. Monitoring of the formation and movement of these light and dark ogives assists in investigations of ice velocity.

At the front of an alpine glacier, some of the ice usually stagnates, and other ice may be forced to override it. The result commonly is sheafs of overthrust ice slabs that carry the debris forward into the terminal moraines. Ice that melts away in these regions produces a considerable amount of meltwater that itself is capable of eroding much sediment from the moraines. Because water is an efficient fluid, it sorts the sediment into different sizes: gravel, sand, silt, and clay. Redeposition of these sediments into layered deposits provides valuable sand and gravel supplies for construction in mountain areas.

Various landforms are produced by glacier meltwater processes. Kame terraces (mounds of sediment dropped from the meltwater) are formed between the glacier ice and the valley wall by the streams and lakes impounded there. Similarly, out in front of the ice, the valley train, or outwash, sediments are spread out into plains of sorted and stratified sediment. Commonly, blocks of ice are stranded in these sediments to melt away later and leave kettle holes and kettle lakes to pockmark the plain. Beneath the flowing ice of the alpine glacier, the bedrock will be abraded by the sediment carried along in the ice. This "glacial

rasp" will groove, striate, and finally polish the bedrock over which it glides. Where blocks of bedrock are frozen onto the base of the ice and plucked out, a roche moutonnée (from the French for sheepskin-wig-shaped, or "curly," rock) is formed, smoothed by abrasion on the up-ice side and rough and broken on the down-ice portion.

After retreat of an alpine glacier upvalley for a long time and wasting away of the ice, the eroded bedrock surface of roche moutonnée and polished and striated bedrock will finally be exposed. In the case of the surface beneath a former icefall, for example, a smoothed "cyclopean," or glacial, stairway will emerge. The many intervening troughs eroded into the bedrock of the alpine valleys commonly become filled with small lakes that are linked by small overflowing streams. The resulting "paternoster" lakes are so named for their

Kinnerly Peak, a horn in Glacier National Park, Montana. (U.S. Geological Survey)

resemblance to beads on a chain. The valley walls themselves, having been undercut and eroded deeply by the ice, become exposed to show the characteristic U-shaped cross-sectional profile of glaciated valleys.

As alpine glaciers melt away entirely, finally the upper cirque basins high on the mountainsides will be exposed. These cirques commonly have steep headwalls, where the old bergschrunds were originally, and an overdeepened floor, where the flowing ice scooped out a basin. Many overdeepened cirque floors fill with water to form a small, steep-sided lake referred to as a tarn. The steep mountain peaks above the cirques are also steepened and undercut by the ice around them and form sharp glacial horns, the characteristic pyramidal peaks of alpine glacial regions. The famous Matterhorn of Switzerland is an example of this landform.

Where cirques have formed on opposite sides of mountain ridges, their headwalls may have merged through back-to-back or headward erosion of the glaciers on opposite sides. If not far advanced when the glaciers melt away, a knife-edged ridge may be the only result. On the other hand, if the glaciation has continued for a long time, the cirques may merge and remove much of the intervening rock mass. After the ice has melted away, a

In Washington's Wenatchee Mountains, looking west, remnants of old alpine glaciation can be seen surrounding Mount Stuart, including U-shaped valleys and four small cirque glaciers in the shadows. (U.S. Geological Survey)

low point, or col, will result. Many of the world's most famous mountain passes are cols that were formed in this fashion.

John F. Shroder, Jr.

CROSS-REFERENCES

Caves and Caverns, 870; Continental Glaciers, 875; Continental Glaciers, 875; Glacial Deposits, 880; Glacial Landforms, 914; Glacial Surges, 887; Ice Ages, 1111; Ice Sheets and Ice Streams, 892; Sand, 2363.

BIBLIOGRAPHY

Bailey, R. H. *Glacier.* Alexandria, Va.: Time-Life Books, 1982. This book, one of several in the Planet Earth series, is a finely written, edited, and illustrated volume. Colin Bull, a well-known glacier specialist, was the main consultant. Because of its exceptionally good presentation, this book is one of the best available on alpine glaciers for the nonspecialist, even though much on continental glaciers is included as well.

Bloom, Arthur L. *Geomorphology: A Systematic Analysis of Late Cenozoic Landforms.* 3d ed. Englewood Cliffs, N.J.: Prentice Hall, 1998. This college-level text covers the basics of geomorphology. Includes three chapters on glaciers and glaciology, as well as an index and bibliography.

Chernikoff, Stanley. *Geology: An Introduction to Physical Geology.* Boston: Houghton Mifflin, 1999. This is a good overview of the scientific understanding of the Earth and of surface processes. Includes sections on glaciers and glaciology, as well as the address of a Web site that provides regular updates on geological events around the globe.

Dolgoff, Anatole. *Physical Geology.* Lexington, Mass.: D. C. Heath, 1996. A comprehensive, well-illustrated guide to the geology of the Earth, including the development of glaciers. Although this is an introductory text for college students, it is written in a style that makes it understandable to the interested layperson. Glossary and index.

Embleton, C., and C. A. M. King. *Glacial Geomorphology.* New York: John Wiley & Sons, 1975. This book is one of the chief sources on all things glacial, but the dominant coverage is of landforms and effects of past glaciation.

Imbrie, J., and K. P. Imbrie. *Ice Ages: Solving the Mystery.* Cambridge, Mass.: Harvard University Press, 1986. This volume is a most readable book that explains the development of the theory of glaciation and examines predictions of future ice ages. While much of the book involves an explanation of past continental glaciers, the orderly exposition of the development of new ideas on general glaciation is most useful. Also, the personalities and inner feelings of many of the scientists involved are described, which adds a human dimension to what can be a dry subject to the nonenthusiast.

Menzies, M., ed. *Modern Glacial Environments: Processes, Dynamics, and Sediments.* Oxford, England: Butterworth-Heinemann, 1996. This multiauthored volume deals with all aspects of the modern study of glaciers. Intended for the specialist. Index and extensive bibliography.

Paterson, W. S. B. *The Physics of Glaciers.* Elmsford, N.Y.: Pergamon Press, 1981. This work is one of the best available references dealing with the basic mechanics of glacier formation, nourishment, structure, flow, and behavior. Although the work contains many pages of rigorous mathematical analyses, a discerning nonprofessional reader can glean a world of useful and reliable information on glacier behavior from the intervening parts of the book.

Sharp, R. P. *Living Ice: Understanding Glaciers and Glaciation.* New York: Cambridge University Press, 1988. This book is one of the most readable and detailed works on glaciers. It is very well illustrated and includes a comprehensive glossary with more than three hundred entries.

CAVES AND CAVERNS

Caves, large natural holes in the ground, are part of the Earth's plumbing system. Groundwater passes through most caves at some point in time, creating many unusual features.

PRINCIPAL TERMS

CALCITE: a common, rock-forming mineral that is soluble in carbonic and dilute hydrochloric acids

GROUNDWATER: water beneath the Earth's surface

LAVA: molten rock extruded from a volcano

LIMESTONE: sedimentary rock, usually formed on the ocean floor and composed of calcite

PHREATIC: a zone in the ground below the level of complete water saturation

SPELEOTHEM: a mineral deposit formed within a cave

VADOSE: a zone in the ground above the level of complete water saturation

FORMATION

Caves, or caverns, are natural cavities in rock large enough for a person to enter. Most caves develop by the process of groundwater dissolving limestone, a common rock deposited on ocean floors. Gypsum, dolomite, and marble (metamorphosed limestone) are other rocks that dissolve readily to form caves. Rain and snow pick up a trace of carbon dioxide as they travel through the atmosphere. Where the ground has a thick layer of decaying vegetation, more carbon dioxide combines with the water, and a dilute solution of carbonic acid forms. (Carbonic acid—carbon dioxide dissolved in water—is also present in soda pop.) The water soaks into the ground and finds its way into cracks in the soluble rock (limestone, dolomite, gypsum, or marble). The acid dissolves the rock in a process similar to that of water dissolving table salt. Groundwater removes what was previously solid rock, and a hole, or cave, remains.

The longest cave in the world is the Mammoth-Flint Ridge Cave System, a solutional cave near Bowling Green, Kentucky. More than 560 kilometers of cave passage have been surveyed and mapped.

In mountainous areas such as the Alps in Europe or the Sierra Madre Oriental in Mexico, groundwater moves hundreds or even thousands of meters downward through cracks in the rocks. The resulting passageways are mostly vertical, with deep shafts. The deepest explored cave in the world is the 1,602-meter-deep réseau Jean Bernard in the French Pyrenees. Scientists have shown that water passes through a 2,525-meter-deep cave in southern Mexico, but explorers have not yet successfully followed much of its course.

Streams flow into or out of the entrances to many actively growing caves. Scientists refer to entrances where water flows into caves as "insurgences." Entrances that have streams or rivers flowing out are called "resurgences." Insurgences and resurgences often mark the boundary between soluble and insoluble rocks. Where water reaches insoluble rock, it flows onto the surface and the cave ends. Similarly, streams flowing over insoluble rocks commonly sink into caves upon reaching a limestone terrain.

In some places, acid-charged water comes from deep in the Earth and not from rainwater. Pockets of carbon dioxide and hydrogen sulfide in the Earth's crust can combine with deep flowing water to form carbonic acid and sulfuric acid, respectively. Caves form when these deep waters rise through cracks in soluble rocks. Water charged with hydrogen sulfide rose up through limestone and dissolved the spectacular Carlsbad Caverns in New Mexico.

Lava flows on the flank of a volcano can create another type of cave, commonly called a "lava tube." As lava flows down the slope of a volcano, the surface cools and solidifies while liquid lava continues to flow under the crust. As the flow cools, self-constructed pipelines under the crust continue to pass fast-moving, hot lava down the slope. Tubes drain when no more lava passes through, and a cave remains. The deepest known cave in the United States, which is 1,099 meters deep, is the 59.33-kilometer-long Kazumura Cave, a lava tube on the island of Hawaii.

A few caves are in insoluble rocks such as granite, sandstone, and volcanic tuff. These features are generally small and have varied histories leading to their development. In many cases, groundwater has carried individual grains of sand, one at a time, from the base of a cliff where a spring emerges. With time, the resulting hole at the cliff base is deep enough to be called a cave. Other caves have formed under blocks of rocks that fell or slid down adjacent hillsides. Pounding waves excavate caves in cliffs along some ocean shorelines. These caves are usually referred to as sea caves.

Ice caves sometimes form under glaciers near their toes. Meltwater flowing under a glacier during summer may enlarge a passageway large enough to form a cave. Ice caves, however, are usually short-lived and constantly change in size and shape.

Solutional caves are often divided into three categories: phreatic caves, vadose caves, and dry caves. Most phreatic caves are still actively forming. Their passages are below the water table and completely filled with water. Vadose caves are above the water table, but water passes through them as rivers and streams. Some caves form under vadose conditions. Other caves in the vadose zone were saturated when they formed. They now provide convenient paths for water to flow through the unsaturated zone. Dry caves are no longer actively enlarging. The water that formed them has withdrawn. The air in a dry cave is usually humid, but the cave does not act as a conduit for water.

Water levels in caves commonly respond quickly to rain. A river may pass through an otherwise dry passage during the spring snow melt. A vadose passage with a small stream can become completely filled with water within a few minutes after a heavy rain commences. Tops of phreatic zones in caves have been observed to rise more than 50 meters within a few hours after the start of a surface deluge.

Once the void forms, nature commonly starts to fill caves. While most of the limestone (or dolomite or gypsum) dissolves, some impurities in the rock always remain as sediments on the cave floor. In addition, sand, mud, and gravel brought into the cave from outside by streams add to these sediments. Over hundreds, thousands, or even millions of years, caves can become completely choked by sediments.

If too much rock is dissolved and the rock is not strong enough to support the void, the ceiling collapses. Failure can occur one small rock at a time or in massive blocks. If the cave is still actively forming, groundwater may eventually dissolve the debris, and the passage will continue to grow upward. However, if the debris is not removed, the passage can completely fill with rubble, ending the existence of the cave.

SPELEOTHEMS

Attractive deposits of minerals, called speleothems, form from supersaturated water in a cave. Supersaturated waters contain more dissolved minerals, usually calcite, than they can maintain in so-

A variety of speleothem formations in Carlsbad Caverns, New Mexico. (U.S. Geological Survey)

lution. Supersaturation can occur when water evaporates and the dissolved minerals stay behind in the remaining liquid water. More commonly, supersaturation happens when cave water releases dissolved carbon dioxide. The less carbon dioxide dissolved in the water, the less acidic the water, and the less mineral the water can hold in solution. Cave water loses carbon dioxide—in the same manner that carbon dioxide bubbles escape from soda pop—when the surrounding pressure on the water drops (like opening a soda can) or when the temperature of the water rises. In both cases, dissolved minerals solidify—a process called precipitation—as speleothems.

One common form of speleothem, stalagmites grow when drops of water from the ceiling of a cave hit the ground, lose carbon dioxide when they splash, and precipitate calcite that builds upward. (U.S. Geological Survey)

The most common speleothems are soda straws, stalactites, stalagmites, and flowstone. Soda straws look like their namesakes and hang from the ceilings of caves. Water is fed from a hole at the top of the soda straw, flows down through the hollow speleothem, and hangs on the end before falling. Calcite precipitates around the edges of the water as it slowly drips from the soda straw's end. Stalactites are typically cone-shaped deposits that hang from the ceiling. Originally soda straws, they grow as water deposits calcite around the outside of the speleothem. Stalagmites grow when drops of water from the ceiling hit the ground, lose carbon dioxide when they splash (like shaking a soda can), and precipitate calcite. They look almost like upside-down stalactites, but their ends usually are more rounded. Stalactites and stalagmites that grow together result in a column. Flowstone, a sheet of calcite coating a sloping wall or floor of a cave, forms under flowing water.

Gypsum, composed of calcium sulfate, is commonly de-

Several types of speleothem in "Violet City," Mammoth Cave National Park, Edmondson County, Kentucky. Columns of stalactites and stalagmites have grown together. (U.S. Geological Survey)

posited within limestone, dolomite, and gypsum caves. Gypsum precipitates in a similar manner as calcite, but the process involves dissolved hydrogen sulfide instead of carbon dioxide. Gypsum speleothem shapes differ from those of calcite speleothems. One type of speleothem, a gypsum flower, looks like clear or white rock flowers growing out of cave walls. They form at the base of the "petals" and extrude earlier-formed deposits away from the wall in a manner similar to the squeezing of toothpaste out of a tube.

Ice forms many of the same speleothems as calcite. Ice stalactites (icicles), stalagmites, and flowstone are displayed in cold caves, particularly in winter and spring. Some caves in the Austrian Alps have moving glaciers and massive ice columns.

ENVIRONMENT

The temperature of most caves is the mean (average) annual temperature of the local area above ground. Temperatures typically fluctuate slightly near entrances and usually are a constant temperature a short way from the entrance. Thus, caves usually seem cool in summer and warm in winter. The moisture in the ground makes most caves very humid. Like temperature, humidity remains nearly constant year-round away from entrances.

Caves will adjust to changes in local barometric conditions. When the outside atmosphere changes from higher barometric pressure to a lower pressure, strong winds blow out the entrances of large, air-filled caves as they also lower their atmospheric pressure. When an area changes from lower pressure to higher pressure, large caves will suck air in from the outside. Just as on the surface, the temperature and air pressure of caves increase at greater depths. The change can be substantial in caves of more than 1,000 meters in depth. "Blowing caves" have entrances at different elevations. A chimney effect causes cold air to drop through the cave and blow out the lower entrance throughout the winter. The effect reverses as air blows out the upper entrance during the summer.

Louise D. Hose

CROSS-REFERENCES

Alpine Glaciers, 865; Aquifers, 2005; Carbonates, 1182; Continental Glaciers, 875; Dolomite, 1567; Glacial Deposits, 880; Glacial Surges, 887; Groundwater Movement, 2030; Groundwater Pollution and Remediation, 2037; Ice Sheets and Ice Streams, 892; Karst Topography, 925; Limestone, 1451; Paleoclimatology, 1131; Water-Rock Interactions, 449; Weathering and Erosion, 2380.

BIBLIOGRAPHY

Bloom, Arthur G. *Geomorphology: A Systematic Analysis of Late Cenozoic Landforms.* 3d ed. Englewood Cliffs, N.J.: Prentice Hall, 1998. This college-level text covers the basics of geomorphology. Includes a chapter on speleology. Index and bibliography.

Courbon, Paul, et al. *Atlas: Great Caves of the World.* St. Louis, Mo.: Cave Books, 1989. Describes and provides maps of the deepest and longest caves in the world. Includes all countries and all types of caves. An essential reference book for understanding the world's greatest caves.

Davies, W. E., and I. M. Morgan. *Geology of Caves.* U.S. Geological Survey, 1991. A brief, inexpensive brochure published by the U.S. government. Explains how most caves form and discusses the common speleothems within them.

Erickson, Jon. *Craters, Caverns, and Canyons: Delving Beneath the Earth's Surface.* Chicago: Facts on File, 1993. Covers structural geology and geomorphology, including caves, at a high school level. A basic explanation of caves.

Exley, Sheck. *Caverns Measureless to Man.* St. Louis, Mo.: Cave Books, 1994. Sheck Exley was the greatest scuba diver ever to explore caves. This book documents his explorations and the water-filled caves he explored. Although the book focuses on his explorations, a good feeling for how caves develop and their significance can be achieved by reading about his adventures.

Gillieson, David. *Caves: Processes, Development, Management.* Oxford, England: Blackwell, 1998. This book describes how caves form, what can be learned from them, and how they can best be managed. Well illustrated and written in an engaging style that makes the material comprehensible to the general reader even though it is intended for specialists. Glossary, bibliography, and index.

Hill, Carol A., and Paolo Forti. *Cave Minerals of the World.* Huntsville, Ala.: National Speleological Society, 1986. The definitive book on speleothems. Describes them and explains how they form; filled with beautiful pictures.

Jagnow, David H., and Rebecca Rohwer Jagnow. *Stories from Stones.* Carlsbad, N.M.: Carlsbad Caverns-Guadalupe Mountains Association, 1992. Describes the geology of the Guadalupe Mountains in New Mexico and gives an excellent description of how the spectacular caves in the area evolved. Carlsbad Cavern and other area caves formed in an unusual manner, and the Jagnows are experts on their origins.

Middleton, John, and Tony Waltham. *The Underground Atlas: A Gazetteer of the World's Cave Regions.* New York: St. Martin's Press, 1986. Describes the major caves and karst areas of nearly every country in the world. The potential for finding caves in countries without presently known caves is also discussed.

Rea, G. Thomas, ed. *Caving Basics: A Comprehensive Guide for Beginning Cavers.* 3d ed. Huntsville, Ala.: National Speleological Society, 1992. A comprehensive book on the geology, biology, archaeology, and exploration of caves written by members of the world's largest organization dedicated to caves. Each chapter was written by a leading expert on the subject. Useful bibliographies accompany each chapter.

CONTINENTAL GLACIERS

Continental glaciers once covered much of northern North America and Europe, but now only Greenland and Antarctica have such huge masses of permanent ice and snow. Because continental glaciers are so large, they affect the climate of large regions outside their boundaries by cooling of air and water temperatures. Continental glaciers of past ice ages have produced a wide variety of erosional and depositional features in northern latitudes.

PRINCIPAL TERMS

ABLATION: the result of processes, primarily melting (evaporation is also involved), that waste ice and snow from a glacier

EQUILIBRIUM LINE: the line or zone that divides a glacier into the upper zone of accumulation and the lower zone of wastage

ISOSTATIC ADJUSTMENT: the adjustment of the crust of the Earth to maintain equilibrium by subsiding when loaded and uplifting when unloaded

PLEISTOCENE ICE AGE: the time from about 2 million years ago to about 10,000 years ago, during which large continental glaciers covered much of northern North America, Europe, and other parts of the world

PRESSURE MELTING TEMPERATURE: the temperature at which ice will melt under a specified pressure; under pressure, water can exist even at temperatures below freezing

FORMATION, STRUCTURE, AND LOCATION

Continental glaciers are ice sheets of huge extent. These continent-sized masses of ice overwhelm nearly all the land surface at their margins. Modern continental ice sheets occur only in Greenland and Antarctica and comprise nearly 96 percent of all glacier ice on the Earth. From about 2 million to 10,000 years ago, during the Pleistocene glacial period of the world's geological history, continental glaciers spread over much of northern North America and Europe.

Continental ice sheets tend to create their own weather that is naturally favorable for glaciers. A huge ice sheet can even have worldwide climatic effects. The greatest part of the world's ice in Greenland and the Antarctic occurs in high latitudes characterized by very low winter temperatures, low summer temperatures, small annual precipitation, and minimal ablation. It does not snow much, but it is so cold that what falls tends to remain for a very long time. As a result of all these factors, such ice masses are relatively inactive and stable. Greater movement activity is noticeable in areas of less extreme cold and moderate precipitation.

Large ice sheets are complex and consist of several domes from which the ice flows radially to the ice margin or to broad interdome saddles, where the ice flow diverges downslope. The location of ice domes and ice saddles determines the flow path of ice on an ice sheet, but such features can change location over time as an ice sheet grows or shrinks in size. Continental glaciers are rarely more than 3,000 meters thick. Ice does not have the strength to support the weight of an appreciably thicker accumulation. If more ice is added by increased precipitation, the glacier simply flows out from the domal centers of accumulation more swiftly. Also, as the pressure at the base increases sufficiently, basal melting occurs that further decreases ice thickness.

Although snow accumulates over much of a continental ice sheet and gradually transforms into granular firn (snow that has survived a melt) and finally to glacial ice, it is from the highest interior domal areas that the main ice streams flow. The ice surface is built up at the interior and slopes outward on all sides. The glacier moves down and out in all directions. At the edge of the sea, if the area is mountainous, such ice sheets will break up into narrow tongues resembling valley glaciers that wind through the mountains to the sea. Otherwise, the ice may end in giant ice ramparts that calve, or break off, icebergs into the ocean, or as floating ice shelves over large continental embayments.

Ice shelves occur at several places along the margins of the Greenland and Antarctic ice sheets, as well as locally in the Canadian Arctic islands. They are nourished by ice streams flowing off the land, as well as by direct snowfall on their surface, and perhaps by freezing of the sea ice on their undersides as well. The largest ice shelves extend hundreds of kilometers seaward from the coastline and can reach a thickness of at least 1,000 meters. The Ross Ice Shelf in Antarctica, for example, is about as large as the state of Texas.

Antarctica is one of two landmasses in the world still overlain by a continental glacier. Greenland is the other. (© William E. Ferguson)

The continental ice sheet of Greenland covers an area of about 1.8 million cubic kilometers, or about 80 percent of the total land area there. The volume of the ice is about 2.8 million cubic kilo-meters. In cross section, the ice has the shape of an extremely wide lens, convex on both the smooth upper surface and the rough lower boundary with the ground. The center of the ice

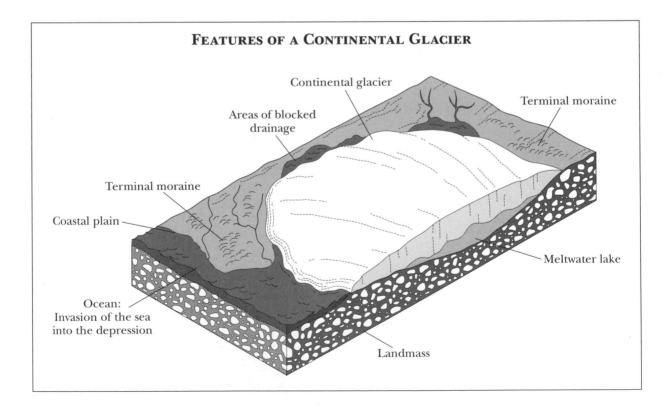

FEATURES OF A CONTINENTAL GLACIER

Continental glacier

Terminal moraine

Areas of blocked drainage

Terminal moraine

Coastal plain

Meltwater lake

Ocean:
Invasion of the sea
into the depression

Landmass

sheet is more than 3,200 meters thick. The greatest area, 18 percent, occurs between 2,440 and 2,740 meters, and 6.5 percent lies between 3,050 and 3,390 meters. The equilibrium line, or boundary between the upper accumulation area and the lower wastage area, is at about 1,400 meters, and 83 percent of the total area lies in the accumulation zone. Measurements of ice velocity in Greenland show that the main ice cap advances at approximately 10 to 30 centimeters per day, but the outlet glaciers near the coast can move as fast as 1 meter per hour. In some places, the ice can actually be seen to move.

The continental glaciers that once covered much of North America, Europe, and elsewhere during the Pleistocene ice age rivaled Antarctica in size and also exerted a widespread effect when they were at their maximum extent. The Laurentide ice sheet of North America at its largest covered an area similar in size to the present Antarctic ice sheet, but the Scandinavian ice sheet of Europe covered only about one-half of this area during the maximum of glaciation.

The Laurentide and Scandinavian ice sheets did not extend to the seas in their southern limits as do the ice caps of Greenland and Antarctica. Instead, these continental glacial systems covered a large part of the northern continents and caused a number of significant peripheral changes of the regional physical setting outside their limits. The weight of the ice depressed the ground surface isostatically (in a process called isostatic adjustment whereby the crust of the Earth maintained equilibrium by subsiding when loaded and uplifting when unloaded), just as it does in Greenland and Antarctica, so that the land sloped toward the glacier. Consequently, glacial lakes formed in the depressions along the ice margins, or arms of the ocean invaded the depressions. The preglacial drainage systems were greatly modified, as the streams that flowed toward the ice margins were impounded to form lakes. Later, as the ice dams melted away and the land again rose isostatically after the weight of the ice was removed, many such lakes and arms of the sea drained away, leaving behind extensive lake and marine clays and silts.

LANDFORMS

Glaciers produce many different erosional and depositional features as a result of their interaction with the ground beneath or at the front of the ice. During the Pleistocene ice age, the great thick continental ice sheets moved over the flat lowlying areas of the northlands, removing the existing soils and eroding up to several meters into the bedrock. As a result, many thousands of square kilometers of northern North America and Europe have little or no soil cover and the effects of the former glaciation are seen everywhere in the polished and grooved fresh bedrock. Continental glaciers are so large that they can produce widespread or massive abrasion and streamlined forms, as well as erode huge lake basins, all of which are exposed after the ice melts away. Large parts of the north-central United States and much of central and eastern Canada have such landforms plentifully displayed.

In many areas that lie near the outer edges of former continental ice sheets, the land surface has been molded into smooth, nearly parallel ridges that range up to many kilometers in length. These

A terminal moraine forming at the foot of Taylor Glacier in the area of McMurdo Sound, Victoria Land, Antarctica. (U.S. Geological Survey)

The shape of this distinctive erosional feature is produced when a stream of glacial ice overrides a bedrock hill, abrading its upstream side into a gradual slope and quarrying or plucking off rock from the downstream side, making a cliff. Geologists use the French term roche moutonnée, *which is descriptive of the resemblance to a sheep's back.* (© William E. Ferguson)

forms resemble the streamlined bodies of supersonic aircraft and offer minimum resistance to glacier ice flowing over and around them. The most common of these forms is the drumlin, which is a smooth, stream-lined hill or ridge consisting of glacially deposited sediment that is elongated parallel with the direction of ice flow. Some drumlins are composed of contemporaneously deposited and smoothed sediment; others are of older sediment that was eroded long after its initial deposition. Bedrock can also be eroded in this fashion. In some places with steeply rising mountains that were overridden by continental ice, the up-ice side of such mountains will be streamlined and smoothed by the ice abrasion, whereas the down-ice side will be plucked and quarried into a rough and jagged cliff. Such asymmetrical mountains are called flyggberg ("flying mountains") topography. Smaller such forms a few meters in height are referred to as roches moutonnées, or "wig-shaped" rocks, after their fancied resemblance to the curls of the smooth and powdered periwigs of the eighteenth century.

Continental glaciers erode by the incorporation of blocks of rocks in the base of the ice and by abrasion with these blocks against the bedrock further along in the ice stream. The process produces large quantities of sedimentary debris. Sediments deposited by continental glaciers can be more than 300 meters thick so that they blanket most of the preglacial topography upon which they rest, thus modifying, disrupting, and obliterating previously established drainage systems.

Most of the rock debris that is transported by glaciers is deposited near the terminus, where melting dominates. The material accumulates as a moraine ridge marking the former front edge of the glacier. As a glacier retreats from an area by backwasting, it may deposit a series of recessional moraines in loops or ridges one behind the other. The sediment of such recessional and terminal moraines is made up of a jumbled mixture of all sorts of rock materials, ranging from clay to boulders, that are collectively referred to as till.

As continental glaciers develop large quantities of meltwater from wastage of the ice, streams of meltwater begin to flow in tunnels within and beneath stagnant ice and carry a large load of sediment in their ice-walled beds. When the ice melts away, such bed loads can be deposited beneath the ice to form a long, sinuous ridge called an esker.

The plentiful meltwater at the terminus of a continental ice sheet also will flow through the terminal moraine landforms and erode away much of the till. This debris will be transported and reworked by the meltwater before being sorted into different sizes and deposited by rivers beyond the ice margins. The resulting layers or strata of sorted sediment can be spread out in broad outwash plains pockmarked with kettle holes where blocks of ice have later melted away. Both the unsorted, unstratified, or unlayered till and the sorted and stratified outwash materials are collectively called drift, the name being a heritage of the time when such materials were thought to have drifted to their present locations during Noah's flood.

John F. Shroder, Jr.

BIBLIOGRAPHY

Andrews, J. T. *Glacial Systems: An Approach to Glaciers and Their Environments.* North Scituate, Mass.: Duxbury Press, 1975. This work contains a wealth of information on glacial classification, formation, mechanics, mass balance, and landforms. Andrews has spent many years in the Arctic working on glaciers and their effects.

Bloom, Arthur G. *Geomorphology: A Systematic Analysis of Late Cenozoic Landforms.* 3d ed. Englewood Cliffs, N.J.: Prentice Hall, 1998. This college-level text covers the basics of geomorphology. Includes three chapters on glaciers and glaciology. Index and bibliography.

Chernikoff, S., and R. Vekatakrishnan. *Geology: An Introduction to Physical Geology.* Boston: Houghton Mifflin, 1999. The authors provide an overview of scientists' understanding of the Earth. Includes the address of a Web site that provides regular updates on geological events around the globe. Contains sections on glaciers and glaciology.

Dolgoff, Anatole. *Physical Geology.* Lexington, Mass.: D. C. Heath, 1996. This is a comprehensive guide to the study of the Earth. Extremely well illustrated and includes a glossary and an index. Although this is an introductory text for college students, it is written in a style that makes it understandable to the interested layperson. Contains a section on the development of glaciers.

Eyles, N., ed. *Glacial Geology.* Oxford, England: Pergamon Press, 1983. This book of edited papers is almost entirely devoted to the effects of glacier erosion and deposition on the land. Engineering applications are stressed and include foundation engineering, road construction, dam and reservoir construction, and groundwater studies in glaciated areas.

Menzies, M., ed. *Modern Glacial Environments: Processes, Dynamics, and Sediments.* Oxford, England: Butterworth-Heinemann, 1996. This multiauthored volume deals with all aspects of the modern study of glaciers. Intended for the specialist. Index and extensive bibliography.

Paterson, W. S. B. *The Physics of Glaciers.* Elmsford, N.Y.: Pergamon Press, 1981. This work is one of the best available references dealing with the basic mechanics of glacier formation, nourishment, structures, flow, and behavior. Despite the considerable amount of rigorous mathematical material, a discerning nonprofessional reader can glean a world of useful reliable information on glaciers from the intervening parts of the book.

Sharp, R. P. *Living Ice: Understanding Glaciers and Glaciation.* New York: Cambridge University Press, 1988. A most readable and detailed book on glaciers, it is very well illustrated. It also includes a comprehensive glossary with more than three hundred entries.

Weiner, Jonathan. "Glacier Bubbles Are Telling Us What Was in Ice Age Air." *Smithsonian* 20 (May, 1989): 78-87. This current article details the use of gas bubbles trapped in glacial ice to find out what the atmosphere was like long in the past. This information is used to understand ice-air interactions in order to better find out what is happening to the present climate of the Earth and what the future holds in store. A very readable and popular account.

GLACIAL DEPOSITS

Glacial deposits are a distinctive association of landforms produced by glacier ice or the water derived from it. An understanding of how these deposits form allows geologists and physical geographers to reconstruct the extent, shape, and other characteristics of former glaciers that covered areas currently free from ice.

PRINCIPAL TERMS

ALPINE GLACIER: a small, elongate, usually tongue-shaped glacier commonly occupying a preexisting valley in a mountain range

CONTINENTAL ICE SHEET: a glacier of considerable thickness that completely covers a large part of a continent, obscuring the relief of the underlying surface

DRUMLIN: a smooth, elongate, oval-shaped hill or ridge formed under a moving glacier

GLACIER: a mass of snow and ice that persists for two or more years and is capable of movement by internal deformation

MELTWATER: water derived from the melting of glacier ice

MORAINE: an arcuate ridge consisting of till, stratified drift, or both, often deformed, deposited at the margin of a glacier

PLEISTOCENE EPOCH: a geologic time period beginning between 2 and 3 million years ago and ending approximately 10,000 years ago, featuring several expansions and contractions of continental ice sheets

STRATIFIED DRIFT: a sorted, layered sediment derived from glacier ice but subsequently reworked and resedimented by meltwater

TILL: an unsorted, unconsolidated sediment consisting of clay to boulder-size particles that are deposited directly by glacier ice without subsequent reworking by meltwater

GLACIAL EXPANSION AND CONTRACTION

Glacial deposits and the landforms produced by them are extremely varied but have a single origin—they are all products of deposition by glacier ice, either through direct deposition from the ice to the land surface or through deposition from water derived from the ice. Glacial deposits are being formed in places such as Greenland and Antarctica, where large glaciers called continental ice sheets still exist. Continental ice sheets are huge accumulations of ice thousands of kilometers in diameter and up to 4 kilometers thick that completely engulf the underlying topography. Glacial deposits are also being formed in some high-altitude mountainous regions where small, tongue-shaped alpine glaciers occupy valleys leading down from upper mountain slopes. The upper slopes of Mount Rainier in Washington State contain twenty such glaciers.

At various times during the past 2 million years, continental ice sheets and alpine glaciers were much more extensive than they are today. This period of glacier expansion and contraction is known as the Pleistocene epoch, which ended only about 10,000 years ago. Whether a particular area was glaciated at some time during the Pleistocene epoch can be determined by examining the glacial deposits left behind. These constitute a distinctive and recognizable association of landscape features, especially if glaciation was relatively recent. Much of the northern United States north of 40 degrees latitude and most of Canada contain these deposits.

Glacial deposits are classified according to mode of origin and environment of deposition. They can be subdivided into unstratified deposits that show little or no evidence of water transport and stratified deposits that have been transported by and deposited in water. These two categories are, in effect, end members of a continuum, with some deposits showing minor effects of water reworking and sorting. Glaciers produce huge amounts of water, referred to as meltwater, especially during the summer months, when air temperatures are at a maximum.

TILL

Glaciers also carry tremendous amounts of eroded rock and soil debris, which get incorporated into the moving ice either by falling onto the glacier surface from adjacent valley sides or, more commonly, by freezing onto the base of the glacier as it moves forward. Once incorporated, the debris is crushed and abraded in transport until it is finally deposited. Unstratified debris deposited directly from glacier ice and not subsequently transported is called till, or ground moraine. Till is characterized by lack of layering or stratification and a wide range of particle sizes from clay (less than 0.002 millimeter) to boulders many meters in diameter. Different types of till may be deposited by a glacier depending on whether the temperature is above or below freezing at the base of the glacier, the amount and distribution of debris within the glacier, and the style of deglaciation (melting of the glacier).

Debris that has been eroded and incorporated

into the glacier may be subsequently deposited beneath it by a plastering-on process as the ice moves forward. This produces lodgment till, a compact, relatively thin deposit (usually fewer than a few meters thick) that may cover thousands of square kilometers. Three other varieties of till are produced when the glacier melts. When debris melts out from a stagnant glacier (no longer moving), it is gradually let down onto the ground surface. If there is no subsequent meltwater reworking and downslope transport of the debris, the resultant deposit is called meltout till. Meltout till tends to be discontinuous but locally quite thick, especially in high-latitude regions where the base of the former glaciers was well below freezing. Debris progressively exposed on the surface of the glacier as it melts may get remixed by slumping and landsliding. The result is a jumble of partially washed debris called superglacial, or ablation, till. It is a very extensive surface deposit in some formerly glaciated regions and often forms hummocky to-

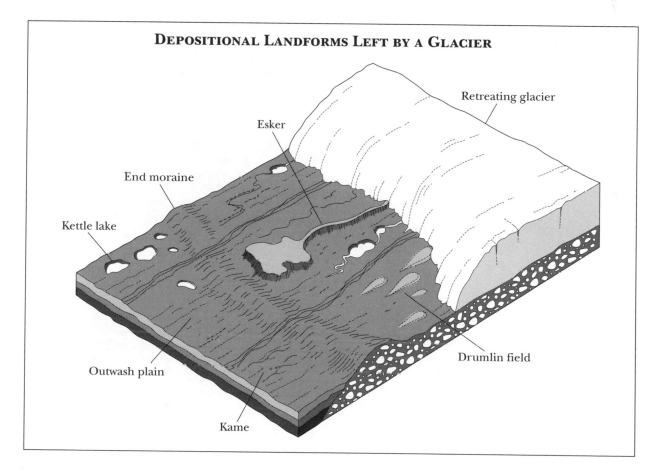

DEPOSITIONAL LANDFORMS LEFT BY A GLACIER

Retreating glacier

Esker

End moraine

Kettle lake

Drumlin field

Outwash plain

Kame

pography such as that found in parts of the Canadian prairies and the Dakotas in the United States.

A fourth type of till, called flowtill or sediment flow diamicton, is deposited by flowage of water-saturated debris from the glacier surface or, less commonly, its base. This thick slurry of mud and debris forms discontinuous, lobate deposits on the land surface. If the flow is into a body of water abutting the front of the glacier, the resulting deposit is termed subaqueous flowtill. Unlike lodgment, meltout, and superglacial till, flowtill shows evidence of water sorting and mass flowage. It is therefore not till in the strict sense, since it has been reworked subsequent to deposition from glacier ice. Each of the four till types may be recognized by its position relative to the other till types, its grain-size distribution, its geometry, its compactness, and its internal features.

MORAINES

In general, the deposition of till produces three types of landforms: a more or less flat surface dominated by lodgment till and often containing elongate, streamlined hills or ridges called drumlins that are all oriented in the direction of glacier flow; a hummocky surface dominated by superglacial till and flowtill; or one or more elongate, arcuate ridges called moraines that are often arranged in a nested series separated by flat areas.

Moraines are usually composed of till but may include other deposits. They tend to have a hummocky or pitted surface. Moraines are formed at the margin of a glacier primarily by a plowing-up of debris in front of the ice combined with deposition of sediment by water flowing off the steep front of the glacier. Moraines form only when the adjacent margin of the glacier is stationary for long periods of time (the longer the stationary position, the larger the moraine). Moraines deposited by continental ice sheets are usually tens of meters high, up to a few kilometers wide, and tens to hundreds of kilometers long. Long Island, New York, is, in part, a large moraine deposited by the most recent continental ice sheet. Moraines formed by alpine glaciers tend to be much steeper and shorter than those formed by continental ice sheets. Although moraines formed by alpine glaciers sometimes reach heights of 200 meters, their lengths are no more than a few kilometers, since the glaciers that produce them are confined by valley walls.

Moraines are classified according to where they form. Ridges of debris accumulated along the sides of alpine glaciers are called lateral moraines, whereas those formed at the foreward margin of any glacier are collectively called end moraines. The end moraine marking the farthest extent of the glacier is called the terminal moraine, while those formed at various distances, usually many kilometers, behind the terminal position are called recessional moraines. Like the terminal moraine, recessional moraines are constructed over long periods of time and require a stationary glacier front for formation. They therefore represent periods of relative climatic stability in which the amount of forward movement of the ice front is

Paired lateral moraines in Lee Vining Canyon east of Tioga Pass in the Sierra Nevada. The outer pair was formed during the Tahoe glaciation; the inner, during the Tioga glaciation. (U.S. Geological Survey)

exactly balanced by the amount of ice melting. The state of Illinois in the United States features a beautiful example of terminal and recessional moraines. The terminal moraine curves across the central part of the state through a distance of more than 400 kilometers. Behind it to the north and east are a series of twelve to fifteen recessional moraines spaced from 3 kilometers to more than 70 kilometers apart.

Kames in Happy Valley, in the Nunatarssuaq region of Greenland. (U.S. Geological Survey)

STRATIFIED DEPOSITS

Stratified glacial deposits, otherwise known as stratified drift, are different from till in that they have been extensively reworked by and deposited in water. They are not as uniformly extensive as till and are confined mainly to lower-elevation areas such as stream valleys, where they form a large variety of individual landforms. Perhaps the most curious of these landforms are eskers, which are discontinuous, steep-sided, sinuous ridges of sand and gravel usually a few meters to tens of meters in height and up to 2 kilometers wide. They tend to snake through the landscape in discontinuous fashion for tens or even hundreds of kilometers. Eskers are actually stream sediments deposited in tunnels partly or wholly within glacier ice. When the bounding ice melts, the esker sediments are left as ridges on the land surface. Unlike normal stream deposits, these sit on the land surface and may locally rise over topographic obstructions. Eskers superficially resemble moraines but are much narrower and steeper, contain very little or no till, and tend to be oriented perpendicular to the margin of the glacier rather than parallel to it as in moraines. Tunnels emanating from the margin of a glacier mark the downstream ends of eskers-in-the-making. Such tunnels can be seen at the front of Malaspina Glacier in Alaska.

Eskers are related to stratified deposits called kames. These are isolated mounds, hills, ridges, deltas, or terraces composed of collapsed meltwater-deposited sediment. Debris falling down crevasses or holes in the glacier or debris deposited by melt-

water streams along the margin between the glacier and a valley side will, upon final ice melting, form kames. Where streams carrying large amounts of sediment drain from the ice into a lake abutting the glacier front, kame deltas may be formed. They resemble ordinary stream deltas except that they have no streams behind them and they may be perched above the surrounding land surface if the meltwater streams feeding them were within or on the surface of glacier ice. In the Appalachian Mountains of southern Quebec, Canada, kame deltas perched on valley sides are often used as cemetery sites, since they are the only flat surfaces to be found. The different types of kames may be recognized by their geometry, coarse grain size, elevation relative to other deposits, and the disruption of layering (stratification) produced by collapse following melting of the bounding ice. Both kames and eskers are termed ice-contact stratified drift, since both form in at least partial contact with glacier ice. Ice-contact deposits are numerous in areas where the glacier stagnated and rapidly melted from the top downward.

OUTWASH AND GLACIOLACUSTRINE DEPOSITS

Some stratified sediments of glacial origin are not deposited in direct contact with ice. Rather, they form in meltwater streams or lakes in front of the glacier. The most abundant of these deposits is outwash, a sorted and stratified layer of meltwater-

Folds in the Malaspina Glacier, with the Mount St. Elias range behind, Alaska Gulf region. (U.S. Geological Survey)

transported sand and gravel deposited as a wide, thin sheet by numerous meltwater streams emanating from the front of a glacier. Outwash is usually associated with moraines, filling valleys in front of them to form what are called valley trains or, where particularly wide, outwash plains. These are forming in Iceland, where small continental-type glaciers called ice caps are in the process of melting. The thickness of outwash deposits is usually measured in meters or a few tens of meters. The top surface is commonly flat but may be riddled with holes called kettles. Kettles form when isolated blocks of ice in front of a glacier become surrounded by outwash and become partially or completely buried by it. When the blocks finally melt, which may take hundreds of years, holes are created in the outwash surface. Kettles commonly are filled with water and form ponds a few meters to tens of meters in diameter. As time passes, the kettles gradually fill with sediment or organic matter and turn into bogs and swamps. The New England region of the United States is famous for its kettles.

Another type of stratified deposit not formed by ice contact is glaciolacustrine (glacial lake) silt and clay. Many glaciers, especially continental ice sheets, block the drainage of streams flowing toward them. Water thus becomes ponded up against the front of the glacier or its moraine. Streams of meltwater flowing from the glacier into the adjacent lake carry huge amounts of fine-grained sediment (silt and clay) that become suspended in the lake water and gradually sink to the bottom, forming a layer that may reach several tens of meters in thickness. When the lake finally drains as a result of melting of the blocking glacier, the glaciolacustrine deposits are left as isolated flat areas. Large areas in North America were occupied by such lakes during melting of the last continental ice sheet. The largest was probably Glacial Lake Agassiz, which covered an area of more than 200,000 square kilometers in the Canadian provinces of Manitoba, Saskatchewan, and Ontario and extended into North Dakota and Minnesota in the United States. It is estimated that this lake was more than 200 meters deep at its maximum extent.

No discussion of glacial deposits is complete without mentioning loess. Loess is a massive (unstratified) accumulation of silt deposited not by ice or water but by wind. It forms a widespread surficial blanket, unless buried by younger deposits, from less than 1 meter to more than 300 meters thick. It occupies an area of more than 200,000 square kilometers in the central United States and more than 800,000 square kilometers in central China. Where present in sufficient thickness (more than a few meters), it forms a distinctive landscape of steep, sometimes vertical bluffs cutting a flat upland surface.

The origin of loess has been a matter of debate since the nineteenth century. The prevailing opinion is that it is a windblown deposit derived from two sources: deserts and outwash surfaces. The latter source appears to be the major one, since loess deposits are thickest near outwash sources and get progressively thinner in a downwind direction. Loess may be thought of as a variety of glacially derived sediment not deposited in direct contact with ice.

STUDY OF GLACIAL DEPOSITS

Geologists and physical geographers who study glacial deposits are primarily interested in reconstructing the geometry, flow directions, ages, and melting patterns of former glaciers. The first step in the process is to identify the different types of glacial deposits in an area. This is accomplished by finding sites where these deposits are exposed and analyzing the exposed deposits for such things as grain size, internal structure and stratification, geometry, and composition. This information allows determination of deposit type. Next, the areal extent of the deposit must be established and its boundaries plotted on a topographic map or aerial photograph, both of which are small-scale, two-dimensional representations of the topography of an area. The spatial relations among deposit types can then be used to establish the geometry of the former glacier at various times during its existence. The shape of the glacier at various times, for example, can be determined by examining the positions and shapes of its terminal and recessional moraines, since they are outlines of its margin.

Ice-flow directions can be established by plotting the orientations of linear features produced by the flowing ice, such as drumlins or striations on bare rock surfaces. In some cases, an investigator wishes to know what kind of terrain a glacier has passed over. Analysis of the distribution of key rock and mineral types in till can yield this information. A sampling grid is drawn on a topographic map, and the investigator attempts to collect at least one sample from each square in the grid. The samples are then analyzed for types of rocks and minerals present and their relative abundance. The spatial change in abundance of these constituents can then be used to establish what terrain the glacier passed over. This method is used extensively by geologists interested in locating valuable ore deposits hidden under a cover of till.

The age of a glacial deposit can be determined by relative age dating and absolute age dating techniques. The former refers to whether one glacial deposit is younger or older than another, whereas the latter actually establishes the age of the deposit

in years. A commonly used relative age dating technique compares the relative amounts of weathering (chemical alteration) of glacial deposits as expressed by depth of soil development or depth of weathering on surface boulders. There are several absolute age dating techniques, the most popular being the carbon 14 method. This method measures the age of organic matter associated with a glacial deposit. Old soils or peat deposits buried by glacial debris can be dated, thereby establishing a maximum age for the overlying deposits and the glacier advance that produced them.

Patterns of ice melting (deglaciation) are determined by analysis of the spatial patterns of till and associated stratified drift deposits. Melting glaciers produce typical sequences of these deposits that can be traced in the direction of glacier movement. For example, a particular valley may contain outwash leading up-valley to a terminal moraine ridge behind which is an esker bordered by kames, both of which are underlain by till. The sequence records a glacial advance to the position of the moraine, followed by stability of the ice front while the moraine forms and outwash is deposited down-valley. Subsequently, the glacier stagnates, and the esker and kames form beneath or on the stagnating ice. Rapid melting of the ice finally exposes the till. Many river valleys in the New England region of the United States contain sequences such as these.

The vertical sequence of glacial deposits at any one spot is often determined by taking core samples using a coring device. Coring devices are long, hollow tubes that are inserted vertically into the ground. When extracted, they contain a continuous vertical sequence of glacial sediments that may be observed by cutting the core tube lengthwise. This is often the only method of studying the history of glaciation in areas where exposures are few.

Philip A. Smith

CROSS-REFERENCES

BIBLIOGRAPHY

Bloom, Arthur G. *Geomorphology: A Systematic Analysis of Late Cenozoic Landforms.* 3d ed. Englewood Cliffs, N.J.: Prentice Hall, 1998. This college-level text covers the basics of geomorphology. Includes three chapters on glaciers and glaciology. Index and bibliography.

Chorley, Richard J., S. A. Schumm, and D. E. Sugden. *Geomorphology.* London: Routledge, Chapman & Hall, 1985. An intermediate-level text covering all aspects of geomorphology, including glacial deposits and processes. Chapters 17 and 19 are especially relevant. Suitable for readers with a basic understanding of geology.

Dolgoff, Anatole. *Physical Geology.* Lexington, Mass.: D. C. Heath, 1996. This is a comprehensive guide to the study of the Earth. Extremely well illustrated and includes a glossary and an index. Although this is an introductory text for college students, it is written in a style that makes it understandable to the interested layperson. Contains a section on the development of glaciers and the types of glacial deposits.

McKnight, Tom L. *Physical Geography: A Landscape Appreciation.* 2d ed. Englewood Cliffs, N.J.: Prentice-Hall, 1987. An introductory physical geography text dealing with the morphology and evolution of landforms. Chapter 19 specifically treats glacial deposits. Suitable for interested readers with no previous background in geology.

Menzies, M., ed. *Modern Glacial Environments: Processes, Dynamics, and Sediments.* Oxford, England: Butterworth-Heinemann, 1996. This multiauthored volume deals with all aspects of the modern study of glaciers. Intended for the specialist. Index and extensive bibliography.

Wright, H. E., and D. G. Fry, eds. *The Quaternary of the United States.* Princeton, N.J.: Princeton University Press, 1965. A collection of articles on the glacial deposits of various regions of the United States. Each article is written by an expert on the glacial geology of a particular region. Suitable for college-level readers with a basic knowledge of glacial geology who are interested in a particular region of the United States.

GLACIAL SURGES

Glacial movement occurs at two points in the glacier: at the base and inside the glacier. Surging glaciers have left an indelible mark on the landscape in the past and will continue to do so in the future.

PRINCIPAL TERMS

BASAL SLIP: glacier movement that is caused when the glacier slides over its floor

DRIFT: an all-inclusive term for any kind of material that is deposited by glaciers and their meltwater

ERRATIC: a glacier-transported rock fragment resting on bedrock unlike that from which the fragment was derived

GLACIATION: the effects of a glacier upon the landscape

GLACIOLOGISTS: scientists who specialize in the study of glaciers and ice

INTERNAL FLOW: occurs as the individual ice crystals are slowly deformed; this motion enables the glacier to flow over irregular surfaces and around curves

MELTWATER: water from the melting of ice and snow

REGELATION: the freezing and thawing of ice as the result of changes in pressure

SNOUT: the terminal end of a glacier

STRIATION: parallel scratches cut in bedrock over which the glacier has passed

TILL: unsorted, unconsolidated glacier-deposited material that was let down directly from the ice and does not include material deposited by meltwater

VARVE: a pair of contrasting layers of sediment deposited over one year's time; the summer layer is light, and the winter layer is dark

ICE-FLOW MECHANISMS

The surging of glaciers, or spontaneous iceflooding, began with the inception of the Ice Age. As a field of study, however, it was a long time before glaciologists gave this phenomenon the attention that it deserves. Glacial surges have not received a great deal of attention because they occur in remote areas and, as a rule, have not been reported. In those cases where the surges have been reported, they are usually over by the time the scientists arrive. In addition, surges are such infrequent occurrences and of such scattered distribution and short duration that glaciologists have had few opportunities to observe them directly.

As a result, glaciologists do not agree on the specific cause of glacier movements. Glaciers that lie at the foot of precipitous slopes may surge as a result of avalanches of snow, but that explanation could not account for all surges. The climatic factor cannot be proven to be the cause of glacial surge either, although climate is the main regulator of a glacier's changing rate of "speed." While some glaciologists have attempted to show that

earthquakes cause glacial surges, most surges occur without the assistance of a big earthquake.

Because the main zone of ice flow is hidden inside the glacier, glaciologists have had to speculate on the mechanism that causes the ice to flow by examining the surface and bottom of the glacier and the degree of deformation in the ice at the snout of the glacier. As a result of these investigations, glaciologists have identified two primary components of movement. The first of these is basal slip, which involves the sliding of the ice over its floor. The major mechanism in this process is regelation, which is produced by tiny irregularities on the rock surface. The freezing point is lowered by any increase in the pressure of the ice mass and is raised by any release of pressure; therefore, when the high-pressure side of the irregularity melts the ice, the meltwater flows around the protuberance and refreezes on the downstream, low-pressure side. This process occurs in a thin zone of ice only 1 or 2 centimeters thick, and it is the constant displacement in this layer that carries the ice along. Although 0.5 millimeter of water is required for this sliding phenomenon, the glacier will move faster if more meltwater is available. Thus, the velocity of a glacier increases during the warm summer months and decreases in the cold winter months. In very cold glaciers, where regelation cannot occur, basal slip is totally absent.

There is also internal flow inside the glacier, which is much slower and much more difficult to study. When snowflakes are buried and pushed together, they become rounded in form. Meltwater that forms at the points of pressure between crystals refreezes and is added only to certain crystals. As a result, some crystals grow at the expense of the others. A glacier is formed as the deepening of the snow and the continued refreezing of meltwater force out the air and convert the individual crystals into a single mass of interlocking crystals. When subjected to steadily applied pressure, such as would result from the weight of the ice itself or from the pull of gravity, the tiny atomic layers of which each crystal consists begin to slide over one another. Recrystallization takes place as the pressure causes the crystals to orient themselves so that their atoms are parallel to the surface of the glacier. After billions of the crystals rearrange themselves in this favorable direction, the glacier begins to move.

Rates and Types of Movement

During a glacial surge, different parts of the glacier move at different rates. There are two basic zones of ice movement. The zone of fracture, which is an upper zone between 30 and 60 meters thick, consists of brittle ice that breaks sharply. In the zone of flow, on the other hand, pressure exerted by the upper layer of ice causes the lower layers to behave like plastic. The plastic deformation of the ice mass allows the ice to flow in the lower zone, which carries the overlying rigid ice along with it.

A less important mode of flow is a shearing type of motion that some glaciers exhibit. This phenomenon occurs because the zone of flow does not proceed at a uniform rate. The more plastic ice can adjust to this differential flow, but the brittle ice cannot and starts to crack and splinter. This breakdown usually happens near the glacier's snout.

Although surge rates of up to 20 meters per day have been recorded, most glaciers move at the rate of only about 1 meter per day. Some over-energetic glaciers have been known to "gallop," or to advance at phenomenal rates of speed. Whenever there are clusters of glaciers, one or more occasionally starts racing ahead of the others. Canada's Steele Glacier, for example, is a galloping glacier that flows at the rate of about 0.6 meter per hour. The Alps, Iceland, and Scandinavia have supplied hundreds of reports over the years of glaciers that snaked into valleys, knocked down orchards, blanketed fields and pastures, and demolished roads and buildings.

Study of Glacial Surges

During the nineteenth century, naturalist Louis Agassiz proved that ice motion is fastest in the center and decreases toward the sides by driving stakes across glaciers. More recently, glaciologists have drilled vertical holes in glaciers and inserted pipes into the holes. After taking careful measurements, they found that as time passed, the glacier bent the pipes. By measuring the bending of the pipes, they found that the deeper the ice below the surface, the slower it flows. Glaciologists also study the movement of glaciers by examining the deposits that they have left behind. These deposits, called drift, take the form of either heterogeneous collections of stony material or uniformly graded material. Drift layers can be distinguished from one another by their composition, color, and degree of weathering. Glaciologists assume that the earliest ones followed a growth pattern similar to the last ones, of which scientists have the clearest picture. Because drift layers overlie one another and have similar shapes, scientists believe that recurring ice sheets grew out of the same source regions and moved in approximately the same directions.

Glaciologists employ several methods to study the direction of ice movement. A composite picture of movement can be formed by studying the striations cut into the bedrock by large, sharp rocks embedded in the base of the moving ice. These striations, some of which are a few meters deep and up to hundreds of meters long, are more pronounced on hillsides and outcrops that faced the glacier. One of the best indicators of the direction of travel is erratic boulders, which are the largest objects deposited by glaciers. Because glaciers cannot move backward, their location leaves no doubt which way the ice went. The distinctive characteristics of a rock can tell a geologist where it came from. For example, if a rock can be identified as coming from an area in the northeast, then the geologist knows that the ice moved in a southwesterly direction. Airplanes are also used to ascertain the ice's direction of travel. Viewed from the air, the paths of the glacier can be clearly seen. In central Canada, for example, elongated hills and valleys and long, narrow lakes have been smoothed in lines parallel to the glacier's path.

Finally, landforms built by the ice illustrate how the ice made its retreat. The manner and direction of the ice's withdrawal are revealed by composition and arrangement of the drift. They show the shape of the ice sheet and where the glacier halted during its retreat. As well as studying the movement of the ice, glaciologists have applied scientific methods to determine the date of these glacier advances. In 1879, Baron Gerard de Geer began his study of varves, the sediment layers in the exposed beds of glacial lakes. Because the lowest varve is laid down when the glacier retreats, Geer was able to count the years since it disappeared. This method is not entirely accurate, however, because some varves have been scattered over a wide range.

Radiocarbon dating, which was developed in 1947 by Willard F. Libby, is a much more reliable method of dating glaciers. This method takes advantage of the fact that the radioactive isotope of carbon, known as carbon 14, changes at a constant rate. By measuring the amount of radioactivity in the remains of a dead plant or animal that was deposited by a glacier, geologists have been able to extend the glacial calendar backward for nearly 70,000 years. The pollen grains embedded in peat bogs, many of which are the remnants of shallow glacial lakes, have enabled scientists using carbon 14 dating to determine the times when various forests covered the landscapes.

Using a related method of dating, chemist and Nobel laureate Harold C. Urey discovered a method of finding the temperature in which a shelled creature lived by analyzing another isotope: oxygen 18. The greater the proportion of oxygen 18, the higher the temperature of the water. Oxygen isotopes tell scientists the temperature of the sediment when they were deposited, which reflects what was happening in the frozen world on land.

Environmental Impacts

In many ways, glaciers are almost as important as air, soil, and water in their effect on human-

Striations, or glacial grooves, were cut into Devonian dolomite near Yule Colorado Quarry by large, sharp rocks embedded in the base of the moving ice. (U.S. Geological Survey)

kind's future. According to the U.S. Geological Survey, more than three-fourths of all the Earth's fresh water is stored in glacial ice. In North America, the volume of fresh water stored in glaciers is far greater than that held in all the continent's rivers, ponds, and lakes combined. Melting mountain glaciers produce fresh water for irrigation in many parts of the world. In fact, glaciers provide much of the water for the world's great river systems. Thus, the availability of this water to humans depends on the condition of the glacier. Whenever glaciers grow and advance, more water is locked up in the ice. When glaciers shrink and retreat, however, the water is released from storage. As one can see, the retreat or advance of a glacier has a tremendous effect on the water resources that depend on it.

Glacier movement also has important implications for mining, because it aids geologists and prospectors in tracking down the source of valuable erratics such as diamonds. Because diamonds can withstand tremendous grinding and travel great distances without disintegrating, however, the volcanic source from which they came may be many miles from the point where they were found. Nevertheless, many geologists and prospectors have hunted for the source of diamonds such as those that have been found in the drift of Wisconsin, Indiana, Ohio, and Michigan.

Many scientists believe that vast sheets of ice will again spread over North America and Europe. This process, taking thousands of years, will transform large-scale agriculture into small-scale subsistence farms. As the ice advances, Chicago, for example, without moving an inch, will become a city of the subarctic within ten thousand years. Even if glaciers had no effect on the Earth, they would be of value simply because of their beauty and majesty. Massive glaciers that dominate the land in many parts of the world have attracted nature lovers and explorers for years. This beauty could be lost forever if mining operations are conducted underneath the glaciers, as some people have suggested.

Although scientists cannot predict with any certainty where and when the great ice masses will move, they do believe that humankind may be hastening their movement. The gradual warming of the earth is being accelerated by the increasing amounts of carbon dioxide that are being created as a result of the burning of mineral fuels, such as coal, oil, and gas. This warming trend could not only result in excessive melting of the ice sheets and produce a dangerous rise in sea level but also enable air masses to carry more snow on the glaciers' feeding grounds and cause them to grow.

Alan Brown

CROSS-REFERENCES

Alpine Glaciers, 865; Caves and Caverns, 870; Climate, 1902; Continental Glaciers, 875; Glacial Deposits, 880; Glacial Landforms, 914; Greenhouse Effect, 1867; Ice Ages, 1111; Ice Sheets and Ice Streams, 892.

BIBLIOGRAPHY

Bloom, Arthur G. *Geomorphology: A Systematic Analysis of Late Cenozoic Landforms.* 3d ed. Englewood Cliffs, N.J.: Prentice Hall, 1998. This college-level text covers the basics of geomorphology. Includes three chapters on glaciers and glaciology. Index and bibliography.

Dolgoff, Anatole. *Physical Geology.* Lexington, Mass.: D. C. Heath, 1996. This is a comprehensive guide to the study of the Earth. Extremely well illustrated and includes a glossary and an index. Although this is an introductory text for college students, it is written in a style that makes it understandable to the interested layperson. Contains a section on the development of glaciers and the types of glacial deposits.

Imbrie, John, and Katherine Imbrie. *Ice Ages: Solving the Mystery.* Short Hills, N.J.: Enslow, 1979. Presents findings by a project sponsored by the National Science Foundation entitled CLIMAP, which studied changes in the Earth's climate over the past 700,000 years. Written by a major scientist and his daughter, it explains what the ice ages were like, why they occurred, and when the next one is due. Contains a good bibliography and appendix but is sparsely illustrated. College level students will find it interesting.

Kurtén, Björn. *The Ice Age.* New York: G. P. Putnam's Sons, 1972. Primarily an introduction to the prehistoric people and animals that existed during the Ice Age. The chapter entitled "The Regimen of Ice" explains how glaciers are born and traces their movements across the globe. Contains many beautiful color illustrations. Useful primarily for secondary school students.

Mathews, William H. *The Story of Glaciers and the Ice Age.* Irvington on Hudson, N.Y.: Harvey House, 1974. Well illustrated with photographs and charts, this book is a good introduction to the study of glaciers and their effects on people's lives. The in-text definitions, nontechnical language, helpful glossary, and large print make this book ideal for junior high and high school students.

Menzies, M., ed. *Modern Glacial Environments: Processes, Dynamics, and Sediments.* Oxford, England: Butterworth-Heinemann, 1996. This multiauthored volume deals with all aspects of the modern study of glaciers. Intended for the specialist. Index and extensive bibliography.

Schultz, Gwen. *Ice Age Lost.* Garden City, N.Y.: Doubleday, 1974. Written mainly for those who have not had formal courses in glacial geology. While this volume explains the scientific processes that created the Ice Age in great detail, it also discusses the ways in which humankind has coped with its frozen environment in the past and present. The photographs and bibliography are good, but the book lacks a glossary.

Sparks, John. *Planet Earth.* Garden City, N.Y.: Doubleday, 1977. The emphasis of this book is on the interplay of all forms of life with their environment. It also documents the history of the Earth. The chapter entitled "The Moving Landscape" briefly explains how glaciers move and places this process in perspective by comparing it to other processes that have altered the Earth. Includes many beautiful color photographs. For high school and college students.

ICE SHEETS AND ICE STREAMS

Ice sheets are extremely large masses of ice that cover most of Greenland and Antarctica. Areas within these ice sheets that move much more rapidly than surrounding areas are known as ice streams. The behavior of ice streams may play an important role in the stability of ice sheets, which could have a profound effect on sea levels.

PRINCIPAL TERMS

GLACIER: a mass of ice, formed from snow, that persists from year to year

ICE AGE: a period of time during which extensive ice sheets exist

ICE CAP: a glacier on a flat area of land

ICE SHELF: a portion of an ice sheet extending into the ocean

MOUNTAIN GLACIER: a glacier in a sloping valley

ANCIENT ICE SHEETS

Throughout the history of the Earth, periods of time have occurred in which large areas of land were covered with relatively permanent expanses of ice formed from snow. These periods of time, known as ice ages, sometimes lasted for millions of years. The ice sheets that were formed during the ice ages had a powerful effect on the physical appearance of the Earth's surface.

The earliest known ice age occurred during Precambrian times, more than 544 million years ago. The most recent and best known series of ice ages occurred during the Pleistocene epoch, lasting from about 2.5 million to 10,000 years ago. During the last 730,000 years of this period, a series of eight climatic cycles, lasting roughly 100,000 years each, occurred. Each cycle alternated an ice age with a warmer period of time. Prior to 730,000 years ago, the climatic cycles were more frequent but of lesser magnitude.

During the most extensive Pleistocene ice ages, more than 45 million square kilometers of the Earth's land area was covered with ice sheets. With the exception of Antarctica, most of this ice was located in the Northern Hemisphere. The largest ice sheet was the Laurentide ice sheet. At its greatest extent, this ice sheet reached from northern Canada to southern Illinois and from the Rocky Mountains to the Atlantic Ocean. The other major ice sheet in North America was the Cordilleran ice sheet, which reached from western Alaska to northern Washington.

In Europe, the Scandinavian ice sheet reached as far as Great Britain, across northern and central Europe, and across northern Russia. Between North America and Europe, the islands of Greenland and Iceland were also covered by ice sheets. Some evidence suggests that other major ice sheets may have existed also. These include the Innuitian ice sheet, north of Canada; the Barents ice sheet, north of Russia; and the Kara ice sheet, in the Baltic region.

The movement of ancient ice sheets had a profound effect on the land left behind when the ice disappeared. The Laurentide and Scandinavian ice sheets carved numerous lakes in many areas of northern North America and Europe. In other areas, including much of the central United States, western Canada, and central Europe, the ice sheets left behind large deposits of sediment. These areas became flat or gently rolling regions with rich soil that was well suited for agriculture.

MODERN ICE SHEETS

The only two ice sheets still found on the Earth are located in Greenland and Antarctica. Masses of ice similar to ice sheets, but much smaller, are known as ice caps or ice fields. Unlike mountain glaciers, which move down sloping valleys in a particular direction, ice sheets and ice caps move outward in all directions.

The Greenland ice sheet is about 1.73 million square kilometers in area, covering about 80 percent of the island's surface. The ice sheet is almost 2,400 kilometers long from north to south, with a maximum width of 1,100 kilometers from east to west at a latitude of 77 degrees north near its northern margin. The average altitude of the ice surface is 2,135 meters. The highest altitudes are found in two elongated domes or ridges. The southern dome reaches a height of almost 3,000 meters at a latitude of 64 degrees north, and the northern dome reaches a height of about 3,290 meters at a latitude of 72 degrees north.

About 8 percent of the freshwater ice on the Earth is found in the Greenland ice sheet. Although most of the surface on which the interior of the ice sheet rests is at about sea level, the margins of the ice sheet occur in more mountainous regions. This prevents the ice sheet from reaching the ocean in most areas. Along a portion of the northwest coast, however, the ice is able to reach the sea, resulting in the creation of numerous icebergs, which may endanger ships in the North Atlantic.

The Antarctic ice sheet, with an area of about 13.8 million square miles, covers nearly all of the continent. Less than 2 percent of the surface of Antarctica is exposed through the ice, in the form of mountain ranges or individual mountains known as nunataks. The ice sheet has an average thickness of about 2,000 meters and contains about 91 percent of the world's freshwater ice.

The western part of the Antarctic ice sheet, from a longitude of about 45 degrees west to a longitude of about 165 degrees east, contains numerous nunataks. It also contains two large ice shelves, regions where the ice sheet extends into the ocean. The Fichner-Ronne ice shelf has an area of about 390,000 square kilometers, and the Ross ice shelf has an area of about 496,000 square kilometers.

The eastern part of the Antarctic ice sheet is separated from the western part by the Transantarctic Mountains, which extend between the two ice shelves. Most of the eastern part of the Antarctic ice sheet consists of a single huge ice dome, reaching a thickness of more than 4,000 meters at the peak.

MOVEMENT OF ICE SHEETS

Ice sheets gain mass by the accumulation of snow, which is transformed into ice. Although a similar process may take only a few years in glaciers located in wet, relatively warm areas, snow is changed into ice much more slowly in ice sheets. In some parts of Antarctica that are extremely cold and dry, this process may take several thousand years.

Ice sheets lose mass through melting or when pieces of the ice sheet break off into the sea. In the Greenland ice sheet, roughly one-half of the mass loss is caused by melting of the surface of the ice, and roughly one-half is caused by the breaking off of icebergs. In the Antarctic ice sheet, very little surface melting takes place. Some melting of the bottoms of the ice shelves occurs, but the majority of mass loss is caused by the breaking off of icebergs, particularly from the ice shelves.

In general, ice sheets move from central high points outward to the sea. This simple movement becomes more complex in areas where the underlying surface is very rugged. The flow of ice in Greenland is generally outward from the two ice domes. In eastern Antarctica, the flow of ice is generally outward from the single dome.

A nunatak encircled by moraine in the Alaska Gulf region. Jefferies Glacier is to the left. (U.S. Geological Survey)

The flow of ice in western Antarctica is more complicated but is generally toward the sea, particularly in the two ice shelves.

The speed of the movement of an ice sheet varies enormously. At the interior, the flow of ice may be only a few meters per year. As it moves outward, the rate increases to tens or hundreds of meters per year. Ice shelves move even more quickly. The outer edge of the Ross ice shelf moves at about 900 meters per year. By comparison, a typical mountain glacier moves at a speed ranging from about 50 to about 400 meters per year.

ICE STREAMS

Certain areas within ice sheets move at rates much faster than the surrounding ice. These regions are known as ice streams. Some ice streams may move as quickly as 1,000 meters per year. The largest ice streams are up to 150 kilometers wide, hundreds of kilometers long, and more than 1 kilometer thick. Six ice streams flow into the Ross ice shelf, and two ice streams flow into the Fichner-Ronne ice shelf. Ice streams are also known to exist in other areas of Antarctica and in Greenland.

Ice streams resemble glacial surges, in which mountain glaciers move forward much more quickly than usual. Glacial surges occur when mountain glaciers slide on a layer of mud made of wet sediment. This fact led scientists to speculate that an ice stream occurs when a part of an ice sheet slides on a similar layer of mud. Evidence for this hypothesis began to appear in the 1980's using a combination of satellite data, seismic data from controlled explosions on the surface of the ice, and data from holes drilled through the ice. Scientists confirmed that ice streams rest on a layer of wet, fine-grained sediment. This slippery mud reduces the friction beneath the ice, resulting in an ice stream. Surrounding areas generally rest on dry, solid bedrock, resulting in more friction and a slower rate of movement.

Some scientists believe that the wet sediment that allows the ice stream to move quickly can exist in areas only where the ice sheet rests on a layer of sedimentary rock. According to this hypothesis, the movement of the ice erodes the rock into sediment. At the same time, the ice traps heat from the interior of the Earth. This heat causes the base of the ice to melt, wetting the sediment and allowing the ice to move more quickly. The faster movement results in more melting of ice, producing wetter mud. This cycle is repeated, allowing the ice to increase in speed until an ice stream is formed.

This simple model of ice stream formation fails to explain some aspects of ice stream behavior. Some ice streams move on a relatively thin layer of sedimentary rock, while ice resting on a thicker layer of sedimentary rock nearby moves much more slowly. Evidence also exists that suggests that the layer of sedimentary rock must be more than 100 meters thick to produce an ice stream. The explanation for these observations was unknown at the end of the twentieth century.

Inside a crevasse in Blue Ice Valley on the Greenland ice sheet, 65 feet below the ice surface. (U.S. Geological Survey)

STABILITY AND SEA LEVEL

The most important question facing scientists who study ice sheets and ice streams is whether the Antarctic ice sheet is stable or whether large portions of it may disappear over a relatively short period of time. The melting of a significant part of the ice sheet, or the breaking off of a large portion of it into the ocean, would raise sea levels by an appreciable amount. In the most extreme case, the complete melting of the Antarctic ice sheet would raise the level of the ocean by about 60 meters, drastically altering the coastlines of the continents. The collapse into the ocean of the western part of the Antarctic ice sheet, believed to be less stable than the eastern part, would raise sea levels by about 5 meters, flooding many coastal cities.

Until fairly recently, the eastern part of the Antarctic ice sheet was generally believed to be very stable, having remained mostly unchanged for 15 million years. In the 1980's, fossils of microscopic sea organisms known as diatoms were discovered in the Transantarctic Mountains. This discovery implied that the ocean had extended into the interior of Antarctica 3 million years ago, suggesting that the ice sheet was much smaller at that time. However, scientists studying rock erosion and deposits of volcanic ash in Antarctica found evidence that the ice sheet had remained unchanged for at least 10 million years. Reconciling this contradictory evidence is a major concern of Antarctic researchers.

Most scientists believe that the western part of the Antarctic ice sheet is more likely to undergo relatively sudden loss of mass than the eastern part. This is because it rests on bedrock that is mostly far below sea level. This means that the ice shelves, floating at sea level, can move much more quickly than the ice sheet, which must move uphill. This effect could cause the border between the ice shelf and the ice sheet, known as the grounding line, to move inland, causing more ice to move into the ocean.

Whether the western part of the Antarctic ice sheet could be reduced in size quickly this way may depend on the behavior of ice streams. Some scientists believe that the starting point of an ice stream may be able to move inland over time, causing ice to move into the ocean at an increasing rate. Whether this is possible depends on whether the sediment on which the ice stream slides extends into the interior of the ice sheet.

On the other hand, some scientists believe that ice streams may actually make the western part of the Antarctic ice sheet more stable than it would be without them. This may happen because the loss of ice from ice streams moving into the ocean could prevent the ice sheet from growing too large and collapsing into the sea under its own weight. In either case, a better understanding of ice streams is needed before scientists will be able to predict the fate of the ice sheets.

Rose Secrest

CROSS-REFERENCES

BIBLIOGRAPHY

Bailey, Ronald H. *Glaciers.* Alexandria, Va.: Time-Life Books, 1982. Full of dramatic black-and-white and color photographs, this book provides interesting statistics concerning ice sheets and glaciers. It also includes a history of the ice ages and a detailed explanation of glacier formation.

Hambrey, Michael, and Jurg Alean. *Glaciers.* New York: Cambridge University Press, 1992. A very detailed discussion of all aspects of glaciers, including statistics and information about ice sheets and ice streams. Many color and black-and-white photographs, diagrams, and charts enhance the text.

Hughes, Terence J. *Ice Sheets.* New York: Oxford University Press, 1998. An exhaustive discussion of the science of ice sheets. Written for the advanced student, this book provides an easy-to-read narrative coupled with charts, maps, diagrams, and the physics and mathematics of ice sheet mechanics.

Livermore, Beth. "Antarctic Meltdown." *Popular Science* 250 (February, 1997): 38-43. Livermore presents both sides of the argument concerning whether an increase in global temperatures would lead to great amounts of ice sheet melting. Provides a good summary of the geological history of Antarctica.

Radok, Uwe. "The Antarctic Ice." *Scientific American* 253 (August, 1985): 98-105. Provides a brief history of Antarctic exploration, along with a summary of discoveries about past climatic events obtained from analysis of ice cores and radar measurements.

Robinson, Andrew. "Avalanches, Glaciers, and Ice Ages." In *Earthshock: Climate, Complexity, and the Forces of Nature.* London: Thames and Hudson, 1993. A dramatic but scientific discussion of ice-related phenomena. Provides an explanation of the necessary conditions for glaciers, as well as the dynamics of ice formation.

Walker, Gabrielle. "Great Rivers of Ice." *New Scientist* 162 (April 17, 1999): 54-58. A short article that nevertheless conveys the nature of Antarctica's ice streams in detail. The researchers' methods, theories of the origin and results of ice streams, and computer models of their formation are all discussed.

8
GEOMORPHOLOGY

GEOMORPHOLOGY AND HUMAN ACTIVITY

Geomorphology is a science that classifies landforms and describes and analyzes the origin and evolution of surface features. It is interdependent with other branches of geology—such as sedimentology and hydrology—that study the processes that act on marine and planetary features. Geomorphology is of immense global importance because of the ever-increasing activity of humans as geomorphic agents. As the human population increases in numbers and complexity, it increasingly affects and is affected by geomorphic processes.

PRINCIPAL TERMS

EROSION: wearing away of soil and rock by weathering, landslides, and the action of streams, glaciers, waves, wind, and underground water

LANDFORM: one of the many features that, taken together, make up the surface of the Earth

LANDSCAPE: broad term for the land setting of an area

TOPOGRAPHY: general configuration of a land surface, including its relief and the position of its human-made features

WEATHERING: destructive processes by which rocks are changed by exposure to the atmosphere at or near the Earth's surface, with little or no transport of the loosened or altered material

LANDFORMS

Geomorphology is the scientific study of topographic surface features and naturally occurring Earth processes. The landform is the basic unit of systematic analysis of geomorphology. Landforms can be considered constructional (coral reefs and volcanic eruptions, for example); however, because of the intensity of atmospheric weathering and erosion, most landforms are erosional. During landscape analysis, a geomorphologist must not only analyze current surface features but also reconstruct rock units and landforms that existed before the present landscape.

Time is a major factor in geomorphology in that landforms are altered by geologic processes over time. A theoretical analysis of landform evolution uses a time range of millions of years. Practical or applied geomorphology uses a time scale of hours to years because of the need to predict changes on a human scale, such as how long an excavated area will remain stable during hot, humid conditions. The use of radioactive isotopes establishes the absolute age of rocks as well as the exposure age of rock surfaces, thus making absolute rates of land surface change possible. The geomorphic application of isotopic dating to land surface changes can establish when an ice sheet retreated from a region.

Central to the geomorphic study of any region is its structure or specific physical properties. Studies of landform structure reveal the type of rock present, its mineral composition, grain size or crystalline structure, and rock strength. Also, studies of landform structure indicate the depositional sequence of the rock material and any naturally occurring stresses that the rock unit has been subjected to, resulting in faults and fractures.

A full range of processes contributes to produce a landscape. The action of wind, water, and ice erodes rocks and transports the eroded material to deposition sites. Gravity pulls down large materials that rise above ground level. Rates and intensity of geomorphic processes vary. Surface weathering rates may be extremely slow, such as a few centimeters per thousand years, or quite rapid, as measured in an avalanche that may travel at 50 meters per second or more. Intensity of geomorphic processes is governed by climate, vegetation present, and elevation of the surface feature.

Solar radiation provides the energy for geomorphic processes at a greater rate than all other energy sources combined. Other energy sources are the gravitational and inertial forces associated with the mass and motion of Earth, the Moon, the Sun, and other planetary objects. Additional energy comes from the outward flow of heat, caused by radioactivity, from the Earth's interior.

The two major geomorphic processes are constructional processes—such as mountain building and volcanic eruptions—and erosional processes—such as weathering, erosion, and the action of wind, rivers, glaciers. Other geomorphic processes are sediment movement and deposition, biologic processes as they pertain to vegetative cover, near-shore and ocean systems, and human-driven processes. There are few spheres of human activity that do not create landforms directly or indirectly.

GLOBAL WARMING

Predicted temperature increases (2 to 6 degrees Celsius per century) and global warming as a result of the buildup of greenhouse gases in the Earth's atmosphere is not a new phenomenon. Although it became a significant issue in environmental management debates only during the last two decades of the twentieth century, the process has operated, with fluctuations, over the past 4.5 billion years of the Earth's history.

Currently, global warming studies focus on the atmospheric, hydrospheric, and biospheric consequences of climate change. Geomorphological consequences of climate change are significant because of potential changes to landscapes and human occupation of these surface areas. Potential geomorphological changes include climatic alterations affecting vegetation cover and agriculture, which could alter soil erosion, surface water runoff, river siltation, and flooding patterns; changes in frequency, magnitude, and geographical extent of tropical storms, which could affect land erosion, especially in coastal zones; changed patterns of rainfall, with excessive rainfall resulting in increased flooding or low amounts of rainfall affecting human water supplies; elevated temperatures at high altitudes and latitudes, which may alter ice and snow distribution and the extent of permafrost; and raised sea level caused by the reduction of glaciers and ice sheets, impacting coastlines through beach narrowing, cliff erosion, and delta formation.

These geomorphic changes may seem catastrophic, but it is helpful to review them in the light of environmental changes in the geologically recent past. Since the end of glaciation, about ten thousand years before the present, sea level has risen approximately 121 meters with periods of rapid sea level rise of 14.6 millimeters per year between 13,000 and 7,500 years ago and 24 millimeters per year between 12,500 and 11,500 years ago. Although these rapid sea-level changes do not compare with current predicted values, they do demonstrate that surface features such as river deltas and coral reefs can continue to exist. Both landforms are capable of quick evolution, river deltas because of sediment supply and reefs by organic structural growth. Both landforms have survived post-Pleistocene rises in sea level and, if not perturbed by further human activity, should continue to do so. The systems that may be most threatened by the projected sea-level rise are land-based human systems such as agriculture.

Accurate geomorphic predictions and changes to landscapes are troublesome because exact weather patterns and distributions of extreme climate events are difficult to assess. Knowledge of average climatic changes does not automatically assign specific locations of weather systems and events. While it is true that landform changes are altered by climatic conditions, additionally local geology, hydrology, topography, and land use play major roles in landform evolution. The climatic, oceanic, and biospheric repercussions of global warming provide the basic ingredients for geomorphic change, but human activity will determine the scale, extent, and rate of the resultant changes.

Human apprehensions about adverse changes to the biosphere and human socioeconomic systems have created the current concerns about global warming. Concern is heightened by the fact that the predicted trends of climate change have never been experienced by humans. The predicted climate changes, if they fully occur, are similar to climatic conditions prior to the last glaciation. Finally, humans now recognize that they are the main influence in climate change and are capable of modifying the rate of this predicted change.

COASTAL ZONE

Major coastal processes are those associated with moving water in near-shore environments. Three forces act upon water to create waves: astronomical, meteorological, and tectonic. Astronomical forces are the driving force for tides, which in turn influence the width of the shoreline and growth patterns of the near-shore flora and fauna. Waves produced by meteorological forces operate to modify the coastline. Storm-generated waves and sea swells, along with ocean currents, operate to erode, transport eroded materials, and deposit debris in coastal areas. Tectonic forces such as earthquakes and volcanic eruptions produce dangerous seismic sea waves known as tsunamis. These three large-scale forces are not affected by human activity, but landforms in the coastal zone have long been altered and used by humans.

Early changes to the coasts by humans were indirect and unintentional because of small populations. Direct and intentional use of the coastal zone began with farming on flat delta land. Lands in these areas were reclaimed or extended by canalization and diking. Reclamation, as a coastal modifier, continues to the present. In the Netherlands, more than 3,800 square kilometers of land have been gained by reclamation in the last eight hundred years. A 1984 study of reclaimed land in Singapore revealed that the country is now 10 percent larger than when it was founded.

Another direct, intentional modification of the coastal zone is the use of shoreline stabilization structures such as seawalls, breakwaters, and jetties. A seawall or breakwater structure is placed parallel to the shore to protect an eroding coastline. Sand-bearing coastal waves are slowed down by seawalls and drop their load of sand up-current from the structure. Down-current from the barrier, the water picks up more sediment to replace the lost load, and the down-current beach is eroded. Jetties built perpendicular to the shoreline also operate to erode and redistribute sand, creating an unnatural scallop shape along the shoreline. Coastal structures placed by humans severely alter the shape and stability of the coastline. Beach erosion is not halted, nor is the shoreline stabilized.

A review of the state of the marine environment for the United Nations Environment Programme in 1990 reported that the open ocean, because of its large diluting capacity, is still relatively clean, despite measurable levels of artificial radioactive material and synthetic organic compounds. Oil slicks from tanker collisions, explosions, navigational or mechanical errors, and offshore oil installations continue to put a human face on ocean pollution. However, pollution from oil slicks is still considered of minor consequence to open ocean marine organisms. Alternatively, coastal areas near large populations are clearly linked to human activity and exhibit detectable increases in phosphate concentrations from sewage and agricultural discharges. The amount of nitrates (a component of fertilizers) found in coastal waters is increasing, as are areas of eutrophication, unusual plankton blooms, and excessive seaweed growth.

RIVER SYSTEMS

Fluvial systems carry water, sediments, and dissolved materials and minerals downstream. Streams, driven by gravity, cut channels and scour the stream bottom and sides with their sediment load. Also, streams deposit sediments in artificial reservoirs and natural freshwater and marine basins. Both stream erosion and deposition create landforms, with stream valleys being one of the most frequent and widespread landforms in North America. Although the portion of water involved in streams constitutes only a small percentage of the total water in the hydrologic cycle, human interference with stream processes perturbs the global water system. The hydrologic cycle is the constant circulation of water from the sea through the atmosphere by evaporation and its return to the land, streams, lakes, glaciers, and the subsurface by precipitation. Water eventually returns to the atmosphere by way of transpiration and evaporation from plants, landforms, and ocean basins.

Human occupation of cities has a profound effect on the hydrological cycle because humans interrupt and rearrange areas where water is naturally stored, such as lakes and groundwater-bearing aquifers. In *Urban Hydrology* (1984), M. J. Hall discusses linkages between hydrology and urbanization. First, the replacement of vegetated soil with impermeable surfaces (asphalt) reduces water storage in the soil horizon, the slow percolation of water into aquifers, and the transpiration mechanism in the hydrologic cycle, which allows moisture to return to the atmosphere. Second, with large amounts of precipitation on solid surface areas, the velocity of water flow is increased, moving the water to stormwater systems rather than natural stream channels, where evaporative processes can occur. Third, construction activity clears the land surface, disturbs the soil layer, reshapes natural slopes with potentially unstable slopes, and leaves a limited vegetation cover or builds additional impermeable surfaces. If overloaded with large quantities of construction-derived debris, streams that routinely carry a small amount of sediment may experience a variety of changes in their physical and biological characteristics. These changes include deposition of sandbars or dunes in the channel. Coarser sediments carried in the water may increase the erosion of channel banks. Bottom-dwelling flora and fauna may be blanketed with sediments, reducing the viability of fish species that feed on such stream organisms.

Dams and reservoirs began to play an important role in water use in the late nineteenth century. The immediate impact of reservoirs is the intentional alteration of flow downstream, usually to increase low flows for year-round water use or to curb floods. Other impacts include the loss of land caused by the filling of the reservoir, increased evaporation from the reservoir surface, groundwater seepage into reservoir walls, channel and bank scouring, and sediment deposition. Inland reservoirs have resulted in major changes to coastlines because of sediment and nutrient starvation. Damming of major streams aggravates beach and delta erosion. Before the Aswan High Dam was constructed in Egypt, the Nile River transported 140 million tons of silt per year to the Mediterranean Sea. Increased erosion of the delta began once Aswan was filled, and the loss of nutrients caused a reduction in fish catches in the Mediterranean Sea.

SOIL

Soil formation is the direct result of physical, biological, and chemical weathering processes acting on rock units and rock-cored landforms. Soil is the active surface layer that mantles most rock masses and supports the growth of rooted plants. It is not just a simple, loose layer with a stock of plants and plant nutrients on the land surface, but rather a specific stratum that regulates biological and geological interactions. Because soil exists between the geological, biological, and atmospheric realms, it is subject to a complicated set of direct

The California Aqueduct, here running through the Mojave Desert. Irrigation in arid regions is one means whereby human activity has shaped the land, both for better and for worse. (©William E. Ferguson)

and indirect links with surface processes, both natural and anthropogenic. Human activities that can degrade soils include improper cultivation, deliberate deforestation, overcompaction by heavy machinery or trampling of grazing domesticated animal herds, and submergence of land because of drainage basin changes.

Globally, only 13 percent of the land surface (1.5 billion hectares) is naturally appropriate for growing crops. An additional 1.7 billion hectares are available for agriculture if proper land management practices are instituted and followed. The remaining vegetated areas are too cold, hot, dry, or low in nutrients or soil cover to produce crops. Soils can be physically removed by wind and water erosion when land is improperly cultivated. The removal of vegetation by clearing the fields for planting increases erosion from water, and fertile soils are washed from the land. With the protective vegetation and topsoils removed, wind erosion can increase soil loss. Tilling land surfaces parallel to the prevailing wind direction exposes soil to wind erosion. Tilling land across contours and valley slopes also degrades soil layers.

Chemical pollution of groundwater can occur as a result of the treatment of land surfaces used for agriculture. Both fertilizers, used to replace soil nutrients, and pesticides, used for raising crop yields, can enter subsurface groundwater zones after irrigation waters percolate through the soil horizons. Groundwater is especially susceptible to this type of degradation because of the slow movement of subsurface waters and the prolonged retention times within the aquifer. Salinization of soils can occur when poor irrigation techniques are used. Overirrigation, poor drainage of irrigated soils, and irrigating with water of high salt content are major causes of this problem. Other issues—such as dam placement and excessive evaporation from rivers in hot, dry climates—complicate soil salinization.

Trees act as a buffer to the erosive processes caused by rain and storms. When trees are removed, much of the rainfall runs directly off the land, taking with it large amounts of soil and sediment that choke stream channels or decrease water quality. Rain forests that are cut and developed for agriculture are particularly susceptible to soil degradation because of the clay-rich layers within tropical soil horizons. Once rain-forest trees are removed, massive amounts of rainfall wash the soil areas, removing the small amounts of nutrients and the fine-grained clay materials. Once this stripping of the soil and nutrients occurs, the area is unsuitable for agriculture or reforestation programs. Another form of deforestation consists of stripping the land for the gathering of fuel wood. Of all the degraded land in the world, 7 percent is the result of firewood gathering and burning, which opens the land to soil erosion by running water and wind erosion.

Mariana Rhoades

CROSS-REFERENCES

Beaches and Coastal Processes, 2302; Dams and Flood Control, 2016; Earth Science and the Environment, 1746; Geomorphology of Dry Climate Areas, 903; Geomorphology of Wet Climate Areas, 909; Glacial Landforms, 914; Global Warming, 1862; Grand Canyon, 919; Karst Topography, 925; Marine Terraces, 929; Permafrost, 933; River Valleys, 936; Soil Erosion, 1513.

BIBLIOGRAPHY

Cvancara, Alan M. *Field Manual for the Amateur Geologist.* New York: John Wiley, 1995. Cvancara's introduction to core geology topics is suitable for high school students. Includes many useful illustrations.

Goudie, A. S., and H. A. Viles, eds. *The Earth Transformed.* Oxford: Blackwell, 1997. A well-illustrated, wide-ranging survey of human activities and their environmental consequences. Includes numerous global case studies.

Mackenzie, Fred T. *Our Changing Planet.* 2d ed. Upper Saddle River, N.J.: Prentice Hall, 1998. Mackenzie covers geologic and human-induced change, in language suitable for high school students. Illustrations and index.

Nisbet, E. G. *Leaving Eden: To Protect and Manage the Earth.* New York: Cambridge University Press, 1991. Nisbet discusses environmental problems and their causes, as well as methods to protect the environment and restore areas that have been damaged.

Simmons, I. G. *Environmental History: A Concise Introduction.* Oxford, England: Blackwell, 1993. A high-school-level introduction to the interactions between nature and humans.

GEOMORPHOLOGY OF DRY CLIMATE AREAS

In dry climates, the great temperature contrasts, strong winds, rare but torrential rains, and limited but powerful water floods produce a diverse geomorphology as well as natural difficulties of existence for those who live in such areas.

PRINCIPAL TERMS

BORAX: a sodium borate mineral that is an ore of boron and occurs as surface crust or large crystals in the muds of alkaline lakes; borax is used in glass, ceramics, cleansing agents, water softeners, and other industrial applications

DESERT (ROCK) VARNISH: a thin, dark, hard, shiny or glazed iridescent (red, brown, black) film, coating, stain, or polish on rocks that is composed largely of iron and manganese oxides and silica formed by weathering of dust films and by microbial action

DIFFERENTIAL WEATHERING: physical and chemical weathering that occurs at irregular or different rates, caused by variations in composition and resistance of a rock or by differences in intensity of weathering, and usually resulting in an uneven surface where more resistant material stands higher or protrudes above less resistant parts

PLUVIAL PERIOD: an episode of time during which rains were abundant, especially during the last ice age, from a few million to about ten thousand years ago

SALTATION: a mode of sediment transport in which the particles are moved progressively forward in a series of short intermittent leaps, jumps, hops, or bounces from a bottom surface

DISTRIBUTION OF DRY CLIMATE AREAS

Dry climate geomorphology is the study of landforms in arid and semiarid regions. Arid lands of various kinds, usually termed deserts, amount to at least 25 percent of the total land area of the world outside the polar regions and up to 43 percent by some definitions. They are generally considered to be those areas with less than 25 centimeters of precipitation or where evaporation amount is twice that of precipitation. Semiarid regions, or steppes, are not quite as extensive as deserts but are still significant. They can be defined as having precipitation between about 25 and 50 centimeters per year, which means that evaporation is about equal to precipitation. The semiarid areas are generally peripheral to the desert areas.

The most extensive arid lands of the world are associated with the two circumglobal belts of dry subtropical air that subsides to the north and south from the rising equatorial air masses, heating and drying in the process. Examples include the Sahara and Kalahari deserts of Africa. A second type of dry area is found in continental interiors far from sources of moisture. The Gobi and Takla Makan deserts of central Asia fall into this category.

A different and more local kind of desert is found on the lee side of mountain ranges. The mountains create a barrier to the flow of moist air, producing a rainshadow effect. As the air rises against the windward slope of a range, it cools and condenses into precipitation. This chain of events removes much of the moisture from the air mass. Then, as the air mass passes over the range and down the other side, it warms in descent and becomes even drier as it moves along. The deserts behind the coastal ranges of California are of this type.

A fourth category of deserts occurs along coastlines. They occur locally along west coasts, where upwelling cold seawater cools passing marine air, thereby decreasing its ability to hold moisture. As the air encounters warm land, its limited moisture condenses and gives rise only to coastal fogs. Deserts of this type occur in Chile, Peru, and southwest Africa.

TEMPERATURE AND RAINFALL

The limited cloud cover in arid areas allows strong thermal contrasts from day to night through intense solar heating during the day and reradiation of heat back out into space at night.

The resulting high temperature gradients result in the strong winds and intensive movement of sediment so characteristic of dry climate geomorphology. No major geologic process, however, is restricted entirely to desert regions. Rather, the same processes operate with different intensities in moist and in arid landscapes. Thus, in a desert, the surface sediments, soils, and landforms show some distinctive differences from those elsewhere.

Although arid areas do not generally have rivers originating in their boundaries, it does rain occasionally, and often the storms are torrential. Most runoff that originates in desert areas never reaches the sea, as the water soon disappears through evaporation, or it soaks directly into the Earth. Rivers such as the Colorado or the Nile begin in high, humid mountains far from the lower deserts through which they flow. Such rivers carry so much water that they keep flowing to the sea, despite great losses where they cross a desert.

FLASH FLOODS

In arid areas where vegetation and water are limited, soils are thin, and bedrock is therefore commonly exposed at the surface. Weathered fragments of rock tend to break off along natural fractures to leave the steep rugged cliffs that are so common in such regions. The plentiful rock fragments can be easily incorporated into flash floods and moved to lower elevations.

Studies show that typical torrential rainstorms in deserts are likely to be accompanied by flash floods that suddenly and swiftly transport large quantities of rock and soil debris. The debris is deposited as sediment that forms alluvial fans at the bases of mountain slopes and alluvial plains on the floors of wide valleys and basins. The debris-laden streams rapidly lose water into the porous and permeable sediment of the desert basins and end as dry washes or gullies.

ALLUVIAL FANS AND PEDIMENTS

Alluvial fans are one of the most characteristic of landforms in dry regions. They are caused by the change from the steep, narrow river channels of upland areas to the lower gradients and the unconfined slopes of desert basins, where the water velocity is checked and the sediment load spread. The fan itself commonly shows a characteristic concave-upward slope profile. Alluvial fans are typically dominated on the upper slopes by coarse fragments and on the lower, gentler slopes by finer sands, silts, and clays. Debris flows—sudden rapid waves of wet, bouldery mud—sometimes rush out of mountains after torrential rains and add to the sedimentation on the fans.

Erosion and sedimentation patterns are responsible for pediments, another of the characteristic landforms of arid zones. These are broad, gently sloping areas of bedrock, spread as aprons around the bases of the mountains. Measured cross sections of a typical pediment and its associated mountain show fairly steep bedrock mountain slopes that abruptly change to the gentler bedrock slopes of the pediment. Pediments are eroded by running water that also builds up a thin apron of sediment below as the water continues to erode back the slopes above.

A pediment meets the upper mountain slope, not at a curve but at a distinct angle. This angularity suggests that mountain slopes in the desert do not become gentler with time, as they would in a humid region where chemical weathering, soil formation, and downslope creeping movement of soil are dominant. Instead, the pediments seem to adopt an angle determined by the resistance of the bedrock and maintain that angle as they gradually retreat. In this way, retreat of the mountain slope should extend a pediment forward at its upslope edge. The growth of the pediment at the expense of the mountain may continue until the entire mountain has been consumed.

PLAYAS AND BADLANDS

In an arid region, water is rarely plentiful enough to flow for long into a basin to maintain a permanent lake on its floor. Instead, the ephemeral streams that flow down from the highlands after an exceptionally large rain can discharge enough water into the basin to produce a temporary shallow lake. These lakes, known as playas, may last a few days or weeks before the water evaporates or soaks into the ground. Many dry playas are white, gray, or tan because of precipitated salts and clays on the surface. The Spanish origin of the term *playa* (meaning "beach") is a reflection of the observation by early explorers of southwestern North America that many old lakes had dried up and left only beaches and dry lake bottoms behind. In fact, these basins were once filled with water during the high-rainfall, or pluvial, periods during the last ice age, when the climate was much wetter and cooler. Now dry, playa basins are scattered throughout the arid areas of the world.

Badlands in dry zones are extremely closely dissected landscapes on weak, impervious rocks or sediments largely devoid of vegetation. Badland topography is a wilderness of steep, smooth slopes, knife-edged or sharply rounded crests, and steep, narrow valleys. Such areas of bare ground erode through rain splash and slope wash in occasional intense rains.

EOLIAN PROCESSES

Although water is the predominant eroding agent in dry areas, the wind is effective in moving large quantities of sand and dust. Contrary to popular belief, however, most deserts are not covered with sand dunes. Nevertheless, landforms resulting from wind erosion or deposition predominate. Both wind activity and the resulting landforms are commonly referred to as eolian (after Aeolus, the Greek god of wind). Eolian processes in the dry regions of the world are responsible for major erosional and depositional landforms as well as significant sedimentary deposits.

Wind erosion acts in two ways: by deflation, the lifting and removal of loose sand and dust particles from the Earth's surface, and by abrasion, the sandblasting action of windblown sand. Deflation causes depressions called blowouts in areas where loose sediment occurs that is of small enough grain size to be moved by wind. These deflation basins can become very large, even below sea level, as long as the underground water table is not reached and as long as large-sized gravels do not accumulate as lag deposits to protect the surface with desert pavements.

YARDANGS AND ERGS

Large landforms produced by wind abrasion are not common, but distinctive linear ridges called yardangs (from a Turkistani word meaning "ridge") occur in some desert regions. Typical yardangs have the form of an inverted boat hull and commonly occur in groups, all aligned with the prevailing wind. This shape and orientation are streamlined by the wind to offer minimum resistance to the moving air. Yardangs develop best in soft sediment that is easily eroded but that is cohesive enough to retain steep slopes.

The most extensive areas of wind-transported sand are the ergs, or great sand seas, that occur in the major deserts of the world. Ergs may cover more than 0.5 million square kilometers: twice the size of the state of Nevada. It has been calculated that 99.8 percent of all windblown sand is in the ergs of the world. The largest are in Africa, Asia, and Australia. One third of Saudi Arabia—about 1 million square kilometers—is covered with ergs in the vast "empty quarter" of the Rub 'al Khali.

SIMPLE DUNES

Individual dunes in an erg form and move as many different mounds or ridges of sand. Generally, dunes form where an obstacle distorts the flow of air. On encountering an obstruction, wind sweeps over and around it but leaves a pocket of slower-moving air directly behind the obstacle. As sand is blown up or around the obstruction and into the protected wind shadow, its velocity is reduced, and deposition occurs. Once a dune is formed, it acts as a barrier itself, further disrupting the flow of the wind and causing sustained deposition downwind. Continued erosion on the windward side and redeposition on the leeward side can produce rates of dune migration of as great as 25 meters per year. As a dune grows larger, the sand grains saltate up the low-angle windward slope to the top, then slip down into the wind shadow. The slip face keeps a more or less constant angle of repose and creates the high-angle crossbedding so typical of dunes. Dunes can range in size from simple ones a few meters high to as much as 250 meters.

In an attempt to make some order out of immense diversity, geologists have lumped dunes into several major types, though there are gradations among them as well as many irregular shapes that are hard to fit into any scheme. The basic classification of dunes is threefold: simple, compound, and complex draa dunes.

Simple dunes consist of arrangements of different dune forms relating to one or more wind directions and different amounts of available sand. Simple dunes may be convex or concave, and they may be transverse or longitudinal to the wind.

COMPOUND AND COMPLEX DRAA DUNES

Compound dunes are combinations of simpler forms without a change of scale or size of the individual component dunes. The solitary crescent-shaped dune, the barchan, which moves over a flat surface of pebbles or bedrock, is the product of a limited sand supply and winds of moderate velocity that blow in a constant direction. Typically, they are small isolated dunes from 1 to 50 meters high. The tips of a barchan point downwind, and sand grains are swept around them as well as up and over the crest. The steep slip face occurs inside the crescent. Transverse dunes typically develop where there is an abundant supply of sand and a constant wind direction. These dunes develop a wavelike form, with sinuous ridges and troughs that extend perpendicular to the prevailing wind.

Longitudinal, or seif (an Arabic word meaning "sword"), dunes are long, parallel ridges of sand; that is, they elongate in a direction parallel to the resultant of several slightly different wind directions. These dunes develop where strong prevailing winds converge and blow over an area having a limited supply of sand. The grains are shepherded into a long ridge by the blowing of the wind, first in one direction across and along the dune and then from the opposite side of the dune. Many longitudinal dunes are less than 4 meters high, but they can extend downwind for several kilometers. In large deserts, they can be more than 100 meters high and 120 kilometers long; they can be spaced 0.5 to 3 kilometers apart.

Parabolic, or blowout, dunes develop out of preexisting, partly vegetated and thereby stabilized dunes. In these dune forms, first a deflation depression develops, and the sand is piled on the downwind edge of the oval depression. As the shallow deflation hollow enlarges, the sand piles up to form a crescent-shaped ridge. In map view, a parabolic dune is similar to a barchan, but the tips of the parabolic dune point upwind around the deflation hollow from which they originate. Where greatly elongated, parabolic dunes have been called hairpin dunes, again with the extended tips pointing upwind.

Complex draa dunes have dimensions of hundreds of meters, spacings of kilometers, and, because of their great bulk, appear to be relatively static. They are thought to be manifestations of atmospheric turbulence on a distinctly larger scale than that manifested by simple dunes, involving interaction between the regional winds and massive, relatively unchanging sand forms. Barchanoid,

transverse, longitudinal, and peaked forms all occur, but in masses that can be more than one hundred times larger than the simple forms. Draa dune fields are characteristic of thick sand accumulations in extremely arid zonal deserts, notably the Sahara and the deserts in Saudi Arabia.

DUST STORMS AND LOESS

Dust storms are a major process in deserts. They can transport thousands of tons of fine sediment high in the atmosphere for hundreds of kilometers. These processes have long been known as sandstorms, but, in fact, far more dust is carried aloft than is sand close to the ground. Great dust storms can reach altitudes of more than 2,500 meters above the desert floor and advance at speeds of up to 200 meters per second. Perhaps 500 million tons of windblown dust are carried out of deserts each year, about the same amount carried annually by the Mississippi River. Large quantities blow out into the oceans from Australia or Africa, and some even move across the Atlantic Ocean from the Sahara to the east coast of South America.

Windblown dust is deposited as loess (a German word meaning "loose") and is defined as wind-deposited silt and clay; loess accumulates slowly and ultimately blankets large areas, commonly masking preexisting landforms. Such deposits may cover as much as one-tenth of the world's land surface and are particularly widespread in semiarid regions along the edges of the world's great deserts. Large deposits can reach 300 meters in thickness. Loess is a quite distinctive sedimentary deposit. Most loess is massive and lacks layering, apparently because grains of different sizes settled progressively from the air and were deposited at random. Where exposed, loess commonly stands in steep cliffs because the molecular attraction between the very fine grains is enough to make the particles quite cohesive. One of the primary differences between loess deposits and sand-dune deposits is that sand continues to be mobile even while it is on the ground, whereas loess dust is cohesive and stable once deposited. Loess also tends to settle out on semiarid grasslands or occasionally on woodland, which are themselves not active areas of further wind erosion. Once deposited, therefore, loess tends to remain in place and can continue to accumulate to great thicknesses over time; loess makes very fertile soil.

STUDY OF PROCESSES AND PHENOMENA

Running water and wind are the primary processes of dry climates, but significant weathering, gravity-driven slope failure, mass wasting, and even cold climate processes occur. With this wide range of landform-controlling factors, study of the relevant processes involves a nearly complete range of geological and geographical techniques. Aside from plentiful climate measurements of different types, sediment movement is monitored, weathering characteristics described and measured, and landform types and distributions mapped.

In order to develop an understanding of the historical dimension of landform development in dry climates, considerable effort is expended by scientists to develop long-term chronologies. This goal is achieved through measurement of landform morphologies that show progressive erosion and deposition development through time, elucidation of the nature of varied sedimentary layers through stratigraphic analysis, and radiometric and relative-age dating techniques to fix the chronologies.

SIGNIFICANCE FOR HUMAN POPULATIONS

Dry climate geomorphology is very important because so much of the world's population lives in such environments and must contend with the extremes of climate and geomorphic process there. Agriculture is particularly difficult in such arid situations. In the United States, much of the dry agricultural grassland and shrubland of the west is characterized by a special suite of arid landforms. Some of the dry desert basins of the southwest are important sources of exotic salts, and their different terrains have special significance to the military and space agencies. For example, playa salt deposits, formed by repeated filling and evaporation of the lakes, can be tens of meters thick and a valuable source of industrial chemicals. Borax, the active ingredient in some scouring powders, is but one example. Dry playas are also extraordinarily flat surfaces and have been used for a long time as huge landing fields for rockets and spacecraft such as a shuttle. In addition, the largely cloud-free and windy dry lands are also important sites for solar-power and wind-power generation.

John F. Shroder, Jr.

CROSS-REFERENCES

Alluvial Systems, 2297; Continental Crust, 560; Continental Growth, 573; Continental Rift Zones, 579; Desert Pavement, 2319; Desertification, 1473; Displaced Terranes, 615; Earth's Crust, 14; Earth's Mantle, 32; Earth's Oldest Rocks, 516; Earthquake Distribution, 277; Evaporites, 2330; Floods, 2022; Geomorphology and Human Activity, 898; Geomorphology of Wet Climate Areas, 909; Glacial Landforms, 914; Grand Canyon, 919; Karst Topography, 925; Marine Terraces, 929; Permafrost, 933; Plate Margins, 73; Plate Tectonics, 86; Precipitation, 2050; River Valleys, 936; Rock Magnetism, 177; Sand Dunes, 2368; Sediment Transport and Deposition, 2374; Sedimentary Mineral Deposits, 1637; Seismic Reflection Profiling, 371; Soil Erosion, 1513; Solar Power, 1674; Stratigraphic Correlation, 1153; Weathering and Erosion, 2380; Wind, 1996.

BIBLIOGRAPHY

Abrahams, A. D., and A. J. Parsons, eds. *Geomorphology of Desert Environments*. London: Chapman and Hall, 1994. The essays in this volume deal with a broad range of topics relating to dry climates. Intended for the specialist but readable by those with some science background.

Bloom, Arthur. *Geomorphology: A Systematic Analysis of Late Cenozoic Landforms*. 3d ed. Englewood Cliffs, N.J.: Prentice Hall, 1998. This college-level text covers the basics of geomorphology. Includes a chapter on arid and savanna landscapes. Index and bibliography.

Chorley, Richard J., S. A. Schumm, and D. A. Sugden. *Geomorphology*. New York: Methuen, 1984. One of the best but more difficult general texts available. Nevertheless, the discussions on arid land phenomena are current, the graphics clear, and the references include later sources. The discerning student of general geomorphology could start with this volume as a guide to the major issues of dry climate geomorphology and then move on to other more detailed, but less rigorous, works.

Lancaster, N. *Geomorphology of Desert Dunes*. London: Routledge, 1995. Lancaster provides an in-depth treatment of desert topography. Technical in some areas. Index and bibliography.

Mabbutt, J. A. *Desert Landforms*. Cambridge, Mass.: MIT Press, 1977. This volume is one of the premier works on desert and desert landforms, with worldwide examples. The emphasis is on landforms and the processes by which they are produced. The style is clear, and the plentiful illustrations are generally well done.

Ritter, Dale F. *Process Geomorphology*. 2d ed. Dubuque, Iowa: Wm. C. Brown, 1986. The author presents useful information about all the processes that act on the Earth's surface, including those that act in dry climate areas. The chapter on wind processes and landforms is clear and well written, but not many additional references are cited.

GEOMORPHOLOGY OF WET CLIMATE AREAS

Wet climate geomorphology is the study of landforms and landforming processes in the humid regions of the Earth. These landforms correspond roughly to tropical rain forest, savanna, and humid-midlatitude climatic regions. Each of these landforms is distinct because of differences in temperature and precipitation.

PRINCIPAL TERMS

CHEMICAL WEATHERING: the chemical alteration of rocks and minerals into new forms that are chemically stable at the Earth's surface

EROSION: the removal of weathered rock and mineral fragments and grains from an area by the action of wind, ice, gravity, or running water

HUMID-MIDLATITUDE: land area with average temperature of the coldest month less than 18 degrees Celsius but at least eight months with average monthly temperatures greater than 10 degrees Celsius; this area has no dry season

LANDFORM: a grouping of genetically related landscapes based on similarity of appearance

MECHANICAL WEATHERING: the physical breakdown of rocks into progressively smaller particles

REGOLITH: the layer of mechanically and chemically weathered rock and mineral matter above fresh, unweathered bedrock

SAVANNA (TROPICAL WET-DRY): land area with monthly temperatures averaging greater than 18 degrees Celsius and with one to six months of average monthly precipitation of less than 6 centimeters

TROPICAL RAIN FOREST: land area with monthly temperatures averaging greater than 18 degrees Celsius and monthly precipitation averaging more than 6 centimeters

CLIMATIC GEOMORPHOLOGY

Geomorphology is a branch of geology and physical geography concerned with the study of the Earth's landforms and the processes that create them. There are several ways of looking at those processes and, as a result, several branches of geomorphology. One is climatic geomorphology, which is the study of the role of climate in shaping the land. Wet climate geomorphology is a component of climatic geomorphology that exclusively addresses land areas having relatively high annual precipitation. This is a natural grouping, since water is instrumental in the wearing down and sculpting of the land.

WEATHERING AND EROSION

Before looking specifically at the landforms and processes of the humid regions, it is first necessary to examine the major natural processes operating on the land, no matter what the climate. These are weathering, transport, and deposition of rock and mineral matter. The first two processes are sometimes collectively referred to as erosion, although a more accurate definition of erosion would be the removal of rock and mineral matter from an area. Weathering may be either mechanical, the physical breakdown of rocks into smaller and smaller particles, or chemical, the chemical alteration of rocks and minerals into new, more stable forms that may become part of a soil. The most important type of mechanical weathering is freeze-thaw, in which freezing water expands within rocks and fractures them. Mechanical weathering, as well as transport of weathered debris, is suppressed in hot, wet climates, where there is little or no freezing and a thick vegetation cover protects the soil from being removed and carried away (eroded). In temperate climates such as Europe and the eastern United States, mechanical weathering is somewhat more important and produces huge amounts of loose rock and mineral particles that can be eroded.

Chemical weathering is most active in tropical wet regions, where both temperature and precipitation are high. It occurs only in the presence of abundant percolating subsurface water and breaks down many common rock-forming minerals into three components: clays, quartz, and dissolved particles (ions). The clay and quartz remain behind, unless eroded, and form a residue from which soils develop, while the ions dissolved in the subsurface water may recombine to form new minerals such as iron oxide. They may also be used by plants or be transported downward in percolating water and eventually deposited into streams and rivers. Some common minerals, and therefore the rocks that contain them, are resistant to chemical weathering and do not erode easily.

Transport of weathered rock and mineral matter occurs primarily by running water, either as sheet flow (when an unchanneled sheet of water flows down a slope) or as channeled flow (in gullies, streams, and rivers). Where the protective cover of vegetation is thick, sheet flow is ineffective as a transporting agent. Where it is thin, sheet flow can erode the underlying soil as fast or faster than it can form. Where very steep slopes exist, landsliding becomes an important transporting agent. Deposition of weathered and transported debris occurs when the velocity of the running water decreases: at breaks in slope or in the floodplains of rivers, for example.

MORPHOCLIMATIC ZONES

Wet climate landforms can be subdivided into three types based on temperature and the seasonal amount and distribution of precipitation. These landforms, or "morphoclimatic zones," are tropical rain forests, where both temperature and precipitation remain high all year long; savannas, or tropical wet-dry regions, where both temperature and precipitation are high but precipitation is restricted to a rainy season; and humid-midlatitude regions, where precipitation is moderate but uniformly distributed throughout the year and temperatures undergo large seasonal fluctuations.

TROPICAL RAIN FORESTS

The tropical rain forests of the world are confined to a belt extending 10 to 15 degrees of latitude north and south of the equator. Examples are the Amazon River Basin in South America, much of central Africa, and the Indonesian Islands. In these areas, high temperatures and rainfall produce a thick vegetation cover even on steep slopes.

As a result, sheet flow is minor, and most rainwater infiltrates the ground, where it promotes chemical weathering. In these areas, mechanical weathering is inhibited, but chemical weathering is intense and produces a deep mantle of weathered rock and mineral grains above fresh bedrock. This mantle of unconsolidated debris is called regolith and may extend more than 100 meters below the land surface. The downward percolation of subsurface water is so intense and constant that the dissolved ions that it carries are completely removed, leaving a residue of clay, quartz, and iron oxides. Soils in tropical wet areas therefore tend to be infertile below the top meter or so.

In mountainous parts of the tropical rain forest, where fast-moving streams have high energy, streams cut rapidly down through the easily eroded regolith and produce deep, narrow valleys with steep, V-shaped slopes separated by narrow ridges. The regolith on these slopes sometimes becomes saturated with water, resulting in landslides. This type of topography is typical of areas such as Papua New Guinea and parts of the Amazon basin.

In contrast with the rugged topography of mountainous regions, lowlands in the tropical rain forest are quite flat. They contain large rivers such as the Amazon in Brazil and the Congo in central Africa. These rivers meander on wide floodplains tens of kilometers wide and thousands of kilometers long. Many rivers have a peculiar stepped appearance, with occasional rapids or even waterfalls interrupting long stretches of sluggish flow. Tropical rain forests contain some of the world's most spectacular waterfalls, such as Angel Falls in Venezuela (900 meters high) and Victoria Falls in Zambia. The waterfalls are produced by resistant rock layers, such as quartzite, that do not chemically weather and that the rivers cannot wear down given the lack of coarse abrasive particles in transport. Another unique landform feature of the tropical rain forest is the occasional presence of large, isolated, haystack-shaped hills devoid of vegetation. These are called monolithic domes, and they present a curious contrast to the surrounding heavily vegetated lowlands. A number of these domes are found around Rio de Janeiro in Brazil. Geomorphologists believe that monolithic domes exist because they are more resistant to weathering than the surrounding rocks. Soils cannot form on them because of steep slopes, landsliding, and lack of the water-retention capacity necessary to promote chemical weathering.

SAVANNA REGIONS

Savanna regions are much larger in area than the tropical rain forests and extend north and south as a wide fringe bordering the tropical rain forests. Savanna regions include parts of southern and northern Brazil, southern Central America, central and southern Africa, and parts of India and Southeast Asia—areas containing more than half the world's population. Savannas have a dry season lasting from three to six months. During the dry season, vegetation dies or becomes dormant and chemical weathering decreases significantly as a result of the lack of percolating water. With a decreased vegetative cover, mechanical weathering and erosion of weathered debris by sheetwash increase significantly, resulting in removal of the regolith faster than it can form.

Savanna landscapes therefore tend to have a stepped appearance, with vast flat or gently sloping plateaus (pediplains) with thin regolith or none, interrupted by abrupt slopes that lead up to the next higher plateau. This style of erosion is called backwasting and is typical of savanna and semiarid regions. Former high plateaus may be reduced by erosion to isolated remnants called inselbergs. The stepped appearance is augmented by the formation of hard, impermeable layers in the soil called crusts or duricrusts. They form progressively during dry seasons, when percolating subsurface water evaporates and chemically precipitates oxides of iron or, less commonly, aluminum. Once formed, these crusts persist and may grow to several meters in thickness. They protect the underlying, less resistant regolith from erosion unless they are breached, in which case erosion proceeds rapidly by undercutting them, sometimes all the way down to bedrock. In this way, the crusts help preserve flat plateau surfaces at various levels, although the surfaces are progressively reduced in area through time.

When stream flow resumes in the wet season, streams are unable to cut down through the crusts and therefore have very shallow, hard-to-recognize channels. Much rainwater runs over the plateau surfaces as sheetwash, carrying with it fine particles of regolith. When flow ceases, areas of shallow ponded water serve as gathering places for animals such as elephants, lions, and wildebeests. The well-known Serengeti Plain in Tanzania, Africa, is an excellent example of this kind of landscape.

HUMID-MIDLATITUDE REGIONS

Landforms in humid-midlatitude regions such as Europe and the eastern United States differ in many respects from the tropical rain forest and savanna regions. Here, the cool winters promote mechanical weathering via freeze-thaw, chemical weathering is moderate because of cooler temperatures, and precipitation is moderate but evenly distributed throughout the year. Streams therefore run all year long and carry higher loads of coarse sediment, such as sand and gravel, that chemical weathering has not reduced to smaller sizes. These mechanically weathered sediments act as abrasive tools and wear down even the harder rocks. Stream courses therefore do not have the stepped appearance seen in the tropical rain forest. They cut valleys whose slopes recede via landslides and, more important, soil creep, the very slow downslope movement of the moist soil.

Soil creep helps to produce the rounded hills and gentle, curved slopes typical of humid-midlatitude areas such as the Appalachian Highlands of the eastern United States or the gently rolling landscapes of parts of Western Europe. Slopes grade uniformly into lowland floodplains, where coarse sediment is stored. Isolated remnants such as monolithic domes or inselbergs are quite rare in humid-midlatitude regions. Since chemical weathering is only moderately active, soils tend to be thin but rich in plant nutrients because the downward percolation of water is not intense enough to cause complete removal of dissolved ions necessary for plant growth. Humid-midlatitude regions are therefore ideal areas for agriculture.

TOPOGRAPHIC MAPS

The primary interest of the geomorphologist is to determine the nature of the important causative processes acting to form the landscape and to construct predictive models of landscape change based on a knowledge of those processes and their rates. A variety of methods is used for this purpose, and specific techniques as well as analytical tools are usually dictated by the problem to be solved. Most geomorphological analyses involve measurement of some landscape feature, such as slope steepness, and observation of the natural processes acting to form it, such as the amount of precipitation, underlying rock type, or type of vegetation cover. The effect of each process must be measured and its relative importance established. Almost all such analyses involve statistical treatment.

Topographic maps, aerial photographs, and geologic maps are essential for any geomorphological study. Topographic maps are two-dimensional representations of the topography of an area. They allow geomorphologists to study landscape features without actually being at a site, and they lend themselves to statistical treatment. For example, the relation between slope steepness and the frequency of landslides can be determined quantitatively because both features have a topographic expression and can be recognized on topographic maps. Such a study may conclude that slopes greater than 40 degrees have 20 percent more chance of landslides than slopes less than 40 degrees. This rather simple example nevertheless has great practical importance to engineers working in wet climate areas.

AERIAL PHOTOGRAPHS

A series of aerial photographs taken at a certain elevation—for example, at 3,000 meters—can be partially overlapped and, when viewed with stereoscopic glasses, can give an exaggerated three-dimensional view of a land area. The photographs therefore present a visual image of the topography with more detail than can be provided by topographic maps.

Photographs taken at different times can be compared in order to determine rapid landscape changes and their rates of change. This method is often used in studies of river-channel migrations across a floodplain. The information can then be used by engineers and land planners to determine future areas of urban expansion.

GEOLOGIC MAPS AND OTHER STUDY TOOLS

Geologic maps are two-dimensional representations of the types and orientations of rocks or surficial materials in an area. They allow geomorphologists to study the relation between the underlying rock or soil and the topography as expressed on topographic maps or aerial photographs. For example, in the humid-midlatitude Valley and Ridge

area of Pennsylvania, elongate ridges stand up to 1,000 meters above intervening valleys. The ridges are composed of resistant, hard sandstone, whereas the valleys are underlain by shale, a weaker and more easily eroded rock.

Other investigative tools have a more specific purpose. These include sieves to measure grain sizes of debris, petrographic microscopes to determine mineral types, and X-ray diffractometers to determine types of clay minerals. In addition, various techniques are used by geomorphologists to determine the age of landscape-forming events. The techniques used are quite sophisticated and include carbon 14 dating, fission track dating, amino acid racemization, and dendrochronology. Finally, computers are essential tools in geomorphological studies. They are used extensively as aids in statistical analyses and for purposes of modeling geomorphic processes and predicting landscape changes.

APPLICATIONS OF GEOMORPHOLOGICAL RESEARCH

The geomorphological study of wet climate landforms has tremendous practical importance through its ability to predict the effects of human-induced changes in the landscape. For example, pressures for development in nations occupying the tropical rain forest have resulted in massive forest-clearing operations that threaten the very existence of these forests. One effect has been a tremendous increase in erosion of the soil and underlying regolith as a result of the high precipitation and stripping of the protective vegetation cover by sheetwash. Parts of the Amazon basin in Brazil now have a badlands type of topography. Crops will not grow in the remaining bare, infertile regolith, because seeds and fertilizer simply wash away with the regolith. The eroded debris rapidly clogs formerly clear-flowing streams and endangers numerous aquatic plant and animal species. Tropical geomorphologists are now involved in projects to reclaim much of this land and to prevent future landscape degradation.

Many applications of geomorphological research in wet climate areas have to do with the dynamics of river channels and river erosion. Engineering works such as reservoir construction and river channelization have had some unanticipated side effects, notably the severe erosion of land downstream from impoundments or increased flood heights downstream from channelized rivers. Geomorphologists have been able to model these river responses to change and have enabled engineers to construct safer and more effective flood control and water supply works.

A new challenge for geomorphologists is the reclamation of disturbed lands such as old landfills and open pit mines. Proper reclamation is based on an understanding of the landforming processes and how they operate in different climatic regions. With this understanding, geomorphologists have had much success in converting potentially hazardous areas to environmentally safe and economically valuable uses such as golf courses, parks, and playgrounds.

SCIENTIFIC VALUE OF CLIMATIC GEOMORPHOLOGY

In addition to its practical value, climatic geomorphology has great scientific value. Because climate exerts strong controls on the landforming processes of weathering, transport, and deposition, climatic geomorphologists can recognize large areas of the Earth where past climates were different from the present one. For example, the currently humid-midlatitude region of central Germany retains remnants of flat plateau surfaces formed under a savanna climate. Many, if not most, land areas of the Earth show such polygenetic landscapes (landscapes having more than one origin). This information is used to establish a history of climate change through the last 70 million years.

A new and rapidly developing area of geomorphology is the study of planetary landforms. Pictures of the Martian surface clearly show valleys similar to those cut by rivers on Earth, yet there is almost no surface water on Mars today. Does this mean that Mars has somehow lost its water, or could some other agent of erosion such as wind have produced the valleys? Current geomorphological research is trying to answer this question, as well as others regarding the formation of planetary landforms.

Philip A. Smith

CROSS-REFERENCES

Aerial Photography, 2739; Climate, 1902; Fission Track Dating, 522; Floodplains, 2335; Geomorphology and Human Activity, 898; Geomorphology of Dry Climate Areas, 903; Glacial Landforms, 914; Grand Canyon, 919; Karst Topography, 925; Marine Terraces, 929; Permafrost, 933; Precipitation, 2050; Radiocarbon Dating, 537; River Valleys, 936; Sediment Transport and Deposition, 2374; Soil Chemistry, 1509; Soil Erosion, 1513; Soil Formation, 1519; Soil Types, 1531; Weathering and Erosion, 2380.

BIBLIOGRAPHY

Bloom, Arthur. *Geomorphology: A Systematic Analysis of Late Cenozoic Landforms.* 3d ed. Englewood Cliffs, N.J.: Prentice-Hall, 1998. This college-level text covers the basics of geomorphology. Includes a section on wet climates, with particular emphasis on weathering. Index and bibliography.

Marsh, William M. *Earthscape: A Physical Geography.* New York: John Wiley & Sons, 1987. An introductory text that can serve as a good starting place for readers who are interested in climatic geomorphology but who have no prior training. Of special relevance are chapters 8 and 9 (climate), 10, 11, and 12 (vegetation), 13 and 14 (soils), 16 (rock structures), and 18, 19, and 20 (weathering and erosion). Suitable for upper-level high school or college-level readers.

Skinner, B. J., and S. C. Proter. *The Dynamic Earth: An Introduction to Physical Geology.* New York: John Wiley & Sons, 1989. A revised physical geology text with an expanded section on landforming processes. It can provide the reader with the necessary geological background to pursue more advanced knowledge in climatic geomorphology. Suitable for upper-level high school or college-level readers.

Thomas, M. F. *Geomorphology in the Tropics: A Study of Weathering and Denudation in Low Latitudes.* New York: John Wiley and Sons, 1994. This detailed and technical text is intended for specialists. Extensive bibliography.

GLACIAL LANDFORMS

Glacial landforms are common in many parts of Canada, the northern United States, northern Europe, Asia, and many mountain ranges of the world. The material composing these landforms, as well as the shapes of some of the landforms, is essential to many human activities.

PRINCIPAL TERMS

ARÊTE: an extremely narrow mountain ridge created between adjacent U-shaped valleys

CIRQUE: a bowl-shaped depression near mountaintops where glaciers originate

DRUMLIN: a streamlined hill formed under actively moving ice

ESKER: a sinuous ridge of stratified drift formed in a tunnel under the ice

KAME: a conical hill of stratified drift

KETTLE: a depression created by the melting of a chunk of ice

MORAINE: landscapes of till, varying from fairly flat terrain to gently rolling hills to long ridges

STRATIFIED DRIFT: material deposited by glacial meltwaters; the water separates the material according to size, creating layers

TILL: a mixture of unsorted, unconsolidated materials deposited by glacial ice

GLACIAL MOVEMENT

Glacial landforms are distinguished primarily on the basis of shape and composition, both of which relate directly to the mode of origin of the glacial features. Some landforms originate by erosion or deposition directly by the ice, while other features are created by erosion or deposition of the meltwaters from the ice. It is important to distinguish whether the ice or its meltwaters produced the landform. Another factor in recognizing glacial landforms is the realization that many glacial features are commonly associated with a particular kind of glacier. Two types of glacier are commonly recognized: continental glaciers and valley glaciers. Although some landform features are common to both types of glacier, others are relatively unique to each type.

An understanding of glacial ice movement is also necessary in order to understand how glacial landforms develop. Whether the ice is moving because of pressures from great thicknesses of continental ice or is responding to gravity down a valley slope, the ice always moves forward. Even when the glacier is melting back, and it looks like it is getting shorter or smaller, internally the ice is still moving forward.

Glaciers can erode the material over which they move, and if they scrape down to solid bedrock, several glacial features can result. If the glacier is carrying small, hard particles along its base, these particles may scratch the bedrock, creating many roughly parallel striations. If large rocks are being transported at the base of the ice, they can gouge out varisized grooves, which can range from a few centimeters to several meters deep. On the other hand, if the ice has ground its load into flour-size particles, called rock flour, the flour can actually polish the bedrock. The orientation of striations and grooves can be used to determine ice-flow direction.

EROSIONAL LANDFORMS

Of all the glacial landforms, perhaps the most spectacular are the erosional products of mountain glaciers. Mountains that have experienced glaciation exhibit some of the world's most breathtaking scenery. Glaciers tend to sharpen peaks and ridges and to steepen valley walls, producing avenues for numerous spectacular waterfalls. As snow accumulates near the tops of mountains, the forming ice tends to carve out a basin-shaped depression on the side of the mountain. This basin is called a cirque and is the home base of the glacier; it is in this basin that more snow will accumulate, thickening the ice, until the ice begins to move out of the cirque and down the valley. If three or more glaciers form around the same mountain peak, the crowding of their

An arête with an elevation of 13,694 feet near Gilpin Peak in the San Juan Mountains, Ouray County, Colorado. (U.S. Geological Survey)

cirques around the peak will erode it into a sharper than usual pyramidal shape. This type of mountain peak is called a horn. Many of the peaks in the Alps have the word "horn" incorporated into their names; the Swiss Matterhorn is probably the most famous glacial horn. If two glaciers are moving down a mountain in adjacent valleys, the ridge between them may be eroded into a narrower, sharper ridge called an arête.

Mountain glaciers tend to widen the base and steepen the sides of the valleys through which they move, creating obviously U-shaped valleys. Large glaciers have more power to erode more deeply, resulting in deeper U-shaped valleys. Small tributary glaciers produce valleys that are not carved as deeply; therefore, when the glaciers retreat from such an area, the floors of the tributary valleys are left high above the floor of the main valley. Such valleys are appropriately called hanging valleys. If streams are present, they become waterfalls that cascade over the edges of the hanging valleys. U-shaped valleys usu-

ally contain strings of lakes that are impounded by irregularities of the valley floor. These lakes are quite picturesque in that the rock flour they contain from glacial meltwaters gives the water a distinctive turquoise color.

Perhaps the most spectacular of the glacial valleys are fjords, most common in Norway, British Columbia, Alaska, and Chile. A fjord is a coastal glacial valley that was carved out when sea level was lower during times of glaciations. Upon retreat of the ice and subsequent rise in sea level, seawater completely inundated the valley. The water in many fjords is often more than 1,000 meters deep. In some areas where fjords extend inland for considerable distances and intersect with other fjords, it may be difficult to distinguish a fjord from a large lake in a U-shaped valley. The fjord, being inundated with marine water, will contain seaweed and experience tidal changes; the lakes will be freshwater and relatively static.

A fjord at the Narrows in Fords Terror, Petersburg district, southeastern Alaska. (U.S. Geological Survey)

GLACIAL DEPOSITS

Glacial deposits are widespread and create interesting landform features. Glaciers deposit till, an unconsolidated mixture of clay, sand, silt, pebbles, cobbles, and boulders. Depending on the area over which the glacier moved, the relative abundance of these components will vary; some till can be very sandy, while other till can be clayey. Abrasion during ice transport causes the particles in till to be usually angular in shape. Because it is the glacial ice itself that deposits this material, the till is unsorted. This lack of sorting results because the materials in till are let down to the surface as the ice melts in the same relative positions they had in the ice. Till can be formed into various types of morainal landforms, depending on how the glacial ice moved. Till can be deposited as ground moraine as the ice retreats. It is laid down under the ice as the ice moves along, resulting in a landscape called a till plain, which usually takes on the appearance of gently rolling, hummocky hills.

When a glacier front becomes stationary, material will continue to be moved forward through the internally forward-moving ice and be deposited along the front of the stationary glacier. It will build up into a ridge called an end moraine. If the glacier then advances, the end moraine will be destroyed by the moving ice. If the glacier retreats, the end moraine will become part of the landscape. The end moraine that is created at the maximum forward position of the ice is called the terminal moraine. Any end moraines that form as the ice periodically retreats are called recessional moraines and usually form concentric bands of ridges in the landscape. If a glacier advances and then retreats without spending any significant length of time in a stationary position, end moraines cannot form; the till will be formed into ground moraine.

Two types of moraines are unique to mountain glaciers. Because these glaciers are restricted to valleys, accumulations of debris tend to gather along the valley walls at the sides of the glaciers. These are called lateral moraines and are left behind as long ridges along the valley walls as the ice melts away. A medial moraine forms when two glaciers join and their lateral moraines are consequently trapped between the two ice rivers, which are now flowing along as one glacier. Several medial moraines can form as more and more glaciers from tributary valleys join, giving the glacier a striped appearance.

MELTWATER

Glacial meltwater is an important agent in creating several landform features. The power of glacial meltwater can be extraordinary, often carving deep gorges. The Channeled Scablands of Washington attest the effects of glacial meltwaters. When glaciers melt, they release not only water but also the huge amounts of debris they carry. The material deposited by glacial meltwater is called stratified drift and differs from till in that it is sorted and layered. The heavier particles drop out first, followed by progressively smaller particles as the water travels farther from the ice, gradually slowing down. Fine clays and silts are usually transported the farthest, sometimes completely out of the area, giving the stratified drift its distinctive sand and gravel characteristics. As the glaciers eventually release less and less water, smaller and smaller particles are deposited over the coarser material deposited previously. Unlike the ice-deposited till, transport in water tends to round the particles of stratified drift.

Outwash plains are huge, flat areas of stratified drift located adjacent to where glaciers existed. If outwash is confined to a river valley, it is called a valley train. Large outwash areas are usually devoid of vegetation cover at first and are subjected to strong winds. Often silt-sized particles in the outwash are transported by wind and deposited in thick layers downwind from the deposit. Wind-blown silt is called loess and is an important ingredient in soils. Soils developed on loess-capped till plains are particularly good for agriculture.

Meltwater sands and gravels can form interesting landscape features if conditions of the melting glacier are just right. If the area near the snout of the glacier becomes too thin, the ice will cease to move internally and become stagnant. When that happens, the stagnant ice may be eroded by its own meltwater as well as that coming off the active ice. Meltwater streams may form on the ice and drill tunnels in and under the ice. The movement of the streams determines what type of landform feature will evolve from the deposition of the sands and gravels transported by the stream. For example, if streams in or on the ice cascade down fairly vertical shafts (called moulins) in the ice or tumble down a fairly steep face of the ice edge, the sand and gravel will spill into a conical pile called a kame, which can be more than 60 meters high. Usually several kames are formed in the same area, dotting the landscape as isolated hills.

Another example of meltwater stream deposition is eskers, which also form in zones of stagnant ice where sand and gravel are deposited within stream tunnels that are in and under the stagnant ice. When the ice melts away, long, sinuous ridges are left in the landscape. New England has some impressive eskers that are more than 40 meters high in places and wind for miles through the countryside.

A kettle, or glacial pothole, near Tuolumne Meadows in Yosemite National Park. (U.S. Geological Survey)

Other features commonly associated with outwash and stagnant ice areas are kettles, which form when chunks of ice break off from the glacier and become buried in outwash or in till as the glacier melts. Gradually the ice chunks melt, leaving depressions in the ground, sometimes in the shape of pioneers' kettles, which is how the name originated. Kettles that intersect the groundwater table or collect precipitation contain water and are called kettle lakes, some of which can cover several acres. An outwash area that contains numerous kettles is called pitted outwash.

OTHER LANDFORMS

Actively moving ice sometimes has the ability to form streamlined shapes through either erosion or deposition. A roche moutonnée is an eroded bedrock form with a gentle up-ice side and a steep, rough down-ice side. Hence, ice-flow direction can be determined by the orientation of roches moutonnées. These features vary in size from 1 meter to large hills in the landscape.

Drumlins are streamlined hills that look like half-buried whales or inverted spoons composed of a variety of materials. The cores can be composed of bedrock or, more commonly, stratified drift and are covered with a veneer of till. The formation process of drumlins is unclear, but agreement on some aspects does exist. They appear to have been formed by actively moving ice near, but not at, the edges of the ice. Drumlins always appear in groups, called drumlin fields, composed of thousands of drumlins. The most famous drumlin fields are in New York, Wisconsin, and New England.

The Great Lakes exist because of the action of continental ice sheets in the Midwest. Tongues of ice gouged out old river valleys, creating the lake basins. Meltwater and precipitation filled the basins. The history of the glacial Great Lakes from the time of the ice sheets to the present day is extremely complicated. As the ice retreated and made minor advances in various parts of the upper Midwest, different sizes and shapes of lakes emerged in each basin. The lakes had different outlets at different times, creating different water levels. Most of the old lake levels can be traced today because old beach and dune deposits exist at various elevations along the present-day Great Lakes shorelines. The old clay-rich lake beds that cover large parts of Ohio, Indiana, and Illinois have some of the richest soils in the United States.

STUDY OF GLACIAL LANDFORMS

Glacial landforms are studied using a variety of methods. The general forms of the features are analyzed using topographic maps and various types of aerial photographs. Such tools not only illustrate the forms of each glacial feature but also show its relationship to other nearby features. Patterns in the landscape can be observed quite easily, and ice-direction movement may be interpreted. Topographic maps are very detailed, showing all natural and human-made features in a small area. These maps also include contour lines that represent elevations above sea level; with some practice, hills and valleys can be identified. All landforms that are large enough to be incorporated by the given contour interval can be shown on the map. All kinds of glacial features can be identified using topographic maps because their shapes are distinctively outlined by the contours.

Aerial photographs can be used in the same way as topographic maps, except that with photographs, the actual feature in the landscape can be seen. There are several types of aerial photographs that can be used for analysis, depending on what features are being studied. Standard black-and-white, color, and a variety of other wavelength bands (from satellite imagery) each emphasize distinct characteristics of the landscape.

Recall that a diagnostic characteristic for many glacial landforms is their composition. The composition of any particular feature is determined by digging into it. Sometimes quarries or roadcuts have already exposed the interior of the structure, making the job of the geologist easier. At other locations, core samples are obtained by drilling. Sometimes well records from already existing water wells may be utilized to determine what kind of material has been deposited in an area. Another clue to the composition of a feature may be obtained by the type of soil that has developed on the material. County soil surveys are useful for this kind of analysis.

On a more detailed scale, geologists may want to understand the mechanics of ice flow when determining the creation of the landform, the direc-

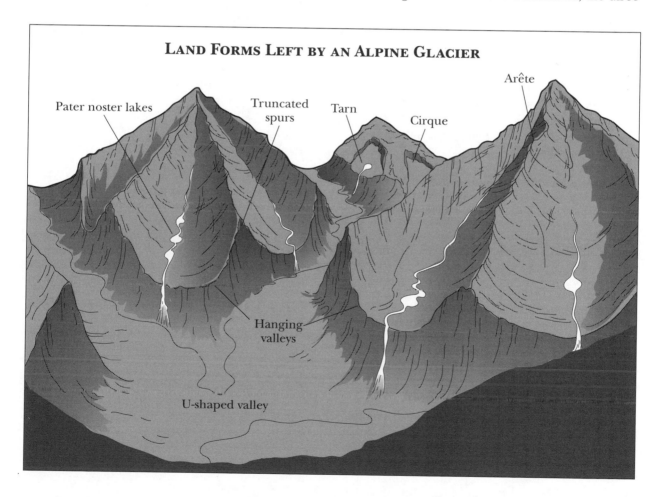

LAND FORMS LEFT BY AN ALPINE GLACIER

Pater noster lakes

Truncated spurs

Tarn

Cirque

Arête

Hanging valleys

U-shaped valley

tion of ice flow, the source area of the material, or the period in which the ice existed in a particular area. These pieces of information can be obtained by analyzing samples taken from the landscape features. Till fabric, pebble lithology and roundness, particle size, radiocarbon dating of organic material, and the amount and type of weathering are all commonly used observations.

Many areas rely on glaciers as a major water resource for either general consumption or power generation. Mountain glaciers are carefully monitored in order to determine the amount of water that can be expected from them each year. The monitoring is done by either digging into the ice at various points and measuring ice layers and rates of melting or, more commonly, by taking precipitation and meltwater stream gauging measurements throughout the entire glacial basin.

Diann S. Kiesel

CROSS-REFERENCES

BIBLIOGRAPHY

Adams, George F., and Jerome Wyckoff. *Landforms.* New York: Golden Press, 1971. This Science Golden Guide covers all aspects of landforms. The section on glaciers and glaciated land is brief but contains excellent diagrams and photographs to illustrate the formation of glacial landscapes. The section on useful publications not only lists general references but also indicates where to obtain regional, state, and local information about geologic features.

Bloom, Arthur G. *Geomorphology: A Systematic Analysis of Late Cenozoic Landforms.* 3d ed. Englewood Cliffs, N.J.: Prentice-Hall, 1998. This college-level text covers the basics of geomorphology. Includes three chapters on glaciers and glaciology. Index and bibliography.

Brunsden, Denys, and John C. Doornkamp, eds. *The Unquiet Landscape.* Bloomington: Indiana University Press, 1974. The introduction to this book contains a good general introduction to geology, maps, and general methods of geological study. A brief bibliography and an index are included. The two chapters explaining glacial erosion and deposition and related landforms are illustrated with excellent photographs. The chapters in the book are interrelated; for example, information about glacial effects on landscape is also explained in the chapters on lakes and changing sea level.

Colby, William E., ed. *John Muir's Studies in the Sierra.* San Francisco: Sierra Club Books, 1960. John Muir was the first to recognize the dramatic impact of glaciation in the Sierra Nevada. His observations and methods of reasoning showed geologists the folly of their previously incorrect interpretations of the development of the Sierra landscape. This book contains Muir's observations and explorations written in his easy-to-understand style. It includes his own drawings as well as several photographs. Because glacial geology is related to other geologic effects, Muir considers all aspects of the Sierra geology.

Dolgoff, Anatole. *Physical Geology.* Lexington, Mass.: D. C. Heath, 1996. This is a comprehensive guide to the study of the Earth. Extremely well illustrated and includes a glossary and an index. Although this is an introductory text for college students, it is written in a style that makes it understandable to the interested layperson. Contains a section on glaciers, glaciology, and glacial deposits.

Forrester, Glenn C. *Niagara Falls and the Glacier.* Hicksville, N.Y.: Exposition Press, 1976. This small book explores all the geology of Niagara Falls, especially the influence of the continental glacier and the development of the Great Lakes. It is well illustrated with maps, photographs, and diagrams and includes a walking tour. Although there is no index, there is a brief bibliography.

John, Brian S. *The Ice Age: Past and Present.* London: Collins, 1977. This book is very thorough in its examination of glacial and meltwater features. Many photographs, maps, and diagrams are used, and both a good index and a bibliography are included.

Matsch, Charles L. *North America and the Great Ice Age.* New York: McGraw-Hill, 1976. Numerous photographs, maps, and diagrams illustrate the chapter on glacial erosional and depositional features. There is also a chapter on lakes and sea-level changes. Brief selected readings and an index are included.

Menzies, M., ed. *Modern Glacial Environments: Processes, Dynamics, and Sediments.* Oxford, England: Butterworth-Heinemann, 1996. This multiauthored volume deals with all aspects of the modern study of glaciers. Intended for the specialist. Index and extensive bibliography.

Moore, W. G. *Glaciers.* London: Hutchinson Educational, 1972. This is a small book of excellent black-and-white photographs with commentaries on glaciers and glacial landscapes from around the world. Although there is no bibliography or index, there is a list of terms as well as questions accompanying each photo.

Price, R. J. *Glacial and Fluvioglacial Landforms.* New York: Hafner Press, 1973. Most of this book treats glacial and meltwater landforms in a more technical, yet nonmathematical, manner, with excellent diagrams, photographs, and maps. The index is good, and the bibliography is very thorough.

Schultz, Gwen. *Glaciers and the Ice Age.* New York: Holt, Rinehart and Winston, 1963. The chapters on glacial landforms are intriguingly named "The Great Detective Story" and "The Artistry of Ice." Both describe erosional and depositional features and discuss how they can be interpreted in the landscape. Illustrated with photographs, maps, and diagrams. Although both the bibliography and the index are brief, there is a glossary.

GRAND CANYON

The Grand Canyon of the Colorado River is considered to be the Earth's most extensive erosional canyon. Besides its awe-inspiring appearance, Grand Canyon provides geologists an opportunity to investigate the uppermost portions of the Earth's crust to examine the interrelationship of igneous, metamorphic, and sedimentary rocks exposed there.

PRINCIPAL TERMS

DIAGENESIS: the alteration of sediment at its initial deposition, which takes place during and after its transformation into sedimentary rock

DIP: the angle between a structural feature (for example, a joint, fault, or bedding plane) and the horizontal, which is usually the Earth's surface

DOLOMITE: a mineral consisting of calcium and magnesium carbonate compounds that often forms from precipitation from seawater

EOLIAN: related to wind deposits or environments

FAULT: a fracture in rock along which an appreciable amount of displacement has occurred

STRATIGRAPHY: the study of sedimentary strata, which includes the concept of time, possible correlation of the rock units, and characteristics of the rocks themselves

UNCONFORMITY: a significant break in a stratigraphic sequence because of nondeposition or erosion of the missing rock layers; much younger rocks are positioned above older rocks

DISCOVERY AND PHYSICAL CHARACTERISTICS

The Grand Canyon of the Colorado River (henceforth referred to as Grand Canyon) is considered to be the most extensive and impressive erosional canyon on the Earth. With an average depth of more than 1,675 meters below the North Rim (1,375 meters below the South Rim) and a length of 350 kilometers, Grand Canyon is a feature that is readily identified even from outer space. Images from the Landsat Earth resource satellites clearly show the enormous extent of the canyon as it winds south and then west across northern Arizona. This view is certainly very different from that experienced in 1540 by a lieutenant of the famous Spanish explorer Francisco Vásquez de Coronado. Don Garcia Lopez de Cardenas, with the help of Hopi guides, was the first non-American Indian to see the canyon as he led a group of Spanish explorers through what is now the southwestern United States.

The South Rim is the best-known portion of the more than 486,000 hectares surrounding the canyon that are included in Grand Canyon National Park. The central area of Grand Canyon was initially designated as a national monument in 1908 by President Theodore Roosevelt. In 1919, Con-

gress proclaimed the area a national park. Several other regions in northern Arizona near the original park boundaries were named as national monuments in 1932 and 1969. Finally, in 1975, the National Park Service reorganized and consolidated these monuments to increase the total area of Grand Canyon National Park to its present expanse of 493,076 hectares. The northernmost extent of the park is at the confluence of the Paria River and the Colorado River near Lees Ferry, Arizona. The Colorado River flows south and then west for 459 kilometers through the park until it empties into Lake Mead. From the confluence of the Little Colorado River, the distance of the Colorado River in the park is 350 kilometers. This is the segment of the park that bears the name "Grand Canyon." The western edge of the park is near the Grand Wash Cliffs in the vicinity of the Arizona-Nevada border. In this area, Grand Canyon National Park is contiguous with Lake Mead National Recreational Area, which extends from southern Nevada into northwestern Arizona. Until 1966, the Colorado flowed unimpeded through the canyon. Once Glen Canyon Dam was completed, the natural balance of river flow and seasonal floods was upset. The flow of the river is now a function of the electrical power needs of the Southwest.

The distance separating the rims on either side of the Colorado River varies from approximately 150 meters in Marble Canyon south of Lees Ferry to 29 kilometers at the downstream portion near Lake Mead. The South Rim is separated from the North Rim by 15 kilometers. It is a drive of 344 kilometers, however, to travel from one rim to the other.

As one looks from the South Rim toward the North Rim, it is evident that the North Rim is higher. The average elevation of the North Rim is more than 2,440 meters, approximately 300 meters above that of the South Rim. A series of uplifted plateaus accounts for this higher elevation. The largest plateau is the Kaibab Plateau, a north-south trending uplifted block that extends westward along the Colorado River from the confluence of the Little Colorado River to Kanab Creek, a major tributary draining the smaller Kanab Plateau to the west. Although the North and South Rims both lie on the Colorado Plateau, the fauna and flora of each area are quite different because of the elevation differences. A number of hiking trails descend from either rim into the inner canyon. Four biological life zones are traversed in making the rim-to-rim hike.

Grand Canyon is situated near the southwestern edge of the Colorado Plateau, a major physiographic province in the western United States, which is characterized by vast amounts of nearly horizontal sedimentary rocks. Although sedimentary rocks are usually deposited in a horizontal manner, it is puzzling to find massive amounts of horizontal strata on this major uplifted area. Geologists expect sedimentary layers to display some significant amount of dip when they are associated with major uplift. Sedimentary units that exist on both the North and South Rims display very little dip. Dip values average about 1 or 2 degrees toward the southwest. As a result, only streams located on the North Rim flow directly into the Colorado River as it passes through the canyon. Drainage on the Coconino Plateau on the South Rim is to the south into the Salt River system of southern Arizona. In the eastern section of the park, the Colorado River drops from an elevation of 852 meters to 544 meters. This vertical distance of 308 meters occurs over a horizontal distance of 169 kilometers. Thus, the gradient or downstream slope of the Colorado River in this area is 1.8 meters per kilometer.

FAULTS AND MONOCLINES

There are a large number of faults present in Grand Canyon. The amount of vertical displacement for these faults ranges up to 300 meters. The majority of these faults have either a northeast or a northwest trend. Perhaps the most visibly evident fault is the Bright Angel fault, which cuts through the Kaibab Plateau. This fault has produced a major zone of weakness that serves as the channel for Bright Angel Creek, a tributary of the Colorado River draining the North Rim. The Kaibab Trail parallels the creek, giving hikers a way of getting from the Colorado River up to the North Rim. Bright Angel fault extends to the South Rim, where it plunges under the surface. Sufficient faulting and mass movement have occurred to produce broken and slumped material.

Grand Canyon in the late afternoon, looking east-northeast from the vicinity of Bright Angel Point. (U.S. Geological Survey)

The Bright Angel Trail descends from the South Rim over this broken, slumped material and heads out across the Tonto Platform down to the river.

Numerous structural features called monoclines pass through this part of northern Arizona. These are rather simple flexures that bend a series of originally horizontal rock layers. Several of the more prominent monoclines are thought to have been formed by the reactivation of Precambrian faults that were later buried by sedimentary strata. Movement along the faults propagated upward and produced a bend in the sedimentary strata overlying the fault. Among the more prominent monoclines in Grand Canyon are the East Kaibab along the eastern border of the park, Grandview-Phantom in the middle portion, and the West Kaibab farther to the west. These monoclines trend basically north-south or northwest-southeast. Because of the flexural nature of these large-scale folds, they are sometimes subtler and hence more difficult to recognize than some of the major faults in the area.

GEOLOGICAL FORMATIONS

The superb exposures of rocks and geomorphic features in Grand Canyon invite all types of geologists to examine them to their further understanding of the history of the Earth. Geologists have conducted countless hours of fieldwork, research, and interpretation over the past century in their quest to understand what Grand Canyon has to offer both the scientist and the casual viewer.

The exposed geology within the canyon changes from place to place. In the easternmost part, the rocks are generally younger. Downriver from Lees Ferry, the sheer cliffs of the Redwall limestone (formed during the Mississippian, 330 million years ago) and the overlying Supai group (formed 300 million years ago) are exposed along the banks of the Colorado. Farther downstream, the river cuts deeper into the rocks, exposing progressively older rocks. In the central portion of the canyon, rock members of the Vishnu group begin to crop out along the riverbanks. These metavolcanics and metasediments (igneous and sedimentary rocks that underwent metamorphism after being formed) are about 1.7 billion years old. These represent the oldest rocks exposed in Grand Canyon. Rocks found in Grand Canyon represent a period of the Earth's history ranging from approximately 1.7 billion years ago to about 200 million years ago. Within this time span are several important missing periods. These unconformities represent either periods of nondeposition of the sediments or periods when rocks that were deposited were later eroded, thus removing their evidence from the geologic record.

Older Precambrian rocks comprise the portion along the riverbanks in the central part of Grand Canyon. Farther west, these rocks form the Inner Gorge. The Vishnu group consists of volcanic debris and sediments that were deposited and later metamorphosed to form the Vishnu schist. These rocks are injected with quartz veins and dikes, along with younger granitic bodies, and are the erosional remnants of a Proterozoic mountain range. The Vishnu group forms the foundation upon which all the other rocks in Grand Canyon rest.

Overlying these rocks is the Grand Canyon supergroup, a very thick sequence of sedimentary rocks and interbedded lava flows. Sedimentary rock types are varied, ranging from siltstone and shales to sandstones. Some portions of the supergroup include metamorphic and igneous rocks. Members of this supergroup include the Bass limestone, Hakatai shale, Shinumo quartzite, Dox sandstone, and Cardenas lavas. The Nankoweap formation and the Chuar group are found in portions of eastern Grand Canyon, which are generally inaccessible to most people.

Three distinct geological formations are representative of the Cambrian period (505-544 million years ago): the Tapeats sandstone, the Bright Angel shale, and the Muav limestone. These three formations constitute the Tonto group. During the early Paleozoic, which began 544 million years ago, northern Arizona was covered by a shallow sea lying to the west. The onset or transgression of this sea, along with migrating coastal sand dunes, produced the first relatively horizontal layer in the lowermost reaches of the canyon—the Tapeats sandstone. This unit ranges in thickness from 30 to 90 meters and consists of medium- to coarse-grained sandstone ranging in color from gray to chocolate brown. The Tapeats sandstone unconformably overlies the Grand Canyon supergroup and is readily recognized from a distance as the first prominent cliff former at the base of the canyon. The Tapeats essentially forms the boundary between the Inner Gorge and the Outer Canyon, two major topographic divisions within Grand Canyon. The Inner Gorge, about 365 meters deep, is a narrow, V-shaped channel cut into the older Proterozoic rocks (primarily the Vishnu schist and Zoroaster granite). The lower reaches of the Outer Canyon are delineated by the Tonto Platform, an extensive bench that is held up by the Tapeats sandstone. The Outer Canyon is roughly 1,220 meters deep and contains all the Paleozoic rocks exposed in Grand Canyon.

Conformably overlying the Tapeats is the Bright Angel shale, the second formation associated with the Paleozoic era. This formation is a distinct slope former, ranging in thickness from 60 to 135 meters. The Bright Angel shale, a mixture of mudstones and fine-grained sandstones, is dominated by a shaly, green mudstone. There are occasional breaks in the slope as low cliffs of more resistant sandstone and siltstones crop out. The rocks have a variegated color in different outcrops within the same canyon. The Bright Angel shale is one of the most fossil-rich units in the canyon, containing numerous traces left by soft-bodied organisms that were not themselves fossilized. The finer-grained rocks of this unit were deposited in quiet, deeper water as the sea that produced the Tapeats sandstone transgressed or moved across the area.

The Muav limestone, a ledge and cliff former composed primarily of mottled gray limestone, overlies the Bright Angel shale. The Muav forms a series of resistant cliffs throughout its 45- to 240-meter thickness. It should be noted that limestones tend to form cliffs and ledges in arid climates. This condition results from the lack of chemical weathering associated with copious rainfall, which is slightly acidic and attacks the calcite present in the limestone. Intermittent recesses on the surface result from relatively thin layers of mudstones and siltstones that have differentially weathered out. The Muav becomes progressively thicker to the west because of the deeper water of the transgressive sea. The occurrence of dolomite in the Muav points to stages of the sea's regression.

Above the Muav limestone is an erosional surface representing the loss of about 160 million years of geologic history (spanning the Late Cambrian to the Early or Middle Devonian). This ma-

jor unconformity produces no discernible interruption in the sequence of horizontally layered strata. Some deposition probably took place, but the rocks that formed were later eroded. Above this unconformity lies the Temple Butte limestone, a prominent cliff former of Devonian age (370 million years ago). In eastern Grand Canyon, this unit consists of intermittent outcrops of channel fill. Two types of dolomite are found in the Temple Butte: the first is a light-colored, thin-bedded dolomite having a porcelain-like texture; the second is sandier and finer-grained. The unit was deposited as a limestone but later underwent diagenesis and was converted into dolomite. The Temple Butte thickens considerably to the west, with outcrops ranging up to 125 meters in thickness. Very few invertebrate fossils are found in the Temple Butte.

Widespread erosion of the Temple Butte limestone took place throughout the region. Above the Temple Butte is an erosional surface that is difficult to recognize in most localities within the canyon. Overlying this erosional surface is the Redwall limestone, certainly one of the most conspicuous layers of rock within Grand Canyon. It forms a sheer cliff more than 150 meters in height. Four members constitute the Redwall limestone: the lowermost Whitmore Wash, the Thunder Springs, the Mooney Falls (thickest of the four), and the uppermost Horseshoe Mesa. All four members are present throughout the canyon. The Redwall limestone is so named because its outer surface is stained by iron oxides that have washed down from the overlying Supai group rocks. For hikers who traverse the many trails in the canyon, their confrontation with the Redwall produces the most difficult portion of their hike. Because of the sheerness of the Redwall, trails often must follow slopes produced by the slumping and sliding of the overlying Supai group or Hermit shale. The Redwall limestone is the product of precipitation of limestone and other carbonates from a warm, shallow sea that transgressed and regressed three times during the Early to Middle Mississippian (330 million years ago). The unit is rich in fossils, such as bryozoans, brachiopods, and crinoids, all indicative of a sea that did not receive much sediment from adjacent continental areas.

The Surprise Canyon formation rests unconformably on the Horseshoe Mesa member of the Redwall. The formation crops out in isolated patches throughout most of the central portion of Grand Canyon. Thicker deposits of channel fill of Surprise Canyon material occur in isolated areas in western Grand Canyon; deposits are much less evident toward the east. Thicknesses can reach 90 meters.

About 25 to 30 million years elapsed before the deposition of the Supai group over the Redwall and intermittent channels of the Surprise Canyon formation. The Supai, which is up to 310 meters thick, is readily evident in the canyon as it is an extensive ledge and cliff former with some slopes separating them. The Supai group is the source of the iron oxide that stained the underlying Redwall limestone. The Supai has been divided into four formations: the basal Watahomigi, Manakacha, Wescogame, and the Esplanade, a distinct sandstone ledge-forming unit. In western Grand Canyon, the percentage of resistant rock increases, resulting in more cliffs being produced in the process of erosion. The formations of the Supai do contain fossils, but they are scarcer to the east. The Supai group, which spans the boundary of the Pennsylvanian and Permian periods (300 million years ago), formed in a combination of coastal plain, eolian, and shallow river environments. The 90-meter-thick, red siltstones of the overlying Hermit shale produce a sloped surface above the Supai. The environment in which it formed was similar to that of the Supai—shallow swamps and lagoons. Cross-bedding present in portions of the unit indicates a sediment source from the north.

Near the top of the canyon, sheer cliffs of the Coconino sandstone overlie the Hermit shale. The Coconino is a very distinctive layer within the stratigraphy of Grand Canyon, as it forms a wide, white band near the upper edge. The Coconino sandstone is a relatively pure, well-sorted quartz sandstone that formed when northern Arizona was covered by a vast sand sea that was driven by the prevailing south and southeast winds. Large cross-bedded dunes are preserved throughout northern Arizona and are exposed within the canyon. The thickness of the unit ranges from 18 to 30 meters in outcrops on the North Rim to more than 90 meters on the South Rim. The unit extends into central Arizona, where thicknesses of more than 245 meters have been recorded. The only fossil record preserved in the sandstone consists of tracks and trails produced by invertebrate and vertebrate animals. Surprisingly, no fossilized remains of the animals have been discovered.

Above the precipitous cliffs of the Coconino sandstone is the Toroweap formation, a ledge and cliff former made of limestone and sandstone layers containing occasional deposits of gypsum. The Toroweap formation interfingers with the Coconino sandstone, which implies that the transgressive and regressive seas that formed the Toroweap inundated parts of the sand sea associated with the Coconino. Because the sea came from the west, layers thicken in that direction within this unit. The Toroweap is moderately fossiliferous, containing corals, bryozoans, brachiopods, and mollusks.

The Kaibab limestone, the uppermost surface of Grand Canyon, is approximately 90 meters thick and is composed of a sandy limestone. The same sea that formed the Toroweap formation returned to deposit the layers of the Kaibab limestone. It was a shallow, warm-water marine environment from which the sediments were deposited. The Kaibab is an extensive deposit ranging from southern Utah to central Arizona. It is fairly rich in invertebrate marine fossils, including mollusks, brachiopods, and crinoids.

Evidence exists at the eastern edge of Grand Canyon that Triassic deposits were laid down in the area. Cedar Mountain, situated about 5 kilometers east of the eastern entrance to Grand Canyon National Park, consists of more than 185 meters of Triassic deposits. This is the only sizable outcrop of Triassic or younger sediments in the area. These sediments were deposited but later eroded from the area, exposing the present-day surface of Kaibab limestone.

COLORADO RIVER

The Colorado River has had a complex history in the area. Several hypotheses have been proposed to explain its present course and how the river cut Grand Canyon. Although these hypotheses differ in their details, they all point to some common points in the evolution of the present river course. Ancestral drainage existed in what is now the upper Colorado River system, which includes the Green and San Juan Rivers, two major tributaries. The Kaibab-Coconino uplift diverted the southward course of the ancestral Colorado River. One theory has the river flowing to the northwest; another has it flowing to the southeast, where extensive lake deposits are found. When the Gulf of California opened about 5 million years ago, headward erosion of the lower Colorado River system began. As the river eroded upstream, it managed to capture small streams situated to the west of what is now Grand Canyon. With increased headward erosion, the lower Colorado River cut its way back through the Grand Wash Cliffs and started to carve western Grand Canyon. Eventually, the lower portion captured the drainage of the upper portion, leading to the drainage of the lake formed by the upper system. This increased water flow enhanced the downcutting of Grand Canyon to its present form. The vast majority of water flowing in the Colorado River now is derived from snowmelt in Colorado and Wyoming. This water, along with that flowing in the San Juan River, passes through the Green and upper Colorado rivers before being impounded behind Glen Canyon Dam. The Little Colorado River drains the arid northeastern Arizona landscape and thus produces a small contribution to the overall flow of the Colorado River as it moves through Grand Canyon.

David M. Best

CROSS-REFERENCES

Biostratigraphy, 1091; Cambrian Diversification of Life, 956; Continental Structures, 590; Diagenesis, 1445; Fossil Record, 1009; Geomorphology and Human Activity, 898; Geomorphology of Dry Climate Areas, 903; Geomorphology of Wet Climate Areas, 909; Glacial Landforms, 914; Karst Topography, 925; Marine Terraces, 929; Paleobiogeography, 1058; Paleozoic Era, 1126; Permafrost, 933; River Valleys, 936; Sedimentary Rock Classification, 1457; Unconformities, 1161.

BIBLIOGRAPHY

Beus, Stanley S., and Michael Morales, eds. *Grand Canyon Geology.* 2d ed. New York: Oxford University Press, 2000. This book consists of a series of articles on the formations and geological history of the Grand Canyon. Intended for the reader with some scientific background, as well as the specialist. Bibliography and index.

Billingsley, George II., and Stanley S. Beus, eds. *Geology of the Surprise Canyon Formation of the Grand Canyon Arizona.* Flagstaff: Museum of Northern Arizona, 1999. A multiauthored volume on a formation recently discovered in the Grand Canyon. The initial overview chapter provides much useful information for the general reader. Bibliography and index.

Frome, Michael. *The National Park Guide.* Skokie, Ill.: Rand McNally, 1987. This book offers a more technical presentation of each of the national parks in the United States. The material is easy to read, however, and is augmented with excellent photographs.

Harris, Ann G., and Esther Tuttle. *Geology of National Parks.* Dubuque, Iowa: Kendall/Hunt, 1983. From the geologic standpoint, this text is unexcelled in its presentation of the geologic evolution and history of the nation's national parks. Location maps along with generalized stratigraphic columns provide the reader with an excellent reference source. College-level readers would benefit the most from this book.

Harris, David V., and Eugene P. Kiver. *The Geologic Story of the National Parks and Monuments.* 3d ed. New York: John Wiley & Sons, 1985. The introductory chapter of this book succinctly covers the basic principles and processes of geology. The reader then is presented with a well-written basic geologic history of national parks and monuments. This well-illustrated book provides an outstanding bibliography for the inquisitive reader. For college-level readers.

Nations, J. Dale, and Edmund Stump. *Geology of Arizona.* 2d ed. Dubuque, Iowa: Kendall/Hunt, 1996. In addition to providing some basic geologic principles, this book discusses the geology of the state in some detail. It gives the overall geologic setting in which the Grand Canyon is located as well as some specifics as to its formation. Suitable for college-level readers. Includes a state geological map.

Powell, John Wesley. *The Exploration of the Colorado River and Its Canyon.* Mineola, N.Y.: Dover, 1961. A reprint of the 1895 edition of *Canyons of the Colorado,* in which Major Powell, the one-armed veteran of the Civil War, detailed the journey he and nine others took along 1,600 kilometers of the Colorado River in 1869 and 1870. Written in diary form, the book places the reader on the river with the first person to explore this part of the western United States. A classic, this text is excellent reading for high school students and above.

Tilden, F. *The National Parks.* 3d ed. New York: Alfred A. Knopf, 1986. Considered to be a classic, this book provides a narrative description of the nation's national parks, monuments, and historical sites and buildings. Aimed at the layperson, it does lack the detail that may be needed to visit an area. Suitable for the generalist.

KARST TOPOGRAPHY

Karst topography is a landform produced by the dissolving action of surface and groundwaters on the underlying bedrock of a region. The landforms produced are unique because they represent internal or underground drainage, forming such features as sinking streams, sinkholes, caves, natural bridges, and springs.

PRINCIPAL TERMS

CARBONIC ACID: a weak acid formed by mixing water and carbon dioxide; it is important in the dissolving of the most common karst rock, limestone

CAVE: an opening or hole in the ground enterable by people; in karst topography, caves have been dissolved out and act (or once acted) as underground conduit flow routes for water

LIMESTONE: a common sedimentary rock containing the mineral calcite; the calcite originated from fossil shells of marine plants and animals

NATURAL BRIDGE: a bridge over an abandoned or active watercourse; in karst topography, it

may be a short cave or a remnant of an old, long cave

SINKHOLE: a hole or depression in the landscape produced by dissolving bedrock; sinkholes can range in size from a few meters across and deep to kilometers wide and hundreds of meters deep

SINKING STREAM: a stream or river that loses part or all of its water to pathways dissolved underground in the bedrock

SPRING: a place where groundwater reappears on the Earth's surface; in karst topography, a spring represents the discharge point of a cave

DISSOLVING BEDROCK

Karst topography is a unique landscape produced by the dissolving of the bedrock of a region, with the consequent development of underground drainage. Most common bedrock materials, such as granite, sandstone, and shale, are resistant to the process of dissolving (also known as dissolution), and landscapes are carved into these bedrock materials by the mechanical action of water, wind, and ice. Some bedrock materials, such as limestone, dolomite, gypsum, and rock salt, dissolve relatively easily. Gypsum and rock salt are soluble in plain water and in most landscapes are chemically destroyed very quickly. Karst landscapes on these two rock types persist only in dryer climates, such as in the American Southwest. Gypsum and rock salt are also mechanically weak rocks and are destroyed rapidly by mechanical erosive activities. Limestone, and its close cousin dolomite, are mechanically strong rocks; therefore, they resist normal erosive activity. Yet calcite, the mineral of which limestone is composed, is especially vulnerable to dissolving by

water that is slightly acidic.

Under normal conditions, the source of such acidity is carbonic acid, a natural combination of carbon dioxide and water. Rainwater absorbs carbon dioxide from the atmosphere and is naturally slightly acidic (modern pollution has accentuated this natural tendency to produce acid rain). In the soil, organic activity can increase the amount of carbon dioxide to levels well above those found in the atmosphere, and water moving through such soils can become very acidic. When rainwater or soil water charged with carbonic acid meets limestone bedrock, it will slowly dissolve the limestone. Limestone, mechanically resistant, supports the development of karst topography well. Dolomite reacts more slowly to acid waters, and generally the karst features found on dolomites are subdued and take longer to form than those on limestones. Discussion of karst topography is therefore, in most situations, a discussion of the dissolving of limestone.

The rate at which limestone dissolves depends on the amount of water in the environment and the amount of carbon dioxide available. Atmospheric levels of carbon dioxide are similar worldwide, but the amount of carbon dioxide in the soil varies greatly. Organic activity controls the amount of carbon dioxide available, and for that reason, the most acid groundwater tends to be found in the warm, wet zones of the tropics, where organic activity is high. Dry, cold climates have erosion rates as little as a few millimeters per one thousand years, whereas warm, wet climates can have rates up to 150 millimeters per one thousand years. These erosion rates are averages of the rate of surface lowering for a region. Over the course of hundreds of thousands of years, significant landforms can be developed.

The dissolving action of carbon dioxide-charged water produces a number of etching patterns on exposed limestone bedrock, from microscopic features to large trough structures more than 1 meter deep and many meters long. Karst topography develops when the water penetrates down pores, cracks, and openings into the limestone. Once inside the rock, the water is capable, over long periods of time, of dissolving voids and passageways in the rock. These openings in the rock become integrated into underground flow networks similar to stream patterns on the Earth's surface. The result is a series of underground passageways called caves, which collect water from sinkholes and sinking streams on the Earth's surface and transmit the water back to the Earth's surface at springs. While the flow pattern of caves is often similar to the pattern of surface streams, caves are tubes in bedrock and may migrate up and down, as well as from side to side, in a manner in which surface streams cannot. As the cave system grows and matures, it can capture larger volumes of surface water and enlarge. Eventually, it will not only carry material in solution but also mechanically transport sediment underground. On the land surface, this underground or internal drainage produces sinking streams and sinkholes. In mature systems, the sinkholes may be kilometers across and hundreds of meters deep, containing many sinking streams.

Depending on the exact nature of the limestone, the climate, and the presence of mountains, a wide variety of karst landscapes may appear. The landscape may be as simple as rolling hills or a flat plain with mostly normal surface drainage and only a few scattered sinkholes, sinking streams, and caves. Other landscapes, however, may have no significant surface drainage and have sinkholes covering the land surface, producing a sinkhole plain. One of the best examples of a sinkhole plain is in the Mammoth Cave area of Kentucky. Under extreme conditions, usually associated with the tropics, the landscape is so altered by dissolution that the sinkholes deepen faster than they widen, producing a landscape of tall limestone towers and pinnacles standing above a flat plain. This landform is called tower karst and is best known from southern China, where thousands of towers several hundred meters high dot the landscape.

The downward erosion by groundwater can produce cave systems at lower eleva-

A sinkhole 350 feet wide and 150 feet deep in central Alabama in December, 1972. (U.S. Geological Survey)

Under extreme conditions, the landscape is so altered by dissolution that the sinkholes deepen faster than they widen, producing a landscape of tall limestone towers and pinnacles standing above a flat plain. China's Shulin (Stone Forest) provides many examples of these so-called tower karst formations. (AP/Wide World Photos)

tions, causing earlier, higher cave systems to become abandoned. These abandoned caves may persist for hundreds of thousands of years, developing complex mineral displays of stalactites, stalagmites, and other deposits. Animals and people may use the cave for shelter, leaving behind important fossil and cultural material. Continuing erosion of the land surface will eventually breach into underlying cave systems to produce new entrances and truncated cave fragments called natural bridges. Sinkholes become natural traps, collecting unwary animals as well as plant remains, soil, and pollen. The transport of material into the subsurface allows the material to be preserved from destruction in the Earth's surface environment.

DEVELOPMENTAL STAGES

Like all landscapes, karst topography can go through a series of developmental stages. Karst de-

velopment begins when the soluble rock is first exposed at the Earth's surface. In some cases, groundwater reaches the soluble rock before it is ever exposed, and cave development can begin before the rock is present on the surface. In either case, the progressive development of a karst landscape by chemical and mechanical erosion will remove the layer of soluble rock responsible for the karst processes. In time, karst processes become less important in the landscape as the soluble bedrock disappears, and the landscape will begin to revert to a more typical, mechanically produced form.

In the United States, limestones form important areas of karst topography in the Virginias, Pennsylvania, Indiana, Kentucky, Tennessee, Alabama, Florida, Missouri, Arkansas, and Texas. Many other states have minor amounts of limestone karst, such as New York, Minnesota, Iowa, Colorado, and New Mexico. Some states have major cave systems located in minor areas of karst, such as Carlsbad Caverns in New Mexico and the Jewel and Wind Caves in South Dakota. Gypsum karst is locally important in Kansas, Oklahoma, and New Mexico, while dolomite karst is found in Missouri. Marble—limestone altered by heat and pressure—forms small karst areas in California, New England, and Oregon. Rock salt karst is very rare in the United States but can be found in Spain and the Middle East. Almost every state and every country has some form of karst development.

The unique features of karst topography can be mimicked by other landscapes in special cases. Such cases are called pseudokarst because they imitate karst topography but are produced by other phenomena. For example, volcanic areas produce eruptive craters and lava tubes that look like sinkholes and caves from karst areas. In deserts, streams can dry up and appear to be sinking, and winds can produce excavated hollows in the sand. Glaciers can produce depressions in the ground that look like sinkholes. In fine-grained clay deposits—which can be seen in the Badlands of South Dakota—caves, bridges, and sinkholes are produced by mechanical flushing of the clay particles, a process called suffusion. Wave activity on rocky coastlines can carve sea caves in a variety of rock materials.

Karst topography is unique because of the chemical manner in which the bedrock is destroyed and because of the internal or underground drainage that results. Most karst is developed in limestone because of this rock's abundance, slow chemical dissolution rates, and great mechanical strength. In certain areas of the world, karst topography is the dominant landscape for thousands of square kilometers. Cave systems in excess of 500 kilometers of surveyed passage exist, and caves have been followed as deep as 1,700 meters beneath the surface.

John E. Mylroie

CROSS-REFERENCES

Aquifers, 2005; Biogenic Sedimentary Rocks, 1435; Carbonates, 1182; Chemical Precipitates, 1440; Drainage Basins, 2325; Evaporites, 2330; Fossilization and Taphonomy, 1015; Freshwater Chemistry, 405; Geomorphology and Human Activity, 898; Geomorphology of Dry Climate Areas, 903; Geomorphology of Wet Climate Areas, 909; Glacial Landforms, 914; Grand Canyon, 919; Groundwater Movement, 2030; Groundwater Pollution and Remediation, 2037; Hydrologic Cycle, 2045; Marine Terraces, 929; Permafrost, 933; Reefs, 2347; River Valleys, 936; Surface Water, 2066; Uranium-Thorium-Lead Dating, 553.

BIBLIOGRAPHY

Beck, B. F., and W. L. Wilson, eds. *Karst Hydrogeology: Engineering and Environmental Applications.* Boston: A. A. Balkema, 1987. This book is the proceedings volume of the Second Multidisciplinary Conference on Sinkholes and the Environmental Impacts of Karst and contains sixty-three papers on a variety of topics concerning human activity and karst, with an emphasis on the high rate of mistakes made as a result of ignorance of karst processes. Accessible to the nonspecialist.

Bloom, Arthur G. *Geomorphology: A Systematic Analysis of Late Cenozoic Landforms.* 3d ed. Englewood Cliffs, N.J.: Prentice Hall, 1998. This college-level text covers the basics of geomorphology. Includes a chapter on the development of karst topography. Index and bibliography.

Chernikoff, Stanley. *Geology: An Introduction to Physical Geology.* Boston: Houghton Mifflin, 1999. This is a good overview of the scientific understanding of the geology of the Earth and surface processes. Includes sections on the development of karst topography. Includes a link to a Web site that provides regular updates on geologic events around the globe.

Gillieson, David. *Caves: Processes, Development, Management.* Oxford, England: Blackwell, 1998. A description of how caves form, what can be learned from them, and how they can best be managed. The first one-third of the book deals with the development of karst and its relation to cave formation. Well illustrated and written in an engaging style that makes the material comprehensible to the general reader. Glossary, bibliography, index.

Jennings, J. N. *Karst Geomorphology.* 2d ed. New York: Basil Blackwell, 1987. A college-level text that emphasizes karst processes overall, as opposed to only their role in cave formation. The text requires some knowledge of the basic sciences. It is interesting to read, as it uses many Australian examples and contains the lifetime knowledge of one of the world's leaders in karst research.

Middleton, John, and Anthony C. Waltham. *The Underground Atlas: A Gazetteer of the World's Cave Regions.* New York: St. Martin's Press, 1987. An easy-to-read guide to the famous long and deep caves of the world, this atlas covers the continents in a general review and then describes the cave resources of the world's countries, from Afghanistan to Zimbabwe. Simple maps are presented to portray the largest and deepest caves. The appendices supply a listing of record-holding caves in terms of length, depth, and volume, as well as data about caving organizations and sources of additional reading.

Sasowsky, I. D. *Cumulative Index for the National Speleological Society Bulletin, Volumes 1 Through 45, and Occasional Papers of the N.S.S., Numbers 1 Through 4.* Huntsville, Ala.: National Speleological Society, 1986. The Bulletin of the National Speleological Society is carried by many libraries and contains papers on the science of speleology and karst processes. This index to that journal allows quick access to published papers on specific topics, by topic, author, or geographical area.

White, William B. *Geomorphology and Hydrology of Karst Terrains.* New York: Oxford University Press, 1988. An advanced college textbook written by one of the leaders in karst research in North America, this volume contains a well-written and comprehensive discussion of karst topography.

MARINE TERRACES

Marine terraces are ancient coastlines that are at elevations well above or well below present-day sea level. Global cycles of sea-level change and tectonic uplift and subsidence of coasts are responsible for creating multiple sets of parallel marine terraces.

PRINCIPAL TERMS

BARRIER ISLAND: a long, low sand island parallel to the coast and separated from the mainland by a salt marsh and lagoon

BIOABRASION: physical and chemical erosion or removal of rock as a result of the activities of marine organisms

GUYOT: an oceanic volcano, presently submerged far below sea level, with a top that has been beveled flat by wave erosion

ISOSTATIC READJUSTMENT: rapid tectonic uplift or subsidence of continental areas in response to the addition or removal of the weight of overlying deposits of glacial ice or seawater

LITHOSPHERE: the outer shell of the Earth, where the rocks are less dense but more brittle and coherent than those in the underlying layer (asthenosphere)

NOTCH or NIP: an erosional feature found at the base of a sea cliff, the result of undercutting by wave erosion, biobrasion from marine organisms, and dissolution of rock by groundwater seepage

PLEISTOCENE EPOCH: the time of Earth history, from about 2 million to 10,000 years ago, during which the Earth experienced cycles of warming and cooling, resulting in cycles of glaciation and sea-level change

STRANDLINE: the position or elevation of the portion of the shoreline between high and low tide (at sea level); usually synonymous with "beach" and "shoreline"

SUBSIDENCE: the sinking of a block of the Earth's lithosphere because of a force pushing it down; coastal areas undergoing rapid subsidence tend to be submerged below sea level

TECTONIC: pertaining to large-scale movements of the Earth's lithosphere

UPLIFT: the rising of a block of the Earth's lithosphere because of a force pushing it up; coastal areas undergoing uplift tend to emerge above sea level

COASTLINES

The coastlines of continents and islands represent a fundamental boundary between the Earth's solid landmass and the constructional and destructional energy of the sea. The landforms in coastal areas are the result of continuous dynamic interaction between these competing geological agents. Marine processes of erosion and sedimentation construct a shoreline profile on the edge of the landmass that defines the strandline, a narrow zone of wave- and tide-washed coast. The coastal strandline usually exhibits a gently sloping platform with its top at an elevation between high and low tides; it is often bounded by adjacent steeper slopes and is a reliable indicator of sea level. Familiar strandline features are beaches, coral reefs, and wave-cut platforms, strandline platforms cut into the bedrock by wave erosion.

If erosion and sedimentation were the only active geological conditions, most coasts would have a single type of strandline landform whose position remained constant through time. The volume of seawater in the world's oceans, however, fluctuates directly with changes in the volume of ice in continental glaciers, causing sea level to rise and fall in response. Glacial ice on the continents melts (sea-level rise) or accumulates (sea-level fall) in response to temperature changes in the Earth's atmosphere. The temperature of the surface ocean reacts in a similar manner: Warming causes expansion of the seawater (sea-level rise), and cooling results in contraction (sea-level fall). It has been estimated from oxygen isotope evidence in deep-sea cores that worldwide sea level has been

130 to 160 meters lower than today several times during the last 2 to 3 million years of Earth history. There is also abundant physical evidence for large-scale sea-level excursions. On the Atlantic coast of the United States, elephant teeth 25,000 years old have been dredged up by fishermen from more than forty locations on the continental shelf, some as far as 120 kilometers offshore. Three-thousand-year-old oak tree stumps and many cultural artifacts used by coastal American Indians have also been found below sea level in this region.

The landmasses are also surprisingly dynamic, rising or falling in elevation in response to a variety of tectonic factors. On some coasts, isostatic adjustment results in uplift and subsidence. The weight of increasing thicknesses of overlying glacial ice or seawater will cause the Earth's lithosphere to be depressed downward; when the weight is removed, the landmass will rebound upward. Volcanic activity will cause expansion (uplift) when the Earth's lithosphere is heated; contraction (subsidence) occurs as the lithosphere cools. An extreme example of this effect is the guyot, an inactive volcanic seamount that subsides far below sea level as the hot lithosphere on which it formed cools and contracts. Guyots have been beveled flat by wave erosion and rimmed with banks of dead coral reef during the period when their tops were in shallow water. Coastal landmasses also undergo uplift and subsidence in response to the forces of plate tectonic movement. Although usually a slow, incremental process, a sudden, rapid coastal uplift occurred in Prince William Sound, Alaska, during the 1964 Good Friday earthquake, which resulted in elevation of a wave-cut platform 400 meters wide.

FOSSIL COASTS

During the Pleistocene epoch, strandline features formed on coasts worldwide that were strongly influenced by these complex and rapidly changing geologic conditions. Most coasts are not tectonically stable (located at one elevation) for a very long period of geologic time, and processes of landscape erosion and decay often act much more slowly than the rates of sea-level and land-level change. Abandoned strandline features, or "fossil coasts," are common to many areas of the world. First recognized in nineteenth century Europe, parallel sets of these distinctive landforms are present on cliffs and coastal hills at elevations high above present-day sea level and on the ocean floor far below present-day sea level. Many contain marine fossils—demonstrating their marine origin—and create a steplike topographic profile of the terrain, leading up the seaward-facing hills. These features are referred to as marine terraces.

Because their formation requires time, strandline landforms usually form when coastal uplift/subsidence and sea-level change are in balance and the position of the strandline is relatively stable for a period of thousands of years, a situation referred to as a sea-level stillstand. In many areas, sea level tends to vary in cycles separated by stillstand events, during which marine terraces can develop. Occasionally, the processes of terrace formation will destroy or obscure an older terrace, so most coasts exhibit only one or two terraces. An unbroken flight of marine terraces climbing inland indicates a coast that is being continuously uplifted. One of the most striking and well-devel-

A marine terrace on the Santa Cruz, California, coast, formed by uplift of former submarine surface. (© William E. Ferguson)

oped sequences of marine terraces in the world is on the seaward slopes of the Palos Verdes Peninsula of Southern California. Here, thirteen distinct terrace levels, from 15 to 448 meters in elevation, form a complete record of coastal uplift and sea-level change spanning several hundred thousand years.

Submerged terraces are usually better preserved than elevated marine terraces because they are less affected by processes of erosion and landscape degradation. Because they are obscured from view and more difficult to study than terraces on land, less is known about them. The most distinctive submerged wave-cut platforms are present off the southwest coast of Great Britain, off northern Australia, and off the Southern California coast and its nearshore islands from Santa Barbara to San Diego. Study of marine terraces has been used to document sea-level variations and the tectonic uplift and subsidence of coastal landmasses through time and to estimate the magnitude of these changes.

COASTAL EROSION

There are several types of marine terraces, differing in their mode of formation. Processes of coastal erosion form marine terraces on coasts where the shoreline is backed by a steep, irregular cliff, where the supply of sediment is too limited to build beaches, and where nearshore waters are so shallow that waves break directly on the sea cliff. The base of the cliff is continuously pounded by the full force of the waves, resulting in very large impact pressures. Beach cobbles and abrasive, sand-laden water are hurled at the base of the cliff. The force with which these stones attack the cliff is difficult to exaggerate; there are reports of beach cobbles thrown 45 meters above sea level.

Wave-cut platforms result from the rapid horizontal cutting away of rock at the base of a sea cliff. Wave erosion quarries a notch into the base of the sea cliff, which undercuts and oversteepens the cliff until landsliding causes the cliff to collapse; this loose rubble is rapidly carried away by the waves and is a source of new rocks to continue wave attack. In some places, notch formation is aided by the bioabrasion activities of marine organisms that live attached to and that feed on rocky shores, and that secrete chemicals to help dissolve the rock or abrade it away with rasping feeding appendages. Dissolution of the rock is sometimes aided by groundwater seeping out of the base of the sea cliff. The net effect of this process is the landward retreat of the sea cliff and formation of a platform that slopes gently seaward. The boundary between the "step" of a wave-cut terrace and the steeper "riser" of the sea cliff that terminates its landward edge is called the inner edge point. The elevation of this point is the position of highest sea level during formation of the feature, and it is used to calculate the height of ancient sea levels in marine terraces.

The platform is continually abraded by the waves transporting sand down the coast, and its width is determined by the amount of landward sea-cliff erosion. The rate of sea-cliff retreat can be surprisingly rapid. Average rates on the California coast exceed 15 centimeters per year for hard rock cliffs and up to 1 meter per year in cliffs composed of soft, unconsolidated sediment. The coast of East Anglia, Great Britain, experiences up to 4 meters of sea-cliff retreat annually. Continuous records are available for the Huntcliff coast of Yorkshire, Great Britain, at a former Ro-

Marine terrace at Natural Bridges Beach State Park near Santa Cruz, California. An emergent marine terrace appears in the distance to the west. (U.S. Geological Survey)

man signal station where the cliff has retreated 30 meters in eight hundred years.

Overlying many elevated wave-cut platforms is a mantle of unconsolidated sediment referred to as coverhead. This material, deposited by a combination of terrestrial and marine sedimentary processes, buries the abraded platform surface and may be as much as 30 meters thick. Typical coverhead deposits include a basal layer of well-sorted beach sand, rounded beach cobbles, and marine fossils—the shoreline sediments left on the platform as sea level retreated. These basal strandline sediments are often covered by a heterogeneous deposit composed of poorly sorted rock debris, soil, and stream gravel deposited by sediment washing or landsliding down from the seacliff slopes above and by coastal streams building their alluvial fans toward the coast.

SEDIMENT DEPOSITION

Marine terraces are also formed by sediment deposition, which usually occurs on relatively flat coasts with a wide continental shelf where the energy of approaching waves is dissipated by friction with the shallow sea bottom. In areas such as the Atlantic and Gulf coasts of North America, rivers transport large quantities of mud and sand into coastal waters. Waves and wind currents pile sand into a long, narrow strandline feature known as a barrier island, a common landform found fringing low-lying coasts worldwide. Once formed during a stillstand of several thousand years, a barrier island will migrate landward as sea level rises. Waves and tidal currents cause barrier sand to wash over the top of the island or around its ends, moving it grain-by-grain landward with the rising sea. When sea level begins to fall, the barrier does not migrate seaward but is left behind inland of the new strandline, where it remains as a record of the former high stand of sea level; this feature is usually referred to as a beach ridge. Subsequent to sea level falling and the strandline moving seaward, river and marsh sediment is often deposited on the seaward side of the abandoned barrier. When sea level again rises to its former position, a new barrier island migrates with it; this island is, in turn, abandoned on the coast at the highest position of sea level. If the new barrier migrates far enough inland to reach the older remnant barrier, it will be welded onto it, forming a wide composite barrier island composed of two or more barrier islands.

Sediment deposition has widened the Atlantic and Gulf coastal plains, and repeated cycles of sea-level rise and fall have formed concentric arcs of abandoned barrier islands stretching inland nearly 50 kilometers from the present-day coast. These beach ridge barrier islands are a type of marine terrace, each recording an ancient high stand of sea level. One of the most prominent is Trail Ridge in southern Georgia, a sand ridge nearly 60 kilometers long that encloses a low swampy area on its landward side known as the Okefenokee Swamp. Submarine beach ridges are less common than their inland counterparts but have been reported from western Brittany.

In tropical areas where the coastal waters are warm all year and clear of suspended mud, corals will flourish and often form massive reefs. Reefs that grow during a long stillstand will become large and well developed, with a gently sloping top that corresponds to sea level. These coastal depositional features are also found stranded above the shoreline when sea level falls or the landmass rises, resulting in reef terraces. This type of marine terrace is most common on island coasts but is also present on the coasts of the Mediterranean and Red Seas.

James L. Sadd

CROSS-REFERENCES

BIBLIOGRAPHY

Hearty, P. J., and P. Aharon. "Amino Acid Chronostratigraphy of Late Quaternary Coral Reefs: Huon Peninsula, New Guinea, and the Great Barrier Reef, Australia." *Geology* 16 (July, 1988): 579-583. A technical article describing the application of the amino acid dating technique to the classic coral reef terraces of the Huon Peninsula and a comparison of the results with uranium-series ages. *Geology* is a monthly publication of the Geological Society of America.

Moore, W. S. "Late Pleistocene Sea-Level History." In *Uranium Series Disequilibrium: Applications to Environmental Problems*, edited by M. Ivanovich and R. S. Hannah. Oxford, England: Clarendon Press, 1982. Half of this chapter in an advanced textbook is a description of state-of-the-art terrace dating techniques written in a technical style. Included is a summary of all known dates for fossil-bearing marine terraces all over the world that is slightly easier to read.

Pethic, J. *An Introduction to Coastal Geomorphology.* Baltimore: Edward Arnold, 1984. Written primarily for the Western European region, this advanced text contains several good sections on the formation of marine terraces, their use in sea-level and tectonic uplift studies, and the correlation of terrace data with other studies.

Plummer, Charles C., David McGeary, and Diane H. Carlson. *Physical Geology.* Boston: McGraw-Hill, 1999. This is a straightforward, easy-to-read introduction to geology intended for those with little or no science background. Includes many excellent illustrations, as well as a CD-ROM.

Press, Frank, and Raymond Siever. *Understanding Earth.* 2d ed. New York: W. H. Freeman, 1998. This comprehensive physical geology text covers the oceans and their mineral resources, as well as the formation and development of the continents. Readable by high school students and general readers. Includes an index and a glossary of terms.

Segar, Douglas. *An Introduction to Ocean Sciences.* New York: Wadsworth, 1997. Comprehensive coverage of all aspects of the oceans and the oceanic crust, including the development of coastal features. Readable and well illustrated. Suitable for high school students and above.

PERMAFROST

Permafrost is a thermal condition existing in any type of Earth material that is perennially frozen. It occurs in ground in which the temperature remains below 0 degrees Celsius for at least two consecutive years. Affecting about 25 percent of the Earth's surface, the condition hampers construction in the far north in regions such as Siberia and Arctic Canada.

PRINCIPAL TERMS

PERIGLACIAL: originally restricted to regions modified by frost weathering such as tundra

PLEISTOCENE: the geologic era between 2 million and 16,000 years ago, characterized by extensive continental glaciation and a colder climate

SUBSIDENCE: a downward vertical movement of the ground, commonly caused by land-cover changes of the surface

TALIK: a layer of unfrozen ground in permafrost; water under pressure frequently occurs in talik layers

THERMOKARST: a group of landforms in flat areas of degrading permafrost and melting ice

FORMATION AND DISTRIBUTION

Permafrost, or permanently frozen ground, is a thickness of Earth material such as soil, peat, or even bedrock, at a variable depth beneath the ground surface, in which a temperature below 0 degrees Celsius exists continuously for a number of years (from at least two years to several thousands of years). Permafrost is thus not a process or an effect but a temperature condition of Earth material on land or beneath continental shelves of the Arctic Ocean. The occurrence of permafrost is conditioned exclusively by temperature, irrespective of the presence of water or of the composition, lithology, and texture of the ground. However, unfrozen water horizons may exist within a body of permafrost as the result of various conditions such as pressure or impurities in the water.

With regard to the climatic history of the Earth, cold conditions and glaciers are the exception, not the rule. Because they are so rare, they are referred to as "climatic accidents." During the Pleistocene, advances of continental glaciers covered vast areas of Europe and North America. Glaciers continue to be significant landscape features, since they lock up as ice some 2.15 percent of the Earth's waters. Permafrost is a relatively new condition introduced with the onset of the Pleistocene glacial events, and it will be maintained as long as the climate remains cold.

As the Earth's climate has warmed during the past 20,000 years and the continental ice has receded, the location, distribution, and thickness of permafrost have changed. Today, the area underlain by permafrost is vast; it has been estimated that about 25 percent of the Earth's surface is underlain by permafrost. One reason that vast areas of the Earth possess permafrost is that the continents widen toward higher latitudes in North America and Eurasia. In the Northern Hemisphere, Mongolia, Russia, and China contain about 12.2 million square kilometers of land underlain with permafrost. North America and Greenland have 8.8 million square kilometers of permafrost. In the Southern Hemisphere, Antarctica accounts for 13.5 million square kilometers of permafrost. During the Pleistocene, perennially frozen ground had a different distribution. Permafrost produces soil and sediment structures that remain after the climate warms. Relict features such as ancient ice wedges can be identified, and former permafrost areas can be mapped. Evidence of preexisting permafrost conditions, and thus a record of climate change, has been extensively documented in Great Britain, southern Canada, and the central United States.

In terms of aerial distribution and continuity, permafrost may be classified as either continuous or discontinuous. At higher-latitude land circling the Arctic Ocean, the lateral extent of permafrost is uninterrupted nearly everywhere, except in areas of recent deposition and under large lakes that do not freeze to the bottom. Discontinuous permafrost includes areas of frozen ground separated by nonpermafrost areas. A transect from the Arctic Ocean southward, into Canada or Russia, reveals more and more permafrost-free areas. Farther south, in central Canada and central Russia, only patchy or sporadic zones of perennially frozen ground are to be found.

Permafrost is a unique condition because it is maintained very close to its melting point. Several thousand square kilometers are warmer than −3 degrees Celsius. If global warming continues, as many scientists predict, most discontinuous permafrost would degrade. Russian scientists, in fact, have

Exposed permafrost in the Katherine River Valley, Torngat Mountains, Labrador. (Geological Survey of Canada)

documented the northward retreat of permafrost near Archangel at a rate of 400 meters per year since 1837. In Canada, permafrost has retreated northward more than 100 kilometers since 1945, as ground temperatures have increased by 2 degrees Celsius in the northern prairie provinces. Changes in temperature may result from a local microclimate change, alteration of land cover, or changes in the atmosphere. Conversely, permafrost not only is degraded by global warming but may indeed contribute to atmospheric change as well. Scientists suggest that thawing of Arctic terrain, which is rich in peat accumulation, may affect the chemistry of the atmosphere. It has been determined that one-quarter of the Earth's terrestrial carbon is stored in organic matter in the permafrost and active layer. Long-term warming would release enormous quantities of greenhouse gases, methane and carbon dioxide in particular, accelerating global warming.

VERTICAL CHARACTERISTICS

A cross section of excavated Earth in the high Arctic reveals the vertical characteristics of permafrost. The thickness of permafrost is variable because of differences in air temperatures, occurrence of water, composition and texture of Earth materials, and many other factors. The permafrost in Siberia may be at least 5,000 meters thick,

whereas in Alaska thicknesses of 740 meters have been recorded. Farther south, thicknesses decrease as discontinuous permafrost becomes more common. The thickness of permafrost is in part determined by a balance between the increase of internal heat with depth and the heat loss from the Earth's surface. If the thickness and depth of permafrost remain unchanged for many years, the heat loss at the surface and the heat at the base of the frozen ground are in equilibrium, and a steady state is maintained. At the surface, the ground seasonally thaws and freezes; this section of the permafrost is known as an "active layer." Its thickness is controlled by numerous variables in addition to seasonal temperature variations, and it exhibits great variability in thickness. Vegetation, snow cover, albedo, water distribution, and human alteration of the surface can all act as insulators and affect the thickness of the active layer. However, in areas not disturbed by people, the ground will typically thaw to a depth of from 1 to 12 meters during the summer.

The deeper boundary separating the active layer from the permafrost is known as the "permafrost table" or "zero curtain." Within the permafrost, unfrozen lenses known by the Russian term *talik* occur. These unfrozen bodies frequently contain water and function as protected water sources or aquifers in the Arctic environment. Talik lenses may be isolated bodies completely enclosed in impermeable permafrost and may result from a change in the thermal regime in permafrost. Conversely, talik may be laterally extensive even in continuous permafrost and is frequently a valuable source of groundwater. Talik lenses are larger in discontinuous permafrost, and the active layer thickens at the expense of the permafrost.

Below the permafrost table, the perennially frozen ground has low permeability. Maps and photographs of the Arctic reveal many lakes, swamps, and marshes, suggesting high precipitation. However, evaporation rates are low because of the cold air temperatures, and standing water occurs because the ground is frozen at depth. Throughout the region, precipitation is probably less than 35 centimeters annually, equivalent to that of marginal deserts. The abundant water and poor drainage result from the occurrence of permafrost at depth, not to high atmospheric precipitation. As seasonal warming and cooling occur, the land surface expands and contracts, particularly in continuous permafrost areas. In essence, the lowering of the ground temperature causes thermal contraction of the ground, and vertical fissures or frost cracks occur. Water filling and freezing in these cracks creates ice wedges in the permafrost that widen with seasonal expansion and contraction. Ice wedges are often 3 to 4 me-

ters wide and extend 5 to 10 meters into the perennially frozen ground below the active layer. In areas of degradation, ice wedges in the thawing permafrost begin to melt, creating thermokarst features characterized by numerous depressions similar to karst plains. As temperatures rise above freezing, the active layer extends deeper and deeper and eventually may intercept an ice wedge. If large ice wedges begin to thaw, the ground subsides, creating steep-sided conical depressions. The perimeter of a depression, often composed of very moist sediment, will slump into the depression, thus making it wider. Water then fills the depression, creating a thermokarst, or thaw lake. Averaging about 3 meters across, the lakes occur in clusters, are elliptical in shape, and are all oriented at right angles to the prevailing wind direction.

C. Nicholas Raphael

CROSS-REFERENCES

Aquifers, 2005; Continental Glaciers, 875; Geomorphology and Human Activity, 898; Geomorphology of Dry Climate Areas, 903; Geomorphology of Wet Climate Areas, 909; Glacial Landforms, 914; Grand Canyon, 919; Groundwater Movement, 2030; Karst Topography, 925; Marine Terraces, 929; Radiocarbon Dating, 537; River Valleys, 936; Soil Liquefaction, 334; Weathering and Erosion, 2380.

BIBLIOGRAPHY

Anderson, Bjørn G., and Harold W. Borns, Jr. *The Ice Age World.* Oslo: Scandinavian University Press, 1994. The well-illustrated text details the Quaternary history of North America and northern Europe over the last 2.5 million years. Contains an extended glossary, reference list, and index. For the general reader as well as the serious student.

Ballantyne, C. F., and C. Harris. *The Periglaciation of Great Britain.* New York: Cambridge University Press, 1994. A good upper-level description of permafrost in Britain. The discussion ranges from basic freeze-and-thaw processes to permafrost conditions and distribution in the geologic past.

Bloom, Arthur. *Geomorphology: A Systematic Analysis of Late Cenozoic Landforms.* 3d ed. Engle-

wood Cliffs, N.J.: Prentice Hall, 1998. This college-level text covers the basics of geomorphology. Includes three chapters on glaciation, cold climates, and permafrost. Index and bibliography.

Chernikoff, Stanley. *Geology: An Introduction to Physical Geology.* Boston: Houghton Mifflin, 1999. This is a good overview of the scientific understanding of the geology of the Earth and surface processes. Includes a section on the development and effects of permafrost. Includes a link to a Web site that provides regular updates on geologic events around the globe.

Clark, M. J., ed. *Advances in Periglacial Geomorphology.* New York: Wiley InterScience, 1988. A series of state-of-the-art reviews of numerous processes and landforms in cold climates. Emphasizes processes of formation; very technical. Suitable for advanced Earth science students as a reference and literature source.

Harris, S. A. *The Permafrost Environment*. Totowa, N.J.: Barnes & Noble, 1986. A primer on the consequences of construction in permafrost regions. Topics include the soil properties and methods of permafrost analysis. Written for contractors and environmental specialists.

Williams, P. J., and M. W. Smith. *The Frozen Earth: Fundamentals of Geology.* New York: Cambridge University Press, 1989. A college-level textbook for science majors, emphasizing the physics of cold regions. Equations are used to illustrate significant physical ideas.

RIVER VALLEYS

The valleys in which streams flow are produced by those streams through long-term erosion and deposition. The landforms produced by fluvial action are quite diverse, ranging from spectacular canyons to wide, gently sloping valleys. The patterns formed by stream networks are complex and generally reflect the bedrock geology and terrain characteristics.

PRINCIPAL TERMS

AGGRADATION: the process by which a stream elevates its bed through deposition of sediment

BASE LEVEL: the theoretical vertical limit below which streams cannot cut their beds

FLUVIAL: of or related to streams and their actions

PROKARYOTIC: an upland area between valleys

STREAM EQUILIBRIUM: a state in which a stream's erosive energy is balanced by its sediment load such that it is neither eroding nor building up its channel

UNDERFIT STREAM: a stream that is significantly smaller in proportion than the valley through which it flows

VALLEY: that part of the Earth's surface where stream systems are established; it includes streams and adjacent slopes

FORMATION OF RIVER VALLEYS

River valleys consist of valley bottoms and the adjacent valley sides. Between valleys are undissected uplands known as interfluves. Valley floors may be quite narrow, as in the case of the Black Canyon of the Gunnison River, or they may be quite wide, as in the case of the Hwang Ho or the Brahmaputra. Similarly, valley sides may have very gentle to rolling slopes, or they may be nearly sheer, as in the case of the Arkansas River's Royal Gorge. In many areas, the interfluves are simply divides between adjacent valleys, but on tablelands such as the Colorado Plateau, they may be tens of kilometers wide in places.

River channels and river valleys are products of the streams that flow through them. As a stream erodes a channel for itself in newly uplifted terrain, it eventually carves a valley whose form is determined by the erosive power of the stream, by the structural integrity of the rock and debris of the valley walls, by the length of time that the stream has been operating on its surroundings, and by past environmental conditions. These past environmental conditions are attested by stream channel and valley profiles that have not entirely erased landforms produced during the most recent episodes of glaciation and climate change. The valley of the Mississippi River, for example, is formed in the complex deposits of what was a much larger, more heavily laden glacial meltwater stream that existed only 15,000 years ago. River channels and valleys may be referred to as palimpsests, a term originally used to describe parchment manuscripts that had been partly scraped clean, then reused. On the fluvial (river-carved or river-deposited) landscape, previous landscape elements are seen, just as old words show through on a recycled piece of parchment.

In many parts of the world, streams are found flowing through valleys that appear to have been formed by a far larger stream. The valley width, amplitude of meanders, and caliber of coarse sediment are proportional to far larger stream courses. Such streams are said to be underfit, and the valleys are largely remnant features from times of wetter and cooler climates that accompanied glaciation, or the streams were glacial meltwater channels during deglaciation.

The fact that stream channels and valleys are not chance features on the landscape was first noted by British geologist John Playfair in 1802. Playfair suggested that instead of being isolated features on the landscape, streams are part of well-integrated networks. More important, these networks are finely adjusted to the landscape and to one another in such a way that tributaries almost always join the main trunk stream at the same level as that stream. Streams that must plunge over

943

waterfalls to join a larger stream are quite rare. Such discordant streams are found primarily in recently glaciated terrain, where the stream-valley system has not yet become fully adjusted, such as in Yosemite Valley. This remarkable consistency of stream accordance is the strongest evidence indicating that streams carve their own valleys.

Streams develop into network patterns that are strongly influenced by bedrock structure. Where the bedrock is relatively uniform and without strong joints and faults, a dendritic pattern of drainage develops. In this pattern, a stream system is branched like a tree. On inclined plains such as the Atlantic coastal plain, the stream pattern is often parallel, with major streams flowing directly down the topographic slope to the sea. Inclined mountain systems such as the Sierra Nevada also produce parallel drainage. Where structural folding of the terrain has produced linear ridge and valley topography, main trunk streams occupy the linear valleys and are quite long, with short, steep tributaries feeding them off the flanks of the hills

or mountains. In areas such as the Great Valley of Virginia, this type of pattern has developed and is known as trellis drainage. Volcanoes often produce a radial pattern of drainage; Mount Egmont, New Zealand, is often cited as an example.

BASE LEVEL

The depth of a river valley is a function of the height of the land above base level, the length of time that has passed since the stream began to erode, the resistance of the bedrock to erosion, the load of sediment that the stream is carrying, and the spacing of adjacent streams. "Base level" refers to the theoretical lower limit to which a stream can cut. Ultimate base level for all streams is sea level, but local base level is significantly higher for many streams. Streams require a minimum slope to transport their sediment to the sea; this limits the depth to which a stream may cut. The upper section of the Hwang Ho near Tibet flows at elevations of more than 3,000 meters, but the river must still flow more than 4,000 kilome-

The Columbia River Gorge. (PhotoDisc)

A dry streambed in eastern Arizona—one of many types of river valley. (© William E. Ferguson)

ters to the sea. In order to carry its heavy load of sediment over that distance, the river must maintain its channel at a rather high elevation.

Valleys that are deep and relatively narrow are called canyons; those valleys that are especially narrow are called gorges. Streams require time to erode great canyons, although the rate of cutting can be quite rapid compared with most geologic processes. The process of downwearing of interfluves is far slower, so young (recently uplifted) landscapes produce the deepest canyons; as downcutting by the main stream ceases, upland weathering and erosion lowers the local relief. The deepest, narrowest canyons are eroded in strong, homogeneous rock such as granite and quartzite. The Royal Gorge of the Arkansas River and the Colca Gorge of Peru are excellent examples; both are cut in recently uplifted masses of resistant rock formations. The Colca Gorge is far narrower than is the Grand Canyon and, at more than 3,000 meters, it is nearly twice as deep.

As a mountain mass or tableland is uplifted, streams often develop that flow directly down the initial slope of the land; such streams are called consequent streams because their course is a di-

rect consequence of the terrain slope. Because streams seek the path of least resistance, their courses often follow the outcrop pattern of weak rocks such as shale, producing what is known as a subsequent stream pattern. In many cases, however, streams seem to ignore the terrain and structural slope of the land entirely, flowing through mountains of quite resistant rock. An excellent example is the Black Canyon of the Gunnison River in Colorado, where the river carves a deep canyon through a high plateau, with its channel cut in resistant gneisses and igneous intrusives. What makes this so surprising is that much lower terrain, underlain by thick sequences of weak shale, lies only 3 kilometers to the west of the head of the gorge, which would seem to provide a much easier path to the sea. The reason for this course—and many other anomalous stream courses—is that the course of the Gunnison was established prior to the uplift of the plateau. As the land rose beneath the river, it maintained its position, carving an ever-deepening gorge in the rocks as they rose. Far older and more extensive is the anomalous course of the New River, which cuts directly across the structure of the Appalachians, flowing a great distance to

A river gorge at Fall Creek, Ithaca, Thompkins County, New York. The gorge wall shows erosion partially controlled by joints. (U.S. Geological Survey)

At any given time, a stream may either erode or aggrade (build up) its bed. This vertical change in stream profile is determined by many factors internal to the stream, as well as outside environmental factors. Over geological time, the tendency of streams is to erode their beds deeper and deeper. Over shorter time intervals, however, a stream may reach a state of equilibrium between its erosive energy and the load of sediment that the stream is carrying. A delicate balance is maintained within the stream channel: If stream energy is increased by an increase in flood volume or frequency, for example, the stream is likely to respond by eroding its bed. Conversely, if stream power remains constant and the sediment load is significantly increased, the stream will respond by aggrading. This channel aggradation increases the channel slope, giving the stream more energy to transport the sediment load, and a new equilibrium is reached.

Michael W. Mayfield

CROSS-REFERENCES

Aerial Photography, 2739; Alluvial Systems, 2297; Alpine Glaciers, 865; Dams and Flood Control, 2016; Drainage Basins, 2325; Floodplains, 2335; Floods, 2022; Geomorphology and Human Activity, 898; Geomorphology of Dry Climate Areas, 903; Geomorphology of Wet Climate Areas, 909; Glacial Landforms, 914; Grand Canyon, 919; Hydrologic Cycle, 2045; Karst Topography, 925; Marine Terraces, 929; Permafrost, 933; River Flow, 2538; Rocks: Physical Properties, 1348; Sediment Transport and Deposition, 2374; Weathering and Erosion, 2380.

the Ohio River rather than the shorter, more direct route to the Atlantic. Once again, this river is one that was established prior to the uplift of the land in which it is now entrenched, in this case the ancient Appalachians. Thus, the river is rather ironically named, considering that its course is older than the Appalachians.

BIBLIOGRAPHY

Bloom, Arthur. *Geomorphology: A Systematic Analysis of Late Cenozoic Landforms.* 3d ed. Englewood Cliffs, N.J.: Prentice Hall, 1998. This college-level text covers the basics of geomorphology. Includes three chapters on fluvial systems and the development of river valleys. Index and bibliography.

Chorley, R. J., S. A. Schumm, and D. E. Sugden. *Geomorphology.* New York: Methuen, 1985. An advanced text in geomorphology that is best used as a reference. Provides detailed citations of the important scholarly works on each subject addressed and is quite thorough. Many diagrams, graphs, and line drawings supplement the material, but there are few photographs. Suitable for college students who have already gained some background in geomorphology.

Dolgoff, Anatole. *Physical Geology.* Lexington, Mass.: D. C. Heath, 1996. This is a comprehensive guide to the study of the Earth. Extremely well illustrated and includes a glossary and an index. Although this is an introductory text for college students, it is written in a style that makes it understandable to the interested layperson. Contains a section on the development of river valleys.

Hunt, C. B. *Natural Regions of the United States and Canada.* San Francisco: W. H. Freeman, 1974. This book is primarily a physiography text, aimed at describing and explaining the pattern of landforms across North America. The author produced it in the light of "the general public's interest in the natural environment and the need for an authoritative account of it in language as nontechnical as possible." In both regards, the book is quite successful. Well illustrated with diagrams, line drawings, and photographs. Introductory chapters provide an explanation of the general phenomena, and the remaining chapters provide a very thorough, understandable coverage of the regional expression of landforms. Suitable for high school students and above.

McKnight, T. L. *Physical Geography: A Landscape Appreciation.* 2d ed. Englewood Cliffs, N.J.: Prentice-Hall, 1987. This physical geography text provides very thorough but understandable coverage of the concepts of stream networks, stream channel history, and fluvial dynamics. Richly illustrated with color diagrams and photographs. Suitable for high school students and above.

Skinner, Brian J., and Stephen C. Porter. *Physical Geology.* New York: John Wiley & Sons, 1987. A consistently excellent text in introductory physical geology, richly and generously illustrated in full color. Very thorough and quite extensive (more than seven hundred pages). Chapter 11, "Streams and Drainage Systems," is one of the best chapters. Suitable for advanced high school students and above.

9
PALEONTOLOGY AND THE FOSSIL RECORD

ARCHAEBACTERIA

Archaebacteria are primitive, one-celled life-forms without a distinct nucleus, different from bacteria in their genetic components. They have been found to be genetically unique and are probably one of the Earth's earliest life-forms.

PRINCIPAL TERMS

EUBACTERIA: minute, prokaryotic organisms that inhabit a range of habitats considerably greater in diversity than those occupied by other organisms; exclusive of archaebacteria, they constitute the majority of monerans

EUKARYOTIC CELL: the cell type present in all animals, plants, fungi, and protists; they have a distinct nucleus and mitochondria, chloroplasts, and other subcellular structures absent in prokaryotic cells

MICROBIALITE: a biogenic sedimentary structure that is found fossilized in sedimentary rock strata of various ages and that is attributed to the life activities of monerans

MONERANS: generally, single-celled organisms that often grow in colonies and that have a prokaryotic cell

PROKARYOTIC CELL: the cell type found in the kingdom monera, characterized by a number of criteria, including the absence of a cell nucleus, mitochondria, and chloroplasts

RIBOSOME: a large multienzyme complex made up of protein and ribonucleic acid (RNA) molecules that carry and process information stored in deoxyribonucleic acid (DNA); this information is also carried by RNA to synthesize proteins

STROMATOLITE: a biogenic sedimentary layered structure produced by sediment trapping, binding, or precipitation as a result of the photosynthesis of microorganisms, principally cyanobacteria (blue-green algae)

EARLY LIFE-FORMS

The nature of the Earth's earliest life-forms has always been an intriguing question for both the Earth sciences (paleontology) and biology. The fossil record shows that one-celled organisms are very ancient, their oldest-known fossils being almost 3.5 billion years old. The long fossil record of prokaryotes consists of both preserved fossil cells and distinctive layered mineral structures called stromatolites and microbialites; these structures were produced from the cell metabolisms of colonies of prokaryotes. From at least 3.5 billion years ago to around 1 billion years ago, microscopic prokaryotes were the Earth's only organisms. They included, as they do today, a diversity of forms commensurate with their long evolutionary history. The appearance of the eukaryotic cell (more than 1 billion years ago) ushered in the age of multicelled organisms (metazoans and metaphytes) some 700 million years ago, and these have become the dominant life-forms on the Earth. What the much earlier prokaryotic organisms were

like and what type of one-celled organisms produced the microbialites is unclear. Usually these oldest of fossils are attributed to the life activities of photosynthetic organisms, particularly the cyanobacteria, referred to in many works as the blue-green algae. A number of other types of one-celled life-forms could have been responsible for some of them, particularly the photosynthetic bacteria and possibly the archaebacteria.

Molecular biologists, utilizing ribonucleic acid (RNA) nucleotide sequencing and other biochemical methods, believe that the nature of early life-forms can be discovered. RNA nucleotide sequences of amino acids can be regarded as a sort of chemical "historical document" that is capable of being "read." The closer two RNA nucleotide sequences are to each other, the smaller is the evolutionary distance between them and the more recent in geologic time did they separate from each other; the further a nucleotide RNA sequence is from another, the greater is the evolutionary distance that separates the two organisms. Utilization of nucleo-

951

tide sequencing can thus produce an evolutionary tree, or phylogeny, for even the most primitive of organisms, and therefore it becomes possible to determine which organisms out of the great variety of primitive life-forms currently living were some of the first to appear. Through information obtained by such sequencing, archaebacteria have been recognized by molecular biologists as some of the most primitive and biochemically unique of organisms. Archaebacteria RNA sequences turn out to be distinctly different from those of other bacteria, even though the various organisms that comprise archaebacteria look like and were previously placed with the bacteria.

A Precambrian stromatolite formation from the Siyeh Formation, Glacier National Park, Montana. (U.S. Geological Survey)

Archaebacteria represent a phylum within the kingdom Monera, a category which includes all those one-celled life-forms that lack a cell nucleus. On the basis of their distinctive biochemistry, the archaebacteria are considered by some life scientists to represent a distinct kingdom, equal in rank to animals, plants, fungi, and protists. In both their tolerance of extreme ecological conditions and their metabolism, the archaebacteria differ from all other monerans, a condition which has led biologists to consider these organisms as particularly well suited to the adverse conditions of the early Earth. The very earliest eras of geologic time may well have been the age of archaebacteria.

MOLECULAR CHARACTERISTICS

On a fundamental molecular level, archaebacteria are different in their biochemistry from the other prokaryotes. Nucleotide RNA sequences and other biochemical differences that exist between the various types of eubacteria (bacteria exclusive of the archaebacteria) are minor when compared with the differences between eubacteria and the archaebacteria. Nongenetic differences include such features as cell walls, those of all eubacteria being composed of a complex polymer called peptidoglycan, which is a sugar derivative. In contrast, cell walls of the various types of archaebacteria are

composed of a variety of other materials, none of which is peptidoglycan. The lipids (fats) in archaebacteria cells are also fundamentally distinct from the lipids in the cells of both eubacteria and eukaryotes. Ribosomal RNA is what ultimately distinguishes the archaebacteria, for it is markedly different in its sequences of bases from any eubacteria. In higher eukaryotic organisms, where the fossil record is good, greater RNA ribosomal differences exist between those organisms that are separated by long periods of geologic time than between those separated by shorter periods of time. Ribosomal RNA differences and the biochemical differences that exist between the eubacteria and the archaebacteria suggest that an evolutionary distance of great magnitude separates them.

The eukaryotic cell has long been observed to be a sort of combination between prokaryotic-type cells, functioning within the cell as chloroplasts and mitochondria, and another cell-type that "ingested" the prokaryotes and incorporated them to become a more complex entity in symbiotic collaboration. Archaebacteria have some genetic characteristics that suggest a link with the eukaryotes, which has led some scientists to propose a predecessor to them both: The "other" cell-type that linked up with prokaryotes underwent substantial further evolution and eventually became the eukaryotic cell. The pre-eukaryotic other cell type is known as a urkaryote. All three of these cell types—the prokaryote, the urkaryote, and the eukaryote—are hypothesized to have arisen from a common cell ancestor, the progenote. The progenote may have been biochemically simpler than any of the three fundamental life-forms that arose from it, an event that might have taken place during the first 1 billion years of Earth history.

THERMOACIDOPHILES

Archaebacteria are represented by three classes: the thermoacidophiles, the extreme halophiles, and the methanogens. The thermoacidophiles occupy hot, acid environments, often rich both in metallic ions and in sulfur compounds, such as hot springs and fumaroles. These organisms are viable under the acidic hot conditions intolerable to other life-forms, with temperatures as high as 75 degrees Celsius. The extreme halophiles, or halobacteria, live only in extremely salty environments. The methanogens are anaerobes that metabolize organic material to form methane; they were the first of the archaebacteria to be discovered.

Igneous activity of various sorts appears from the geologic record to have been much more intensive and widespread in the early Earth (between 2 and 4 billion years ago) than during more recent geologic times or at present. A terrestrial geologic record for the first 1 billion years of the Earth's history is unknown, as the widespread igneous and tectonic activity during this time seems to have destroyed the evidence; an actual record begins at nearly 4 billion years ago. During the next 1.5 billion years (the Archean eon), igneous phenomena and massive tectonism were still dominant. Hot-spring and fumarolic activity would have been more commonplace during these early times than during later geologic time, and these environments are favored by the thermoacidophiles. Although the fossil and sedimentational record of the Archean eon does not negate the possi-

bility that archaebacteria were some of the most widespread and dominant life-forms of that time, determination that a particular organic—or presumed organic—structure of the early Earth was produced by archaebacteria or by some other moneran is quite difficult and may well be impossible. A number of puzzling structures, seemingly of biogenic origin, have been reported from Archean strata. Some of these stromatolite-like or microbialite-like structures are associated with what appear to be hot-spring deposits. These structures may well represent minerals deposited as a consequence of life activities of thermoacidophiles associated with geothermal activity. Like so many stromatolite-like and microbialite-like structures, unequivocal proof as to their biogenic origin is difficult to obtain. Structures similar to them, however, are produced today in hot springs, in the hottest waters of which live communities of thermoacidophile archaebacteria.

Sometimes hot-spring deposits and structures contain carbon-rich sediments or graphite. In addition, some hydrothermal veins of various ages have associated carbonaceous or graphitic material. It is possible that such material might have originated from thermophyllic archaebacteria. Stratified metallic element deposits are known from Archean strata, some of which have been thought to be of biogenic (possibly archaebacterial) origin. Some

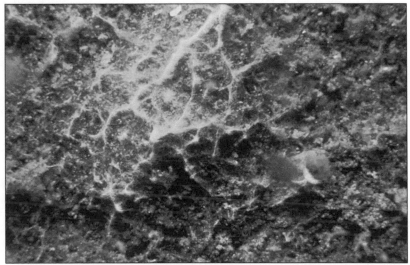

Bacterial mats form on the seafloor, where oxygen is low, and may provide insight into the earliest life-forms on Earth. (National Oceanic and Atmospheric Administration)

of these deposits yield microbialites exhibiting distinctive dome, finger, or layered structures containing metallic oxides or carbonates. Today, such structures are produced by the cyanobacteria and by the photosynthetic bacteria, but these younger stromatolites lack components like the oxidized metals. Other Archean stromatolites or microbialites and stromatolite-like structures associated with geothermally active environments have a distinctive "signature" different from later forms, and their origin by thermoacidophile archaebacteria cannot be ruled out.

EXTREME HALOPHILES

The second group of archaebacteria, the extreme halophiles, requires an intensely saline environment. Shallow, marginal marine areas, evaporite basins, and salt flats are the niches in which these organisms generally flourish. Physiologically, the extreme halophiles are photosynthetic; however, the photosynthetic pigment is not chlorophyll, but is rather a light-sensitive red pigment, bacterial rhodopsin. The cell walls of the extreme halophiles differ from those of other bacteria in the presence of compounds that prevent destruction of the walls in the high salt concentrations under which they live. The chemical similarities of ribosomes and lipids of both the extreme halophiles and the methanogens suggest a common origin.

Again, the fossil record of these organisms is difficult to interpret; some biologists suggest that the halophiles were more prevalent early in the Earth's history than they are today. Fossil rod-shaped bacterial cells have been found as far back as the mid-Archean (3.2 billion years ago); however, as the gross morphology of archaebacteria differs little from that of eubacteria, evidence remains inconclusive.

Peculiar and distinct microbialites of Archean age that are associated with radial sprays of gypsum crystals were described from western Ontario in the 1910's by Charles Doolittle Walcott. Walcott, a pioneer North American paleontologist who concentrated on the early (Precambrian and Cambrian) fossil record, made many finds of peculiar structures resembling fossils in Precambrian strata, many of which remain a mystery. Walcott thought that the radiating gypsum crystals were the rays, or spicules, of a type of spongelike organism he called

atikokania. Associated with Walcott's atikokania are distinctive microbialites that contain "lenses" of gypsum that almost certainly originated in a very saline environment. These microbialites could possibly represent the product of physiological activity of the extreme halophiles when, during the process of photosynthesis, they locally removed carbonic acid from the saline water. The white lenses that characterize these distinctive microbialites are gypsum fillings between the black calcium carbonate bands, possibly precipitated by photosynthesis of halophilic archaebacteria.

METHANOGENS

Methanogens produce their metabolic energy either from the breaking down of organic compounds incorporated into sediments or from the reduction of carbon dioxide in the presence of elemental hydrogen, with the consequent release of methane. Were it not for the methanogens, organic carbon would eventually become incorporated into the sediments of the Earth's crust, where it would accumulate and could not be recycled back into the biosphere; the methanogens facilitate this recycling of carbon. Methanogens, like the other archaebacteria, have biochemical features distinct from all other bacteria, suggesting that they evolved separately from them; like the other archaebacteria, methanogens differ from other prokaryotes in the sequences of nucleotides that make up the RNA in their ribosomes and protein. Fossil methanogens are more difficult to distinguish from the geologic record than other archaebacteria, however, as they leave no distinctive chemical "footprint," as the others can. The abundance of black, carbon-rich sediments in strata of the Archean eon suggests that the oxygen-free, anaerobic environment in which the methanogens flourish was commonplace during that time. The methanogens' biochemical uniqueness, and thus presumed great geologic age, along with the anaerobic Archean Earth environment, suggests that they may have been a dominant part of the Archean biosphere and not restricted as they are today.

Bruce L. Stinchcomb

CROSS-REFERENCES

Archean Eon, 1087; Biogenic Sedimentary Rocks, 1435; Cambrian Diversification of Life, 956; Car-

BIBLIOGRAPHY

Kandler, Otto, and Wolfram Zillig, eds. *Archaebacteria Eighty-five: Proceedings of the EMBO Workshop on Molecular Genetics of Archaebacteria.* Forestburgh, N.Y.: Lubrecht and Cramer, 1987. A proceedings volume on molecular genetics, biology, and biochemistry of archaebacteria. Two papers are concerned with the geologic and paleontologic record of archaebacteria: "Traces of Archaebacteria in Ancient Sediments," by J. Hahn and Pat Haug, and "Morphological and Chemical Record of the Organic Particles in Precambrian Sediments," by H. D. Pflug. This latter paper illustrates a wide variety of microstructures from Precambrian sediments and discusses possible pathways by which archaebacteria and other prokaryotes could have been responsible for the concentration of many metallic ore deposits throughout various parts of the Precambrian. Other papers such as "Archaebacterial Phylogeny: Perspectives on the Urkingdoms," by C. R. Woese and G. J. Olsen, present biochemical reasons substantiating the uniqueness of archaebacteria from both the eubacteria and the eukaryotes. Other papers probe the biochemical similarities with eukaryote cytoplasm and peculiar substrate requirements.

McMenamin, Mark. *Discovering the First Complex Life: The Garden of the Ediacara.* New York: Columbia University Press, 1998. This entertaining study of the earliest complex life-forms on the planet details the author's work on these organisms. Written for the interested student but understandable by the general reader.

Schopf, J. William, ed. *Major Events in the History of Life.* Boston: Jones and Bartlett, 1992. An excellent overview of the origin of life, the oldest fossils, and the early development of plants and animals. Written by specialists in each field but at a level that is suitable for high school students and undergraduates. Although technical language is used, most of the terms are defined in the glossary.

Woese, Carl R. "Archaebacteria." *Scientific American* 244 (June, 1981): 98. One of the most comprehensive articles available on the archaebacteria. Distinctive attributes characteristic of the archaebacteria as determined through molecular biology are enumerated. The author was one of the workers originally involved in the discovery of the biochemical uniqueness of archaebacteria.

CAMBRIAN DIVERSIFICATION OF LIFE

The abrupt appearance of a great variety of animal fossils about 544 million years ago and about 3 billion years after the origin of life is termed the Cambrian diversification. Study of this spectacular development provides insights into the processes of evolution.

PRINCIPAL TERMS

COELOMATE: an organism possessing an internal cavity termed the coelom

EUKARYOTE: a cell that has a nucleus surrounded by a well-defined membrane; the type of cell present in the metazoans

METAZOAN: a grade of organization of living organisms in which the cells are specialized for various functions and cooperate for the good of the whole organism

PHANEROZOIC: the time period from 544 million years ago to the present, during which

sediments accumulated containing obvious and abundant remains of animals and plants

PRECAMBRIAN: the time period that includes nearly 90 percent of geologic time, ranging from 4.6 billion years ago, when the Earth formed, to 544 million years ago, when the Cambrian period started

TRACE FOSSILS: traces of animal activity, such as burrows or trackways, preserved in the sediment

APPEARANCE OF COMPLEX ORGANISMS

Although the remains of microscopic organisms are known from rocks more than 3 billion years old, the remains of complex animals are not known before about 680 million years ago, while plants and animals with preserved hard parts did not become established until about 544 million years ago. The abrupt appearance of a great variety of animal fossils at the end of the Precambrian (the time period ranging from about 4.6 billion years ago to the start of the Cambrian period, 544 million years ago) is termed the Cambrian diversification and was a spectacular radiation heralding the start of the Phanerozoic (the time period ranging from the Cambrian to the Quaternary period). By the middle of Early Cambrian time, most of the major invertebrate animal groups had appeared in the oceans, and this rapid appearance is a striking feature of the fossil record. The diversification has been documented and shows a period of about 80 million years during which the variety of life increased exponentially, suggesting that each group originated by the simple splitting into two of an ancestral group. Unfortunately, although there is fossil evidence of early metazoans (complex animals), it is sparse, and reconstruction of the events during which simpler animals

groups gave rise to more advanced animal groups is based primarily on comparative anatomy and embryology of modern forms.

The earliest record of metazoans occurs in the Ediacaran interval, which is a stratigraphic unit deposited during the last 100 million years of the Proterozoic. Prior to that there is only evidence of unicellular organisms. Initially these were prokaryotic (single cells lacking nuclei), but about 1.75 billion years ago, the first eukaryotic organisms (cells with a nucleus) had developed. The earliest metazoans were probably loose aggregates of eukaryotic cells that were not differentiated by function; however, the organisms found in the Ediacaran interval are already more complex. The fauna was first described from the Pound quartzite in the Flinders Ranges of South Australia. The fossils are impressions, left as the animals were stranded on mud flats and subsequently covered by sand, and they are interpreted as having been made by metazoans lacking hard parts. Many of the impressions are circular in outline with concentric or radial striations and have been considered to be jellyfish, though none can be tied with confidence to living organisms. In addition, there are large (up to 1-meter-long), elongate, leaf-shaped forms considered to be related to modern sea pens, which are frondlike representatives of one of the coral groups. Other impressions are regarded as having been made by polychaete worms (bristle worms), while much rarer are impressions of supposed arthropods and animals of uncertain affinities.

Whether these impressions should be interpreted in terms of living jellyfish, corals, and worms has been questioned, but they are clearly evidence of the existence of early metazoans. This fauna significantly predates the earliest Cambrian faunas and is now known from five continents. The presence of the same or similar forms as far apart as South Australia and northern Russia is striking evidence of the widespread distribution of this early shallow-marine fauna. In this fauna, scientists can glimpse a fleeting stage in metazoan diversification, a stage that may have given rise to the more diverse and abundant invertebrates of the Cambrian and later periods but that ultimately became extinct before the end of the Proterozoic.

Trace fossils also provide evidence of the early development of metazoans and can be particularly valuable, as impressions left by animals in sediment may be preserved when the animals themselves may not be. Trace fossils do not become abundant or diverse until near the Precambrian/Cambrian boundary, and most of these early traces probably resulted from the burrowing activity of soft-bodied infaunal worms. Trace fossils take the presumed origin of mobile metazoans further back into the Precambrian than do the Ediacara fauna but only to about 100 million years before the start of the Cambrian. As the Ediacara fauna indicates that a number of major groups had already developed, the initial radiation of Metazoa had clearly taken place earlier. Study of the relationships of modern metazoan groups can help shed light on the course of this diversification.

LEVELS OF ORGANIZATION

Several grades or levels of organization are recognized in metazoans living at present. The simplest forms are ones in which the cells are not separated by function; some modern sponges can serve as examples of this grade. Most simple metazoans have cells that are separated by function, however, and the simplest grade is represented by the Cnidaria (corals and jellyfish), in which the body wall is separated into two layers: the outer

ectoderm and the inner endoderm. The next grade, which includes all higher organisms, consists of forms in which the body wall is in three layers, the mesoderm being sandwiched between the ectoderm and endoderm. In more advanced invertebrate animals, the mesoderm forms a lining to the ectoderm and overlies a fluid-filled cavity termed the coelom, which may have originally functioned as a hydrostatic skeleton but in higher organisms has also been used as space for the placement of internal organs.

The coelomates (organisms with a coelom) can be divided into several major groups on the basis of embryology, indicating the early subdivisions of the basic stock. These include a lophophorate group, encompassing Brachiopoda and Bryozoa, in which a food-gathering apparatus termed a lophophore is present; a large group including both Echinodermata and Chordata (the group that includes vertebrates); and a further group that includes Arthropoda and Annelida (segmented worms). The larvae of marine annelids are very similar to those of Mollusca, and the mollusks may, therefore, have split off early from the main line of descent of segmented animals. Although the time

Trilobite and brachiopod fossils from the Bonanza King Formation, Death Valley, California. (U.S. Geological Survey)

of origin of these groups is unknown, it can be concluded that coelomates were present already in the Late Precambrian and that the origination of arthropods was closer to the age of the Ediacara fauna than that of the annelids or the cnidarians that form the majority of the population.

The Ediacara fauna vanish from the sedimentary record before the end of the Precambrian, although the presence of trace fossils indicates that metazoans were still present. The base of the Cambrian is recognized as the appearance of fossils of animals that secreted a skeleton, and this basal stage of the Cambrian, the Tommotian stage, contains a diverse fauna of small shelly organisms, many tubelike and composed of calcium phosphate. This fauna also includes sponges, brachiopods, gastropods, primitive mollusks, and hyolithids (conical shells usually considered to be mollusks, though their soft parts cannot be reconstructed with any degree of confidence). The best known sequences through the Precambrian-Cambrian boundary are in Siberia, where sedimentation occurred without a major interruption through this period. In these sequences, small shelly fossils occur only rarely in the Precambrian, although they are abundant in the Tommotian, pointing to an explosive development of life in the earliest Cambrian. In fact, all the modern phyla (except the Bryozoa) have a fossil record that starts in the Cambrian.

Above the Tommotian stage, a number of groups of larger animals with hard skeletons appear. They are characteristic of later Cambrian faunas, appear very abruptly, and are fully organized and differentiated on their first appearance. One of the most conspicuous groups is the trilobites, many-legged arthropods that crawled along the seafloor, often forming conspicuous trace fossils. Brachiopods, two-valved filter-feeding organisms, are also common, and echinoderms were represented by a remarkable variety of types. Although the animals with hard skeletons were very diverse, there is also evidence that the soft-bodied faunas were equally diverse.

One indication of this diversity is provided by the fauna preserved in the Middle Cambrian Burgess Shale of British Columbia and the slightly older Chengjian fauna from China. The black shale of the Burgess Shale accumulated in an oxygen-free environment, which prevented the destruction of the animals that were washed in with it. The majority of organisms in this fauna are nontrilobite arthropods, but there are also numerous polychaete and priapulid worms, which were already highly diversified and which might have caused the burrows found in rocks of latest Precambrian age. In addition, there are many animals of unknown affinity, often forms found only in the Burgess Shale. This fauna indicates the presence of complex communities composed of highly organized animals; the diversity in the range of feeding adaptations, for example, is quite as varied as that found in modern animals. This complex fauna is quite unlike that of the earliest Cambrian or of the Ediacaran and demonstrates the explosiveness of the early radiation.

DIVERSIFICATION THEORIES

A number of explanations of the sudden diversification in the Cambrian have been put forward. Environmental factors were certainly very important, and one suggestion is that the diversification may have been related to an increase in suitable environments at the end of the Precambrian. That would have been caused partly by the breakup of continental areas, which would have provided a greater extent of coastline and, therefore, a rapid increase in the availability of habitats for the shallow-marine organisms. In addition, the end of an extensive period of glaciation that took place at the end of the Precambrian would have resulted in the flooding of coastal areas, providing increased shallow-marine environments. The end of the glacial period would also have resulted in a warming trend that would have opened up new marine environments and possibly helped to trigger the expansion of marine diversity.

It has also been suggested that an increase in oxygen levels may have contributed to the sudden development of complex metazoans. Large and thin animals such as those present in the Ediacara fauna may have been adapted to respire by diffusion in oxygen concentrations as low as 8 percent of present levels. Oxygen levels of up to 10 to 15 percent of present levels may have been present in the earliest Cambrian, which may have been sufficient to allow the increased diversification of invertebrate organisms.

Further evidence for oxygen levels is also available in the appearance of hard skeletons at the base of the Cambrian. External skeletons are useful protection, suggesting their development in response to predation; however, they are also useful as solid surfaces for the attachment of muscles and as supports to lift organisms above the bottom. Calcium and phosphate ions are both essential to the processes within metazoan cells, and it may be that skeletal parts originated as reserves of these materials. Calcium phosphate is a hard tissue commonly found in earliest Cambrian faunas; organisms using calcium carbonate did not become common until the end of the Cambrian.

At present, an atmospheric oxygen level of at least 16 percent is required before marine organisms can secrete calcium carbonate skeletons, suggesting that increased oxygen levels in the Cambrian may have contributed to the development of diverse organisms at that time. It appears, therefore, that a variety of environmental factors—including increase of shallow-marine areas, a warming trend, and increasing oxygen levels—may all have contributed to the explosive Cambrian diversification of invertebrate organisms.

STUDY OF FOSSILS

The evidence for the early evolution of complex organisms rests on the fossils present in the sediments. Studies of these remains give scientists insights into the way in which early life developed. Preservation of organisms is an unusual occurrence, however, and particularly rare when the organisms are completely soft-bodied. Preservation potential is enhanced by the presence of some hard parts, skeleton or shell, that will resist erosion and decay. The potential for preservation can be additionally enhanced if the organism lives in an environment in which sedimentation is taking place, thus improving the chances of incorporation in the sediment. Because of these constraints on preservation, scientific knowledge of shallow-marine faunas through time is much greater than knowledge of faunas in other environments.

The earliest metazoan faunas are those from the Ediacaran. These organisms were soft-bodied, and their preservation is therefore an extremely uncommon event. The fossils consist of natural molds and casts formed when the animals were covered by drifting sands in nearshore environments. The specimens can be studied directly by scientists, as they are normally clearly exposed on the bedding planes of the sediments when the rocks are split. In some cases where molds only are present, a latex cast may be made to facilitate detailed study. Preservation of the early Cambrian fauna in the Burgess Shale is rather different, however, and a greater variety of techniques can be used in its study. The specimens are compressed into thin films of carbon and are exposed on the bedding planes of the shale when it is split. In many cases, the splitting results in breakage through the specimen so that parts of the original adhere to two separate rock slabs (termed the part and the counterpart).

Such compression means that in complex animals such as arthropods, delicate preparation may be necessary to remove surfaces so that other surfaces and structures may be examined. Such preparation is conducted under a binocular microscope, and small engraving tools and needles are used to remove the matrix. Interpretation of the specimens involves producing drawings and photographs, and the Burgess Shale material provides special problems here, as the specimens are preserved as black carbon films on a black shale. Ultraviolet light may be used to enhance the reflectivity of the specimens relative to the surrounding matrix; they may also be photographed under a liquid (water or ethyl alcohol), as this process will also enhance differences between the specimen and the matrix. Drawings are normally made using a camera-lucida attachment on a binocular microscope. This attachment allows the scientist looking through the microscope to see both the specimen and his or her drawing hand and pencil, thus enabling the tracing of the specimen and production of an accurate illustration.

Study of the data as a whole requires analysis of large amounts of information so that evolutionary trends can be assessed and periods of rapid diversification or higher than normal extinction can be recognized. These studies are dependent on the original description and identification of the fossils, but manipulation of the data is accomplished by computers.

David K. Elliott

CROSS-REFERENCES

BIBLIOGRAPHY

Glaessner, M. F. *The Dawn of Animal Life: A Biohistorical Study.* New York: Cambridge University Press, 1984. A rather detailed account of the Ediacara fauna and its significance. Suitable for college-level students.

Gould, Stephen J. *Wonderful Life: The Burgess Shale and the Nature of History.* New York: W. W. Norton, 1989. An extremely readable text, aimed at the general reader, that also manages to provide an enormous amount of information about the Burgess Shale, the Cambrian explosion, and the nature of evolution.

Prothero, Donald R. *Bringing Fossils to Life.* Boston: McGraw-Hill, 1998. This well-illustrated and entertaining text covers a broad range of paleontological topics, including the Cambrian explosion. Glossary, bibliography, and index.

Stanley, S. M. *Earth and Life Through Time.* 2d ed. New York: W. H. Freeman, 1989. A general text on paleontology and historical geology that provides considerable background on the evolution of the Metazoa in its environmental context. Suitable for high school or college students.

Stearn, C. W., and R. L. Carroll. *Paleontology: The Record of Life.* New York: John Wiley & Sons, 1989. A general paleontology text that provides an extremely clear section on the early development and diversification of the Metazoa. Suitable for high school or college students.

Whittington, H. B. *The Burgess Shale.* New Haven, Conn.: Yale University Press, 1985. A very readable and well-illustrated account of the history of study of the Burgess Shale, its fauna and significance. Suitable for college-level students.

CATASTROPHISM

Historically, catastrophism was the doctrine that a series of sudden and violent events caused widespread or even global effects, producing the differences in fossil forms and other features found in successive geological strata. More recently, a new school of catastrophism has arisen, prompted by growing evidence that much of the geological record, including mass extinctions of living organisms, has been greatly influenced by rare events of large magnitude, such as widespread flooding, volcanic activity, and asteroid impacts.

PRINCIPAL TERMS

ASTEROID: one of the numerous small rocky bodies bigger than about 10 meters in size orbiting the Sun, mostly between Mars and Jupiter; some, however, cross the Earth's orbit

CRATER: an abrupt circular depression formed by extrusion of volcanic material or by the impact of an asteroid or meteorite

CRUST: the outermost solid layer or shell surrounding the Earth

FOSSIL: naturally preserved remains or evidence of past life, such as bones, shells, casts, and impressions

GEOLOGICAL COLUMN: the order of rock layers formed during the course of the Earth's history

METEORITE: a fragment of an asteroid that survives passage through the atmosphere and strikes the surface of the Earth

STRATIGRAPHY: the study of rock layers to determine the sequence of layers and the information this provides on the geological history of a region

UNIFORMITARIANISM: the doctrine that geological events are caused by natural and gradual processes operating over long periods of time

VOLCANISM: the processes by which magma is transferred from the Earth's interior to produce lava flows on the surface and the ejection of gases and ash into the atmosphere

NEPTUNISM AND VULCANISM

Although the term "catastrophism" is usually associated with the work of Georges Cuvier near the beginning of the nineteenth century, most theories of Earth history before that time involved various ideas of catastrophism, emphasizing sudden and violent events rather than gradual processes. Based on the biblical account of creation in six days and genealogies of the descendants of Adam, most writers assumed that the Earth was only about six thousand years old. Early theories of the Earth's surface features were based on the biblical account of Noah's flood. In early eighteenth century England, Thomas Burnet and John Woodward used the idea of a universal flood to explain geological phenomena such as the formation of mountains and valleys, irregularities in strata, and the existence and location of fossils. These ideas stimulated the collection of fossils as evidence of biblical veracity.

In Italy, where volcanoes were active, Venetian priest Anton Moro suggested in 1740 that Noah's flood was a more localized event and that rock strata formed from a series of violent volcanic eruptions that entombed plants and animals, forming fossils in the rocks. These catastrophic ideas were sometimes viewed as complementary; by the late eighteenth century, however, they led to a controversy between the Neptunists, who stressed the role of water and floods, and the Vulcanists, who emphasized fire and heat.

In 1749, Georges Leclerc, Comte de Buffon, keeper of the King's Gardens in Paris, suggested a speculative natural history of the Earth with a vastly expanded time scale. Instead of a recent six-day creation, he proposed seven epochs of development over a span of about seventy-five thousand years. Using calculations devised by Sir Isaac Newton for estimating the cooling of comets, Buffon experimented with the cooling of a red-hot globe of iron; he extrapolated his findings for a mass the size of the Earth, arriving at an estimate that it would have taken 74,832 years for the Earth to cool to its present temperature. By the time he finished his *Épochs de la Nature* (1779), Buffon had divided his history of nature into seven "epochs" as metaphors of the seven "days" of creation.

Although his treatise did not refer to catastrophism, Buffon's epochs included catastrophic events of both fire and water. In the first epoch, the Earth formed out of matter ejected from a collision of a comet with the Sun. As the Earth solidified in the second epoch, its crust wrinkled to form the mountain ranges. In the third epoch, vapors condensed as the Earth cooled, covering the Earth with a flood in which fish flourished and sediments formed, enclosing fossils and organic deposits such as coal. The fourth epoch began after further cooling produced subterranean openings, causing a rush of waters, earthquakes, and volcanoes that produced dry lands. Land animals and plants appeared in the fifth epoch, and the continents moved apart in the sixth after migrations of animals had separated various species. Finally, wrote Buffon, humans appeared in the seventh epoch.

By the end of the eighteenth century, the Neptunists and Vulcanists became more sharply divided. The British geologist Sir William Hamilton developed in more detail the implications of Vulcanism from the action of volcanoes. He identified basalt and other rocks found near volcanoes as products of lava flows. He argued that volcanic action played a constructive role in uplifting new land from the sea, shaping the landscape, and providing a safety valve for excess pressure below the crust.

ABRAHAM WERNER AND JAMES HUTTON

A purely Neptunist school was established by the German mineralogist and geologist Abraham Werner. He accepted the idea of geological succession in sedimentary deposits but did not develop its historical implications, since he classified rocks by mineral content rather than by fossils. Secularizing earlier theories based on the biblical flood, Werner held that rock strata formed from a universal primeval ocean, which produced four types of rocks by sequential processes. Primitive rocks, such as granite, crystallized out of the primeval ocean and contained no fossils. Transitional rocks, such as micas and slate, contained only a few fossils. Sedimentary rocks such as coal and limestone were next and were rich in fossils. Derivative rocks such as sand and clay formed from the other three by processes of weathering. Werner believed that volcanoes resulted from the burning of underground coal and were not an important geological force.

Although Werner's theory about the origin of sedimentary rocks was largely upheld, most other rocks were eventually shown to have an igneous origin from a molten state. This idea was developed by the Plutonist school of geology, which stressed the geological activity of the internal heat of the Earth, in addition to the volcanic eruptions of the Vulcanists. This view was developed by the Scottish amateur geologist James Hutton in his *Theory of the Earth* (1795). Hutton believed that the geological forces seen in the present operated in the same way and at the same rate in the Earth's past and that this principle should be the basis of geological explanation: The present is the key to the past.

Hutton's "uniformitarian principle" contrasted with Werner's idea of a primeval ocean, which involved catastrophic events confined to the past and unobservable in the present. Hutton carefully observed the slow and steady erosion of the land as rivers carried silt into the sea. He examined the weathered beds of gravel, sand, and mud brought down by the rivers, as well as the crystalline granites of the Scottish mountains. He concluded that sedimentary rocks formed from beds of mud and sand compressed by overlying seas and heat pressure from below, while crystalline rocks came from molten material inside the Earth brought to the surface by volcanic action.

Developing the idea that the interior of the Earth is molten, Hutton suggested that molten rock pushes into cracks beneath the Earth's crust, tilting up sedimentary strata and solidifying to form granites. Thus mountains were built with a crystalline core and sedimentary surface. This principle of injection was apparent in some granite intrusions into crevices in sedimentary rocks above, indicating that the granite was younger. The existence of granites of differing ages was contrary to Werner's assumptions. In some cases, Hutton found horizontal sedimentary strata covering tilted strata near the base of mountains, suggesting long periods of time since the strata had tilted. The age of the Earth appeared to be so long that catastrophic events did not seem to be necessary to account for its surface features.

CUVIER'S THEORY INTRODUCES CATASTROPHISM

Although these uniformitarian ideas found some early support, it was not enough to overcome religious objections to a theory that required such an ancient Earth, delaying its eventual acceptance. In France, Georges Cuvier opposed Hutton's idea of slow geological processes with his theory of catastrophism in the introduction to his *Researches on Fossil Bones* (1812). Since there was no apparent continuity between successive strata and their fossils, he believed that a series of catastrophic floods must have occurred—rather than continuous forces—with each flood wiping out many species and eroding the Earth. These catastrophes also tilted strata left by earlier floods, ending with Noah's flood some six thousand years ago. Cuvier's catastrophism applied Neptunism to the vast time scale of Hutton. His influence delayed the acceptance of both biological and geological ideas of evolution in France for several decades.

In England and France, the study of strata was made easier by many well-exposed horizontal layers rich in fossils. These were systematically studied in France by Cuvier and in England by William Smith, who discovered in 1793 that each stratum had its own characteristic form of fossils. Their work revealed that strata near the surface were younger than those farther down, and a history of life-forms could be worked out from their fossils. Further stratigraphic studies by Adam Sedgwick and Roderick Murchison identified the Cambrian series of strata with the oldest fossil-bearing rocks, the Silurian series with the earliest land plants, and the Devonian series dominated by fish fossils.

The discoveries of this "heroic age of geology" (1790-1830) were summarized by the Scottish geologist Charles Lyell. Reviving Hutton's uniformitarian ideas, Lyell published three volumes entitled *The Principles of Geology: Being an Attempt to Explain the Former Changes of the Earth's Surface by Reference to Causes Now in Operation* (1830-1833). Assuming indefinitely long periods of time, he insisted that geological forces had always been the same as they are now. Yet few of Lyell's contemporaries accepted his ideas before Charles Darwin developed them in his theory of organic evolution. The combined influence of Lyell and Darwin caused many scientists to shift away from Cuvier's catastrophism.

One of Cuvier's later associates, the Swiss-American naturalist Louis Agassiz, helped to modify the extreme uniformitarianism of Lyell. From the distribution of boulders and the grooves scratched on solid rock in the Swiss Alps, he showed in 1837 that Alpine glaciers had once stretched from the Alps across the plains to the west and up the sides of the Jura Mountains. In 1847, he accepted a position at Harvard University, and in North America, he found evidence that glaciers had also overrun that continent's northern half. Agassiz's Ice Age theory gradually won acceptance over more catastrophic flood theories, and evidence for several long ages of advancing and retreating ice over millions of years was eventually found.

STUDY OF EARTH HISTORY

By the end of the nineteenth century, uniformitarian logic had become the primary method of studying the history of the Earth's surface. Early in the twentieth century, the development of radioactive dating techniques confirmed the enormous age of the Earth, revealing some 4.6 billion years during which the same slow processes of erosion, eruption, sedimentation, and ground movement visible today could account for everything from the Grand Canyon to marine fossils in the Alps. Yet this very method has identified discontinuities and anomalies that reveal the importance of catastrophic events in Earth history. It now appears that the planet and its life-forms have been shaped by more than gradual processes still operating today. Evidence has accumulated that many sudden and violent events in the past had widespread consequences, including torrential flooding, massive volcanic activity, and huge asteroid impacts causing global disasters. A "new catastrophism" uses uniformitarian methods to show that past catastrophes may be the key to understanding the present.

An early attempt to revive catastrophism was made by J. Harlen Bretz in 1923 to explain certain landscape features. He suggested that some of the world's largest floods poured down the Columbia River Gorge from melting glaciers into the Pacific Ocean, scouring much of the Columbia Plateau down to bedrock and creating the Channeled Scablands of eastern Washington. His ideas were finally vindicated from aerial and satellite photographs. It was then shown that glacial Lake Missoula in Montana produced as many as seventy floods from about 15,000 to 12,000 years ago, matching legends of several Native American tribes in the Pacific Northwest. At that time, a glacial ice dam impounded a body of water some 400 kilometers long. Repeated emptying of Lake Missoula occurred when the ice dam floated and broke, releasing large amounts of water over the course of at least forty hours. This deluge, which was some ten times the combined flow of all the world's rivers, removed soil as deep as 300 meters, inundating 7,800 square kilometers as deep as 106 meters.

More obvious catastrophic events are associated with volcanic activity. The most devastating volcanic eruption in recorded history was that of the Tambora volcano on Sumbawa Island, Indonesia, in 1815. The explosion killed twelve thousand people, and another eighty-two thousand died of starvation and disease. Tambora ejected so much volcanic ash into the stratosphere that Europe and North America experienced a year without a summer. Snow blanketed New England in June, and frosts blighted crops throughout the growing season.

Much larger volcanoes changed the landscape in prehistoric times. Volcanic activity in the Yellowstone region began about 2 million years ago as the continental crust moved westward. Subterranean melting of the crust produced a large underground reservoir of magma that resulted in three major explosions over a period of 1 million years. The first produced one of the largest eruptions to occur on the Earth. After the roof of the magma reservoir collapsed, it subsided hundreds of meters, producing a caldera (giant crater) that covered an area of nearly 3,000 square kilometers. The volcanic ash canopy from the last eruption annihilated life over much of the western United States, and its fallout is preserved in strata from California to Kansas.

Other discoveries have revealed that asteroid or comet impacts are the probable source of even more widespread annihilation of life and may be associated with extensive volcanic activity. In 1958, the Estonian astrophysicist Ernst Öpik suggested that a sufficiently large asteroid collision could

penetrate the continental crust, triggering the formation of huge areas of lava floods such as the Deccan Traps in western India and the Columbia River Plateau in the Pacific Northwest. In 1973, the American chemist Harold C. Urey proposed that a comet collision could cause sufficient heating of the biosphere to explain global extinctions. More recently, University of Montana geologists have shown that an asteroid impact 17 million years ago could account for the immense lava flows of the Columbia River Plateau, spreading as far as 500 kilometers from their source to form the largest volcanic landform in North America. As the North American continental plate moved westward, the resulting hot spot shifted to form the Yellowstone volcanic region.

In 1980, a team of physicists and geologists headed by Luis Alvarez and his son Walter discovered that the thin global sediment that separates the end of the Cretaceous era (age of dinosaurs) from the Tertiary era (age of mammals) contained anomalous quantities of the element iridium, rare on the Earth but common in meteorites. They suggested that this K-T boundary layer, dated at 65 million years ago, was evidence of an asteroid collision that ejected enough matter into the atmosphere to produce a "cosmic winter," killing the dinosaurs and many other species. Their estimate of at least a 10-kilometer asteroid was confirmed in 1990, when a 200-square-kilometer crater was identified in the Yucatán region of Mexico, dating from 65 million years ago. Such a collision could have produced shock waves that came to a focus on the opposite side of the Earth, explaining the 65 million-year-old eruptions that formed the Earth's largest lava fields in the Deccan Traps. Satellite surveys have revealed at least one hundred large but weathered craters on the Earth known as "astroblemes."

Joseph L. Spradley

CROSS-REFERENCES

BIBLIOGRAPHY

Chapman, Clark, and David Morrison. *Cosmic Catastrophes.* New York: Plenum Press, 1989. A good review of classical catastrophism and uniformitarianism. Includes an authoritative discussion of the "new catastrophism" associated with asteroid collisions. The book is well illustrated and contains a good glossary.

Clube, Victor, and Bill Napier. *The Cosmic Winter.* Oxford, England: Basil Blackwell, 1990. A good history of meteorite collisions and their effects, including a chapter on assessing the risk from meteorite and asteroid collisions.

Harris, Stephen L. *Agents of Chaos.* Missoula, Mont.: Mountain Press, 1990. A good geological description of catastrophic events that have shaped the Earth's crust, including earthquakes, volcanoes, floods, and asteroids. Many interesting illustrations and a good glossary are included.

Huggett, Richard. *Catastrophism: Systems of Earth History.* London: Edward Arnold, 1990. A careful geological assessment of the rise and fall of classical catastrophism and its modern neo-catastrophist revival. Includes an extensive bibliography.

Lewis, John S. *Rain of Iron and Ice.* New York: Addison-Wesley, 1996. A survey of impact cratering in the solar system and the implications for bombardment of the Earth by comets and asteroids. Includes a dozen photographs.

Lucas, Spencer G. *Dinosaurs: The Textbook.* 3d ed. Boston: McGraw-Hill, 2000. This book provides good coverage of the extinction of the dinosaurs. Intended for university undergraduates but readable by anyone with an interest in dinosaurs. Glossary, bibliography, and index.

Raup, David M. *The Nemesis Affair: A Story of the Death of Dinosaurs and the Ways of Science.* New York: W. W. Norton, 1986. An excellent history of catastrophism and its revival in the light of asteroid-collision evidence, including a discussion of possible sources of periodic extinctions and the nature of scientific controversies.

Sharpton, Virgil L., and Peter D. Ward, eds. *Global Catastrophes in Earth History.* Boulder, Colo.: Geological Society of America, 1990. A compilation of fifty-eight papers on various aspects of catastrophic extinction and its causes. Although written for specialists, it is accessible to readers with some scientific background. Bibliography and index.

Steel, Duncan. *Rogue Asteroids and Doomsday Comets.* New York: John Wiley & Sons, 1995. An interesting discussion of asteroid and comet impacts, past and future, including speculations about their historical influence and precautions that could be taken in the future. Contains a good glossary and bibliography.

COLONIZATION OF THE LAND

The advent of animals and plants on land during the Ordovician period added new complexity to preexisting ecosystems of microbes. The newly increased mass of vegetation on land served to stabilize soils against erosion and promoted the weathering of their nutrient minerals. Arthropods, too, found a place in this early ecosystem of nonvascular plants on land.

PRINCIPAL TERMS

ARTHROPODS: a group of animals lacking backbones but with rigid, proteinaceous, and mineralized external skeletons that are jointed at the limbs and at other points of movement; includes crabs, spiders, and insects

CHAROPHYTE: a kind of freshwater green alga with filamentous branches arranged in whorls around a threadlike central stem, characterized by egg cells that are large compared with those of other algae

CUTICLE: an outer, thin, waterproof, waxy, and proteinaceous cover to the bodies of land plants and arthropods

MILLIPEDE: a kind of arthropod with an elongate body of many, nearly identical segments, each with a pair of limbs on each side

NEMATOPHYTE: an extinct kind of nonvascular land plant consisting of flat sheets, stems, or trunks, supported by densely interwoven, microscopic, proteinaceous tubes

PHOTOSYNTHESIS: the process by which plants and some microbes create organic matter and oxygen from carbon dioxide and water, using the energy of the Sun and a catalytic pigment

RHIZOME: an organ of vascular plants with the anatomy of a stem rather than of a root but running along or under the ground more like a root than a stem

RHYNIOPHYTE: an extinct kind of vascular land plant with rhizomes and simple spore-bearing organs but lacking true roots or leaves

SPORE: a reproductive propagule with a tough, acid- and desiccation-resistant, external, proteinaceous coat in land plants lacking seeds

STOMATE: an opening in the surface of the green parts of vascular land plants that can be closed by the flexing of adjacent cells in order to control gas exchange between the atmosphere and the plant interior

VASCULAR PLANTS: those plants with elongate, woody, water-conducting, tubular cells (tracheids) in their stems or veins of leaves; now includes ferns, conifers, and flowering plants

APPEARANCE OF LIFE ON LAND

The appearance of animals and plants on land by the Middle Ordovician period, some 450 million years ago, was a major event in the evolution of terrestrial ecosystems. Nevertheless, they probably were not the Earth's first inhabitants; there is a fossil record of blue-green algae and other microscopic life well back into Precambrian time, as much as 3.5 billion years ago. Indeed, it is doubtful that plants and animals visible to the naked eye could have lived on land without preexisting microbial ecosystems, which served to stabilize minerals in soils, to decompose and circulate organic matter of dead organisms, and to oxygenate the atmosphere by photosynthesis. The increased mass of more complex animals and plants on land during Ordovician time further stabilized soils, invigorated the recycling of organic matter, and boosted atmospheric oxygenation. In addition, large plants provided greater depth and structure to terrestrial ecosystems than was possible with microbes and so may have promoted photosynthetic efficiency, biological diversity, and perhaps also resistance to disturbance by floods and storms. This self-reinforcing boost to terrestrial productivity firmly established life on land.

Because there are marine fossils of plants and animals visible to the naked eye in Precambrian rocks (at least 600 million years old), it has commonly been assumed that the earliest creatures on land during the Ordovician and Silurian periods invaded from the sea. Reasons advanced to explain why the land was unavailable for marine creatures for more than 200 million years include the lack of available oxygen, the poverty of terrestrial microbial photosynthetic productivity, and an unpredictable land surface of flash floods and erosional badlands. This view of an invasion from the sea has been used to explain the origins of earliest land animals, which probably were arthropods, such as millipedes and spiders. A tremendous variety of fossil arthropods have been found in Cambrian, Ordovician, and Silurian deposits of shallow seas and estuaries. Like modern marine crabs, these creatures may have ventured out to a limited extent on land, and some may have become more fully adapted to more difficult conditions there. The external skeletons of arthropods, important for defense in the sea, also are effective for support, movement, and preventing desiccation on land.

On the other hand, millipedes and spiders are not very closely related either to any known fossil or to living aquatic arthropods. A reassessment of the earliest fossil scorpions, formerly regarded as possible early land animals, has shown that they had a breathing apparatus that would have been effective only in water. Substantial evolution on land must have occurred to produce the earliest spiders and millipedes, perhaps from microscopic early microbial feeders that have left no fossil record.

IMMIGRANT VERSUS INDIGENOUS EVOLUTION

The idea of invasion of the land by marine and freshwater algae is supported to some extent by the close biochemical similarities between modern land plants and charophytes (a kind of pond weed commonly called stonewort because of its calcified egg cells). Charophytes, however, are very different from land plants, and it is unlikely that such soft-bodied aquatic algae in the geological past were any more successful in colonizing the land than are the mounds of rotting seaweed now thrown up on beaches by storms. Land plants differ from stoneworts and seaweeds in many ways: They have a waxy and proteinaceous outer coating (cuticle) to prevent desiccation and to allow the plant body to remain turgid through internal water pressure; they have small openings (stomates)

surrounded by cells that can open and close the opening in order to control loss of water and oxygen and intake of carbon dioxide; they have internal systems of support and water transport, which include tubular thick-walled cells (hydroids) in nonvascular plants, such as mosses and liverworts, and elongate cells with helical or banded woody thickenings (tracheids) in vascular plants; they have roots, unicellular root hairs, or rootlike organs (rhizoids) that gather water and nutrients from soil; and they have propagules (spores) protected from desiccation and abrasion by proteinaceous envelopes. To some botanists, the coordinated evolution of all these features from aquatic algae is extremely unlikely, notwithstanding the impressive diversity of algae today. This consideration, plus the simple nature of the earliest fossil land plants, has led to the argument that land plants evolved on land from microscopic algae already accustomed to such conditions.

While immigrant versus indigenous evolutionary origins of the earliest land creatures remains a theoretical problem, there is fossil evidence of very early land ecosystems. In Late Ordovician rocks are found the earliest spores of land plants. Most of them are smooth and closely appressed in groups of four, somewhat similar to spores of liverworts and mosses today. This is not to say that they belonged to liverworts and mosses; no clear fossils of land plants visible to the naked eye have yet been found in rocks of this age. Early moss and liverwort ancestors are found in Silurian rocks, but so are extinct nonvascular plants, such as nematophytes. These early experiments in the evolution of land plants had tissues supported by densely interwoven proteinaceous tubes. In life, they had the rubbery texture of a mushroom and a variety of bladelike and elongate forms similar to those of some living algae.

Although the botanical affinities of the earliest spores of nonvascular land plants remain unclear, there is evidence that they grew in clumps. Buried soils of Late Ordovician age have been found with surficial erosion scours of the kind formed by wind around clumps of vegetation. The clumps are represented by gray spots from the reducing effect of remnant organic matter. Burrows also have been found in Late Ordovician buried soils as an indication of animals in these early land ecosystems. The fossil burrows are quite large (2 to 21 millimeters).

They are similar in their clayey linings, backfill structures, and fecal pellets to the burrows of modern roundback millipedes. The buried soils are calcareous and strongly ferruginized—indications that they were nutrient-rich, periodically dry, and well drained, as are modern soils preferred by millipedes. Actual fossils of millipedes have not yet been found in rocks older than Late Silurian, so all that can be said at present is that these very early animals on land were in some ways like millipedes.

DIVERSIFICATION OF LIFE ON LAND

By Silurian time (some 438 million years ago) there was a considerable diversification of life on land. Spores of fungi and of vascular land plants have been found fossilized in Early Silurian rocks. During Mid-Silurian time, there were small, leafless plants with bifurcating rhizomes and photosynthetic stems terminated by globular, spore-bearing organs. These matchstick-sized fossil plants have been called *Cooksonia*. Although not so well preserved as to show their water-conducting cells, they have been regarded as the earliest representatives of the extinct group of vascular plants called rhyniophytes. In Devonian rocks (some 408 million years old), some well-preserved rhyniophytes are known to have been true vascular plants, but there are other plants similar in general appearance that had simpler thick-walled conducting cells like those of nonvascular plants. By Devonian time, there were also vascular plants with spore-bearing organs borne above lateral branches (zosterophylls), plants with true roots and spore-bearing organs borne in clusters (trimerophytes), and spore-producing plants with woody roots and tree trunks (progymnosperms). The evolution of the earliest vascular plant cover on land, and of the first forests, involved different kinds of plants now extinct.

To fossil millipedes of Silurian age were added during Devonian time spiders, centipedes, springtails (Collembola), and bristletails (Thysanura). The earliest vertebrates on land are known from bones of extinct amphibians (Ichthyostegalia) and from footprints of Devonian age, some 370 million years old.

This great Silurian and Devonian evolutionary radiation promoted environmental changes similar to those initiated by the first colonization of land by plants and animals, as well as some new changes. For example, the formation of charcoal from wildfires in woodlands and the accumulation of peat in swamps were ways of burying carbon that otherwise might have decayed or been digested into carbon dioxide in the atmosphere. Removal of carbon dioxide in this way allowed increased oxygenation of the atmosphere. Oxygenation was kept within bounds by increased flammability of woodlands when oxygen reached amounts much in excess of the present atmospheric level.

Late Devonian ecosystems were very different from modern ones. Major ecological roles, such as insect-eating large animals on land, were still being added. More changes were to come, but the world at that time would have seemed a much more familiar place than the meadows of *Cooksonia* during the Silurian, the patchy cover of Ordovician nonvascular plants, and the red and green microbial earths of earlier times.

Gregory J. Retallack

CROSS-REFERENCES

BIBLIOGRAPHY

Gordon, M. S., and E. C. Olson. *Invasions of the Land: The Transition of Organisms from the Aquatic to Terrestrial Life*. New York: Columbia University Press, 1995. This is a detailed text, but it is written in a very readable style. It covers all aspects of the invasion of life onto land.

Little, C. *The Colonization of the Land*. Cambridge, England: Cambridge University Press, 1983. This extended account of the invasion of land by a variety of animal groups is aimed at the level of a high school graduate or beginning university student. Includes an extended bibliography of technical works.

Schopf, J. William, ed. *Major Events in the History of Life*. Boston: Jones and Bartlett, 1992. An excellent overview of the origin of life, the oldest fossils, and the early development of plants and animals. Includes a chapter on the origin and evolution of the earliest land plants. Written by specialists in each field but at a level that is suitable for high school students and undergraduates. Although technical language is used, most terms are defined in the glossary.

Schumm, Stanley A. *The Fluvial System*. New York: Wiley-Interscience, 1977. This book summarizes many years of theoretical and experimental work on river channels and includes interesting speculations on the effects of early land plants on styles of deposition in and around rivers. Some parts of the book are highly mathematical, but most of it is suitable for senior high school and college readers.

Stanley, Steven M. *Earth and Life Through Time*. 2d ed. New York: W. H. Freeman, 1989. This textbook for entry-level college students in geology includes a short section on colonization of the land, with illustrations of a few of the pertinent fossils. Also useful is a concise treatment of basic ideas in the study of fossils and sedimentary rocks.

Stebbins, G. L., and G. J. C. Hill. "Did Multicellular Plants Invade the Land?" *American Naturalist* 115 (1980): 342-353. The short answer to the title of this well written article is no. The argument is theoretical but well reasoned. Because it challenges conventional views, this article would be a useful basis for debate among college or senior high school students.

Wright, V. P., and Alfred Fischer, eds. *Paleosols: Their Recognition and Interpretation*. Princeton, N.J.: Princeton University Press, 1986. This collection of articles aimed at the senior college student or geological professional has two chapters that discuss the interpretation of fossil soils (paleosols) thought to have supported early land plants. Other chapters introduce various other aspects of the study of paleosols, and all have copious references to technical articles.

Zimmer, Carl. *At the Water's Edge: Macroevolution and the Transformation of Life*. New York: Free Press, 1998. Zimmer discusses the development of vertebrates onto land and the return path later taken by turtles and aquatic mammals. Intended for the general reader as well as the college student.

DINOSAURS

Dinosaurs were one of the most successful early life-forms, thriving for more than 150 million years before becoming extinct by the end of the Cretaceous period. The reasons behind their longevity and their sudden extinction hold important implications for humankind's own survival.

PRINCIPAL TERMS

ANKYLOSAURS: a group of later ornithischians characterized by heavy armor

CEROTOPSIANS: a group of later ornithischians characterized by a beaked snout and a bony frill on the back of the head

ORNITHISCHIANS: one of the two orders of dinosaurs; it comprises the "bird-hipped" dinosaurs

ORNITHOPODS: the early, bipedal ornithischians

SAURISCHIANSIANS: one of the two orders of dinosaurs; it comprises the "reptile-hipped" dinosaurs

SAUROPODS: the herbivorous, quadrupedal saurischians

STEGOSAURS: a group of later ornithischians characterized by a row of plates down the back

THECODONTS: an order of Triassic reptiles that were the ancestors of dinosaurs, birds, and crocodiles

THEROPODS: the carnivorous, primarily bipedal saurischians

SAURISCHIANS

"Dinosaur," which is derived from the Greek term for "terrible lizard," is the popular name for a group of extinct land-dwelling reptiles. They were the dominant vertebrate animals during most of the Mesozoic era, which began 225 million years ago and ended 65 million years ago. Among the dinosaurs were the largest animals that ever walked the Earth, although some of the earliest dinosaurs were very small.

The Mesozoic era is divided into three periods, Triassic, Jurassic, and Cretaceous, of approximately equal length. Dinosaurs first appeared in the later third of the Triassic period. Experts believe that dinosaurs developed from a group of archosauromorph reptiles such as *Marasuchus*, which was a lightly built flesh eater about 1.3 meters long. It was clearly a biped, running on its hind legs, and the long tail was presumably used as a balancing organ.

Dinosaurs are divided into two separate orders, depending on the arrangement and shape of the hip bones, which determine the way an animal walks and holds its body. The saurischians, or "reptile hips," as they are commonly called, arose in the early part of the Late Triassic; the ornithischians, or "bird hips," arose toward the end of the Triassic period.

The earliest dinosaurs were saurischians, which are best known from the Ischigualasto Formation of Argentina. The order Saurischia may be divided into two major suborders: the theropods, or "beast-footed dinosaurs," and the sauropods, or "reptile-footed dinosaurs." The theropods, which were more primitive than the sauropods, were primarily bipedal, although many of them probably used all four feet when walking or resting. The hind legs were strong and bore birdlike feet, while the forelimbs bore sharp, curved claws for seizing and holding prey. All theropods had long tails that functioned as stabilizers. The head was large, and the jaws of most of the theropods contained sharp teeth.

The theropods are divided into two major groups. A basal group, the ceratosaurs, includes such dinosaurs as *Coelophysis*, a small, agile carnivore with a long, narrow skull represented by many hundreds of specimens from the Ghost Ranch Quarry in New Mexico. However, larger dinosaurs, such as *Ceratosaurus*, are included within this group. The remaining theropods, termed tetanurans, include the Maniraptoriformes, which share many advanced characteristics with birds. The largest of these is *Tyrannosaurus rex*, from the Late Cretaceous period of North America, which grew to a weight of 4,500 kilograms, a height of 6 meters, and a length of 15 meters.

The sauropods, which appeared slightly later in the Triassic than the theropods, have come to stand as a symbol of gigantism in land animals. They were all quadrupeds and vegetarians. They had small skulls, long necks and tails, large barrel-shaped bodies, padded feet, and large claws on the innermost toe of the forefoot and the innermost toe of the hind foot. The ancestral stock of the sauropods were the prosauropods, which were much smaller than the sauropods. Like most prosauropods, *Plateosaurus* had blunt, spatulate teeth, was an herbivore, and was quadrupedal, although it was capable of bipedal posture and gait.

The later sauropods had longer necks, and their skulls were relatively small. The limb bones became solid and pillarlike to support their great weight. This category contained the largest of the dinosaurs, *Brachiosaurus*, which is estimated to have weighed 73,000 kilograms. The best known sauropods are *Brontosaurus* and *Diplodocus*, from the Late Jurassic period of North America. Although it was once assumed that these huge beasts had to live in swamps where the water could support their great weight, it is now clear that they were terrestrial animals that used their long necks to eat from trees.

ORNITHISCHIANS

The sauropods reached their zenith in the Late Jurassic; the ornithischians replaced them as the dominant herbivores in the Cretaceous period. The expansion of this group was associated with the advent of the flowering plants during the Cretaceous period. Characteristically, a horny beak was developed at the front of the mouth, and the toes ended in rounded or blunt hooves instead of claws.

The earliest ornithischians were the ornithopods. A typical example is *Hypsilophodon*, a small, swift dinosaur with a long, slender tail and long, flexible toes. The most specialized of the ornithopods were the "duck-billed dinosaurs," also known as hadrosaurs. Although they had flat beaks and no anterior teeth, the cheek region had rows of grinding teeth. The various types of duck-billed dinosaur can be distinguished by modifications of the bones associated with the nostrils. Some were molded into hollow, domelike crests, bizarre swell-

ings of the nasal region, or long, projecting tubular structures that were used to warm the air or to produce sounds. The remaining three groups of ornithischians presumably evolved from the primitive ornithopods.

The earliest of these three groups of highly specialized quadrupeds was the "plated dinosaurs," or stegosaurs, which first appeared early in the Jurassic period. This large dinosaur was more than 6 meters long. In comparison to its body size, its head was extremely small. *Stegosaurus* had an average of twenty plates arranged alternately in two parallel rows down the back. The plates were originally thought to have been used for protection, but scientists now believe that the plates could have been used for thermoregulation. *Stegosaurus* died out in the Early Cretaceous period.

The "armored dinosaurs," or ankylosaurs, are not very well known, even though their remains have been found over much of the world. Their armor consisted of a mosaic of studs over the body, spikes that protected the legs, and, in some cases, spikes on the tail. They protected themselves by crouching and drawing in their head and legs.

The last dinosaurs to develop were the "horned dinosaurs," or ceratopsians. The skull was characterized by a beaked snout and a bony frill that extended from the back of the head. The ceratopsians were also distinguished from others by various patterns of horns. The skull of *Triceratops*, for example, had three sharp horns, one on the snout and one above each eye. The best known of the small ceratopsians was *Protoceratops*, which was a small, hornless dinosaur from the Gobi Desert in Mongolia.

Studies of dinosaur eggs, nests, trackways, and bone structures have shown that smaller dinosaurs probably had a warm-blooded, or endothermic, metabolism similar to mammals. This is supported by the discovery of small theropods in China that show a covering of feathers, presumably for insulation. Large dinosaurs, such as sauropods, would have been more efficient as ectotherms, similar to most modern reptiles.

EXTINCTION THEORIES

Several theories regarding the dinosaurs' extinction were first proposed in the late nineteenth and the early twentieth centuries. According to one popular theory, dinosaurs were wiped out because early mammals of the Cretaceous period ate their eggs. Yet the eggs of many modern reptiles have faced the same threat and have survived, primarily because reptiles lay so many eggs. Another theory suggested that the same animals ate the plants on which the dinosaurs depended. Although that is possible, virtual plagues of mammals would have been required to eradicate the dinosaurs. Some early scientists also believed that the dinosaurs became too big for their environment; that is unlikely, however, because gigantic dinosaurs had been successful for millions of years. Changes in the physical environment also occurred in the Late Mesozoic. Evidence indicates that the sea levels fell. Geologic evidence shows, though, that drastic environmental changes had occurred many times during the dinosaurs' reign without any detrimental effect.

A theory proposed in early 1979 by Luis Alvarez and Walter Alvarez suggests that the iridium that has been found in several samples of sedimentary layers between the rock of the Cretaceous and Tertiary periods came from an asteroid that struck the Earth at that time. Such a catastrophic event could have caused an enormous cloud of dust to circle the Earth and cut off the sunlight, destroying the plants and the dinosaurs that depended on them. This theory, however, fails to explain

why so many other animals, such as the mammals, managed to survive.

Another modern theory places the blame on the greenhouse effect. It has been argued that the reduction of the seas that occurred during the Cretaceous period caused a reduction of marine plants. As a result, the amount of carbon dioxide in the air increased, trapping heat from the Earth's surface. A similar theory suggests that the eruption of a tremendous volcano produced a fatal amount of carbon dioxide. Neither theory, however, explains why other animals, especially heat-sensitive reptiles, survived.

The main alternative to the extraterrestrial catastrophist explanation is a gradual ecosystem change model. Declines in many groups of organisms that started well before the Cretaceous-Tertiary boundary are seen as being caused by long-term climatic change, as lush tropical environments were replaced by strongly seasonal, temperate climates. The best explanation may be a combination of the two main theories.

STUDY OF DINOSAURS

Scientists study dinosaurs by examining fossils, which are animal remains that have turned to stone. If a dinosaur died near a river or in a swamp, it stood an excellent chance of being preserved. Its body might sink into the mud, or floodwaters might float it downstream, where it would end up on a sandbar, on the bottom of a lake, or even in the sea. After the flesh decayed, the bones would be covered by sediments, such as mud or sand. The weight of accumulated layers of sediment would compress the remains and turn them into rock: mud into shale, sand into sandstone, limy oozes into limestone or chalk.

The way a fossil is studied is determined by the category to which it belongs. The first category is petrified fossils. They may be preserved in two ways. In replacement, minerals replace the original substance of the animal after water has dis-

Paleontologists excavate dinosaur bones in a fossil quarry at Dinosaur National Monument, Colorado. (© William E. Ferguson)

A model at the Museum of Natural History, Vernal, Utah, of a 90-foot-long Diplodocus, *a plant-eating dinosaur that flourished during the Jurassic age, about 150 million years ago.* (© William E. Ferguson)

solved the soft body parts. In permineralization, minerals fill in the small air spaces in bones or shells, thereby preserving the original bone or shell. The second group of fossils is composed of natural molds that form when the bodies dissolve. Scientists make artificial casts of these molds by filling them with wax, plastic, or plaster. The third type is prints, which are molds of thin objects, such as feathers or tracks. Sometimes, even skin is preserved. Prints are formed when the soft mud in which they are made turns to stone. Scientists can determine the length and weight of the dinosaur that made a set of footprints by studying the depth, size, and distance between them.

Most fossils are found in sedimentary rocks, which lie beneath three-fourths of the Earth. The best collecting areas are places where the soil has worn away from the rocks. Areas in Colorado, Montana, Wyoming, and Alberta, Canada, have been especially rich in fossils. Most of the finds consist of no more than scraps of limb bones, odd vertebrae, loose teeth, or weathered lumps of rock with broken bone showing on the surface. Once a scientist has discovered a few fossilized fragments, he or she combs the area to find the rest of the animal. If the skeleton is embedded, it is extracted with the help of a wide variety of tools, ranging from picks and shovels to pneumatic drills. Loose

fragments are glued back into place, and parts that are too soft or breakable are hardened by means of a special resin solution that is sprayed or painted on.

As the fossil is uncovered, it is encased in a block of plaster of paris. (A more modern method uses polyurethane foam instead of plaster.) After the entire surface is covered, the fossil is rolled over, and another layer of plaster is added. After the fossil has been transported to the museum, the plaster is removed. The "development" stage involves the removal of the rock around the bones. The oldest way is by hand, using tools such as hammers and chisels; a more modern technique uses electrically powered drills similar to dentists' drills. Sandblasting and chemicals may also be employed. After the fossil is cleaned, it is ready for mounting. The bones are fastened to a steel framework that makes the skeleton appear to stand by itself.

LIFE-EARTH INTERACTION

From the dinosaurs, scientists are learning new lessons about the physiology of such beasts, their relationship to the world in which they lived, their distribution and the bearing of that distribution on the past arrangements of the continents, various aspects of evolution, and the reasons that they became extinct. The dinosaurs played a major part in the shaping of the natural world. Birds, for example, are probably their descendants, as evidenced by the intermediary species *Archaeopteryx,* a primitive bird that lived during the Late Jurassic period; although its beak contained teeth, *Archaeopteryx* also had feathers and could fly.

The disappearance of a species that seemed to rule the world for more than 100 million years brings into question the notion of a "dominant" species. Most people believe that mammals are now the dominant form of life; however, dinosaurs did not "rule," and neither do mammals. If one were to list the biological organisms whose influence on the planet is such that their removal would produce chaos, then that list would be headed by microorganisms so small that they can be seen only through powerful microscopes. The list would also include the green plants and the fungi.

The extinction of the dinosaurs also brings into question the ability of humans to destroy the world. All species, from the simplest microorganism to the largest plant or animal, modify their immediate surroundings. They cannot avoid doing so. The success of one group, however, does not imply the failure of the groups it exploits. The complexity of individual organisms may increase, but the simpler forms do not necessarily disappear. Life continued after the demise of the dinosaurs and would probably continue to do so if humankind were destroyed.

Alan Brown

CROSS-REFERENCES

Archaebacteria, 951; Cambrian Diversification of Life, 956; Catastrophism, 961; Colonization of the Land, 966; Ediacaran Fossils, 976; Eukaryotes, 981; Evolution of Birds, 985; Evolution of Flight, 990; Evolution of Humans, 994; Evolution of Life, 999; Fossil Plants, 1004; Fossil Record, 1009; Fossilization and Taphonomy, 1015; Fossils of the Earliest Life-Forms, 1019; Fossils of the Earliest Vertebrates, 1024; Geoarchaeology, 1028; Life's Origins, 1032; Mammals, 1038; Mass Extinctions, 1043; Microfossils, 1048; Neanderthals, 1053; Paleobiogeography, 1058; Paleobotany, 1062; Paleosols, 1144; Paleozoic Era, 1126; Petrified Wood, 1058; Prokaryotes, 1071; Stromatolites, 1075; Systematics and Classification, 1079.

BIBLIOGRAPHY

Allaby, Michael, and James Lovelock. *The Great Extinction*. London: Secker and Warburg, 1983. This book posits that the dinosaurs died as the result of some sort of catastrophic occurrence, such as the collision of an asteroid with the Earth. The authors support their contention by examining the effects that modern catastrophes, such as the eruption of Mount St. Helens, have had on wildlife. The conclusion discusses the implications that the extinction of the dinosaurs has for humankind's possible extinction. It is well written and well indexed. Suitable for college students.

Bakker, Robert T. *The Dinosaur Heresies*. New York: William Morrow, 1986. Bakker's book paints a revolutionary picture of dinosaurs as dynamic, intelligent, hot-blooded creatures that dominated the Earth for more than 150 million years. Bakker suggests, for example, that *Brontosaurus* was not a sluggish, swampbound creature but rather a nimble landdweller. He also theorizes that the dinosaurs died as the result of diseases transmitted across the land bridge by foreign species. This highly personal account is interesting throughout and written with both the general reader and the college student in mind.

Benton, Michael J. *Vertebrate Paleontology*. 2d ed. London: Chapman and Hall, 1997. This comprehensive text provides a complete outline of the history and development of vertebrates. Designed for college courses but also suitable for the informed enthusiast. The first three chapters deal with the origin of vertebrates and their early development.

Charig, Alan. *A New Look at the Dinosaurs*. New York: Avon, 1983. Charig begins by answering the question "What were the dinosaurs?" and then proceeds to answer other intriguing questions, such as "Were dinosaurs warmblooded?" "Did birds evolve from dinosaurs?" "Were dinosaurs too heavy to walk on land?" and "Why did dinosaurs suddenly die out 65 million years ago?" Charig explains all the known facts and theories with the aid of black-and-white drawings, maps, charts, photographs, and several beautiful watercolor plates.

Colbert, Edwin H. *Dinosaurs: An Illustrated History*. Maplewood, N.J.: Hammond, 1983. This volume, beautifully illustrated with more than one hundred color photographs and paintings, is essentially a condensed version of the following source, although Colbert discusses all the theories regarding the extinction of the dinosaurs in the conclusion. A fascinating book written for the general reader.

Currie, P. J., and Kevin Padian. *Encyclopedia of Dinosaurs*. San Diego, Calif.: Academic Press, 1997. This is an enormous compendium of information on dinosaurs. Includes 275 articles by experts in the field. An extremely useful and user-friendly publication.

Lambert, David. *A Field Guide to Dinosaurs*. New York: Avon, 1983. This book discusses the physical characteristics, behavior, and evolution of all known dinosaur genera, one in five of which has been named since 1970. Also covers the various theories regarding the extinction of dinosaurs, the process of fossilization, and the discovery and display of fossils. Richly illustrated and nontechnical enough for junior high and high school students.

Lucas, Spencer G. *Dinosaurs: The Textbook*. 3d ed. Boston: McGraw-Hill, 2000. This book provides good coverage of the extinction of the dinosaurs. Intended for university undergraduates but readable by anyone with an interest in dinosaurs. Glossary, bibliography, and index.

Witford, John Noble. *The Riddle of the Dinosaurs*. New York: Alfred A. Knopf, 1985. Although Witford relates the history of paleontology in the beginning of this book, his primary purpose is to demonstrate how discoveries made since the 1960's have revolutionized dinosaur theory. He proposes, for example, that some dinosaurs were warm-blooded, quick moving, and good parents. He also discusses the most radical change in dinosaur theory: the idea that they disappeared because of a massive catastrophe, such as global floods, asteroid collisions, or exploding stars. Filled with photographs and color drawings, the book is accessible to the general reader but technical enough for the researcher.

EDIACARAN FOSSILS

Ediacaran fossils provide evidence of the oldest remains of animal life with moderately complex body structures. These organisms, known as metazoans, lacked internal skeletal structures and lived in ancient oceans, either as bottom dwellers or as primitive swimmers and floaters. Ediacaran fossils are now known from widely scattered locations around the world and represent important data relative to the interpretation of ancient plate tectonic activity.

PRINCIPAL TERMS

CORRELATION: the establishment of the fact that an event in one place occurred at the same time as a similar event in another place

EARLY PALEOZOIC: that part of geologic history that is somewhat younger than about 550 million years before the present

EDIACARAN FOSSILS: fossils of marine animals with moderately complex body structures (metazoans) that lived some 570 to 670 mil-

lion years ago; the term "Ediacaran" was derived from the well-known Ediacara Hills fossil locality in south Australia

LATE PRECAMBRIAN: that part of geologic time from about 550 million years to 1 billion years before the present

STRATIGRAPHIC UNIT: any rock layer that can be easily recognized because of specific characteristics, such as color, composition, or grain size

DISCOVERY OF EDIACARAN FOSSILS

The Ediacaran fossil assemblage is a group of animal remains that represent the oldest metazoan life-forms of significant size and diversity known from the fossil record. The Ediacaran assemblage is composed of a number of different fossils that occur in a part of the Earth's stratigraphic record that dates from about 565 million to 543 million years ago. Fossils that are now considered part of the Ediacaran assemblage were originally described in 1933 from a location in southwest Africa (Namibia), and from south Australia in 1949. The age and significance of these fossils were not realized, however, until the mid- to late 1960's, because until that time the particular rock units that include the Ediacaran assemblage were not of interest to most geologists. Since then, geologists in many parts of the world have been conducting research on both the Ediacaran fossils and the rocks that contain them. These ancient metazoans, the Ediacaran assemblage, are now recognized from stratigraphic units in several geographically isolated areas, including North Carolina, Newfoundland, British Columbia, England, Russia, China, Africa, and Australia. Significant numbers and varieties of individual fossils, however, have been found only in southwest Africa and south Australia.

Rocks of the appropriate age and type that could contain Ediacaran fossils belong to what geologists define as the Late Precambrian eon on the geologic time scale or calendar. Because the crust of the Earth is constantly undergoing change, rocks of Late Precambrian age are not commonly preserved at the surface of the Earth, and where preserved, they are generally unfossiliferous or have undergone such intense alteration or metamorphism that any fossil evidence has been destroyed. Also, locations of preserved rocks of this age occur in such widely separated areas that geologists working in one location frequently do not have the opportunity to become familiar with the geology in other areas. Thus, the discovery of this unique and easily identified fossil assemblage from widely scattered locations provides geologists with a correlation tool to help unravel the history of the Earth during the Late Precambrian.

The Ediacaran fossils described to date represent remains of a number of distinct soft-bodied animals. These organisms were very different in appearance from modern creatures, and consequently geologists have a difficult time establishing evolutionary connections between the ancient and living organisms. In 1983, A. Seilacher of the University of Tübingen proposed that the Ediacaran organisms represented distinct taxa unrelated to the modern groups that appeared in the Cambrian, thus constituting a "failed experiment." Some of the Ediacaran organisms were bottom-dwelling, or benthic, creatures with bilateral symmetries. Some of the fossils have symmetries and shapes analogous to modern swimming and bottom-dwelling medusoids (jellyfish-like organisms). Other Ediacaran fossils display a body geometry similar to that of living flatworms and were apparently swimmers or crawlers on the ocean floor. Many of the Ediacaran fossils have the appearance of an elm leaf and were originally mistakenly classified to the biologic class Petalonia because of this appearance. Fossils of the Ediacaran assemblage, however, do represent animal remains and not plant remains. Fossil sizes for organisms represented by the Ediacaran assemblage range from several millimeters in length or diameter to several centimeters and, in some cases, several tens of centimeters.

One of the unusual qualities of this assemblage is that its components appear to have left little or no direct evidence of evolutionary descendants in the fossil record. This situation, which could be the result of the incomplete nature of the geologic record, makes it difficult to compare the Ediacaran organisms with either modern creatures or those preserved in the early Paleozoic stratigraphic record.

DISKLIKE AND FRONDLIKE FOSSILS

There are two main body geometries of the Ediacaran fossils. One of the most common shapes is discoidal, similar to that of many living jellyfish. One sees Ediacaran fossil remains preserved in the rock in the form of platelike, or disklike, impressions similar to those left in beach sand by a jellyfish that has washed up onto a beach. This platelike or disklike impression is commonly marked by a number of concentric rings, narrow ridges, or grooves. These features can be variously located near the rim of the disk or more or less evenly spaced across the surface of the disk or concentrated at intervals in different places across the disk surface. Some of the discoidal fossil impressions have suggestions of tentacles on the margin of the disk. In addition, many of the Ediacaran discoidal fossils have radial ornamentation and various bumps and depressions located near the center of the disk. These latter have been interpreted as either feeding or reproductive organs.

The fossil record is not sufficiently detailed to be able to establish an evolutionary line from these metazoans to fossils with similar features preserved in younger rock units or to modern medusoids.

There are, however, sufficient similarities between the Ediacaran fossils and living medusoids to regard the Ediacaran disklike fossils as ancient medusoids. The occurrence of these impressions in rocks of this particular age argues for an extensive evolutionary history prior to this time frame.

The other common shape of fossils in the Ediacaran assemblage is the generally flattened, bilaterally symmetrical shape that is overall very similar to that of a leaf from an elm tree. This frondlike or leaflike shape is preserved in many of the fossil localities around the world and was apparently common to several different benthic Ediacaran organisms.

A cluster of Ediacaran fossils of the soft-bodied Beltanelliforis brunsae *on the underside of a siltstone bed found in the Blueflower Formation in the Wernecke Mountains, Yukon Territory.* (Geological Survey of Canada)

These bottom-dwellers were somewhat inflated, saclike creatures whose exterior appears to have had a supple, leathery texture. The creatures lived attached to the bottom in a manner very similar to that of a modern sea pen. The attachment mechanism was a bulbous expansion of the lower end of a basal stalk. This expanded portion of the stalk was anchored within the bottom sediment and is best preserved in the form *Charnodiscus*, described from England. Some variations, such as *Pteridinium*, exhibited a triradiate symmetry from a central stalk. The frondlike organisms apparently fed by filtering food materials from the surrounding water.

The several varieties of frondlike and disklike Ediacaran organisms are preserved as impressions in shaly sandstone layers, as a footprint is in mud. (There are no actual remains or even altered remains of the Ediacaran organisms known.) These impressions indicate that the organisms lacked an internal skeleton and that the external body covering was probably leatherlike. This leathery exterior and inflated habit have created difficulties in classifying many of the remains. When the organisms died and fell into the soft muds of the ocean bottom, sediments would fill the open sac in a haphazard and incomplete manner.

As the sediment compacted and squeezed the organic remains during the fossil-making process, a variety of preservation styles from the same type of organism resulted. Consequently, a particular frondlike organism may be represented by multiple styles of preservation. For example, body geometries from opposite sides of the creature can be superimposed on each other, giving the impression of a very different creature. Also, there are situations in which internal and external body geometries are superimposed on each other, and situations even exist in which internal geometries from opposite body walls appear as a single fossil. Thus, several different names have been assigned to the same organism.

There is evidence that some of the bilaterally symmetrical Ediacaran fossils are remains of swimming organisms or perhaps swimming or crawling flatworms. Similarities to living annelids are sufficiently convincing that some Ediacaran fossils—such as the fossil *Spriggina*—have been tentatively classified as annelids.

The Ediacaran assemblage of fossils is preserved in rocks whose compositions, textures, sedimentary structures, and positions in sequence indicate that the original sediments were deposited in an ancient ocean, near the shoreline. Preserved sedimentary features in the rocks suggest that the depositional environment was generally marine, probably in nearshore shallow water and intertidal situations on relatively low-energy sandy bottoms. Thus, when Ediacaran fossils are recognized in previously undescribed rocks, it is possible to interpret the original depositional environments.

STUDY OF FOSSILS

Fossils are one of the essential pieces of geologic data used to determine time equivalence

On March 17, 1995, geologist Mark McMenamin discovered one of the world's oldest fossils, a jellyfish or sea anemone from an Ediacaran biota in northwestern Mexico. (AP/Wide World Photos)

(correlation) of stratigraphic units that are separated by great distances, such as between fossil-bearing rocks in North Carolina and South Australia. Fossils are also of importance to the evolutionary biologist or paleontologist who is interested in the history and development of life-forms on the Earth throughout geologic time. Fossil evidence is routinely employed by the geologist interpreting ancient depositional environments. The general assumptions are that fossils are indicative of the depositional environment preserved in the rock in which the fossils are found and that identical fossils found in different areas of a state or on different continents are of essentially the same age.

The fundamental techniques in the utilization of fossils are critical observation, accurate collection of data, synthesis of data, knowledge of the biology of living organisms, and knowledge of modern sedimentary processes. The latter two items can be gained from reading, while the former three require practice. The geologist or paleontologist makes detailed observations of fossils and the enclosing rock characteristics before the fossils are collected for study, noting such things as position of fossil relative to the probable living positions and searching for any indications that the fossil was transported after the animal died. The geologist also looks for relationships to other fossils and rock types and relative abundances of particular fossils. After the fossils are collected, they are carefully prepared for laboratory study, where detailed observations and accurate descriptions of fossil morphology, or shape, are made. These analyses form the basis for biological comparisons with other fossils and with living organisms.

The occurrence of Ediacaran fossils in both South Australia and North Carolina, as well as other parts of the world, has been established by careful fieldwork and descriptive work. These data tell the geologist that the enclosing rock units are the same age even though they are not now connected in any way. In those parts of the stratigraphic column that are unfossiliferous, such interregional or intercontinental correlations are possible only by using expensive techniques that employ radioactive elements or materials. These materials are in such low concentrations that they are detectable only with very sensitive instruments. Thus, the discovery of metazoan fossils in stratigraphic units previously thought to be unfossiliferous has been of tremendous value to geologists working in different parts of the world.

Some geologists are very interested in the biologic aspects of the fossil record. The working hypothesis is that creatures with similar anatomical features or creatures living in similar environments can be used as indicators for the physical surroundings and to determine biologic affinities. Based on this hypothesis and the hypothesis of uniformitarianism (the notion that the present is the key to the past), geologists seek to find similarities between preserved or implied anatomical structures of fossils and similar features in living organisms. Also, the geologist examines the rock record and compares that with modern depositional environments, attempting to reconstruct the ancient conditions. In the case of the Ediacaran fossils, geologists have been able to establish ancient depositional environments based on enclosing rock characteristics and from these environments to interpret how the organisms lived. By comparing the preserved anatomical features of the fossils with apparent living analogs, geologists then make interpretations about the organisms themselves. The geologist thus compiles many lines of evidence to support a hypothesis.

Gail G. Gibson

CROSS-REFERENCES

BIBLIOGRAPHY

Gibson, G. G., S. A. Teeter, and M. A. Fedonkin. "Ediacaran Fossils from the Carolina Slate Belt, Stanly County, North Carolina." *Geology* 12 (July, 1984): 387. This article describes some of the better preserved Ediacaran fossils known. Nontechnical.

Glaessner, M. F. *The Dawn of Animal Life: A Biohistorical Study.* New York: Cambridge University Press, 1985. A semitechnical synthesis of the origin and evolution of early life-forms on the Earth, including a discussion of the distribution and variations of the Ediacaran assemblage.

Hofmann, H. J., W. H. Fritz, and G. M. Narbonne. "Ediacaran (Precambrian) Fossils from the Wernecke Mountains, Northwestern Canada." *Science* 221 (1983): 45. A semitechnical discussion of the geologic occurrence of Ediacaran fossils within rock units of the Canadian Rocky Mountains.

McMenamin, Mark. *Discovering the First Complex Life: The Garden of the Ediacara.* New York: Columbia University Press, 1998. This entertaining study of the earliest complex life-forms on the planet details the author's work on these organisms. Written for the interested student but understandable by the general reader.

Prothero, Donald R. *Bringing Fossils to Life.* Boston: McGraw-Hill, 1998. This well-illustrated and entertaining text covers a broad range of paleontological topics, including early life. Glossary, bibliography, and index.

Schopf, J. William, ed. *Major Events in the History of Life.* Boston: Jones and Bartlett, 1992. An excellent overview of the origin of life, the oldest fossils, and the early development of plants and animals. Written by specialists in each field but at a level that is suitable for high school students and undergraduates. Although technical language is used, most terms are defined in the glossary.

EUKARYOTES

All the commonly seen organisms on the Earth are eukaryotes, or organisms built up of eukaryotic cells. These organisms evolved from a prokaryotic ancestor and have developed over the last 1.4 billion years into the extremely diverse and successful groups of invertebrates, fish, amphibians, reptiles, and mammals. Today, eukaryotes live in a vast array of different environments in almost all areas on the surface of the Earth.

PRINCIPAL TERMS

CHLOROPLAST: the organelle in eukaryotes in which photosynthesis is performed by algae and green plants

DEOXYRIBONUCLEIC ACID (DNA): a stable molecule that contains most of the genetic information of a cell

ENDOSYMBIOTIC THEORY: the concept that eukaryotes arose from prokaryotes by incorporating free living microbes into symbiotic relationships

EUKARYOTIC CELL: a cell that contains a nucleus and other membrane-bounded organelles

MITOCHONDRION: the eukaryotic organelle in which energy is generated by aerobic respiration

ORGANELLES: subcellular membrane-bounded units that perform specific functions within the eukaryotic cell

PRECAMBRIAN: the interval of geologic time from the formation of the Earth (4.6 billion years ago) to the beginning of the Cambrian period (544 million years ago)

PROKARYOTIC CELL: a cell that does not contain a nucleus or other membrane-bounded organelles

CHARACTERISTICS

Eukaryotes are organisms whose constituent cells are characterized by the presence of a nucleus and other membrane-bounded organelles such as mitochondria and, in plants, chloroplasts. prokaryotic cells do not have a nucleus or membrane-bounded organelles. Biologists now recognize that the greatest discontinuity in life is not between animals and plants but between the prokaryotes and the eukaryotes. All organisms on the Earth except the viruses, bacteria, and cyanobacteria (blue-green algae) are eukaryotes, and the vast majority of fossil species of the last 700 million years have also been eukaryotes. The fossil record of the Precambrian was dominated by the prokaryotes, and the transition from prokaryotic to eukaryotic cells represents one of the greatest steps in the development of life on the Earth. This transition probably occurred between 2 billion years ago and 1.4 billion years ago, and it led to the evolution of complex multicellular plants and animals. The evolution of highly successful groups such as fish, reptiles, and mammals was dependent upon the formation and elaboration of the eukaryotic cell.

The major distinctive trait of the eukaryotic cell is the nucleus, an organelle that houses the deoxyribonucleic acid (DNA). In prokaryotic cells, the DNA molecules are arranged in a single loose strand within the cytoplasm of the cell. Prokaryotes generally reproduce by simple splitting (binary fission), whereas the reproduction of the cells and organisms is much more highly organized in eukaryotes. Eukaryotic cells reproduce by the complicated processes of mitosis and meiosis. In mitosis, the DNA is copied, and each new cell receives an exact copy of the original DNA. In meiosis, however, a second division of the genetic information occurs, forming sperm and egg cells. These sex cells from separate organisms then can fuse to form an offspring that has a combination of genetic information from each parent. This sexual reproduction, which is not found in prokaryotes, has led to a high level of diversity among the eukaryotes.

Eukaryotic cells also contain other organelles that perform specific metabolic functions in a more tightly organized manner than they occur in prokaryotes. Mitochondria are small organelles found in almost all eukaryotic cells and are the sites of aerobic respiration. In aerobic respiration, carbon-rich molecules are broken down into smaller molecules, and chemical energy is released and captured by the cell. This process requires oxygen. The amount of energy released in aerobic respiration is much greater than is the energy released by fermentation, a process that occurs in prokaryotes.

Algae and green-plant cells also have chloroplasts, organelles that are the sites of photosynthesis. During photosynthesis, sugars are produced by the combination of carbon dioxide and water, with the release of oxygen (from the water molecule). The energy to drive this reaction is in sunlight captured by pigments located in the chloroplasts. Some prokaryotes are photosynthetic, but the process is more highly organized on the membranes within the chloroplasts of eukaryotes. Eukaryotic photosynthesis is efficient enough to supply most of the energy for all other eukaryotes on the Earth. Many other organelles are found within the eukaryotic cell, and they perform such functions as waste removal, transport of materials, and movement of the cells.

The advantage given to eukaryotes by these organelles is the spatially ordered arrangement of sequential biochemical reactions that allows very efficient metabolic processes. The increased efficiency permitted the development of larger and more complex multicellular organisms. Although a few prokaryotes are loosely congregated into multicellular organisms, it is in the eukaryotes that well-defined separation of functions between cells is found.

EUKARYOTE FOSSILS

The origins of eukaryotes are largely unknown as a result of the notorious selectivity of the fossil record. Advanced eukaryotes such as clams, fish, birds, and mammals have shells, bones, or teeth that are commonly preserved in sedimentary rocks. These eukaryotes have left behind a decipherable, if sporadic, fossil record over the last 700 million years. The older single-cell eukaryotes were rarely fossilized, and the evidence for the first eukaryotes probably will never be found. It is clear, however, that the eukaryotes arose after the first accumulation of significant amounts of oxygen in the atmosphere at about 2 billion years ago and the appearance of the first multicellular animals at about 680 million years ago.

However rare an occurrence, single cells are occasionally fossilized in the rocks. The oldest fossils are from sedimentary rocks in the Pilbara Shield of Australia and are dated at about 3.5 billion years old. All these fossils and all the other Archean and early Proterozoic fossils are prokaryotes, showing that only prokaryotes lived during the first two-thirds of the Earth's history.

Fossil eukaryotic cells are rarely preserved with intact organelles, but they may be distinguished from prokaryotes by size. Eukaryotic cells range in size from 5 microns to 1 millimeter (one micron equals 0.001 millimeter), whereas prokaryotes are generally 1 to 20 microns in diameter. Relatively large fossilized cells have been found in the Precambrian rocks in the Death Valley region of California and in Australia, and these rocks have been dated at about 1.4 billion years old. Although some paleontologists are not convinced that these fossils are eukaryotes, a statistical analysis of more than eight thousand fossil cell sizes from Precambrian rocks throughout the world shows that the eukaryotic stage of evolution had probably been achieved by about 1.75 billion years ago. Large sheets of algae have been found in some of the sedimentary rocks of the Northwest Territory of Canada and have been dated at about 1 billion years old, and almost certainly eukaryotic cells have been found in the 900-million-year-old rocks of Australia. It appears that eukaryotes developed in the interval between 2 billion years and the appearance of definite eukaryotes at about 1 billion years ago, or about 1.75 billion years ago. The manner in which complex eukaryotic cells arose from a prokaryotic ancestor could not be recorded in the fossil record, and theories attempting to describe the evolution of the early eukaryotes must be based upon the biochemical relationships between present-day organisms.

EVOLUTION THEORIES

Two general theories have been proposed to explain the evolution of eukaryotes. The traditional view, direct filiation, suggests that the nucleus, mitochondria, chloroplasts, and other organelles arose by mutations with the prokaryotes. Although most mutations are deleterious, some may have been beneficial to the ancestral prokaryotes and were retained within the organisms. Through sequential accumulation of mutations, each of the organelles developed over a long interval in the Precambrian eon. This view is supported by the very complex interrelationships between the organelles within the eukaryotic cell.

A competing theory was proposed in the early twentieth century by the Russian biologist R. C. Mereschkowsky and has been revived and revised by some modern biologists, particularly Lynn Margulis of Boston University. This theory, known as the endosymbiotic theory, suggests that mitochondria, undulipodia (an organelle for mobility), and chloroplasts were sequentially incorporated into symbiotic relationships within the ancestral prokaryotes. The mitochondria were, in this theory, originally free-living aerobic bacteria that arose after the initial oxygenation of the atmosphere. These bacteria were capable of normal living processes, but they invaded a large prokaryotic cell and continued to live and respire within the host cell. Both the invader and the host cell benefited from this arrangement. The host provided food to the invader in exchange for some of the energy derived from that food. In addition, the invader may have supplied some protection from oxygen to the host cell. The two continued to live together and, over a long period of time in the later Precambrian, became closely related until they grew completely dependent upon each other. The same general scenario is used to explain the origin of the undulipodia from spirochetes and chloroplasts from cyanobacteria.

There is biochemical evidence to support the endosymbiotic theory as interpreted by Margulis and others. First, separate and different DNA has been found in mitochondria, chloroplasts, and the connecting sites for the undulipodia. This DNA is not coated with proteins as is the nuclear DNA, and it is usually found in a single strand as it is in the bacterial cells. Second, some mitochondria can replicate independently of the main cell. Finally, the ribonucleic acid (RNA) in some mitochondria is more closely related to bacterial RNA than to the RNA of the main cell. All these data suggest an endosymbiotic origin of some of the organelles in eukaryotic cells.

However the earliest eukaryotic cells arose, they rapidly diversified. This diversification in the eukaryotes is in the arrangement and the organization of the cells, not in the internal biochemistry of the cells, which remains remarkably consistent throughout all the eukaryotes. The eukaryotes joined together and differentiated into tissues, organs, and systems to form multicellular organisms. Soft-bodied metazoans were present by 680 million years ago, and all the major phyla of animals were present by 500 million years ago. The eukaryotes have developed different forms, abilities, and behaviors over the last 500 million years and are able to live in most of the environments of the world. Eukaryotes are found in polar to tropical regions, from the depths of the oceans to the tops of mountain ranges, and from rain forests to deserts. It is estimated that there are between 3 and 10 million living species of eukaryotes today and perhaps one hundred to one thousand times as many eukaryotic species that have lived in the geologic past. All those organisms are based upon the eukaryotic cell that developed about 1.75 billion years ago from some prokaryotic ancestor.

Jay R. Yett

CROSS-REFERENCES

BIBLIOGRAPHY

Cloud, Preston C. *Oasis in Space.* New York: W. W. Norton, 1988. A history of the Earth and life written by one of the major geologic researchers. Contains very well written sections on the development of life throughout the Precambrian eon. Suitable for college students.

McMenamin, Mark. *Discovering the First Complex Life: The Garden of the Ediacara.* New York: Columbia University Press, 1998. This entertaining study of the earliest complex lifeforms on the planet details the author's work on these organisms. Written for the interested student but understandable by the general reader.

Margulis, Lynn. *Early Life.* Boston: Jones and Bartlett, 1982. An excellent introduction to prokaryotes, eukaryotes, and the endosymbiotic theory. The characteristics of prokaryotes and eukaryotes are presented in a relatively nontechnical way.

_____. *Symbiosis in Cell Evolution: Life and Its Environment on the Early Earth.* New York: W. H. Freeman, 1981. A detailed, technical account of the development of eukaryotes from prokaryotes by one of the major proponents of the endosymbiotic theory. This text is recommended for college students interested in the background data for this theory. Contains information on the biology and fossil record of eukaryotes.

Margulis, Lynn, and Karlene V. Schwartz. *Five Kingdoms.* 2d ed. New York: W. H. Freeman, 1988. A complete text on the phyla of the organic world. A very good introduction to the diversity of eukaryotes. Short sections describe the most important biologic facts about each phylum, and a short discussion of the fossil record of the group is also presented. Suitable for high school and college students.

Schopf, J. William, ed. *Major Events in the History of Life.* Boston: Jones and Bartlett, 1992. An excellent overview of the origin of life, the oldest fossils, and the early development of plants and animals. Written by specialists in each field but at a level that is suitable for high school students and undergraduates. Although technical language is used, most terms are defined in the glossary.

Starr, Cecie, and Ralph Taggart. *Biology: The Unity and Diversity of Life.* 4th ed. Belmont, Calif.: Wadsworth, 1987. An excellent, comprehensive introduction to biology, written for the interested high school and college student.

EVOLUTION OF BIRDS

Birds evolved from small carnivorous dinosaurs by the development of feathers into flight structures. The earliest known bird is the Late Jurassic Archaeopteryx, *but almost immediately birds became extremely diverse and numerous, with many of the modern bird groups being present by the Late Cretaceous period.*

PRINCIPAL TERMS

ARCHAEOPTERYX: the oldest known bird, fossils of which were first discovered in the Upper Jurassic Solnhofen Limestone of Bavaria in 1860

CLADISTICS: a method of analyzing biological relationships in which advanced characters of organisms are used to indicate closeness of origin

CRETACEOUS PERIOD: a period of time that lasted from about 146 to 65 million years ago, the end of which was marked by the extinction of the dinosaurs

ENANTIORNITHES: a diverse group of Cretaceous birds that filled many of the avian niches now inhabited by modern bird groups

JURASSIC PERIOD: a period of geological time that lasted from about 208 to 146 million years ago, during which birds originated

THEROPODS: the group of carnivorous dinosaurs from which birds developed

THEORIES OF BIRD ORIGINS

The major groups of living birds have fossil records that can be traced back to the Early Tertiary period (later than 65 million years ago) and in some cases to the Late Cretaceous period (up to 75 million years ago). However, the first known fossil bird, *Archaeopteryx*, is from the Late Jurassic period (about 150 million years ago), showing that almost one-half of the history of bird evolution had taken place before the modern groups appeared.

Interest in bird origins was sparked by the discovery in 1860 of a solitary bird feather in the Solnhofen Limestone of Bavaria, which is about 150 million years old (Late Jurassic). This was one year after the publication of Charles Darwin's *On the Origin of Species by Means of Natural Selection* (1859) and thus came at an opportune time in the study of evolution. Shortly after this, several skeletons were discovered in the same deposits, and the presence of feather impressions made it clear that these were the remains of true birds. The animal was named *Archaeopteryx lithographica* and is still the most primitive known member of the birds. Although it had feathers and wings and clearly could fly, it also had a number of unbirdlike features such as a toothed jaw and a long, bony tail.

Thomas Henry Huxley was the first person to make a connection between birds and dinosaurs. In 1870 he noted similarities between the hind legs of the theropod (carnivorous dinosaur) *Megalosaurus* and those of the ostrich and concluded that they were closely related. This was contested on the grounds that both *Megalosaurus* and the ostrich were large and bipedal and that the similarities in leg structure might be caused by a similar mode of life. It was also pointed out that dinosaurs were even larger than ostriches and that none of them had been able to fly, so they seemed rather unlikely bird ancestors.

In 1926 Gerhard Heilmann's book *The Origin of Birds* renewed interest in the idea. He showed that birds were anatomically closer to theropods than to any other group, except for the lack of clavicles in theropods—the structures that fuse in birds to form the wishbone. Since other reptiles had clavicles and theropods apparently did not, he concluded that they had become lost secondarily in theropods. Therefore, they could not be the ancestors of birds, and the similarities must be the result of the fact that they were both bipedal and shared a similar lifestyle. Surprisingly, information was available at that time to show that dinosaurs did have clavicles, as this structure had been illustrated in a description of the theropod *Oviraptor,* although it had been misidentified.

Other views on the origin of birds were put forward during the next forty years, including the ideas that they had been derived from crocodiles, that they had been derived from an early archosaur (the group from which both dinosaurs and crocodiles are derived), and even that they were closely related to mammals. However, idea of a bird-dinosaur relationship was revived by John H. Ostrom of Yale University in the 1960's. He had described a sickle-clawed theropod, *Deinonychus*, a predator from the Early Cretaceous period (115 million years ago) of Montana. The skeletal anatomy of *Deinonychus* showed a number of features that were shared with birds—including *Archaeopteryx*—and other theropods but not with other reptiles. On the basis of these findings, Ostrom concluded that birds are descended from small theropod dinosaurs.

At the time that Ostrom was putting forward his evidence for the theropod origin of birds, a new method of analyzing relationships between organisms was being developed. Called "cladistics," this method groups organisms entirely on the presence of shared characters that are particularly informative. Thus, it is thought that as evolution proceeds, new heritable traits emerge and are passed on; hence, two groups of animals sharing such new traits will be more closely related to each other than to groups that share only the original traits. Analyses of these characters are presented as treelike diagrams called "cladograms" showing the order in which the new characters, and thus the new creatures, evolved. Each branching point in the diagram reflects the emergence of an ancestor that founded a group having advanced characters not present in groups that developed earlier.

Although Ostrom did not use cladistics during his analysis of bird-dinosaur relationships, a study carried out ten years later by Jacques A. Gauthier of the University of California at Berkeley showed that birds were most closely related to small carnivorous dinosaurs and particularly to sickle-clawed theropods such as *Deinonychus*.

DEVELOPMENT OF BIRDLIKE CHARACTERISTICS

Analysis of the features considered to be "birdlike" has shown that many of them appeared early in the history of bird ancestors and then were adapted later to support flight and an arboreal way of life. Some of these were already pres-

ent in the earliest dinosaurs; for example, bipedality (the ability to walk on two limbs) and an upright stance were already present in the immediate ancestors of dinosaurs. In addition, dinosaur ancestors had a hingelike ankle joint and foot bones that were elongated so that they walked on their toes and placed their feet in line when walking. The earliest theropods had hollow bones and a lightened skull, features that provided a light skeleton as in birds. They also had long necks and held their backs horizontally as birds do. Many birdlike features of the limbs developed during the evolution of the theropods, including the reduction of the number of digits from five to three. *Archaeopteryx* also showed this feature, although in the later history of birds, the fingers fused together. The legs of early theropods were long, with a short thigh and long shin as is seen in birds, and the foot was birdlike, with only the three inner toes used for walking.

During theropod evolution, the arm became progressively longer, and the wrist became more flexible, features that were useful for prey capture

Archaeopteryx, the oldest known bird, fossils of which were first discovered in the Upper Jurassic Solnhofen Limestone of Bavaria in 1860. (© William E. Ferguson)

A small, feathered dinosaur, Sinosauropteryx prima, *shown in this 1997 photo, is believed to be between 120 and 140 million years old and is thought by some scientists to be the earliest bird in the world. Other scientists have noted that some features of this animal suggest it is not closely related to birds.* (AP/Wide World Photos)

but were also an advantage later in the development of flight. The shoulder girdle was also strengthened by elongation of the scapula (shoulder blade) and the coracoid (which forms part of the shoulder joint), as well as by the fusion of the clavicles (the collar bones) in the midline to form the wishbone. The breastbone, which was originally made of cartilage, formed a bony plate in advanced theropods. These features provided a stronger skeleton for the dinosaurs but would later form a solid support for the flight muscles in birds. It is also clear that the tail became shorter and stiffer in the theropods and that muscles that once attached between the leg and the tail later connected the pelvis to the tail, thus becoming ideally positioned for tail control during flight in birds.

Behavioral traits also seem to have developed first in the ancestors of birds. One of the most vivid examples of this is seen in a small theropod from the Gobi Desert called the *Oviraptor* ("egg stealer"), which was so named because it was thought to be a predator on the eggs of other dinosaurs. However, one specimen was found sitting on a nest full of eggs, apparently brooding them at the time that it was overwhelmed by sand and died.

ORIGINS OF FEATHERS AND FLIGHT

A major diagnostic feature of birds is the presence of feathers. It is thought that these developed originally from long scales in the reptilian ancestors of birds. They may originally have served as heat shields to reduce heat flow into the body and then, after fraying along the edges, as an insulating covering to retain heat in these small, probably endothermic animals (animals whose internal temperature is maintained by food consumption). Evidence for the presence of feathers in dinosaurs was provided in 1996 and 1997 by Ji Qiang and Ji Shuan of the National Geological Museum of China. They published reports on fossils from Late Jurassic or Early Cretaceous rocks in Liaoning Province, one of which, named *Sinosauropteryx*, bears filamentous fringes along its back and on its body that could have consisted of precursors of feathers. However, *Sinosauropteryx* has short arms and other characteristics indicating that it is not a close relative of birds. A second animal, *Protoarchaeopteryx*, has short true feathers on its body and longer ones on its tail. Skeletal features indicate that it is probably a theropod closely related to birds.

There are two main views on how flight developed in the ancestors of birds. The arboreal hypothesis suggests that ancestral birds started by gliding from branches with the help of incipient flight feathers and then graduated to flapping flight as the feathers enlarged. Although this is an intuitively attractive idea, it is not supported by any obviously arboreal adaptations in theropods or in *Archaeopteryx*. The cursorial hypothesis suggests that, as small dinosaurs chased prey along the ground, they developed lift from feathers along their arms as they stretched or jumped. As the feathers enlarged, lift would have increased incrementally until sustained flight was achieved. This hypothesis is supported by the terrestrial mode of life of theropods and by the fact that the structure of the arm in these dinosaurs, developed to grab prey, is preadapted to create the flight stroke in birds.

LATER EVOLUTION OF BIRDS

Little was known of ancestral birds other than *Archaeopteryx* until the late 1970's, when members of a new group of birds, the Enantiornithes, was found in northwestern Argentina. The name means "opposite birds" because their ankles fused at the opposite end from those of modern birds. They are now known from Spain, China, Mongolia, Madagascar, Argentina, North America, and Australia. They varied in size from vultures to sparrows and had teeth and claws on their wings in some cases, but they were all extremely capable fliers. A specimen from Spain with feather impressions shows the presence of an alula, a tuft of feathers on the first digit that is a key to maneuverability at slow speeds and is found in all modern birds. This 115-million-year-old bird, named *Eoaluluavis*, or "dawn alula bird," represents the oldest record of modern flight.

Although Enantiornithes became extinct at the end of the Cretaceous (65 millions years ago), at least four major lineages of living birds, including ancient relatives of shorebirds, seabirds, loons, ducks, and geese, were already present before the end of the Cretaceous; evidence from calculated divergence times using deoxyribonucleic acid (DNA) studies suggests that others were also present. Thus, a picture is developing of a major radiation of birds through the Cretaceous at the same time that dinosaurs were already thriving in terrestrial environments. Although some bird groups did become extinct at the end of the Cretaceous (most notably the Enantiornithes), the overall picture of bird extinctions is spotty because of the low potential for preservation of small, fragile bird skeletons. Birds that frequent aquatic environments, where suitable environments for preservation exist, are better represented in the fossil record than other groups, and it will take many more discoveries to complete the picture of bird evolution.

SIGNIFICANCE

Birds—an important and diverse group of organisms—are currently the most completely known of modern vertebrates. They pose a number of intriguing questions, including those concerning the origins of flight as well as their relationship with other major groups. The earliest known bird, *Archaeopteryx*, from the Late Jurassic of Germany, was clearly already able to fly, as demonstrated by the presence of flight feathers on the wings. However, it also still retained a set of unbirdlike characters that link it closely to theropod dinosaurs. Discoveries from China have shown that at about the same time, small feather-covered dinosaurs were present, and further discoveries from South America and Spain have shown that birds were diverse and numerous shortly afterward. The method by which flight developed in birds is still debated, however, as is the exact group from which they developed and the time of divergence.

David K. Elliott

CROSS-REFERENCES

BIBLIOGRAPHY

Benton, Michael J. *Vertebrate Palaeontology.* 2d ed. London: Chapman and Hall, 1991. This general vertebrate paleontology text devotes one chapter to the birds and discusses the history of ideas about their evolution and origins. The treatment is detailed, but it is not over-complicated and is very readable. Bibliography and index.

Morell, Virginia. "The Origin of Birds." *Audubon* (March/April, 1997): 36-45. Good illustration of the Chinese feathered dinosaur material together with a discussion of the opposing views of the origin of birds. Suitable for high school readers.

Norell, Mark, Luis Chiappe, and James Clark. "New Limb on the Avian Family Tree." *Natural History* 9 (September, 1993): 38-42. A short article on a bizarre possible early bird from the Gobi Desert by the team that collected the fossils.

Padian, Kevin, ed. *The Origin of Birds and the Evolution of Flight.* Memoirs of the California Academy of Sciences 8. San Francisco: California Academy of Sciences, 1986. A collection of four articles, one of which deals with the origin of birds, while the other three deal with the mechanical constraints on flight and the opposing theories of how flight started. An excellent preface by Padian puts the work in context. A list of relevant research articles is provided.

Padian, Kevin, and Luis M. Chiappe. "The Origin of Birds and Their Flight." *Scientific American* (February, 1998): 28-37. A clear, concise overview of the state of knowledge on bird origins and diversification. Good reconstructions of early birds and illustrations of important anatomical features. Suitable for high school readers.

EVOLUTION OF FLIGHT

The evolution of flight has provided an enormous selective advantage to those organisms that have mastered it. True flight has developed independently on four separate occasions throughout geologic history, beginning with the insects and followed by pterosaurs, birds, and bats. By studying how these animals have mastered flight, humans have learned to fly via machines and have also learned about the evolutionary process.

PRINCIPAL TERMS

ARCHAEOPTERYX: a now-extinct transitional animal between dinosaurs and birds but generally regarded as the first bird because it flew on feathered wings

DERMOPTERAN: a gliding squirrel-sized mammal probably most closely related to primates; the so-called flying lemur of Southeast Asia

DRAG: the component of force that slows the speed of a moving wing as a result of friction and turbulence of the air

INSECTIVORE: any animal that eats insects, such as shrews, bats, and many birds

LIFT: the component of air force that acts on a wing in an upward direction

PTEROSAUR (formerly PTERODACTYL): a now-extinct flying animal closely related to the dinosaurs; it had scales and wings made of a membrane attached to its "fingers"

THEROPOD: a branch of carnivorous dinosaurs, thought to be ancestral to birds, that had large heads on long necks and a deep, compact body and that ran on two strong, bird-like hind limbs

VERTEBRATE: any animal with a backbone, such as birds, dinosaurs, and fish; insects are invertebrates

DEVELOPMENT OF FLIGHT

As much as humans would like to imagine that they have mastered the technique of flying with their machines, they still have much to learn from the animal world. Insects have dominated the air for about 300 million years and remain the most successful of all animals in terms of abundance and diversity. Following insects into the air were the flying dinosaurs, or the pterosaurs, now extinct. They were rapidly followed by those wonderfully successful flyers, the birds. About 50 million years ago, the last group of flying animals joined the birds and insects: bats.

Although these animal groups are capable of true powered flight, there are many additional species that can also be said to "fly": flying squirrels (the marsupial flying phalangers of Australia) and the dermopterans, or "flying lemurs." There are also at least two dozen species of flying lizards, frogs, and snakes. All these animals are really gliders, not flyers. These creatures can leap from a branch, extend a skin membrane outward to increase their surface area, and coast to a lower branch or the ground rather like a toy paper airplane. In the case of the flying squirrels and dermopterans, the gliding membrane consists of a skin fold extending between the fore and hind limbs, sometimes also enclosing the tail. The living dermopteran of the genus *Cynocephalus* has been recorded gliding a horizontal distance of 136 meters with a fall of only 11 meters. Gliding lizards can extend a membrane from their sides that is not connected to the limbs; gliding frogs use their webbed feet as a membrane. The Indian parachute snake expands its ribs and draws in its belly after leaping from a branch, parachuting downward to a soft landing.

Flying fish use their abilities to escape submerged predators by launching themselves into the air. The fish have highly enlarged pectoral fins, which fan out into gliding surfaces in the air while the fish vigorously sways its tail just below the water. In this way, it can reach launch speeds of up to 75 kilometers per hour and then glide freely in the air for a few hundred meters. One family of freshwater flying fish in South America actively flaps with its winglike pectoral fins, making a buzzing sound, although these fish cannot remain airborne for more than a few meters.

In order for an animal to develop true powered flight, it must be able to lift itself upward, propel itself forward, and overcome the friction of the air (drag) as it moves. Birds, pterosaurs, and bats have streamlined their wings in such a way that the upper surface of the wing is more curved than is the lower surface. This shape produces an acceleration of air above the wing, which generates upward suction, or lift. Flying animals can lift themselves only when moving forward so that air flows over the wing, which induces friction, or drag. Thus, the wings must be flapped vigorously to maintain forward motion and to replace the energy lost to friction, and the body must be as streamlined as possible.

The structures that the flying animals have evolved are designed to maximize lift and minimize drag. Insects have developed true wings, but the flying vertebrates (animals with backbones, such as birds, pterosaurs, and bats) have modified their forelimbs for this purpose. The forelimb bones must be exceptionally strong, lightweight, and powered by large, well-anchored muscles, which can make up to one-fourth the body weight in some birds. In addition, all flying vertebrates must consume a large amount of energy in order to overcome drag. Therefore, they must have highly efficient hearts, lungs, and metabolisms, which means that they must be warm-blooded, or able to generate their own internal energy. Flying vertebrates also need greater brainpower (larger skulls) in order to coordinate their flying and to regulate their metabolism.

Insects and Pterosaurs

The first and most successful animals to develop flight were insects. Because insects lack bones or teeth, they do not become preserved readily in sediments, and therefore their fossil record is very poor. The first insects, which were wingless, probably crawled up from the sea onto land around 380 to 400 million years ago, following the colonization of the land by primitive plant life. The first winged insect appears in the fossil record about 300 million years ago, by which time flight was already highly developed in insects. Thus, scientists can only speculate about these first flying animals by studying the fossils available and by observing those living insects whose characteristics seem "primitive." It is likely that insect wings first developed for some other purpose such as gills, gill plates, camouflage, or armor plating as protection from predators. These structures probably then became useful as gliding surfaces, allowing the insects to launch into the wind and to drift to a new location or escape predators. Insects that could fan their protowings could prolong the length of their glides and, eventually, develop true flight. The advantages of true flight to an insect were (and are) enormous in terms of migration, finding food and mates, and escaping predators.

The first vertebrate animal to share the skies with insects, the pterosaurs, appeared suddenly in the fossil record about 200 million years ago during one of those creative spurts in the evolutionary process that also produced dinosaurs, crocodiles, turtles, primitive mammals, and many other animals. Pterosaurs, of which there are about eighty-five known species, flew on wings made of a skin membrane connected to the side (or in some, the hind leg) and stretched out over the

True flight developed independently on four separate occasions throughout geologic history, beginning with the insects and followed by pterosaurs, birds, and bats. This is a fossil of Rhamphorhynchus phyllurus, *a pterosaur, from Eichstadt, Germany. (© William E. Ferguson)*

greatly elongated fourth digit. (The thumb is digit one, index finger number two, and so on.) The membrane was reinforced with stiff fibers of connective tissue. When not flying, the pterosaurs could neatly tuck their wings next to their scaly bodies and scramble about on all fours, using their three-clawed front digits for walking, climbing trees, or grasping. Most had long toothy snouts and large heads on long necks; some had long tails, and several had a head crest. Pterosaurs are thought to have mostly fed on fish, like shorebirds, and in fact probably strongly resembled modern pelicans (without feathers). Modern investigation has shown that pterosaurs were agile, active flyers and possibly endothermic (warm-blooded).

The head, neck, ankle, and shoulder anatomy all indicate that pterosaurs evolved from a primitive archosaurian reptile such as *Lagosuchus*, but because they appear so suddenly in the fossil record, there is no evidence to show how the forelimbs evolved into wings. It is not the pterosaurs but the first birds that shed some light on this dilemma. The first birds, in the form of the dinosaur-bird *Archaeopteryx*, appear in the fossil record about 150 million years ago. Birds shared the skies with pterosaurs for about 80 million more years, after which the pterosaurs (as well as the dinosaurs) became extinct.

ARCHAEOPTERYX

Unlike insects, pterosaurs, and bats, which appear suddenly and already as highly developed flyers in the fossil record, *Archaeopteryx* provides the ideal transitional animal, in both time and structures, between reptiles and modern birds. Like birds, *Archaeopteryx* had feathers on streamlined wings. The feathers, which are modified reptilian scales, vastly increased the lift and surface area of the wing while adding little weight. The wing itself consisted of elongated digits two and three, tightly connected by tissue in *Archaeopteryx* and fused solidly together in modern birds. Also like modern birds and unlike most reptiles, *Archaeopteryx* was warm-blooded and had fused collarbones (the wishbone). Unlike modern birds, *Archaeopteryx* lacked the large-keeled breastbone and the ability to fold its wings tightly. Also more like reptiles, *Archaeopteryx* had a long, toothy snout; a long, bony, feathered tail; and claws on its digits. In fact,

if its feathers had not been preserved in such exquisite detail in the fossil record, *Archaeopteryx* would most likely have been classified as a small theropod. Theropods are a branch of dinosaurs that ran on their strong, birdlike hind limbs, were carnivorous, and had deep, compact bodies with long necks. They were not related to pterosaurs. If a theropod were warm-blooded and small enough (*Archaeopteryx* was pigeon-sized), its scales could have evolved into feathers to help keep it warm.

From its other anatomical details, *Archaeopteryx* was clearly a competent flyer in spite of its lack of keeled breastbone to anchor flight muscles, although it probably lacked the maneuverability of modern bird flight. Traditionally, it was thought that *Archaeopteryx* developed the ability to fly via a gliding stage (the trees-down hypothesis), but additional research has indicated that this scenario is unlikely. Instead, perhaps *Archaeopteryx* ran along the ground, pursuing insects, and used its wings as aerodynamic stabilizers to balance itself after leaping into the air to catch its prey in its mouth (the cursorial hypothesis). Another possibility is that *Archaeopteryx* scrambled among the trees and used its wings to balance itself as it leapt between branches (the arboreal hypothesis).

BIRDS AND BATS

Whatever the means of development, modern birds have clearly adapted flight and feathers to a wide variety of lifestyles and environments, from the tiny tropical hummingbird to the penguin and ostrich. The huge success of insectivorous birds has no doubt placed strong selective pressure on many insects to become nocturnal, as almost all birds are inactive at night. This adaptation allowed the evolution of the fourth and final group of flying animals, the bats.

Bats are the only flying mammals. Like other mammals, they are warm-blooded, have fur, bear live young, and nurse their young with milk. Bat wings consist of a skin membrane stretched over the elongated digits two through five, like an umbrella. Digit one is short, free of the membrane, and bears a claw for grasping. The large wing membrane, which is also attached to the animal's side, encloses the weak hind leg and, in many species, even the long tail. When not flying, bats can scramble over the ground on all fours, but generally they prefer to hang upside down from their hind limbs. Most are tropical, nocturnal insectivores and use echolocation (sonar), not sight, to find their prey.

Fossil bats are quite sparse, but the oldest fossil is around 50 million years in age. *Icaronycteris* appears almost indistinguishable from modern bats, so scientists are unable to explain how the bats learned to fly. The teeth of *Icaronycteris* bear some resemblances to those of fossil shrews and moles, which were also nocturnal insectivores (some of which also use a rudimentary form of echolocation), so it is likely that the bats split from this family some time before 50 million years ago.

Sara A. Heller

CROSS-REFERENCES

Archaebacteria, 951; Cambrian Diversification of Life, 956; Catastrophism, 961; Colonization of the Land, 966; Dinosaurs, 970; Ediacaran Fossils, 976; Eukaryotes, 981; Evolution of Birds, 985; Evolution of Humans, 994; Evolution of Life, 999; Fossil Plants, 1004; Fossil Record, 1009; Fossilization and Taphonomy, 1015; Fossils of the Earliest Life-Forms, 1019; Fossils of the Earliest Vertebrates, 1024; Geoarchaeology, 1028; Life's Origins, 1032; Mammals, 1038; Mass Extinctions, 1043; Mesozoic Era, 1120; Microfossils, 1048; Neanderthals, 1053; Paleobiogeography, 1058; Paleobotany, 1062; Petrified Wood, 1058; Prokaryotes, 1071; Stromatolites, 1075; Systematics and Classification, 1079.

BIBLIOGRAPHY

Alexander, R. McNeill. *Exploring Biomechanics: Animals in Motion.* New York: Scientific American Library, 1992. This text explores the entire range of animal motion but concentrates on the development of flight in three of eight chapters. Extremely well illustrated and well referenced.

Bakker, Robert T. *The Dinosaur Heresies.* New York: William Morrow, 1986. A well-illustrated, engaging, and easy-to-read book that advances the claim that dinosaurs were warm-blooded, active, quick, and intelligent creatures. Chapter 13 describes the anatomical characteristics and likely life habits of the pterosaurs; chapter 14 discusses the characteristics of *Archaeopteryx* and reevaluates its evolutionary relationship to the dinosaurs. Suitable for the high school reader or interested layperson.

Benton, Michael J. *Vertebrate Paleontology.* 2d ed. London: Chapman and Hall, 1997. This comprehensive text provides a complete outline of the history and development of vertebrates. Designed for college courses but also suitable for the informed enthusiast. Deals extensively with the development of flight in vertebrates.

Lewis, Roger. *Thread of Life: The Smithsonian Looks at Evolution.* Washington, D.C.: Smithsonian Books, 1982. This easy-to-read book describes with clear text and illustrates with color photographs and paintings the history of life on the Earth. A chapter entitled "Up from the Sea" contains a short discussion of the rise of insects; the chapter "Life on the Wing" describes the origin of birds and pterosaurs. Suitable for the high school reader.

Paul, Gregory S. *Predatory Dinosaurs of the World.* New York: Simon & Schuster, 1988. This up-to-date and well-illustrated book provides a general introduction to the life and evolution of the predatory dinosaurs, followed by a detailed catalog of same. Chapter 9 discusses the beginnings of bird flight and postulates that flight originated from the leaps of tree-dwelling dinosaurs rather than from gliding. Suitable for the college-level reader.

Tributsch, Helmut. *How Life Learned to Live.* Cambridge, Mass.: MIT Press, 1982. This short paperback book shows how plants and animals have used physics and engineering principles to their advantage in the struggle to survive. Chapter 3, "Locomotion," provides a detailed but not too technical description of the aerodynamic principles of flight and how various animals are adapted to different flying life-styles. Suitable for the college-level reader.

Wellnhofer, P. *Illustrated Encyclopedia of Pterosaurs.* London: Salamander, 1991. A beautifully illustrated and comprehensive treatment of the pterosaur, an extinct flying reptile.

EVOLUTION OF HUMANS

The biochemical and behavioral evidence for an evolutionary relationship between humans and apes is amply confirmed by the fossil record of human ancestors, mainly from Africa. The evolutionary tree of humanity can be traced back some 33 million years, when ancestral apes and monkeys had evolved. The earliest erect-walking humans evolved some 5 to 10 million years ago from the stock that gave rise to gorillas and chimpanzees. The earliest morphologically modern humans, from which all living races arose, date back only about 120,000 years.

PRINCIPAL TERMS

ACHEULIAN: a cultural tradition distinguished by stone tools characterized by large (about 15-centimeter), crudely fashioned, leaf-shaped hand axes

APES: mammals somewhat like monkeys but lacking a tail and with five low cusps on their molars instead of four sharp ones; they include gibbons, orangutans, chimpanzees, and gorillas

AUSTRALOPITHECINES: erect-walking early human ancestors with a cranial capacity and body size within the range of modern apes rather than of humans

CRANIAL CAPACITY: the internal volume (in cubic centimeters) of the braincase of a skull

DRYOPITHECINES: extinct kinds of apes that walked and climbed like monkeys but that lacked the four-cusped molars of monkeys,

the tree-swinging specializations of gibbons and orangutans, and the knuckle-walking specializations of chimpanzees and gorillas

KNUCKLE WALKING: a form of locomotion found in gorillas and chimpanzees whereby they walk on all fours, with feet facing forward but hands curled backward so that the knuckles are placed on the ground

OLDOWAN: a cultural tradition distinguished by stone tools characterized by rounded pebbles crudely chipped to a sharp edge on one side

RAMAPITHECINES: extinct kinds of apes in which the males had prominent canines as in other apes but the females had smaller canines as in humans; otherwise, these creatures were like dryopithecines or orangutans

EARLY APES AND MONKEYS

The fossil record of human evolution in Africa can be traced back to the Oligocene and the Eocene epochs, some 30 to 50 million years ago, in the al-Fayyūm depression of Egypt. Some of the creatures fossilized here may have been ancestral to modern monkeys. Others, such as *Aegyptopithecus*, had the low-crowned, five-cusped molars of modern apes and humans. *Aegyptopithecus*, a small creature with an average weight of 6 to 7 kilograms, was monkeylike in body and may have moved on all fours through trees in search of fruit and leaves. The fossils of similar, weakly differentiated monkey-apes have been found in rocks of Eocene age in Burma (*Pondaungia* and *Amphipithecus*) and of Oligocene age in Bolivia (*Branisella*). The South American forms evolved in isolation into New World monkeys (howlers, marmosets, and capu-

chins). Evolution took a different course in Africa and Asia, culminating in the Old World monkeys (vervets, colobuses, and baboons), as well as apes and humans.

By the Early Miocene, some 23 to 17 million years ago, there was a variety of early apes and monkeys in East Africa. They are particularly well known as fossils from Rusinga Island and nearby localities in southwestern Kenya. The fossils of Old World monkeys (*Victoriapithecus*) have been found, as have the fossils of a much larger group of monkeylike apes referred to as dryopithecines (*Proconsul, Rangwapithecus, Dendropithecus, Limnopithecus*). The monkeylike group had four ridged cusps on their molars, which may reflect a preference for eating leaves. The dryopithecines were apelike in having five low cusps, which suggests that their diet included more fruit. The dryopithecines were nevertheless monkeylike in the way they walked on all fours and climbed trees. Their habitat was dry tropical forest, like that covering large areas around the margin of the jungles of Zaire. These forests were less lush than are rain forests, and they offered more local and seasonal variation in food and shelter.

By the Middle Miocene, some 14 million years ago, there appeared a new group of apes. These ramapithecines are distinguished primarily by canines in females reduced almost to human proportions. The males, however, were strongly built and had well-developed canines like other apes. In other respects, they can be pictured as monkeylike ancestors to orangutans. An early African ramapithecine (*Kenyapithecus*), known from fossils in Kenya, is found along with early monkeys and dryopithecine apes in almost equal abundance. By this time, dryopithecines had waned in abundance and diversity, as monkeys became more prominent. Great environmental changes occurring then would have favored leaf-eating monkeys over fruit-eating dyropithecines. Fossil soils and grasses in Kenya indicate the appearance of wooded grassland. This open vegetation, often called savanna, developed in response to a drier climate than had existed earlier during the Miocene. By this time, both dryopithecines and ramapithecines had dispersed into Europe and Asia. The European apes became extinct 10 million years ago, but ramapithecines evolved into the orangutan and dryopithecines into gibbons in Southeast Asia.

DEVELOPMENT OF THE ERECT STANCE

In Africa, meanwhile, monkeys continued to evolve and diversify, while ramapithecines and dryopithecines declined in importance. A small fragment of a possible ancestor of the gorilla has been found in rocks about 8 million years old in the Samburu Hills of Kenya. Gorillas and chimpanzees share a distinctive series of anatomical features in their hands useful for knuckle walking. These great apes probably had evolved from an African ramapithecine. By 5 million years ago, this same evolutionary lineage had given rise to the first australopithecines. These erect-walking creatures were short (1 to 1.5 meters), and, by human standards, their brains were small (cranial capacity of only 400 to 500 cubic centimeters). They can be

pictured as erect-walking chimpanzees. Early australopithecines are well understood from a spectacular series of skeletal remains (*Australopithecus afarensis*) from Harar, Ethiopia, and a set of trackways from Laetoli, Tanzania, both about 3 to 4 million years old. Both localities also have yielded fossilized soils, pollen, and mammals of a mosaic of woodland and wooded grassland similar to that still seen in many East African game parks.

Erect stance appears to be the most important evolutionary innovation separating humans from apes, and its origin continues to attract speculation. The earliest evidence of this is a thigh bone from the Awash Valley of Ethiopia dated at between 4.0 and 3.5 million years ago. According to a series of martial hypotheses, erect stance was an adaptation for hunting big game in open savanna. Not only did it leave hands free to throw spears, but it also enabled early humans to see potential prey and predators over the tall grass. A second set of hypotheses emphasizes the innovative use of new resources in open habitats. For example, a dietary shift toward small seeds would require precise manipulation, and the need to locate carcasses and to scavenge would require greater mobility. Erect stance would help to meet both these needs. A third set of hypotheses emphasizes the nurturing side of human nature, such as the need to carry food and other materials back to a home base. Nurturing hypotheses are preferred by many paleoanthropologists over the other hypotheses because, for one thing, the earliest erect-walking australopithecines had curved fingers and toes and so retained a facility for climbing trees. In addition, savanna habitats were a part of the African environment and were exploited by a great variety of monkeys for at least 10 million years before the advent of erect stance.

DEVELOPMENT OF TOOL USE

From 2.5 to 1 million years ago, at least two distinct kinds of australopithecines coexisted in savanna mosaic vegetation then covering many parts of Africa. Some were strongly built and weighed 60 kilograms on the average. These robust forms (*Australopithecus robustus*, *A. boisei*, or *Paranthropus*) had large grinding molar teeth and heavily muscularized skulls. They may have eaten large quantities of tough vegetable material, such as seeds and tubers of open savanna, or low-grade browse of woodland and forest. Others were smaller and lightly built, weighing an average of 30 kilograms. These slight, slender (gracile) forms (*Australopithecus africanus, Homo habilis*) had smaller teeth, indicating a softer diet that included meat, fruits, and vegetables. Crudely fashioned pebble tools of the oldowan culture are found as old as 2.6 million years. Although their makers are not known with certainty, they are widely attributed to the lightly built species *Homo habilis*, in large part because of that group's cranial capacity (about 650 to 680 cubic centimeters) compared with that of the robust forms (about 500 to 530 cubic centimeters) that lived at the same time. This difference in brain size—and presumed intelligence—is all the more striking because of the greater body size of the robust forms.

About 1.5 million years ago, a new suite of stone tools of the acheulian tradition appeared, along with a new and more human species, *Homo erectus*. At about this time, earlier australopithecine species declined in abundance and became extinct. The new humans were taller (1.8 meters) and heavier (55 kilograms), with markedly enlarged brains (cranial capacity of 750 to 1,225 cubic centimeters). Their molar teeth were even more reduced in size than in their gracile australopithecine ancestors. Their tools and butchery sites indicate organized big-game hunting. Fruit and vegetables still may have played a large role in the diet, as in modern hunter-gatherer societies. *Homo erectus* may have used fire as long as 1.4 million years ago in Africa, and certainly used fire by 500,000 years ago. They spread out of Africa into Spain, Europe, China, and Indonesia. Archaic humans of the species *Homo sapiens* evolved from this stock by about 500,000 years ago in North Africa or the Middle East. A distinctive local race of archaic *Homo sapiens*, called the Neanderthalers, evolved during the Ice Age in northern Europe.

APPEARANCE OF HOMO SAPIENS

Remains of the earliest morphologically modern humans (*Homo sapiens*) are found in African deposits ranging back in age to about 120,000 years. These skulls have the prominent chin and highly inflated brain case of modern humans (cranial capacity of 1,276 to 1,400 cubic centimeters). The remains indicate that these humans used new and more finely tooled stone implements; they

also appear to have been hunters and gatherers. They diffused into Europe and Asia, extending to Australia by 40,000 years ago and into North and South America perhaps as long ago as 30,000 years. All modern races of humans are thought to have descended from this African stock, which replaced the earlier forms (*Homo erectus* and archaic *Homo sapiens*) throughout their range. It was this new wave of African emigrants of 30,000 years ago that brought to the world the first art, in the form of cave paintings and portable statuettes. They may also have practiced a crude form of animal husbandry, by controlling horse herds in a manner comparable to the shepherding of reindeer by the Lapps of Finland. Since that time, there has been some limited human evolution, particularly a reduction of tooth size in response to softer cooked and processed foods. Physical evolution has been surpassed in the modern era by dramatic and increasingly interracial cultural evolution.

STUDY OF HUMAN EVOLUTION

The most important evidence of human evolution is fossil remains. Many thousands of such fossils have been found over the years, but scientific excitement still is aroused by the discovery of an especially complete skull or skeleton. The bones are compared with one another and with bones of living apes and humans in order to reconstruct the modifications to bone shape that have evolved over geological time and to determine what they may indicate concerning the lifestyle of early humans. The use of X rays can reveal bone structures and evidence of stress or disease not visible on the surface of the bones. Medical computerized axial tomography (CAT) scans also are being used to explore the interior spaces of fossil skulls. Study under the scanning electron microscope can distinguish scrapes on bone surfaces made by cutting with tools from those made by gnawing. Images of microscopic grooves and pits on the surface of teeth can be a clue to the hardness and grit content of the diet of human ancestors.

Also important to the study of human evolution is an evaluation of the geological circumstances of the fossils: their distribution in the rocks; the origin of the entombing ancient soil, stream or lake deposits, or volcanic ash; and the affinities of associated fossilized pollen, leaves, snails, or mammals. These separate lines of evidence are useful for understanding the lifestyles of early humans and for reconstructing the stages of death, burial, and fossilization of human remains.

Consistent methods of determining geological age are essential. The traditional methods of dating deposits by their fossil content (biostratigraphically) or by stone tools have been widely used but lack fine resolution. Correlation of lavas and ashes by distinct differences in chemical composition and their dating by isotopic methods, such as potassium-argon and rubidium-strontium dating, have become especially important for calibrating human evolution.

The methods of modern molecular biology show the evolutionary relationships and distances between humans, apes, and monkeys. Large complex molecules such as deoxyribonucleic acid (DNA) and ribonucleic acid (RNA) have evolved in time as small parts of them changed. The degree of similarity of such molecules in different organisms has been shown to be proportional to the geological time that has elapsed since they had a common ancestor. Monkeys and humans are quite different in overall DNA compared to humans and chimpanzees. These evolutionary distances can be used to estimate geological time since evolutionary divergence. The fossil record shows that a common ancestor of humans and monkeys existed no more than 33 million years ago. Using this datum to calibrate the DNA data, the common ancestor of chimpanzees and humans, not yet known from the fossil record, probably lived at least 7 million years ago. Similar kinds of evidence confirm that modern humans originated in Africa about 120,000 years ago.

Gregory J. Retallack

CROSS-REFERENCES

Archaebacteria, 951; Cambrian Diversification of Life, 956; Catastrophism, 961; Colonization of the Land, 966; Cretaceous-Tertiary Boundary, 1100; Dinosaurs, 970; Ediacaran Fossils, 976; Eukaryotes, 981; Evolution of Birds, 985; Evolution of Flight, 990; Evolution of Life, 999; Fossil Plants, 1004; Fossil Record, 1009; Fossilization and Taphonomy, 1015; Fossils of the Earliest Life-Forms, 1019; Fossils of the Earliest Vertebrates, 1024; Geoarchaeology, 1028; Life's Origins, 1032; Mammals, 1038; Mass Extinctions, 1043; Microfossils, 1048; Neanderthals, 1053; Paleobiogeography, 1058; Paleobotany, 1062; Petrified Wood, 1058; Prokaryotes, 1071; Stromatolites, 1075; Systematics and Classification, 1079.

BIBLIOGRAPHY

Benton, Michael J. *The Rise of the Mammals.* New York: Crescent, 1991. A beautifully illustrated book, suitable for high school students, that explains when and how various mammal families developed. Includes an extensive section on the evolution of humans.

_____. *Vertebrate Paleontology.* 2d ed. London: Chapman and Hall, 1997. This comprehensive text provides a complete outline of the history and development of vertebrates. Designed for college courses but also suitable for the informed enthusiast. The last one-third of the book deals with the rise of mammals, their diversification, and the evolution of humans.

Day, Michael H. *Guide to Fossil Man.* 4th ed. Chicago: University of Chicago Press, 1986. This classic reference work includes basic information on locality, geological age, morphology, evolutionary significance, and other aspects of the most important early human fossils. The book is aimed at the professional scientist but is also a well-illustrated and organized overview of the data base of human evolution that should interest the layperson.

Johanson, Donald C., and Maitland A. Edey. *Lucy.* New York: Simon & Schuster, 1981. This enormously popular account of the personalities and their adventures studying human evolution is at the same time a sound guide to some basic principles of ape and human anatomy and of geological dating. Written in the style and pace of a good detective story, it can be enjoyed without any training in science.

Leakey, Richard E. *The Making of Mankind.* New York: E. P. Dutton, 1981. This book-length exposition of the author's popular television series of the same name deals with the whole sweep of human and cultural evolution from apes to atom bombs. It not only covers a lot of scientific ground at an easily understandable level but also offers personal reflection on the philosophical significance of research into human origins.

Poirier, Frank E. *Understanding Human Evolution.* Englewood Cliffs, N.J.: Prentice-Hall, 1987. This concise guide to the evolution of monkeys, apes, and humans is a widely used senior-level university textbook. Its special strength is explaining the varied lines of scientific evidence that bear on the subject.

Tobias, Philip V. "Major Events in the History of Mankind." In *Major Events in the History of Life*, edited by J. William Schopf. Boston: Jones and Bartlett, 1992. An excellent synthesis of the evolution of hominids from their emergence in the Miocene to the development of modern *Homo sapiens*. Although written in a technical style, the article is profusely illustrated and includes a glossary.

EVOLUTION OF LIFE

Evolution is change in species through time. From a one-celled ancestor, many billions of species have evolved. Evolution is the cause of all life's diversity.

PRINCIPAL TERMS

EVOLUTIONARY TRENDS: statistical directions of evolutionary changes; major trends have been toward increased body size, brain size, and complexity

GENE: the basic unit of heredity; composed of deoxyribonucleic acid (DNA), genes are located in the cell nucleus on chromosomes

GRADUAL EVOLUTION: the theory that evolution occurs throughout much of a species' existence, mostly at slow rates

MUTATION: a spontaneous change in a gene; the ultimate source of variation on which natural selection acts

NATURAL SELECTION: the main process of biological evolution; the production of the most offspring by individuals with the most adaptive traits

PUNCTUATED EVOLUTION: the theory that evolution occurs mainly in rapid "spurts"

SPECIES: a group of individuals that can successfully interbreed only among themselves

INHERITANCE AND NATURAL SELECTION

"Evolution" comes from the Latin word meaning "to unroll." In a general sense, it refers to any change through time, but it is often restricted to biological change. For most of human history, the universe was thought to be unchanging. Then, in the eighteenth century, expeditions to new continents and the discovery of extinct fossil animals convinced many people that the biological world was not as unchanging as had been thought. There could be no proof for this hypothesis, however, until an explanation of how such change occurred could be found. In 1809, Jean-Baptiste Lamarck became the first to propose an explanation; his theory was based on the inheritance of acquired traits. According to this theory, giraffes, for example, obtained long necks because individual giraffes stretched their neck muscles more and more to reach ever higher leaves, and the longer necks were passed on to the offspring. This idea was quickly shown to be false by experiment. Traits acquired during an individual's lifetime (such as larger muscles acquired through weightlifting) are not passed on to offspring.

It was 1859 when Charles Darwin proposed what is now known to be the actual process by which evolution occurs: natural selection. This process can be divided into three steps: Individuals in a species vary in their traits; some individuals will have more offspring than others, depending on how advantageous their particular traits are; advantageous traits will increase in a species, and disadvantageous traits will be lost through time and as the environment changes. For example, climatic change may cause a forested environment to become a snowy tundra. Creatures with dark coats are best off in a forest because of the concealment of the dark, shadowy environment, but as the lighter, snowy environment becomes dominant, lighter individuals will have an advantage because they are more easily concealed. Over a long period, enough changes will accumulate in a group that an observer might say that a new species had been created. This process has often been called "survival of the fittest," but the "fittest" organisms are not always the fastest, fiercest, or even most competitive. For example, animals that cooperate with other animals or are the most timid and conceal themselves readily may survive more often and produce more offspring.

When the whole species changes at once, "nonbranching" evolution occurs. Many species, however, have wide ranges and occur in many different geographic areas, so often only some populations of a species are subjected to environmental changes. That is an important point because it explains how so many species can be created from one ancestral species. "Branching" evolution is especially common when one of the populations becomes cut off from the others by a barrier of some kind. For example, a new river may form. This river prevents interbreeding and allows each population to form its own pool of traits. In time, differences between the two environments will cause the two populations to become so distinct that they form two different species.

Species have hitherto been described as groups that are visibly distinct enough to be distinguishable from one another; however, there is a much more objective definition of species, based on the criterion of interbreeding. To a biologist, members of a species can produce fertile offspring only when they breed with other members. Therefore, a new species has evolved not when it "looks" sufficiently different from its ancestors or neighboring populations but when it can no longer successfully interbreed with them. This definition of reproductive isolation is important because many closely related species look quite similar yet cannot successfully interbreed.

Evolution by natural selection explains not only how species have changed but also why they are so well adapted to their surroundings: The best-adapted individuals have the most offspring. Further, it explains some crucial aspects of basic anatomy, such as why vestigial, or "remnant," organs exist: They are in the process of being lost. For example, the now-useless human appendix was once an important part of the human digestive tract. Also, it explains why many organisms have similar organs that are used for different purposes, such as five-fingered hands on humans and five-digit organs on bat wings. Such "homologous" organs have been modified from a common ancestor. This modification also explains why many organisms pass through similar embryonic stages; human embryos, for example, have tails and gills like a fish.

LAWS OF HEREDITY

After Darwin proposed natural selection as the process of evolution, it was readily accepted, and most scientists have accepted it ever since. The explanation was incomplete, however, in one major area: Darwin could not explain how variation was produced or passed on. The laws of heredity were discovered by Gregor Mendel in 1866 while Dar-

win was wrestling with this problem. Mendel's work lay unnoticed until the early twentieth century, when other scientists independently discovered the gene as the basic unit of heredity. Genes are now known to be molecular "blueprints" that are repeatedly copied within each cell. They contain instructions on how to build the organism and how to maintain it.

Genes are passed on to the offspring when a sperm and an egg cell unite. The resulting fertilized egg consists of one cell that contains all the genes on strands, or chromosomes, in the cell nucleus. The chromosomes occur in pairs such that one member of each pair is from the father and one is from the mother. As growth occurs, certain genes in each cell will be biochemically "read" and will give instructions on what happens next. Genes are composed of the molecule deoxyribonucleic acid (DNA), which is shaped like a twisted ladder and is copied when the "ladder" splits in half at the middle of the "rungs." Once the instructions are copied, they are carried outside the cell nucleus by messenger molecules, which proceed to build proteins (such as enzymes and muscle tissue) using the rungs as a blueprint.

With this added knowledge of genes as the units of heredity, evolutionists could see that natural selection acting on individuals selects not only traits but also the genes that serve as blueprints for those traits. Therefore, as well as being a change in a species' traits through time, evolution is also often defined as a change in the "gene pool" of a species. The gene pool is the total of all the genes contained in a species. Individual variation in a gene pool originally arises via mutations, errors made in the DNA copying process. Usually, mutation involves a change in the DNA sequence that causes a change in the genetic instructions.

Most mutations have little effect, which is fortunate because those that are expressed generally kill or handicap the offspring. That occurs since any organism is a highly integrated, complex system, and any major alterations to it are therefore likely to disrupt it. Nevertheless, rare improvements do occur, and it is these that are passed on and become part of the breeding gene pool. Although mutations provide the ultimate source of variation, the sexual recombination of genes provides the more immediate source. Each organism has a unique combination of genes, and it is the fitness of this combination that determines how well those genes survive and are passed on. Although brothers and sisters have the same parents, they are not alike because genes are constantly shuffled and reshuffled in the production of each sperm and egg cell.

RATES AND PATTERNS OF EVOLUTION

A major area of debate is how fast evolution occurs. Some scientists believe that most evolution occurs rapidly. This view has been called "punctuated" evolution. Another group argues that evolution is more often gradual, as Darwin originally proposed. To some extent, this disagreement is a matter of different perspectives. A geneticist working with flies in a laboratory would see the evolution of a new species in ten thousand years as very slow. To a paleontologist, however, who often deals with fossil species lasting millions of years, ten thousand years is brief indeed. Nevertheless, there is more to the debate than perspective alone. Punctuationists argue not only that evolution is rapid but also that species have such tightly integrated gene pools that virtually no change at all occurs during most of a species' existence.

In contrast, gradualists view species as being much less integrated, so that change can be a continuous process. The fossil record at first glance seems to support the punctuated view. The majority of species show very little change for long spans of time and then either disappear or rapidly give rise to another species. The fossil record is very incomplete, however, being full of gaps where no fossils were deposited. As a result, it is often impossible to tell whether the "rapid" change in species is real or only a gap in what was actually a gradual sequence. Also, fossils represent only part of the original organism—usually only the hard parts, such as shells, bones, or teeth. Therefore, any changes in soft anatomy, such as tissues or biochemistry, are lost, making it impossible to say with certainty that no change occurred. Whatever the outcome of the debate, all scientists agree that evolutionary rates vary.

In addition to the rate of evolution, much has also been written about the patterns produced by evolution since life arose about 3.5 million years ago. Evolutionary trends are directional changes seen in a group. The most common trend, found in many groups, is an increase in size. Another trend, seen mainly in mammals, is an increase in brain size. Life as a whole has shown an increase in total diversity and complexity. These trends, however, are only statistical tendencies. They are not inevitable "laws," as many have misinterpreted them in the past. Often, groups do not show them, and in those that do, the change is not constant and may reverse itself at times. Finally, trends are often interrupted by mass extinctions. At least five times in the past 600 million years, more than 50 percent of all the species on the Earth have been wiped out by catastrophes of different kinds, from temperature changes to impacts of huge meteorites.

STUDY OF FOSSILS

Fossils, the remains of former life, provide the only record of most evolution, because more than 99 percent of all species that have ever existed are now extinct. Paleontology is the study of fossils. Such study begins with identification of the remains—usually hard parts, such as bones—and ends with measurement of fossil size, shape, and abundance. The extreme incompleteness of the fossil record is a major obstacle to this method, since only some parts are preserved, and these are from strictly limited periods of the evolutionary past. Nevertheless, many evolutionary lineages can be traced through time. Indeed, refined measurements of rate and direction of anatomical change are often possible when used in conjunction with dating techniques.

The study of living organisms permits observation of the complete organism. Comparative anatomy reveals similarities among related species and shows how evolution has modified them since they separated from their common ancestor. For example, humans and chimpanzees are extremely similar in their organ and muscle anatomy. This method is not limited to comparison of adults but includes earlier stages of development as well. Comparative embryology often shows anatomical similarities, such as those between humans and other vertebrates. For example, the human embryo goes through a stage with gills and a tail, resembling stages of an amphibian embryo. Comparative biochemistry is also very useful, revealing similarities in proteins and many other molecules. Such comparisons are based on differences in molecular sequences, such as amino acids. Molecular "clocks" are sometimes calculated in this manner.

More distantly related species are thought to have more differences. The accuracy of such clocks, however, is hotly debated.

A major technique is DNA sequencing, whereby the exact genetic information is read directly from the gene. This method will greatly add to knowledge of evolutionary relationships, although it is expensive and time-consuming. Biogeography, or the distribution of organisms in nature, is a method that Darwin used and that is still important today. This technique often reveals populations (races) within a species' overall range that differ from one another because they inhabit slightly different geographic areas. These populations give the scientist a "snapshot" of evolution in progress. Given more time, many of these races would eventually become different species.

Artificial breeding is a method of directly manipulating evolution. The most widely used experimental organism for this purpose is the fruit fly, which is used in part because of its exceptionally large chromosomes; they make the genes easy to identify. A common experiment is to subject the flies to radiation or chemicals that cause mutations and then to analyze the effects. The gene pool is then subjected to extreme artificial selection as the experimenter allows only certain individuals to breed. For example, only those with a gene for a certain kind of wing may reproduce. Although such experiments have often altered the organisms' gene pools and created new varieties within the species, no truly new species has ever been created in the laboratory. Apparently, more time is needed to produce a new species. Outside the laboratory, artificial breeding has been done for thousands of years. Food plants and domesticated animals have had much of their evolution controlled by humans. Analysis of the effect of this breeding on the organisms' gene pools is the most complete and direct method of studying evolution.

SIGNIFICANCE

The study of fossils has been a major tool in understanding the Earth's history. This understanding has allowed more efficient exploitation of Earth resources. For example, petroleum and coal provide the major energy resources today and were both formed by organisms of the past. Petroleum comes from the biochemicals of marine organisms, and coal comes from fossilized plants.

Most paleontologists are employed in the costly search for these "fossil fuels," and knowing the evolutionary history of these groups helps to determine the most productive places to search. Fossils also form nonenergy resources. Limestone is used in many processes, from making cement to making steel. Most limestone is composed of the fossilized remains of seashells and other marine skeletons. Phosphate minerals, essential for fertilizers in almost all forms of agriculture, come from marine fossil deposits as well.

Darwin's theory of natural selection caused a violent reaction throughout much of the world when it was applied in social contexts (to which Darwin himself disagreed) as "social Darwinism." The notion that humans evolved from lower lifeforms such as the ape was truly revolutionary. Instead of creatures of a divine plan, humans were now seen as products of natural, sometimes "random," processes. The impact of this realization on ethics, the arts, and society in general is still being felt. Evolution, however, does not necessarily conflict with religion, as is often thought. Science seeks to find out only how things happen, not the ultimate reasons why they happen. Therefore, most major religions have reconciled their tenets with the fact of evolution by viewing natural selection as simply a mechanism employed by God to meet his ends.

Michael L. McKinney

CROSS-REFERENCES

BIBLIOGRAPHY

Dawkins, Richard. *The Blind Watchmaker: Why the Evidence of Evolution Reveals a Universe Without Design*. New York: W. W. Norton, 1988. Perhaps the best popular book available on evolution. It is clearly written and well argued, and the author is a well-qualified evolutionary biologist. The book specifically addresses major questions that have bothered many critics of evolutionary theory, in a nontechnical manner. The most unique aspect is a simple computer model that actually "produces" evolution. Readable by any interested person and readily available.

Futuyma, Douglas J. *Evolutionary Biology*. Sunderland, Mass.: Sinauer Associates, 1986. This book has been widely praised by evolutionary scientists. It is more theoretical than most but includes interesting discussion on the history of evolutionary thought and social issues of evolution. An advanced book, suitable mainly for college students.

McNamara, K. J. *Shapes of Time: The Evolution of Growth and Development*. Baltimore: The Johns Hopkins University Press, 1997. McNamara explores evolution from the perspective of heterochrony (changes in the timing of embryological development) and shows how this phenomenon has affected many aspects of evolution. Each chapter has a bibliography. Indexed.

Prothero, Donald R. *Bringing Fossils to Life*. Boston: McGraw-Hill, 1998. This well-illustrated and entertaining text covers a broad range of paleontological topics, including the Cambrian explosion. Glossary, bibliography, and index.

Schopf, J. William, ed. *Major Events in the History of Life*. Boston: Jones and Bartlett, 1992. An excellent overview of the origin of life, the oldest fossils, and the early development of plants and animals. Written by specialists in each field but at a level that is suitable for high school students and undergraduates. Although technical language is used, most terms are defined in the glossary.

Stanley, Steven M. *The New Evolutionary Timetable: Fossils, Genes, and the Origin of Species*. New York: Basic Books, 1981. A popular review of evolutionary theory. The book is especially useful for its discussion of Darwin and the history of evolutionary ideas. It also provides a view of ideas about evolution that have been raised by evolutionary scientists since 1980, such as the widely discussed "punctuated" patterns of evolution. For general audiences.

Stebbins, G. Ledyard. *Darwin to DNA, Molecules to Humanity*. New York: W. H. Freeman, 1982. A philosophical and personal review of major concepts of evolution. Unlike many books on evolution, this one discusses its social implications and is not afraid to speculate about what the theory might mean for the human species and civilization.

FOSSIL PLANTS

The rise of land-dwelling animals paralleled the rise of plants, which have always been the basis for animal life. Fossil plants are a valuable source of information regarding such phenomena as changes in climate, ancient geography, and the evolution of life itself.

PRINCIPAL TERMS

CATKIN: a dense cluster of scalelike flowers

MESOZOIC ERA: the era that began 245 million years ago and ended 66 million years ago; it includes three periods: the Triassic, the Jurassic, and the Cretaceous

MOTILE: having the ability to move spontaneously

PALEOZOIC ERA: the era that began about 543 million years ago and ended 245 million years ago; it includes six periods: the Cambrian, the Ordovician, the Silurian, the Devonian, the Carboniferous, and the Permian

PHYLUM: one of the broad categories used in the classification of organisms

PISTIL: the female organ of flowers, which receives the pollen

POLLEN: the male cells of fertilization

STAMEN: the male organ of flowers, which bears the pollen

SYMBIOSIS: a condition in which organisms live together in close association, especially when the relationship is mutually beneficial

VASCULAR: possessing a series of vessels that form a conducting apparatus for food and sap

THALLOPHYTES

The earliest fossil plants are represented by a phylum called the thallophytes. The geological record of the thallophytes is incomplete. Of the seven large groups, only a few are represented by fossils. Although doubtful records from the Paleozoic era have been found, the earliest identifiable specimens are found in the Jurassic period. The dearth of fossils from this group of plants can be attributed to their minute size and the fragile nature of their remains. The thallophytes are the most primitive plants, lacking roots, stems, leaves, and conducting cells. The simplest thallophytes are the subphylum autophytic thallophytes, which include blue-green algae diatoms and algae. All these plants produce chlorophyll. Blue-green algae made up a class of unicellular plants occurring in colonies held together by a jellylike material. Related to green and brown algae were the diatoms, one-celled plants enclosed in a wall consisting of two overlapping valves. The next class, simply called algae, consists of several different types of seaweed, such as chara, or stonewort, which secretes lime with which it encrusts its leaves and is responsible for many freshwater limestones of the past. Many fossils that have been described as algae were actually molds of burrows or tracks of animals.

The second subphylum of the thallophytes is called the heterophytic thallophytes. These plants are distinguished by the absence of chlorophyll; as in animals, their principal source of energy is organic. The heterophytic thallophytes are subdivided into three classes. Bacteria, one-celled plants without definite nuclei, are the chief agents of the decomposition of organic matter; without bacteria, more prehistoric plant and animal remains would have been preserved. The next class, slime fungi, are sticky masses enclosing many nuclei but without cell walls. Slime fungi have never been found as fossils. The final class is fungi, which are composed essentially of a branching mass of threads called the mycelium, which penetrate the cell walls of their "host"—plant or animal—and live upon its substance. Fungi are also rarely preserved as fossils, although mycelial threads of a fungus have been detected under the bark of trees from the Pennsylvanian coal beds. Lichens are a type of fungi made up of a fungus and an algae living together in symbiosis. Fossil lichens have been recognized only from very recent formations.

BRYOPHYTES AND PTERIDOPHYTES

The next phylum to emerge, the bryophytes, exhibits a distinct advance over the thallophytes. Bryophytes adapted more successfully to the terrestrial environment. They were able to take water and other necessary substances from the soil by means of rootlike hairs called rhizoids. The most distinct advance of the bryophytes over the thallophytes is in their method of reproduction. The spores produced by these plants germinate by sending out a mass of green threads, the protonema. The simplest bryophytes are the liverworts. The mosses, which are more abundant today than the liverworts, possess leaves consisting of many small chlorophyll-bearing cells. Because the ancient members of the bryophyte group were more delicate than the modern forms, they have been preserved only under exceptional conditions, such as those provided by the silicified peat beds at Rhynie, Scotland, which contain fossils from the Devonian period.

The pteridophytes were much more advanced than the bryophytes. While the structure of the bryophytes was primarily cellular, that of the fern plant is vascular. Unlike the bryophytes, the pteridophytes originate from the fertilized egg and produce spores. Pteridophytes are well represented by the ferns, which have existed from the Devonian period. Another class of pteridophytes is the horsetails (Equisetales), which have existed from the Devonian to the present. Equisetales include the calamites, trees of the Permian period, with lance-shaped leaves attaining heights of 30 meters, and annularia, a smaller plant with a stem of 5 to 8 centimeters in diameter, which was abundant in the Pennsylvanian period.

The third class of pteridophytes is the club mosses, which are largely creeping, many-branched plants with numerous tiny mosslike leaves spirally arranged on the stem. One member of this class, *Lepidodendron*, was a lofty tree of the later Paleozoic, appearing in the Lower Devonian and dying out in the Permian. The leaves were arranged in a convex outline to form a "leaf cushion." The stigmaria consisted of spreading, rootlike underground stems and were common in the Pennsylvanian period. *Sigillaria* was a tree that resembled *Lepidodendron* in general appearance except for the fact that its leaf cushions were hexagonal in outline. The final class of pteridophytes, Sphenophyllales, consisted of slender plants with jointed stems and leaves in whorls. These climbing plants are known from the Devonian to the Permian periods and are represented by sphenophyllum, a small branching plant with slender, ribbed stems.

SPERMATOPHYTES: GYMNOSPERMS

The fourth phylum, the spermatophytes, are distinguished by the production of seeds, although the lower groups have the same alternation of the vegetative (asexual) and reproductive (sexual) generations as is seen in the pteridophytes. The chief distinguishing characteristics of the spermatophytes are the formation of a pollen tube and the production of seeds. The first class, the gymnosperms, are typified by the pines, mostly evergreens. One order of gymnosperms, Cycadofilicales, were fernlike in habit but were not actually ferns. Because the leaf and stem remained practically unchanged, it is very easy to mistake the early seed plants for ferns. One of the most familiar of the fernlike fronds of the Pennsylvanian coal deposits is *Neuropteris*, which had large, compound leaflets. *Pecopteris* had fronds

Fossil leaves from sandstone of the Gerome Andesite, Northwest Uranium Mine, Stevens County, Washington. (U.S. Geological Survey)

like those of *Neuropteris*, but the leaflets were attached to the stalk by their whole width. *Lyginodendron* also had stems varying in diameter from 3 millimeters to 4 centimeters. Its leaves were very large, divided, and fernlike in appearance.

The second order of gymnosperms, Cycadales, or the cycads, is an extinct family of trees or shrubs. The stem in most forms was thick and short and covered with an armor of leaf bases. The method of fertilization in the cycads was also primitive. The male cells were motile, swimming actively to the ovule after the rupture of the pollen tube. By virtue of its lateral branching and hairlike scales covering the leaf bases, Cycadales also recalls the ferns. Nevertheless, it represents an advance over previous plants in that it had a true flower because both male and female organs were borne on the same axis and were arranged in the manner of later flowering plants. Thus Cycadales is an intermediary in the line of development of the angiosperms from their fern ancestors. This order formed the dominant vegetation of the Mesozoic, ranging from the Triassic into the Lower Cretaceous.

The next order of gymnosperms, Cordaitales, is an extinct group of tall, slender trees that thrived throughout the world from the Devonian to the Permian period. The leaves of these trees were swordlike and distinguished by their parallel veins and great size, reaching up to 1 meter. The stem resembles that of the conifers except for the very large pith, which recalls the cycads. The reproductive organs of *Cordaites* were small male and female catkins. The Cordaitales were the dominant members of the gymnosperm forests during the Devonian period. The fourth order of gymnosperms, Ginkgoales, resembles the conifers in general appearance. The leaves, however, are fanlike and are shed each year. Like the cycads and ferns, the male cells are motile in fertilization.

The order Coniferales includes mostly evergreen trees and shrubs, with needle or scalelike leaves and with male and female cones. Derived from Cordaitales of the Paleozoic, Coniferales possesses fewer primitive characters than Ginkgoales. The yews, which are comparatively modern, have fruit with a single seed surrounded by a scarlet, fleshy envelope. Another family, Pinaceae, having cones with woody or membranous scales, are represented by *Araucaria*, which is very common in the Petrified Forest in eastern Arizona. The Abietae, one of the more common families of evergreens, includes pines, cedars, and hemlocks dating back to the Lower Cretaceous. One of the most extraordinary members of the conifers was the family Taxodiaceae. Taxodiaceae includes the genus *Sequoia*, which is represented today only by the redwood and the *Sequoia gigantea*. The trunk attains a height of more than 90 meters and a diameter of more than 9 meters. *Sequoia*'s twigs, cones, and seeds were abundant in the Lower Cretaceous of North America. Finally, the family Cupresseae includes the junipers and is known from the Jurassic.

SPERMATOPHYTES: ANGIOSPERMS

The second class of spermatophytes is the angiosperms. The angiosperms contain the plants of the highest rank. This group comprises more than one-half of all known living species of plants. The members of the angiosperms are commonly known as the flowering plants. The typical flower is composed of an outer bud-covering portion, the stamens, and the pistil. Each part of the flower is a specially modified leaf. When the wind or an insect brings the pollen into contact with the pistil, the pollen is held in place by a sugary solution. After the pollen penetrates an ovule, the nucleus divides several times. This fusion is called fertilization. The embryo, consisting of a stem with seedling leaves, is called a seed.

Both subclasses of the angiosperms first appeared in the upper part of the Lower Cretaceous. Dicotyledoneae is a primitive subclass that begins with two seedling leaves that are usually netted-veined. The stem is usually thicker below than above, with the vascular bundles arranged to form a cylinder enclosing a pith center. As growth proceeds, new cylinders are formed. One of the most primitive of the dicotyledones is the American tulip tree, appearing first in the Upper Cretaceous. Sassafras is the last representative, flourishing throughout North America and Europe since the Lower Cretaceous. The poplar is known to have lived during the Lower Cretaceous period in Virginia and Greenland.

The second subclass, Monocotyledoneae, descended from the dicotyledones and is distinguished by the fact that the plant begins with a single leaflet, or cotyledon. The veins of the leaves are parallel, the stem is cylindrical, and the roots are fibrous. This subclass, which is represented by the grasses and grains, is especially important to humans today. Fossils from this subclass date back to the upper part of the Lower Cretaceous of eastern North America. The fossil record of the palm goes back to the mid-Cretaceous.

SPERMATOPHYTES: A SUCCESS STORY

The evolution of plants is the story of their struggle to adapt themselves to land. One of the changes necessary in the development of land flora was the change from a cellular structure to a vascular one, which opened up possibilities for increase in size and laid the foundation for the trees. In order to adapt to land, plants also had to develop a resistance to the dehydrating quality of the air. The earliest plants, the thallophytes, were closely tied to water. One of the first examples of flora adapting to land were the freshwater algae. The change from a cellular to a vascular structure led to the development of roots; the pteridophytes were the first plants to take this step. The mosses and ferns adapted to land but still required rain or dew for the union of the gametes. It is only the spermatophytes that developed a device that freed them from the necessity of external water for fertilization to occur. This ability permitted the spermatophytes to proliferate throughout the Earth.

Alan Brown

CROSS-REFERENCES

Archaebacteria, 951; Biostratigraphy, 1091; Cambrian Diversification of Life, 956; Catastrophism, 961; Colonization of the Land, 966; Dinosaurs, 970; Earth Resources, 1741; Ediacaran Fossils, 976; Eukaryotes, 981; Evolution of Birds, 985; Evolution of Flight, 990; Evolution of Humans, 994; Evolution of Life, 999; Fossil Record, 1009; Fossilization and Taphonomy, 1015; Fossils of the Earliest Life-Forms, 1019; Fossils of the Earliest Vertebrates, 1024; Geoarchaeology, 1028; Geologic Time Scale, 1105; Life's Origins, 1032; Mammals, 1038; Mass Extinctions, 1043; Microfossils, 1048; Neanderthals, 1053; Paleobiogeography, 1058; Paleobotany, 1062; Petrified Wood, 1058; Prokaryotes, 1071; Stromatolites, 1075; Systematics and Classification, 1079.

BIBLIOGRAPHY

Andrews, Henry N. *The Fossil Hunters: In Search of Ancient Plants.* Ithaca, N.Y.: Cornell University Press, 1980. A comprehensive and detailed work about the major figures in the field of paleobotany. The book traces the growth of paleobotany over much of the globe, from the late 1600's to the present. Scientists are classified according to the country in which they did most of their research. Andrews is especially concerned with the work accomplished by several naturalists in the late seventeenth and the eighteenth centuries. Intended primarily for students of paleobotany, botany, and geology.

Gaylord, George. *Life of the Past.* New Haven, Conn.: Yale University Press, 1968. One of the world's leading paleontologists explains the major historical biological principles that have been deciphered from fossilized clues. The author also promotes the broader significance of paleontology regarding life and its evolution. The appendix contains a useful summary of the subphyla of ancient plants. This book will appeal to the general reader and to the scientist alike.

Kirklady, J. F. *Fossils in Colour.* Dorset, England: Blandford Press, 1980. Beautifully illustrated with color plates and numerous sketches, the section on plants discusses the morphology, geological history, and the occurrence of fossil plants. Useful for college-level students and specialists in paleobotany.

Murray, Marian. *Hunting for Fossils.* New York: Macmillan, 1967. Beginning with an introduction to paleontology, the author explains how and where to find fossils. The chapter on plants explains how old ideas regarding the "primitive" plants have been revised. For the general reader and the high school student.

Niklas, K. J. *The Evolutionary Biology of Plants.* Chicago: University of Chicago Press, 1997. A text on the biology of plants with particular emphasis on the evolutionary history of morphology. Suitable for readers with a background in botany and paleobotany.

Ransom, Jay Ellis. *Fossils in America*. New York: Harper & Row, 1964. In nontechnical style, the author explains how to prospect and prepare fossils. The book contains a useful glossary, a bibliography, an index, an appendix of reference libraries, and more than one hundred drawings and photographs.

Taylor, T. N., and E. L. Taylor. *The Biology and Evolution of Fossil Plants*. London: Prentice-Hall, 1993. This detailed textbook on paleobotany is intended for the specialist. However, it is an excellent resource for anyone interested in the subject.

Thompson, Ida. *The Audubon Society Field Guide to North American Fossils*. New York: Alfred A. Knopf, 1983. The section of fossil plants is beautifully illustrated with color plates and discusses the adaptation of plants to land, structural support, reproduction, and fossilization. The book also discusses the various divisions of fossil plants. Recommended for college-level students or specialists in the field.

White, M. E. *The Greening of Gondwana*. Sydney: Reed Books, 1996. This very well illustrated book has color and black-and-white photographs of Australian fossil plants, with an emphasis on the Gondwana flora.

Zim, Herbert S. *A Golden Guide to Fossils*. New York: Western Publishing, 1962. Discusses the process of fossilization and the collection and exhibition of fossils. The short section on plants classifies them and discusses them in easily understood terms. Also contains helpful color illustrations.

FOSSIL RECORD

The fossil record provides evidence that addresses fundamental questions about the origin and history of life on the earth: When did life evolve? How do new groups of organisms originate? How are major groups of organisms related? This record is neither complete nor without biases, but as scientists' understanding of the limits and potential of the fossil record grows, the interpretations drawn from it are strengthened.

PRINCIPAL TERMS

BODY FOSSIL: the petrified remains of a plant or animal

FOSSIL: evidence of organic activity, usually preserved in sedimentary rock strata

GRADUALISM: an explanation of how evolution works involving slow, constant change through time

LAGERSTATTE: an assemblage of exceptionally well-preserved fossils

MACROFOSSIL: a fossil that is large enough to study with the unaided eye, as opposed to a microfossil, which requires a microscope for examination

MORPHOLOGY: the appearance (shape and form) of an organism

PALEONTOLOGIST: a scientist who studies ancient life; invertebrate paleontologists study fossil invertebrate animals, vertebrate paleontologists study fossil vertebrates, and micropaleontologists study microfossils

PUNCTUATED EQUILIBRIA: an alternative model of how evolution works, by rapid speciation events that involve major changes in morphology

SPECIATION: the evolutionary process of species formation; the process through which new species arise

SPECIES: a group of similar, closely related organisms

TAPHONOMY: the study of the sequence of events that led to the burial and preservation of fossils

TRACE FOSSIL: indirect evidence of an organism's presence through tracks, trails, and burrows

OVERVIEW

The term "fossil" originally referred to any object dug up from the Earth and included minerals as well as the petrified remains of once-living organisms. The term is now used in a restricted sense to describe the preserved remains of organic life, both plant and animal. Probably the most familiar kinds of fossils are body fossils, which are the fossilized remains of the actual organisms, but there is also indirect fossil evidence of organic activity, such as footprints as evidence of walking. Collectively, these tracks, trails, and burrows are termed "trace fossils."

The term "fossil record" refers to the sum total of fossils preserved in geological strata on the Earth. The fossil record extends back in time to rocks 3.5 billion years old. The first entries in the fossil record are single-celled, plantlike organisms. There are about 250,000 known fossil species of plants and animals. That seems to be a large number until one compares it with the approximately 4.5 million species of plants and animals that are alive today. The entire fossil record of ancient life amounts to 5 percent of the total number of modern species. What is the reason for the paucity of fossils compared with the abundance of modern species? Does this difference in number of fossils and recent species reflect true differences in diversity, or does it reflect limitations in the compilation of the fossil record? In other words, were there fewer species in the geological past than in the recent past, or is the fossil record incomplete?

Charles Darwin, famous for his contributions to evolutionary theory, came to the conclusion that the fossil record is incomplete. Darwin was one of the earliest naturalists to publish on what he termed the "imperfection" of the fossil record. In his book *On the Origin of Species by Means of Natural Selection* (1859), Darwin compared the preserved record of life on the Earth to a set of books in which several volumes are missing, and from the remaining volumes of which chapters are missing, and from the remaining chapters of which pages are missing, and from the surviving pages of which words are missing. Darwin and others concluded that not every kind of organism that once lived is fossilized and that the fossil record is in fact biased toward preservation of some forms over others.

RAPID BURIAL AND ANOXIA

The biases inherent in the fossil record stem from the fact that fossilization of organic material is the exception, not the rule, and very specific and relatively rare conditions must be met for an organism to become fossilized. Fossilization favors organisms with hard parts—for example, an exterior shell (exoskeleton) or internal skeleton (endoskeleton). Fossilization also favors organisms living in certain environments. Two particular environmental conditions favor fossilization: rapid burial and anoxia (lack of oxygen). Rapid burial protects organic remains from predators or scavengers and physical reworking by tides and waves. Oxygen supports bacteria and decomposition of organic material. Burial in an oxygen-free (reducing) environment insulates organic material from decay and thus favors fossilization.

The most exceptional fossils known are from environments in which one or both of these two environmental conditions were met. German paleontologists call exceptionally preserved fossil assemblages fossil "lagerstatten," or mother lodes. Famous fossil lagerstatten include the Mazon Creek fauna, from Pennsylvanian-period strata (300 million years old) in Illinois, in which insects, crustaceans, and previously unknown soft-bodied organisms are preserved in ironstone concretions; the Burgess Shale fauna, from Middle Cambrian (500-million-year-old) strata in British Columbia, famous for the discovery of a great variety of unusual arthropod and annelid-like animals; the Solnhofen Limestone, from Jurassic strata (200 million years old) in Bavaria, in which *Archaeopteryx* (a reptile with feathers) was discovered; insects preserved in amber (fossilized tree sap), of the Oligocene epoch (40 million years old) in Germany; and the La Brea Tar Pits, from the Pleistocene epoch (10,000 years old) in California, in which a variety of animals, including saber-toothed tigers and mastodons, were ensnared and preserved in natural asphalt springs.

From this brief survey, it is clear that fossil lagerstatten occur in a variety of geological and geographical settings and through a wide range of geologic time. These lagerstatten share the characteristic of rapid burial or burial in a biologically inert environment. They provide information on the morphology of previously unknown groups or a better record of known groups, but do such assemblages reveal anything about how the organisms lived? Some of these lagerstatten represent environments in which the animals died rather than the environments in which they lived; the La Brea animals, for example, certainly did not live in the tar pits. Lagerstatten, however, can be used to reconstruct paleocommunities, to create a picture of organisms that lived contemporaneously, even though that picture of paleocommunity structure or paleoecological relationships might not be complete. Each different fossil lagerstatten provides an exceptional view of a geologically unique situation.

PRESERVATIONAL BIASES

Because of the preservational biases inherent in the fossil record, most fossilized species represent only a few major groups—the numerically abundant and well-skeletonized organisms that lived in or near an anoxic environment or an environment that was subjected to rapid, episodic influxes of sediment. The environment with the best fossil preservation potential on the Earth is the shallow marine shelf: Most marine life lives in the shallow shelf; the shelf is subject to rapid influxes of sediment (via storms and rivers, for example), and many marine invertebrates have exoskeletons. Thus, hard-shelled invertebrates from shallow marine environments constitute the bulk of the fossil record. The major marine invertebrate groups that dominate the fossil record include corals, bryozoans, brachiopods, mollusks (clams, cephalopods, and snails), arthropods (especially trilobites), and echinoderms (starfish and their relations).

The fossil record of other groups, including marine and terrestrial vertebrates and plants, is neither as abundant nor as complete as the marine invertebrate record. Leaves and flowers are rare as fossils, often preserved as imprints in sediments deposited in ancient inland lakes. Woody

tissue of trees is often preserved by a process called replacement, in which mineral-
rich water percolates through
the porous wood and minerals
(especially quartz) precipitate
from the water, filling the voids
in the plant structure. The logs
of the Petrified Forest in Arizona are examples of replacement by silica. Pollen spores
and seed pods are important
constituents of the paleobotan-
ical record because these structures are abundant and often
have tough outer coverings. Paleobotanists face special problems in identification and classification of plant fossils because
entire plants are rarely found
as fossils. Stems, roots, leaves,
flowers, and seeds are described
separately as they are found.
Consequently, a single plant

Ammonites, descended from a single group of shelled cephalopods surviving into the Early Triassic, quickly evolved into many species, so much so that the Mesozoic is known as the age of ammonites as well as the age of dinosaurs. Sometimes the evidence for ancient ecological relationships is striking: Ammonites were closely related to the modern genus Nautilus. *(© William E. Ferguson)*

species is likely to have separate species names given to each of these structures.

Quality and quantity of fossil material are restricted for the vertebrate paleontologist, compared to the wealth of fossil material available to
the invertebrate paleontologist, because vertebrates have an endoskeleton that is more easily
broken apart (disarticulated) after death of the
organism than the well-calcified exoskeletons of
marine invertebrates. Terrestrial vertebrates also
live and die in environments in which their remains are subject to destruction by predators,
scavengers, and bacterial decomposition. The vertebrate fossil record is dominated by teeth, which
contain the mineral apatite, which is hard and resistant to weathering. Often, entire fossil vertebrate species are defined on the basis of solitary
bits of jaws, teeth, and bone.

ORIGIN AND EVOLUTION OF LIFE

The biases associated with fossilization do not
diminish the importance of the fossil record for
scientists' understanding of the origin and evolution of life. The fossil record not only enables scientists to reconstruct the morphology of long--
extinct individuals but also provides evidence from

which ecological relationships—for example, interactions between different organisms and the
structure of ancient communities of organisms, or
paleocommunities—can be inferred. Sometimes
the evidence for ancient ecological relationships is
striking, as in the find of an ammonite (related to
the modern genus *Nautilus*) with circular holes in
the shell associated with circular teeth of a mosa-
saur (marine reptile)—evidence of a predator-prey
relationship.

Trace fossils contribute their own unique information to scientists' understanding of ancient life.
They are often found in rocks devoid of body fossils and provide the only evidence of biological activity in these ancient sediments. Unlike body fossils, trace fossils cannot be transported by wind or
currents. Traces were made where they are found,
and there is little fear of erroneous interpretations
based on mixing trace fossils from different environments. Trace fossils also provide important
evolutionary information: The earliest evidence of
multicellular organisms is small vertical burrows
in rocks 2 billion years old. There are no body fossils in these rocks; the trace-makers were undoubtedly soft-bodied organisms that were poor candidates for fossilization. The presence of the bur-

rows in these ancient rocks suggests that the evolutionary transformation from unicellular to multicellular life took place in a shallow marine, nearshore environment. The diversity of trace-fossil forms reveals a variety of animal behaviors: burrowing, crawling, walking, feeding, and grazing. For example, the gait of some extinct arthropods and details of their appendages can be determined from the pattern of their fossilized footprints, even though body fossils of these animals are rarely found with intact appendages. The spacing of dinosaur footprints has led to the idea that many dinosaurs were able to move rapidly and were not the slow-gaited, lumbering giants formerly imagined.

The fossil record documents the fact of evolution and provides data from which scientists can infer the processes by which evolution works. Darwin believed that evolution was a slow and gradual process of small changes, accumulating over time to transform one fossil species into another. This gradualistic view of evolution, however, is not strongly supported by the fossil record. Gradualism predicts that numerous intermediate forms (transitional in appearance between the old species and the new species) should exist. In fact, many fossil groups have left no record of these intermediate forms. Darwin attributed the absence of transitional forms to the incompleteness of the fossil record. This view of the fossil record as incomplete provided a convenient explanation for gradualists, and gradualism was widely accepted as an explanation of how evolution works for many years.

The gradualist view was challenged by paleontologists who saw a different pattern in the geological record of some groups of organisms. Many fossil groups appear suddenly in the fossil record and persist largely unchanged in appearance through most of their geologic history. This pattern suggests that evolution (as measured by change in appearance of an organism) happens very quickly and involves a major change in appearance, as the old form "jumps" to the new form. According to this alternative view, the sudden appearance of new species in the fossil record and the absence of intermediate fossils is real and not an artifact of an imperfect fossil record. This alternative model of evolution, termed "punctuated equilibrium," holds that evolution is not the gradual, constant process

envisioned by Darwin but rather a rapid process involving major changes in morphology. Speciation in this punctuated model of evolution does not require the hypothetical (and problematic) transitional intermediates proposed by Darwin. Both gradualistic and punctuated modes of evolution have been documented for different groups of organisms. The evidence suggests that different groups of organisms evolved through different evolutionary pathways, some gradual, some punctuated.

STUDY OF THE FOSSIL RECORD

Study of the fossil record begins in the field, where paleontologists collect fossils and record observations on the orientation of the fossils and the character of the sediment in which the fossils are found. The specific technique used in collecting fossils depends on the kinds of fossils sought, the nature of the research, and time constraints. A major problem faced by paleontologists in collecting fossils in the field is separating the fossils from the surrounding sediment. Certain microfossils that are commonly found in limestone are collected by dissolving many kilograms of the rock in hydrochloric acid. The microfossils themselves are insoluble in weak acid and are easily recovered from the acid bath. Other microfossils may be picked from deep-sea cores that consist of unconsolidated mud. Macrofossils can be removed from the outcrop with the aid of hammers and chisels, although vertebrate paleontologists sometimes use brushes in excavating fragile vertebrate material. Unless the fossils have weathered naturally from the outcrop, they probably need to be cleaned of adhering sediment. Macrofossils can be cleaned by boiling in solvents, by sandblasting with a miniature air-abrasion unit, or by meticulous, time-consuming hand cleaning with small probes (old dental instruments make good fossil-cleaning tools).

Strategies for collecting fossils from outcrop exposures include collecting fossils exposed at the surface from weathering, and "mining," digging down to or along targeted fossiliferous bedding surfaces. First-time collectors usually begin by collecting everything in sight. Once the novelty has worn off, they select only the best, or most complete and well-preserved, specimens and toss aside broken or deformed fossils. Perfect specimens make good museum exhibits, but experienced paleontologists realize that important information is often revealed by the broken or imperfect specimen. The pattern of shell wear or breakage may reveal important information about the transport and burial history of the fossil after its death. The study of these postmortem processes is called taphonomy. Taphonomic studies require detailed observation about the preservation of the fossils and the original orientation of the fossils in the outcrop.

In the laboratory, fossils are examined with a binocular microscope, or they may be ground down to translucent slices and examined under a petrographic microscope. This thin-section technique permits microscopic examination of shell structure. The scanning electron microscope is used to examine surficial features of the shell, at magnifications on the order of 30 to 5,000 times. The composition of fossil shells might be determined by microprobe or isotopic analysis.

Much paleontological study concerns description of fossil species. Paleontologists seek not only to differentiate one species from another but also to discover the relationship between different species. This is accomplished by describing and comparing the key morphologic characters of different species. Computers and digital imaging equipment enable paleontologists to measure many morphologic features quickly and process large sets of numerical data rapidly. Relationships between species can be represented numerically through a variety of mathematical techniques and the results presented in graphical form in computer printouts.

Not all paleontological discoveries are made in the field. There is a wealth of fossil material residing in museum collections, awaiting future study. For example, a new dinosaur species was discovered when crates of fossils collected one century earlier were unpacked and examined in detail for the first time.

Danita Brandt

CROSS-REFERENCES

BIBLIOGRAPHY

Briggs, Derek E. G., and Peter R. Crowther, eds. *Paleobiology: A Synthesis.* Oxford: Blackwell Scientific Publications, 1990. This multiauthored text deals with all aspects of paleontology in a series of short articles by experts in each field. Intended for university students but also useful to high school students and interested readers.

Darwin, Charles. *On the Origin of Species: A Facsimile of the First Edition.* Cambridge, Mass.: Harvard University Press, 1975. Darwin's seminal 1859 work is worth reading to appreciate the depth and breadth of evidence that he marshaled for his theory of evolution by natural selection. Of special interest is Darwin's struggle to reconcile the evidence from the fossil record (the absence of transitional fossils) with his gradualist theory of evolution and his description of the imperfection of the fossil record. Although the Victorian-era writing style requires patience on the part of the reader, Darwin's book is accessible to a general audience.

Gould, Stephen J. *Hen's Teeth and Horse's Toes: Further Reflections in Natural History.* New York: W. W. Norton, 1983. This is one of several volumes of essays in which Gould interprets paleontological principles for a lay audience (other volumes are *The Panda's Thumb: More Reflections in Natural History, Ever Since Darwin: Reflections in Natural History,* and *The Flamingo's Smile: Reflections in Natural History*). A recurring theme in many essays concerns the mechanism of evolution; Gould is a leading proponent of punctuated equilibria. These essays are eminently readable, often humorous, and always enlightening.

Hamblin, W. Kenneth. *The Earth's Dynamic Systems.* Minneapolis, Minn.: Burgess, 1975. This textbook gives an overview of physical geology. Chapter 5 contains a clear discussion of the principle of faunal succession, with particular emphasis given to William Smith's work. Useful illustrations and a glossary are included.

Laporte, Leo F., ed. *The Fossil Record and Evolution.* San Francisco: W. H. Freeman, 1982. This volume comprises sixteen articles reprinted from *Scientific American,* united by the theme of the fossil record. Included are articles on the origin and earliest fossil record of life, the Burgess Shale fossil lagerstatte, and a reinterpretation of dinosaur behavior from fossil evidence. These articles are suitable for a high school audience and are characterized by numerous well-constructed illustrations. An index and a bibliography for further reading are included.

Prothero, Donald R. *Bringing Fossils to Life.* Boston: McGraw-Hill, 1998. This well-illustrated and entertaining text covers a broad range of paleontological topics, including the fossil record. Glossary, bibliography, and index.

Rudwick, Martin J. S. *The Meaning of Fossils: Episodes in the History of Palaeontology.* London: Macdonald, 1972. Rudwick traces the history of humankind's understanding of the fossil record, from the time of the ancient Greeks, when "fossil" referred to any object dug up, through the Dark Ages, when fossils were viewed as tricks of the Devil, to the modern understanding of fossils as evidence of past life. Accessible to a general audience, the book includes a glossary, an index, and a bibliography.

Thompson, Ida. *The Audubon Society Field Guide to North American Fossils.* New York: Alfred A. Knopf, 1982. One of the most comprehensive and useful books of its kind, the Audubon guide is published in a slim, soft-cover format that is meant to accompany the reader into the field. Primarily a guide to the fossil record of marine invertebrates of North America; vertebrates and plants receive coverage proportional to their fossilization potential. More than five hundred full-color plates of the most common North American fossils are accompanied by detailed descriptions of each. Also included is a chapter on where to find fossiliferous rocks in North America.

FOSSILIZATION AND TAPHONOMY

Taphonomy is the subfield of paleontology that addresses the processes by which the evidences of once-living organisms pass from the biosphere into the lithosphere, where they are fossilized and eventually are discovered. In all but the rarest situations, the resulting fossil is much different from the original organism from which it was derived. The paleontologist unravels the taphonomic history of a fossil in order to gain an accurate impression of the organism when it was living.

PRINCIPAL TERMS

CAST: a fossil that displays the form of the original organism in true relief

DISTILLATION: the driving of carbon dioxide and water out of tissues, leaving only free carbon

EPIFAUNA: organisms that live on the seafloor

FOSSIL: any remains or evidence of a once-living organism

FOSSILIZATION: the processes by which the remains of an organism become preserved in the rock record

INFAUNA: organisms that live in the seafloor

MOLD: a fossil that displays the form of the original organism in negative relief

PERMINERALIZATION: the filling of pores and cells with minerals without changing the material surrounding the pores or cells

REPLACEMENT: molecule-by-molecule substitution of the original material of the organism by a different mineral material

STEINKERN: an internal mold that preserves the interior form of an organism in negative relief

TAPHONOMY: the study of all the processes that take place between the death of an organism and its discovery as a fossil

EVIDENCE OF PAST LIFE

Of the many millions of kinds of living things that have inhabited the Earth over the past several billion years, only a small fraction have left any evidence of their existence. Paleontologists call these evidences of past life fossils and use them to try to reconstruct the appearance and lifestyle (for example, ecology) of the original organism. From the time an organism dies until its remains are discovered by the paleontologist, however, many things may occur that cause a loss of information and make the task of reconstruction very difficult. Therefore, understanding how fossils are formed becomes essential to their interpretation. A separate subfield of paleontology called taphonomy has arisen, which is devoted to the study of how organisms become fossils. Taphonomy is traditionally divided into two parts. The first concerns the many destructive events that take place between the death of an organism and its final burial. The second concerns the many changes that can take place following burial and prior to discovery. It is this second part that traditionally has been called fossilization.

To illustrate the role of events that take place between death and burial, it is convenient to imagine a scenario involving a typical vertebrate organism—for example, a deer. When the organism dies—by predation, disease, or accident—the body of the animal begins to change, and the information that was contained in the living animal begins to decrease. Even the mode of death can be destructive, since predators can crush and tear parts of the deer, often leaving telltale marks on the skeleton. Should the death itself be quiet, however, the body will quickly be attacked by scavengers that will tear at the flesh and will tend to dismember the animal, sometimes taking parts (for example, a leg) far from the site for consumption. Thus, almost immediately, remains of the vertebrate would no longer be complete at the death site. With the majority of the flesh stripped off the skeleton, the smaller scavengers will then remove the last of the tendons and ligaments holding bones together, and the entire skeleton will then be separate bones. Once again, many of the smaller scavengers will chew on bones and antlers, weakening them or consuming them com-

1015

pletely. Because most natural waters are acidic (a result of dissolved atmospheric carbon dioxide), the rain that falls on the isolated bones will have a tendency to dissolve them. Eventually, the last part of the skeleton remaining will be the teeth, as they are the hardest and most compact parts of the animal. Thus, a moderately large animal, such as a deer, may be reduced to only a few teeth prior to burial. This explains the fact that the majority of vertebrate fossils, particularly of smaller vertebrates, tend to be teeth.

PRESERVATION OF ORGANIC MATERIAL

A scenario similar in broad outline to what happens to land animals could be written for marine invertebrates. What is significant is that the many different processes that are common to the environment of an organism are able to reduce a complex living individual to a few fragments. Clearly, then, one of the major factors that will determine the survivability of a fossil is the durability of its skeletal material. Second, the environment in which the organism lives will be of great significance. For the deer in the example, the wooded terrestrial environment is not a place where sediments accumulate rapidly to bury animal remains quickly, and preservation is far less probable. For many organisms living on or in the floor of the sea, the chances of preservation are much greater, as sediment accumulates in these environments at much greater rates than on land. Organisms living in the seafloor (infauna) are especially prone to burial because they are already within soft sediment; those living on the seafloor (epifauna) are also well positioned for such preservation. Thus, the fact that marine invertebrates are far better represented in the fossil record than are terrestrial vertebrates becomes readily understandable. It will always be easier to preserve a clam than a squirrel.

Once buried, the remains are still not guaranteed preservation. What is recovered will depend upon the mode of fossilization. In the best of situations, permitting the greatest amount of information retrieval, no alteration of the original material will have occurred, so that what is recovered is compositionally and morphologically identical to what was buried. Typically, unaltered fossils are relatively recent, as the probability of change increases with geological age. Good examples are

the insect and plant remains that become entombed in tree sap that hardens to produce amber. Enclosing the organisms in the thick, sticky sap effectively excludes oxygen and protects against putrefaction and the scavenging of other organisms. The chitinous exoskeleton of the insect is likely to be preserved without any change. Spectacular preservation of ants, wasps, and other insects has been reported from the famous Baltic amber of the Eocene epoch (about 65 to 55 million years ago). Similarly, many of the bison bones that have been found in prehistoric sites in the western United States are unaltered. Plant material is often slower to be modified during fossilization, and some wood from the Cretaceous period (about 140 to 65 million years ago) appears unchanged, even under the microscope.

PROCESSES OF FOSSILIZATION

Fossils are often said to be "petrified." This is a nonscientific term referring to three different processes: permineralization, replacement, and distillation. Permineralization is the infilling of pores and cells by the precipitation of mineral matter. The pore walls and cell walls remain unchanged. The mineral involved is typically opaline silica; many other minerals have been found, although less commonly. Replacement refers to the substitution of some other mineral for the substance of the fossil. These substitutions may take place on a molecule-by-molecule basis, leaving a nearly perfect copy of the original, with much of the detail intact. While opaline silica is again very common, many other minerals are also common, including pyrite, hematite, and calcite.

Distillation, unlike the other two processes of fossilization, involves the soft tissues. These tissues are composed of compounds of carbon, hydrogen, and oxygen. Under burial pressures and temperatures and in the absence of oxygen, destructive distillation takes place that liberates water and carbon dioxide until only free carbon remains. The carbon forms a thin, black film in the rocks. The outline of the organism is often preserved, sometimes around its skeleton, giving clues to the fleshy shape of the organism, which could not be determined from the skeleton alone. Plants are often preserved this way. The three-dimensional form of the plant may be completely distorted. In some cases, however, the surface textures are molded into the surrounding sediment before distillation, leaving very thin carbon films lining the molds, which preserve the details of veins and stems. Many fern and fernlike fossils are preserved in this manner.

Perhaps the most common form of preservation is as molds and casts. A mold is a fossil that preserves the surface features, which may be either external or internal to the organism, in negative relief—that is, with raised areas on the original appearing as depressions on the fossil. To obtain an accurate impression of the original, paleontologists often pour rubber into the mold; when hardened, it will have the surface features of the original in true relief, that is, with projections now appearing as projections of rubber. The rubber reproduction is termed a cast. Nature may also produces casts by filling a mold in rock with minerals or sediment. Because shelled organisms such as clams have large internal cavities that may become sediment-filled, it is easy to imagine this material as a cast, yet in fact it is a mold of the interior of the shell, known as a steinkern. An indentation on the interior of the clam shell will appear as a raised area (that is, in negative relief) on the steinkern. Internal molds are common in large bivalved organisms, and spiral cones of sediment may be all that remains of a coiled gastropod, having filled the interior of the shell. When the shell was subsequently dissolved away, only the solidified filling was left.

A subtler change may take place if the skeletal material was originally aragonite, a variety of calcium carbonate often used by marine organisms. Aragonite is chemically unstable at or near the Earth's surface and will recrystallize as calcite. Calcite has a slightly different crystal shape and size. As a result, when the aragonite crystallizes, the fine details of the microstructure of the skeleton are often obliterated, even though the gross external form appears unchanged.

TRACES OF ORGANIC ACTIVITY

Tracks, trails, and traces of organisms are another category of fossil that records the activity of the animals that made them. These trace fossils are extremely useful in revealing behavioral habits that cannot be determined from skeletal remains.

The best known of these are the dinosaur footprints found in numerous localities around the world. Less impressive, but more commonly found, are the burrows and feeding marks of various marine worms. While a particular trace can seldom be assigned to a specific animal, comparison with the habits and markings of modern organisms has allowed paleontologists to make general identifications. Because none of the original substance of the organism is involved, the process of preservation of trace fossils depends entirely upon preserving the texture of the sediments in which they were formed, both before and after burial. Where there are abundant organisms living on and in the seafloor, their activities continually churn and disturb the sediment, usually wiping out most traces before they can be fossilized. Rapid sedimentation, however, can often bury the traces below the reach of most infaunal organisms, which tend to live near the surface, thus preventing the subsequent disruption of the traces.

Finally, regardless of the type of fossil preservation, many fossils are destroyed when the sediments containing them are compacted, thus crushing and distorting any fossils within. Moreover, if the pressures and temperatures of burial become great enough, the rock will become metamorphosed, and all the original features, including the fossils, may be lost, especially if the rock undergoes recrystallization.

John J. Ernissee

CROSS-REFERENCES

BIBLIOGRAPHY

Behrensmeyer, Anna K., and Andrew P. Hill. *Fossils in the Making: Vertebrate Taphonomy and Paleoecology.* Chicago: University of Chicago Press, 1980. An advanced version of the Shipman book mentioned below. Consists of a series of papers presented at a symposium, which are reproduced here with few illustrations and in typescript: a major collection of works by the principal researchers in the field. The bibliography is exceptional and constitutes, at least for vertebrates, a nearly exhaustive listing up to the date of publication. Available in the libraries of universities with active departments of geology.

Briggs, Derek E. G., and Peter R. Crowther, eds. *Paleobiology: A Synthesis.* Oxford: Blackwell Scientific Publications, 1990. This multiauthored text deals with all aspects of paleontology in a series of short articles by experts in each field. Intended for university students but also useful to high school students and interested readers.

Donovan, Stephen K., ed. *The Process of Fossilization.* New York: Columbia University Press, 1991. A multiauthored text that provides a modern survey of taphonomy, the scientific study of the processes and patterns of fossilization. Intended for the specialist. Includes an index and bibliography for each chapter.

Fenton, C. L., and M. L. Fenton. *The Fossil Book.* Garden City, N.Y.: Doubleday, 1958. This long-out-of-print book remains one of the clearest and most enjoyable general introductions to fossils ever published. Still likely to be found in most school and public libraries. The abundant line drawings are superb and have not been equaled since this book was published. Fossilization is briefly discussed in chapter 1. Suitable for junior high school students and above.

Müller, A. H. "Fossilization (Taphonomy)." In *Treatise on Invertebrate Paleontology.* Boulder, Colo.: Geological Society of America, 1979. A comprehensive treatment of fossilization and taphonomy of the invertebrates, written by one of the pioneers of the field. An advanced work, intended for specialists. Moreover, it suffers from being an English translation of a German-language manuscript, with all of the attendant difficulties in readability. The illustrations are excellent, as is the bibliography, up to the year of publication.

Prothero, Donald R. *Bringing Fossils to Life.* Boston: McGraw-Hill, 1998. This well-illustrated and entertaining text covers a broad range of paleontological topics, including extinctions. Glossary, bibliography, and index.

Shipman, Pat. *Life History of a Fossil: Introduction to Taphonomy and Paleoecology.* Cambridge, Mass.: Harvard University Press, 1981. One of the first introductory textbooks on taphonomy, but unfortunately restricted to the vertebrates. Although intended for upper-level college students and above, the writing is exceptionally clear and lucid throughout and is likely to be accessible to well-read high school students. Features a glossary and has a good bibliography, although limited to vertebrate taphonomy, which is current as of the date of publication.

Whittington, H. B. *The Burgess Shale.* New Haven, Conn.: Yale University Press, 1985. A detailed account of the taphonomic and paleoecologic interpretation of one of the most spectacular fossil localities in the world—the Cambrian Burgess Shale of British Columbia. Here were found an enormously diverse assemblage of superbly well preserved fossils notable because of the large number of soft-bodied organisms represented. This book details the organisms, many of which are strange and fascinating, and the mode of fossilization that preserved them so perfectly. Well written and reasonably well illustrated. Both the Royal Ontario Museum in Toronto, Canada, and the U.S. National Museum of Natural History in Washington, D.C., have excellent collections of materials from this site, which are normally on display. Should be read as a prelude to visiting these collections. Suitable for college students, but accessible to well-read high school students.

FOSSILS OF THE EARLIEST LIFE-FORMS

Identification and study of the earliest fossils have expanded the time frame during which life is known to have existed on the Earth, revealing information that has important implications for the origins of life and thus for human evolutionary ancestry.

PRINCIPAL TERMS

ARCHEAN EON: the early Precambrian period, from about 2.5 to 4 billion years ago, roughly corresponding to the earliest known fossils

CAMBRIAN PERIOD: the period from about 544 to 505 million years ago, marked by the appearance of hard-shelled organisms

CYANOBACTERIA: previously called blue-green algae, these oxygen-producing, photosynthetic microbes often live in matlike clusters

ISOTOPES: atoms of the same element that differ in mass

MICROFOSSIL: the characteristic imprint left by a microscopic organism in a geological formation such as a stromatolite

PALEOBIOLOGY: the study of the most ancient life-forms, typically through the examination of microscopic fossils

PHOTOSYNTHESIS: the biological process of using sunlight to form energy-rich compounds, usually with the production of free oxygen

PROTEROZOIC ERA: the late Precambrian era, from about 600 million to 2.5 billion years ago, before the proliferation of macroscopic life

STROMATOLITE: a structure produced by the trapping or binding of sediment as a result of the growth and metabolic activity of microorganisms

WARRAWOONA SEQUENCE: a geological formation in Western Australia, dated at 3.5 billion years before the present, where "North Pole" stromatolites were discovered

DISCOVERY OF PRECAMBRIAN FOSSILS

The search for the earliest fossils has focused upon the remains of microscopic, single-celled organisms. The study of these microfossils has yielded a better appreciation of the antiquity of life. Fossils visible to the naked eye have been found in rocks dating back 600 million years. This record corresponds to the Cambrian diversification, sometimes called the Cambrian "explosion," a proliferation of hard-shelled animals that left their imprint in the geological ledger, and the preceding Ediacaran, in which soft-bodied organisms are preserved. The Cambrian period heralded the arrival of complex organisms that had skeletons and that were composed of many cells adapted to specialized functions. Prior to 1950, little was known about the Precambrian history of life, principally because there were no fossils to provide the necessary information. Did organisms suddenly appear after the Earth endured 4 billion years of a lifeless existence? Scientific intuition suggested that simple, unicellular creatures must have preceded the arrival of macroscopic organisms, yet no evidence of this development had been found.

In 1954, Stanley Tyler and Elso Barghoorn published a report of bacteria-like fossils found in the Gunflint iron formation of Canada, dated at nearly 2 billion years before the present. Through the Gunflint research and related studies on early life, Barghoorn emerged as a leading figure in the growing discipline of paleobotany—the study of ancient plants—until his death in 1984. In the words of his colleagues, "Elso will undoubtedly be most clearly remembered as the man who pushed the history of life back sixfold."

The Gunflint fossils have been extensively studied in the years since their discovery. The geological formations that house them are known as stromatolites, a term that refers to a layered structure formed by successive generations of sticky microbial mats that trap sediment and gradually build a dome-shaped structure up to several meters high. Such mat communities are known to be associated with photosynthetic cyanobacteria and certain other microorganisms that occur widely on the modern Earth. The singular advantage of stromatolites from a paleontological viewpoint is that, after the organic material decomposes, the sedimentary structure provides a permanent record of the morphology of the organisms that formed it. The Gunflint rocks are composed of chert, a form of quartz that is especially resistant to compression and thus helps to preserve the fossils. The stromatolites that were studied by Barghoorn and Tyler revealed the presence of six distinct organisms, indicating a highly diversified biological community from 2 billion years ago.

A stromatolite formation from Ross River Canyon in Australia. (U.S. Geological Survey)

PRECAMBRIAN PALEOBIOLOGY RESEARCH GROUP

The documentation of the Gunflint fossils from the Proterozoic era (about 600 million to 2.5 billion years ago) spurred the search for more ancient remains of life. The African continent, considered by many scientists to be the cradle of human evolution, has also provided a rich source of microbial fossils. In the 1960's, Barghoorn and his student J. William Schopf studied a set of 3.1-billion-year-old rocks known as the Fig Tree Cherts from the border region between South Africa and Swaziland. Microscopic analysis revealed two unknown microbes, to which Barghoorn and Schopf assigned characteristic Latin names in the naturalist's tradition: *Eobacterium isolatum* for a rod-shaped relic, and *Archaeosphaeroides barbertonensis* for a spheroidal fossil. The latter resembles modern cyanobacteria and, like the Gunflint study, suggests the operation of photosynthesis at an early stage in biological evolution. More significantly, the Fig Tree Cherts established that life has existed on the Earth for a majority of its history. Hence the origin of life had to be placed in the Archean eon, the period from about 2.5 to 4 billion years before the present.

In an effort to codify and extend the evidence of early life, Schopf initiated an organized project in the later 1970's known as the Precambrian Paleobiology Research Group (PPRG). With support from the National Aeronautics and Space Administration (NASA), Schopf and a team of twenty-three scientists from every relevant field compiled an exhaustive study of ancient fossils with the goal of characterizing the earliest biosphere of the Earth. A major site of the PPRG effort was a set of fossils from a region of Western Australia known as North Pole, a name given by early prospectors to this remote location. Following the discovery of stromatolites in 1977 by an Australian graduate student named John Dunlop, Schopf's group conducted extensive fieldwork and laboratory research to establish the age and authenticity of these microfossils. Dating techniques used by the Geological Survey of Western Australia confirmed the age of this formation, called the Warrawoona sequence (or Warrawoona group), at 3.5 billion years. The PPRG scientists identified a variety of round and wormlike structures that provide strong evidence for the existence of life only 1 billion years after the Earth formed. Isotopic analysis also supports a biological origin for these structures, some of which correspond in shape to cyanobacteria and other modern organisms but also including others that have no modern counterpart. Most researchers thus accept the North Pole stromatolites as the oldest evidence of life yet discovered.

CLASSIFICATION

The classification of the organisms that created the Archean stromatolites has generated considerable debate. If the fossils in the structures represent the ancestors of modern cyanobacteria, as suggested by their general size and shape, then photosynthesis had presumably become an important metabolic pathway at this stage in evolution. Geologists have argued, however, that the photosynthetic production of oxygen must have been a later development, since the geological record shows little evidence of free oxygen during the Archean eon. Stromatolites can also be formed by bacteria that are not photosynthesizers, so the existence of such formations in the Archean does not require the existence of photosynthesis. Schopf, who led the study of the North Pole fossils, believes that the biological production of oxygen did not emerge until 1 billion years after these stromatolites were formed.

Other paleobotanists, such as Andrew Knoll (like Schopf, a student of Elso Barghoorn), proposed that photosynthesis had already developed 3.5 billion years ago but that such organisms were so limited by the small continental shelf and by the harsh ultraviolet radiation on the early Earth that they did not significantly alter the atmosphere. An alternative way to reconcile the postulated early photosynthesis with the geological record is to consider that the biologically produced oxygen might have disappeared by combining with organic molecules in the ocean. The similarity between carbon isotope ratios in stromatolites and modern cyanobacteria also suggests an Archean beginning for photosynthesis. Nevertheless, the limited evidence prevents any definitive conclusions regarding the metabolism of the organisms that made up the North Pole communities.

Rocks that are older than the Warrawoona sequence have been discovered by geologists, but the existence of fossils in these formations is more problematical. Much attention has centered on a set of rocks from Greenland known as the Isua deposits, dated at 3.8 billion years before the present. Unlike the North Pole fossils, these rocks have undergone extensive deformation as a result of the extreme temperatures and pressures to which they have been subjected throughout geologic time. While fossils may once have existed, their presence is difficult to establish because of the lack of pristine evidence. The longer a rock has been on the Earth, the higher the probability that it will have been deformed in the manner of the Isua deposits. Thus, attempts to push the frontier of life beyond the 3.5-billion-year mark set by the Australian discoveries will undoubtedly encounter this dilemma, though the search for more ancient fossils will continue.

STUDY OF CELLULAR FOSSILS

The two primary techniques that are employed to characterize cellular fossils are microscopic examination and chemical analysis. The former method is important for studying the shape of the organism and thus relating it to contemporary cells with a similar form. Thin sections of the rock are examined first under a light microscope using reflected and transmitted white light. In addition, impressions of the fossils are examined by electron microscopy, which reveals even finer details. Since the organisms themselves have long since decayed, such an inspection of the sedimentary imprint provides only the outlines of the cellular structure, from which inferences must be drawn regarding its relationship to modern biology. Special care must be taken to ensure that the suspected fossil is not merely a bubble or other artifact and that a true biological structure is not a result of contamination by a modern organism.

Chemical analysis of the fossils is important both to establish their age and to deduce the nature of their previous biochemistry. For both these objectives, measurements of specific isotopes (atomic species with a characteristic mass) play a major role. The decay of radioactive isotopes is the best technique for learning how old a sample is, and a variety of methodologies have been developed for this purpose. Fortunately, nature has provided many different radioisotopes, each with a distinct rate of disintegration. Carbon dating, for example, is usually employed in archaeological research because the half-life (the time required for 50 percent of the element to decay) of carbon 14 is only 5,730 years. For fossils that are billions of years old, a longer-lived isotope is required: Two that are commonly utilized are potassium 40, with a half-life of 1.27 billion years, and rubidium 87, with a half-life of 48.9 billion years. By comparing the ratios of the radioisotopes with their more stable products (argon 40 in the case of potassium and strontium 87 in the case of rubidium), scientists can establish the age of sediments to within an uncertainty of a few percentage points.

The analysis of organic material in fossils can provide clues to the biological function of ancient organisms. Although one cannot usually determine which compounds were originally present in the living microbe, the distribution of carbon, hydrogen, and oxygen isotopes provides a partial record of the metabolic activity of the fossilized cell because organisms selectively incorporate one isotope in preference to another form of the same element. In general, the lighter forms are enriched in living organisms, as enzymatic reactions proceed faster than with the heavier isotopes, which, for carbon chemistry, means that cells contain a higher percentage of carbon 12 than would be expected in the absence of biological activity. The extent of isotopic enrichment in microfossils can provide a marker for the nature of such metabolic activity.

Another direct marker of metabolic productivity consists of rocks deposited during the period when a particular organism flourished. Geologists have focused especially on the presence of minerals that are formed by reactions with molecular oxygen, a by-product of green plant photosynthesis. Analysis of one such mineral, uraninite, indicates that atmospheric oxygen at the end of the Archean eon (about 2.5 billion years ago) was only about 1 percent of the present level. Another possible marker of atmospheric oxygen is the so-called banded iron formations, (BIFs), alternating layers composed of the minerals magnetite and hematite. Although the abundance of iron minerals has been correlated with relative amounts of atmospheric oxygen, similar deposits can also be formed by reactions that do not require oxygen, and thus the interpretation of the banded iron formations is difficult. The key point on which geological studies concur is that free oxygen was nearly absent during the Archean eon, when the earliest microbes lived.

William J. Hagan, Jr.

CROSS-REFERENCES

Archaebacteria, 951; Cambrian Diversification of Life, 956; Catastrophism, 961; Colonization of the Land, 966; Dinosaurs, 970; Ediacaran Fossils, 976; Eukaryotes, 981; Evolution of Birds, 985; Evolution of Flight, 990; Evolution of Humans, 994; Evolution of Life, 999; Fossil Plants, 1004; Fossil Record, 1009; Fossilization and Taphonomy, 1015; Fossils of the Earliest Vertebrates, 1024; Geoarchaeology, 1028; Life's Origins, 1032; Mammals, 1038; Mass Extinctions, 1043; Microfossils, 1048; Neanderthals, 1053; Paleobiogeography, 1058; Paleobotany, 1062; Petrified Wood, 1058; Prokaryotes, 1071; Stromatolites, 1075; Systematics and Classification, 1079.

BIBLIOGRAPHY

Hartman, H., J. G. Lawless, and P. Morrison, eds. *Search for the Universal Ancestors: The Origins of Life*. Reprint. Palo Alto, Calif.: Blackwell Scientific Publications, 1987. A general report by a group of scientists active in the study of the origin and early evolution of life.

Kutter, G. Siegfried. *The Universe and Life: Origins and Evolution*. Boston: Jones & Bartlett, 1987. Intended for use as a college textbook, this book is divided equally between cosmic and biological evolution.

McMenamin, Mark. *Discovering the First Complex Life: The Garden of the Ediacara*. New York: Columbia University Press, 1998. This entertaining study of the earliest complex life-forms on the planet details the author's work on these organisms. Written for the interested student but understandable by the general reader.

Margulis, Lynn. *Early Life*. Boston: Jones & Bartlett, 1982. A succinct overview by a leading authority in the field of early evolution. Written on the level of *Scientific American*, the book includes a glossary to help the reader with technical terms.

Margulis, Lynn, and Dorion Sagan. *Microcosmos: Four Billion Years of Microbial Evolution*. New

York: Summit Books, 1986. A provocative and lively account for the general reader, this book stresses the role of microorganisms throughout biological history.

Prothero, Donald R. *Bringing Fossils to Life*. Boston: McGraw-Hill, 1998. This well-illustrated and entertaining text covers a broad range of paleontological topics, including early life. Glossary, bibliography, and index.

Schopf, J. William, ed. *Earth's Earliest Biosphere: Its Origin and Evolution*. Princeton, N.J.: Princeton University Press, 1983. As the official report of the Precambrian Paleobiology Research Group, this collection of articles is intended for readers with an advanced knowledge of science. Nevertheless, because of the interdisciplinary nature of the project, the authors have been careful to define specialized terms, and a glossary is also included.

_____, ed. *Major Events in the History of Life*. Boston: Jones & Bartlett, 1992. An excellent overview of the origin of life, the oldest fossils, and the early development of plants and animals. Written by specialists in each field but at a level that is suitable for high school students and undergraduates. Although technical language is used, most terms are defined in the glossary.

FOSSILS OF THE EARLIEST VERTEBRATES

Fossils of the earliest vertebrates are present in rocks dating back to 470 million years ago. Although they are rare and fragmentary, they provide the only contact with the planet's early life-forms and help establish an understanding of how vertebrates developed.

PRINCIPAL TERMS

AGNATHA: a class of vertebrates that includes all forms in which jaws are not developed; the group to which the earliest vertebrates belong

CLADISTICS: a method of determining relationships in which shared advanced characteristics exhibited by the organisms are used

EXOSKELETON: a bony armor covering the outside of the animal

ORDOVICIAN: a time period covering the interval from 490 to 430 million years ago; follows the Cambrian, which covers the interval from 543 to 490 million years ago

VERTEBRATE: a subphylum of the Phylum Chordata, distinguished by the presence of a brain with paired sensory organs and the development of bone

EARLY EVOLUTION OF VERTEBRATES

The record of the early evolution of vertebrates is of particular interest to humankind as members of that subphylum, yet the evidence of evolution is tantalizingly incomplete. The vertebrates constitute a subphylum of the phylum Chordata, a group of organisms characterized by the presence of a notochord (a longitudinal stiffening rod), a dorsal hollow nerve cord, gill slits, and segmented muscle masses. The subphyla other than the vertebrates are often referred to as invertebrate chordates (because they lack vertebral elements); modern examples of these forms include the Hemichordata (pterobranchs and acorn worms) and the Cephalochordata (*Branchiostoma*, formerly known as *Amphioxus*). Fossil evidence for the history of invertebrate chordates is very limited. *Pikaia* from the Middle Cambrian Burgess Shale (500 million years ago) is a vaguely fishlike animal that appears to resemble the modern cephalochordate *Branchiostoma*.

Vertebrates differ from the invertebrate chordates in the possession of a brain, specialized paired sensory organs (hearing, sight, smell), and bone. The absence of fossils showing the earliest stages of vertebrate evolution has allowed little to be known for certain about how this took place. It is assumed that the earliest vertebrates and their immediate ancestors were soft-bodied animals that would leave no fossil traces. They would not be capable of preservation until the development of bone had taken place, a capacity that does not seem to have occurred until well after the early radiation of the vertebrates. As relevant fossils are not present for the earliest stages of development, a picture of an idealized early vertebrate has been developed based on features common to primitive members of the modern fish groups and on evidence from the earliest fossil vertebrates.

The picture developed is of a small fishlike animal with a fusiform body and a head that is not clearly differentiated. Anterior jaws were not present, but the animal may have fed by engulfing small prey whole through a flexible mouth opening. Gill slits were present, perforating the wall of the pharynx and associated with gills, thin lamellar tissues that provide a surface for gas exchange and allow the animals to breathe. The dorsal hollow nerve cord expanded anteriorly and was associated with specialized paired sensory structures (smell, sight, and balance). Lateral line organs extending over the body surface detected vibrations in the water, as they do in modern fish. The notochord formed a supporting rod extending the length of the animal, and anchored to it were the segmental swimming muscles (the myotomes) that extended along both sides of the body. No paired fins were present, but a tail and possibly a dorsal fin were. This picture of a primitive vertebrate is very similar to the modern cephalochordate *Branchiostoma* and also to the fossil cephalochordate *Pikaia*, implying that the cephalochordates may be the closest relatives of the vertebrates within the phylum Chordata.

DISCOVERY OF VERTEBRATE FOSSILS

Although bone does not appear to have been present in the earliest vertebrates, it is present in the first fossil forms found in rocks of Ordovician age (490 to 430 million years ago) from North America, Australia, and Bolivia. The fossils consist mostly of disarticulated fragments of bony exoskeleton, although several specimens from North America are sufficiently well preserved to have been used as the basis for reconstructions, and finds in Bolivia have included complete animals. Because the evidence is usually so fragmentary, a variety of fossils from the Cambrian (543 to 490 million years ago) and Ordovician periods have been identified as early vertebrates during the past century.

These determinations have not always been accurate, and it is now clear that many of these fragments do not show vertebrate morphology or histology (internal structure). In particular, a number of fragments of bony armor have been reported from the Upper Cambrian of North America and the Lower Ordovician of Spitsbergen, Norway. These are earlier than the accepted dates for vertebrate occurrences; however, there is as yet no definite proof that these fragments are vertebrate. Although they appear to be composed of hydroxyapatite, as are vertebrate skeletons, this is not in itself proof of vertebrate origin, as a number of invertebrate groups (such as arthropods, conodonts, and brachiopods) have hard tissues of a similar composition. As these fragments are microscopic in size, nothing can be deduced about the gross morphology of the animal that produced them. It may be that these fragments represent parts of the exoskeleton of early arthropods, but as yet this is still an area of hot debate, and there is clearly insufficient evidence to designate these fragments as vertebrate.

Remains that are still accepted as vertebrate were first described from the Ordovician Harding Sandstone of Colorado in 1892. The two species

described, *Astraspis desiderata* and *Eriptychius americanus*, consist of numerous fragments of bony plates and scales that covered the animals with a flexible armor. Histology of the individual elements and gross morphology of the few partially articulated specimens prove conclusively that these animals were vertebrate. A reconstruction of *Astraspis desiderata* shows it to have been roughly 130 millimeters long and covered anteriorly by ornamented bony plates that were often fused to each other peripherally to form a continuous shield. Posteriorly, the tail was covered by imbricating scales and terminated in a small, scale-covered caudal fin. Laterally, there were eight pairs of gill openings. The animal had no fins other than the caudal fin, suggesting that it was a poor swimmer. Nothing is known of the internal skeleton of these animals, and it is assumed that it consisted of cartilage, a material that is not normally fossilized. Some calcified fragments from the head region are known, but it has not yet been possible to determine what their original position might have been. Nothing is known of the mouth area of these vertebrates, but they are presumed to have been Agnatha, vertebrates in which jaws are not developed. Armored jawless vertebrates of a similar type are common in more recent rocks of Silurian (430 to 395 million years ago) and Devonian (395 to 345 million years ago) age, and modern jawless forms (lampreys and hagfishes) do exist, though they are not common and are no longer armored.

Vertebrate fragments of a somewhat similar type were described from the Ordovician Stairway Sandstone in Australia in 1977. Enough material of one species, *Arandaspis prionotolepis*, is known for a reconstruction to have been made, and this suggests an animal approximately 150 millimeters in length in which anterior dorsal and ventral armor plates were separated by a slanting row of small square plates that probably protected the gill openings. The tail was covered by long, thin, flexible scales. This picture has been improved by the discovery in Bolivia in 1986 of complete specimens of a closely related species, *Sacabambaspis janvieri*. These specimens show that the animal had eyes in the front of the head and a mouth that was a flexible opening not supported by jaws, thus confirming that these early vertebrates were Agnatha.

ORIGIN OF VERTEBRATES

Although they differ in details of their morphology, indicating that a considerable amount of evolution had already taken place, these Ordovician vertebrates are clearly closely related and indicate the presence of diverse and widespread early vertebrate faunas at this time. The environments in which they lived and their mode of life have been the subjects of some dispute over the years because of their bearing on the environment in which the vertebrates originated. It was originally suggested that vertebrates originated in fresh water, an interpretation supported at that time by the supposed freshwater environment of deposition of the Harding Sandstone.

These ideas have since been questioned, and fresh interpretations of the sediments in which vertebrate fragments occur show that they were marine. It now appears that in all cases the environments in which early vertebrates were living were shallow, near the shore, and sandy. The animals probably lived as benthonic deposit feeders in these sandy nearshore environments, which is very similar to the mode of life of the modern cephalochordate *Branchiostoma*. Unlike *Branchiostoma*, however, they probably did not bury themselves in the sediment to feed. Once they died, the carcasses would have been disarticulated by scavengers and distributed across the current-swept nearshore environment, explaining why complete skeletons are so rare. The present wide distribution of early vertebrate occurrences (Australia, Bolivia, and North America) is unfortunately not improved when paleogeographic reconstructions taking account of plate tectonic movements are made for the Ordovician. At that time, South America was far south and Australia and North America, though both near the equator, formed part of different continental masses. It is very difficult, therefore, to relate the distribution of early vertebrate occurrences in a meaningful way, and it is to be hoped that additional occurrences, as yet undiscovered, will aid scientists in their interpretations.

STUDY OF FOSSIL REMAINS

Although the fossil remains of early vertebrates have been studied for almost one hundred years, the techniques used have changed very little. The fragments of bone are removed from the matrix (surrounding sediment) by mechanical or chemical means initially. Mechanical preparation involves the removal of the matrix by small needles, either hand-held or in small electrical engraving tools. By this means, the rock is slowly chipped and flaked away from the bone until it is entirely exposed and can be studied. Chemical preparation exploits the fact that the calcium phosphate of which bone is composed will resist some acids that can break down carbonate matrices. This means that if the vertebrate fragments are preserved in a carbonate rock (limestone) or in one that has a carbonate cement, the vertebrate fragments can be dissolved out. If the fragments are very fragile, they are often backed with plastic first, so that they do not disintegrate when the supporting matrix has been removed. The acids most commonly used in this technique are acetic acid and formic acid. Both mechanical and chemical techniques are slow and painstaking and require a high level of skill.

Once the material has been prepared, it is studied optically using microscopes. The internal structure can be studied by making thin sections of bone fragments and then viewing them by transmitted light. The thin sections are made in a similar way to thin sections of rock. A thin slice of the bone is cut, and one side is ground and polished and then attached to a glass slide. The bone is then ground down until only a thin film is left attached to the glass slide. The bone is so thin that light can shine through it, and the details of the internal structure of the bone can be seen when it is viewed through a microscope. The external morphology of the bone can be studied using a light microscope also, but more recently bone fragments have been studied using a scanning electron microscope. This microscope uses a beam of electrons that are generated and focused on the specimen. As the electrons rebound from the specimen, they pass through a phosphorous disk, and the energy produced is seen as a scan line on a cathode-ray tube, thus producing a picture that can be observed. This microscope can produce very high magnifications, thus enabling minute detail of surface structures to be seen.

Studies on the relationships of the earliest vertebrates have relied heavily on a methodology termed "phylogenetic systematics" or, more commonly, "cladistics." This method uses the characteristics of organisms to develop a picture of the way in which they are related. Cladistics is distinguished from other taxonomic methods (taxonomy is the study of interrelationships) by the fact that it is a rigorous system in which shared advanced characteristics alone are used to show relationships. These relationships are expressed as branching diagrams termed cladograms (*klados* is Greek for "branch," hence the name cladistics). Studies using this methodology have shown that modern Agnatha are composed of two disparate groups, so widely separated that one is in fact more closely related to jawed vertebrates. The exact relationship of the earliest fossil vertebrates to the modern forms is still the subject of intense debate.

David K. Elliott

CROSS-REFERENCES

Archaebacteria, 951; Atmosphere's Evolution, 1816; Cambrian Diversification of Life, 956; Catastrophism, 961; Colonization of the Land, 966; Dinosaurs, 970; Ediacaran Fossils, 976; Eukaryotes, 981; Evolution of Birds, 985; Evolution of Flight, 990; Evolution of Humans, 994; Evolution of Life, 999; Fossil Plants, 1004; Fossil Record, 1009; Fossilization and Taphonomy, 1015; Fossils of the Earliest Life-Forms, 1019; Geoarchaeology, 1028; Life's Origins, 1032; Mammals, 1038; Mass Extinctions, 1043; Microfossils, 1048; Neanderthals, 1053; Paleobiogeography, 1058; Paleobotany, 1062; Paleoclimatology, 1131; Petrified Wood, 1058; Prokaryotes, 1071; Radioactive Decay, 532; Stromatolites, 1075; Systematics and Classification, 1079.

BIBLIOGRAPHY

Benton, Michael J. *Vertebrate Paleontology*. 2d ed. London: Chapman and Hall, 1997. This comprehensive text provides a complete outline of the history and development of vertebrates. Designed for college courses but also suitable for the informed enthusiast. The first three chapters deal with the origin of vertebrates and their early development.

Carroll, R. L. *Vertebrate Paleontology and Evolution*. New York: W. H. Freeman, 1987. This is a most up-to-date text on vertebrate paleontology and is intended for the college student. Chapters 2 and 3 concern the origin of vertebrates and the diversity of jawless forms, but chapter 1 also provides much useful information on evolution and the uses of taxonomy.

Elliott, D. K. "A Reassessment of *Astraspis desiderata*, the Oldest North American Vertebrate." *Science* 237 (July, 1987): 190. This provides an overview of knowledge of early vertebrates together with a description of the earliest North American forms. It contains an illustration of a North American Ordovician vertebrate. *Science* is a weekly scientific magazine that publishes articles of interest to all branches of the scientific community. Suitable for college-level readers.

Gagnier, P.-Y. "The Oldest Vertebrate: A 470 Million Year Old Jawless Fish, *Sacabambaspis janvieri*, from the Ordovician of Bolivia." *National Geographic Research* 5 (Spring, 1989): 250. An article on a discovery of early vertebrates that provides good illustrations of the appearance of these animals. This journal publishes articles in a variety of scientific disciplines for the general reader as well as the scientist. Suitable for high school and college-level students.

Janvier, Philippe. *Early Vertebrates*. Oxford: Clarendon Press, 1998. This book is written by the foremost expert in the field. Although intended for experts, it is accessible to the interested reader with some scientific background. Extremely well illustrated.

Maisey, John G. *Discovering Fossil Fishes*. New York: Henry Holt, 1996. A broad coverage of the field aimed at the interested reader who may not have a scientific background. Well illustrated but lacks a bibliography.

GEOARCHAEOLOGY

Archaeology is rapidly becoming a markedly more scientifically based field, a trend started with the "New Archaeology" of the 1970's. Archaeological geology covers the wide range of geological sciences that are applied to archaeology during excavation and postseason, or postexcavation, sorting, classifying, and analyzing.

PRINCIPAL TERMS

ABSOLUTE DATE: a date that gives an actual age, though it may be approximate, of an artifact

CURIE POINT, or CURIE TEMPERATURE: the temperature at which materials containing iron oxides lose their magnetic pattern and align with the Earth's magnetic field

RELATIVE DATE: a date that places an artifact as older or younger than another object, without specifically giving an age for it

REMOTE SENSING: any of a wide variety of techniques, such as aerial photography, used for collecting data about the Earth's surface from a distance

STRATIGRAPHY: the deposition of artifacts in layers or strata

DATING METHODS

Archaeological geology is the application of geological methods and techniques to archaeology. The two disciplines have become so closely intertwined at times that some have spoken for a new term to describe their partnership: "geoarchaeology." A term aimed more specifically at the contributions of the physical and chemical sciences associated with archaeology (such as potassium-argon dating) is "archaeometry." Without such scientific methods, archaeology becomes guesswork at best; dating of finds, for example, should be derived from empirical data, or information that can be proven through experiment and observation, rather than deduced from theory without corroboration. Archaeology is, basically, deductions made about an artifact from the context in which it is found; geology helps to define and date that context, thereby providing the empirical information from which speculation about the artifact can be derived.

The principle of superposition was probably one of the first geological methods that archaeology utilized. This law states that a layer superimposed on another layer should be younger, having been laid down after the lower, or older, layer. This study of stratigraphy is a keystone to archaeological dating but constitutes relative rather than absolute dating. One of the better-known methods of absolute dating is carbon 14 dating. Carbon 12 and carbon 14, an isotope of carbon 12, are elements that exist in all living organisms. Once the organism dies, whether it is plant or animal, the input of carbon 14 from the environment stops, and the remaining carbon 14 begins to decay. The amount of carbon 14 left relative to the amount of carbon 12 is used to calculate the amount of time that the organism has been dead, using the known half-life of carbon 14 (5,730 years).

Archaeomagnetism (the term is the archaeological equivalent of geology's "paleomagnetism") is another dating method, but one that uses changes in the Earth's magnetic field as recorded by archaeological artifacts such as kilns. The Earth experiences continual changes in the intensity and polarity of its magnetic field. If a clay artifact has been heated to its Curie point, or temperature, its magnetic particles will align in the direction of the polarity of the Earth's magnetic field in which they cool. Geological identification of the polarity pinpoints the time of firing. Thermoluminescence dating is also a method used on clay. A piece of pottery is heated to 500 degrees Celsius, and the ensuing emission of light is measured as an indication of the length of time since its firing, as the energy of the thermoluminescence increases as it is stored up over time. Another method for dating inorganic objects, in this case volcanic rock, is potassium-argon dating. When volcanic rock is newly solidified, it contains no argon, but it does contain potassium 40. As the potassium 40 decays, it becomes argon 40. Because the half-life of potassium 40 is known (1 billion years), the amount of argon 40 gives the absolute age of the rock.

REMOTE SENSING AND PETROLOGY

Geological techniques are also used for locating and recording archaeological sites. Remote sensing is a technique by which data are collected on a site in a "remote" rather than a hands-on way. Photographic images are a major part of remote sensing. Aerial photography, for example, is used to photograph the landscape from airplanes or satellites. Images and patterns that record electromagnetic radiation provide another source of remote-sensing data, as does soil resistivity surveying. Soil resistivity surveying is a method used to map buried features by finding electrical conductivity differences between the features and the soil around them. The ease or resistance with which the current penetrates the soil is the basic principle. To test the resistivity, four electrodes are inserted into the ground. Two generate the current, and two measure the drops in voltage; an equation taking into account the distance between the electrodes, the amperage, and the drop in voltage then gives the total resistance to conductivity.

Petrology, or the study of different aspects of rocks, is a standard feature of archaeological geology. Most techniques focus on the study of thin sections or powdered samples of rock—for example, through the use of scanning electron microscopes and X-ray diffraction, respectively. In addition, archaeologists also utilize geological studies of cryoturbation (freeze-thaw cycles in soil), argilliturbation (shrinking and swelling cycles in clays), aeroturbation (disturbances by gas, wind, and air), aquaturbation (disturbances from the movement of water), and seismiturbation (disturbances by earthquakes). Study of these different types of environmental disturbance helps identify the different site-disturbing processes at work.

APPLICATION OF GEOARCHAEOLOGICAL METHODS

Many aspects of the geological sciences can be applied to archaeological sites. It is the nature of the site that determines the appropriate method to use. For example, a historical archaeological site, or one that is dated to within the parameters of recorded history, would not probably be a site at which radiocarbon dating would be useful: The artifacts would not be old enough for a dating method that yields figures in thousands of years. On the other hand, a prehistoric site such as the possibly Iron Age Caer Cadwgan, a probable hillfort in Wales, would benefit primarily from radiocarbon dating: The site is probably about three thousand years old, and charcoal and bone, excellent types of samples for carbon 14 testing, are the main elements found in excavation at this site. This site has also yielded small glass beads, items that would perhaps be datable by thermoluminescence.

Historical sites in general, however, are not the prime candidates for archaeological geology that prehistoric sites are. Prehistoric archaeology, on the other hand, depends completely on geological analysis for some conclusions because it predates any written records; there are no fortuitously preserved documents to fall back on for verification. Archaeological geology can accomplish much, but it is most useful for four principal processes during an excavation: locating a site, recording and analyzing the features of a site, and dating a site.

The archaeological use of soil resistivity for locating sites or individual features was first applied to locate prehistoric stone monuments just after World War II. The differences in ground conductivity are used to locate anomalies, which can range from buried ditches to stone walls. The method was first developed in geology to locate ore deposits, faults, and sinks (sunken land where water can collect). Remote sensing can be a useful method for several different goals. Remote sensing can be used to locate sites, monitor changes in the archaeological record, reveal the distribution of archaeological sites, or map sites. Aerial photography can be used to locate sites, and it can then be used to map a site and record its features for planning an excavation. On some sites in North Africa, the camera will be sent up in a balloon to take the photographs, as the site may be too small or the budget of the excavation too limited for airplane-carried camera work.

Archaeology is a destructive process; once a site is excavated or even surface collected, it cannot be restored. Remote sensing can, sometimes, take the place of excavation in what is termed nondestructive archaeology. It can also be a useful substitute for excavation when there is not time for a full-scale dig to be mounted, as during times when a formation may be temporarily visible (for example, winter snows revealing significant gradations in the land) but conditions may not be right for excavation. Remote sensing can also preserve at least an image of a site that must be destroyed, as during construction of a road. With this use, it is an invaluable tool in rescue archaeology.

APPLICATION OF DATING METHODS

Petrology is a useful tool in general for archaeology. Petrology can be used on many different types of sites, from prehistoric to historic. It can be used to give a provenance, or origin, for building and sculptural stones as old as Stonehenge or as young as the classical Greek marbles. Archaeomagnetism can be used not only to date artifacts but also to distinguish sources for substances. Obsidian sources, for example, have been found to each have different magnetization strengths. It is possible then that the source of ore for coins, which may contain trace amounts of iron, may also be able to be discovered. Archaeomagnetic dating of lava has been used to date the end of the Minoan civilization (about 1500 B.C.E.), which was destroyed by the volcanic eruptions at Santorin (or Thera) in Greece. It is used mainly for early human sites, or for sites up to 10,000 years old.

Thermoluminescence can also be used to date lava as well as burnt flint (implements heated by accident or on purpose to improve certain qualities), burnt stones (heated on a fire and then placed in a food container as a "pot-boiler"), glass (volcanic glass especially), sediments (buried soils), and ceramics. This method is popular because of its absolute dates for a wide range of ages: 50 to 500,000 years old. Potassium-argon dating is also used for inorganic samples but only ones that predate humankind. Samples can only be as young as 100,000 years old. Another limitation, however, is that because it is used to date lava, in order to be of any use the site must be connected to a particular volcanic eruption, and sites such as

Santorin and Pompeii are not common (and too young anyway). The usefulness of potassium-argon dating to archaeology is its ability to fix dates for reversals of the Earth's magnetic field, which in turn are used to date archaeological sites through archaeomagnetism.

Using petrology to establish the provenance of a rock artifact is of great importance in prehistoric archaeology. The evaluation of rock types found can help to identify the mining or quarrying skills of a culture, its determination of the usefulness of various kinds of rocks, and even some of the places to which people may have traveled. If a certain type of rock is not native to the area in which it is found yet occurs in large quantities, it may be inferred that perhaps significant trade or travel was taking place. The value that a culture may have placed on a particular rock may also be determined, judging by its use (whether ceremonial or practical, for example). Use-wear as opposed to earth-moving processes that have changed the shape of the rock are another object of study. Petrological methods such as X-ray diffraction can be used to identify the origin of rocks on a site. Studying the rocks from a number of different sites can be a way of tracing the trade routes of a culture.

J. Lipman-Boon

CROSS-REFERENCES

BIBLIOGRAPHY

Brothwell, Don, and Eric Higgs, eds. *Science in Archaeology*. With a foreword by Grahame Clark. 2d ed. New York: Praeger, 1970. Despite the date of the book, it is helpful because it has chapters on many of the scientific methods that are in use, including thermoluminescence, potassium-argon, and radiocarbon (carbon 14) dating. This volume is useful for seeing the wide range of scientific methods that are applied in archaeology and the beginning of New Archaeology. Suitable for undergraduate college-level students.

Butzer, Karl W. *Archaeology as Human Ecology: Method and Theory for a Contextual Approach*. New York: Cambridge University Press, 1982. Butzer sets out the principles and objectives of geoarchaeology as a subdiscipline of archaeology. His emphasis is on the environmental sciences in general as necessary for the best empirical data in archaeology.

Kelley, Jane H., and Marsha P. Hanen. *Archaeology and the Methodology of Science*. Albuquerque: University of New Mexico Press, 1988. Kelley and Hanen have prepared this volume as a way to reconcile the philosophies of science and archaeology. They include case histories that target the methodological problems of interpreting even sound empirical data. Useful for college-level students seeking to discover how the sciences fit into archaeology.

Kempe, D. R. C., and Anthony P. Harvey, eds. *The Petrology of Archaeological Artifacts*. New York: Clarendon Press, 1983. A very useful book for different techniques of archaeological geology and their various applications, including specific examples. As the title indicates, petrology is the main focus. It is technical and therefore best suited to advanced college-level readers.

Lasca, N. P., and J. Donahue, eds. *Archaeological Geology of North America*. Boulder, Colo.: Geo-

logical Society of America, 1990. This multiauthored volume covers the paleoarchaeology of North America. A useful introductory chapter reviews the entire subject. Bibliography and index.

Parkes, P. A. *Current Scientific Techniques in Archaeology*. New York: St. Martin's Press, 1986. This book includes detailed information on archaeometry. Technical but well written. Suitable for college-level readers.

Ryan, William, and Walter Pitman. *Noah's Flood: The New Scientific Discoveries About the Event that Changed History*. New York: Simon & Schuster, 1998. The author uses an interesting combination of geologic, archaeologic, and linguistic information to conclude that the biblical Flood myth derives from the catastrophic flooding of the Black Sea. Written for the general reader but with plenty of scientific information.

Schiffer, Michael. *Formation Processes of the Archaeological Record*. Albuquerque: University of New Mexico Press, 1987. This book includes chapters on the Earth processes that can disturb and change the archaeological record, such as earthquakes and freeze-thaw cycles. Contains photographs and is suitable for the general reader.

_____, ed. *Advances in Archaeological Method and Theory*. 11 vols. San Diego, Calif.: Academic Press, 1978-1987. This series is a comprehensive review of changes in scientific archaeological methods and new applications of those methods. Each chapter is by a different contributor, and many include specific examples of studied sites.

West, Frederick H., ed. *American Beginnings*. Chicago: University of Chicago Press, 1996. This multiauthored volume deals with the geological and archaeological evidence for the movement of humans across the Bering Strait land bridge. Well illustrated. Extensive bibliography.

LIFE'S ORIGINS

At some time after the formation of the Earth some 4.5 billion years ago and before the earliest known microfossils some 3.5 billion years old, life evolved from widely available organic compounds into unicellular creatures similar to modern blue-green algae. Plausible schemes for their evolution from abiotically produced organic matter remain elusive. Much research into the origin of life examines the environment of early life, comparing the relative merits of the ocean, submarine volcanic vents, and soil as possible places where life began.

PRINCIPAL TERMS

AMINO ACID: an organic acid consisting largely of carbon, oxygen, and hydrogen but with an amino group and sometimes also sulfur, phosphorus, and iron

CHERT: a hard rock of minutely crystalline, and often partly hydrous, silica

CHONDRITE: a kind of meteorite that has conspicuous globular grains (chondrules) of high-temperature minerals such as pyroxene and olivine

CLAY: a hydrous aluminosilicate mineral with sheetlike crystal structure

CYANOBACTERIA: commonly known as blue-green algae, a group of microscopic unicellular organisms that lack a membrane dividing their polynucleotides from the rest of the cell and that have a photosynthetic metabolism

ENANTIOMER: a particular version of the same kind of asymmetric chemical compound, such as sugars and amino acids, that may be left- or right-handed and so polarize light in a clockwise (D enantiomer) or counterclockwise (L enantiomer) direction, respectively

FERMENTATION: a metabolic process found in microbes such as yeast, whereby complex organic molecules are partially broken down to alcohol and carbon dioxide, thus releasing energy for work

PHOTOSYNTHESIS: a metabolic process found in plants and microbes whereby the energy of the Sun and the catalytic properties of a pigment (bacteriophyll or chlorophyll) are used to reduce carbon dioxide to organic matter

POLYNUCLEOTIDE: a high-molecular-weight compound formed from many units of a sugar joined to an organic base and to a phosphate group; includes ribonucleic acid (RNA) and deoxyribonucleic acid (DNA), both of which carry genetic information in all living cells

PROTEIN: a high-molecular-weight compound that is a long chain or aggregate of amino acids joined by hydrogen bonds

RESPIRATION: a metabolic process found in animals and microbes whereby complex organic molecules ("food") are oxidized to carbon dioxide, thus releasing energy for work

ORGANIC COMPOUNDS

Life on the Earth is probably no older than the solar system, which accreted, or built up, some 4.5 billion years ago. Early stages in this accretionary process are revealed by meteorites, most of which have been radiometrically dated to about this age. A particular kind of meteorite called a carbonaceous chondrite is characterized by the high-temperature globular mineral grains called chondrules, mixed with low-temperature iron-rich clays and organic compounds. Most of the organic compounds are dispersed among the clays, but some of them form organic lumps and globules.

These latter have been mistakenly identified as microfossils, but the organic compounds of carbonaceous chondrites have some diagnostic features of organic matter formed abiotically in solution from mixtures of gases including carbon dioxide.

Many organic molecules, such as sugars, are asymmetric. When synthesized in the laboratory, there are about equal amounts of left-handed and right-handed versions (each is known as an enantiomer) that polarize light in a clockwise or counterclockwise direction, respectively. Organic matter behaves similarly in carbonaceous chondrites, whereas organisms are highly selective for a particular enantiomer of a compound, including, for example, only D enantiomers of sugars. Organic compounds that also are presumed to have formed abiotically have been detected from the characteristics of light spectra reflected from asteroids, comets, and interstellar dust. Thus, simple organic compounds used by living creatures are widely available in the universe and have been for at least 4.5 billion years.

DEVELOPMENT OF COMPLEX ORGANISMS

It is an enormous step from a tarry mixture of simple organic compounds to a fully functioning organism, with its long polynucleotide instructions and messengers and its complex molecular assembly line and controlling compounds of protein, all bound up within a system of membranes and cell wall. This degree of complexity, seen in some of the simplest living forms of unicellular life, was attained by about 3.5 billion years ago. Microscopic fossils of unicellular cyanobacteria have been found permineralized in black cherts of this age in the Pilbara region of western Australia. These are among the oldest little-deformed sedimentary

rocks in which such fossils are likely to be found. Highly deformed and metamorphosed sedimentary rocks as old as 3.8 billion years crop out around Isua in southwestern Greenland. These rocks have not yet yielded convincing microfossils, but the isotopic composition of carbon in these rocks appears to be a metamorphosed version of compositions found in younger rocks with this isotopic signature of photosynthetic life. Older fossil evidence of life on the Earth is unlikely. There is, however, a younger fossil record from cherts and shales of increasingly complex microfossils in rocks ranging from 3.5 billion years ago to the present. By 700 million years ago, there were also complex multicellular creatures. Fossil plants and animals become increasingly familiar in rocks of younger geological age.

The vast gulf between primordial organic compounds and functioning cells revealed in the rock and meteorite record also is apparent from experimental studies in organic chemistry designed to investigate how life originated. It has been shown many times that organic compounds including amino acids are produced readily within water in

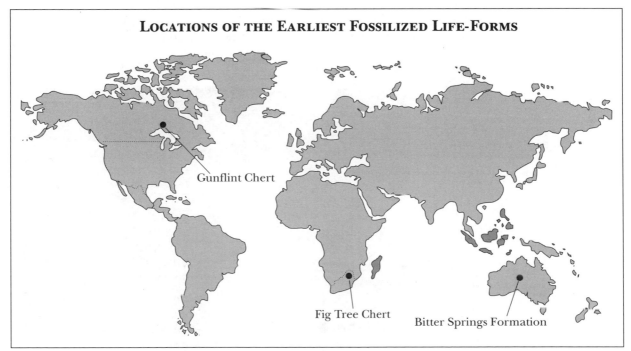

LOCATIONS OF THE EARLIEST FOSSILIZED LIFE-FORMS

Gunflint Chert

Fig Tree Chert   Bitter Springs Formation

Cherts are hard, well-cemented sedimentary rocks produced by recrystallization of siliceous marine sediments buried in the seafloor. These cherts—such as the 3.1-billion-year-old Fig Tree Chert, the 2-billion-year-old Gunflint Chert, and the 0.9-billion-year-old Bitter Springs Formation—contain microfossils of early bacteria and other, more complex, life-forms.

sealed flasks containing reducing gases such as carbon dioxide energized by electrical discharges, ultraviolet light, or even shock waves. The exact gas mixture or energy source is not critical to organic matter production, although yields are limited in gas mixtures including oxygen or with weak energy sources. Clay, siderite, and other minerals added to these reaction flasks have been shown to promote both the production of simple organic compounds and their assembly into larger compounds. How they accomplish it is not certain: They may act as a template to hold molecules in a favored orientation for combination, either as a catalyst participating in reversible reactions that tend to promote a particular combination or as a chemostat in buffering solutions from extreme acidity. Thus the origin of organic compounds such as those in carbonaceous chondrites can to some extent be duplicated in a chemistry laboratory.

Chemical experiments fall far short, however, of creating such complex organic molecules as ribonucleic acid (RNA) and deoxyribonucleic acid (DNA). In some cases parts of nucleotides have been synthesized and nucleotides have been aggregated into polynucleotides, but it remains a problem how there could arise a nucleotide-rich environment of the correct chemical nature under abiotic conditions. Nevertheless, it is clear from much experimental work in molecular biology that polynucleotides such as RNA and DNA are the basis for a system of information storage and transfer in all known life-forms on the Earth.

The odds of tarry abiotic organic matter becoming organized into a functioning cell by pure chance are astronomically unfavorable. It has been argued that, given the long period of time available, even the rarest of chances becomes a certainty. This argument now has little appeal, because geological and meteoritic evidence constrains the origin of life to within the first 1 billion years of the Earth's history

and because studies in molecular biology have revealed astounding complexity in even the simplest of organisms. One escape from this dilemma is to postulate seeding of the Earth with some form of extraterrestrial life, either by long-distance transport of spores across space or by the technological enterprise of an extraterrestrial civilization. These views have not proven very productive, because they are difficult to test and, in addition, do not solve the problem. Even the simplest life-forms are well suited to Earth and are made of materials commonly available here. Thus, if life is extraterrestrial, it originated under broadly Earth-like conditions elsewhere in the universe. It remains a useful exercise to speculate on how life might have arisen from natural causes here on Earth.

ORIGIN-OF-LIFE THEORIES

The most widely held hypothesis for the origin of life until recently has been that life arose in the "primordial soup" of the world ocean. The early atmosphere of the Earth probably included little or no free oxygen, like that of most planets. Under these conditions, organic matter would have been produced in the sea by lightning strikes, ultraviolet light from the Sun, and shock waves from earthquakes or meteorite impacts. This vast organic nutrient solution is viewed as the source of ever more complex molecules, some of which at-

Chemosynthesis, rather than photosynthesis, may have given rise to life: Minerals venting from the seafloor supply bacteria with chemical sustenance, and the same may have been true for the earliest life-forms. (National Oceanic and Atmospheric Administration)

tained the critical innovation of self-replication. These early versions of RNA and DNA then evolved increasingly complex mechanisms for self-preservation, such as cell organelles, cells, and bodies. By this view, the primitive molecular eco-system fed on abiotically produced organic com-pounds, and the earliest organisms may have been fermenters. It was only when primordial organic compounds of the ocean were depleted that there arose more sophisticated cellular machinery for the biological production of organic matter by photosynthesis.

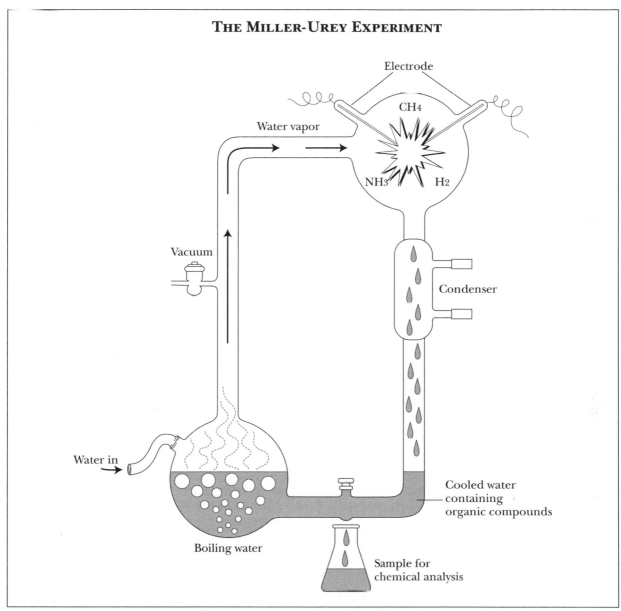

THE MILLER-UREY EXPERIMENT

Electrode

CH4

Water vapor

NH3   H2

Vacuum

Condenser

Water in

Cooled water containing organic compounds

Boiling water

Sample for chemical analysis

It has been shown many times that organic compounds, the beginnings of life, including amino acids, are produced readily within water in sealed flasks containing reducing gases such as carbon dioxide energized by electrical discharges, ultraviolet light, or even shock waves. The most famous of these experiments, depicted here, was conducted by Stanley L. Miller and Harold C. Urey in 1953. Since then, however, scientists have come to believe that hydrothermal vents, openings along mid-ocean ridges where heated water is vigorously expelled into the sea, act as flow reactors capable of stripping oxygen from carbon dioxide to produce organic matter.

The most serious problem of this hypothesis is the gap between abiotic organic matter and organized self-replicating molecules. The ocean is, and probably was, too dilute a solution, too low in dissolved phosphorus, and too uniform in its alkalinity and oxidation state for synthesis and maintenance of complex organic molecules. The laboratory synthesis of DNA requires many different reactions under very different chemical conditions: acidic and alkaline, oxidizing and reducing, wet and dry. Such fluctuating conditions can be imagined during the intermittent evaporation of seawater ponded on ice floes or in tidal flats, but low temperatures and the formation of salts would retard production of complex organic molecules.

The discovery of deep-sea volcanic vents prompted a second group of hypotheses about the origin of life. These "black smokers" are openings along mid-ocean ridges where internal water of the Earth as well as seawater percolating through seafloor basalt has been heated by contact with hot volcanic rocks and is vigorously expelled into the sea at temperatures of up to 380 degrees Celsius. The water gushing from the vents is also strongly acidic and depleted in oxygen. The "black smoke" billowing forth, as well as the vent chimneys, consists of dark sulfide minerals. The idea that ocean vents are flow reactors capable of stripping oxygen from carbon dioxide to produce organic matter is one of the arguments that they may prove pertinent to the origin of life. In addition, the complex system of fractures and vesicles in volcanic rocks around vents would have encouraged a variety of simultaneous and very different chemical reactions, in contrast to the tendency for chemical uniformity in the open ocean. Furthermore, experimental studies have shown that proteins aggregate into spheres and sheets at high temperatures, and such structures can be imagined as crude membranes and cell walls. If life did evolve in oceanic vents, then it would initially have lived by fermentation or respiration of abiotically produced organic compounds, and photosynthesis would have arisen with more complex organic systems of those few vents exposed to sunlight on land or in shallow water.

Despite some appealing features of this hypothesis, it remains to be demonstrated whether the minute amounts of organic matter detected in submarine volcanic vents were produced abiotically or are remains of the distinctive fauna of tube worms, clams, and crustaceans that now colonize the seafloor around vents. The high temperatures and strongly acidic nature of vent water are incompatible with the production and preservation of complex proteins and nucleotides. These problems are reduced considerably even a short distance away from the vent, but then there remain the problems already mentioned for an oceanic origin of life.

A third set of hypotheses, that life evolved in soil, can be taken to include ideas involving clay minerals. Soils are, and probably were, the principal sites where clay is formed on the Earth. Soil hypotheses for the origin of life combine many of the desirable features of ocean and vent hypotheses. Soil water could have been enriched in organic matter like the early ocean, and soils are in a sense flow reactors driven by the energy of the Sun at the surface. Soil water is compartmentalized and evaporates into a complex array of films around mineral grains and other particles. Soils enjoy moderate temperature and mild chemical conditions.

The most critical advantage of the soil hypothesis is that the production of clay by weathering or of organic matter by abiotic synthesis would have been encouraged because clayey and organic soils are less easily eroded than are sandy, inorganic soils. This matter of advantage can be imagined as a crude form of "natural selection" in which the most clayey and organic soils survived erosion for the longest periods of time. Thus, even the least efficient mechanisms of abiotic weathering and organic synthesis could result in soils as organic and clayey as carbonaceous chondrites. Increasingly complex compounds such as sugars and proteins would be preserved preferentially to the extent that they could bind the soil and maintain it within the zone of materials and energy transfer at the surface. In this view, the primary production of organic matter would initially have been fueled by light from the Sun at the surface, and photosynthetic organisms may well have preceded fermenting and respiring creatures. The idea that life arose in soils reduces some of the difficulties of forming functioning cells from tarry organic matter, but it cannot be claimed to eliminate them.

Gregory J. Retallack

CROSS-REFERENCES

Aerial Photography, 2739; Archaebacteria, 951; Cambrian Diversification of Life, 956; Catastrophism, 961; Colonization of the Land, 966; Dinosaurs, 970; Earth's Magnetic Field, 137; Earth's Magnetic Field: Secular Variation, 137; Ediacaran Fossils, 976; Electron Microscopy, 463; Eukaryotes, 981; Evolution of Birds, 985; Evolution of Flight, 990; Evolution of Humans, 994; Evolution of Life, 999; Fossil Plants, 1004; Fossil Record, 1009; Fossilization and Taphonomy, 1015; Fossils of the Earliest Life-Forms, 1019; Fossils of the Earliest Vertebrates, 1024; Geoarchaeology, 1028; Geologic Time Scale, 1105; Mammals, 1038; Mass Extinctions, 1043; Microfossils, 1048; Neanderthals, 1053; Paleobiogeography, 1058; Paleobotany, 1062; Paleosols, 1144; Petrified Wood, 1058; Petrographic Microscopes, 493; Potassium-Argon Dating, 527; Prokaryotes, 1071; Radiocarbon Dating, 537; Remote Sensing and the Electromagnetic Spectrum, 2802; Sediment Transport and Deposition, 2374; Stromatolites, 1075; Systematics and Classification, 1079; X-ray Powder Diffraction, 504.

BIBLIOGRAPHY

Cairns-Smith, A. G. *Genetic Takeover.* Cambridge, England: Cambridge University Press, 1982. This masterpiece of logic develops the startling theory that clay not only encouraged the origin of organic-based life but also can be considered as part of a quasi-living system itself. This book assumes a university-level background in science, but care is taken to explain technical lines of evidence in various scientific specialties. Also offered is a comprehensive bibliography.

Crick, Francis. *Life Itself: Its Origin and Nature.* New York: Simon & Schuster, 1981. In this slim volume, a Nobel laureate in biochemistry explains, in terms easily understood by the layperson, why life is too complex a phenomenon to have developed on the Earth within the limited geological time available. In arguing that the Earth was deliberately seeded with microbial life by alien civilizations, he also reviews a variety of other hypotheses for the extraterrestrial origin of life.

Hartman, H., P. Morrison, and J. G. Lawless, eds. *Search for the Universal Ancestors: The Origins of Life.* Palo Alto: Blackwell Scientific, 1987. This short book for a popular audience outlines the scientific quest to understand the origin of life and introduces ideas on how life may have evolved.

McMenamin, Mark. *Discovering the First Complex Life: The Garden of the Ediacara.* New York: Columbia University Press, 1998. This entertaining study of the earliest complex life-forms on the planet details the author's work on these organisms. Written for the interested student but understandable by the general reader.

Nisbet, Evan G. *The Young Earth: An Introduction to Archaean Geology.* Winchester, Mass.: Unwin Hyman, 1987. This colorful account of the basic data and controversies of geologists researching the oldest rocks on the Earth is aimed at an audience with at least high school training in science. It is an excellent guide for the primary geological literature on these rocks and especially for the idea that life originated in and around submarine volcanic vents.

Schopf, J. William, ed. *Major Events in the History of Life.* Boston: Jones and Bartlett, 1992. An excellent overview of the origin of life, the oldest fossils, and the early development of plants and animals. Written by specialists in each field but at a level that is suitable for high school students and undergraduates. Although technical language is used, most terms are defined in the glossary.

MAMMALS

Mammals are four-legged backboned animals distinguished by a number of unique characteristics, including hair, constant warm body temperature, mammary glands, and specialized teeth that are replaced only once. Mammals first descended from a group of animals known as synapsids, or mammallike reptiles, about 200 million years ago but remained small (mouse-sized) and unspecialized until the dinosaurs died off about 65 million years ago. They then underwent a huge evolutionary radiation—that is, they spread into different environments and diverged in structure.

PRINCIPAL TERMS

CENOZOIC ERA: the period of geologic time from about 65 million years ago to the present

GENUS (plural, GENERA): a group of closely related species; for example, *Homo* is the genus of humans, and it includes the species *Homo sapiens* (modern humans) and *Homo erectus* (Peking Man, Java Man)

MAMMARY GLANDS: the glands that female mammals use to nurse their young

MARSUPIALS: mammals that give birth to a premature embryo that then finishes its development in a pouch

MONOTREMES: primitive mammals, such as the platypus and spiny anteater, that lay eggs and have other archaic features

ORDER: a group of closely related genera; in mammals, orders are the well-recognized major groups, such as the rodents, bats, whales, and carnivores

PLACENTALS: mammals that carry the embryo in the mother until it is born in a well-developed state; it is nourished in the womb by a membrane (the placenta)

SYNAPSIDS: the mammallike reptiles that lived from about 300 to 200 million years ago and evolved into mammals

THE AGE OF MAMMALS

Mammals are a group of vertebrates (animals with backbones) that have been the dominant animals on land and sea since the dinosaurs died out about 65 million years ago. Indeed, the present period of geologic time, the Cenozoic era, is often called the age of mammals. About 4,170 species of mammals are alive currently, but at least five times that many are now extinct. About 1,010 genera of mammals are living, but, according to a 1945 tabulation, an additional 2,000 genera were extinct, and that number has greatly increased since 1945.

Mammals have been very successful in occupying a great variety of terrestrial and aquatic ecological niches. They include terrestrial meat-eaters and plant-eaters, tree dwellers, burrowing forms, and even aquatic forms. The largest terrestrial mammals today are elephants, but the extinct hornless rhinoceros *Paraceratherium* was much larger, reaching 6 meters at the shoulder and weighing about 20,000 kilograms. The largest mammals, however, are whales; they can weigh up to 150,000 kilograms in the case of the blue whale, which is larger than even the largest dinosaurs.

Living mammals are easily distinguished from all other vertebrates by a number of unique evolutionary specializations. Unlike other vertebrates, mammals have hair, are equipped with mammary glands to nurse their young, and bear live young (except for the most primitive egg-laying mammals, the platypus and spiny anteater). Mammals maintain a constant, relatively high body temperature. They have a four-chambered heart and a very efficient digestive and respiratory system, which includes a diaphragm in the chest cavity to aid in breathing. Mammals develop rapidly as juveniles and then stop growing when they reach adult size, unlike other animals, which grow continuously through life.

Fossil mammals are known only from their fossilizable parts, which are mostly their teeth and bones. Their fossil record can be traced back to their ancestors, the synapsids, or mammal-like reptiles, which are known as far back as 300 million years ago. Fossil mammals are usually distinguished from their ancestors by a number of skeletal features. These features include specialized teeth that are replaced only once (comparable to our "baby teeth" and adult teeth), a jaw joint between the dentary and squamosal bones, and a middle ear that is composed of three bones (the hammer, anvil, and stirrup).

MAJOR MAMMAL GROUPS

As the large synapsids were being replaced by the dinosaurs about 210 million years ago, mammals evolved as tiny, mouse-sized animals that fed on insects. Mammals remained in this state for almost 150 million years, living in the nooks and crannies of the world of the dinosaurs, which dominated the planet. A number of different "experimental" groups of insectivorous (insect-eating) mammals lived during this time, but all became extinct. About 100 million years ago, the three major groups of living mammals evolved. These include the monotremes, the marsupials, and the placentals.

The monotremes are the egg-laying mammals, the platypus and the spiny anteater of Australia and New Guinea. The more advanced mammals bear live young. One group, the pouched mammals, or marsupials, give birth to a premature, partially developed embryo. The embryo then climbs to the mother's pouch and fastens to a nipple, where it finishes its development. The most familiar marsupials are the kangaroo, koala, Tasmanian devil, and opossum, although there have been many other types of marsupials in the past, and many are still alive today in Australia and South America. Where marsupials lived in isolation with no competition from placental mammals, they have evolved into many different body forms, which converge on the body forms of their ecological equivalents in the placentals. In Australia presently, there are marsupial equivalents of cats, wolves, mice, flying squirrels, rabbits, moles, tapirs, and monkeys. Among extinct marsupials of South America were the equivalents of lions and of saber-toothed cats. As similar as these animals look to their placental equivalents in their external body form, they are not related to true cats, wolves, or the rest, as they are all pouched mammals.

Unlike marsupials, placentals must carry the embryo in the womb through its full development. To allow this, the embryo is nourished by an extra membrane surrounding it in the womb. This membrane, the placenta, is shed when the baby is born and is part of the "afterbirth." This mode of reproduction makes the placental baby less vulnerable than a marsupial baby, but it means that the mother is more vulnerable, as she must carry a larger embryo for a longer time. A female marsupial can also reproduce faster, as she can carry one baby in a pouch and be pregnant with another.

APPEARANCE OF MAMMALS

After the extinction of the dinosaurs about 65 million years ago, the planet was ready for a new group of large animals to evolve and take over the vacant ecological niches. Placental mammals underwent a tremendous diversification until they occupied many ecological niches, and some reached the size of sheep. Most, however, were no larger than a cat. The placentals split into the edentates (anteaters, sloths, armadillos, and their relatives) and the rest of the mammalian orders (groups of genera). Like marsupials, edentates had their greatest success in isolation in South America, although presently the armadillo is successfully spreading northward. The most successful placentals in the Northern Hemisphere in the early Cenozoic were the archaic ancestors of the insectivores (moles, shrews, hedgehogs, and their extinct relatives), primates (lemurs, monkeys, apes, humans, and their extinct relatives), the carnivores (meat-eating mammals), and a great number of extinct, archaic orders that have no living descendants.

About 54 million years ago, in the Eocene epoch, some of the modern orders of mammals began to appear. They were still small and unspecialized and would not be easy to recognize today. Some of these groups included the rodents (mice, rats, squirrels, guinea pigs); the even-toed hoofed mammals, or artiodactyls (pigs, camels, deer, cattle, antelopes, giraffes, and their relatives); and the odd-toed hoofed mammals, or perissodactyls (horses, rhinos, tapirs, and their extinct relatives). These groups lived along with many archaic groups that are now extinct, so the Eocene world had a strange mixture of modern and archaic mammals. During the Eocene, the first bats began to fly, and the first whales to swim.

About 38 million years ago, the world's climate got much cooler and more extreme. The tropical rain forests that had dominated the world in the Eocene were replaced by more mixed vegetation, including open grasslands. As a consequence, most of the archaic groups of mammals died out, and they were replaced by a great diversification of mammals from living groups. The archaic forms were mostly leaf eaters that could not survive the loss of the tropical forest or tree-dwelling forms, such as the ancient primates. At this time, the first true dogs, cats, elephants, rhinos, tapirs, camels, pocket gophers, rabbits, and shrews appeared. Although still very archaic representatives of their respective orders, the mammalian communities began to take on a more modern look. From this point on, environments (like the East African savanna today) were dominated by perissodactyls, artiodactyls, carnivores, rodents, rabbits, and insectivores.

CLIMATE CHANGES

Since the end of the Eocene, the world's climate continued to get cooler and more extreme. More and more forests were replaced by open grasslands, and the mammals evolved in response to the changing environment. Most of the larger plant eaters (perissodactyls and artiodactyls) had to develop ever-growing molars to chew tougher grasses and long legs to escape their predators by running. This development can be seen not only in the evolution of the horse, but also in the rhinos, camels, and the many artiodactyls that chew their cud (including antelopes, deer, sheep, and cattle). Carnivores became more specialized into ambush hunters (cats), pack hunters (dogs), bone-crushing scavengers (hyenas), and a specialized type, the sabertooths, which evolved four different times independently. Rodents became the most common mammals of all, and they dominate the burrowing, ground-foraging, and tree-dwelling ecological niches.

During the last 5 million years, the Earth's climate became increasingly severe as the ice ages developed. Many mammals became extinct, while others became adept at migrating away from ice sheets or developed thick coats of hair (such as the woolly mammoth and woolly rhino) in order to live in glaciated regions. The most severe extinction of large mammals happened only 10,000 years ago, when the ice sheets retreated and the present interglacial period began. Whether this extinction was attributable to the change in climate or to the severe overhunting by prehistoric humans is still disputed. Since that time, many mammals have been wiped out by overhunting, by the destruction of habitat, or by competition from mammalian "weeds" (such as rats, rabbits, and goats) that accompany human habitation. The extinction of mammals has reached its worst levels during the last two centuries, as human populations have exploded to more than 5 billion. Only a small remnant of the once-rich assemblage of mammals that dominated the Earth for 65 million years is left presently, and much of it is endangered in the wild.

STUDY OF MAMMAL REMAINS

Because only the hard parts of the mammals commonly become fossils, mammalian paleontologists specialize in studying bones and teeth. Ideally, the paleontologist would study complete skeletons, but these are rarely found, as scavenging and stream erosion break up and scatter them. Most paleontologists thus make do with teeth, jaws, and skulls. Teeth, in particular, are valuable, as they are the hardest and most durable tissue in the body and thus are the most resistant to breakdown by erosion and stream abrasion. In addition, teeth are highly diagnostic of species and are influenced by the diet of the animal. Therefore, the paleontologist can not only identify a mammal by its teeth but also learn much about its ecology.

A paleontologist starts searching in sedimentary beds of the appropriate age and environment (river-channel sandstones and floodplain mudstones), particularly those that have produced fossil mammals in the past. The best results occur in areas with desert climates and badlands exposures, as there are few plants to cover the rocks, and the erosion is fast enough that new fossils are exposed each year. The most successful method is to prospect, or to walk head-down with the eyes "fastened" to the ground, for as many hours and days as it takes until bones or teeth are seen. When fossils are found, they may not be worth collecting if they are too fragmentary or are unidentifiable. If the fossil is worthwhile, the paleontologist collects not only the fossil itself but also exact data about where it occurred and at what stratigraphic level (so that it can be used for dating). The most common mistake made by amateurs is

A cave bear skeleton, reconstructed from Pleistocene-age fossils found in a cave in Germany. (© William E. Ferguson)

ten collected by finding a rich, bone-producing level from which bags of fossil-rich material are then filled. This material is later placed in a wooden box with a window-screen bottom and allowed to soak in a stream. The water washes out all the silt and clay, leaving a concentrate of pebbles and fossils. This concentrate can then be dried and spread out on a table to be sorted through by hand.

Donald R. Prothero

CROSS-REFERENCES

Archaebacteria, 951; Atmosphere's Evolution, 1816; Cambrian Diversification of Life, 956; Carbonaceous Chondrites, 2634; Catastrophism, 961; Chondrites, 2660; Clays and Clay Minerals, 1187; Colonization of the Land, 966; Dinosaurs, 970; Earth's Age, 511; Ediacaran Fossils, 976; Eukaryotes, 981; Evolution of Birds, 985; Evolution of Earth's Composition, 386; Evolution of Flight, 990; Evolution of Humans, 994; Evolution of Life, 999; Fossil Plants, 1004; Fossil Record, 1009; Fossilization and Taphonomy, 1015; Fossils of the Earliest Life-Forms, 1019; Fossils of the Earliest Vertebrates, 1024; Geoarchaeology, 1028; Geochemical Cycle, 412; Life's Origins, 1032; Mass Extinctions, 1043; Meteorite and Comet Impacts, 2693; Meteors and Meteor Showers, 2711; Microfossils, 1048; Neanderthals, 1053; Paleobiogeography, 1058; Paleobotany, 1062; Paleosols, 1144; Petrified Wood, 1058; Prokaryotes, 1071; Soil Formation, 1519; Solar System's Origin, 2607; Stromatolites, 1075; Systematics and Classification, 1079.

to collect a fossil without carefully noting this information.

Occasionally, mammal fossils are large enough or fragile enough that special methods are necessary. In this case, the paleontologist digs around the exposed fossil and clears the rock around it until it is almost free, resting on a thin pedestal of rock. Then it is encased in a "jacket" of burlap and plaster of paris, much like the plaster cast used to set broken limbs. When the plaster cast is dry and hard, the fossil is removed from the pedestal, and more plaster and burlap seal it in the cast. When it reaches the lab, technicians can open the cast and carefully clean away the excess rock, treating the fossil with hardener and preservative as they expose it.

Fossils of small mammals (particularly rodents, rabbits, primates, and insectivores) are too tiny to see while prospecting. These fossils are more of-

BIBLIOGRAPHY

Benton, Michael J. *The Rise of the Mammals.* New York: Crescent, 1991. A beautifully illustrated book, suitable for high school students, that explains when and how various mammal families developed.

_____. *Vertebrate Paleontology.* 2d ed. London: Chapman and Hall, 1997. This comprehensive text provides a complete outline of the history and development of vertebrates. Designed for college courses but also suitable for the informed enthusiast. The last

one-third of the book deals with the rise and diversification of mammals.

Carroll, Robert L. *Vertebrate Paleontology and Evolution*. New York: W. H. Freeman, 1988. A complete, well-illustrated textbook on fossil vertebrates. Includes several chapters on fossil mammals. Although written at the college level, some parts can be understood by the general reader.

MacDonald, David W., ed. *The Encyclopedia of Mammals*. New York: Facts On File, 1984. An excellent, colorful review of the living mammals, with some very good sections on extinct mammals as well.

Savage, R. J., and M. R. Long. *Mammal Evolution: An Illustrated Guide*. New York: Facts On File, 1986. A beautifully illustrated account of mammal evolution for the general reader. It groups the mammals by their ecology, however, rather than by their normal classification, and contains some inaccuracies.

Woodburne, Michael O., ed. *Cenozoic Mammals of North America: Geochronology and Biostratigraphy*. Berkeley: University of California Press, 1987. A valuable review of the dating and correlation of strata by fossil mammals. Intended for the advanced college student and the paleontologist.

MASS EXTINCTIONS

Mass extinctions demonstrate that the evolution and extinction of life are not smooth but rather are interrupted by mass dyings as a result of as yet poorly understood causes. Mass extinctions seem to be an integral part of the overall pattern of life processes on the Earth.

PRINCIPAL TERMS

ECOLOGY: the study of the relations between organisms and their environments

ENVIRONMENT: the conditions affecting the existence, growth, and overall status of an organism

ERA: one of the major divisions of geologic time, including one or more periods

GEOLOGIC RECORD: the history of the Earth and its life as recorded in successive layers of sediment and the fossil specimens they contain

NICHE: in an ecological environment, a position particularly suited for its inhabitant

PERIOD: the fundamental unit of the geologic time scale

PLATE TECTONICS: the theory and study of the formation and movement of the sections that make up the outer portion of the Earth, which move independently over the interior

STRATIGRAPHIC TIME SCALE: the history of the evolution of life on the Earth broken down into time periods based on changes in fossil life in the sequence of rock layers; the time periods were named for the localities in which they were studied or from their characteristics

PATTERNS OF EXTINCTION

The extinction of animal species appears to be as much a part of the pattern of life on the Earth as is the evolution of new species. Indeed, the two processes go hand in hand, as new species fill the environments vacated by dying ones. Extinction is so common that more than 99 percent of the species that have ever lived are now extinct. At periods throughout the Earth's history, extinctions of numerous and widely varied species have occurred within a relatively narrow range of time. Climatic changes, worldwide catastrophic events, and changing sea levels lead the list of possible causes of these mass extinctions. A species cannot be considered separately from the environment in which it lives. A change in the environment may occur faster than the organism can adapt and evolve along with it. This species may become extinct, while another species will develop to fill the environmental niche left behind. Thus, it appears that the ability of a species to survive may depend upon its ability to adapt quickly to changing conditions.

The pattern holds true for both plant and animal life. Mass extinctions of plant and animal life, however, do not seem to coincide. There have

been three revolutions in principal plant forms. Each evolved quickly, and once a major group of plants was established, it continued for millions of years. Animal life is marked by frequent extinctions; mass extinctions of animal life appear successively throughout the geologic record. In fact, it is these very extinctions that have, in many cases, provided the reference points that separate one geologic period from another. Catastrophic extinctions and revolutionary changes in animal life have occurred during at least six and perhaps as many as nine distinct junctures in the history of the Earth. Hundreds of minor extinctions have occurred as well.

PERIODS OF MASS EXTINCTION

The earliest known mass extinction occurred some 650 million years ago toward the end of the Precambrian eon. Animal and plant life was primitive at this time, and it is possible that chemical changes in the atmosphere were occurring as the oxygen-producing plants developed. At the time of the mass extinction, more than 70 percent of the forms of algae disappeared. The end of the Cambrian period, about 500 million years ago, is generally seen as the first of the major extinctions of life. More than one-half of the existing species of animal life vanished at this time. At the end of the Devonian period, 360 million years ago, many types of fish became extinct, along with 30 percent of the animal life on the planet. The largest mass extinction in the Earth's history happened some 248 million years ago at the end of the Permian period. This extinction also marked the end of the Paleozoic era. One-half of all the families of animal life were exterminated. This included 75 percent of all the amphibian families and 80 percent of the reptiles. Marine life, however, was hit hardest. It has been estimated that 95 percent of shallow marine organisms became extinct. It took 15 million to 20 million years for animal life to recover to anywhere near the previous variety. Another mass extinction took place at the end of the Triassic period some 213 million years ago. Up until this time, primitive amphibians and reptiles had dominated the land. By the end of the period, they had dropped out and were replaced by the early dinosaurs. It could appear that competition with the dinosaurs led to the demise of these creatures, but other extinctions took place at this time, particularly in the oceans, that point to a more general cause.

The most studied and best-known mass extinction occurred at the end of the Cretaceous period some 65 million years ago, at a point called the Cretaceous-Tertiary boundary. It is known mostly for the disappearance of the dinosaurs, but it also marks the extinction of many other forms of life, including many ocean reptiles, shelled creatures called ammonites, and forms of microscopic plankton. In all, 25 percent of all the known families and 75 percent of the known species of animals were eliminated. In this extinction, as is true with previous extinctions, there was a gradual decline in diversity among animal life for tens of millions of years before the final extinction at the end of the period.

The latter part of the Tertiary period, 10,000 years ago, saw the extinction of the giant land mammals. From the woolly mammoth and the mastodon to the saber-toothed cats, all perished. There is some evidence that implicates humankind in the extinction of these giant mammals. With organized hunting and the use of fire, it is conceivable that early humans played a role in the decimation of animal populations. Remnants at ancient kill sites show that thousands of animals were killed at a single hunt. Even in historical times, it has been seen that with the arrival of primitive people to a new area, extinctions of indigenous life-forms follow.

Although it has not been proven that humankind is responsible for the extinctions of the giant mammals toward the end of the Tertiary period, there is little doubt as to its role in the most recent mass extinction of life. The twentieth century witnessed the beginning of a widespread extinction of life-forms resulting from the explosive spread of the human species. While human demand for space increases, available habitats for other animals diminish. Remote areas that were once the havens of wildlife are now being penetrated by hunters, farmers, lumbermen, fishermen, and developers. Those areas devoid of humans are visited by the products of civilization—acid rain, air and water pollution, and destruction of the ozone layer. Thus, humankind is directly or indirectly responsible for the extinctions of more than 450 species of animals, and many more are endangered.

CAUSES OF MASS EXTINCTION

Current extinctions aside, why should a large percentage of life on the Earth die out suddenly? The question can be misleading. For example, "suddenly" is a relative term when applied to geologic events. Changes happen over thousands of millions of years. An abrupt change in layers of sediment may indicate a change that took place over a period of a few days or a few million years. It is difficult to determine from the geologic evidence how sudden "suddenly" was.

Although it was long known that mass extinctions occurred from time to time throughout the Earth's history, the causes were thought to be unknowable for lack of definitive evidence until high levels of the element iridium were found at the boundary marking the end of the Cretaceous period. With the possibility of testable evidence, the study of mass extinctions was opened again in earnest, and new relationships among available data began to be formed. As mass extinctions were documented exhaustively, what appeared to be a cyclic pattern emerged. According to one study, exterminations of life occurred about every twenty-six thousand years. The periodicity remains in question, and theories range from an orbiting death star to the fact that random extinction events create an intrinsic periodicity.

Something in the Earth's environment definitely changes at the time of a mass extinction. Many theories begin, then, by determining the changing conditions at the time of the extinction. Although valid, a weakness in this method is that the Earth's environment is constantly in a state of flux. In whatever time period one examines, changes are bound to be found. It is therefore merely speculation whether any particular change or set of changes was sufficient to cause a mass dying. In addition, it seems that mass extinctions cannot be studied from the perspective of a single species or even a single extinction event. Only by studying them as a whole can scientists determine patterns that may explain them. In general, mass extinctions kill species on the land and in the oceans at the same time. They strike hardest at large animals on the land. Freshwater animals are generally left unscathed. Plants are not affected as severely as animals. Any theory of mass extinction must therefore address these issues.

TWO MASS EXTINCTION THEORIES

Two theories have emerged that seem to handle the criteria. Neither theory is new, but each has received renewed interest with the emerging evidence. One popular theory invokes a global catastrophe with a celestial origin. Impacts by meteorites or comets are proposed to be responsible for at least the mass extinction that killed the dinosaurs, if not others. A 1979 study by Walter Alvarez and Luis Alvarez showed that the presence of glass microspherules, shocked quartz, and the rare element iridium at the Cretaceous-Tertiary boundary provides evidence for the impact of a bolide (a meteorite or comet that explodes upon striking the earth). However, the effect of this on the extinction event has been hotly debated.

The other theory has to do with the changing sea levels of the Earth, a theory from the 1920's that was revived during the 1960's and 1970's in the light of the theory of plate tectonics. As the sea level drops, warm, shallow seas recede. The habitat of some marine animals is eliminated, while chemical changes in the deeper ocean water cause other extinctions. On land, bridges appear between areas that were once separated by water,

and the migration of animals follows. Competition for available habitats is generated among species that were previously separated, also leading to extinctions. The drifting together and breaking apart of continents would dramatically affect habitats. It has been shown that in general, the arrangement of continents affects the variability of climate, the variety of species on land, and the nature of habitats along the continental shelves in the ocean. Changes in these factors over long periods of time would lead to extinctions of animal species.

There is no single or simple answer to the cause of mass extinctions. The pivotal question seems to be whether they occurred instantaneously, requiring a single catastrophic event, or over an extended period. In a period of several million years, many changes will occur in the environment, some of them catastrophic. Volcanoes erupt, meteorites strike, climates change, and continents drift. There is no doubt that catastrophes occur, but how they affect life on the Earth is as yet poorly understood. All things considered, it is most likely that mass extinctions occur as a result of a combination of many factors, including catastrophic ones that happen over long periods of time. Perhaps a cyclic pattern exists, or perhaps species survival is a matter of playing the odds on a changing planet. Yet, if the Earth is considered to be a living organism, it is no wonder that as the planet ages and evolves, the life it supports also changes.

STUDY OF THE GEOLOGIC RECORD

The major clues to unravel the pattern of life on the Earth come from the study of the geologic record. The sequential layering of rocks, their deformation and erosion, allows scientists to construct a history of the processes the rocks have undergone. By analyzing the chemical composition of layers of rock and sediment and studying the nature of the fossil remains in them, scientists can begin to understand what forms of life existed during which time periods and when (and sometimes how) they became extinct. It was discovered that the layers could be dated by the decay of radioactive elements and that, in this way, the ages and periods of geologic time could be dated accurately.

Of all the fossils in the geologic record, fossils of tiny marine animals (microfossils) give the clearest picture of mass extinctions. Produced prolifically in the world's oceans, they settled among the sediment on the ocean floor and fossilized in layers. These layers tell the story of how the creatures developed, diversified, flourished, and finally when they ceased to exist. Found in diverse areas of the world, from desert to mountainous regions, the fossils also tell the story of the movement of continents. Certain of them form such a complete record that their forms and types can be used to date sediments all over the world. While fossil remains of land animals can also be dated and related to specific time periods, these larger fossils are much scarcer and therefore do not present as complete a picture.

By studying traces of magnetism preserved in rock along with fossil deposits, geologists can chart the drifting of the continents in the distant past. The emergence of the theory of plate tectonics has provided scientists with a framework for understanding why the Earth's environment and all its features are in a constant state of change. This changing environment is a key clue to periodic mass extinctions.

Besides the rise and fall of fossil species, other characteristics of the geologic layers can be measured. The relative abundances of certain elements indicate the processes that may have been occurring at the time the deposits were laid down. An example is the iridium found at the site of the extinction of the dinosaurs. Iridium is uncommon on the Earth but is found in certain types of meteorites, which has led many scientists to believe that the extinction of the dinosaurs was caused by the impact of a comet or meteorite. Particles of soot

were also found in the layers from that time period, suggesting that a large part of the Earth may have been burned. Shocked quartz, indicating violent activity, also supported the theory.

The story of life found in the geologic record leaves scientists with myriad clues but few answers. It is as if they had millions of minute pieces to an enormous puzzle; collecting the pieces does not solve the puzzle. The various clues must be interpreted, seen in the light of their historical significance, and put together in fresh ways. As patterns emerge in the study of mass extinctions, the clues are beginning to fall into place, and an understanding of the dynamics of life on the Earth is emerging.

Divonna Ogier

CROSS-REFERENCES

BIBLIOGRAPHY

Alvarez, W. *T. Rex and the Crater of Doom.* Princeton, N.J.: Princeton University Press, 1997. A history of the discovery of the data that has been put forward to support an impact-caused extinction of dinosaurs. For the general reader.

Archibald, J. D. *Dinosaur Extinction and the End of an Era.* New York: Columbia University Press, 1996. A thorough review of dinosaur extinction and advocacy of a prolonged event caused by falling sea levels culminated by the

impact of an asteroid. Intended for general readers.

Bakker, Robert T. *The Dinosaur Heresies.* New York: William Morrow, 1986. An expert on current theories concerning the dinosaurs explains them in an easy-to-read and interesting way. Dinosaurs are studied in terms of their place in geologic history, and their extinction is explained as part of a larger pattern. The pattern is compared with other mass extinctions in the history of the Earth.

Colbert, Edwin H. *Dinosaurs: An Illustrated History.* Maplewood, N.J.: Hammond, 1986. Portrays the natural history of the dinosaurs, their beginnings, evolutionary development, how they influenced the environment in which they lived, and their extinction. Well illustrated.

Erwin, Douglas H. *The Great Paleozoic Crisis.* New York: Columbia University Press, 1993. Examines the events recorded at the Paleozoic-Mesozoic boundary, where the most important extinction event in the Earth's history took place. Written for the specialist but accessible to students. Glossary and extensive reference list.

Press, Frank, and Raymond Siever. *Understanding Earth.* 2d ed. New York: W. H. Freeman, 1998. This comprehensive physical geology text covers the formation and development of the Earth. Contains extensive sections on plate tectonics and its effect on the distribution of continents through time and thus on extinction events. Readable by high school students, as well as by general readers. Includes an index and a glossary of terms.

Wilford, John Noble. *The Riddle of the Dinosaur.* New York: Alfred A. Knopf, 1985. A well-written, sparsely illustrated account of the life and death of the dinosaurs. More in-depth than most books about dinosaurs, the book deals with each area of the study of the dinosaurs, discusses the various and sometimes controversial ideas concerning them, and comes to a synthesis without espousing a particular conclusion.

MICROFOSSILS

Micropaleontology is the study of plant and animal fossils that are too small to be seen without magnification. These microscopic objects provide valuable information about the evolution of life on the Earth and about changes that have occurred on the Earth's surface through time. They also have great value as index fossils and as indicators of ancient environments, data useful in the search for oil and natural gas.

PRINCIPAL TERMS

FORAMINIFERA: single-celled, amoeba-like animals

INDEX FOSSILS: indicators of geologic age; they have short geologic (time) ranges and broad geographic distribution

ISOTOPES: atoms of an element that differ in weight because different numbers of neutrons are present in their atomic nuclei

MORPHOLOGIC EVOLUTION: changes in the body or skeleton shape of organisms through time

PARTHENOGENIC: organisms in which unfertilized females produce viable, fertile offspring without copulation with males of the species

MICROPALEONTOLOGY

When people think of fossils, they envision items such as the bones and teeth of dinosaurs, the petrified bones of Neanderthals, or the shells of oysters. Micropaleontologists study the preserved remains of organisms too small to be seen with the unaided eye. It is convenient to divide micropaleontology into two subfields: animal micropaleontology and plant micropaleontology. Animal micropaleontology encompasses the study of a wide variety of fossils. In most cases, the material studied is actually the shell constructed by the once-living animal. Shells are commonly composed of calcium carbonate (the common mineral calcite), calcium phosphate (the mineral apatite, a constituent of human teeth and bones), and opaline silica (quartz with water molecules in its crystal structure). Other microscopic animals build shells from mixes of sand grains, shell fragments, and volcanic ash, all glued together with organic or mineral cement. Thus, shells are referred to as calcareous, phosphatic, siliceous, or agglutinated.

Microfossils vary in size as well as in shell composition. As a rule, few are smaller than 0.05 millimeter in diameter; most are in the range of 0.75 to 2.0 millimeters, and a very few are as large as 5 centimeters. Among the larger microfossils are those called nummulites, which are major constituents of the limestone used in the Egyptian pyramids. In about 450 B.C.E., Herodotus described them, erroneously, as mummified lentils (the food of pyramid construction crews).

Systematic study of microfossils did not begin until the early years of the nineteenth century. Alcide Dessalines d'Orbigny published a paper describing some microfossils that he thought to be microscopic cephalopods related to the chambered nautilus. Subsequently, it was realized that d'Orbigny's microfossils were types of protozoans (single-celled animals) called foraminifera. Another pioneer was Christian Gottfried, who was the first to treat micropaleontology as a field of study. As is the case for larger fossils, microfossils can be used to determine the age of sedimentary rocks; European micropaleontologists used them for this purpose as early as 1874 and, by 1930, micropaleontology was aiding in petroleum exploration worldwide.

FORAMINIFERA AND RADIOLARIANS

Many types of animals can produce microscopic shells or other hard parts that may be preserved in sedimentary rocks. Among these are protozoans, gastropods (snails), worms, crustaceans (crabs, lobsters), sponges, echinoderms (starfish, sea urchins), and fish. Protozoans contribute significantly to the microfossil record. Two are particularly useful as indicators of geologic age and of ancient environments: the foraminifera and the radiolarians. Most living foraminifera are found in the ocean, where their distribution is controlled by water temperature, salinity, depth, turbulence, light intensity, bottom conditions, availability of food, predators, parasites, and other biologic factors. Foraminifera, or forams, usually have calcareous or agglutinated shells. The largest number of species are bottom dwellers (benthic), but one group, the globigerinids, evolved to live as passive floaters (or plankton) in the surface waters of the oceans. Benthic forams have a long geologic history extending back 500 to 600 million years. For the last 300 million years, their shells have been significant rock formers. The rock that forms the White Cliffs of Dover is composed largely of the shells of foraminifera. Planktonic forams evolved more recently, first appearing in rocks about 175 million years old. Since planktonic species live in surface waters, ocean currents may carry them thousands of miles. This wide geographic distribution coupled with their rapid morphologic evolution makes planktonic foraminifera especially valuable as index fossils.

Radiolarians are also planktonic protozoans, but they differ from foraminifera in their soft-body-part anatomy and in their shell construction. Radiolarian shells are composed of opaline silica, are usually in the 0.05 to 0.5 millimeter size range, and display a bewildering array of shapes. Most are variations on three shapes: spheres, cones, or disks. Some radiolaria secrete their shells as spongy masses, others build shells of perforated sheets of silica, and still others construct latticework shells of great delicacy and beauty. Radiolaria are found in rocks almost 600 million years old, giving them the longest geologic range of planktonic microfossils. Mesozoic and Cenozoic radiolarians are better known than are Paleozoic ones.

OSTRACODS

Microscopic crustaceans called ostracods (or ostracodes) have a 500-million-year-long fossil record and are abundant in many sedimentary rocks. Unlike the familiar macroscopic crustaceans, ostracods encase their minute, shrimplike bodies in a pair of tiny, calcareous, bean-shaped shells 1 to 5 millimeters long. Outer surfaces of these shells may be smooth, or they may bear spines, wartlike bumps, grooves, ridges, flanges, and pores. Hinge structures and muscle scars, found on the inner surfaces of the shells, are also useful features in distinguishing different ostracods. Although some species are planktonic, most ostracods are benthic. Living ostracods can be found from the deepest ocean floor to the shoreline and landward into lakes and streams. A few species have adapted to live on land in moist ground litter. Playa lakes in desert regions often are inhabited by ostracods. Species living in these harsh environments are often parthenogenic: Unfertilized females lay fertile eggs, which lie dormant in the muddy bottom of a dried-up lake; the next time the lake fills, the eggs hatch into a new generation of females, and the cycle continues. Dormant periods of a decade or more have been reported.

Two different kinds of toothlike structures are frequently found in assemblages of microfossils. Annelid worm jaws, called scolecodonts, are composed of a resistant organic material (chitin) and have been found in rocks up to 600 million years old. The second type are called conodonts ("cone-

Cretaceous-period chert from the Franciscan Formation, composed of microscopic siliceous shells of radiolarians. (© William E. Ferguson)

toothed" animals). These fossils are 0.2 to 6.0 millimeters in greatest dimension, are composed of calcium phosphate, and occur as isolated specimens or in clustered assemblages. The composition suggests that they were produced by an animal, and their distribution in sedimentary rocks suggests that these "conodont animals" (ordinal name Conodontophorida) were marine and planktonic. They first appeared in Cambrian time, flourished through the rest of the Paleozoic, and became extinct during the Triassic period. Specimens of conodont animals are known from the Carboniferous of Scotland and the Ordovician of southern Africa. They appear to have been eel-like animals in which the conodonts themselves formed a grasping structure in the throat. Although the affinities of this animal remain obscure, it is generally thought to have been a chordate of some kind.

STUDY OF MICROFOSSILS

While microfossils can be recovered from many types of marine sedimentary rocks, they are most abundant in fine-grained rocks such as shale. Samples of the rocks can be obtained from surface outcrops or from subsurface boreholes in the form of cores or cuttings. In collecting samples for study, care must be taken to eliminate any contamination, and accurate records of sampling localities must be maintained.

In the laboratory, a variety of techniques are used to extract the microfossils from the host rock and to clean them. Composition of the fossils, composition of the host rock, and the kind of study to be done dictate the separation methods used. Some foraminifera, for example, can be studied profitably only in thin section. To prepare them, fossiliferous samples are glued to microscope slides and are ground and polished until the rock is paper thin and the internal features of the shell can be seen. Acids are used to dissolve sedimentary rocks effectively, in order to release insoluble siliceous or phosphatic microfossils. Acetic acid will dissolve limestone without damaging conodonts, and radiolarians can be freed best by dissolving the host rock in hydrochloric acid.

Calcareous-shelled ostracods and foraminifera are usually extracted from shales in the following manner. A clean, dry, crushed (to pea size) sample is soaked for thirty to sixty minutes in kerosene. The kerosene is poured off, filtered, and saved for reuse. The sample container is filled with water, and a wetting agent is added. This mixture is boiled for twenty to thirty minutes, and then it is poured through a 200-mesh screen. Fine clay particles pass through, leaving the microfossil residue on the screen. Usually it is necessary to repeat the boiling and screening process several times to yield a good, clean residue. After drying, the residue is ready for "picking." A one-grain-thick layer of residue is spread on a picking tray, which is then placed on the stage of a binocular microscope. Magnification of thirty to forty times is required. A picking brush is also needed. The micropaleontologist slides the picking tray back and forth so that all the residue is scanned. Once a microfossil is found, the picking brush and a steady hand come into play: The brush is moistened and guided to the fossil, which sticks to the brush while being transferred to a microscope slide for future study. Routinely, several hundred specimens will be picked from each sample; each specimen will be identified and tallied in the sample census. Specimens can be repicked for further study using the greater magnification of the scanning electron microscope (SEM). Thin coatings of a conductive metal are vacuum-plated on the specimens, which are then ready for SEM analysis. In a vacuum chamber, a beam of electrons is focused on the specimen; reflected electrons are collected and are converted electronically into an image of the specimen. Magnification of fifty thousand times or more is possible, and the electronic image has a three-dimensional appearance.

BIOSTRATIGRAPHY

Probably the single most important application of microfossils is in biostratigraphy, or the use of fossils to determine the age of rocks. Fundamental to biostratigraphy has been the establishment of reference sections. These are sequences of rock that have been precisely dated and whose fossils have been described in considerable detail. Fossils from a rock sequence of unknown age are compared with fossil sequences from reference sections and the best match is obtained. Because of their rapid evolution and their broad geographic distribution, fossils of planktonic organisms are particularly well suited for this work. For example, radiolarian assemblages from Japan, western Texas, and the Ural Mountains of Eurasia can be compared directly with one another.

Using a picked collection of foraminifera, a number of other studies can be carried out. If the assemblage contains both planktonic and benthonic species, the planktonics-benthonics ratio gives a measure of water depth at the time of deposition. Another depth indicator is the ratio of calcareous-shelled benthonics to agglutinate-shelled benthonics. The diversity of the assemblage may also provide useful information. In a tropical assemblage, many species are usually present, but only a few individuals of each species occur (high diversity); in a high-latitude assemblage, only a few species are present, but each is represented by a large number of individuals (low diversity). Diversity measures must be used carefully, as similar effects can be seen in relation to water depth—low diversity in very shallow or very deep water and higher diversity at intermediate depths.

Calcareous-shelled forams can also be analyzed to give a direct measure of temperature of the water in which they lived. Oxygen atoms in two forms (isotopes) occur in seawater. The ratio of oxygen 18 to oxygen 16 is a function of water temperature. When forams extract calcium carbonate from seawater and build their shells with it, the oxygen isotopes are incorporated as well. In the laboratory, the shells are converted to carbon dioxide, and the ratio of oxygen 18 to oxygen 16 is determined. The measured ratio is then plugged into a mathematical formula to calculate water temperature. A second temperature-measuring technique can be used with planktonic forams. As the animals grow, they add chambers to their shells in a spiraling pattern. In warm water, the spiral is clockwise (right-handed); in cold water, it is counterclockwise (left-handed). The coiling ratio for planktonic forams in a sample thus gives an indication of water temperature. Although this technique is less precise than isotopic measurements, it is faster and less expensive.

Studies of evolution can also be conducted using microfossils. Because planktonic animals seem to evolve faster than do benthonic ones, plankton are often preferred in these studies. Ideally, one needs closely spaced samples from a core in an area where insignificant environmental change occurred while deposition of the planktonic shells was continuous. Some lineages of Cenozoic radiolarians have been traced for several million years, during which time small morphologic changes accrued so that the descendant species are quite different from their ancestors in morphology. In some of these studies, hybridization between different but related species seems to have been recorded by the fossils.

William C. Cornell

CROSS-REFERENCES

BIBLIOGRAPHY

Anderson, O. R. *Radiolaria.* New York: Springer-Verlag, 1983. This volume considers the biology of living radiolarians and includes discussion of morphology, ultrastructure, physiology, ecology, and evolution. A glossary and appendix are included, as is an index. College level.

Brasier, M. D. *Microfossils.* Winchester, Mass.: Allen & Unwin, 1980. This concise volume treats fifteen groups of microfossils, emphasizing their features as seen in a microscope. Illustrated with high-quality line drawings. An appendix summarizes sampling and preparation techniques. The bibliographic entries are chiefly twentieth century articles and books. Both a general index and a systematic index are included. College level.

Briggs, Derek E. G., and Peter R. Crowther, eds. *Paleobiology: A Synthesis*. Oxford: Blackwell Scientific Publications, 1990. This multiauthored text deals with all aspects of paleontology in a series of short articles by experts in each field. Intended for university students but also useful to high school students and interested readers.

Feldmann, Rodney M., Ralph E. Chapman, and Joseph T. Hannibal, eds. *Paleotechniques*. Knoxville, Tenn.: Paleontological Society, 1989. This extremely useful volume discusses specific techniques for collecting, preparing, and studying microfossils. Extensive bibliography.

Hag, Bilal, and Anne Boersma, eds. *Introduction to Marine Micropaleontology*. New York: Elsevier, 1978. Articles by specialists on fourteen groups of microfossils. Illustrations are outstanding; indexing is thorough. A helpful glossary of terms is included. Each chapter includes lists of general references and of cited references. College level.

Pokorný, Vladimír. *Principles of Zoological Micropaleontology*. Translated by K. A. Allen. 2 vols. Elmsford, N.Y.: Pergamon Press, 1963-1965. Translated from the German, this 650-page book presents an excellent overview of animal micropaleontology. Includes discussions of micropaleontological methods and of the major groups of animal microfossils. Profusely illustrated with high-quality photomicrographs and line drawings. References are grouped by chapters and include good coverage of literature published outside of North America. Fossil and subject indexes are given. College level.

Prothero, Donald R. *Bringing Fossils to Life*. Boston: McGraw-Hill, 1998. This well-illustrated and entertaining text covers a broad range of paleontological topics, including micropaleontology. Glossary, bibliography, and index.

NEANDERTHALS

Neanderthals are the best-known extinct members of the human lineage. It is generally agreed that they were close relatives of modern humans, but the nature of the relationship is vigorously debated. They have been assumed to be a direct ancestor, a diseased member of the species, or an extinct side branch of the family tree.

PRINCIPAL TERMS

DEOXYRIBONUCLEIC ACID (DNA): the chemical that carries the instructions for all living things; closely related organisms have very similar DNA

GENUS: the first part of the scientific name of an organism; members of the same genus but different species are closely related, but cannot mate and produce fertile offspring

MITOCHONDRIA: subcellular structures containing DNA used to estimate the relationships between groups of organisms; the more similar the DNA, the more closely related the groups

SPECIES: the second part of the scientific name of an organism; members of the same species can mate and produce fertile offspring

SUBSPECIES: the third part of a scientific trinomial, assigned to one of two groups that can mate and produce fertile offspring, but that have some strikingly different characteristics

TAXONOMY: the science of classifying and naming living and fossil organisms, or the classification and scientific name of a living or fossil group

HISTORY AND CHARACTERISTICS

The fossil that gave the Neanderthals their name was found in a cave being quarried for limestone in Germany's Neander Valley in 1856. At least two Neanderthal fossils were discovered before the Neander Valley individual; however, neither was recognized as a member of an extinct human group until after the name "Neanderthal" was assigned. Many similar fossils have been found in scattered locations all over Europe and the Middle East since the Neander Valley discovery. Dates assigned to the various fossils indicate that the Neanderthals originated late in the Ice Age and became extinct a few thousand years before the last glacial retreat (from about 200,000 years ago to about 30,000 years ago). Thus the Neander Valley specimen lent its name to a fossil relative of modern humans that occupied Europe and the Middle East late in the Ice Age.

Though the Neanderthals were very similar to modern humans, they had several distinctive characteristics. Neanderthals were short and exceptionally stout-bodied with broad, supportive bones and joints. This body form suggests a life filled with intense physical effort. Perhaps the compact body also helped them cope with cold stress under Ice Age conditions. Their brains were somewhat larger than modern human brains. That size may have compensated for the more massive total body size of the Neanderthals, since large-bodied organisms generally have larger brains. Their foreheads sloped up from their exceptionally heavy eyebrow ridges, their jaws extended forward beyond the plane of the face, and their chins were weakly developed. These and several other characteristics are used to define a fossil find as a Neanderthal.

STRUCTURE AND BEHAVIOR

Rudolph Virchow's initial interpretation of the Neander Valley fossil as a diseased human was popular for a time. Virchow held that the fossil was a modern human whose unique features were the result of disease. However, as more fossils with the same characteristics were discovered all around Europe and the Middle East, this explanation became untenable. Later, misinterpretation of the characteristics of Neanderthal fossils led Marcellin Boule and others to interpret Neanderthals as stooped, bent-kneed, apelike subhumans with an animal nature to match.

Additional fossil discoveries, including evidence for toolmaking and burials, sometimes with flowers placed in the grave, caused anthropologists to rethink the presumed animal nature of the Neanderthals. Although the evidence for flowers has been challenged, the evidence for burials, presumably accompanied by mourning, is accepted by many anthropologists. In addition, fossils showed that some Neanderthals lived much of their lives with deformed limbs and other disabilities, which would have made it difficult or impossible for them to fend for themselves. Yet they apparently lived many years in that condition, suggesting the support of other members of a social group. Such behavior was not in keeping with Boule's picture of the Neanderthals as nonhuman animals.

Reinterpretation of the anatomic evidence also suggested that, instead of a bent-kneed, stooped posture, the Neanderthals walked on two rather straight legs and had hands capable of manipulating materials and making tools, much as modern humans do. All this indicated that the Neanderthals were more like modern humans than Boule's interpretation, and they came to be thought of in that light.

TAXONOMIC RELATIONSHIP TO MODERN HUMANS

Neanderthals have always been recognized as close relatives of modern humans, but the specific taxonomy of the relationship is still a point of contention. They are placed in the same genus (*Homo*) as modern humans by almost all anthropologists, but researchers debate whether they were members of our species, *Homo sapiens*, or belonged in their own species, *Homo neanderthalensis.*

The discussion of the structure and behavior of Neanderthals bears directly on this question. If the Neanderthal characteristics were the result of disfigurement caused by disease, Neanderthals were simply aberrant humans and not especially interesting from the perspective of human evolution. However, if they were stooped, bent-kneed, and animal-like in behavior, they were probably a separate species, perhaps ancestral to modern humans, and therefore more interesting from the evolutionary perspective. On the other hand, if they were upright in stature, were skilled toolmakers, were supportive of their handicapped and elderly, and buried their dead with mementos such

as flowers, they might earn the designation *Homo sapiens* and take on an even greater interest to the more modern members of that species.

Such arguments are part of the practical taxonomy of the Neanderthals, but the real key to species identification and species separation is (at least theoretically) interbreeding. If the members of two groups can mate with each other and produce fertile offspring, and if these offspring can produce fertile offspring, the two groups are generally considered to be members of the same species. Therefore, the real taxonomic question becomes: Could Neanderthals and early-modern humans interbreed?

Because it is very difficult to determine whether fossil groups interbred with one another, Neanderthal taxonomy has been primarily determined by anatomic and presumed behavioral characteristics, such as those already discussed. That taxonomy has vacillated with changing interpretations of those characteristics. Neanderthals have been placed in their own species (*Homo neanderthalensis*) for much of their history, but they have been identified as a human subspecies (*Homo sapiens neanderthalensis*) at other times. The latter designation implies that the Neanderthals and modern humans (*Homo sapiens sapiens*) were members of the same species and therefore could interbreed.

Determination of the Neanderthals' taxonomic position is an integral part of arguments over the mechanism of the origin of modern humans. There are two main hypotheses for that origin: the replacement hypothesis of Christopher Stringer and the multiregional hypothesis vigorously supported by Milford Wolpoff. The replacement hypothesis is also designated "out-of-Africa" because it assumes that a population of African origin expanded throughout Africa, Europe, and Asia and rapidly replaced the more primitive humanlike species living there, including the Neanderthals. Whether this replacement was by competition or by more direct and violent means is undetermined. The multiregional hypothesis suggests that the widespread, more primitive humanlike populations evolved into modern humans rather than being replaced by new immigrants. Both hypotheses hold that the more primitive populations also originated in Africa and spread to Europe and Asia at a much earlier date.

Because the Neanderthals are the best known and best understood early human group, an understanding of the Neanderthal relationship is critical to an understanding of the evolutionary history of humanity. A Neanderthal contribution to modern human ancestry would support the multiregional hypothesis, and the lack of such a contribution would be consistent with the replacement hypothesis.

ADVANCES OF THE 1990'S

By the 1990's, the Neanderthals were well established as a group related to modern humans, but the questions remained: How close was the relationship? Did the two groups interbreed? Were Neanderthals a part of the evolutionary heritage of modern humans? During the 1990's, improved techniques and additional fossil discoveries led to greater understanding of the Neanderthals but little consensus on these questions. A few examples will illustrate the situation.

In a 1996 study, Jean-Jacques Hublin and several coworkers determined that Neanderthals found at an archeological site in France made bone tools and wore decorative emblems on their bodies, behaviors not uncovered with older Neanderthal fossils. They concluded that the Neanderthals were influenced by early-modern humans who lived in the same area at the same time and that a reasonably elaborate cultural exchange must have occurred between the two groups. However, based on the strikingly different anatomy of the two groups' inner ears, they also concluded that the Neanderthals and modern humans did not interbreed. The investigators reasoned that if interbreeding had occurred, the two groups would have shared a common ear structure.

In 1997, Matthias Krings, Svante Paabo, and their colleagues isolated and engineered deoxyribonucleic acid (DNA) from the mitochondria of Neanderthal bones and compared it to DNA from modern human mitochondria. They found the Neanderthal DNA to be quite different from that of modern humans and concluded not only that the two groups were different species but also that Neanderthals were not ancestral to modern humans.

In 1998, Daniel Lieberman proposed that a reduction in the length of the sphenoid bone during embryology can explain most differences between the two groups' skulls. The sphenoid is a bone in the skull of both Neanderthals and mod-

ern humans, and Lieberman showed it to be shortened in modern humans but not in Neanderthals. He hypothesized that the impact of shortening the sphenoid resulted in the modern human skull characteristics, while the longer sphenoid resulted in the Neanderthal skull. Based on the fundamental nature of the change, he concluded that Neanderthals do not belong to the same species as modern humans and were probably not ancestral to modern humans.

In 1999, Cidàlia Duarte, Erik Trinkaus, and several colleagues discovered the buried remains of a four-year-old child in southern Spain. The skeleton was estimated to be about 24,500 years old, and they interpreted its anatomy to be a mixture of modern human and Neanderthal characteristics. Most anthropologists agree that southern Spain supported Neanderthal populations longer than other parts of the world, perhaps as late as 27,000 years ago, and that modern humans and Neanderthals coexisted in the region. Duarte, Trinkaus, and their group suggested that the skeleton they found demonstrated that the two groups did interbreed and that Neanderthals were part of the ancestry of *Homo sapiens*.

SIGNIFICANCE

Consideration of this short list of studies in the 1990's demonstrates the state of knowledge about the Neanderthals' place in human evolution. Viewed alone, each study seems to clinch the position of its authors. In fact, the first three reinforce one another so well that Neanderthals would seem to be eliminated from direct participation in the evolution of modern humans. However, Duarte and Trinkaus's study would seem to clinch the opposite position, that Neanderthals were direct participants in the evolution of modern humans. This situation symbolizes the absence of consensus in the field. There are also established scientists with alternative viewpoints for each of these studies. Lieberman himself is a coauthor of a letter that criticizes his own conclusions about the sphenoid and points to the need for a better understanding of the development of primate skulls to help clarify the situation.

A number of anthropologists have pointed out that Krings and Paabo's conclusions are extrapolated from a single, short segment of the mitochondrial DNA and that more extensive studies, including studies of DNA from the nucleus, are necessary before definitive conclusions can be drawn. In fact, nuclear DNA studies of modern humans have suggested that modern human DNA comes from a number of sources rather than a single African source as in the out-of-Africa hypothesis. Clearly, extensive DNA comparisons would be helpful; however, DNA from fossils is difficult to find and difficult to work with, so an extensive collection of such studies is not likely to accumulate.

Ian Tattersall, who rejects the Neanderthals as direct contributors to modern human evolution, has criticized Duarte and Trinkaus's data and their interpretation of the data. The verbal exchange has been bitter, not an unusual circumstance for disagreements in this field.

Although anthropologists have learned an enormous amount about the Neanderthals, their relationship to modern humans continues to escape consensus. This is, without question, a result of the difficulty of the problem and the tentative nature of the evidence. Most agree that the Neanderthals were a successful group closely related to modern humans. Everyone's hope is that more fossils, improved technology, and fresh insight will clarify the question because understanding the Neanderthals is likely to contribute to an understanding of humanity.

Carl W. Hoagstrom

CROSS-REFERENCES

Archaebacteria, 951; Biostratigraphy, 1091; Cambrian Diversification of Life, 956; Carbonate Compensation Depths, 2101; Catastrophism, 961; Cenozoic Era, 1095; Colonization of the Land, 966; Deep-Sea Sedimentation, 2308; Dinosaurs, 970; Ediacaran Fossils, 976; Electron Microscopy, 463; Eukaryotes, 981; Evolution of Birds, 985; Evolution of Flight, 990; Evolution of Humans, 994; Evolution of Life, 999; Fossil Plants, 1004; Fossil Record, 1009; Fossilization and Taphonomy, 1015; Fossils of the Earliest Life-Forms, 1019; Fossils of the Earliest Vertebrates, 1024; Geoarchaeology, 1028; Geologic Time Scale, 1105; Life's Origins, 1032; Mammals, 1038; Mass Extinctions, 1043; Mesozoic Era, 1120; Microfossils, 1048; Paleobiogeography, 1058; Paleobotany, 1062; Paleozoic Era, 1126; Petrified Wood, 1058; Prokaryotes, 1071; Stratigraphic Correlation, 1153; Stromatolites, 1075; Systematics and Classification, 1079.

BIBLIOGRAPHY

Akazawa, Takeru, Kenichi Aoki, and Ofer Bar-Yosef, eds. *Neandertals and Modern Humans in Western Asia.* New York: Plenum, 1998. A scholarly but understandable group of papers with contributions from many of the major workers in Neanderthal evolution and biology.

Ciochon, Russell L., and John G. Fleagle, eds. *The Human Evolution Source Book.* Englewood Cliffs, N.J.: Prentice Hall, 1993. The broad spectrum of human evolution is covered in this book, but a large section titled "The Neanderthal Question and the Emergence of Modern Humans" covers Neanderthals. Several of the most fundamental questions are dealt with in a clear and interesting fashion.

Fox, Richard G. "Agonistic Science and the Neanderthal Problem." *Current Anthropology,* supp. 39 (June, 1998). All articles in the supplement concern "The Neanderthal Problem and the Evolution of Human Behavior," the supplement's title. The interaction between Neanderthals and modern humans in Europe and in the Middle East and an archaeological consideration of the out-of-Africa model are the subjects of three other articles.

Shreve, James. *The Neanderthal Enigma.* New York: William Morrow, 1995. Written by a science writer rather than a scientist, this is an interesting account of the history of Neanderthals and Neanderthal studies.

Stringer, Christopher, and Robin McKie. *African Exodus.* New York: Henry Holt, 1996. A small, interesting book on the out-of-Africa hypothesis written by an anthropologist intimately involved with the development of the hypothesis (Stringer) and a science writer.

Tattersall, Ian. *The Last Neanderthal.* New York: Macmillan, 1995. An extensively illustrated account of what is known about Neanderthals and how it was learned. The prologue outlines two interesting, if imaginative, characterizations of the "last Neanderthal."

Tattersall, Ian, and Jeffrey H. Schwartz. "Hominids and Hybrids: The Place of Neanderthals in Human Evolution." *Proceedings of the National Academy of Science* 96 (June, 1999): 7. This commentary is a reply to the paper by Duarte, Trinkaus, and their colleagues (pages 7604-7609 of this same issue) in which they report the description of a 24,500-year-old fossil with characteristics of both Neanderthals and early-modern humans. The two positions are clearly laid out in the article and the commentary.

Trinkaus, Erik, and Pat Shipman. *The Neanderthals: Changing the Image of Mankind.* New York: Alfred A. Knopf, 1992. An interesting and well-written account of the Neanderthals and the scientists who study them.

PALEOBIOGEOGRAPHY

Paleobiogeography concerns the study of the geographic distribution of past life-forms. Paleobiogeographic studies, coupled with reconstructions of ancient plate configurations, serve as tools for reconstructing a detailed history of the Earth's changing geography and environments through time and the concurrent evolution of the organisms living in these environments.

PRINCIPAL TERMS

BARRIER: any sort of physical, chemical, or biological obstacle that prevents the migration of a species into or out of a given geographic area

BIOGEOGRAPHIC PROVINCE: a geographic region distinguished by a unique set of endemic organisms that live in the region

CONVERGENT ORGANISMS: evolutionarily unrelated organisms that come to resemble one another through adaptation to the same life habit

COSMOPOLITAN: having a broad, essentially worldwide distribution

DIVERSITY: the variety of life, usually described in terms of the number of species present

ENDEMIC: restricted to a specific, limited geographic region

EPICONTINENTAL SEAS: shallow seas that periodically covered large portions of the continents in the geologic past

LIMITING FACTORS: the physical, chemical, and biological factors that govern the distribution of a given organism

PALEOBIOGEOGRAPHY: the study of the geographic distribution of past life-forms

PLATE TECTONIC THEORY: the concept that the Earth's exterior is divided into an interlocking mosaic of moving plates

DISTRIBUTION OF LIFE

Life is distributed on the Earth's surface in complex patterns, governed by local environmental conditions and distributed in specific geographic regions. Very few living species of plants or animals are cosmopolitan in their distribution. *Homo sapiens* (modern humans) is one of the few cosmopolitan species. Most organisms evolve, exist, and eventually become extinct all in the same limited geographic region; these organisms are said to be endemic in their distribution. Endemic species may be restricted to a continent (the African elephant in sub-Saharan Africa), a portion of a continent (the snow leopard in the Himalaya), an island (the orangutan in Borneo), or even smaller geographic areas (the giant sequoia redwood in Northern California).

Most organisms are adapted to a specific set of narrowly defined environmental requirements that must be met if the organism is to exist in a given area. If only one of these environmental requirements or limiting factors is not present in the right proportion, the species will be unable to exist in a particular environment. These limiting factors include all the physical, chemical, and biological properties that define a given environment. For all organisms, an adequate food supply or source of nutrients is basic for continued existence. Factors controlling the distribution of organisms in terrestrial (land) environments include the availability of water, the temperature, and the altitude above sea level; humidity, precipitation, soil type, and sunlight intensity are limiting factors for plants, and types of vegetation are limiting factors for animals. For aquatic organisms, factors such as water depth, salinity, temperature, dissolved oxygen content, current flow, clarity, and bottom conditions become important. These various environmental factors in a particular geographic area are critical to determining the kinds of animal and plant life that can exist there.

Each environment within a given geographic region supports flora and fauna that are adapted to the specific conditions of that environment and are known collectively as biota. These geographically restricted biotas can be used to define bio-

1058

geographic provinces, specific geographic regions distinguished by their endemic floras and faunas. The boundaries between biogeographic provinces are marked by the development of barriers to the free movement of species from one area to another. These barriers might be physical, chemical, or biological in nature. For example, land areas would be barriers to the dispersal of marine organisms, open oceans would be barriers to the dispersal of terrestrial organisms, and deep oceanic waters might be barriers to the dispersal of shallow-water, nearshore organisms. Many barriers differentiating biogeographic provinces today are climatic in nature, marking zones of changing atmospheric or water temperature.

MAJOR BIOGEOGRAPHIC PROVINCES

At the present, the major continental land areas can be divided into seven major biogeographic provinces, based primarily on the occurrence of distinctive groups of large mammals. These provinces are delineated by climate and vegetation, which, in turn, control the distribution of these animals. These land areas and the terrestrial organisms they support are isolated by large oceans. The greater the isolation of these provinces (that is, the more effective the barriers between provinces), the more distinctive the flora and fauna characteristic of each province. Similarities between faunas in North America and Eurasia result from similar climatic conditions in these land areas and the development of a "corridor," the Bering Strait land bridge, within the past 2 million years; this corridor allowed the unobstructed interchange of land animals between the two continents. Australia, on the other hand, until the nineteenth and twentieth centuries, maintained its geographic isolation and therefore developed distinctive marsupial mammalian fauna. The fossil record shows that a similar marsupial fauna existed in South America until the Pliocene epoch (about 5 million years ago), when the Central American isthmus was uplifted, allowing for the invasion of South America by more advanced placental North American mammalian forms. This endemic South American fauna is now largely extinct; survivors include the opossum and armadillo.

The great diversity of animal and plant life in the world at present is largely the result of the present-day configuration of the continents. The continents are spread all over the globe in rough north-south alignments that cross the entire spectrum of latitudinal climatic belts, from the Arctic Circle to the equator and from the equator to Antarctica. These climatic belts, as already noted, delineate many of the presently defined biogeographic provinces. The north-south arrangement of continents separated by large ocean basins allows cold-water polar currents to flow nearly unimpeded toward the equator, generating cooler climatic conditions on a global scale. In addition, large, deep oceanic basins separate the continents, forming formidable barriers to the dispersal not only of land animals and plants but also of shallow-water marine life.

The development of these biogeographic barriers is an important evolutionary mechanism, promoting increases in species diversity through the creation of new species by geographic isolation. The more effective these barriers are in preventing the interchange of species between provinces, the more genetically isolated faunas and floras in these provinces become, promoting greater differences between organisms living in these areas. As a result, similar environments in different geographic regions will support totally separate but often convergent faunas and floras. If these barriers did not exist, the same organisms would occupy the same environments in all geographic regions.

At present, nearly all the large continents are emergent, or above sea level. Their emergence resulted from a draw-down of worldwide sea level caused by relatively recent continental glaciation in the Northern Hemisphere (Pleistocene epoch, about 2 million to 12,000 years ago). The formation of this enormous mass of ice tied up much of the available surface water, which would otherwise have flowed into the ocean basins. Thus, sea level was depressed, and shallow seas were drained off most of the continents. Shallow epicontinental seas had previously covered portions of the continents and had had a climatically ameliorating effect, for bodies of water tend to hold heat longer than exposed land areas. Their disappearance contributed to the development of extreme climatic conditions. The development of emergent continents and intervening deep-water ocean basins also isolated shallow-marine faunas living in the narrow continental shelf regions ringing the continents. In this way, numerous marine biogeographic provinces have developed in these coastal areas, and the diversity of endemic shallow-marine faunas in these areas has increased.

PLATE TECTONICS AND SPECIES DIVERSITY

The lateral motions of the continents, because of plate tectonic activity, have been a powerful force in reshaping the distribution of biogeographic boundaries in time and space. According to the plate tectonic theory, the Earth's exterior is divided into an interlocking mosaic of moving plates composed of low-density crustal rocks and underlying, denser mantle rocks. These plates are created at spreading centers, where dense volcanic rocks well up from below and flow away laterally. Continental crust at these spreading centers is rifted apart, allowing for the formation of a new ocean basin dividing what was once a continuous continent. Such is the case in the Near East, with the formation of the Red Sea between the rifting African continent and the Arabian plate.

Plates can also move together, colliding with each other. At the point of impact, the crust shortens, thickens, and is deformed, leading to the formation of a new mountain belt that welds the colliding continents together into one larger continent. An example of this continent-to-continent collision is the collision between India and Eurasia, upthrusting the Himalaya. Plates also slide past each other along enormous linear cracks in the crust, termed strike-slip or transform faults. The San Andreas fault in California is such a fault, with California west of the fault moving north relative to the rest of North America. Such fault systems serve to displace continental crust horizontally along plate boundaries. All these plate tectonic activities cause the continents to change their shape and geographic position with time as well as result in the rise and fall of topographic features on the Earth's surface.

Plate tectonic reconstructions for the major continents indicate major geographic realignments of these continents over the past 600 million years. These reconstructions show a pattern of cyclic periods of plate divergence (plates moving away from each other) followed by plate convergence (plates converging and eventually colliding). Plate divergence causes the formation of large ocean basins

separating the continents. Convergent cycles culminate in the development of supercontinental masses composed of multiple, sutured continental plates. These "supercontinents" exist for several tens of millions of years, only to rift apart with the initiation of the next divergent cycle, which would fractionate the continents and disperse them over the globe. Detailed studies of the fossil organisms preserved in the sedimentary rocks on these continents indicate that the large-scale evolution of life over the past 600 million years parallels these plate tectonic events, with the changing geographic patterns having a profound effect on the evolution and extinction of organisms.

An example of this relationship is the Early Paleozoic interval of geologic time (Cambrian and Ordovician periods, about 600 to 450 million years ago). This was a time of plate divergence, with the continents aligned into rough east-west belts. The various continents that now make up the Northern Hemisphere were situated as isolated masses along the equator, in tropical latitudes. The present-day Southern Hemisphere continents were sutured into one large supercontinent, which scientists call Gondwanaland, located at higher, cool-temperate latitudes south of the equator. The continents were eroded down to flattened platforms and were flooded (more than perhaps 95 percent of their surface area) by shallow, epicontinental seas. Sea level was comparatively high because of the absence of continental collisions (which would elevate the continents through mountain building) and a lack of glaciation through much of the period. World climate was uniformly mild. The continents were separated by deep ocean basins and, as a result of the flooding of the continents by these epicontinental seas, shallow-marine environments were abundant.

This Cambro-Ordovician plate tectonic scenario coincides with a seemingly explosive evolution of shallow-marine animal life, especially the major groups of shell-secreting invertebrates (corals, brachiopods, snails, clams, nautiloids, trilobites, and echinoderms). These shallow marine faunas were highly provincial for much of this time interval because of the divergent nature of the major continental masses, with the intervening deep-water oceanic basins forming formidable barriers to the dispersal of these shallow-water marine animals. The highly endemic nature of these faunas and the mild climatic conditions promoted high levels of diversity. Terrestrial floras and faunas are unknown for this time interval, with the earliest land plants and animals being known from the succeeding Silurian period (about 425 million years ago).

Robert C. Frey

CROSS-REFERENCES

Archaebacteria, 951; Biostratigraphy, 1091; Cambrian Diversification of Life, 956; Catastrophism, 961; Cenozoic Era, 1095; Colonization of the Land, 966; Dinosaurs, 970; Dolomite, 1567; Ediacaran Fossils, 976; Eukaryotes, 981; Evolution of Birds, 985; Evolution of Flight, 990; Evolution of Humans, 994; Evolution of Life, 999; Fossil Plants, 1004; Fossil Record, 1009; Fossilization and Taphonomy, 1015; Fossils of the Earliest Life-Forms, 1019; Fossils of the Earliest Vertebrates, 1024; Geoarchaeology, 1028; Life's Origins, 1032; Mammals, 1038; Mass Extinctions, 1043; Mesozoic Era, 1120; Microfossils, 1048; Neanderthals, 1053; Paleobotany, 1062; Paleozoic Era, 1126; Petrified Wood, 1058; Prokaryotes, 1071; Stromatolites, 1075; Systematics and Classification, 1079.

BIBLIOGRAPHY

Briggs, Derek E. G., and Peter R. Crowther, eds. *Paleobiology: A Synthesis.* Oxford: Blackwell Scientific Publications, 1990. This multiauthored text deals with all aspects of paleontology in a series of short articles by experts in each field. Intended for university students but also useful to high school students and interested readers.

Dott, Robert H., Jr., and Donald R. Prothero. *Evolution of the Earth.* 5th ed. New York: McGraw-Hill, 1994. This basic textbook on historical geology is aimed at students of geology. However, it is very readable by anyone with a background in science. Presents an up-to-date account of the Earth's history from the viewpoint of plate tectonics. Includes a glossary.

Hallam, A., ed. *Atlas of Palaeobiogeography.* New York: Elsevier, 1973. One of the first volumes to be devoted exclusively to paleobiogeography and to incorporate plate tectonic information. Includes chapters by individual experts detailing the paleobiogeography of individual groups of fossil organisms for specific geologic periods. An important source of paleobiogeographic information. Most articles are well written and illustrated. Suitable for the informed layperson and college-level students with a basic knowledge of paleontology.

Press, Frank, and Raymond Siever. *Understanding Earth.* 2d ed. New York: W. H. Freeman, 1998. This comprehensive physical geology text covers the formation and development of the Earth. Readable by high school students, as well as by general readers. Includes an index and a glossary of terms.

Prothero, Donald R. *Bringing Fossils to Life.* Boston: McGraw-Hill, 1998. This well-illustrated and entertaining text covers a broad range of paleontological topics, including biogeography. Glossary, bibliography, and index.

Stanley, Steven N. *Earth and Life Through Time.* San Francisco: W. H. Freeman, 1986. A college-level historical geology textbook, this volume contains information on many aspects of plate tectonics, paleogeography, and evolution. Plate arrangements and the distribution of fossil organisms and environments are well illustrated for each geologic period.

PALEOBOTANY

Only a small percentage of the plants that ever lived left a record of their existence. Others survive as fossils: mineralized wood, flowers in amber, leaf imprints in coal, or other indicators of life in an earlier era. Paleobotanists document this fossil record and use it to interpret the past evolution of plants.

PRINCIPAL TERMS

AMBER: hardened resin from prehistoric trees; some amber contains intact, preserved organisms

FOSSILIZATION: the gradual process by which the remains of a once-living organism are naturally preserved, usually in sedimentary rock

MICROORGANISM: a very small, usually one-celled, organism such as algae, bacteria, and protozoans

ORGANISM: any particular living thing, whether a microorganism, plant, or animal

SEDIMENTARY ROCK: a rock that forms from sediment deposited on the bottom of lakes and the ocean

TAXONOMY: the branch of biology concerned with naming and classifying organisms

VASCULAR PLANT: a plant with vascular conducting tissues

IMPORTANCE OF PLANT FOSSILS

Paleontology (or paleobiology) is the science concerned with fossils, the physical evidence of prehistoric life—including plants, animals, and microorganisms—on the Earth. Paleobotany focuses on plant fossils, including algae, fungi, and related organisms, as well as mosses, ferns, and seed plants. As most organisms decompose rapidly after death, their preservation in nature is a rare event. Most individuals are not represented in the fossil record, and even many species that must have existed have vanished without a trace.

As a branch of botany, paleobotany is of importance primarily because the record of fossil plants helps scientists understand the long process of plant evolution. Especially since the 1940's, fossil evidence has helped to explain the origin of major groups of lower plants, such as algae and fungi. Researchers now also have evidence for the origin of the earliest vascular plants and the formation of reproductive structures, such as cones of gymnosperms (evergreen trees and relatives) and flowers of angiosperms (flowering plants).

The location of fossils, including both their temporal (age) and their spatial (geographical) arrangement, is used to determine past climates. The climates of the world have changed continuously as continents have shifted over the Earth's surface. For example, the location of coal deposits (which are the remains of giant tree ferns) in what is now Pennsylvania indicate the warmer climate that must have existed then.

Although perhaps most of the contributions to paleobotany have been made by professionally trained scientists with a solid background in geology, botany, and related sciences, amateurs have also made significant discoveries. Many valuable specimens of university and museum collections were made by people interested in paleobotany as a hobby.

HOW FOSSILS FORM

As already pointed out, the formation of a fossil is an exceptional event, one that requires a special combination of favorable environmental conditions. The most common scenario is a fossilization process in which the plant becomes covered by a soft sediment, which then hardens to form a sedimentary rock. This type of rock forms gradually over long periods of time as particles produced by erosion are compacted on the bottom of the body of water.

The large-scale process by which plant parts become impregnated with minerals produces what has traditionally been called "petrified wood." The modern term for this process is "permineralization." Soluble carbonates, silicates, and other compounds infiltrate plant cells and the spaces between them. Eventually, the mineral deposits may completely replace the naturally occurring organic matter, preserving the details of the plant's microscopic architecture. Well known are the petrified forests of western United States, many of which are protected within national parks, such as Petrified Forest National Park in Arizona.

Being trapped in a sedimentary rock does not automatically guarantee that the organism will be preserved. The environment must be an anaerobic one—that is, one in which oxygen is excluded—thus preventing the decay that would otherwise result. The process may be interrupted by the action of waves or other erosive forces which re-expose the developing fossil before the process of fossilization is completed. Even after the process is completed, the well-preserved specimen may become distorted or altered in appearance because of the combined effects of time, pressure, and high temperatures that convert sedimentary rocks into metamorphic ones.

As one would expect, the harder cells and tissues of plants are more likely to be preserved as fossils than are softer ones. For example, the thick-walled cells of wood and bark (called xylem) are more often preserved than are those of the pith (center of a stem) or cortex (found in stems and roots beneath the bark or outer covering). Other cells that are often fossilized are pollen grains and spores, both of which have outer shells that are highly resistant to decay.

Limestone and dolomite are among the most common types of rocks that form in such a way that they trap plants and form fossils. Coal, a combustible sedimentary rock, is formed in much the same way as others but is distinctive because the sediment involved is of plant rather than mineral origin. Within this matrix of plant-derived material is often embedded a variety of plant parts.

SPECIAL TYPES OF FOSSILS

Two special kinds of rock that may contain plant and animal fossils are diatomite and amber. Diatomite is a rock that forms from the silica cell walls of a group of unicellular algae known as diatoms. Since silica is the same material that sand and quartz are composed of, it is unusually permanent. Diatoms are found in both fresh and salt water in great numbers and diversity. When they die, their cell walls accumulate on the bottom of the

water and become compacted over time into diatomite. The rock, itself formed by fossilization, may have fossil remains of various kinds of plants and animals preserved within it.

Amber, considered a semiprecious stone by gemologists and valued because of its beauty and distinctive appearance, is also of interest to paleobotanists. Amber is basically the fossilized resin produced by ancient cone-bearing evergreen trees. Sticky resins oozed from trees in response to injuries. Before such resins harden, various small animals, floral parts, pollen grains, fungal spores, and other plant parts may become trapped and be preserved intact. Deposits of amber valued for their use as jewelry and as fossils are recovered from two world areas: the Baltic region of northern Europe and the Dominican Republic in the Caribbean Sea.

Paleobotanists are sometimes challenged by puzzling specimens. Outright fakes are sometimes presented by pranksters, but more common are various mineral structures that bear a superficial resemblance to a plant. Such specimens are called pseudofossils. Mineral deposits called dendrites found in rock crevices bear a resemblance to fern leaves. A coprolite (fossilized feces) from the upper Cretaceous of Alabama was initially mistaken for the cone of a conifer (cone-bearing evergreen tree); these specimens may be referred to as pseudo-plant fossils, since they are true fossils of animals. During the formation of flint, bands are sometimes formed that suggest fossil mollusks or coral. Suspicious specimens require careful analysis by a specialist. In general, plants and animals, and therefore their remains, possess details and a characteristic regularity of form absent in pseudofossils.

NAMING AND CLASSIFYING FOSSILS

In order to prevent confusion, fossils, like living species, need to be named and classified in a consistent, systematic fashion recognized by paleobiologists throughout the world. The branch of biology devoted to the naming and classification of organisms is called taxonomy. Fortunately, the same system is applied to both living and fossilized plants.

According to binomial nomenclature (naming), each species is given a scientific name consisting of two parts: the genus name followed by the trivial, or species, name. The former is capitalized,

The ginkgo tree is considered a "living fossil," persisting today as a commonly planted shade tree harvested for the medicinal properties of extracts from its leaves. (© William E. Ferguson)

cause they are (or were) vascular plants, are assigned to the Division Tracheophyta. Within that division, the redwood and ginkgo, because they produce uncovered seeds, are placed into the Class Gymnospermopsida (naked seeded plants), whereas the palm, a flowering plant, is assigned to the Class Angiospermophytina.

The plant fossil record is often used to establish natural relationships among various extant plant species and other taxa (taxonomic categories) at higher levels. This is especially true of the vascular plants. In fact, the division Tracheophyta was established by A. J. Eames in 1936 to show the natural relationship between seed plants and ferns. The basis for this new category was the discovery, earlier in the twentieth century, of Devonian fossils of a group of primitive vascular plants known as "psilophytes." They were recognized as ancestral to both ferns and seed plants. Previously, ferns and seed plants had been assigned to a separate division of the plant kingdom.

As the plant fossil record becomes more complete, further revision of the classification system becomes necessary to allow the system to more nearly reflect the true or natural relationships among the various categories. This is, at least, the common goal of both paleobotanists and those who study modern plants.

Thomas E. Hemmerly

CROSS-REFERENCES

Archaebacteria, 951; Cambrian Diversification of Life, 956; Catastrophism, 961; Colonization of the Land, 966; Dinosaurs, 970; Ediacaran Fossils, 976; Eukaryotes, 981; Evolution of Birds, 985; Evolution of Flight, 990; Evolution of Humans, 994; Evolution of Life, 999; Fossil Plants, 1004; Fossil Record, 1009; Fossilization and Taphonomy, 1015; Fossils of the Earliest Life-Forms, 1019; Fossils of the Earliest Vertebrates, 1024; Geoarchaeology, 1028; Geologic Time Scale, 1105; Ice Ages, 1111; Life's Origins, 1032; Mammals, 1038; Mass Extinctions, 1043; Microfossils, 1048; Neanderthals, 1053; Paleobiogeography, 1058; Petrified Wood, 1058; Prokaryotes, 1071; Stromatolites, 1075; Systematics and Classification, 1079.

whereas the latter is written in lower case; both are italicized. Often a name follows that belongs to the person who assigned that name. As example, the scientific name of an extinct redwood tree is *Sequoia dakotensis* Brown, while that of a stemless palm is *Nipa burtinii* Brongniart. The ginkgo tree, *Ginkgo biloba L.*, is considered a "living fossil." Known from the fossil record, it persists as a commonly planted shade tree. The initial following its scientific name is that of Carolus Linnaeus, the Swedish botanist who established this binomial system of nomenclature in 1753.

Species, living or dead, are classified using a hierarchical system (also by Linnaeus), which reflects degrees of similarity or dissimilarity to other species. All three trees already mentioned, be-

BIBLIOGRAPHY

Bengtson, Stefan, ed. *Early Life on Earth.* New York: Columbia University Press, 1994. A collection of serious articles that relate early biological evolution to conditions on the primitive Earth. Perhaps of most interest is the article written by Stephen Jay Gould, "Introduction: the Coherence of History."

Cvancara, Alan M. *Sleuthing Fossils: The Art of Investigating Past Life.* New York: John Wiley & Sons, 1990. This interesting introduction to fossils is written in the first person for the amateur. A good place for beginners to start.

Golding, Roland. *Fossils in the Field.* New York: John Wiley & Sons, 1991. In lab manual format, this book is invaluable for giving specific directions for field and laboratory methods. Written for the more advanced amateur. Numerous interesting and instructive drawings.

Roberts, D. C., and W. G. Hodson. *A Field Guide to Geology: Eastern North America.* Boston: Houghton Mifflin, 1996. Like other Peterson Field Guides, this is a small volume that is useful both in the field and as a reference.

Stewart, W. N., and G. W. Rothwell. *Paleobotany and the Evolution of Plants.* 2d ed. Cambridge: Cambridge University Press, 1993. Probably the most comprehensive and authoritative textbook on the subject. Of special interest to the beginner are chapters 1 through 3, which give the scientific foundation for paleobotany.

Walker, Cyril, and David J. Ward. *Fossils.* New York: Dorling Kindersley, 1992. The nearly three hundred pages of full-color illustrations are the most striking features of this very useful book. The introduction contains an excellent orientation to fossil study.

PETRIFIED WOOD

Petrified wood is the fossilized remains of ancient forests that were saturated with mineral-filled water, which converted the woody tissues into stone. Such preserved trees are studied by scientists interested in learning about prehistoric plants and their environments.

PRINCIPAL TERMS

MIOCENE: geological epoch of the Tertiary period in the Cenozoic era, beginning about 26 million years ago

OLIGOCENE: geological epoch about 38 million years ago in the Tertiary period of the Cenozoic era

PLIOCENE: geological epoch in Tertiary period of the Cenozoic era that began about 12 million years ago

PTERIDOPHYTE: plants, such as ferns, that have leaves, stems, and roots but no seeds or flowers

SILICIFICATION: a process in which molecules of silica that are dissolved in groundwater replace the organic material in fallen tree trunks and branches

STRATA: layer of Earth where petrified fossils are located according to the time period in which the organism lived

TRIASSIC: period of time about 225 to 195 million years ago at the beginning of the Mesozoic era, when dinosaurs lived

XYLEM: plant cells through which water and nutrients move

PETRIFYING ANCIENT FORESTS

The Latin word *petros*, meaning rock, is the source for the scientific term "petrification." Petrified wood is actually stone remnants from prehistoric trees. During the Triassic era, gymnosperms—seed-producing trees without flowers, such as gingkos and conifers—grew over much of the Earth's landmass. Volcanic eruptions triggered tremors, lightning, and heavy rains, which washed trees from higher elevations down to swampy valleys. As they were pushed downhill, the trees were stripped of their bark, branches, and roots from the force of the water's impact, and they also broke into pieces. Usually, trees soaking in deep, muddy water would decay, but silt rapidly and completely covered these trees, preventing exposure to oxygen and inhibiting aerobic decomposition. Volcanic ash in the floodwater consisted of inorganic compounds such as magnesium carbonate and iron sulphide, and the trees also absorbed silicon dioxide (called silica) that had dissolved in groundwater. The minerals filled the spaces between cells in the tree trunks and branches. Molecules of these inorganic materials replaced molecules of organic tissues. During the next millions of years, wood gradually became stone in the process of silicification. Assisted by extreme pressure and temperatures, the silica that was lodged in the wood was transformed into quartz. Plants that have undergone petrification are also referred to as being permineralized.

The trees remained preserved under the soil for millions of years until soil erosion and shifting plates exposed them. Manganese, lithium, copper, and iron created patterns of bright colors as wood fossilized. Some petrified wood displays varying rings of vivid colors, resembling agates. Other pieces are brown and look like driftwood. More significant than its beauty, petrified wood reveals information about the history of plants on the Earth. Unlike other fossils that are seen as an impression or compression, petrified wood is a three-dimensional representation of its organic material that preserves its external shape and internal structure. The preserved tree trunk sections also indicate the size of the Triassic forest. Scientists have even seen chromosomes and stages of nuclear division in petrified cells. A termite nest was discovered in one petrified log, offering clues about that insect's communal evolution.

IDENTIFYING PETRIFIED WOOD

Geologists and paleobotanists analyze petrified wood samples to specify the type of ancient tree that became fossilized. A piece of petrified wood from a hardwood tree is carved into a small cube approximately 2.5 centimeters square. The cell structure of hardwood fossils is more diverse than that of softwood trees, causing its source to be more easily identified. Scientists choose pieces that exhibit an intact cell structure. Using a saw and grinding wheel as needed, they flatten the cube's surfaces and sand them to preserve the natural rings of the tree and to remove any marks made while shaping the cube. Researchers then examine the specimen with a microscope at ten times, forty times, and one hundred times magnification to scrutinize how the wood's bands and pores are arranged, both of which are crucial identifying characteristics. Softwood samples require greater magnification (four hundred times) of thin slivers only one or two cells thick in order to permit enough light to shine through when mounted on a slide.

Wood anatomists then describe the sample's cell structure, which they compare to records of previously identified petrified wood and existing trees. North Carolina State University and the International Association of Wood Anatomists created a computer program known as the General Unknown Entry and Search System (GUESS), which efficiently checks its databases for matching cell patterns. In addition to databases that contain information about existing hardwood and softwood trees, another database for fossil hardwoods has information on at least 1,356 types of trees of more than 1,200 species that have been identified. Codes direct researchers to scientific journal citations for detailed illustrations.

Early identification of petrified wood relied on seeking similarities with existing trees even though the petrified sample may have represented an extinct tree type. Because most petrified wood lacks fruit or leaves to aid categorization, paleobotanists had to focus on unique aspects of cell structure to distinguish fossil trees. As petrified wood samples were described in journals, scientists developed models for comparison to determine the geological age and geographical source of fossil wood. Identification remained difficult, however, because the cell structures of some trees resembled those of other species more than trees in their own families.

Three types of petrified wood are found in the Tertiary strata: nondescript silicified wood, petrified palm wood, and massive silicified wood. Nondescript silicified wood has undergone silicification but still appears to be woody structurally. Difficult to identify because of its generic structure, nondescript silicified wood requires expert authentication for accurate labeling. Petrified palm wood has rod structures that reinforced the tissue strength before the wood grain became silicified and that look like spots or lines when the wood is cut. Popular among rock collectors, this type of petrified wood is also Arizona's state fossil. Massive silicified wood is difficult to recognize because the tree's grain was destroyed during silicification, thus making identification reliant on awareness of the area in which the tree was located and comparison to other petrified wood in adjacent territory.

PETRIFIED WOOD IN THE UNITED STATES

Arizona's Petrified Forest National Park is one of the best-known areas with petrified wood. This wood originated from a forest of giant conifer-like trees that grew from Texas to Utah. Located in northeastern Arizona, the park's 93,533 acres are home to one of the largest and most varied assortment of petrified wood in the world. The desert is dotted with stone log fragments. Although some visitors expect to see rock trees standing in clumps similar to a natural forest, these petrified trees rest where they fell individually or in groups.

The park features four primary petrification zones. At Blue Mesa, petrified logs have emerged from an area of blue- and gray-colored clay. The Jasper Forest consists of a valley scattered with petrified wood. At one time it was covered with logs, but nineteenth century prospectors collected the stones to sell. Nearby, a petrified tree trunk called Agate Bridge crosses a stream. The Rainbow Forest contains the largest amount of petrified wood in the national park, as well as some of the largest logs, some of which are several meters long. The Agate House is a hut that sixteenth century Native Americans constructed completely from petrified wood. The Crystal Forest's petrified wood represents what was once a vast landscape of logs sparkling with quartz and purple amethyst crystals. By the twentieth century, prospectors and tourists had taken the most beautiful pieces; as a result, local residents sought government protection against further theft. In 1906, the petrified wood site was declared a national monument, finally achieving national park status in 1962.

The National Park Service tries to protect petrified wood by preventing the nearly 1 million annual visitors from seizing samples. Rangers patrol sites and ask tourists to report any thefts they witness. Despite these precautions, several tons of petrified wood disappear annually from the Arizona park. As a deterrent, local superstitions have been encouraged to warn people that illegally acquired fossils will bring them bad luck. Privately owned sites adjacent to the park offer collectors opportunities to search for petrified wood without any restrictions. Stores sell small pieces that have been artistically cut and polished at affordable prices, which people use to decorate their homes, gardens, and even aquariums. Petrified wood is both Arizona's state fossil and its state rock.

Because the ancient forest lived simultaneously

A petrified tree trunk, c. 1890, on Specimen Ridge, Yellowstone National Park. (U.S. Geological Survey)

with the dinosaurs, archaeologists look for dinosaur fossils near petrified wood in the Arizona park. Although collection and thefts have greatly reduced the number of petrified logs, authorities believe that some areas of the park may shelter petrified wood buried as much as 100 meters beneath the surface.

Petrified wood has been discovered in many other regions of the United States, especially in areas in the West where volcanic activity occurred, such as Yellowstone National Park, and in areas where rivers and streams deposited large amounts of sand, such as Louisiana and Texas. Washington State is home to the Ginkgo Petrified Forest State Park, which contains petrified logs that began fossilizing during the Miocene epoch. This petrified wood is unique because it includes petrified ginkgo, an indigenous tree that no longer grows naturally there. An unusual type of petrified wood that resembles pebbles and that originated in the Chehalis Valley is sometimes seen on the state's beaches.

The Calistoga, California, petrified forest is considered one of the best sources of Pliocene fossils similar to existing redwoods. Measuring more than 2 meters in diameter, the fossil logs reveal grey stone veins of quartz. Petrified wood in New Mexico's Bisti Badlands is not as colorful as neighboring Arizona's fossil wood. In Utah, petrified wood has been found near the Escalante River and the Coyote Buttes region near the Paria River. Petrified wood has also been located in the southern states of Mississippi and Alabama.

Petrified Wood Worldwide

Scientists have also found petrified forests in other parts of the world. A petrified forest on the Greek island of Lesvos was named a protected site by presidential decree. A museum was established to encourage scientific research about this Greek forest's origins. Scientists have determined that the petrified wood began fossilizing during the Late Oligocene to Lower-Middle Miocene epochs. Unlike other sites of petrified forests, the Lesvos fossil trunks are erect and still have intact roots penetrating into fossilized soil, branches, leaves, cones, and seeds.

Scientists realize that these trees were fossilized where they grew, offering insight about the environment and climate of ancient Lesvos. The ancient trees of Lesvos were well preserved during the petrification process, and such details as rings indicating age and growth patterns are visible. Paleobotanists have determined that both gymnosperms and angiosperms (flowering trees) grew in the ancient forests, along with pteridophytes such as ferns. Many of these ancient species no longer grow in the Mediterranean. Instead, they are found in Asian and American tropical and subtropical regions, indicating that Lesvos's petrified forest presents information about how the Earth evolved as the continents moved apart.

Additional sites of fossilized wood also reveal details about the planet's development. Argentina's Petrified Forest on the Central Steppes was created after the formation of the Andes Mountains. Larger than American wood fossils, Argentinean petrified wood includes pieces 27 meters long and 3 meters in diameter. The petrified forest in Namibia on the African continent contains giant tree trunks as long as 30 meters. Petrified wood has also been discovered in Australia, India, England, Turkey, and Switzerland.

Petrified wood of a conifer, showing annual rings, from the Eocene epoch. (© William E. Ferguson)

PETRIFIED WOOD'S LEGACY

Paleobotanists research petrified wood to determine how the Earth and the plants that have grown on its surface have changed since ancient geological periods. They study how plants are related and descended from similar ancestors. In addition to compiling lists and descriptions of thousands of modern tree species, scientists have noted existing trees that are represented by petrified wood and have cataloged types of ancient trees identified solely by fossil samples. Paleobotanists realize that during the Earth's existence many more tree species have existed than are documented and that many species, existing and extinct, await discovery. They note that pieces of fossil wood are often the oldest known specimens of a tree species and might be the predecessors of living trees.

Petrified wood helps researchers comprehend how trees have evolved to adapt to environmental and climatic conditions. For example, researchers have hypothesized that the cell structure of tree xylem has not changed as much as fruit and leaves. Fossil wood also sometimes reveals the reason for the demise of extinct trees. The study of petrified wood has altered how scientists perceive both ancient geological ecosystems and modern environments by offering perspective on which plants lived on the Earth and survived, adapted, or died according to changes in the atmosphere and crust.

Elizabeth D. Schafer

CROSS-REFERENCES

Archaebacteria, 951; Biostratigraphy, 1091; Cambrian Diversification of Life, 956; Catastrophism, 961; Colonization of the Land, 966; Dinosaurs, 970; Ediacaran Fossils, 976; Eukaryotes, 981; Evolution of Birds, 985; Evolution of Flight, 990; Evolution of Humans, 994; Evolution of Life, 999; Fossil Plants, 1004; Fossil Record, 1009; Fossilization and Taphonomy, 1015; Fossils of the Earliest Life-Forms, 1019; Fossils of the Earliest Vertebrates, 1024; Geoarchaeology, 1028; Gondwanaland and Laurasia, 599; Ice Ages, 1111; Life's Origins, 1032; Mammals, 1038; Mass Extinctions, 1043; Microfossils, 1048; Neanderthals, 1053; Paleobiogeography, 1058; Paleobotany, 1062; Paleoclimatology, 1131; Prokaryotes, 1071; Stromatolites, 1075; Systematics and Classification, 1079.

BIBLIOGRAPHY

Agashe, Shripad N. *Paleobotany: Plants of the Past, Their Evolution, Paleoenvironment, and Application in Exploration of Fossil Fuels.* Enfield, N.J.: Science, 1997. A discussion of how ancient trees became petrified and how they are beneficial to environmental research. Illustrated with photographs of fossil chromosomes and diagrams of cell structure. Also includes an index and bibliographical notes.

Andrews, Henry Nathaniel. *The Fossil Hunters: In Search of Ancient Plants.* Ithaca, N.Y.: Cornell University Press, 1980. Contains biographical information about paleobotanists who have studied petrified wood. Illustrated and indexed, with bibliographical references. Useful reference for high school students wishing to supplement more technical discussions.

Daniels, Frank J., Richard D. Dayvault, and Brooks B. Britt, eds. *Petrified Wood: The World of Fossilized Wood, Cones, Ferns, and Cycads.* Grand Junction, Colo.: Western Colorado, 1998. This elaborately illustrated book covers petrified wood found around the world. The text explains how wood petrifies, its scientific applications to paleobotany and geology, and how to collect petrified wood. Contains an index and bibliographical sources. Suitable for readers of all ages.

Lubick, George M. *Petrified Forest National Park: A Wilderness Bound in Time.* Tucson: University of Arizona Press, 1996. Lubick discusses the scientists and custodians who have studied and protected fossil wood in Arizona's Petrified Forest National Park. Illustrations, maps, index, and bibliography. Suitable for high school readers.

Prothero, Donald R. *Bringing Fossils to Life: An Introduction to Paleobiology.* Boston: McGraw-Hill, 1998. Prothero explains how fossil wood offers clues to the evolution of existing trees from ancient species. Includes photographs and drawings, bibliographical sources, and index. An excellent reference for students.

Stewart, Wilson N., and Gar W. Rothwell. *Paleobotany and the Evolution of Plants.* 2d ed. New York: Cambridge University Press, 1993. This text provides fundamental information about the science of studying plant fossils to identify species past and present and to determine how they are related. Bibliographical references and index.

PROKARYOTES

Prokaryotes are primitive, one-celled organisms that have left an extensive fossil record in the form of sedimentary structures produced by physiological activity of cell communities. For 80 percent of the Earth's history, communities of prokaryotes made up the biosphere of the Earth. They are a well-defined group of organisms and occupy a highly diverse variety of habitats.

PRINCIPAL TERMS

AEROBES: prokaryotes (usually bacteria) that live in the presence of elemental oxygen

ANAEROBES: prokaryotes that can live only in an atmosphere that is free of elemental oxygen

DEOXYRIBONUCLEIC ACID (DNA): a molecule made up of two strands of nucleotides arranged in a double helix; the molecular basis of heredity

GRAZING and CROPPING EUKARYOTES: single-celled (protists) or multicelled (metazoans) eukaryotes (cells with a definite nucleus), which appear in the fossil record during the close of the Precambrian eon, about 1 billion years ago

NUCLEOTIDE: a molecule made up of a series of amino acids that, when linked together, are capable of carrying genetic information

PRECAMBRIAN EON: the first 3.5 billion years of the geologic record; it is followed by the Paleozoic era of the Phanerozoic eon

RIBONUCLEIC ACID (RNA): a complex compound made up of nucleotide bases that acts as a template, or "messenger," in the replication of DNA

RIBOSOME: a large multienzyme found associated with cell nuclear materials, composed of RNA and protein molecules

CHARACTERISTICS OF PROKARYOTES

From about 3.5 to 1 billion years ago, life on Earth, as determined from the fossil record, consisted entirely of one-celled organisms that have a cell morphology and a metabolism different from those of all other life-forms. These organisms, the prokaryotes, are characterized by their lack of a cell nucleus, their lack of sexual reproduction (meiosis), the small size of the prokaryotic cell, and their distinctive biochemistry. Prokaryotes are neither plants nor animals, although the aerobic photosynthetic forms, often called blue-green algae, have in the past been placed with the plants. Eukaryotes, organisms with a cell nucleus and a larger, more complex cell, make up the animals, plants, fungi, and protists. Prokaryotes are thus quite separate from all other life-forms in terms of their cell biology. Prokaryotes and eukaryotes are the two most basic categories of living things, exhibiting basic differences in their biologic processes that are greater than those that exist between animals or plants or between any of the other kingdoms.

Prokaryotes are mainly single-celled organisms, usually found living together in "colonies" consisting of immense numbers of cells. Their deoxyribonucleic acid (DNA) is distributed throughout the cell, not, as in the case of the eukaryotes, localized in a cell nucleus surrounded by a nuclear membrane. The prokaryotic cell is smaller by a factor of ten than the average eukaryotic cell. It lacks chloroplasts and mitochondria and consequently is considered primitive when compared with the eukaryotic cell.

Prokaryotes constitute the kingdom Monera, one of five kingdoms in modern taxonomy. (The other kingdoms are the protists, fungi, animals, and plants, all of which have more complex eukaryotic-type cells.) Phyla, or categories, within the kingdom Monera include the bacteria, the cyanobacteria (or blue-green algae), the archaebacteria, and the prochlorophytes. The bacteria, as well as the other moneran phyla, are further subdivided into a number of classes. Bacterial classes of the Monera include the eubacteria, photosynthetic bacteria, myxobacteria (slime bacteria), actinomycetes (moldlike bacteria), and other groups, each characterized by its own distinctive metabolism and biochemistry. The bacteria consist of obligate or strict anaerobes and facultative anaerobes; the former include the photosynthetic bacteria, which differ from the cyanobacteria not only in their ability to function, if required, under anaerobic conditions and low light levels but also in their different photosynthetic pigment.

The archaebacteria are considered by some to be the most primitive and ancient of the monerans. Archaebacteria have a number of biochemical and metabolic characteristics that allow them to live under very adverse conditions—conditions such as those that appear to have existed during the early history of the Earth. The archaebacteria are defined from their ribosomal ribonucleic acid (RNA), which in sequencing is quite different from that of all other monerans. The archaebacteria differ fundamentally from the other bacteria classes in structural and biochemical aspects as well.

GEOLOGICAL SIGNIFICANCE

Fossil prokaryote cells of great antiquity have become widely known from the fossil record. They were first reported in the 1910's by C. D. Walcott from 1.5-billion-year-old strata of western Montana (Belt series); however, the authenticity of these fossils was doubted until the discovery, in 1954, of one of the oldest known paleontological "windows" on life of the past, the 2-billion-year-old Gunflint biota. Since then, many occurrences of prokaryote cell fossils have been reported, most from very fine textured flinty cherts associated with stromatolites of the Proterozoic (latter part of the Precambrian) eon and dating from as far back as 3.5 billion years.

The geologic significance of the prokaryotes is great: Not only do they (at present as well as in the geologic past) play an important part in the recycling of many chemical elements, but they also have a role in basic geologic processes such as weathering and other alteration of rocks. For example, prokaryotes are involved in the formation of stromatolites. Stromatolites are layered organosedimentary structures, frequently found fossilized in rock strata of many different geologic ages. Stromatolites come in a considerable variety of shapes and sizes; the different types have often been given Linnaean biological names because, when originally discovered, they were thought to

be fossil organisms like corals or sponges. Most stromatolites are dome-shaped, finger-shaped, or laminar structures that have a characteristic "signature"; they can form significant parts of rock strata, particularly in limestone and dolomites. Stromatolites are found in rock strata as ancient as the Archean (former part of the Precambrian) eon and are particularly diverse and abundant in strata of the Proterozoic; locally, they can be quite common in early Paleozoic marine strata as well.

The origin of stromatolites was debated for many years; as late as the 1950's, many paleontologists seriously doubted their biogenic origin. This doubt stemmed, at least in part, from the fact that stromatolites occur so much further in the geologic past than do any other fossils. Through thousands and thousands of meters of Precambrian strata, they are the only fossils that can be found. Early workers on stromatolites, such as Walcott, suggested a cyanobacterial origin for them. The discovery of the well-preserved cells of prokaryotic type in digitate (fingerlike) stromatolites of the Gunflint Chert of Ontario in the 1950's led to a gradual acceptance by most geologists and paleontologists of the organic origin of the majority of stromatolites. It became clear that, under the right conditions, small, fragile cells could be preserved in very ancient strata.

During the 1970's and 1980's, studies on Precambrian stromatolites and the prokaryotic organisms responsible for them became widespread. Stromatolite occurrences going as far back as 3.5 billion years have been documented. These ancient stromatolites yield not only morphological information but also carbon isotope ratios indicative of a biogenic origin. They sometimes supply biochemical information in the form of hydrocarbons, amino acids, and porphyrins (the latter is apparently a degradation product of original photosynthetic pigment). In 1999, studies of Proterozoic rocks from western Australia confirmed the presence of chemicals that could only have been synthesized by cyanobacteria.

The morphology of a prokaryotic organism is simple. Unlike fossils of eukaryotic organisms, fossils of most prokaryotes provide little specific information about the actual living organism. Prokaryotic cells can be single coccoid (spherical) forms, or they can be elongate chains of cells, as with the filaments, or trichomes, of the cyanobacteria.

STUDY OF PROKARYOTES

A standard petrographic thin section mounted on a glass slide is the common mount for observing cells preserved in a stromatolite. Oil immersion is usually required if fossil prokaryote cells are to be observed. Thin slivers of a stromatolite can also be examined under oil immersion; however, the best results are generally with well-made thin sections. Often considerable trial and error is involved in finding stromatolites that preserve cells and then in actually locating those cells; different parts of a particular stromatolite specimen usually have varying degrees of cell preservation. Very fine grained sediments, such as those that occur with stromatolites preserved by black cherts or finely crystalline limestones, generally give the best results.

In this section under high optical magnification, a stromatolite may exhibit fossil cyanobacterial cells as either filaments or rod-shaped forms. If preservation of these small prokaryote cells is excellent, as in the stromatolites of the Gunflint formation, the biogenic origin of the cells will be clear, and distinct cell types can be observed. When most stromatolites are examined in thin section under high magnification, however, the biogenicity of the small objects seen is usually not so certain. Often, small black globules of carbon, suggestive of macerated cells, are evident, but their origin usually cannot be proved. Contaminants such as spores, pollen grains, bacteria cells, and fungi fragments can be a problem, particularly in examination of suspected fossiliferous rocks when thin sections are not used. Even with most thin sections, the unequivocal verification of a biogenic origin for fossil cells is rare. In the case of the Gunflint prokaryotes, the detail preserved in these fossil cells is highly remarkable; some of them show internal cell structure and cells in the process of division.

The earliest stromatolites that yield these fossil cells are generally either broad domes or laminar forms. Associated cells either are single-cell coccoid forms or consist of probable chains of photosynthetic bacteria. Chains of cells of filamentous cyanobacteria generally first appeared about 2.3 billion years ago, and this appearance of filaments agrees fairly well with the first appearance of branched or digitate stromatolites, for which filamentous cyanobacteria seem to be responsible.

CHEMICAL SIGNATURES

Often more significant than single-cell morphology or the megascopic morphology of a stromatolite is the chemical signature left by a group of prokaryotes as a consequence of their metabolic activity. Prokaryotes are classified according to their type of metabolism; some prokaryotes have a metabolism that enables them to occupy a wider variety of ecological niches than do eukaryotic organisms. Anaerobic and aerobic forms are the two fundamental forms of prokaryotic metabolism. In these two categories are the autotrophs and the heterotrophs; heterotrophic prokaryotes require previously formed organic material on which to live, while autotrophs do not. The autotrophs obtain their energy from their environment either in the form of sunlight (photoautotrophs) or through chemical reactions such as oxidation, as in the sulfur-oxidizing bacteria; such bacteria are called chemoautotrophs. This type of metabolism is unique in the organic world, for all other life-forms obtain their energy from photosynthesis or through utilization of the chemical energy contained in previously formed organic compounds. The cyanobacteria are photoautotrophs and are responsible for the formation of the various types of stromatolites. The process of photosynthesis changes the microenvironment around the photosynthesizing prokaryote; the mineral precipitation that results is responsible for the formation of stromatolite layers.

Some stromatolites contain oxidized manganese, cobalt, or other "transitional" elements, possibly incorporated into these fossil communities by oxidative metabolism of bacteria. Chemoautotrophic prokaryotes, which are various types of bacteria, may leave a chemical signature in the form of these oxides and precipitate their production of a layered stromatolite-like structure containing these oxidized metals. A number of bacteria oxidize manganese to higher oxidation states so that it is precipitated; deep-sea manganese nodules presently being formed are believed to have such an origin. Sectioning of these nodules shows a finely layered, stromatolite-like structure. Some of the heavy-metal-bearing stromatolites of the early Precambrian may reflect a similar chemoautotrophic metabolism. Analysis of organic residues present in many stromatolites in small quantities can sometimes shed light upon the specific organisms responsible for forming them. This technique, however, has met with only limited success, although degradation products of the photosynthetic pigment present in cyanobacteria have been identified, supporting the cyanobacterial origin of many ancient stromatolites.

The earliest stromatolites, those of Archean age (about 3.5 billion years old), exhibit certain distinctive morphological and chemical aspects. Some of these early stromatolites may be products of anaerobic assemblages of photosynthetic bacteria rather than of cyanobacteria communities. Geochemical evidence suggests that the atmosphere in the Archean may have been anoxygenic (oxygen-free) and that the photosynthetic bacteria, not being obligate aerobes, would have been favored by such an environment.

Bruce L. Stinchcomb

BIBLIOGRAPHY

Broadhead, T. W., ed. *Fossil Prokaryotes and Protists: Notes for a Short Course.* Knoxville: University of Tennessee, 1987. Information on a broad range of fossil prokaryotes and protist microfossils is presented. This work is the text of one of a series of "short courses" sponsored by the Paleontological Society, but it can be useful to anyone interested in the various fossil groups covered.

McMenamin, Mark. *Discovering the First Complex Life: The Garden of the Ediacara.* New York: Columbia University Press, 1998. This entertaining study of the earliest complex lifeforms on the planet details the author's work on these organisms. Written for the interested student but understandable by the general reader.

Margulis, Lynn, and Karlene V. Schwartz. *Five Kingdoms.* 2d ed. New York: W. H. Freeman, 1988. A concise, useful, and readable examination of five kingdoms: the prokaryotes, protists, fungi, plants, and animals. The prokaryotes are presented within a framework that easily permits one to compare their taxonomic diversity and evolutionary radiation with those of the other kingdoms.

Nisbet, Evan G. *The Young Earth: An Introduction to Archean Geology.* Boston: Unwin Hyman, 1987. A highly comprehensive coverage of geologic phenomena of the Earth's earliest geologic time span, the Archean eon. Included in this work is information on both crustal evolution and biosphere. The book's sections vary considerably in technical coverage and terminology, some parts being readily comprehensible to the lay reader while others are quite technical and require considerable background in trace element geochemistry, isotope geochemistry, and petrology.

Schopf, J. William, ed. *Major Events in the History of Life.* Boston: Jones and Bartlett, 1992. An excellent overview of the origin of life, the oldest fossils, and the early development of plants and animals. Written by specialists in each field but at a level that is suitable for high school students and undergraduates. Although technical language is used, most terms are defined in the glossary.

STROMATOLITES

Stromatolites are the most common megascopic fossils contained within ancient rocks, dating to 3.5 billion years in age. In both the living and fossil form, they are created by the trapping and binding of sediment particles and the precipitation of calcium carbonate to the sticky surface of matlike filaments grown on a daily cycle by blue-green algae. Modern stromatolites are found throughout the world; they are of particular use in the creation of hydrocarbon reservoirs, in geologic mapping, and as indicators of paleoenvironments.

PRINCIPAL TERMS

ALGAE: primitive, one-celled, chiefly aquatic plantlike organisms lacking stems, roots, and leaves

BLUE-GREEN ALGAE: any of the algae classified within the division *Cyanophyta*

CYANOBACTERIA: an algaelike bacteria commonly known as "blue-green algae"

FOSSIL: the remains, trace, or imprint of any plant or animal that lived during the geologic past

ONCOLITE: an organosedimentary rock structure, concentrically laminated, formed by blue-green algae and smaller in size than a stromatolite

ORGANOSEDIMENTARY STRUCTURE: a sedimentary rock feature developed by the life processes of blue-green algae

PROKARYOTE: an organism characterized by simple protoplasmic structure and lacking a nucleus

STROMATOLITE: an organosedimentary structure formed by the trapping, binding, or precipitation of sediment upon the laminae surface of blue-green algae

THROMBOLITE: a stromatolite-like fossil characterized by an obscurely clotted, unlaminated internal structure

STROMATOLITE-PRODUCING ALGAE

Stromatolites ("layered rock" in Greek) are organosedimentary structures associated with certain types of the sedimentary class of rock; they develop through the metabolic processes of plant microorganisms. Stromatolites can also be considered to be fossils; in this sense, they do not represent the remains of actual organisms but rather of material deposited and collected by a living organism in a manner such that the original size, shape, and morphology of the organism are preserved.

The organisms principally responsible for the development of stromatolites are photosynthetic cyanobacteria, commonly referred to as "blue-green algae." Structurally, blue-green algae are among the most primitive of this class of life, lacking true stems, roots, and leaves. They grow on a daily cycle, in response to the rising and setting of the Sun. In response to this cycle, a sticky, filamentous, organic surface is produced in the form of wavy, matlike laminae. This mucus-coated mat traps and binds fine and coarse-grained sediment

to its surface during daylight hours, increasing the thickness of the laminae. During evening hours and nighttime, new algae growth penetrates the sediment-coated layer, producing a new gelatinous filament that in turn will trap and bind more sediment during the following daylight period. As this process continues, a layered, or laminated, structure is built.

Blue-green algae are prokaryotic (prenuclear) microorganisms, as they possess neither a cell nucleus nor a specialized cellular organelle (that portion of a cell that functions as an organ). Blue-green algae are also asexual and thus restricted in degree of variability, resulting in many living species that are almost indistinguishable from species that lived more than 1 billion years ago. Because of this relative lack of evolution over geologic time, much of what is known regarding ancient stromatolites has been gained by the study of living stromatolites produced by blue-green algae. The mineral matter associated with extant stromatolites is primarily composed of calcium carbonate, although a few species are associated with siliceous material, as evidenced by stromatolites studied in Yellowstone National Park, Wyoming. In the fossil form, the original calcium carbonate content is sometimes altered through replacement by magnesium, iron, and silica.

Modern blue-green algae have a wide geographic distribution and grow in diverse aquatic environments, including salt water, fresh water, and even moist soils. Some blue-green species thrive in hot springs and geysers with temperatures only a few degrees below the boiling point, while others have been collected in Arctic and Antarctic regions. Regardless of their aquatic environment, all modern blue-green algae depend upon sunlight for growth and survival. Because of this need for light, it was formerly thought that marine varieties were restricted to shallow waters near shore. Today, however, it is known that the photic zone—that region of the ocean penetrated by sunlight—extends to depths approaching 150 meters, allowing growth of deep-water varieties.

Stromatolite-producing algae are found in aquatic environments that are generally hostile to grazing and burrowing invertebrates. Sharks Bay in western Australia, a classic site for the study of modern (living) stromatolites, is conducive to blue-green algal growth because the hypersaline environment limits the activity of grazing snails. In Yellowstone National Park, alkaline waters up to

59 degrees Celsius in temperature present an environment supporting the development of unique siliceous stromatolites produced by bacteria rather than by algae. In the shallow marine environment of the Bahama Banks in the Caribbean, blue-green algae are abundant in supertidal channels with current flow too strong for the effective colonization of stromatolitic-grazing invertebrates such as gastropods and ostracods.

DISTRIBUTION OF STROMATOLITES

In the geologic record, stromatolites are the most abundant of fossils found in rocks dating to the Precambrian era, that period of time from the origin of Earth (approximately 4.6 billion years ago) up to 544 million years ago. The oldest fossil stromatolites are contained in the 3.3-billion- to 3.5-billion-year-old Warrawoona group of rocks in Australia. Close in age are the 3.4-billion-year-old stromatolites of the Swaziland group of South Africa; somewhat less removed are those associated with the 2.5-billion- to 2.8-billion-year-old Bulawayan Limestone, also in Africa. All these examples originated in the Archean eon, the first recorded period of geologic history, extending by definition from approximately 4 billion to 2.5 billion years ago. During the Proterozoic eon, immediately following the Archean eon and extending to 544 million years ago, stromatolites became prolific. This Proterozoic expansion is probably reflective of the initial development of continental landmasses and associated warm, photic continental shelf regions, as plate tectonics became a controlling process in the early development of Earth's crust.

Throughout the Archean and Proterozoic eons, blue-green algae underwent a steady and progressive state of biologic evolution, recognized today as the singular, common megascopic fossil of the Precambrian time period (the Archean and the Proterozoic eons). For this reason, the Precambrian is often referred to as the "age of algae," or, more specifically, the "age of blue-

A stromatolite formation from Grand Canyon National Park, Coconino County, Arizona. (U.S. Geological Survey)

green algae." Throughout the Phanerozoic eon, defined as 544 million years ago to the present, blue-green algae underwent minimal evolution, probably because its evolutionary state had become adapted to a variety of environments, reducing the need for further diversification.

Stromatolitic-building algae maintained their dominance of the aquatic world during the Early Phanerozoic eon (544 million to about 460 million years ago). With the rather abrupt appearance, however, of shelled, grazing, and cropping invertebrates in the early Phanerozoic eon, blue-green algae began to decline in significance. Today, as compared to their Precambrian domination, they have, on a relative scale, become endangered.

Geographically, fossil stromatolites are ubiquitous on every continent, especially within sedimentary carbonate rock sequences older than 460 million years. On southeastern Newfoundland, stromatolites built by blue-green algae of the genus *Girvanella* are found in conglomerate and limestone strata of the Bonavista Formation (approximately 550 million years in age). In the Transvaal region of South Africa, delicately banded stromatolitic structures compose one of the most widespread of early Proterozoic shallow-water carbonate deposits in the world, extending over an area exceeding 100,000 square kilometers. In nearby Zambia, algal stromatolites are closely associated with rock sequences containing economic levels of copper and cobalt. Upper Permian (250 million years ago) stromatolite horizons can be traced over an area of northern Poland exceeding 15,000 square kilometers. Miocene age (15 million years old) algae of the species *Halimeda* compose the limestone-forming rocks of the island of Saipan in the Mariana Islands of the Pacific Ocean. In North America, fossil algal-bearing rocks include the 2-billion-year-old Gunflint (Iron) Formation of Ontario, Canada, and the well-developed stromatolitic horizons of Early Paleozoic era age (450 million years ago) composing the Ellenberger Formation of Oklahoma and Texas.

CLASSIFICATION AND IDENTIFICATION

The classification and identification of stromatolites are often concluded on the basis of overall morphology, particularly the size, shape, and internal construction of the specimen. The relevant literature makes use of a variety of morphological terms, including the adjectives "frondose" (leaflike), "encrusting," "massive," "undulatory" (wavelike), "columnar," "laminar," "domed," "elliptical," and "digitigrade" (divided into fingerlike parts). Through the study of modern blue-green algae, it is suggested that three environmental criteria are of importance in stromatolite geometry development. These are direction and intensity of sunlight, direction and magnitude of water current, and direction of sediment transport. As an example, the extant elliptical stromatolites of Shark Bay, western Australia, are oriented at right angles to the shoreline as the result of strong current-driven wave and scour action. Under certain environmental conditions, cyanobacteria growth surfaces are not preserved, producing fossil algal structures characterized by a lack of laminae. These structures are termed "thrombolites," in contrast to laminar-constructed stromatolites.

While stromatolites are generally described as megascopic in size, discussion of specific dimensions relates both to laminae thickness and to overall size. Stromatolite laminae of the Precambrian-aged Pethei Formation, an outcropping along the shores of Great Slave Lake in the Northwest Territories of Canada, are both fine and coarse in dimension. The coarse-grained layers, formed of lime-mud pellets and calcium and magnesium carbonate rhombs, are principally less than 5.0 millimeters in thickness. The fine-grained laminae, composed of calcium carbonate clay and silt-sized particles, are, on average, only 0.5 millimeters thick.

In size, individual stromatolites can range from centimeters up to several meters. Fossils of the common Precambrian genus *Conophyton* occur in a range of sizes, from pencil-sized shapes to columns up to 10 meters in diameter. Subspherical varieties of stromatolite-like structures, formed by the accretion of successive gelatinous mats of blue-green algae and generally less than 10 centimeters in diameter, are termed "oncolites." Stromatolitic complexes in the Great Slave Lake district measure 80 meters long by 45 meters wide by 20 meters in thickness and can be continuously traced for distances exceeding 160 kilometers.

Albert B. Dickas

CROSS-REFERENCES

Archaebacteria, 951; Archean Eon, 1087; Atmosphere's Evolution, 1816; Biogenic Sedimentary

BIBLIOGRAPHY

Dott, R. H., Jr., and D. R. Prothero. *Evolution of the Earth.* 5th ed. New York: McGraw-Hill, 1994. A well-illustrated textbook on the subject of the geologic history of Earth. Contains several sections on stromatolites, their significance to Precambrian history, and their role in the understanding of very early forms of life.

Gebelein, C. D. "Biologic Control of Stromatolite Microstructure: Implications for Precambrian Time Stratigraphy." *American Journal of Science* 274 (1974): 575-598. This lengthy paper discusses the various environmental factors, including water and wind currents, that affect the detailed structure of modern stromatolitic organisms. Such knowledge is useful in worldwide mapping of stromatolite horizons. Written for the undergraduate college student.

Levin, H. L. *The Earth Through Time.* 5th ed. New York: Saunders College Publishing, 1996. This college-level textbook discusses fossil stromatolites of the Proterozoic eon and their importance as heliotrophic rock structures. Chapters 6 and 7 place these fossils within the context of the evolution of prokaryotic organisms. An easy-to-read source complete with numerous photographs.

McMenamin, Mark. *Discovering the First Complex Life: The Garden of the Ediacara.* New York: Columbia University Press, 1998. This entertaining study of the earliest complex life-forms on the planet details the author's work on these organisms. Written for the interested student but understandable by the general reader.

Schopf, J. William, ed. *Major Events in the History of Life.* Boston: Jones and Bartlett, 1992. An excellent overview of the origin of life, the oldest fossils, and the early development of plants and animals. Written by specialists in each field but at a level that is suitable for high school students and undergraduates. Although technical language is used, most terms are defined in the glossary.

Stewart, W. N. *Paleobotany and the Evolution of Plants.* London: Cambridge University Press, 1983. The first four chapters discuss the preservation, preparation, and age determination of fossil plants, the geologic fossil record, and the Precambrian eon, when stromatolites were common throughout the aquatic world. Suitable for college-level readers.

Walter, M. R., ed. *Stromatolites.* New York: Elsevier, 1976. Contains papers on zonation, paleoenvironments, environmental diversity, morphologies, origin, geologic significance, and distribution of fossil stromatolites. Several papers discuss modern algal stromatolites and the organisms that build them. A valuable reference for the serious student.

Walter, M. R. "Understanding Stromatolites." *American Scientist,* September-October, 1977, 563-571. An in-depth paper on stromatolites, thrombolites, and oncolites. Discusses means of development of differing morphologies, modern analogues, and uses in stratigraphic mapping of stromatolitic-bearing rocks. Contains a complete reference section. Written for the high school science reader.

SYSTEMATICS AND CLASSIFICATION

The modern classification of both fossil and living organisms attempts to organize life-forms in a logical and uniform manner and to illustrate the evolutionary relationships between different groups of animal and plant life. Only 250,000 fossil species have been described, while it has been estimated that some 500 million species have existed over the past 3.4 billion years. Thus, there are enormous numbers of fossil species still awaiting discovery and classification.

PRINCIPAL TERMS

BINOMIAL SYSTEM: the current system of classifying organisms, which gives each organism a dual name consisting of the genus and the species

GENUS: a group of closely related species that share a common ancestry and similar morphological characteristics

KINGDOM: one of the five large subdivisions of life, differentiated on the basis of gross body plan (single-celled versus multicellular) and method of obtaining food or nutrients (produces own food or obtains it from other organisms)

NOMENCLATURE: the names and terms used in a classification system

PHYLOGENY: the study of the evolutionary relationships between organisms

PHYLUM: a major grouping of organisms, distinguished on the basis of basic body plan, grade of anatomical complexity, and pattern of growth or development

SPECIES: a group of actually or potentially interbreeding organisms; the basic taxonomic unit

TAXON: one of several systematic groups or categories into which a particular organism may be placed

TYPE SPECIMENS: the original fossil specimens upon which a species was first erected

BINOMIAL SYSTEM

The system of classification utilized by biologists and paleontologists to classify both living and fossil organisms is based on the premise that a similarity in morphology or structure provides evidence of a biological kinship between organisms. The biological classification employed by biologists and paleontologists not only expresses the similarities and differences between groups but also summarizes or conveys a vast amount of information concerning these groups, especially the general phylogenetic or evolutionary relationships between groups.

"Systematics" is the study of taxonomy, which is concerned with the identification and classification of fossil organisms into morphologically and evolutionarily related groups. These groups or taxonomic categories are referred to as taxa. The system of classification in use worldwide is the binomial system of nomenclature, proposed by the Swedish naturalist Carolus Linnaeus in the 1700's. The fundamental unit, or taxon, in this classifica-

tion is the species. A species can be defined as a group of morphologically similar individuals whose ecological demands and physiologic functions are the same and whose members interbreed to produce fertile offspring. Groups of closely related species that share a common ancestry and possess a set of similar morphological characteristics are termed "genera" (singular, genus). Under the binomial system, each plant or animal is given two names, the genus name first, the species second. The human species, for example, is referred to as *Homo* (genus) *sapiens* (species). The names of both the genus and the species are printed in italics. Species names are given in Latin, as originally this was the universal language of science in the Western world, and naturalists were looking for a standardized nomenclature recognizable to scientists from all countries. The Latinized names of species have remained universal names, recognizable to biologists and paleontologists worldwide.

There are also taxonomic categories above the genus level. Similar genera are grouped into a single family; groups of related families are placed into an order; similar orders form a class; and closely related classes form a phylum. The more than thirty-five phyla are placed into one of five large groups termed kingdoms. These include the Monera (single-celled organisms with no cell nucleus), the Protista (single-celled organisms with a cell nucleus), the multicellular Plantae (organisms that produce their own food via photosynthesis), the multicellular Animalia (organisms that gain nutrients from the ingestion of other organisms), and the Fungi (multicellular organisms that absorb nutrients from their environment). Whereas the species represents a real, natural taxonomic unit, all taxonomic categories above the species level are artificial and subjective. The classification of the carnivorous dinosaur *Tyrannosaurus rex* would be as follows: kingdom, Animalia; phylum, Chordata; class, Reptilia; order, Saurischia; family, Tyrannosauridae; genus, *Tyrannosaurus;* species, *Tyrannosaurus rex.* The basic categories of higher taxa are sometimes supplemented by additional categories with the use of a sub-prefix or super-prefix.

HIERARCHICAL CLASSIFICATION

This type of classification is referred to as a hierarchical system. In such a system, each fossil specimen belongs to one "lower" taxonomic category or species and one of each of the "higher" or superspecific categories. From the highest rank (kingdom) to the lowest rank (species), these categories become progressively narrower in scope. For example, 1.5 million or more species form the kingdom Animalia, but only one species constitutes the species *Homo sapiens.* The criteria used to differentiate taxonomic groups also becomes narrower and more trivial as one goes from higher to lower taxa.

The binomial system attempts not only to divide organisms into structurally differentiated groups but also to illustrate the phylogenetic, or evolutionary, relationships between these groups. This classification scheme serves as a phylogenetic outline for the evolution of life on Earth and as an indicator of the diversity of basic body plans available to organisms through time. This is often diagrammatically depicted in the rough form of a branching tree, with the more primitive, structurally simpler organisms at the base and the more advanced, structurally more complex forms in the crown. In such figures, one can see how animals evolved from simple, single-celled protozoans through increasingly more complex multicellular organisms with successively more sophisticated grades of development. These evolutionary pathways are determined from studies of the comparative anatomy of living organisms and by information provided by the fossil record.

The identification of fossils and the integration of fossil organisms into the system of biological classification allows for reconstructing the history of life on Earth. The present records just one instant in the long history of life on this planet—a "snapshot" of life as it exists today. The fossil record, however, extending back some 3.4 billion years, provides the whole epic motion picture. Although some frames are missing or blurred by the effects of time, fossils provide the only record of life on Earth and therefore offer the sole source for the factual documentation of the evolution and extinction of organisms through time. The systematic study of fossil organisms provides four types of evolutionary information: It allows scientists to reconstruct the phylogenetic relationships that exist between various taxonomic groups; it reveals the times of major evolutionary events, such as major evolutionary radiations and so-called mass extinction events; it gives some idea as to the rates of evolutionary change; and it reveals patterns of evolutionary change.

FOSSIL REMOVAL AND DESCRIPTION

Determining the identity of a particular fossil provides an outline of its biological affinities, evolutionary pedigree, anatomical complexity, and also provides some idea of how the organism lived in its environment. The fossil then becomes more than a curiously shaped stone; it becomes the remains of a once-living organism that, in itself, is a time capsule for information about that interval of geologic time during which the organism lived and eventually died.

The basic tools of the systematic paleontologist traditionally have been the various implements used to remove the fossils from the rock matrix and prepare them for description, calipers to measure morphological dimensions, and an adequate reference library to aid in determining the identity and relationships of the fossils under study and compare them with those already described.

Today, computers can store and compare the dimensions of large suites of specimens and statistically analyze large sets of data in order to determine the possible relationships between fossil taxa. Studies of shell microstructure and geochemistry have also provided new approaches to fossil identification and classification, beyond the study of the external features of a fossil specimen.

The basic approach to fossil identification, however, remains essentially the same as it has been for the past two centuries. A fossil is discovered and collected, with the details of its occurrence—that is, the rock layer or stratum and geographic locality where it was found—being determined and recorded with the fossil. Ideally, associated fossil species should also be noted, as well as the orientation of the fossil specimens in the rock and the nature of their preservation. This may help the paleontologist to infer something about the life habits and habitats of the fossils. A single specimen of a particular fossil taxon is not enough; the larger the number of specimens of the fossil available for collection from the site, the better for the purposes of statistical analysis of the species.

The fossil specimens are removed from the enclosing rock matrix, and rock matrix adhering to the specimens is removed using preparatory equipment ranging from hammers and chisels to air-abrasive equipment and vibra-tools. This preparation is necessary to expose the main morphological features of the shell or bone, features critical in determining the fossil's identity. Paleontologists occasionally use sonic cleaners, weak acids, or strong detergents to remove more stubborn matrix adhering to the fossils. Fortunately, many fossils weather cleanly from soft sedimentary rocks like shales, marls, and poorly cemented sandstones and require little cleaning and preparation.

Following this, a detailed description of the specimen is formulated, including the dimensions of the specimen (measurements of length, width, height, thickness, and so on), the general arrangement of the parts that compose the fossil, and a description of the external and internal morphological features that distinguish the specimen. For certain fossil groups, especially corals, bryozoans, and nautiloid cephalopods, it is necessary to cut open specimens with rock saws, both longitudinally and tangentially, to expose critical internal structures and the microstructure of the shell if it

is preserved. If more than one specimen was collected of the fossil type under study, each specimen is described and then compared with its associates in order to determine any morphological variation present within the fossil population or to determine whether one or more distinct species are present in the collection.

FOSSIL CLASSIFICATION

To classify the fossil specimen, the fossil at hand is compared with fossil forms already described in the scientific literature. The usual process for identifying a fossil specimen is to determine its phylum first and then proceed from the general to the specific. Information to aid in the identification of fossils can be found in a number of reference books, textbooks, and popular field guides on fossils. Invertebrate fossils (remains of animals that lack backbones) constitute the majority of fossils found. These primarily marine animals are also the easiest to identify, largely a consequence of the possession of hard parts consisting of mineralized external or internal shells that usually are composed of a minimal number of parts that are preserved intact.

The fossil remains of vertebrate animals and plants are much less common and more difficult to identify and classify. This is attributable, in part, to the fact that these organisms are most common in terrestrial environments where the rapid and complete burial necessary for preservation as a fossil is not a common process. The internal skeletons of vertebrates consist of a multitude of individual bones held together by tendons, cartilage, and fleshy tissue. Upon the death of the organism, these soft parts decay, and the skeleton readily disaggregates into a pile of loose bones that, upon transport by stream action or scavengers, are spread over a broad area and are difficult to reassemble into the original skeleton. Plant remains lack mineralized hard parts and consist of multiple parts (roots, stems, leaves, seeds) that also are easily separated from one another in the process of burial.

After comparing the fossil specimens with species that have already been described, one of two conclusions can be made. Either the fossil specimens can be identified as those of an existing species, or they constitute a species new to science. If the specimens are identified as a preexisting species, they should be closely compared with the type specimens of the known species to ascertain if any variation exists between the two suites of specimens. The "type specimens" are the original specimens for which the fossil species was first erected. They are usually preserved for posterity in collections held at major natural history museums. The age of the rock unit and the geographic locality of the collecting site should be compared with that of the existing species to determine whether these new specimens extend the range of the species in time or space. If it is determined that the fossil specimens are sufficiently distinct from existing species, then a new species is created for these specimens.

DESCRIPTION OF NEW SPECIES

The description of a new species must follow a specific format that has been specified in the Code of Zoological Nomenclature established by the International Commission on Zoological Nomenclature in the early 1900's to bring a measure of uniformity to animal taxonomy. A comparable set of procedures has been established for plants. This code specifies the choice of the species name, publication of the name, description of the new species, and the designation of one or more type specimens.

The name given to a new species must be binomial, consisting of the genus name and the species name. The name used for the species cannot already be in use with the genus name with which it is to be associated. The names of the species and the genus must be Latin words or words that have been Latinized. Species names can be Latinized place-names, the names of people, or descriptive words. A species cannot be named anonymously, and the author's name is part of the official species name. For example, the official name for the living species of the chambered nautilus from the southwest Pacific would be *Nautilus pompilius* Linnaeus, Carolus Linnaeus being the original author of the species.

For a new species to be officially recognized, the name and a description of the distinctive features that identify it must be published in an approved medium; that is, it must be in print, published in quantity, and circulated to libraries to ensure that the announcement of the new name is readily available to scientists around the world.

The description of a new species should include a diagnosis listing the characteristics distinguishing the new species from similar species as well as a fuller description of the various morphological aspects of the species, along with its geographic locality, geologic occurrence, and a reference to the museum where the designated type specimens for the species are stored. It is also strongly recommended that an illustration of the new species be included in this description.

The International Code also specifies that each new species description must be accompanied by the designation of a type specimen or set of type specimens. These are the actual original fossil specimens upon which the species is based. The code requires that type specimens be clearly labeled as such and that measures be taken to ensure their preservation and accessibility to interested scientists. This usually means that specimens are deposited in a major natural history museum, where such facilities are available. If the new species is based on a single specimen, this specimen is designated the type specimen, or "holotype." If several specimens serve this purpose, they are designated the "syntypes" for the species.

The naming of new species is best left to a professional paleontologist who is familiar with the rules of zoological nomenclature and the accepted format for publication. Nevertheless, amateur fossil collectors, who typically find the bulk of new species in the field, are often recognized for their contributions by being named a coauthor of the new species or having the new species named for them.

CLADISTICS

Reconstruction of the phylogenies, or evolutionary family trees, of fossil organisms operates in a manner opposite that of fossil identification and classification. A paleontologist starts at the species level, reconstructing evolutionary relationships between species, and then works up the taxonomic hierarchy, from the narrow to the broad. Phylogenies are reconstructed using a system called "cladistics," in which organisms are grouped according to the presence of shared characteristics that are particularly informative. This system assumes that as evolution progresses, new, heritable traits emerge and are passed on; two groups of animals sharing such new traits will be more closely related to each other than to groups that only share the original traits. Analyses of these characters are presented as treelike diagrams called "cladograms," which show the order in which the new characters, and thus the new species, evolved. Each branching point in the diagram reflects the emergence of an ancestor that founded a group that exhibited advanced characters not present in the groups that had developed earlier. Through detailed studies, or "monographs," of groups of related species and genera, the phylogenies of most of the major animal groups have been mapped out. These reconstructions, however, are constantly undergoing change, primarily as the result of the description of newly discovered fossil taxa and the restudy of older, established species.

Robert C. Frey

BIBLIOGRAPHY

Boardman, R. S., A. H. Cheetham, and A. J. Rowell, eds. *Fossil Invertebrates.* Oxford: Blackwell Scientific, 1987. Up-to-date college-level text on invertebrate paleontology, with individual sections that each describe a major phylum and its component classes and orders. Very good illustrations of the major invertebrate groups and their characteristic morphological features.

Briggs, Derek E. G., and Peter R. Crowther, eds. *Paleobiology: A Synthesis.* Oxford: Blackwell Scientific Publications, 1990. This multiauthored text deals with all aspects of paleontology in a series of short articles by experts in each field. Intended for university students but also useful to high school students and interested readers.

Clarkson, E. N. K. *Invertebrate Paleontology and Evolution.* London: Allen & Unwin, 1986. A college invertebrate textbook up-to-date with good description of the major invertebrate phyla, classes, and orders. Good illustrations of the major morphological features that distinguish each group. A brief general description of taxonomy and fossil systematics.

Kitching, Ian J., Peter L. Forey, Christopher J. Humphries, and David M. Williams. *Cladistics: The Theory and Practice of Parsimony Analysis.* 2d ed. Oxford: Oxford University Press, 1998. An introduction to the study of systematics using the method of phylogenetic analysis. Intended for the college student.

Moore, R. C., ed. *Treatise on Invertebrate Paleontology.* 27 vols. Lawrence: University of Kansas Press, 1953-1975. This series of volumes (with some volumes currently being revised) includes descriptions and illustrations of all fossil invertebrate groups, down to the genus level. Each volume concentrates on a particular phylum or class and includes general chapters on evolution, ecology, and classification. The standard reference for the identification and classification of fossil invertebrates. Some basic knowledge of biology or paleontology required.

Prothero, Donald R. *Bringing Fossils to Life.* Boston: McGraw-Hill, 1998. This well-illustrated and entertaining text covers a broad range of paleontological topics, including systematics and classification. Glossary, bibliography, and index.

Radinsky, Leonard B. *The Evolution of Vertebrate Design.* Chicago: University of Chicago Press, 1987. A very good introductory text to the vertebrates. Well written and illustrated, with a unique approach to vertebrate classification, relating important morphological features in each group to their function. Highly readable and understandable to the college-level reader.

Raup, D. M., and S. M. Stanley. *Principles of Paleontology*. San Francisco: W. H. Freeman, 1978. An excellent college-level text emphasizing paleontological concepts, including thorough sections on fossil classification, the description of new species, and reconstruction of fossil phylogenies. Suitable for college-level readers with some background in biology or paleontology.

Smith, Andrew B. *Systematics and the Fossil Record*. Oxford: Blackwell Press, 1996. An overview of approaches to systematics and biostratigraphy, with particular emphasis on their application to fossil organisms. For college students.

10
EARTH HISTORY AND STRATIGRAPHY

ARCHEAN EON

The Archean eon is the earliest on the geologic time scale. Within this time, the basic structure and chemical composition of the Earth evolved. Because the rocks of this period are complex and fragmentary, their history remained opaque until twentieth century advances in technology allowed scientists to study them with a greater degree of accuracy.

PRINCIPAL TERMS

ACCRETION: the gradual accumulation of matter in one location, typically because of gravity

BASALT: a fine-grained, dark mafic igneous rock composed chiefly of plagioclase feldspar and pyroxene

BASIN: a regionally depressed structure available for the collection of sediments

BRECCIA: a rock formed by the amalgamation of various rock fragments

GNEISS: a coarse-grained metamorphic rock that shows compositional banding and parallel alignment of minerals

GRANITE: a coarse-grained, light-colored igneous rock composed of three types of minerals: two types of feldspars, quartz, and variable amounts of darker minerals

GREENSTONE: a field term used to describe any altered basic igneous rock that owes its color to the presence of various green minerals

MARIA: Latin plural meaning "sea" used to describe the Moon's dark areas; the light-colored areas of the Moon are called the highlands

PLATE TECTONICS: the study of the movement and deformation of large segments or plates of the Earth's surface over the underlying mantle

SILICATE: a substance whose structure includes silicon surrounded by four oxygen atoms in the shape of a tetrahedron

---

EVOLUTION OF EARTH

The Archean eon is the earliest in the history of the Earth and accounts for about 54 percent of all geologic time. The time of its exact beginning is unclear, but an estimate of 4.5 billion years ago is generally accepted; it continued until approximately 2.5 billion years ago. The Archean was a time of major evolution in the Earth's chemical and physical structure, which gave the planet its basic character. The current paradigm, first enunciated in 1905 by the American geologist Thomas C. Chamberlin and astronomer Forest R. Moulton at the University of Chicago, describes the Earth and the solar system in the beginning as a gas cloud rotating around a center point. Shock waves from two nearby and independent supernovas caused the cloud to collapse as the rate of rotation increased. With greater rotation, the cloud progressively flattened into a disk shape.

The dominant physical process at this time was the condensation of tiny particles consisting mostly of silicate and nickel-enriched iron. The resulting rotational eddies concentrated the particles, and they clustered at discrete distances from the protosun. The particles literally fell together under their own mutual gravitational attraction, creating larger particles, which, in turn, grew through impact with other masses. This process continued until planetesimal bodies (kilometers in size) accreted and acted as gravitational "dust mops" sweeping through space, collecting more mass. Astronomers believe that once accretion through gravitational attraction began to create density centers, the centers reached their present masses rather quickly, requiring something on the order of ten thousand years. The impact-accretion action resulted in the fragmentation and heating of the protoearth, differentiating the preplanetary material. A segregated interior began to develop, and the process accelerated as larger masses retained more and more of the impacting fragments. When the protoearth reached a sufficient mass, a segregation by density occurred similar to the overall density segregation of the solar system.

At or near this time, the Sun ignited. Its gravitational influence, temperature gradient, and solar wind produced a strong chemical/density segregation among the planets. The "rocky" planets—Mercury, Venus, Earth, and Mars—formed close to the Sun, while the frozen gas planets—Jupiter, Saturn, Uranus, and Neptune—formed in the outer orbits. By 1954, Harrison Brown had proposed an impact hypothesis to account for the initial accretion and differentiation of the Earth into core, mantle, and crust. Following a large number of gravitational impacts, the Earth accreted as a homogeneous mixture of silicates and iron-nickel. Radioactive heating caused the dense iron-nickel to melt and sink to the center of the Earth, where it formed the core. The remaining lighter silicates formed the mantle and the crust. A slightly different impact hypothesis, described by Robert Jastrow in 1963, requires the existence of a dense iron-nickel condensate phase at the Earth's core; a mantle and crust of silicates accreted around the core by gravitational impact. The mechanism of accretion and differentiation probably incorporates features from both hypotheses.

METEORITES

Meteorites may aid understanding of the Archean eon. Their chemical composition suggests that the condensation or accretion from the nebular disk was not homogenous, falling instead into three general groups. These groups—called iron, stony-iron, and stony—appear to represent an early crystallization of two distinct chemical phases. Iron meteorites consist chiefly of iron with 4 to 20 percent alloyed nickel and small amounts of other elements such as chromium. The stony irons, as the name suggests, consist of roughly equal amounts of rock and iron. Stony meteorites are largely silicate minerals. The marked difference in the two earliest known groups (4 to 4.5 billion years old) found in the primordial solar system increases the likelihood that the Earth accreted as a partially differentiated body.

A comparative study of Earth's moon strengthens the impact model of Earth's early history. Because of the close ratio of the two masses compared to other planet-satellite systems and the fact that Earth and the Moon revolve around a common center, astronomers often describe the Earth-Moon relation as a dual planet system rather than as a planet-satellite system. They assume that Earth and the Moon formed contemporaneously and in close proximity to each other. The Moon, however, lacks the destructive erosional and tectonic forces found on Earth; therefore it functions as a time capsule that mirrors an earlier Earth phase. During the later stages of the Moon's evolution, a molten stage appeared. The heat energy that caused the melting came from the impacts of the planetesimals and the decay of radioactive elements. Shortly after its crust cooled, a period of intense meteoric bombardment left the Moon with numerous craters, resulting in highland-type terrain. Near the end of this cratering period, several asteroid-sized objects (100 kilometers in diameter) struck the lunar surface, breaking through the thin crust. This allowed the darker-colored basaltic lava to escape to the surface, flooding the low-lying areas; these darker areas are the maria.

Equally large meteoroids must have struck Earth in the same fashion and during this time period. All maria basalts sampled so far have ages in the range of 3 to 4 billion years and are contemporaneous with the oldest dated Earth rocks. In 1958, radioactive decay studies by L. T. Aldrich and G. W. Wetherill showed that the oldest surviving relics of terrestrial rocks date from about 3.8 billion years. Two explanations proposed in the late 1970's may account for the age differences of terrestrial rocks, lunar rocks (highlands) at 4.1 billion years, and meteorites at 4.5 billion years. According to J. V. Smith, the Earth's mantle was so cool after accretion (about 4.5 billion years ago) that it did not heat up sufficiently from gravitational pressure to produce magmas for another 700 million years. The second explanation is B. M. Jahn's and L. E. Nyquist's plate tectonic model of subduction. It requires the crust's continuous generation and destruction and its recycling into the mantle via convection currents until some of the crustal material became stable. Meteor bombardment may have destroyed some of this early thin crust, but tectonics was probably the dominant factor.

IMPACT ESTIMATES

Geologists have experienced difficulty in estimating the number of meteors that struck Earth. They are also not certain whether these meteors were of similar size and energy as those that formed the lunar craters in the early Archean time period. R. A. F. Grieve and H. Frey have extrapolated much of the physical and statistical modeling for the Archean from the Moon, whose early Archean crust equivalent is preserved. Maria lava dates obtained from the Apollo missions indicate an age of 3.1 to 3.8 billion years for these forty well-defined maria basin structures, suggesting that perhaps up to three thousand basins might have existed on Earth. These statistical models of the early 1980's project the formation of at least two thousand—and more likely twenty thousand—basins on Earth with ages between 3.9 and 4.4 billion years. Through time, the frequency of these impacts should have decreased nearly exponentially.

Meteors with diameters greater than 100 kilometers are the most significant geologic agents of the early Archean. Geologists estimate that several thousands of 100-kilometer meteors impacted Earth's early Archean surface and converted 30 to 50 percent of the crust into impact basins. These impacts could produce walls 3 kilometers above the surrounding terrain, with depths of 10 to 12 kilometers. An early ocean basin probably had this type of topography. The energy expended upon impact fractured the thin Archean crust to a depth of 25 kilometers and allowed molten material from the mantle to escape to the surface and flood the basin. The resulting structure became an ideal trap for the accumulation of sediments. Its stratigraphy would have included several kilometers of impact melts, crustal breccia, volcanics, and highland sediments. With the passage of time, the basins subsided, underwent a second partial melting, and produced a new generation of magmas. If the recycling continued, the magmas could have produced rocks higher in silicon and aluminum, more like the continental cores. Geologists believe that once the basins became tectonically stable, their stratigraphy included a mixture of metamorphosed rock intruded by granites and capped by crustal sediments. Such an interpretation can explain the formation of the protocontinent nuclei.

Further tectonic development of the early crust led to the partial aggregation of nuclei, which, with the evolution of the greenstone belts, produced the familiar Archean Shields. Greenstone belts are unique to the Archean eon. They consist largely of volcanic rocks and sedimentary rocks derived primarily from volcanics. Their stratigraphy is often metamorphosed, producing the mineral chlorite, which has the characteristic green color from which the belts derive their name. Because the composite features of the greenstone belts are without a counterpart in modern geology, the geologic conditions of their formation were very different from what is observed today. The Canadian Shield demonstrates the greenstone stratigraphy and tectonics. Characteristically, it exhibits alternating linear belts (compressed basins) of greenstone-granite and gneiss. It also contains a series of elliptically shaped basins such as the Abitibi. While no large continents existed during the Archean eon, the nuclei necessary for their formation were present as protocontinents. These protocontinents were separated by numerous marine basins that accumulated lava and volcanic sediments. They later became greenstone belts. The thin Archean crust often broke under the active tectonic forces in the mantle and interjected magma into the protocontinents.

Geologists have had difficulty unraveling the true nature of the Archean stratigraphy. The two main problems are that the greenstone-granite terrains contain extensive metasedimentary sequences and that the combined igneous-metasedimentary successions grade laterally and vertically from intermediate- to high-grade metamorphic rocks. Large-scale magmatic intrusion and the structural response to the meteor impacts characterize the tectonics and stratigraphy of the Archean. The stratigraphic succession—the product of these tectonic forces—suggests that a deep crustal fracture system controlled the geology of this time and that meteorite impact produced the surface topography.

STUDY OF THE ARCHEAN EON

Scientists have used a diverse range of analytical techniques to study the Archean eon. Because tectonics or erosion has destroyed much of Earth's original Archean material, scientists look to the Moon to sample and observe this early stage of planetary development. The imagery from various lunar orbiter missions of the 1970's yielded a clarity and perspective of the Moon previously unknown. Few outside the field of geology realized that these images also functioned as a snapshot of the early Earth. Voyager photos of September 18, 1977, provided the first look at Earth and the Moon as a dual planet system. Moon rock samples obtained by Apollo astronauts during the missions of the early 1970's indicated chemical compositions and histories similar to the oldest of Earth's rocks and provided the first evidence supporting a parallel history.

Geologists have a variety of tools and techniques to unravel the Archean story. These range from viewing thin sections of rock samples under a microscope to the use of remote-sensing, Earth-orbiting satellites beginning in July, 1972. Because fossils are virtually nonexistent in the Archean, geologists rely on radioactive dating techniques for sequencing its events. Field mapping and sample collecting are their primary geological tools.

Anthony N. Stranges and Richard C. Jones

CROSS-REFERENCES

Biostratigraphy, 1091; Cenozoic Era, 1095; Cretaceous-Tertiary Boundary, 1100; Dinosaurs, 970; Ediacaran Fossils, 976; Evolution of Humans, 994; Evolution of Life, 999; Fossil Plants, 1004; Fossil Record, 1009; Fossilization and Taphonomy, 1015; Geologic Time Scale, 1105; Ice Ages, 1111; Land Bridges, 1116; Mammals, 1038; Mesozoic Era, 1120; Paleoclimatology, 1131; Paleogeography and Paleocurrents, 1135; Paleoseismology, 1139; Paleosols, 1144; Paleozoic Era, 1126; Proterozoic Eon, 1148; Stratigraphic Correlation, 1153; Transgression and Regression, 1157; Unconformities, 1161; Uniformitarianism, 1167.

BIBLIOGRAPHY

Briggs, Derek E. G., and Peter R. Crowther, eds. *Paleobiology: A Synthesis.* Oxford: Blackwell Scientific, 1990. This text deals with all aspects of paleontology in a series of short articles by experts in each field. Intended for university students but also useful to high school students and interested readers.

Dott, Robert H., Jr., and Donald R. Prothero. *Evolution of the Earth.* 5th ed. New York: McGraw-Hill, 1994. This basic textbook on historical geology is aimed at students of geology. However, it is very readable by anyone with a background in science. Presents an up-to-date account of the Earth's history from the viewpoint of plate tectonics. Includes a glossary.

Kauffmann, William J., III. *Universe.* New York: W. H. Freeman, 1985. The chapters on the Moon, Earth, and modern cosmology provide appropriate background for the general reader interested in the Archean eon. This freshman textbook is well written and includes excellent visuals.

Press, Frank, and Raymond Siever. *Understanding Earth.* 2d ed. New York: W. H. Freeman, 1998. An excellent freshman-level textbook in geology. A comprehensive physical geology text that covers the formation and development of the early Earth. Readable by high school students, as well as by general readers. Includes an index and a glossary of terms.

Scientific American 249 (September, 1983). The entire issue deals with the study of Earth. There are many excellent and readable articles accompanied by superb illustrations.

Stanley, Steven. *Earth and Life Through Time.* 2d ed. New York: W. H. Freeman, 1989. A freshman-level textbook for the Earth sciences suitable for the interested amateur.

BIOSTRATIGRAPHY

Biostratigraphy is that branch of the study of layered rocks—stratigraphy—that focuses on fossils. Its goals are the identification and organization of strata based on their fossil content. Biostratigraphy thus investigates one of the principal bases of the geologic time scale of Earth history.

PRINCIPAL TERMS

CORRELATION: the determination of the equivalence of age or stratigraphic position of two strata in separate areas, or, more broadly, determination of the geological contemporaneity of events in the geologic histories of two areas

FOSSILS: remains or traces of animals and plants preserved by natural causes in the Earth's crust

INDEX FOSSIL: a fossil that can be used to identify and determine the age of the stratum in which it is found

SEDIMENTARY ROCKS: rocks formed by the accumulation of particles of other rocks or of organic skeletons or of chemical precipitates or some combination of these

STRATIGRAPHY: the study of layered rocks, especially of their sequence and correlation

STRATUM (pl. STRATA): a single bed or layer of sedimentary rock

BIOSTRATIGRAPHIC ZONES

Biostratigraphy is the method of identifying and differentiating layers of sedimentary rock (strata) by their fossil content. Strata with distinctive fossil content are termed biostratigraphic units, or zones. Zones vary greatly in thickness and in lateral extent. A zone may be a single layer a few centimeters thick and of very local extent, or it may encompass thousands of meters of rocks extending worldwide. The defining feature of a zone is its fossil content: The fossils of a given zone must differ in some specific way from the fossils of other zones.

Zones are usually recognized after fossils have been collected extensively over the lateral and vertical extent of a rock sequence or at many sequences over a broad region. The positions of the fossils in the strata are carefully recorded in the field. Fossils that co-occur in a single layer are noted, as are fossils found isolated in the strata. In the laboratory, the biostratigrapher, usually a paleontologist, then tabulates the vertical and lateral ranges of the fossils collected. It is from these ranges that the paleontologist recognizes zones. Different types of zones are recognized depending on the way in which the fossils in the strata prove to be distinctive. Assemblage zones are strata distinguished by an association (assemblage) of fossils. Thus, not one type but many types of fossils are used to define an assemblage zone. All dinosaur fossils, for example, can be thought of as defining an assemblage zone that encompasses Earth history from about 220 to 66 million years ago.

Range zones are strata that encompass the vertical distribution, or range, of a particular type of fossil. Thus, one fossil type, not many, is used to define a range zone. In contrast to the example just given, one type of dinosaur, *Tyrannosaurus rex*, lived only between 68 and 66 million years ago. Its fossils thus define a range zone that corresponds temporally to this 2-million-year interval.

Acme zones are rock layers recognized by the abundance, or acme, of a type (or types) of fossil (or fossils) regardless of association or range. Horned dinosaurs (*Triceratops* and its allies) reached an acme between 70 and 66 million years ago; that is, during this period they were most diverse and most numerous. This acme zone thus overlaps the *Tyrannosaurus rex* range zone and represents a small portion of the dinosaur assemblage zone.

Finally, interval zones are recognized as strata between layers where a significant change in fossil content takes place. For example, the mass extinctions that took place 250 and 66 million years ago bound a 184-million-year-long interval zone that is popularly referred to as the "age of reptiles."

DEVELOPMENT OF BIOSTRATIGRAPHY

Biostratigraphy developed independently in England and France just after 1800. In England, William Smith, a civil engineer, worked in land surveying throughout the country. From his vast field experience, he recognized that a given stratum usually contains distinctive fossils and that the fossils (and the stratum) could often be recognized across a large area. Smith's work culminated in his geological map of England (1815), based on his tracing of rock-fossil layers across much of the country.

Meanwhile, in France, Georges Cuvier and Alexandre Brongniart studied the succession of rocks and fossils around Paris. They too discovered a definite relationship between strata and fossils and used it to interpret the geological history of the rocks exposed near Paris. In this history, Cuvier saw successive extinctions of many organisms coinciding with remarkable changes in the strata. To him, these represented vast "revolutions" in geological history, which Cuvier argued were of worldwide significance. It is now known that Cuvier was mistaken, but the discovery that a particular fossil type (or types) was confined to a particular stratum became the basis for biostratigraphy. This allowed geologists to identify strata from their fossil content and to trace these strata across broad regions of the Earth's crust.

Almost simultaneous with the development of biostratigraphy was the development of biochronology. Biochronology is the recognition of intervals of geologic time by fossils. It stemmed from the realization that during Earth's history, different types of organisms lived during different intervals of time. Thus, the fossils of any organism represent a particular interval of geologic time. (Such fossils are called index fossils because they act as an "index" to a geologic time interval.) Biochronology thus identifies intervals of geologic time based on fossils. These time-distinctive fossils are the fossils by which zones are defined, which is to say that each zone represents, or is equivalent to, some interval of geologic time.

The time value of zones made them more useful in tracing strata and deciphering local geological histories. Biostratigraphy now became one of the central methods of stratigraphic correlation. With the aid of fossils, it became possible to determine the ages of strata and thus demonstrate the synchrony or diachrony of these strata in different areas. Through its use in stratigraphic correlation, biostratigraphy became one of the bases for constructing what is called the relative geological time scale of Earth history composed of eons, eras, periods, epochs, and ages. This time scale is the "calendar" by which all geologists temporally order their understanding of the history of the Earth.

APPLICATION OF BIOSTRATIGRAPHY

Biostratigraphy is generally used as a method of stratigraphic correlation, the process of determining the equivalence of age or stratigraphic position of layered rocks in different areas. Stratigraphic correlation by biostratigraphy is extremely important in deciphering geological history; it reveals the sequence of geological events in one or more regions. Understanding geological history is of interest for its own sake to scientists and laypersons alike. It is crucial to the discovery of mineral deposits and energy resources within the Earth's crust. In addition, it provides insight into the biological events that have taken place on this planet for the last 3.9 billion years.

A good example of the use of biostratigraphy in this last regard comes from the study of dinosaur extinction. When dinosaurs were first discovered in England in 1824, and when the term "dinosaur" was coined by the British anatomist Sir Richard Owen in 1841, nobody realized that dinosaurs had lived on Earth for only 150 million years and that their extinction had taken place rather rapidly about 66 million years ago. By 1862, however, enough dinosaur fossils had been collected around the globe that a biostratigraphic pattern was beginning to emerge. In that year, the American geologist James Dwight Dana, in his classic *Manual of Geology*, noted that all dinosaurs disappeared before the end of the Mesozoic era, which is now considered as the interval of Earth history between 250 and 66 million years ago. This biostratigraphic generalization was possible because geologists noticed that many Mesozoic rocks (but no older or younger rocks) were full of dinosaur fossils, and thus the Mesozoic came to be termed "the age of reptiles." It might just as well be referred to as the "dinosaur zone," except of course for the first 30 or so million years of the Mesozoic, during which dinosaurs apparently did not exist.

More than a century of research has confirmed

Dana's biostratigraphic generalization and considerably refined it. Scientists now generally agree that the last dinosaurs disappeared worldwide approximately 66 million years ago. It is also known that dinosaurs first appeared about 220 million years ago. Thus, scientists are able to recognize a dinosaur zone and erect many types of zones based on the ranges and acmes of specific types of dinosaurs. This biostratigraphy of dinosaurs is the basis for informed discussion of the sequence and timing of events during the evolution of the dinosaurs. For example, scientists are now confident that *Stegosaurus* lived long before *Tyrannosaurus* and that stegosaurs as a group of dinosaurs became extinct long before the end of the Mesozoic.

Although discussion here has relied heavily on dinosaurs for examples of biostratigraphy at work, the fossils of these giant reptiles are not ideal for use in biostratigraphy because it is not easy to identify most dinosaur fossils precisely and because most dinosaurs were not animals with broad geographic ranges. Indeed, the fossils of most use in biostratigraphy, index fossils, are those that are easy to identify precisely and that represent organisms that had wide geographic ranges, enjoyed broad environmental tolerances, and lived only for a brief period of geologic time.

Usually an entire skull or skeleton is needed to identify a dinosaur fossil precisely; the isolated bones most often found are not enough, although they do indicate the fossil is that of a dinosaur. Most dinosaurs (there are some notable exceptions) seem to have lived in one portion of one continent; indeed, fossils of the horned dinosaur *Pentaceratops* (a cousin of *Triceratops*) have been found only in New Mexico. There is strong evidence that some dinosaurs preferred coastlines, whereas others preferred dry areas. Thus, many, if not most, dinosaurs did not live in a wide range of environments. Finally, although many dinosaurs apparently lived for only brief intervals of geologic time, the fossil record of most of these giant reptiles is not extensive enough to pin down their exact interval of existence

The factors that mitigate the use of most dinosaur fossils in biostratigraphy are quite different for microscopic fossils of pollen grains and the shelled protozoans known as foraminiferans. These microscopic fossils fit well the four criteria that identify fossils most useful in biostratigraphy. Indeed,

such "microfossils" (studied by micropaleontologists) are some of the mainstays of biostratigraphy.

SIGNIFICANCE

Biostratigraphy, the recognition of strata by their fossil content, is a cornerstone of stratigraphic correlation. By using fossils to identify bodies of rock, they can be traced over broad areas, and their sequence in distant areas can often be determined. Stratigraphic correlation by biostratigraphy is critical to deciphering geological history; without it, the search for mineral deposits and energy resources would be considerably more difficult. Furthermore, understanding the history of geological disasters—earthquakes, volcanic eruptions, meteorite impacts, and the like—and thereby being able to predict future disasters, relies on knowledge of the sequence and timing of geological events, knowledge often derived from biostratigraphy. Deciphering the history of life on this planet, including the myriad appearances, changes, and extinctions of Earth's biota during the last 3.9 billion years, largely depends on the sequence and timing established by biostratigraphy.

Biostratigraphy has also given rise to biochronology, the recognition of intervals of geologic time based on fossils. As a result, scientists have been able to construct a relative global geologic time scale, and it is within the context of this time scale that all geological and biological events in Earth history have been placed.

Spencer G. Lucas

CROSS-REFERENCES

BIBLIOGRAPHY

Ager, Derek V. *The Nature of the Stratigraphical Record.* 2d ed. New York: Halsted Press, 1981. A witty and unabashed look at stratigraphy; some of the discussion centers on biostratigraphy. An extensive bibliography, index, and a few well-chosen illustrations illuminate the text.

Barry, W. B. N. *Growth of a Prehistoric Time Scale.* Rev. ed. Palo Alto, Calif.: Blackwell Scientific Publications, 1987. Largely devoted to the history of how the global geologic time scale was formulated, much of this book is a history of biostratigraphy. Well illustrated, with a good bibliography and an index.

Brenner, R. L., and T. R. McHargue. *Integrative Stratigraphy Concepts and Applications.* Englewood Cliffs, N.J.: Prentice-Hall, 1988. Chapter 11 of this college-level textbook provides a detailed look at biostratigraphic concepts, methods, and applications. Well illustrated, with extensive reference lists and an index.

Dott, Robert H., Jr., and Donald R. Prothero. *Evolution of the Earth.* 5th ed. New York: McGraw-Hill, 1994. This basic textbook on historical geology is aimed at students of geology. However, it is very readable by anyone with a background in science. Presents an up-to-date account of the Earth's history from the viewpoint of plate tectonics. Includes a glossary.

Hedberg, H. D., ed. *International Stratigraphic Guide.* New York: John Wiley & Sons, 1976. The international "rule book" for stratigraphy. It sets procedures and standards to be met when naming stratigraphic units. It also defines many terms used in stratigraphy and has an extensive bibliography. Chapter 6 is devoted to biostratigraphy.

Press, Frank, and Raymond Siever. *Understanding Earth.* 2d ed. New York: W. H. Freeman, 1998. This comprehensive physical geology text covers the formation and development of the Earth. Readable by high school students, as well as by general readers. Includes an index and a glossary of terms.

Prothero, Donald R. *Bringing Fossils to Life.* Boston: McGraw-Hill, 1998. This well-illustrated and entertaining text covers a broad range of paleontological topics, including biostratigraphy. Glossary, bibliography, index.

Stanley, S. M. *Exploring Earth and Life Through Time.* New York: W. H. Freeman, 1992. An excellent introductory-level college textbook on historical geology. It reviews the history of life and the many fossil forms found in strata in the Earth's crust. Chapter 5 includes a discussion of biostratigraphy. Lavishly illustrated, with extensive references, glossaries, appendices on fossil groups, and an index.

CENOZOIC ERA

During the Cenozoic era, the last 66.4 million years of Earth history, all aspects of the Earth's modern environment developed. Through study of the geologic record from this era, scientists are able to distinguish between environmental changes caused by a normal progression of geologic phenomena and those changes that are related to human activity.

PRINCIPAL TERMS

EPICONTINENTAL SEA: any body of marine water that is present on the continents; epicontinental seas were more extensive in the past than at present

EPOCH: a relative time unit and a subdivision of a period

GLACIERS: systems of moving ice that can occur at any time in high elevations and periodically will occur on the continents in the high latitudes during periods known popularly as ice ages

PERIOD: a relative time unit and a subdivision of an era

REGRESSION: a very slow fall in sea level that may result in the exposure of the continental shelves

TECTONICS: the general term for deep-Earth geologic phenomena such as mountain building, volcanism, earthquakes, continental collisions, and seafloor spreading

TRANSGRESSION: a very slow rise in sea level that usually results in the flooding of the continents

MODERN EARTH ENVIRONMENT

The Cenozoic era began approximately 66.4 million years before the present and extends into modern time. It is subdivided into two periods, the Tertiary and the Quaternary, and these two periods are subdivided into seven epochs. The subdivisions of the Cenozoic era are not equal in duration; rather, the periods and epochs, as well as the Cenozoic era itself, are relative time units and do not have a fixed time value, as do absolute time units such as hours, days, and years. Relative time units are based on geologic events and phenomena such as fossils. Their position in geologic history is determined by the relative position of these events and phenomena in the rock record. The values in years are determined by radiometric methods after the relative units are recognized.

The Cenozoic era represents a time in Earth history when the modern Earth environment began to develop. The geologic setting, geography, atmosphere, climate, oceans, and fauna and flora all began to exhibit a decidedly modern appearance. In previous eras of Earth history, many environmental conditions were very different from those of the present, as were the fauna and flora. Despite the differences, there is a progression in

geologic history that can be followed. The Earth of the Cenozoic era most closely resembles the present-day Earth, because the Cenozoic era is closer in time and contains the recent epoch of Earth history. The further back one goes in time, the greater the differences. In order to understand fully and appreciate the development of the Earth through the Cenozoic era, a brief survey of the conditions and phenomena of the epochs of the Cenozoic era from the oldest to the youngest is necessary.

PALEOCENE EPOCH

The Paleocene epoch (between 66.4 and 57.8 million years ago) is the first epoch of the Cenozoic era. The transition from the preceding Cretaceous period of the Mesozoic era is one of the most pronounced in the geologic record. Great physical and accompanying biological changes took place. Whatever the cause of such changes, the setting of the Paleocene was very different from that of the Cretaceous. Worldwide, the beginning of the Paleocene is marked by a regression of the seas, a fall in sea level. This change resulted in reduced shallow seas for the dwelling of marine organisms and a decrease in humid conditions on the continents. The geographic setting was also changing with the continued enlargement of the Atlantic Ocean Basin and the northerly drift of North America. This northerly drift resulted in the increase of the temperate zone in North America at the expense of the subtropical zones, causing a cooling of the climate from the tropical conditions of the Cretaceous period.

Another important geographic change in the Paleocene epoch was the decrease in size of Tethys, the worldwide warm-water ocean roughly parallel to the equator. In North America, mountain building that began in the late Cretaceous continued in the Rocky Mountains region. Although sea level fell worldwide in the earliest Paleocene, a large transgression, or sea-level rise, soon began. This transgression was large enough that the sea reached all the way into the present-day High Plains of North and South Dakota. It is estimated that sea level was more than 250 meters above its present level during this episode. When this transgression and the following regression ended, the sea never again rose to this level. Since the end of this transgression-regression interval, sea level has never risen above the coastal plains.

Life of the Paleocene epoch was very different from that of the Cretaceous period. Marine life was not as diverse in the Paleocene, and many prominent forms of marine invertebrates that were common in the Cretaceous were extinct by the Paleocene. Many of the marine organisms looked very much like modern marine invertebrates and, in fact, are the ancestors of modern biota. On land, a fauna that had been previously dominated by the dinosaurs was now dominated by the mammals. Most of these mammals were small and did not closely resemble many of the modern mammals. Others, such as the rodent *Paramys*, which resembled a modern squirrel in appearance, are clearly the ancestors of modern mammal faunas. The flora of the Paleocene was dominated by the flowering plants.

EOCENE EPOCH

The Eocene epoch (between 57.8 and 36.6 million years ago) contained two major tectonic events of global significance. The first major event was the beginning of the closure of Tethys. This was caused by the collision of two continents, India and Asia. Although mountain building did not

begin in the area at this time, the two continents were in close proximity to each other. Mountain building did begin in the Mediterranean region during the Eocene as a result of the African plate moving relative to the Eurasian plate. Another significant tectonic development was the separation of Australia from Antarctica. This event allowed the development of a cold current around the Antarctic continent and a climatic isolation of this landmass. In North America, volcanic activity was extensive throughout the West.

Environmental conditions of the Eocene showed a continued cooling and drying trend from the Paleocene. Large basins between the mountains that were built during the Paleocene were filled with river and lake sediments during the Eocene. Sea level remained higher than at present, but the epicontinental seas in North America were confined to the coastal plain regions.

One of the major innovations in marine life during the Eocene epoch was the whale. The whales developed from land carnivores that had adapted to a marine existence. On land, the most significant development was the appearance of the grasses. These plants enabled the widespread development of savannas (semiarid grasslands) and a wide variety of grazing animals. Included among these are the earliest members of the horse, elephant, and rhinoceros lineages. Also present were the first very large mammals, the titanotheres, and early ancestors of the camels. The Eocene also saw the first development of the dog, cat, and weasel families and the existence of numerous birds.

OLIGOCENE AND MIOCENE EPOCHS

The Oligocene epoch (between 36.6 and 23.7 million years ago) saw a continuation of the tectonic and environmental conditions that were present in the Eocene. An important change occurred in the Earth's sea level at about the middle of the Oligocene. Sea level fell radically to a point well below present sea level. This change appears to be related to the development of continental glaciers in Antarctica and resulted in a further cooling and drying of the continents.

Life of the Oligocene is characterized by the success of the large land mammals such as titanotheres and rhinoceros. One member of the rhinoceros family, *Indricotherium*, was the largest mammal to walk the Earth, standing about 5.5 meters tall at the shoulder. The mammal fauna of the Oligocene continued to become more modern in appearance; monkeys and other apes began to develop at this time.

The Miocene epoch (between 23.7 and 5.3 million years ago) represents a time of marked change in Earth history. Although throughout the Cenozoic era the fauna and flora have been very modern in appearance, the Miocene marks a time when fauna and flora began to resemble modern ones closely. In the marine realm, invertebrates resembled modern ones, and, in fact, many modern species of invertebrates trace their origins to the Miocene epoch. The expansion of the whales in the Miocene was the greatest marine-related change. On land, the flora began to be dominated by small, nonwoody plants. These plants were well adapted to life under somewhat dry, cool conditions. Many of the modern common families of wildflowers began their rapid expansion in the Miocene. Small mammals, such as the rodents, and the perching birds became more common. Large mammals were present, but very large forms, such as the titanotheres, were extinct by the Miocene. In the Miocene, two major additions to the carnivores, the bear and hyena families, developed.

The tectonic and environmental conditions of the Miocene are closely akin to modern processes. One major worldwide tectonic event was the beginning of the uplift of the Himalaya as India began to move beneath the Asian continent. Another important event was the closure and evaporation of the entire Mediterranean Sea. Whether this event, known as the Messinian Event, was caused by tectonic processes or was related to a worldwide drop in sea level, or some combination of the two, is still debated among scientists. No matter what the cause, such a phenomenon represents a major change in the Earth's environment. As a result of this major change, development of widespread continental glaciation began in the Southern Hemisphere. Global conditions produced by this process were lower sea level and dryer conditions in the Northern Hemisphere.

PLIOCENE, PLEISTOCENE, AND HOLOCENE EPOCHS

The Pliocene epoch (between 5.3 and 1.6 million years ago) represents a time of equable climate over much of the Earth, partly because of a rise in sea level that began at this time. Other major events that began in the Pliocene were the renewed uplift of the Rocky Mountains and the uplift of many of the other mountain ranges throughout the American West. Two new locations for seafloor spreading began to develop. The Red Sea began to expand at this time, as did the Gulf of California. Continued volcanism between North and South America caused a land bridge to develop between the two landmasses. A sharp climatic change began in the late Pliocene with the widespread development of continental glaciers. Life of the Pliocene resembled the life of the Miocene in many ways. One important development was the first well-preserved fossils of the hominids, the family to which the human species belongs.

The Pleistocene epoch (between 1.6 million years ago and 10,000 years ago) is a time in Earth

During the Pleistocene epoch, a subdivision of the Cenozoic era, the last of the truly large mammals became extinct. This model of an imperial mammoth is shown mired in a tar pit in modern Los Angeles. (© William E. Ferguson)

history dominated by glaciers. The Pleistocene is commonly referred to as the "ice age," but there were many advances and retreats of the great continental ice sheets during this epoch. Much of the modern Earth surface acquired its present appearance as features such as the Great Lakes and many present-day river systems were formed. During the Pleistocene epoch, the last of the truly large mammals became extinct. Forms such as mammoths, mastodons, and giant sloths disappeared from the fauna. Some of these extinctions appear to be related to the expansion and success of the human species. A part of this expansion includes the arrival of humans in the Americas more than 30,000 years ago, possibly over a land bridge between Siberia and Alaska, which was the result of lower sea level in the present-day Bering Sea.

The Holocene epoch began about 10,000 years ago. This date is approximately the time that the last glacier retreated from the temperate latitudes. The Holocene epoch is not recognized by all geologists as different from the Pleistocene. The question is whether the present conditions on the Earth mark a fundamental change in the Earth's climate or whether they simply mark another warm period between glacial episodes. One thing the Holocene does characterize is the importance of humans as a geologic agent and the ability of humans to reshape the environment. For this reason, many geologists recognize the Holocene as a unique epoch in Earth history.

Richard H. Fluegeman, Jr.

CROSS-REFERENCES

Archean Eon, 1087; Biostratigraphy, 1091; Cretaceous-Tertiary Boundary, 1100; Fossil Record, 1009; Geologic Time Scale, 1105; Ice Ages, 1111; Land Bridges, 1116; Mesozoic Era, 1120; Microfossils, 1048; Paleoclimatology, 1131; Paleogeography and Paleocurrents, 1135; Paleoseismology, 1139; Paleosols, 1144; Paleozoic Era, 1126; Proterozoic Eon, 1148; Stratigraphic Correlation, 1153; Transgression and Regression, 1157; Unconformities, 1161; Uniformitarianism, 1167.

BIBLIOGRAPHY

Cooper, John D., Richard H. Miller, and Jacqueline Patterson. *A Trip Through Time: Principles of Historical Geology.* Columbus, Ohio: Merrill, 1986. A basic introduction to historical geology. Excellent discussions of the Cenozoic history of western North America, illustrations that are clear, and a text that is fairly easy to read. Designed as an introductory text for college students, but the readership level is such that any interested individual can gain information from this book. A glossary is included.

Dott, Robert H., Jr., and Donald R. Prothero. *Evolution of the Earth.* 5th ed. New York: McGraw-Hill, 1994. This basic textbook on historical geology is aimed at students of geology. However, it is very readable by anyone with a background in science. Presents an up-to-date account of the Earth's history from the viewpoint of plate tectonics. Includes a glossary.

Lane, N. Gary. *Life of the Past.* 2d ed. Columbus, Ohio: Merrill, 1985. A basic introduction to the study of paleontology, fossils, and life of the past. An excellent text for the amateur or student interested in the animals of geologic history. In a clear style, the characteristics of the different fossil groups are explained. Illustrations of the different groups are included. Each chapter is followed by a series of key words, and an extensive glossary is included at the end of the text. Written for a general audience with a high school education, but more advanced students will find the book useful.

Levin, Harold L. *The Earth Through Time.* 3d ed. Philadelphia: W. B. Saunders, 1988. A thorough, well-illustrated text on historical geology. The text is well organized and proceeds in a logical manner. Illustrations are abundant, and selected ones are in full color. Although the book is written for first-year college students, the general reader will find much of interest. Two of the four appendices are especially helpful: One is a classification of living things, and the other is a summary of important rock sections of various ages, including the Cenozoic. A glossary and an index are included.

Press, Frank, and Raymond Siever. *Understanding Earth*. 2d ed. New York: W. H. Freeman, 1998. This comprehensive physical geology text covers the formation and development of the Earth. Includes sections on the Cenozoic. Readable by high school students, as well as by general readers. Includes an index and a glossary of terms.

Prothero, Donald R. *Bringing Fossils to Life*. Boston: McGraw-Hill, 1998. This well-illustrated and entertaining text covers a broad range of paleontological topics, including biostratigraphy and the use of fossils in the study of Cenozoic rocks. Glossary, bibliography, and index.

Stanley, Steven M. *Earth and Life Through Time*. 2d ed. New York: W. H. Freeman, 1989. This book is a beautifully illustrated and thorough treatment of Earth history. Provides a detailed account of the evolution of the Earth and its life-forms. A major feature of this text is the worldwide approach used to discuss major events. Examples from outside North America are discussed. Also, very clear illustrations of the positions of the continents through time are included, as well as illustrations of animals and plants in their environments. Although a bit more difficult to read than other texts on historical geology, it can be understood by college students, and the illustrations make this book worthwhile for any interested person.

Wicander, E. Reed, and J. S. Monroe. *Historical Geology*. St. Paul, Minn.: West Publishing, 1989. Intended as a basic introduction to historical geology for first-year college students. Beautifully illustrated, containing many full-color diagrams and photographs. Also contains many case histories written in clear, nontechnical language. Although designed for college students, the writing style is accessible to the general reader.

CRETACEOUS-TERTIARY BOUNDARY

The Cretaceous-Tertiary boundary, 66 million years ago, is the junction between the Mesozoic and Cenozoic eras. This boundary coincides with a major extinction of marine and terrestrial organisms, the most conspicuous of which were the ammonoid cephalopods in the sea and the dinosaurs on the land. A 10-kilometer-diameter bolide that collided with Earth at this time has been invoked by some as the cause of these extinctions.

PRINCIPAL TERMS

BOLIDE: a meteorite or comet that explodes upon striking Earth

CENOZOIC ERA: the youngest of the three Phanerozoic eras, from 66 million years ago to the present; it encompasses two geologic periods, the Tertiary (older) and the Quaternary

CRETACEOUS PERIOD: the third, last, and longest period of the Mesozoic era, 144 to 66 million years ago

ERA: a large division of geologic time composed of more than one geologic period

EXTINCTION: the disappearance of a species or large group of animals or plants

FAMILY: a grouping of types of organisms above the level of a genus

MESOZOIC ERA: the middle of the three eras that constitute the Phanerozoic eon (the last 544 million years), which encompasses three geologic periods—the Triassic, the Jurassic, and the Cretaceous—and represents Earth history between about 250 and 66 million years ago

STRATUM (pl. STRATA): a single bed or layer of sedimentary rock

TERTIARY PERIOD: the earlier and much longer of the two geologic periods encompassed by the Cenozoic era, from 66 to 1.6 million years ago

MASS EXTINCTION

The Cretaceous-Tertiary boundary is a point in geological time located 66 million years before the present. It corresponds to the junction between the geological eras known as the Mesozoic, of which Cretaceous is the youngest subdivision, and the Cenozoic, of which Tertiary is the oldest subdivision. This boundary coincides with (and, using fossils, is recognized by) a major extinction of marine and terrestrial organisms. This extinction is not the most massive extinction in the history of life; the Paleozoic-Mesozoic extinction, 250 million years ago, holds that honor. The extinction at the end of the Cretaceous is, however, the most talked about extinction in Earth history because it was during this time that the dinosaurs disappeared.

When British paleontologist John Phillips coined the terms Mesozoic and Cenozoic in 1840, he already knew that they represented time intervals in Earth history characterized by very different types of organisms. It was not until the beginning of the twentieth century, however, that paleontologists recognized the full significance of the boundary between the Mesozoic and Cenozoic eras. By 1900, about a century of scientific collecting and study of fossils demonstrated that many types of organisms had become extinct at or just before the Cretaceous-Tertiary boundary. This extinction thus ended what is popularly termed "the age of reptiles," setting the stage for the appearance and proliferation of the types of organisms that have inhabited Earth for the last 66 million years, or what is popularly called "the age of mammals."

In examining the extinctions that took place in the seas at the end of the Cretaceous, scientists have learned that about 15 percent of the families (or approximately one hundred families) of shelled invertebrates became extinct. Particularly hard-hit groups were the ammonoid cephalopods, relatives of living squids and octopi, who suffered total extinction; clams and gastropods (snails), who endured significant losses; and the marine reptiles, the mosasaurs (giant marine lizards) and plesiosaurs (long-necked reptiles), who vanished altogether. Major changes also occurred in the marine plankton, and the foraminiferans (microscopic shelled protozoans) also suffered heavy losses. On land, the flying reptiles (pterosaurs) and the dinosaurs became extinct, many types of marsupial mammals disappeared, and a few types of flowering plants, especially broad-leafed forms and those living in low latitudes, died out.

After the extinction, the land surface was populated by many placental mammals, which rapidly diversified during the early Tertiary; by turtles, crocodiles, lizards, snakes, and other reptiles little affected by the extinction; and by birds and flowering plants, groups not seriously impaired by the extinctions. In the sea, the most conspicuous Mesozoic denizens—ammonoids, mosasaurs, and plesiosaurs—were gone, as were some types of clams, especially the reef-building rudists and the platelike inoceramids. However, many other clams survived, as did representatives of the other hard-hit invertebrate groups. The plankton and bony fish recovered, and sharks remained unscathed by the extinctions.

PROBLEMS OF EXTINCTION CRITERION

The Cretaceous-Tertiary boundary is almost always identified by the extinctions that took place at that time. Thus, in the sequence of strata, certain fossil groups (for example, dinosaurs) are present in Cretaceous rocks but are absent in Tertiary rocks. Using the criterion of extinction, however, to identify the Cretaceous-Tertiary boundary produces two significant problems.

The first of these problems stems from the inherent diachrony of extinction—in other words, the fact that an extinction almost always does not occur simultaneously across the geographic range of an organism. Thus, hippopotamuses have been undergoing extinction for thousands of years and disappeared from Europe and Asia a few thousand years ago. They are now restricted to small areas in Africa, where they will probably suffer extinction within the next few thousand years unless human intervention saves them. With the exception of a possible pervasive global catastrophe at the Cretaceous-Tertiary boundary, why should not the extinction of many Cretaceous organisms have taken place in the same diachronous fashion as the ongoing extinction of the hippopotamus? Indeed, some paleontologists believe that there is evidence that dinosaurs became extinct in South America after their extinction in North America. If this is correct, then what is identified as the Cretaceous-Tertiary boundary in North America is older than what is identified as the boundary in South America. This presents a serious problem for placing the Cretaceous-Tertiary boundary, which, ideally, should represent the same point in time everywhere.

The second problem faced when using extinctions to identify the Cretaceous-Tertiary boundary is the circularity of reasoning that can result; that is, if one identifies the Cretaceous-Tertiary boundary by the extinction of dinosaurs, one must be careful in saying that dinosaurs became extinct at the Cretaceous-Tertiary boundary. What if, as some believe, dinosaurs survived longer in some parts of the world than in others? To determine if this was the case, another criterion (usually another group of fossils) must be used to determine the age of the youngest dinosaur fossils.

An outcrop exposing a K-T boundary ejecta layer near Beloc, Haiti. Cretaceous limestone is overlain by the ejecta layer, composed of tektites pseudomorphed by clay. Darker Tertiary limestone overlies the ejecta layer. (Geological Survey of Canada)

EXTINCTION THEORIES

Without question, the most intriguing aspect of the Cretaceous-Tertiary boundary is what caused all of the extinctions. In order to answer this question, the timing of these extinctions must be determined. Did they occur simultaneously and suddenly? If so, then a major catastrophe of global proportions apparently was their cause. If, however, the extinctions were not simultaneous, and if some groups of organisms were already in decline prior to the Cretaceous-Tertiary boundary, then a single catastrophe alone cannot explain the extinctions.

In 1979, Nobel physics laureate Luis Alvarez, his geologist son Walter Alvarez, and two nuclear chemists, Frank Asaro and Helen Michel, proposed that a bolide (a comet or meteorite) 10 kilometers in diameter collided with Earth 66 million years ago and caused the extinction of the dinosaurs and other groups of organisms that died out at the end of the Cretaceous. They initially based this proposition on chemical analysis of a clay layer at Gubbio in northern Italy. This clay layer was deposited at the bottom of the sea 66 million years ago, and the chemical analysis revealed that it contains an unusually large concentration of the platinum-group metal iridium. Such a high concentration of iridium, reasoned Alvarez and his colleagues, could not be produced by known terrestrial mechanisms and thus must have settled in the dust produced by a huge bolide impact. They later identified a site at Chicxulub on the Yucatán Peninsula in southern Mexico as the point of the bolide impact.

Geological studies at other localities worldwide where 66-million-year-old rocks are preserved have confirmed the Alvarez team's proposition of a bolide collision with Earth 66 million years ago. Their claim that the bolide impact is linked directly to the Cretaceous-Tertiary-boundary extinctions has not fared as well. Indeed, the fossil evidence indicates that many groups of organisms in the sea (for example, the ammonoids and inoceramid clams) and on the land (dinosaurs) were declining millions of years before the Cretaceous-Tertiary boundary. Furthermore, some groups of organisms (rudist clams are an example) became extinct 1 million or more years before the boundary. Also, there is some evidence, hotly debated, that a few types of dinosaurs may have survived into the earliest Tertiary. Nevertheless, the fossil evidence is not without its detractors, since many fossils remain to be discovered, and the suddenness and synchrony or diachrony of some extinctions still is subject to debate.

A dispassionate reading of the existing fossil evidence does not support a single, mass extinction at the Cretaceous-Tertiary boundary. Instead, it suggests that, as a result of changing climates and sea levels, a period of extinction beginning 3 to 5 million years before the Cretaceous-Tertiary boundary was culminated by the final disappearance of several groups of organisms at (or perhaps just after) the end of the Cretaceous. Perhaps the bolide impact at the Cretaceous-Tertiary boundary is best interpreted as the last piece of bad luck encountered by a Mesozoic biota already doomed to extinction.

STUDY OF THE CRETACEOUS-TERTIARY BOUNDARY

Research on the Cretaceous-Tertiary boundary first must focus on locating the boundary in strata of a given region. To facilitate this, there are two places—Stevns Klint in Denmark and Gubbio in Italy—where by international agreement the position of the Cretaceous-Tertiary boundary is fixed in the strata. Identifying the boundary elsewhere on Earth has thus been reduced to a problem of stratigraphic correlation, the method by which the equivalence in age or position of strata in disparate areas is determined. The goal of the fieldworker then has to be identifying criteria (usually fossils) by which correlation with the Cretaceous-Tertiary boundary in Denmark or Italy can be demonstrated.

Since the Cretaceous-Tertiary-boundary rocks in Denmark and Italy were deposited at the bottom of the sea 66 million years ago, it is sometimes difficult to identify good criteria for stratigraphic correlation in 66-million-year-old rocks that were deposited on land. In these rocks, the youngest dinosaur fossils usually are believed to mark the Cretaceous-Tertiary boundary until other evidence demonstrates otherwise. This other evidence sometimes comes from fossil pollen grains, numerical ages, or other geophysical techniques, such as studying the magnetic properties of the rocks in order to determine their age.

Once the boundary has been placed with confidence, other aspects of studying the Cretaceous-Tertiary boundary are even more complex. They focus on the extinctions themselves and their potential causes. Data and techniques from many fields are brought to bear here, including paleontology (the study of fossils), sedimentology (the study of how sediment is transported and deposited), and geochemistry (the study of rock chemistry). At its simplest, in a given sequence of strata that encompasses the Cretaceous-Tertiary boundary, the goal of research is to collect and document the vertical ranges of all fossils, their relative abundances, and how their ranges and abundances correspond to environmental changes indicated by the sediments and rock chemistry.

One of the problems these studies face is the incompleteness of the fossil record. For example, when paleontologists think that they have found the youngest dinosaur fossil in a local sequence of strata, how can they be sure? Maybe younger dinosaurs lived in the area and their fossils were not preserved, or, if they were preserved, the fossils may not have yet been found. This caveat makes it difficult, especially in rocks deposited on land, where fossil occurrence often is very spotty, not only to be certain of the position of the Cretaceous-Tertiary boundary but also to be confident of the correspondence between fossil range, fossil abundance, and environmental changes indicated by sediments and rock chemistry. The potential for new fossil discoveries always exists. This is only one reason that research on the Cretaceous-Tertiary boundary continues at a fast pace and that the cause of the extinctions at and around this boundary remains a subject of heated debate.

DINOSAUR MYSTERY

One of the most interesting aspects of the extinctions at the Cretaceous-Tertiary boundary is the disappearance of the dinosaurs. Dinosaurs included the largest land animals of all time and dominated Earth's surface for 150 million years.

Why such large and seemingly successful reptiles died out has captured the imagination of scientist and layperson alike for more than a century. More important, understanding extinctions in the past, such as those that took place at the Cretaceous-Tertiary boundary, may allow humankind to understand the causes and effects of massive extinctions. This understanding may, in turn, help humans avoid extinction in the future. Insight into these past extinctions would also provide some basis for understanding the potential effects of the ongoing extinction of species in the tropical regions of the globe.

Finally, there is seemingly incontrovertible evidence that a large bolide impacted Earth 66 million years ago. The effects of this impact have been likened to the "nuclear winter" that might result from a global thermonuclear war. Such a nuclear winter would be a period of intense cold when all incoming sunlight is blocked out by the smoke accumulated in the atmosphere from continent-wide forest fires. Analogous conditions may have existed on Earth during the first ten to one hundred years that followed the bolide impact at the Cretaceous-Tertiary boundary. Studying the effects of this impact thus provides insight into a global disaster of horrific proportions and, if nothing else, is an inducement to the human species to avoid such a cataclysm.

Spencer G. Lucas

CROSS-REFERENCES

BIBLIOGRAPHY

Alvarez, Luis. "Mass Extinctions Caused by Large Bolide Impacts." *Physics Today* 40 (July, 1987): 24-33. This very polemical article presents a strong argument for the bolide impact at the Cretaceous-Tertiary boundary having caused sudden and simultaneous mass extinctions. It also relates a very readable chronology of the Alvarez team's work on the iridium-rich clay layer and the other lines of evidence and arguments that ensued. This is the late Luis Alvarez's last written word on the subject. Well illustrated and referenced.

_____. *T. Rex and the Crater of Doom.* Princeton, N.J.: Princeton University Press, 1997. A history of the discovery of the data that has been put forward to support an impact-caused extinction of dinosaurs. For the general reader.

Archibald, J. D. *Dinosaur Extinction and the End of an Era.* New York: Columbia University Press, 1996. A thorough review of dinosaur extinction and advocacy of a prolonged event caused by falling sea levels culminated by the impact of an asteroid. Intended for general readers.

Hsü, Kenneth J. *The Great Dying: A Cosmic Catastrophe Demolishes the Dinosaurs and Rocks the Theory of Evolution.* New York: Harcourt Brace Jovanovich, 1986. Hsü extensively reviews and accepts the ideas of the Alvarez team. He then argues against typical notions of Darwinian evolution to support the idea that major crises (extinctions) are the driving force of evolution. Some very debatable ideas are wrapped up in this well-written, novel-like book. Indexed but lacks illustrations and references.

Lucas, Spencer G. *Dinosaurs: The Textbook.* 3d ed. Boston: McGraw-Hill, 2000. This book provides good coverage of the extinction of the dinosaurs. Intended for university undergraduates but readable by anyone with an interest in dinosaurs. Glossary, bibliography, and index.

Officer, C., and J. Page. *The Great Dinosaur Extinction Controversy.* Reading, Mass.: Addison-Wesley, 1996. This text critiques the impact hypothesis at length and advances the view that massive volcanism was the cause of the extinction of the dinosaurs. For the general reader with some scientific background.

Stanley, S. M. *Extinction.* New York: Scientific American Books, 1987. A very readable, extensive treatment of the subject of extinction. Chapter 7 reviews the extinctions at the end of the Cretaceous and elegantly reduces the welter of data and viewpoints to explain why the fossil record does not support a single, massive extinction at the Cretaceous-Tertiary boundary. Well illustrated and indexed.

Ward, Peter. "The Extinction of the Ammonites." *Scientific American* 249 (October, 1983): 136-141. A very readable and extensively illustrated article that presents the evidence that ammonoids were declining well before the end of the Cretaceous. Ward sees this decline as a losing battle against more mobile, shell-crushing predators. No references.

GEOLOGIC TIME SCALE

Geologic science has contributed to modern thought the realization of the immense time involved in the Earth's history. So vast is this time span that the term "geologic time" is used to distinguish it from other kinds of time.

PRINCIPAL TERMS

BRACHIOPOD: a bivalved filter feeder; clams and oysters are the modern equivalents

BRYOZOAN: a colonial marine animal very much like modern sponges

FAUNAL SUCCESSION: the sequence of life-forms, as represented by the fossils within a stratigraphic sequence

GEOLOGIC MAP: a map illustrating the age, structure, and distribution of rock units

HOLOTYPE: the definitive example of a specimen, used to compare all others

LITHOLOGY: the mineral composition and texture of a rock

PALEONTOLOGY: the science of ancient life-forms and their evolution as studied through the analysis of fossils

STRATIGRAPHIC SEQUENCE: a set of rock units

that reflect the geologic history of a region

TONGUE STONES: an ancient colloquial term used to describe what is now recognized as fossil sharks' teeth; if viewed from the convex side, a large shark's tooth might resemble a tongue turned to stone

TRILOBITE: a many-legged arthropod named for its three symmetrical lobes; the principal index fossil for the Cambrian period

UNCONFORMITY: a surface that separates two strata; represents a gap in time in which no geologic records remain

UNIFORMITARIANISM: the general principle that the Earth's past history can be interpreted in terms of what is known about present natural laws, as these processes differ neither in degree nor in kind

DEVELOPMENT OF HISTORICAL GEOLOGY

The ancient civilizations were very indefinite regarding time periods. For Strato and Eratosthenes, invertebrate fossils were evidence only of ancient seas having existed. Herodotus associated vertebrate fossils with Greek mythology, concluding that large fossilized bones were remnants of battles between giants and their gods. During the mid-seventeenth century, the science of historical geology began to branch from the trunk of natural philosophy and develop its own identity. The first step toward that development was a new understanding about fossils. In the fall of 1666, fishermen fishing off the west coast of Italy caught a great white shark. Word of this unusual fish spread to the Medici court in Florence, where the Grand Duke Ferdinand II ordered the head cut off and brought for examination to Neils Stensen (known as Nicolaus Steno), a young Danish doctor serving the court. Steno recognized the strong resemblance of the shark's teeth to tongue stones. He made the intuitive leap that contrary to common belief, tongue stones did not grow in the ground but had their origin in the heads of sharks. The problem was to account for the transposition of the teeth from the shark's head into the solid rock that enclosed them. Through a series of critical observations and deductions, Steno arrived at three basic tenets of modern geology: Layered rocks result from sediments settling out of water, the oldest strata are on the bottom, and the strata originally are deposited in an essentially horizontal position. Steno published these conclusions in *Prodromus*, his great work of 1669.

Robert Hooke gave Steno's ideas a wide hearing in lectures he presented before the Royal Society of London in 1667-1668. Hooke's main contribution to the budding science of historical geology was his support of the fossils' organic origin, the extinction of species, the change within species over time, and the theory that subterranean forces have caused the continents to rise and fall with respect to the sea. Thomas Burnet's very controversial *Sacred Theory of the Earth* (1681-1689) sparked further debate about the origin and changes in the Earth's surface features. Burnet called his theory sacred because it would justify by reason the biblical doctrines, specifically the Fall and the Universal Deluge. Burnet held that the Earth was around 4,000 years old and that the Universal Deluge occurred about 1,600 years later. Benoît de Maillet, a French diplomat, proposed in *Telliamed* (1748) a theory that put the age of the Earth at more than 2 billion years. Maillet based his theory on the Sun's life expectancy and on observations of the fall of sea level. In this work, he supported two fundamental ideas: the belief that terrestrial life originated in the sea and Aristotle's "infinite age of the Earth," which became Maillet's "vast amount of time" to build mountains layer by layer from strata once submerged.

Another description of the Earth's age appeared in the efforts of a second Frenchman, Georges-Louis Leclerc, the Comte de Buffon. Beginning in 1749, Buffon published a comprehensive multivolume work with the modest title *Natural History*, in which he divided the history of the Earth into seven epochs. Buffon's contribution to the question of age was an empirical study of the cooling rates of iron. In his own foundry, he heated to incandescence and then cooled iron balls of different diameters. He recorded the time for each ball to cool, extrapolating from his results the time it would take for a ball the size of the Earth to cool to the current level: 96,670 years and 132 days. Buffon generated a second timetable based on sedimentation rates observed in oceans. The variable deposition rates reported revealed a considerable range of time. The longest estimate placed the Earth's age at nearly 3 million years and had life appearing between 700,000 and 1 million years. These ideas circulated widely because of Buffon's strong influence on the intelligentsia of his time, which included correspondence with Benjamin Franklin and Thomas Jefferson. Buffon contributed two important ideas that helped to build a sense of time. First, he expanded the age of the Earth to millions of years. Second, he showed that scientists could understand past geologic events by observing the causes of change that are in operation today.

UNCONFORMITIES AND FAUNAL SUCCESSION

In 1785, James Hutton published "Theory of the Earth" in the *Transactions of the Royal Society of Edinburgh*. The ideas expressed in this essay and his later elaborations of them are the beginning of modern geology. He concluded that if humans could measure the rate at which erosion destroys lands, that rate could be used to calculate the time needed to form the strata observed in the field. The rates Hutton observed were so small, however, that he questioned the possibility of measuring them in a year or even a lifetime.

The vastness of the time required to describe the Earth's history came to Hutton as he observed what modern geologists call an unconformity (a gap of time in the geologic record). Hutton observed this sequence of strata at Siccar Point near St. Abbs, Scotland, where a sequence of horizontal strata rested on a sequence of vertical strata. He concluded that if all strata began as horizontal, a vast amount of time must have elapsed to produce the present configuration. According to Hutton, first the sediments had to be deposited in a marine environment, where they solidified. Next, as a result of the Earth's internal heat, the horizontal sediments rose above sea level to a vertical position and eroded. Then they were submerged in their vertical position, and a new deposition placed horizontal strata on top of the vertical strata. Finally, the whole structure again rose and eroded to expose the structure. Accounting for this history required a time factor of enormous scale. In Hutton's closing essay he suggested an infinite time frame: "The result, therefore, of this present inquiry is, that we find no vestige of a beginning, no prospect of an end."

In the late eighteenth and early nineteenth centuries, a new principle for determining the geologic ages of fossiliferous strata emerged from field studies in England and France. This was the principle of faunal succession. In a simplified form it stated that within sequences of strata, different kinds of fossils succeed one another in a definite order. The Englishman William Smith most graphically demonstrated this principle. Smith was a self-taught surveyor and civil engineer working on the construction of canals in England in 1794. In the course of these excavations, he discovered that each of the formations revealed a distinctive fossil species. Extrapolating this knowl-

THE GEOLOGIC TIME SCALE

MYA	Eon	Era	Period	Epoch		Developments
0.01	Phanerozoic Eon (544 mya-present)	Cenozoic (65 mya-today)	Quaternary (1.8 mya-today)	Holocene (11,000 ya-today)		Ice Age ends; humans begin to impact biosphere
1.8				Pleistocene (1.8 mya-11,000 ya)		Glaciation leads to Ice Age; modern humans evolve
5			Tertiary (65-1.8 mya)	Pliocene (5-1.8 mya)		Cooling period leads to Ice Age
23				Miocene (23-5 mya)		Erect-walking human ancestors
38				Oligocene (38-23 mya)		Primate ancestors of humans
54				Eocene (54-38 mya)		Intense mountain building: Alps, Himalaya, Rockies. Modern mammals: rodents, hoofed animals
65				Paleocene (65-54 mya)		Cretaceous-Tertiary event (?) leads to dinosaurs' extinction c. 65 mya
165		Mesozoic (245-65 mya)	Cretaceous (146-65 mya)			Birds arise; breakup of super-continents into present form
208			Jurassic (208-146 mya)			Earliest mammals
245			Triassic (245-208 mya)			Dinosaurs develop
286		Paleozoic (544-245 mya)	Permian (286-245 mya)			Permian extinction
325			Carboniferous (360-286 mya)	Pennsylvanian (325-286 mya)		Supercontinent Pangaea forms
360				Mississippian (360-325 mya)		Reptiles
410			Devonian (410-360 mya)			Amphibians; vascular plants; diverse insects
440			Silurian (440-410 mya)			Early land plants, insects
505			Ordovician (505-440 mya)			Life colonizes land; earliest vertebrates appear in fossil record
544			Cambrian (544-505 mya)	Tommotian (530-527 mya)		Cambrian diversification of life
900	Precambrian Time (4500-544 mya)	Proterozoic (2500-544 mya)	Neoproterozoic (900-544 mya)	Vendian (650-544 mya)		Earliest invertebrates Marine plants, animals
1600			Mesoproterozoic (1600-900 mya)			
2500			Paleoproterozoic (2500-1600 mya)			Transition from prokaryotic to eukarotic life leads to multicellular oganisms, c. 2 bya
3800		Archaean (3800-2500 mya)				Microbial life as early as 3.5 bya
4500		Hadean (4500-3800 mya)				Earth forms 4.5 bya

NOTES: mya = millions of years ago; bya = billions of years ago.

SOURCE: Data on time periods in this version of the geologic time scale are based on new findings in the last decade of the twentieth century as presented by the Geologic Society of America, which notably moves the transition between the Precambrian and Cambrian times from 570 mya to 544 mya.

edge to other regions allowed him to identify and predict the stratigraphic sequence. In 1814, Smith consolidated his findings in what geologists often consider the first geologic map. The next year he published his major work, *Delineation of the Strata of England and Wales, with Part of Scotland*. Smith was recognized in his own time as the father of English geology. His contribution to dating was to utilize the fossils in determining the relative ages of strata.

Already at the beginning of the nineteenth century, European geologists had begun a systematic classification of fossiliferous strata into coherent units or periods. The first of the geologic time periods appeared in 1799 with the work of Alexander von Humboldt, who applied the name Jurassic to a coherent sequence of fossiliferous limestone strata found in the Jura Mountains of Switzerland and France. Just to the west of these mountains, the limestones and their associated fossils dip under a dominantly chalky sequence studied by the Belgian geologist Omalius d'Halloy. He gave it the name Cretaceous in 1822. The pattern of naming rock units, based on the lithology and associated fossils, for the geographic region in which they were first described continued through the nineteenth century.

DEFINITION OF PERIOD BOUNDARIES

The original boundaries of these periods were neither distinct nor easily translated outside the holotype regions. Methods for defining the boundaries resulted from a dispute in England over a sequence of strata that Adam Sedgwick and Sir Roderick Impey Murchison described in 1835. Murchison based his chronological sequence on the fossil order that he observed and named it the Silurian. Concurrently, Sedgwick had relied on lithology to establish a sequence of strata that lay below the Silurian and was therefore older. He called it the Cambrian. Initially, the two periods seemed to create order for these strata; however, under closer examination the periods overlapped, and dispute developed between Sedgwick and Murchison over the commonly held strata. The answer appeared in 1879 after their deaths, when Charles Lapworth separated the systems based on fossils. Lapworth collected evidence that illustrated that the Cambrian-Silurian sequence actually contained three distinct fossil assemblages

and resolved the dispute by removing the lower Silurian from its previous classification and renaming it the Ordovician. The discrimination of fossil assemblages provided the key to distinguishing other increments of geologic time, such as the Devonian and Permian periods.

By the middle of the nineteenth century, geologists recognized that the rock units they studied in one location were not universal geographically. Their systems were highly variable in lithology and thickness from region to region. The distinctiveness of the fossils within each group enabled them to translate from one geographic area to another. The power of this approach was that the order of the sequence was not random but predictable; for example, bryozoans and corals characterize the Ordovician period the world over. Each has its own distinctive suite of fossils. Recognition of the sequencing of life-forms through time had a profound effect on Charles Darwin. Indeed, Darwin's paleontological investigations led him to focus on the idea of a species changing through time. Later, in his *On the Origin of Species by Means of Natural Selection* (1859), Darwin pointed out that natural selection was a viable concept if and only if enough time had elapsed for its operation.

In 1862, Lord Kelvin made the first attempt to calibrate the geologic time scale. Working with the thermodynamic laws of heat production and radiation, he calculated the cooling rates of Earth and the Sun. Kelvin's calculations set a "natural" upper limit to their ages and indicated that Earth had solidified from the original molten state between 20 million and 400 million years ago. This was much less than the time scale advocated by the uniformitarians. Geologists thus found themselves no longer in a position to assume unlimited time; their theories had to fit the time interval established by Kelvin's thermodynamic studies. Resolution of the time interval came in 1896 with the discovery of radioactivity by Antoine-Henri Becquerel, Pierre Curie, and Marie Curie. This was the beginning of a chain of events that revolutionized science and expanded the geologic time range into thousands of millions of years.

RADIOACTIVE DATING

In 1902, Pierre Curie announced that radioactive minerals constantly radiate heat. Two years later, Ernest Rutherford established that the amount of heat they radiate is proportional to the number of alpha particles they emit. John Joly provided additional support for Rutherford's discovery in his 1909 publication, *Radioactivity and Geology*. Joly demonstrated that the heat from radioactive decay within the Earth could alter the Earth's actual cooling rate to make it appear younger than it really was. Kelvin's calculations of 20 to 400 million years did not include the masking effect of internal radioactive heat, and thermodynamics alone therefore no longer established the boundaries. Geologists and biologists legitimately could claim a longer time interval for the evolutionary process. Continued investigations revealed that radioactive minerals might serve not only as sources of heat but also as clocks to date the rocks that contained them.

In 1905, Lord Rayleigh and Sir William Ramsay calculated an age of 2,000 million years for a specimen containing uranium. The American physical chemist Bertram Boltwood, noting that uranium ores always contain lead, speculated that lead might be the end product of a uranium decay series. By comparing the ratio of lead to uranium in forty-three minerals, he calculated their ages and obtained results ranging from 400 to 2,200 million years. Boltwood's results were the first quantitative proof of the Earth's age. Then, in 1913, Frederick Soddy and, independently, Kasimir Fajans demonstrated that radioactive elements can have the same chemical properties but slightly different atomic masses. Their discovery of isotopes was the next step in establishing intervals on the time scale. This development enabled researchers to measure the decay rate or half-life of a radioactive isotope.

By the late 1920's, Francis William Aston had begun to use mass spectrometry to make significant improvements in isotopic analysis. Alfred Otto Carl Nier continued to improve the mass spectrometric technique in the 1930's and 1940's, providing the most accurately determined values of uranium-lead ratios found in naturally occurring uranium minerals. Earlier, in 1913, the English geologist Arthur Holmes had made the first attempt at a quantified time scale using Boltwood's calculations of radioactive decay. In a 1948 paper, he developed an expanded time scale based on Nier's more recent values. By the 1960's, J. L. Kulp was continuing Holmes's investigations, introducing the rubidium-strontium decay series to date rock samples.

Mass spectrometric investigations continued throughout the twentieth century, and by the mid-1970's, the discovery and dating of Precambrian rocks in Greenland, South Africa, Australia, and Canada yielded ages of about 3.7 billion years. The Earth appears to be even older. Its age seems bracketed somewhere between these ancient terrestrial rocks and the lunar rocks, which date from 4.6 billion years. Isotope analysis of meteorites supports this same age range.

Anthony N. Stranges and Richard C. Jones

CROSS-REFERENCES

Archean Eon, 1087; Asteroids, 2640; Biostratigraphy, 1091; Cenozoic Era, 1095; Cretaceous-Tertiary Boundary, 1100; Fossil Record, 1009; Ice Ages, 1111; Land Bridges, 1116; Mammals, 1038; Mass Extinctions, 1043; Mesozoic Era, 1120; Meteors and Meteor Showers, 2711; Paleoclimatology, 1131; Paleogeography and Paleocurrents, 1135; Paleoseismology, 1139; Paleosols, 1144; Paleozoic Era, 1126; Proterozoic Eon, 1148; Stratigraphic Correlation, 1153; Transgression and Regression, 1157; Unconformities, 1161; Uniformitarianism, 1167.

BIBLIOGRAPHY

Albritton, Claude C., Jr. *The Abyss of Time: Unravelling the Mystery of the Earth's Age.* San Francisco: Freeman, Cooper, 1980. A very readable history emphasizing the people who shaped the concept of geologic time. Suitable for college freshmen and the interested reader.

Berry, William B. N. *Growth of a Prehistoric Time Scale, Based on Organic Evolution.* New York: W. H. Freeman, 1968. A good introduction to the history of the geologic time scale. Well written and illustrated. Suitable for college-level students.

Briggs, Derek E. G., and Peter R. Crowther, eds.

Paleobiology: A Synthesis. Oxford: Blackwell Scientific Publications, 1990. This multiauthored text deals with all aspects of paleontology in a series of short articles by experts in each field. Intended for university students but also useful to high school students and interested readers.

Dott, Robert H., Jr., and Donald R. Prothero. *Evolution of the Earth.* 5th ed. New York: McGraw-Hill, 1994. This basic textbook on historical geology is aimed at students of geology. However, it is very readable by anyone with a background in science. Presents an up-to-date account of the Earth's history from the viewpoint of plate tectonics. Includes a glossary.

Prothero, Donald R. *Bringing Fossils to Life.* Boston: McGraw-Hill, 1998. This well-illustrated and entertaining text covers a broad range of paleontological topics, including biostratigraphy and the development of the geologic time scale. Glossary, bibliography, and index.

ICE AGES

Several periods of Earth's history were marked by major glacial episodes. Possible causes of these episodes include the movement of Earth's continental areas into higher latitudes, changes in atmospheric composition and motion, and changes in Earth's orbit. Effects of the ice ages included substantial changes in sea level and coastal topography, alteration of lake and river drainage, and the shifting of plants and animals accompanying the changing climates.

PRINCIPAL TERMS

ALBEDO: the amount of solar energy reflected by the Earth's surface back into space; an increase in albedo is believed to lead to lower temperatures and stimulates the expansion of glaciers

GLACIATION: a major formation of land ice and the period in which it occurs

GLACIER: an accumulation of ice that flows viscously as a result of its own weight; it forms when snowfall accumulates and recrystallizes into a granular snow (firn, or névé), which

becomes compacted and converted into solid, interlocking glacial ice

INTERGLACIAL PERIOD: the interval of milder climate between two major glacial episodes

PLEISTOCENE: the epoch of Earth history characterized by the presence of large ice sheets in the higher latitudes of the Northern Hemisphere, approximately 2 million to 10,000 years before the present

SCABLAND: a region characterized by rocky, elevated tracts of land with little soil cover and by postglacial dry stream channels

PERIODS OF WIDESPREAD GLACIATION

There have been several periods in Earth's history in which large glaciers have covered substantial portions of Earth's surface. These "ice ages" were created during times in which more snow fell during the winter than was lost by ablation (melting, evaporation, and loss of ice chunks in water) during the summer. Approximately 2 billion years ago, during the Proterozoic eon, large ice sheets covered substantial portions of North America, Finland, India, and southern Africa as indicated by glacially deposited sediments termed tillites. Such glacial deposits have also been found in rocks of late Ordovician and early Silurian age in northern Africa, approximately 440 to 420 million years before the present. Glacial episodes intensified on the Gondwanaland continent during the Carboniferous (Mississippian and Pennsylvanian periods of North American classification) and Permian periods, as tillites and glacially scoured areas indicate the presence of vast ice sheets over large areas within the southern portions of South America, Africa, India, and Australia. Antarctica was probably almost completely covered with ice sheets at this time, and there are indications that

the ice sheets expanded and retreated at intervals within these periods, spanning some 360 to 245 million years before the present.

Although these earlier glacial episodes may represent substantial cooling periods for portions of Earth's surface, the term "ice age" has almost become synonymous with the last great glacial episode at the end of the Cenozoic era. There is evidence that within South America, this glaciation began somewhere between 7 and 4.6 million years ago. In most areas, the major glacial episodes spanned the latter part of the Pliocene and the Pleistocene epochs, a period between 3 million and 10,000 years before the present. During this interval, ice sheets with thicknesses of 3 kilometers or more accumulated over much of the higher latitudes of the Northern Hemisphere.

Traditionally, these late Cenozoic ice ages have been separated into four subdivisions. In the European Alps, these included, from oldest to youngest, the Gunz, Mindel, Riss, and Würm glacials. These periods were believed to be separated by warmer interglacial periods termed the Gunz-Mindel, Mindel-Riss, and Riss-Würm. In North America, a fourfold subdivision was also utilized, which included (from oldest to youngest) the Nebraskan, Kansan, Illinoian, and Wisconsin glacials. These were divided by the Aftonian, Yarmouth, and Sangamon interglacials, respectively. This fourfold subdivision seemed to be represented in other global regions as well. By the late 1970's, however, studies on deep-sea sediments indicated that the Alpine glacial sequence covered at least eight glacial cycles rather than four or five. Because severity of glaciation depends on both latitude and altitude as well as on local climatic factors, it has become apparent that such a simplistic classification for the last great ice age is untenable. Interdisciplinary studies are better establishing these glacial episodes and more precisely dating the glacial-interglacial cycles.

CAUSES OF ICE AGES

The causes of the ice ages may be quite varied. One possible explanation seems to be related to the distribution of continental areas as a result of plate tectonics. According to this theory, Earth is divided into a series of rigid plates that shift their position relative to one another because of the movement of underlying ductile material. Calculations on the position of Gondwanaland during the Carboniferous-Permian glaciation indicates that the southern portion of the supercontinent was situated over the South Pole. Such a position stimulated the growth of large ice sheets. The position of the continents during the last great ice age may also have caused the growth of ice sheets, as the most prominent glacial areas were at higher latitudes. Plate tectonics cannot, however, entirely account for late Cenozoic glaciation, as several glacial-interglacial periods have been recorded during a relatively brief period in which Earth's plates could not have been repeatedly repositioned in "colder" and "warmer" latitudes. Therefore, other explanations need to be sought to determine the specific causes of the last ice age.

Another explanation for the initiation of ice ages is that at certain intervals, the Sun's luminosity decreases. Proposed mechanisms have been the movement of the Sun through a dense interstellar cloud of gases and particles or fluctuations in the energy output of the Sun. Other suggestions have concerned Earth's albedo, or the amount of solar energy reflected back into space. Twice during the nineteenth century, eruption of volcanoes in Indonesia resulted in colder winter temperatures worldwide as the result of a blockage of the Sun's radiation by particles of floating volcanic ash. Short intervals of declining temperatures following these volcanic eruptions led to the hypothesis that severe climate changes could have been caused by more severe volcanism. It has also been suggested that the emergence of land areas during a relative drop in sea level also increases albedo because of the greater reflectivity of terrestrial surfaces. Such episodes may lead to a decrease in the relative temperature at Earth's surface. The greenhouse effect, in which an increase in carbon dioxide content in the atmosphere leads to greater temperatures, may also indirectly result in increased albedo. The higher temperatures may cause greater evaporation rates, and therefore more clouds would form. In turn, the tops of the clouds would reflect the Sun's energy, possibly leading to a drop in temperature and initiating glaciation. A decrease in carbon dioxide content, however, could lead directly to the ice ages, because a decrease in this heat-trapping gas would cause a concomitant decrease in temperature. Another hypothesis suggests that mountain building along continental margins, where Earth's lithospheric plates collide, may have resulted in the formation of substantial mountain glaciers at the higher altitudes. The increased albedo resulting from the presence of these highly reflective ice surfaces may have resulted in cooler temperatures and more ice buildup, setting off a chain reaction leading to larger continental glaciers and major glacial episodes. There is no direct evidence, however, that volcanic eruptions, interstellar clouds and solar phenomena, emergence of landmasses, creation of mountain chains, or carbon dioxide fluctuations have led to glacial episodes.

Two hypotheses seem to fit the evidence better as pertains to the initiation of glaciation. Approximately 3.5 million years ago, the isthmus of Panama emerged, apparently resulting in the strengthening of the Gulf Stream's northward flow. This strengthening may have fed more moisture to the northern high latitudes, therefore creating more snowfall and thus a buildup in glaciation. One of the most widely accepted theories concerning the origin of the ice ages was proposed by a Yugoslavian mathematician, Milutin Milankovitch. During the 1920's and 1930's, Milankovitch studied the possible effects of variations in Earth's orbit upon the timing of glacial episodes. These effects include the angle of the ecliptic (axial tilt), the precession of the equinoxes, and the eccentricity of Earth's orbit. At present, Earth's axial tilt is approximately 23.5 degrees from the perpendicular to the plane of Earth's orbit around the Sun. According to Milankovitch's calculations, Earth's axial tilt would vary from approximately 22.1 to 24.5 degrees every 41,000 years. As the angle of the ecliptic is the primary factor producing the seasons, such cyclicity may lead to significant climate change.

The second possible cause of climate change involves a variance in the eccentricity of Earth's orbit. A complete cycle between times of maximum orbital eccentricity occurs at approximately 93,000-year intervals, which closely corresponds to the twenty cold-warm cycles recorded in deep-sea cores from the last ice age. The final cycle theorized by Milankovitch involved the precession of the equinoxes: The movement of Earth's axis would approximate the wobbling of a spinning top, with a periodicity of 21,000 years, which would cause a slow shift in the position of the solstices and equinoxes through time. As Earth's climate may be affected by each of these three cycles, the combination of their effects may at times create significant climate changes. As evidence from deep-sea cores seems to support the theory that Milankovitch's cycles correspond to glacial periodicity during the last ice age, this theory has become especially popular for explaining the origin of the major glacial periods.

Whatever caused the initiation of the ice ages, once large glaciers began forming, their presence may have stimulated further ice buildup. With more ice, Earth's albedo would increase, with resulting lower temperatures. Also, a drop in sea level accompanying glacial buildup would likely have led to variation in oceanic and atmospheric circulation patterns, along with a relative rise in altitude of the landmasses. Such a cycle may have been self-perpetuating, with more ice buildup creating more severe glaciation.

EFFECTS OF GLACIATION

The effects of widespread glaciation were varied and profound. One result was a relative decrease in sea level, as more moisture became locked within the glacial ice. Estimations vary as to the amount of sea-level drop, although many scientists believe that it was 75 meters or more. This drop resulted in many changes in shoreline topography. Land bridges were formed between the British Isles and Europe, as well as between Asia and North America across the Bering Strait. The ice ages also greatly affected regional drainage patterns. Prior to Pleistocene glaciation, the northern portions of the Missouri and Ohio Rivers drained toward the northeast. With incursion by the great ice sheet, called the Laurentide Ice Sheet, drainage patterns changed, eventually resulting in the modern southward drainage patterns. Because of this ice sheet's great weight, the continental crust beneath the glacial areas was depressed as much as 300 meters. As the glaciers melted and receded, the Great Lakes were formed within these basins.

Even outside glacial areas, large lakes were formed because of the increased rainfall in certain areas. One of the largest of these pluvial lakes was Lake Bonneville, the remnant of which constitutes the Great Salt Lake of Utah. Glaciers also created large lakes by damming watercourses. In the Pacific Northwest, glacial Lake Missoula was a glacially dammed lake covering an area almost 8,000 square kilometers in extent. The disastrous collapse of the dam occurred during glacial retreat, with the ensuing catastrophic flooding creating the Channeled Scablands of eastern Washington. Another effect of glaciation is downcutting and subsequent erosion by rivers as a result of the lowering of sea level. Once the glaciers melt, sea level again rises, and rivers deposit large amounts of sediment in their floodplains.

Even in areas not covered by glacial ice, climates were affected. Studies indicate that glacial periods in the higher latitudes corresponded to drier (interpluvial) periods in temperate and tropical regions. Pluvial periods were essentially equivalent to the interglacials of higher latitudes.

Another possible effect of changing climates within and between glacial periods may be the selective extinction of animal species. At the end of the last great ice age, especially in North America, many types of large mammals became extinct. These extinctions have been theorized as resulting from the direct or indirect influence of climate change or, alternatively, as the result of the inva-

THE LAURENTIDE ICE SHEET

sion of North America by Paleo-Indians across the Bering Strait land bridge. A seemingly less disastrous change that occurred during the Pleistocene—but one that is just as crucial for understanding the climate changes during the past glacial-interglacial periods—involves the displacement of plants and animals. The distributions through time of a wide variety of Pleistocene invertebrates, vertebrates, and plants have been documented. During periods of glaciation, forests diminished, and desert areas, steppes, and grasslands increased in size. The peculiar presence of seemingly cold- and warm-climate species of mammals within the same deposits have led some scientists to speculate that some areas in front of the glaciers had a moderate climate. The winters were warmer (allowing the warm-climate species to live

within the area), and the summers were cooler (enabling the cold-climate species also to become established in the same region). The fossil distributions of plants and animals during the past ice age indicate that they were often quite different from those observed at present.

Phillip A. Murry

CROSS-REFERENCES

BIBLIOGRAPHY

Anderson, Bjørn G., and Harold W. Borns, Jr. *The Ice Age World.* Oslo: Scandinavian University Press, 1994. The well-illustrated text details the Quaternary history of North America and northern Europe over the last 2.5 million years. Contains an extended glossary, reference list, and index. For the general reader as well as the serious student.

Bloom, Arthur G. *Geomorphology: A Systematic Analysis of Late Cenozoic Landforms.* 3d ed. Englewood Cliffs, N.J.: Prentice Hall, 1998. This college-level text covers the basics of geomorphology. Includes three chapters on glaciers and glaciology. Index and bibliography.

Chernikoff, Stanley. *Geology: An Introduction to Physical Geology.* Boston: Houghton-Mifflin, 1999. This is a good overview of the scientific understanding of the geology of the Earth and surface processes. Includes sections on glaciers, glaciology, and glacial deposits. Includes a link to a Web site that provides regular updates on geologic events around the globe.

Dolgoff, Anatole. *Physical Geology.* Lexington, Mass.: D. C. Heath, 1996. This is a comprehensive guide to the study of the Earth. Extremely well illustrated and includes a glossary and an index. Although this is an introductory text for college students, it is written in a style that makes it understandable to the interested layperson. Contains a section on the development of glaciers and the types of glacial deposits.

Lowe, J. J., and M. J. C. Walker. *Reconstructing Quaternary Environments.* London: Longman Group, 1984. This volume outlines the techniques for analyzing the environments and ecology of the Quaternary period. Included are discussions of landforms and characteristic sediments of the Quaternary, as well as dating techniques and analyses of plants and animals. Although the text would most easily be read by students with introductory training in geology and biology, all terms are thoroughly defined, and it could thus be comprehended by a general audience.

Nilsson, T. *The Pleistocene: Geology and Life in the Quaternary Ice Age.* Stuttgart: Ferdinand Enke Verlag, 1983. One of the most thorough accounts of the ice ages available. The bulk of the text concerns continental reviews as to the extent and characteristics of Pleistocene glaciation, as well as the mammal faunas and human fossils found. Emphasis is on the European record, although chapters are devoted to each of the other continents. Although all terms are thoroughly defined, the text is primarily designed for upper-level students of biology, paleontology, and geology.

Sutcliffe, A. J. *On the Track of Ice Age Mammals.* Cambridge, Mass.: Harvard University Press, 1985. This volume is written for a general audience with no formal training in geology or paleontology. Well written, it avoids unnecessary scientific jargon and is profusely illustrated. Several chapters give a detailed summary of the chronology and general features of the ice ages. Other chapters cover the general characteristics of Pleistocene mammals, modes of preservation, and regional accounts of Pleistocene mammalian faunas.

LAND BRIDGES

The theory that Earth's continents were once connected by land bridges that accounted for the migration of flora and fauna was considered viable by many leading geologists during the last half of the nineteenth century and for the first three decades of the twentieth century.

PRINCIPAL TERMS

CENOZOIC ERA: the geologic era dating from the present to about 65 million years ago

CONTINENTAL DRIFT: the theory that landmasses separated and drifted apart in prehistoric times

CRETACEOUS ERA: a geologic period ending some 65 million years ago, during which seas covered much of North America and the Rocky Mountains were formed

EOCENE EPOCH: part of the Cenozoic era, dating to about 37 million years ago

GLACIATION: the process of being covered by an ice sheet or glacier

ISTHMIAN LINKS: chains of islands between substantial landmasses

LAND BRIDGES: narrow land formations that connect landmasses

MESOZOIC ERA: the geologic era spanning the period from about 245 million years ago to about 65 million years ago

CHANGING EARTH

For geologists, Earth has long been a huge, complicated jigsaw puzzle waiting to be reconstructed. The difference between this jigsaw puzzle and the ones with which most people are familiar is that this one is constantly changing. Earth changes in small ways—and sometimes in larger ways, such as when a volcano erupts—on a daily basis. It changes more drastically over longer periods of time. It is now thought to be changing as global warming slowly melts the polar ice caps, with a resulting increase in the volume of ocean waters and a corresponding decrease in the shorelines adjacent to the rising oceans.

It has long been acknowledged that drastic climatic changes have occurred on Earth over millions of years. The Arctic and Antarctic, now solidly frozen, once had tropical and subtropical vegetation, as is evident from impressions of plants found in mineral deposits in these regions and from other paleontological evidence. Now-moderate regions were, during the Ice Age, much colder. Glaciation pushed debris unrelentingly from one region to another as the glaciers moved toward the equator.

Long baffled by the existence of similar flora and fauna in areas seemingly unrelated to each other climatically—Africa, Australia, and Antarctica, for example—geologists arrived at various conclusions regarding the phenomenon. It must be remembered that geologists talk in terms of hundreds of millions of years more often than they do in terms of a century or a millennium. Geological change often occurs so slowly as to be virtually unnoticeable within such geologically abbreviated time spans as thousands of years.

LAND-BRIDGE THEORIES

Among the theories promulgated to explain the presence of similar flora and fauna in widely disparate areas was that of land bridges. Some land bridges obviously existed, such as the one across the Bering Strait that linked Siberia to what is now Alaska. It is apparent that two continents, Asia and North America, were at one time linked by this broad bridge. Similarly, some geological conjecture also has it that land bridges existed between present-day Ireland and Scotland and between present-day Gibraltar and North Africa.

A Viennese geologist, Edward Suess, proposed an extensive land-bridge theory in his influential book *The Face of the Earth* (1885). Suess coined the term "Gondwanaland," named after Gondwana in east-central India, an area with rocks that revealed unique fossil plants and showed indisputable signs of glaciation dating from the late Paleozoic to the early Mesozoic eras.

As Suess pieced together the discrete elements of his geological puzzle, he found that separate continents had the sorts of fossil plants and glacial rocks that he found in Gondwana. He attributed these occurrences in widely separated areas to land bridges rather than to continental drift. Suess, and the scores of notable geologists who accepted his thesis, contended that the land bridges connecting the continents sank beneath the ocean at some time in prehistory.

Other geologists were trying to solve the same riddles with which Suess was confronted. In 1908, F. B. Taylor published a privately printed pamphlet in which he suggested that present mountain ranges occurred when enormous landslides advanced slowly and steadily from the polar regions toward the equator. He conceived of the world's great continents as originally consisting of huge sheets of rock that were torn apart by glaciation as gigantic fields of ice moved toward zero latitude.

Taylor advanced two fundamental ideas, both of which have been opposed by other geologists to the point that these two hypotheses have been discredited. First, he speculated that mountain ranges were products of thousands of miles of lateral movement; second, he believed that the Moon became associated gravitationally with Earth during the Cretaceous period, about 100 million years ago. In this period, there was a considerable increase in the variety of flora and fauna found on Earth. Taylor suggested that in the earliest times, the Moon created phenomenal tidal forces that slowed the rate at which Earth rotated and pulled the continents away from the poles. Both of these theories were subsequently disproved.

H. H. Baker, an American geologist, considered the supercontinent of Earth's earliest years to have been split along the Atlantic and Arctic Oceans as the Miocene period neared its end some 7 million to 8 million years ago. He asserted that great tidal

A low coastal plain that is part of the former, and now mostly submerged, Bering Strait Land Bridge. (© William E. Ferguson)

action had torn a huge piece from Earth. The resulting void, according to Baker, became the Pacific Ocean. The portion that was catapulted into space, he speculated, became the Moon.

In his landmark 1915 book *Die Entstehung der Kontinente und Ozeane* (*The Origin of Continents and Oceans*, 1924), Alfred L. Wegener proposed that Earth had at one time consisted of a single land mass that he called "Pangaea," the Greek term for "all Earth." He concluded that during the Mesozoic era, the single landmass was torn apart, with the southern continents (Africa, Australia, South America, Antarctica) being pulled away first. Looking at the globe as though it were a jigsaw puzzle, one can imagine that the east coast of South America would fit easily into the declivities in the west coast of Africa.

EVIDENCE FOR LAND BRIDGES

As this investigation into Earth's origins was ongoing, the fossil record uncovered during the 1870's and 1880's led to the inevitable conclusion that land links, either land bridges or isthmian links, had existed between Africa, India, Australia, South America, and Antarctica. These areas of vastly differing climates in modern times (geologically speaking) appeared to share a fossil plant record that suggested a close affinity among them.

F. H. Knowlton, speaking in 1918 about Mesozoic floral relations, supported the notion that land bridges from Antarctica once linked most of the southern landmasses. Writing in 1947, however, after the land-bridge theory had been largely discredited, Theodor Just suggested that "the needs of animal and plant geographers . . . vary sufficiently with their respective interests and so do the land bridges assumed by them." Just discounted land links as deciding factors in the migration of flora.

During the Eocene epoch, the southeastern United States experienced an influx of tropical flora, suggesting to some that a land bridge or isthmian link had once existed between South America and what is now the United States, possibly between South America and southern Florida. Modern researchers suggest that this tropical foliage actually came to the southeastern United States from an intermediate region such as Central America or the Antilles.

During the Oligocene epoch, a considerable exchange of plant life appears to have taken place between present-day Panama and the Antilles. Petrified wood found on Antigua reflects several forms of flora found in both Panama and southeastern America. While the land-bridge or isthmian-link theory might explain this coincidence, other, more viable theories exist to explain it. Among these is the theory that certain flora simply drifted on the open sea, perhaps hitchhiking on debris that eventually washed ashore in a place where the climate could support its germination and growth.

During the Pleistocene epoch, North and South America were connected. The Isthmus of Panama did not exist, and the region of what is now the Antilles was considerably higher than it is today, possibly with a land bridge to South America, less than a hundred miles away, and certainly with isthmian links to that continent. Vegetation common to the rain forests of the Amazon and Orinoco basins occurred in southern Florida, but most geologists consider it improbable that a land bridge existed between that region and the Antilles or South America.

On the other hand, those who dispute the existence of a land bridge between North and South America are on shakier ground than those who dispute a land bridge between Florida or the Antilles and South America. Vertebrate paleontologists have made a convincing case for animal migration (including human) in this area, as they have for animal migration between Asia and North America by way of the land bridge that is known to have existed between Siberia and present-day Alaska. In 1917, Edward W. Berry postulated that the Antilles were once a part of South America but that continental drift separated the region from that continent. Such a theory would explain much of the coincident flora and fauna in both places.

Numerous catastrophic theories regarding continental drift have been used to explain the separation of Earth's landmass into continents. Most of these theories have to do with the separation of the Moon from Earth, accompanied by the violent tidal action that followed this separation. As late as 1932, Charles Schuchert supported the notion that land bridges connected parts of an enormous supercontinent and that these bridges subsequently sank beneath the deep waters of the Atlantic, Pacific, and Indian Oceans, as suggested several decades earlier by Edward Suess.

The major portions of the jigsaw puzzle on which geologists have worked for the past century are now firmly in place. Other portions do not yet fit into any reasonable pattern. The land-bridge theory, although it explains certain animal migrations in a limited geographical range, has been displaced by other theories relating to diverse flora and fauna in such widely separated areas as India, Antarctica, Africa, and South America. The major importance of the land-bridge hypothesis is that it demanded testing, and this testing led to valuable insights about Earth's earliest history.

R. Baird Shuman

CROSS-REFERENCES

BIBLIOGRAPHY

Carey, S. Warren. "A Tectonic Approach to Continental Drift." In *Continental Drift: A Symposium. Being a Symposium on the Present Status of the Continental Drift Hypothesis, Held in the Geology Department of the University of Tasmania, in March, 1956.* Hobart: University of Tasmania, 1958; revised 1959. Carey presents one of the most balanced overviews of the continental-drift/land-bridge controversy in this somewhat specialized article. Important, but not easily accessible to the beginner.

Fuentes, Carlos. *The Buried Mirror: Reflections on Spain and the New World.* New York: Houghton-Mifflin, 1992. Celebrating the five hundredth anniversary of Christopher Columbus's first trip to the New World, Fuentes traces with considerable skill the migration patterns that accounted for the presence on the American continent of a large population of people whose ancient ancestors were of Asian origins and who probably arrived in America by way of the Bering Sea land bridge. A most engaging book.

Press, Frank, and Raymond Siever. *Understanding Earth.* 2d ed. New York: W. H. Freeman, 1998. This comprehensive physical geology text covers the formation and development of the Earth. Discusses the idea of land bridges and why such a theory was not accepted. Readable by high school students, as well as by general readers. Includes an index and a glossary of terms.

Ross, Charles A., ed. *Paleobiogeography.* Stroudsburg, Penn.: Dowden, Hutchinson, and Ross, 1976. Ross has gathered in one compendious volume the salient works of most of the geologists and paleontologists who worked toward evolving the land-bridge/isthmian-link theory, as well as the work of those who questioned and eventually discredited the theory.

Shea, James H., ed. *Continental Drift.* New York: Van Nostrand Reinhold, 1985. The section entitled "Gondwana" and Alfred Wegener's chapter "The Origins of Continents" are particularly valuable to those interested in the land-bridge theory. The presentations are clear and informative.

Suess, Edward. *The Face of the Earth.* Oxford, England: Oxford University Press, 1909. This translation of Suess's *Das Anlitz der Erde* (1885) is essential reading for anyone seriously interested in this field. Suess was the major early proponent of the land-bridge/isthmian-link theory. He presents his salient arguments, since discredited, in this book.

Sullivan, Walter. *Continents in Motion: The New Earth Debate.* 2d ed. New York: American Institute of Physics, 1991. Sullivan presents considerable information about land bridges and isthmian links, offering a comprehensive overview of the theory and its eventual loss of credibility. The presentation is well balanced and accessible to people not specifically trained in the field.

Wegener, Alfred. *The Origin of Continents and Oceans.* New York: Dover Books, 1966. This translation of Alfred Wegener's landmark book *Die Entstehung der Kontinente und Ozeane* (1915), in combination with Suess's *The Face of the Earth*, helps one to understand the considerable controversy that Suess's hypothesis evoked and some of the reactions to it. Fortunately, this relatively recent translation of the original work is available in many college and university libraries.

MESOZOIC ERA

The Mesozoic era was a major episode in Earth history during which a primitive flora and fauna and physical environment changed and became progressively more familiar. During this era, continents and ocean basins nearly achieved their present configuration and dinosaurs and other large reptiles flourished.

PRINCIPAL TERMS

EPICONTINENTAL SEA: a sea covering part of a continental block; such seas generally are less than 200 meters deep and are the depositional site of most exposed sedimentary rocks

GONDWANALAND: an ancient, large continent in the Southern Hemisphere that included Africa, South America, India, Australia, and Antarctica

LAURASIA: an ancient, large continent in the Northern Hemisphere that included North America and Eurasia

PANGAEA: the supercontinent containing all continental crust that existed at the beginning of the Mesozoic

PERIOD: a unit of geological time forming part of an era and subdivided, in decreasing order, into epochs, ages, and chrons

RIFTING: a process of faulting and basaltic intrusion occurring where crustal plates separate during continental drift; may cause mountains

SUBDUCTION: a process by which one crustal plate rides over another, which descends and melts, generating molten rock that then intrudes the deformed plate above; it causes mountains

SYSTEM: the rocks deposited during a period, which is defined by age of the rocks making up its system

TETHYS: a seaway embayed into Pangaea between the southeast corner of Asia, the western end of the Mediterranean, and the southeast end of Pangaea

BREAKUP OF PANGAEA

The Mesozoic era and system are, respectively, a major subdivision of geologic time and the rocks of that age. The Mesozoic began about 225 million years before the present and ended about 65 million years before the present. It was preceded by the Paleozoic era and followed by the Cenozoic era and is divided into three periods: Triassic (about 225 to 180 million years before the present), Jurassic (about 180 to 144 million years before the present), and Cretaceous (about 144 to 65 million years before the present).

All the continents were gathered in a single large landmass, the supercontinent Pangaea, at the beginning of the Triassic period. South America, Africa, Antarctica, Australia, New Zealand, Arabia, and Peninsular India previously formed the supercontinent Gondwanaland during the Carboniferous (about 350 to 285 million years before the present), while North America, Greenland, Eurasia

(less Peninsular India), and Borneo formed Laurasia. Laurasia and Gondwanaland merged during the Permian (about 285 to 225 million years before the present) by welding northwest Africa and South America to the south and east margin of North America. At this time, the Atlantic Ocean and Gulf of Mexico did not exist, but a large wedge-shaped seaway, Tethys, separated the former Laurasia and Gondwanaland blocks between the Mediterranean and opened to the east. The equator transected Mexico, the Sahara Desert, and northern India, continuing eastward to divide the Tethys seaway. The South Pole was slightly offshore of the western base of the Palmer Peninsula, and the North Pole was in eastern Siberia.

During the Late Triassic, Pangaea began to break up as North America moved away from Europe and northwest Africa. The resultant narrow North Atlantic Ocean and Caribbean Gulf of Mexico widened in the Jurassic and Cretaceous, reaching something near the present size and shape by the end of the period. Starting in the Jurassic, North America pulled away from South America, leaving a Caribbean Gulf of Mexico seaway connecting the Atlantic and Pacific. The South Atlantic opened later in the Cretaceous, resulting in a relatively narrow strait between Africa and South America by the end of the Mesozoic. The Late Triassic also was the time during which India, Australia, and Antarctica began moving away from Africa as a single block. India separated from the Australian-Antarctica block as movement continued, and, in the Late Jurassic, Madagascar split from Africa. Cenozoic movement then divided Antarctica and Australia and fused Peninsular India to Asia, eventually resulting in the present configuration of lands surrounding the Indian Ocean.

By the end of the Cretaceous, the equator passed just north of South America, crossed Africa at the southern margin of the Sahara, clipped the southwest corner of Arabia, traversed India along a north-south line through western India, and crossed Sumatra and Java. The North Pole was in the Arctic Ocean off western Siberia, and the South Pole lay within Antarctica.

Movement of the continents and opening of ocean basins caused compressional deformation of the Earth's crust accompanied by massive igneous intrusion. The Andes and the mountains of western North America resulted from North and South America moving westward over oceanic crust in the Pacific Basin. Thrust faulting, folding, and batholithic (igneous) intrusion began in the Late Jurassic from California and Nevada northward to British Columbia and Alaska. This process continued during the Cretaceous, spreading to South America, the Palmer Peninsula, and northwest Antarctica. Similar deformation associated with subduction also took place along the northern margin of the Tethys seaway from the Mediterranean through the Balkans, Caucasus, and into Indonesia. Rifting along the East Coast of North America from Nova Scotia to Florida caused fault-block mountains and major basaltic flows as the Atlantic Ocean began opening in the Triassic and early Jurassic. Similar rifting with mountain making and basaltic flows occurred in South Africa, India, and Australia as the eastern part of Gondwanaland fragmented during the Late Triassic through the Cretaceous.

SEDIMENTARY DEPOSITION

Triassic sedimentary deposits are characterized by terrestrial red beds laid down on the supercontinent, Pangaea, under arid climatic conditions. These conditions occur between paleolatitudes (lines of latitude shown on the present pattern of land and sea where those lines existed in the past) from 10 to 30 degrees north and south. Thus, red beds were deposited north of the equator, from the north coast of Africa through Spain and Germany to the Ural Mountains and northwestward into Britain and Scandinavia. Similar rocks in East Greenland and southwestern North America are part of the same belt, as well as red beds in the rift valley system between Nova Scotia and Florida. Red beds and evaporites (salt and associated minerals precipitated from evaporating water) also accumulated in a parallel southern belt in Brazil, South Africa, Madagascar, eastern India, Australia, and Antarctica. Triassic marine rocks are sparse because Pangaea stood well above sea level during most of the Triassic, but fairly complete sequences do occur between the Alps and southern China along the northern margin of the Tethys seaway. Similarly, marine Triassic sediments occur in the Alps and western Cordillera of North America, which then were on the continental side of a subduction zone. Epicontinental seas encroached on both the Americas and Eurasia from these bordering areas, leaving marine rocks interbedded with the great red bed accumulations on their western and southern margins.

Growth and expansion of spreading centers as Pangaea began breaking up probably caused the advance of epicontinental seas in the Jurassic and through the Middle Cretaceous. Resultant marine sedimentary rocks are widespread in North America from the Rocky Mountains westward and in western Europe. In addition, growth of mountain ranges along continental margins, as a result of either subduction or collision, led to extensive alluvial (river or stream) deposition. Also, relatively deep sea deposition continued in subduction zones bordering the Pacific and the northern shore of Tethys. Epicontinental seas also encroached on the western Americas prior to the Late Jurassic and were responsible for extensive deposits in western Europe. As the North Atlantic opened, Late Jurassic deposits first filled the Nova Scotia to Florida rift zone and then overlapped the continent in the Middle Jurassic and into the Cretaceous. Also, a great evaporite basin in the Gulf of Mexico area accumulated Jurassic salt deposits. Rising mountains from California to western Canada shed extensive alluvial deposits eastward as far as central Kansas during the latest Jurassic. Widespread alluvial deposits associated with basalt flows in India, Australia, Antarctica, Madagascar, and New Zealand. Terrestrial, coal-bearing rocks also border the northern and western shores of marginal seas from Iran to Siberia.

Processes responsible for Jurassic depositional events accelerated in the Cretaceous, with maximum expansion of epicontinental seas in the middle of the period. Drying and filling of most of these seas followed as a result of expanded mountain making toward the end of the Cretaceous. Thus, deep-water, continental margin deposition continued around the Pacific perimeter and on the southern margin of Eurasia until these basins were engulfed by growing mountain chains. Epicontinental seas expanded over Europe west of the Urals, western and northern Africa, central Australia, western and northern South America, and on the Atlantic coast of North America. In addition, a new seaway opened between the Arctic Ocean and the Gulf of Mexico, separating the rising mountains of the western Cordillera from the eroded stumps of the Appalachians. Widespread chalk deposits characterize these seas in areas remote from actively rising mountains, including the notable chalk cliffs of Dover and Normandy. The rising western North and South American mountains, however, shed abundant detritus to their east in the form of alluvial plains, deltas, and muddy marine deposits, which very nearly filled the North American midcontinent seaway by the end of the Mesozoic. Most, but not all, of the other Cretaceous epicontinental seas either dried up or were greatly constricted at the end of the period or early in the Tertiary.

PLANT LIFE AND DINOSAURS

Early Mesozoic plants resembled those of the preceding Paleozoic. Primitive Carboniferous coal swamp plants persisted in restricted moist environments in the Northern Hemisphere but died out at the end of the Triassic. Cycadeoids, gingkoes, conifers, and ferns dominated Triassic through Early Cretaceous floras, with the cycads, primitive

conifers—including the auracariaceans of the Petrified Forest—and gingkos first expanding in the Triassic. This plant life was joined by more modern conifers and early relatives of cypress in the warmer, moister Jurassic and Early Cretaceous. Abruptly, in the middle of the Cretaceous, modern broad-leaved trees appeared and quickly dominated the forests. Thus, Late Cretaceous forests closely resembled those of the present.

Large amphibians, along with an expanding contingent of reptiles, constituted the terrestrial vertebrate fauna at the beginning of the Triassic period. Phytosaurs (large crocodile-like reptiles) were the dominant reptiles, but the first dinosaurs appeared in the Late Triassic, along with the first ichthyosaurs (porpoiselike reptiles), nothosaurs (ancestors of plesiosaurs), and the first turtle. The first frogs also were Triassic. Mammal-like reptiles (reptiles ancestral to mammals), living mostly in the Southern Hemisphere, declined during the Triassic but probably gave rise to the first mammals late in the period.

Rapid dinosaur diversification in the Jurassic led to bipedal (two-footed) and quadrupedal (four-footed) forms as well as both herbivores and carnivores. *Brachiosaurus*, a quadrupedal herbivore, was probably the largest land animal of all time. Other, bizarre, nondinosaurian reptiles also appeared in the Jurassic. Ichthyosaurs and the first plesiosaurs (large animals with long necks and tails and four paddles for limbs) became abundant in the seas, and the first batlike pterosaurs invaded the skies. The only mammals were small, inconspicuous, and generalized. *Archaeopteryx*, from the Upper Jurassic of Germany, generally is considered to be the first bird.

Dinosaurs continued to dominate terrestrial Cretaceous faunas but began a slow decline culminating in extinction at the end of the period. Large sauropods (quadrupedal, herbivorous saurischians, saurischians being "reptile-hipped" dinosaurs) declined throughout and disappeared before the end of the period. Theropods (bipedal saurischians with short forelimbs), including *Tyrannosaurus*, the largest known terrestrial carnivore, also declined, but a few persisted to the end of the period. Ornithischians ("bird-hipped" dinosaurs) include a number of quadrupedal herbivores. One of these, *Triceratops*, with its distinctive three horns and fringed neck, persisted to the very end of the Cretaceous. Other, bipedal, ornithischians include the duckbilled dinosaurs. Pterosaurs also declined to extinction at the end of the period, but one of them, *Quetzalcoatlus*, was a giant, having a wingspan of 10 meters. Ichthyosaurs declined, but plesiosaurs were abundant, along with giant lizardlike mosasaurs in the marine fauna. All, however, became extinct at or before the end of the Cretaceous. The Cretaceous also saw the rise of turtles, snakes, lizards, and crocodilians—all currently prominent reptiles. Mammals of the time included both marsupials (mammals retaining the newborn in a pouch on the mother's abdomen) and placentals (mammals in which the young are not nurtured in a pouch). All of them were small, resembling modern shrews and insectivores, but include the ancestors of all modern mammals. Cretaceous birds are rare, but a diversity of toothed birds are known, including the flightless, marine, diving bird *Hesperornis*. More modern birds appear in the very latest Cretaceous.

MARINE AND INSECT LIFE

Chondrostian fish (ray-finned, but with largely cartilaginous skeletons) diversified greatly in the Early Triassic as sharks declined and more primitive fish characteristic of the Paleozoic persisted in low numbers and variety. Cartilaginous fish, including sharks, continued to be dominant in the Jurassic, but holostean fish gave rise to the first teleosts (bony fish) in the Late Jurassic. This group continued to expand and diversify in the Cretaceous and persists to the present.

Primitive insects occur sporadically in Triassic through Middle Cretaceous fossil faunas but were reduced to a minor role by an explosive appearance of very modern-appearing insects accompanying the appearance of broad-leaved forest trees in the Middle Cretaceous.

Marine invertebrates suffered extensive extinction in the Permo-Triassic transition. The Paleozoic corals, trilobites, and graptolites completely disappeared, and other groups, such as brachiopods, bryozoans, and crinoids, were decimated. By Late Triassic, however, new organisms had occupied the environments inhabited by the extinguished animals. Modern corals, not at all closely related to Paleozoic corals, already were constructing reefs. The old bryozoans were replaced by new

At Alaska's Chulitna Terrane, Triassic and Jurassic rocks are exposed in an overturned (tilted) fold: Black and white banded rocks are limestone and pillow basalt overlain by red beds (Triassic age); above these bands are Jurassic sandstone and argillite with marine fossils. (U.S. Geological Survey)

bryozoan types, and the reduction in brachiopods was matched by expansion among the pelecypods (clams, oysters, and scallops). Ammonites, descended from a single group of shelled cephalopods surviving into the Early Triassic, quickly evolved into many species, so much so that the Mesozoic is known as the age of ammonites as well as the age of dinosaurs. Other characteristic organisms appeared during the Jurassic and Cretaceous, including redistid clams (giant pelecypods that built reefs), a wide variety of echinoids (sea urchins and sand dollars), globigerinids and other planktonic foraminifera (single-celled animals secreting shells), and coccoliths (single-celled, planktonic algae secreting minute carbonate plates), along with crabs, shrimp, and lobsters.

Globigerinids and coccoliths became sufficiently abundant in the early part of the Late Cretaceous to cause worldwide deposition of chalk and chalky limestone.

By the end of the Mesozoic, many organisms had become extinct, including dinosaurs and ammonites. These extinctions, though not as extensive as those at the end of the Paleozoic, are the reason for recognizing separate Mesozoic and Cenozoic eras. Not all the characteristic Mesozoic animals that failed to survive into the Cenozoic, however, became extinct at the very end of the period. In addition, very little change occurred in the flora, as the major change took place in the middle of the Cretaceous.

Close examination of the rocks at the Mesozoic-Cenozoic boundary as defined by occurrences of planktonic foraminifera, coccoliths, ammonites, and dinosaurs has uncovered the widespread occurrence of a thin layer containing anomalous concentrations of iridium and, more questionably, sooty material and minerals of the sort associated with meteoritic impacts. This evidence has been used to advance a theory that the end-of-Cretaceous extinctions were caused by meteoritic impact, perhaps through drastic cooling caused by vast quantities of dust and smoke thrown into the atmosphere. Physical evidence of an actual impact has not convinced all researchers, however, and many paleontologists do not feel that the gradual disappearance of the Mesozoic fauna can readily be explained by an instantaneous event, even though it might be synchronous with the final extinctions.

Mesozoic rocks are rich in mineral fuels. Coal is sparse in the Triassic, reflecting the aridity of the time. Much coal, however, occurs in the Jurassic of Asia, deposited on alluvial plains north of the Tethyan orogenic belt, and even more extensive deposits in the Cretaceous of North America and northern South America are in a similar position relating to the growing western American mountains. Jurassic and Cretaceous oil and gas are very important, especially around the Gulf of Mexico and in the Persian Gulf region. Triassic ore deposits are rare because of little mountain building and igneous intrusion, but Jurassic and Cretaceous metallic ores are very extensive. The gold of the California Mother Lode is in Jurassic rocks, and many of the precious and base metal reserves of the Rocky Mountains and the Andes are Cretaceous. Similarly, the mountains formed elsewhere on the Pacific Rim and on the northern margin of the Tethys are rich in metallic ores. Finally, the diamond pipes of southern Africa are Cretaceous.

Ralph L. Langenheim, Jr.

CROSS-REFERENCES

Archean Eon, 1087; Biostratigraphy, 1091; Cenozoic Era, 1095; Climate, 1902; Continental Drift, 565; Continental Glaciers, 875; Continental Rift Zones, 579; Continental Shelf and Slope, 584; Continental Structures, 590; Continents and Subcontinents, 595; Cretaceous-Tertiary Boundary, 1100; Earth's Oldest Rocks, 516; Earth's Structure, 37; Geologic Time Scale, 1105; Ice Ages, 1111; Land Bridges, 1116; Paleoclimatology, 1131; Paleogeography and Paleocurrents, 1135; Paleoseismology, 1139; Paleosols, 1144; Paleozoic Era, 1126; Proterozoic Eon, 1148; Stratigraphic Correlation, 1153; Transgression and Regression, 1157; Unconformities, 1161; Uniformitarianism, 1167.

BIBLIOGRAPHY

Arkell, W. J. *Jurassic Geology of the World.* Edinburgh: Oliver and Boyd, 1956. A summary description of the Jurassic rocks of the world, their location and age, as determined in detail by fossil content. A professional work with an extensive bibliography.

Dott, Robert H., Jr., and Donald R. Prothero. *Evolution of the Earth.* 5th ed. New York: McGraw-Hill, 1994. This basic textbook on historical geology is aimed at students of geology. However, it is very readable by anyone with a background in science. Presents an up-to-date account of the Earth's history from the viewpoint of plate tectonics. Includes a glossary.

Levin, Harold L. *The Earth Through Time.* 3d ed. Philadelphia: Saunders College Publishing, 1988. A history of the Earth, emphasizing North America. Especially good account of the fossil record. A beginning college text.

Press, Frank, and Raymond Siever. *Understanding Earth.* 2d ed. New York: W. H. Freeman, 1998. This comprehensive physical geology text covers the formation and development of the Earth, including information on the Mesozoic. Readable by high school students, as well as by general readers. Includes an index and a glossary of terms.

Prothero, Donald R. *Bringing Fossils to Life.* Boston: McGraw-Hill, 1998. This well-illustrated and entertaining text covers a broad range of paleontological topics, including biostratigraphy and the use of fossils in the study of Mesozoic rocks. Glossary, bibliography, and index.

Stanley, Steven M. *Earth and Life Through Time.* 2d ed. New York: W. H. Freeman, 1989. All aspects of the Earth's history, including the Mesozoic, are covered in this book. Especially comprehensive for the history of life; good coverage of Europe. A beginning college text, easy to read but not superficial.

PALEOZOIC ERA

Paleozoic stratigraphy concerns the study of rock sequences that date from 544 to 245 million years before the present. The primary method by which the stratigraphic relationships of the Paleozoic are established is stratigraphic analysis of distinctive types of fossils.

PRINCIPAL TERMS

BIOSTRATIGRAPHY: defining rock layers on the basis of their fossil content

CORRELATION: matching rock units of equivalent age

INDEX (GUIDE) FOSSIL: the remains of an ancient organism that are useful in establishing the age of rocks; index fossils are abundant and have a wide geographic distribution, a narrow stratigraphic range, and a distinctive form

SERIES: a time-rock unit representing rock deposition during a geologic epoch

STAGE: a time-rock unit representing rock deposition during a geologic age

STRATIGRAPHY: the study of rock layers (strata)

SYSTEM: a time-rock unit representing rock deposition within a geologic period

CAMBRIAN SYSTEM

The Paleozoic is the oldest era of the Phanerozoic eon, ranging from approximately 544 to 245 million years before the present. In North America, it is divided into seven periods. Beginning with the oldest, these are the Cambrian, Ordovician, Silurian, Devonian, Mississippian, Pennsylvanian, and Permian. Outside North America, the Mississippian and Pennsylvanian are combined to form the Carboniferous period. In stratigraphy, strata laid down during a particular period are referred to as a system. Systems are subdivided into series (the time-rock unit equivalents of epochs), which may be further subdivided into stages (the time-rock unit equivalents of ages). These subdivisions are very important in subdividing time-rock units of the Paleozoic and other eras of Earth history.

The Cambrian is the oldest system in the Paleozoic, ranging from approximately 544 to 505 million years before the present. It was named by Adam Sedgwick in 1835 on the basis of rock exposures in northern Wales. The Cambrian system is subdivided globally into a Lower (Early), Middle, and Upper (Late) series. The earliest rocks of the Phanerozoic eon are stratigraphically difficult to interpret. One reason is that although most rocks of the Paleozoic are subdivided primarily on the basis of their included fossils, the record of Pre-cambrian and Early Cambrian life is poorly known. By Middle Cambrian times, the composition of the fossil communities had changed greatly; this proliferation of forms preserved in the rock record has enabled paleontologists to establish fairly well the stratigraphy of the Middle and Upper Cambrian series. The primary tool used in the establishment of Middle and Upper Cambrian stages is the stratigraphic distribution of trilobites, marine arthropods that somewhat resembled modern pill bugs. Acritarchs, tiny planktonic algae that form a resistant, fossilizable covering, were another group that have become useful index fossils for the Proterozoic (Upper Precambrian) through Devonian. Another common Cambrian fossil group is the brachiopods. These two-valved suspension feeders were of relatively simple form during the Cambrian, a fact that lessens their utility as stratigraphic indicators. During the Late Cambrian, a series of events caused many types of trilobites to become extinct. These events of extinction and evolutionary radiation are especially useful in precise definition of Cambrian stratigraphy. The last of these extinction episodes, which occurred at the very end of the Cambrian, eliminated numerous trilobite species. After this event, the trilobites never rebounded; in post-Cambrian strata they cannot be widely used as index fossils.

ORDOVICIAN, SILURIAN, AND DEVONIAN SYSTEMS

The second Paleozoic system is the Ordovician. It ranges from approximately 490 to 438 million years before the present. The Ordovician was named in 1879 by Charles Lapworth, who combined portions of the Cambrian and Silurian systems as first defined in Wales, thus ending a debate that had developed concerning those sequences of strata. The Ordovician may also be subdivided into Early, Middle, and Late series, and for these the stages are much better established than in the Cambrian. Strata of Ordovician age are often characterized by an abundance of carbonate (limestone and dolomite) rocks, which were deposited upon shallow seaways extending over many of the continents. Although trilobites declined in number throughout the Ordovician, many other animal groups became abundant or appeared for the first time. Especially important for biostratigraphy are graptolites, primitive hemichordate animals that lived in colonies; often they are preserved as black marks that resemble tiny hacksaw blades on rocks. Also of biostratigraphic importance are the acritarchs and the nautiloids. The latter group were predatory mollusks that at this time typically had straight shells, with tentacles emanating from one end. The chitinozoans are another group used in biostratigraphy for rocks of Ordovician through Devonian age. Although they are often classified as algae, the precise relationships of these microscopic, typically vase-shaped organisms are unknown. Finally, the evolution of more complex, hinged-shell groups of brachiopods with distinctive form enables stratigraphers to utilize this group in biostratigraphy from Late Ordovician through Early Pennsylvanian series rocks. The end of the Ordovician is marked by another mass extinction event, in which approximately one hundred families of marine animals were wiped out.

The Silurian system was named by Roderick Murchison in 1835 on the basis of rock exposures in southern Wales. The age of the Silurian ranges from 438 to 408 million years before the present. Worldwide, an Early and a Late series are recognized. In the best-known exposures of Silurian rocks, however, various series designations are used. Silurian rocks are common on all continents and were deposited in continental or marine environments. The most common rocks of the Silu-

rian are limestones and dolomites, although many dark, clay-rich marine rocks containing graptolites are present. As in the Ordovician, this group is a useful index fossil for the Silurian. Brachiopods, acritarchs, and chitinozoans are other important fossils used in determining the age of Silurian strata.

The fourth system of the Paleozoic is the Devonian, ranging from 408 to 360 million years before the present. The Devonian system was named by Murchison and Sedgwick in 1839 for rock exposures in Devon and Cornwall, in southwest England. Globally, the Devonian is subdivided into Early, Middle, and Late series. Devonian rocks are very common; they are present on all continents, where they formed in a great variety of environments. Thick sequences of Devonian-age sediments accumulated along the edges of ancient continents and record a time of changing climate and environments. On land, red bed sequences, colored by oxidation of iron-bearing minerals, are widely distributed for the first time. Of these units, the Old Red Sandstone of the British Isles is best known. As in the Ordovician and Silurian, graptolites, brachiopods, and chitinozoans are important marine index fossils for the Devonian. The proliferation of different groups of organisms also adds to the list of fossils used in stratigraphic correlation. Here, for the first time, pollen and spores can be used to define strata; they are especially important in correlating Devonian continental rocks. Also of biostratigraphic utility are the strange, tiny, toothlike conodont fossils (from animals of unknown affinity) and ostracods (minuscule, two-valved crustaceans). In addition, for the first time, ammonoids (cephalopods with typically coiled shells) are of widespread use in biostratigraphy. Goniatite ammonoids are of major importance in defining rocks of the Devonian through Permian age.

MISSISSIPPIAN, PENNSYLVANIAN, AND PERMIAN SYSTEMS

The term "Carboniferous" was first applied to rocks within the area of north-central England by William Conybeare and William Phillips in 1822. Yet no type section (the area in which rock units are first described) was ever designated. The Carboniferous lasted from approximately 360 to 286 million years before the present. Although the system is named for coal-bearing strata, only its upper portion contains large amounts of coal; the lower portion typically consists of carbonates. In North America, a different classification is employed, based largely on these rock differences. In 1870, Alexander Winchell proposed the name "Mississippian" for Lower Carboniferous exposures in the Mississippi River valley between southeastern Iowa and southern Illinois. Rocks of the Mississippian range from 360 to 320 million years before the present. In 1891, Henry Williams named Upper Carboniferous exposures the Pennsylvanian, after the widespread coal-bearing strata in that state. The Pennsylvanian ranges from 320 to 286 million years before the present. The widespread application of these names to Carboniferous-age rocks in North America led to the formal recognition of the Mississippian and Pennsylvanian as systems by the U.S. Geological Survey in 1953.

The Mississippian system (Lower Carboniferous series) is characterized by the widespread distribution of limestone and dolomite formed by marine incursions over many continents. Cyclic successions are developed in many places within Mississippian-age strata, especially within the upper portion of the system, in which sequences of limestones, sandstones, and shale were repetitively deposited one upon the other. Goniatites, brachiopods, spores, and pollen are all utilized in the Mississippian-age strata for correlation purposes.

Within North America, there is a widespread unconformity (a gap in the rock record) between the Mississippian and Pennsylvanian (Upper Carboniferous). The Pennsylvanian was a time of extensive coal deposition within low-lying swamps within North America, Europe, and many portions of Asia. Coal deposits are often found capping cyclical sequences consisting of limestone, shale, sandstone, and more shale (with the associated coal). Within the marine sequences, goniatite ammonoids are utilized for determining age and, at the base of the system, brachiopods are used for correlation. In addition, fusulinid foraminifers (small fossils resembling grains of rice but giants of this one-celled group) become important for Upper Pennsylvanian correlation. With the abundance of coal-producing plants, fossil spores and pollen serve as valuable index fossils for continental sediments of the Upper Carboniferous.

The Permian system was named for the region of Perm in the western Ural Mountains of Russia.

The system, first proposed by Murchison in 1841, roughly corresponds to a period of Earth history between 286 and 245 million years before the present. The Permian may be subdivided into a Lower and Upper series. Permian deposits are thick and widespread in many parts of the world and indicate a time of climatic complexity. Permian stratigraphic sequences are quite varied. Coal was still present but was no longer being extensively formed in North America or central Europe. Continental deposits of the Permian are primarily known for extensive red bed sequences created through the oxidation of iron-rich sediments. Extensive evaporites are also present, especially in Upper Permian-age strata, with thick layers of gypsum and salt. Large reefs are found, and, on the Gondwana continents, tillite deposits give evidence for extensive Permian glaciation. Major index fossils for the Permian include goniatite ammonoids and fusulinids for the marine sequences and spores and pollen for continental deposits. The end of the Permian is marked by an extinction event of major significance. Fusulinids, tabulate and rugose corals, and trilobites became extinct; brachiopods, bryozoans, and echinoderms sustained huge losses, and substantial numbers of bivalve and gastropod mollusks also became extinct. On land, many groups of the larger coal-swamp trees became extinct, and several major groups of vertebrates also died out. These huge losses in plant and animal types mark the stratigraphic boundary between the Paleozoic era ("the era of ancient life") and the subsequent Mesozoic era.

STUDY OF PALEOZOIC STRATIGRAPHY

Studies of Paleozoic stratigraphy primarily utilize analyses of rock units, paleontological studies, and radiometric dating techniques. Lithostratigraphy defines strata (rock layers) on the basis of rock, or lithologic, characteristics. Distinctive rock layers are very useful in determining the position and relative age of local outcroppings of Paleozoic strata. Most Paleozoic systems tend to be dominated by specific rock types. For example, carbonate rocks (limestones and dolomites) are characteristic of many Ordovician, Silurian, and Mississippian sequences. Clastic rocks, including sandstones and mudrocks with their included coals, are diagnostic of Pennsylvanian-age deposits. Red bed deposits, consisting of oxidized iron-rich sediments, are often associated with clastic sequences in Devonian- and Permian-age strata. Unconformities—gaps in the rock record representing periods of erosion or nondeposition—often separate the Paleozoic systems and form natural boundaries for their separation. A major unconformity also typically separates Precambrian rocks from those of the earliest Paleozoic.

Biostratigraphy defines rocks on the basis of their fossil content. The basic unit of biostratigraphic classification is the biozone, in which strata are characterized by the occurrence of a certain fossil or fossils. These index or guide fossils are abundant, distinctive forms with a wide geographic distribution but narrow stratigraphic range and are therefore useful in correlation and in determining the age of strata. The fossils most often used are planktonic (floating) or nektonic (swimming) forms. Of special importance for the study of Paleozoic stratigraphy are trilobites, brachiopods, and graptolites. For certain portions of the Paleozoic, acritarchs, conodonts, chitinozoans, fusulinids, pollen, and spores, as well as nautiloid and ammonoid cephalopods, are widely utilized in biostratigraphic studies.

Lithostratigraphic and biostratigraphic studies can establish only the relative ages of strata. Determination of the absolute age of Paleozoic rocks is done primarily through radiometric dating techniques. These studies utilize the known decay rates of radioactive isotopes to establish the age of rocks in terms of years. Isotopes used for studies of Paleozoic rocks are uranium 235, uranium 238, and thorium 232, all of which decay to lead. Also used are potassium 40, which decays to argon 40, and rubidium 87, whose daughter isotope (product of decay) is strontium 87. Igneous and metamorphic rocks are generally utilized in radiometric dating techniques.

SIGNIFICANCE

Probably the most important application of Paleozoic stratigraphy is in the search for natural resources. Without knowing the age of the rocks being evaluated or the economic products potentially included within them, the exploration geologist's job would be impossible. Of particular significance is the widespread presence of coal in rocks of Pennsylvanian and Permian age. Most of the coal produced in North America and central Europe has been mined from Pennsylvanian-age rocks. As these coal beds often occur within the same portion of a cyclical sequence of sediments, the search for them is made easier through knowledge of Paleozoic stratigraphy. Permian-age coal sequences are found in Australia, China, the former Soviet Union, India, and South Africa. There are also great quantities of oil and natural gas within rocks of Paleozoic age. Significant finds of hydrocarbons have been made in Devonian- and Carboniferous-age rocks, and discoveries of huge oil and gas fields in the Permian Basin of west Texas were a result of exploration of Upper Paleozoic marine strata. Arid conditions, especially common during the Permian, resulted in huge deposits of sodium and potassium salts as well as gypsum and anhydrite. Mountain-building events during the Paleozoic also created tremendous deposits of both precious metals and metals that are important for industry. Without an understanding of the stratigraphic occurrence of these valuable resources, the recovery of these materials would be impossible.

Phillip A. Murry

CROSS-REFERENCES

BIBLIOGRAPHY

Bruton, David L., ed. *Aspects of the Ordovician System.* New York: Oxford University Press, 1984. A symposium volume from the Fourth International Symposium on the Ordovician system, held in Norway in 1982. The introductory paper and a few of the papers on the stratigraphic framework and sea-level changes of the Ordovician would be understandable to a college-level student with some training in geology. The sections on Ordovician environments and their included faunas are written by and for specialists in those areas.

Dineley, D. L. *Aspects of a Stratigraphic System: The Devonian.* New York: John Wiley & Sons, 1984. A well-written and well-illustrated book covering many aspects of Devonian stratigraphy, sedimentology, and fossils. Because it is a single-authored text, it possesses much more continuity than do most symposium volumes. The text would be understandable and very readable for a college student with some training in geology.

Dott, Robert H., Jr., and Donald R. Prothero. *Evolution of the Earth.* 5th ed. New York: McGraw-Hill, 1994. This basic textbook on historical geology is aimed at students of geology. However, it is very readable by anyone with a background in science. Presents an up-to-date account of the Earth's history from the viewpoint of plate tectonics. Includes a glossary.

Levin, Harold L. *The Earth Through Time.* 3d ed. Philadelphia: Saunders College Publishing, 1987. An introductory college-level text that covers aspects of historical geology. A large section is devoted to Paleozoic-age rocks. Well written and profusely illustrated.

Press, Frank, and Raymond Siever. *Understanding Earth.* 2d ed. New York: W. H. Freeman, 1998. This comprehensive physical geology text covers the formation and development of the Earth. Readable by high school students, as well as by general readers. Includes an index and a glossary of terms.

Prothero, Donald R. *Bringing Fossils to Life.* Boston: McGraw-Hill, 1998. This well-illustrated and entertaining text covers a broad range of paleontological topics, including biostratigraphy and the use of fossils in the study of Paleozoic rocks. Glossary, bibliography, and index.

Stanley, Steven M. *Exploring Earth and Life Through Time.* New York: W. H. Freeman, 1993. This introductory college-level text is well written and well illustrated and contains very few technical errors. Would be accessible to high school students with some training in geology.

Whittington, H. B. *The Burgess Shale.* New Haven, Conn.: Yale University Press, 1985. Whittington reviews the beautifully preserved animals from the Middle Cambrian of the Rocky Mountains in Canada and gives a detailed look at the often strange animals from that famous locale. A review of the stratigraphy and sedimentology of the locality is given, along with comments on the evolutionary significance of the Early Paleozoic animals. The work is written for general audiences.

PALEOCLIMATOLOGY

Climate is the average of weather elements over long periods of time and over large areas. The study of ancient climates, termed paleoclimatology, utilizes sedimentologic, paleontologic, and geochemical data to reconstruct ancient temperature, wind patterns, precipitation, and evaporation.

PRINCIPAL TERMS

ATMOSPHERIC CIRCULATION: the movement of air as a result of regional pressure differentials

CLIMATE: the accumulative effects of weather; its basic elements include radiation, temperature, atmospheric moisture and precipitation, evaporation, and wind

INSOLATION: incoming solar radiation; differences in global insolation at various places on the Earth's surface create weather and climate patterns

WEATHER: the condition of the atmosphere at a given moment and at a given place

CLIMATIC PROCESSES

The basic elements of climate include radiation, temperature, atmospheric moisture and precipitation, evaporation, and wind. The features of ancient climates that have been most studied are temperature, wind patterns, amount of precipitation, and evaporation. In order to understand ancient climates, it is necessary to understand the fundamentals of modern climatic processes.

Solar radiation is the electromagnetic radiation emitted by the Sun's surface; this radiation is primarily within the infrared, visible, and ultraviolet ranges. The incoming solar radiation is termed global insolation. The amount of energy received by the Earth at any one time is essentially constant, although paleoclimatic evidence indicates that it has not always been so. The amount of global insolation received depends on the output energy of the Sun, the distance between the Earth and the Sun, the angle at which the Sun's rays strike the Earth's surface, the duration of daylight, and atmospheric composition. Therefore, global insolation depends primarily on the latitude and the seasons: Regions within the higher latitudes receive the least amount of insolation, primarily because of the angle at which the Sun's rays strike the Earth, and they have the greatest seasonal variation. The result is an energy deficit poleward of 40 degrees north and 40 degrees south latitude, respectively; the equator has the least variation, so that between these degrees of latitude there is an energy surplus. Therefore, a continuous horizontal exchange of energy between these regions occurs, which is the cause of atmospheric circulation and weather patterns.

The unequal heating of the Earth's surface creates pressure differences. Since gases move from areas of high pressure to areas of low pressure, this differential heating and pressure variation causes the wind to blow. If the Earth were not rotating, its general circulation pattern would be one of ascending air at the equator and descending air at the poles, with continuous horizontal flow between. The Coriolis effect, however, changes this ideal pattern, and free-moving objects are deflected from a straight-line path in response to the Earth's rotation: In the Northern Hemisphere, the deflection is to the right, and in the Southern Hemisphere, it is to the left. Ocean surface circulation patterns are therefore typically clockwise in the Northern Hemisphere and counterclockwise in the Southern, yet atmospheric patterns are in fact more complex because air parcels are ascending and descending as well as moving horizontally. General atmospheric circulation patterns from the equator poleward include the doldrums, the trade winds, the horse latitudes, the westerlies, and the easterlies. In the geologic record, general atmospheric circulation patterns are difficult to determine as a result of the scarcity of data points necessary for accurate calculation of such large-scale phenomena. Wind patterns, however, may be calculated for some areas on the basis of sedimentary structures, and general atmospheric circulation patterns may be inferred from models based on modern atmospheric circulation and the distribution of the Earth's land and water areas through time.

In order for precipitation to occur, there must be atmospheric instability. An air parcel will rise until it reaches an altitude where the surrounding air is of equal temperature; this process may be enhanced where intense solar heating creates lower pressures, where the air mass is heated by a warm surface below it, where sloping terrain such as mountains forces air to ascend, or where cool air acts as a barrier over which warmer, lighter air rises. As the unstable air parcel moves vertically, it cools as a result of the expansion of gases (the adiabatic process). Condensation will occur when the dew point temperature is reached, which varies according to the initial relative humidity of the air parcel. Provided that there is sufficient moisture and that nuclei are present around which the moisture may accumulate, precipitation may occur. The major types of precipitation include rain, freezing rain, sleet, snow, and hail. The type of precipitation that occurred in ancient times is difficult to determine, although occasional examples of raindrop prints are found as casts in sediments, and ancient glacial deposits indicate the accumulation of frozen precipitation. In some situations, the amount of precipitation may be inferred, typically through the utilization of fossil plants. In other cases, the amount of precipitation versus evaporation may be indicated, primarily through the utilization of sedimentary mineral types such as evaporites.

CLIMATE CHANGES

The result of the interaction of these climatic processes is the formation of regional climates. On the modern Earth, a wet equatorial belt lies within about 20 degrees of the equator. The adiabatic cooling of very moist and warm air results in heavy precipitation within these regions, yielding a wet tropical climate. These air parcels then begin descending, adiabatically heat up through compression, and the now-dry air encounters the Earth's surface. The result is the formation of tropical deserts, centered at approximately 25 to 30 degrees north and south latitude. At the middle latitudes (approximately 35 to 65 degrees), the interaction of air masses between the polar and tropical regions and the variation in global insolation throughout the year result in large variations in temperature and seasonality. Polar regions are characterized by cold and dry conditions, because the Sun's rays are typically at low angles and cold air parcels in those regions are unable to accumulate large amounts of moisture.

One phenomenon that seems to affect climate, at least for relatively short cycles of Earth history, is sunspot activity. Although the data are incomplete, the increase and decrease in these solar disturbances seem to suggest that they create variations in global insolation and subsequent warming and cooling trends. The cooling of the Earth's climate during the early-to-middle portion of the second millennium has been linked to variations in sunspot activity. Another explanation of causes of Earth-climate cyclicity was put forth in the twentieth century by Milutin Milankovitch. Milankovitch suggested that Earth's orbit around the Sun may vary from more elliptical (with more pronounced differences in global insolation during the summer versus the winter) to more circular (with less seasonality). Such a cycle would take about 100,000 years to complete. Other Milankovitch cycles proposed include changes in the tilt of Earth's axis in relationship to the Sun (a 40,000-year cycle) and the wobble of Earth upon its axis (a cycle of 21,000 years). Studies on ocean-floor sediments for the Pleistocene (within the past 2 million years of Earth history) and in lake basin sediments of the Triassic and Jurassic (approximately 200 to 180 million years before the present) of the East Coast of the United States indicate that such cycles may indeed exist.

Another possible cause of short-term changes in climate may be asteroid impacts. Scientists have theorized that such impacts would increase the particulate levels in the atmosphere to such an extent that a phenomenon similar to "nuclear winter" would occur. Overall global insolation would decrease dramatically, and temperatures would plummet. Suggestions have been made that this phenomenon could account for the periodic extinctions that take place on the Earth.

Another possible cause of changes in the Earth's atmosphere involves changes in carbon dioxide levels: a heating trend caused by increasing levels (the greenhouse effect) or a cooling trend caused by decreasing levels. Such changes may be caused by changes in the overall metabolism of organisms or by decreases and increases in volcanic activity. Increases in volcanic activity could also cause cooling trends because of a decrease in global insolation as a result of volcanic particles suspended in the atmosphere.

Ocean current patterns tremendously influence weather patterns and regional climates, and it has been suggested that the atmosphere and oceans can be viewed as a single, interacting unit. Changes in oceanic circulation would certainly modify climates, although the geologic evidence of a definite link between the two is equivocal. Stronger evidence for causes of variation may be related to plate tectonics. The movement of continental plates over polar areas may certainly account for the buildup of ice during certain periods of Earth history, such as occurred on the southern continents during the Permian period (about 290 to 240 million years before the present). In part, such shifting of plates may also ac-

count for the last ice ages in Europe and North America, although the presence of several periods of glaciation and interglacial episodes indicates that each episode of glaciation during the Pleistocene cannot be explained this way.

Phillip A. Murry

CROSS-REFERENCES

BIBLIOGRAPHY

Boggs, Sam, Jr. *Principles of Sedimentology and Stratigraphy.* Westerville, Ohio: Charles E. Merrill, 1987. As many of the paleoclimatological methodologies rely on sedimentological features, any detailed study of paleoclimates will require some knowledge of sedimentology. This text is one of the best books on the subject in terms of readability and accuracy. The chapters on sedimentary textures, sedimentary structures, and sedimentary environments are especially useful for understanding the sedimentological results of paleoclimatological processes. Designed for junior- or senior-level geology students.

Cox, C. B., and P. D. Moore. *Biogeography: An Ecological and Evolutionary Approach.* 4th ed. Palo Alto, Calif.: Blackwell Scientific Publications, 1985. This book contains a wealth of information on many aspects of biogeography, but certain sections may be too detailed for the general college-level reader. Discussions of climate and other atmospheric limits on biogeography and coverage of ecology

and plate tectonics serve as a useful overview for analysis of paleoclimate conditions and influences through time.

Dodd, Robert J., and Robert J. Stanton, Jr. *Paleoecology: Concepts and Applications.* New York: John Wiley & Sons, 1981. A comprehensive treatment of many aspects of paleoecology. Provides detailed coverage of various plant and animal groups as concerns their utilization in analysis of ancient ecologies and climates as well as detailed analysis of the use of isotope studies in paleoecology and paleoclimatology. Also included are discussions of ecosystems and communities in paleoecology and paleobiogeography. Written at the introductory college level.

Dott, Robert H., Jr., and Donald R. Prothero. *Evolution of the Earth.* 5th ed. New York: McGraw-Hill, 1994. This basic textbook on historical geology is aimed at students of geology. However, it is very readable by anyone with a background in science. Presents an up-to-date account of the Earth's history from the viewpoint of plate tectonics. Includes a glossary.

Sharpton, Virgil L., and Peter D. Ward, eds. *Global Catastrophes in Earth History.* Boulder, Colo.: Geological Society of America, 1990. A compilation of fifty-eight papers on various aspects of catastrophic extinction and its causes. A large number of the essays deal with climatic changes associated with the boundary event. Although written for specialists, it is accessible to readers with some scientific background. Bibliography and index.

PALEOGEOGRAPHY AND PALEOCURRENTS

Paleogeography is both the geography of a past geologic time and the science of its determination. Paleocurrents are currents existing within a paleogeographic environment or region.

PRINCIPAL TERMS

AZIMUTH: degrees of arc measured clockwise from the north

CORRELATION: the determination of identity in age, fossil content, or physical continuity of rocks observed in different areas

ENDEMIC SPECIES: species confined to a restricted area in a restricted environment

EPICONTINENTAL SEA: a sea overlapping continental crust, as opposed to a sea underlain by oceanic crust

ISOTOPES: forms of an element having the same atomic number but differing atomic weights

PLANETARY WIND SYSTEM: a global atmospheric circulation pattern, as in the belt of prevailing westerlies

REMANENT MAGNETISM: the magnetic field imposed upon a rock at the time of its formation in accord with the global magnetic field

ANCIENT GEOGRAPHY

Paleogeography is the geography or the study of the geography of the past. Paleogeographic interpretations or maps may refer to any geographic phenomenon or group of phenomena, so there are many special types of paleogeographic maps. Paleobiogeographic maps show distribution of fossil species, ecological assemblages, floras, or faunas. Paleoclimatological maps plot ancient climatic realms, and paleophysiographic maps depict ancient landforms. Paleotectonic maps outline regions of deformation, including crustal compression, crustal extension, persistent uplift or depression, and stable areas. Paleobathymetric maps describe the depths of ancient bodies of water. Maps defining past patterns of circulation in the sea, a lake, or stream are called paleocurrent maps. Any geographic property, in fact, may be the subject of a specialized paleogeographic map. Paleogeographic maps are, however, interpretive maps; that is, they are not maps of existing conditions or objects but describe conditions inferred from the rock record. They cannot be verified by direct observation in nature. Paleogeographic maps and interpretations may illustrate the geography of the entire globe, of whole continents, or merely of specific regions or features.

Global maps generally show continents and ocean basins as they are inferred to have existed during a major interval of time, such as a geologic period. The continents shown on these maps may be parts of present-day continents or may be composed of the merged parts of several modern continents. Furthermore, the ancient continents are not likely to be located in their present position in reference to latitude and longitude. On many of these maps, ancient continental landmasses and archipelagos are plotted over the whole Earth, with the outline of present continents and parts of continents superimposed. A series of such maps thus records the manner in which segments of continental crust have continually divided, split, merged, and moved about throughout geologic time. The maps for the Precambrian, which are least reliable, record the appearance of many "protocontinents," coalescing blocks of continental crust, lighter rocks of the composition of granite that form an incomplete outermost layer of the rocky Earth.

These protocontinents eventually gathered to form perhaps five ancestral continents. During the Early Paleozoic, these gathered in two large continents: Laurasia in the Northern Hemisphere and Gondwanaland in the Southern Hemisphere. During the Late Paleozoic, Laurasia and Gondwanaland coalesced in a single continental mass, Pangaea. A large embayment, the Tethys Sea, indented the eastern side of Gondwanaland and is

1135

the remote ancestor of the Mediterranean Sea. During the Mesozoic and Cenozoic, this single large continent progressively separated into the modern suite of continents: North America, South America, Eurasia, Africa, Australia, and Antarctica. These continents gradually moved to their present positions relative to the poles and the equator. Global paleogeographic maps of this sort first appeared in the middle twentieth century, as the theory of plate tectonics and paleomagnetic concepts were discovered and accepted.

PAST POSITIONS OF CONTINENTS AND OCEAN BASINS

Knowledge of the global position, size, and interconnections of the continents and ocean basins is required to place past geologic events in a geographic context. It also is needed for full understanding of present global geologic relationships. The past position of crustal blocks, continents or parts of continents, and ocean basins may be inferred in part from the remanent magnetism of rocks of known age. Remanent magnetism is a magnetic field imprinted in a rock and conforming to the world's magnetic field existing at the time the rock was formed. This remanent magnetic field remains fixed, even if the rock is subsequently rotated or moved. It is, therefore, a permanent record of the original latitude and orientation of the rock. Thus, the azimuth direction of the remanent magnetism in a rock indicates the position of the pole with respect to the locality of a rock when it was deposited or intruded. The polarity of the remanent field reveals the polarity, normal or reversed, at the time the rock formed. The inclination of the vertical component of the remanent field gives the latitude at which the rock formed. Also, regions in which the remanent magnetic fields in rocks of common age are oriented uniformly may be considered parts of the same continental blocks or fragments. The remanent field is susceptible to destruction only by heating the rock to near-melting temperatures. The longitudinal position of a rock cannot be determined by paleomagnetic analysis, so other phenomena must be called upon to place ancient continental crustal blocks in their proper positions on the globe.

Former continuity between separated continental blocks may be inferred from their shape, as in the apparent match between the eastern coastline of the Americas and the western coasts of Europe and Africa. This pattern was noted and a relationship inferred as early as the first half of the nineteenth century. Such "jigsaw puzzle" techniques are a significant aid in continental placement. It should be noted, however, that the land area of a continent is not what is being matched; much of the continent is submerged, so the location of the continental margin and slope is essential.

Apparent continuity of ancient geologic features on the margin of separated continental masses suggested former connections before the advent of plate tectonics, however, and still reinforces paleogeographic reconstructions. For example, the apparent continuation of the Appalachian orogenic belt and the band of carboniferous coal deposits from Texas to Newfoundland, the British Isles, and Western Europe supports the idea of former continuity between the landmasses involved. Common occurrence of glacial deposits, coal beds, similar volcanic extrusives, and disjunct occurrences of fossil organisms incapable of crossing open oceans in Australia, India, Antarctica, South America, and Africa all support reconstruction of the Gondwanaland continent. Before the advent of plate tectonics, now-submerged continent-sized landmasses in the South Atlantic and Indian Oceans or long isthmian land bridges were postulated as crossing those oceans to explain the present distribution of the Gondwanaland fauna and flora.

PALEOGEOGRAPHIC MAPS

Paleogeographic maps of continental masses may show the distribution of ancient geographic features on modern-day continents, or they may refer to continents as they existed at some former time. Maps of the distribution of the land, sea, and ancient mountain ranges on present-day continents have been produced since the mid-nineteenth century. These maps will show the epicontinental seas, seas consisting of marine waters overflowing continental crust, as opposed to seas in deep ocean basins. Such maps also show the continental shelf or inundated continental margin, the relatively steep continental slopes at the very edge of the continental crust, and dry-land areas. In addition, major belts of deformation, broad areas of uplift, and broad areas of subsidence, or basins, are frequently shown on these maps. More detailed, and generally more specific, maps will show such things as deltas, ancient volcanic tracts, coastal lagoons, estuaries, reefs, dune fields, mountains, and glaciers.

Regional maps may show the paleogeography of areas defined by geologic interest, or they may be bounded arbitrarily. Examples of the first would include a map of the Late Paleozoic Ancestral Rocky Mountains. A set of paleogeographic maps for a state would exemplify the second. Regional maps of smaller areas generally are more detailed than continental or global maps and are more directly derived from local geology and paleontology. Latitude and pole position may, however, be shown and are determined by the same techniques employed in producing global or continental scaled maps. At the regional level, paleogeography may be concerned with the distribution of land and sea, mountain ranges, volcanic regions, lakes, flora and fauna, and the climate.

Paleogeographic maps are constructed by correlating and establishing the extent of rocks of the age in question. The conditions under which contemporaneous rocks were deposited are then determined. In this way, areas of uniform or nearly uniform environment may be outlined and mapped. Areas lacking rocks of the age under consideration were regions of nondeposition or intrusion or were areas subsequently eroded to remove rocks of that age. Areas undergoing deformation and areas suffering igneous intrusion or extrusive volcanism also are defined and mapped.

Ralph L. Langenheim, Jr.

Cross-References

Archean Eon, 1087; Atmosphere's Global Circulation, 1823; Atmosphere's Structure and Thermodynamics, 1828; Biogenic Sedimentary Rocks, 1435; Biostratigraphy, 1091; Cenozoic Era, 1095; Chemical Precipitates, 1440; Climate, 1902; Cretaceous-Tertiary Boundary, 1100; Evaporites, 2330; Fossil Plants, 1004; Fossil Record, 1009; Geologic Time Scale, 1105; Glacial Deposits, 880; Greenhouse Effect, 1867; Ice Ages, 1111; Land Bridges, 1116; Mesozoic Era, 1120; Microfossils, 1048; Paleobiogeography, 1058; Paleoclimatology, 1131; Paleoseismology, 1139; Paleosols, 1144; Paleozoic Era, 1126; Proterozoic Eon, 1148; Sediment Transport and Deposition, 2374; Stratigraphic Correlation, 1153; Transgression and Regression, 1157; Unconformities, 1161; Uniformitarianism, 1167.

Bibliography

Dott, Robert H., Jr. and Donald R. Prothero. *Evolution of the Earth.* 5th ed. New York: McGraw-Hill, 1994. A geologic history, centering on North America. Includes a set of global paleogeographic maps, paleogeographic maps of North America for Precambrian through Cenozoic times, and regional paleogeographic maps of areas of particular interest. Maps are supported by summaries of the data upon which they are based. Paleocurrents are shown on many of these maps. Comprehensive index and a useful glossary of technical terms. Designed for use as a beginning college geology text.

Press, Frank, and Raymond Siever. *Understanding Earth.* 2d ed. New York: W. H. Freeman, 1998. This comprehensive physical geology text covers the formation and development of the Earth. Contains extensive sections on plate tectonics and its effect on paleogeography. Readable by high school students, as well as by general readers. Includes an index and a glossary of terms.

Seyfert, Carl K., and Leslie A. Sirkin. *Earth History and Plate Tectonics: An Introduction to Historical Geology.* 2d ed. New York: Harper & Row, 1979. Contains a set of global paleogeographic maps from the Precambrian through the present, along with more detailed maps of North America for the geologic periods. Discusses how paleogeographic maps are compiled using paleomagnetic, geologic, biologic, and climatic data. Written for beginning students of geology.

Shell Oil Company. Exploration Department. *Stratigraphic Atlas, North and Central America.* Houston: The Department, 1975. Maps of the distribution and lithologic character of rocks belonging to the major subdivisions of each geologic system. Also paleogeologic maps, isopachous (thickness) maps, and lithofacies (regional changes in rock character) maps. Directly interpretable as paleogeographic maps.

Smith, A. G., D. G. Smith, and B. M. Funnell. *Atlas of Mesozoic and Cenozoic Coastlines.* New York: Cambridge University Press, 1994. Global paleogeographic maps showing past positions of Mesozoic and Cenozoic epicontinental seas.

Wanless, H. R., and Cynthia R. Wright. *Paleoenvironmental Maps of Pennsylvanian Rocks, Illinois Basin and Northern Midcontinent Region.* Boulder, Colo.: Geological Society of America, 1978. Detailed regional paleogeographic maps for individual subunits of the coal-bearing rocks of middle America.

PALEOSEISMOLOGY

Paleoseismology is the study of the evidence of past earthquakes. By studying the physical features of previous earthquakes, scientists gain knowledge of the forces operating within the Earth, knowledge that may be used to develop a means of predicting future large earthquakes.

PRINCIPAL TERMS

EARTHQUAKE: a shaking or trembling of the Earth caused by a break or rupture in the rocks of the Earth's outer crust

ELASTIC REBOUND: the process of the buildup and release of geologic stress; the release of the strain in the rocks results in earthquakes

FAULT: a fracture of the Earth's crust along which rocks move

FAULT OFFSET: a feature such as a road, creek bed, or tree line across a fault that separates and becomes misaligned by the fault movement

FAULT SCARPS: a steep cliff or slope created by movement along a fault

PLATE TECTONICS: the theory that the Earth's hard outer crust (lithosphere) is composed of sections, or plates, that are in constant motion

RECURRENCE PERIOD: the range of time between successive earthquakes

SEISMIC GAP: a fault region known to have had previous earthquakes but not within the area's most recent recurrence period

TSUNAMI: a large ocean wave caused by an earthquake in the ocean floor

ANCIENT EARTHQUAKES

Seismology is the science that deals with earthquakes and movement within the Earth. Seismologists and geologists use historic and prehistoric information about earthquakes in order to make better calculations of the probability of earthquakes occurring in a particular area. However, historical evidence from earthquakes—that is, information acquired from observation and instrumental measurements—is short and incomplete. Paleoseismology is a relatively new branch of earthquake science that uses archaeological techniques to find physical evidence of large ancient earthquakes and thus to expand the seismic history of the Earth.

In the 1960's, the theory of plate tectonics was introduced. According to this theory, the Earth's outer crust is composed of sections, or plates, that are in constant motion. When these plates collide or spread apart, the resulting forces produce deformations, or tectonics, of the Earth's surface that are identified with earthquakes. When plates collide and one plate moves beneath the other, the resulting fault is called a "subduction zone." Subduction zones are under water and cause changes in land at

the shoreline. Earthquakes at the subduction zones often cause tsunamis that are devastating to coastal areas. When the plates rub against one another, the resulting fault is known as a "plate interaction." The San Andreas fault in California is a well-known example of a plate-interaction fault.

The idea of elastic rebound is used to explain how earthquakes originate. Under normal circumstances, the rocks at a fault boundary move a few centimeters per year. If the rock cannot move, the force of the normal movement becomes stored as "strain energy." When the force becomes too great, the fault slips to release energy. As the plate boundaries shift or rupture, earthquakes occur. By studying the historic record, scientists have found that the plate boundaries rupture in segments until the entire section is broken, and the cycle then begins again. Scientists believe that future earthquakes are more likely to occur in places where the segments have not recently ruptured. These places are called "seismic gaps."

With the discovery of plate tectonics, scientists believed that long-term and short-term prediction of earthquakes would be possible. Long-term prediction is based on the probability that an earthquake will occur. Paleoseismic evidence of past earthquake recurrence periods allows scientists to determine when a large earthquake is "overdue" for a particular area.

Short-term prediction involves specifying the future place, magnitude, and time of the earthquake. Short-term prediction relies on data collected from monitoring equipment on known faults as well as precursory signals such as an increase in fault movement, change in deep-well water levels, changes in surface elevations, an increase in minor earthquake activity, the occurrence of a moderate foreshock, or an increase in the release of radioactive radon gas. Since other environmental factors can cause some of these signals, there are no definitive predictors of an imminent earthquake.

The advantages of knowing when, where, and how strong an earthquake will be are numerous. The cost of an earthquake in terms of loss of human life and property can be tremendous. In addition to the damage caused by the actual ground shaking from the earthquake, there may be loss of life and property caused by structural collapse and secondary effects such as landslides, fires, and tsunamis. The September 19, 1985, earthquake that struck Mexico City killed more than eight thousand people, injured thirty thousand, destroyed or severely damaged more than one thousand multistoried buildings, and caused an estimated $5 billion in damage.

Study of Earthquakes

Scientists use several methods to study earthquakes. One of the methods is based on recorded observations. The earliest known scientifically collected observational data about earthquakes is a survey taken after a strong earthquake, known as the All Saints Day earthquake, struck Lisbon, Portugal, on November 1, 1755. Local officials sent a list of questions to the affected areas asking about the duration of the shock, the times and intensities of aftershocks, the number of people killed, the structures destroyed, and the extent of damage from related fires. Firsthand accounts of death and structural damage, as well as obvious changes in the physical features of the land, allow scientists to judge the severity of an earthquake. This information is useful in identifying seismic gaps. Probability indicates that an area is more likely to have a

large earthquake if a large earthquake has occurred there before.

Another method for studying earthquakes is through instrumentation. Tiltmeters and strainmeters are used to measure changes in the crustal movement in earthquake areas. Hidden faults are located using a device that relies on how sound waves are reflected. Satellites, aerial photographs, and surveying equipment also provide clues to the location of underground faults. Seismic activity is monitored worldwide with seismographs, which record the occurrence, direction, severity, and duration of earthquakes. Modern seismographs are extremely sensitive and can detect ground motions of very small earthquakes. Scientists estimate that approximately eighty thousand small tremors and fifteen to twenty severe earthquakes occur around the world annually.

In the late 1970's, Kerry Sieh, a geologist at the California Institute of Technology, conducted the pioneering work in paleoseismology at Pallet Creek on the San Andreas fault. Scientists could now gather additional information about earthquakes to supplement and expand the existing observational and instrumentation knowledge. With the prehistoric evidence of earthquakes, scientists can better calculate the probability of another large earthquake.

Paleoseismic techniques may involve digging a trench across a fault and examining the rock strata, or layers, for evidence of previous ruptures. Scientists also look for evidence of flooded forests or sheets of sand along coastlines, which would indicate tsunami activity. The evidence can be direct, such as fault offsets, or indirect, such as landslides or sand deposits, which occur with strong ground shaking. By measuring the distance separating the two rock segments in a fault offset, scientists can determine how much the fault has slipped. The samples can be radiocarbon dated in order to determine when the earthquakes occurred. When the time gap between the earthquakes is determined, the recurrence period and average interval can be established. The recurrence period is a range of time between earthquakes; the interval is the average time between earthquakes. For example, if earthquakes happened in a particular region in 1700, 1775, 1825, and 1925, the recurrence range is fifty to one hundred years, with an average interval of seventy-five years.

PALEOSEISMIC DATA

Seismologists need information about past earthquakes in order to study and prepare for future earthquakes. Paleoseismology has been very useful in North America because of the short historical record. Additionally, the techniques have helped establish the accumulation of stress in faults all over the world.

Paleoseismic data enables scientists to determine higher levels of earthquake hazard in a geographic area. Engineers and public officials use the information from scientists to revise building codes and redraw maps of earthquake-shaking hazard zones. The U.S. Uniform Building Code, which includes nationwide standards for designing structures, defines six levels of earthquake-shaking hazard. By convincing people to reinforce existing structures and adopt earthquake-resistant construction standards, loss of life and property can be greatly reduced.

In California, a 1971 earthquake in the San Fernando Valley killed sixty-five people and caused an estimated $500 million in damages. California has had earthquake-resistant building criteria since 1933, which have been updated several times. In contrast, a 1972 earthquake of similar magnitude in Managua, Nicaragua, killed more than five thousand people and caused economic losses equal to the country's gross national product for that year. Managua had few buildings designed under earthquake-resistant standards.

Paleoseismology helps answer questions about faults, folds, secondary effects of prehistorical earthquakes, length of an earthquake cycle, the cycle's regularity or irregularity, and similarity of characteristics of an area's earthquakes. Scientists believe that earthquake processes of the past will probably be the same as those in the near future. The longer the record of earthquake occurrences in a particular area, the greater the predictive value of the knowledge. The ability to predict the likelihood and possible effects of future earthquakes enables people to take precautions to protect themselves against injury, death, and economic losses.

STUDY OF SHORELINES

Paleoseismology initially studied active faults to help establish the rates of movement. The technique was then applied to interpreting earthquake crustal deformations or shaking effects. Although paleoseismology is usually used to examine surface faults, the technique is also used to examine subduction-zone earthquakes indirectly. Since subduction zones are underwater faults, scientists examine changes in the land at the shoreline for evidence of past activity. The sudden lowering of a shoreline or a series of terraced beaches are associated with major earthquakes.

In Alaska, scientists examined terraced beach lines near an island in Prince William Sound. This area was the site of the March 27, 1964, "Good Friday" earthquake, the second-strongest earthquake recorded. The earthquake killed 114 people and caused $350 million in property damage. The coastlines of nearby landmasses bowed by as much as 2.5 meters, flooding them with seawater and killing many square kilometers of forest. Scientists discovered that large earthquakes have occurred in the area an average of once every 850 years during the past 5,000 years. Because of this investigation, scientists believe that it is unlikely that a major earthquake will occur in this area in the near future.

Until the mid-1980's, scientists believed that the area off the coast of Washington, Oregon, and Northern California had a low probability of a massive earthquake, since there was no historical record of such an earthquake. Paleoseismic studies, however, show evidence of a sudden lowering of the shoreline associated with major earthquakes. In addition, scientists also found sheets of sand deposited by tsunamis and sand-filled cracks caused by strong earthquake shaking. Researchers have thus concluded that three massive earthquakes have struck this area in the past three thousand years; the most recent great earthquake struck about three hundred years ago. Based on this knowledge, measures have been taken to reinforce existing structures and toughen earthquake design standards. Without the paleoseismic data, a powerful earthquake would have taken the area by complete surprise.

Scientists have been gathering much information about faults at plate boundaries, since an estimated 95 percent of all earthquakes occur at these boundaries. The probability of occurrence for some earthquakes is more difficult to predict, because they do not occur near plate boundaries. Often, these inner-plate earthquakes are devastating, because they occur so infrequently that there may be no historical record of an earthquake in that area.

On September 30, 1993, ten thousand people in Killari, India, died when an earthquake devastated the area. Prior to the earthquake, the area was included in the lowest level of the earthquake-hazard map of India. The seismologists were wrong at Killari because the hazard map relied mostly on the historical record of earthquakes. In India, this record is only about 150 years old. Investigation following the earthquake suggested that the area had been seismically quiet for 65 million years or more. Scientists believe that the earthquake may have been triggered by an artificial, or human-constructed, reservoir that changed the stress on a fault. Understanding how such construction affects the stress on the Earth's crust that triggers earthquakes allows more considered and intelligent decisions about building nuclear power plants, dams, reservoirs, and cities.

In the United States, the area east of the Rocky Mountains has faults that are buried deep and accumulate stress slowly. Earthquake risk estimates for the eastern United States are not as advanced as those for the West, where the faults break the surface and earthquakes occur more often. Eastern faults may be quiet for hundreds, thousands, or millions of years.

One of the most destructive series of earthquakes in American history, however, occurred in 1811 and 1812 in New Madrid, Missouri. At least one earthquake per day took place in the area for more than a year, with three extremely powerful ones; these quakes were felt over approximately 2.6 million square kilometers and caused tremendous physical changes to the land. Destruction

was severe, but casualties and economic losses were low because there were few settlements in the affected area. The earthquakes triggered landslides, submerged islands, opened fissures in the ground, flooded lowlands, and caused areas of land to rise or sink.

Large earthquakes have also occurred in Charleston, South Carolina, in 1886; Newfoundland, Canada, in 1929; and New York State in 1988. Numerous small earthquakes have occurred over the past two hundred years, with at least twenty damaging earthquakes in the central Mississippi Valley alone. Eastern earthquakes are puzzling because they do not generate visible surface faults. Because of this lack of physical evidence and the infrequent occurrences of eastern earthquakes, determining the probability and location of another large earthquake in the region is very difficult.

Virginia L. Salmon

CROSS-REFERENCES

Archean Eon, 1087; Atmosphere's Global Circulation, 1823; Biostratigraphy, 1091; Cenozoic Era, 1095; Continental Growth, 573; Continents and Subcontinents, 595; Cretaceous-Tertiary Boundary, 1100; Displaced Terranes, 615; Geologic Time Scale, 1105; Gondwanaland and Laurasia, 599; Ice Ages, 1111; Land Bridges, 1116; Mesozoic Era, 1120; Oceans' Origin, 2145; Paleobiogeography, 1058; Paleoclimatology, 1131; Paleogeography and Paleocurrents, 1135; Paleosols, 1144; Paleozoic Era, 1126; Plate Motions, 80; Proterozoic Eon, 1148; Sediment Transport and Deposition, 2374; Stratigraphic Correlation, 1153; Supercontinent Cycles, 604; Transgression and Regression, 1157; Unconformities, 1161; Uniformitarianism, 1167.

BIBLIOGRAPHY

Adams, John. "Paleoseismology: A Search for Ancient Earthquakes in Puget Sound." *Science* 258 (December 4, 1992): 1592-1593. Introduces a five-article series on paleoseismology fieldwork in the Pacific Northwest. Explains what scientists look for and how they interpret physical evidence of prehistoric earthquakes at coastlines. Briefly discusses the history of paleoseismology. Easy to follow.

Bolt, Bruce A. *Earthquakes and Geological Discovery.* New York: Scientific American Library, 1993. Explains why earthquakes occur and how they are measured. Also discusses some of the most famous earthquakes and what can be done to minimize their destructive effects. Suitable for anyone interested in earthquakes.

Harris, Stephen L. *Agents of Chaos: Earthquakes,*

Volcanoes, and Other Natural Disasters. Missoula, Mont.: Mountain Press, 1990. The first third of this book discusses earthquake hazards in the United States. Looks at historic earthquakes and explains why future earthquakes will occur in various geographic areas. Well-written and suitable for high school readers. Includes a glossary and a lengthy bibliography.

Lay, Thorne, and Terry C. Wallace. *Modern Global Seismology.* New York: Academic Press, 1995. A complete text discussing seismology and its various applications. Includes a chapter dedicated to the earthquake cycle. Contains only brief information specifically on paleoseismology, but the overview of seismology is excellent.

McCalpin, James P., ed. *Paleoseismology.* New York: Academic Press, 1996. Includes an introductory chapter on paleoseismology and discusses field techniques and interpretation of data. Includes a bibliography and index. Suitable for more advanced readers.

Main, Ian. "Long Odds on Prediction." *Nature* 385 (January 2, 1997): 19-20. Brief article discussing the possibility of reliable earthquake prediction. Discusses why earthquakes are difficult to predict and suggests that efforts should concentrate on hazard mitigation.

Moseley, Charles. "Slow Progress in Earthquake Prediction." *Editorial Research Reports* 2 (July 15, 1988): 354-363. Well-written and easily understood overview of earthquake prediction and the role of paleoseismology. Includes examples of major earthquakes and a clear explanation of the need for and problems associated with prediction. Includes a bibliography.

Worsnop, Richard L. "Earthquake Research." *CQ Researcher* 4 (December 16, 1994): 1105-1128. A balanced discussion of the state of earthquake research. Discusses the issues of earthquake prediction and hazard-reduction preparedness; also gives background on faults and current risk areas and includes an explanation of how earthquake strength is measured. Includes a bibliography and suggestions for further reading. Suitable for high school readers.

Yeats, Robert S., Kerry Sieh, and Clarence R. Allen. *The Geology of Earthquakes.* New York: Oxford University Press, 1997. Detailed information about the geological explanations for and effects of earthquakes. Includes a discussion of tectonics and faults as well as an assessment of seismic hazards. Suitable for college-level readers. Includes a bibliography and informative illustrations.

PALEOSOLS

Paleosols are ancient soils that have been buried. Although natural acids (largely dissolved carbon dioxide) are almost entirely supplied by the atmosphere during modern soil formation, that does not seem to have been the case with all paleosols, particularly the most ancient paleosols and those that are associated with ore deposits. Paleosols that were produced by ancient atmospheric gases may record the environmental conditions of the ancient Earth.

PRINCIPAL TERMS

CALICHE: a type of soil or paleosol that contains a high proportion of calcium carbonate, calcium sulfate, or both

CLAY MINERAL: a type of mineral that is the most common product of soil formation; it is composed of silicon, oxygen, hydrogen, usually aluminum, and possibly other elements

PARTIAL PRESSURE: the proportion of a gas mixture (for example, the atmosphere) that a particular type of molecule (for example, carbon dioxide) comprises

SOIL: all material that has been substantially altered at the Earth's surface by interaction with the atmosphere, living things, or both, and that has not been laterally displaced subsequent to that alteration

SOIL HORIZON: a distinct layer in a soil

ANCIENT SOILS

Soil is one of the best-known yet least precisely defined geologic entities. Many definitions would not exclude beds of graded sediment; other definitions would not exclude sediment that has experienced only mild physical and chemical alteration. Additionally, many soil science textbooks restrict "soil" to that lying within the depth of plant roots, but such a definition is inappropriate for Precambrian and Early Paleozoic soils, which formed before the evolution of rooted plants.

The rock record of the Earth covers the past 3.9 billion years (out of a total Earth history of 4.6 billion years). Throughout the known rock record, soil formation (weathering) has involved more energy than other geologic processes, such as mountain building. Mountain building is driven by the internal energy of the Earth (for example, by natural radioactivity), whereas soil formation is driven by solar energy and by chemical reactions between acidic atmospheric gases and exposed rock. Solar energy fuels photosynthetic plants, which concentrate the most abundant atmospheric acid (carbon dioxide), and subsequent decay of these plants releases concentrated carbon dioxide into soil waters, which thereby become acidic and dissolve minerals. The downward solar flux of energy, which enhances weathering, currently is about 7,500 times greater than the upward flux of internal energy. The internal energy of the Earth has progressively decreased with time as radioactive elements decay; solar radiation has progressively increased. In addition, the input of solar energy has exceeded the output of internal energy by at least a factor of 1,000 since the beginning of the rock record.

The collective volume of soils produced throughout Earth history may have been comparable to the present volume of the continental crust, but only an insignificant volume of these soils has become preserved by burial within the crust. Paleosols now constitute a smaller proportion of continental crust than most other well-known rock types. The proportion of paleosols appears to be particularly small in the oldest rocks. In these rocks, the preserved portion is so small and the conditions for preservation of paleosols appear to have been so peculiar that it is dangerous to make sweeping interpretations of ancient Earth environments based on paleosols.

PALEOSOLS AND SEDIMENTARY ROCK

Virtually all old soils have been eroded to become sediment rather than buried to become paleosols. Intense weathering causes a high proportion of rock to dissolve, and the remnant soil becomes rich in insoluble elements such as aluminum. Erosion of this soil produces aluminum-rich sediment, which usually accumulates in the shallow ocean as a clay-rich rock (mudrock), such as shale. Sedimentary rocks therefore may be environmental indicators, and the vastly greater volume of sedimentary rock has resulted in most interpretations of ancient environments coming from study of sedimentary rock rather than from study of palcosols. Environmental study of paleosols is less constrained, however, than that of sedimentary rock because the unweathered parent of the weathered material (soil) is observable beneath a paleosol, whereas the parent for a weathered sediment generally cannot be deduced with certainty; the parent rock for any given sediment may lie thousands of miles from where the sediment accumulates—for example, the Andean Mountain parent for much of the Amazon deltaic sediment.

Paleosols are known from all major divisions of Earth history, from the Archean eon to the present epoch. The proportion of paleosols to other rock types roughly increases with time, consistent with the theory of continental growth through Earth history; an increase in the area of exposed continents would lead to an increase in the volume of soil. Although the area of exposed continents probably was smallest in the Archean eon, well-preserved Archean paleosols are known from Canada and from South Africa, where Archean continental crust never has been so deeply buried that regional metamorphism could destroy evidence of the weathering processes which produced the paleosols. The area of exposed continents apparently increased dramatically from the Archean to the Proterozoic, so Proterozoic paleosols are correspondingly more abundant.

The role that plant life may have played in the weathering of these Proterozoic soils is unclear. Plants currently concentrate so much carbon in the upper portion of a typical soil that oxidation of this carbon provides more carbon dioxide to the underlying soil than does diffusion of carbon dioxide directly from the atmosphere. Much of this carbon dioxide comes from roots, but roots did not evolve until after the Proterozoic. Plant life in Precambrian (Archean plus Proterozoic) soils was limited to bacteria. Mats of photosynthesizing cyanobacteria (also called blue-green algae) lay at the surface, and other bacteria occurred deeper in the soil. Modern soil contains abundant bacteria below the surface, but the soil surface generally is occupied by complex photosynthesizing plants—for example, trees—instead of by cyanobacteria. A partial image of Precambrian soils may be obtained by studying cyanobacterial mats that grow in environments too harsh for more complex plants—for example, on salt flats and around geysers. These environments are not only harsh but also highly variable. The chemical composition of water on a modern salt flat may vary from hypersaline to nonsaline following a thunderstorm. The water temperature around a geyser may vary by several tens of degrees within a few minutes. The ability of cyanobacteria to withstand such variations may indicate that the environmental conditions of Precambrian soils either were consistently harsh or were more variable than those of more recent soils.

GEOGRAPHIC VARIATION

Ancient paleosols generally are too scarce for study of the geographic variation of soil types.

The middle layer, a paleosol once a soil, was baked by the lava layer that flowed on top of it. (© William E. Ferguson)

Geographic variations in rainfall and temperature are the two prime variables that control the distribution of modern soils because the atmosphere mixes so rapidly that any local variation in the partial pressure of oxygen or carbon dioxide rapidly becomes globally homogenized. Molecular oxygen and carbon dioxide also would have been evenly distributed in ancient atmospheres, but their proportions of the atmosphere may have varied throughout Earth history. The oldest paleosols therefore are studied to examine variation in the Earth's atmosphere through Earth history, whereas the youngest paleosols are studied to learn about geographic variation in climate. Paleosols and other indicators record dramatic variation in climates on Earth during the past 2 million years, related to the growth and melting of enormous continental glaciers.

Tropical climates produce distinct soils and thus distinct paleosols. Both wet and dry tropical climates characteristically produce yellow-to-red soils because of oxidation and retention of iron in the soil. Wet tropical conditions can produce soils that are so rich in aluminum that these soils may be mined profitably. In such soils, the more soluble elements have been leached away by groundwater draining lush vegetation. Decay of the vegetation produces acids that attack even quartz, leaving only aluminum-rich minerals. Aluminum-rich paleosols are among the oldest (Archean) paleosols on Earth, so wet tropical conditions appear to have existed long ago, despite the fact that solar radiation should have been much smaller during the Archean eon.

Warm, dry climates may produce little soil of any kind. For example, more characteristic of the Sahara than the "sand seas" that are commonly illustrated in documentaries are vast stretches of bare rock. Caliche is a characteristic soil and paleosol produced in such a climate. Caliche generally forms by precipitation of calcium carbonate from upwardly moving groundwater as the groundwater approaches the Earth's surface. The upward decrease in pressure may allow carbon dioxide to be released, just as carbon dioxide is released upon opening a bottle of soda pop. Release of carbon dioxide favors precipitation of calcium carbonate; this precipitation may be aided by evaporation of the groundwater—which points to a dry climate for the formation of an ancient

caliche. A caliche paleosol generally may be interpreted to record an ancient dry climate, even if the precipitation of calcium carbonate was not related to evaporation within soil but was simply the result of release of carbon dioxide. The original excess of carbon dioxide could have been provided by the escape of gases to the Earth's surface during metamorphism of carbon-bearing sedimentary rocks at great depth. The abundant infiltration of rainwater in a wet climate would dissolve calcium carbonate from soil, whatever its origin, so the preservation of calcium carbonate in a caliche paleosol generally records a dry climate, even if the precipitation of calcium carbonate were induced by deep crustal processes independent of climate.

ELEMENTAL COMPOSITION AND MINERALOGY

Interpretation of the elemental composition and mineralogy of a paleosol generally is controversial, especially for paleosols older than 544 million years (Precambrian paleosols), because paleosols potentially have experienced substantial modification (diagenesis) after burial. One of the most consistent chemical peculiarities of Precambrian paleosols is that they contain extreme ratios of potassium to sodium, unlike modern clayey soils. No known weathering process could fractionate sodium from potassium so severely. Precambrian paleosols commonly contain more than ten times as much potassium as sodium, whereas these two elements generally behave similarly under modern weathering conditions. This potassium in Precambrian paleosols mostly occurs in fine-grained, aluminum-rich mica. The potassium either is a record of pervasive diagenetic alteration of Precambrian paleosols or indicates that, unlike modern soils, they did not form as a result of atmospheric acid-forming gases. The majority of Precambrian paleosols could represent alterations on the ancient land resulting from exhalation of acid-containing mud from deep in the Earth. Although paleosols that are older than 544 million years generally have the greatest ratios of potassium to sodium, paleosols that are 245 to 544 million years old (Paleozoic paleosols) also are more potassium-rich than are modern soils. In these Paleozoic paleosols, the potassium-bearing mineral typically is a clay mineral called illite. In illite-bearing paleosols, some investigators attribute the high potassium content to peculiar weathering conditions, whereas others attribute it to precipitation of potassium from through-flowing groundwater long after burial of the soil.

Michael M. Kimberley

CROSS-REFERENCES

BIBLIOGRAPHY

Bloom, Arthur. *Geomorphology: A Systematic Analysis of Late Cenozoic Landforms.* 3d ed. Englewood Cliffs, N.J.: Prentice Hall, 1998. This college-level text covers the basics of geomorphology. Includes a section on the development of soils. Index and bibliography.

Brady, N. C., and R. R. Weil. *Nature and Properties of Soils.* 11th ed. Englewood Cliffs, N.J.: Prentice Hall, 1996. This advanced text requires some background in geology and geomorphology. However, it contains a wealth of information on the development of soils.

Chernikoff, Stanley. *Geology: An Introduction to Physical Geology.* Boston: Houghton-Mifflin, 1999. This is a good overview of the scientific understanding of the geology of the Earth and surface processes. Includes a section on the development of paleosols. Includes a link to a Web site that provides regular updates on geologic events around the globe.

Kimberley, M. M., and D. F. Grandstaff. "Profiles of Elemental Concentrations in Precambrian Paleosols on Basaltic and Granitic Parent Materials." *Precambrian Research* 32 (1986): 133-154. All papers in this issue are devoted to Precambrian paleosols. This particular paper compares Precambrian paleosols that developed on basalt with those that developed on granitic rocks.

Reinhardt, J., and W. R. Sigleo. *Paleosols and Weathering Through Geologic Time: Principals and Applications.* Special Paper 216. Boulder, Colo.: Geological Society of America, 1988. Several paleosols in the United States (from Pennsylvania, Kentucky, Georgia, Alabama, Colorado, and California) are described in this volume, along with field methods and theoretical models for the development of ancient paleosols.

Retallack, G. *Laboratory Exercises in Paleopedology.* Eugene: University of Oregon Press, 1985. Practical exercises are outlined for the study of paleosols. Examples of paleosols are described from several localities around the world.

Samama, J. C. *Ore Fields and Continental Weathering.* New York: Van Nostrand Reinhold, 1986. Some paleosols are ore deposits for aluminum or nickel; part of this book reviews the characteristics and metal-concentrating processes which produce such paleosols. Another part reviews a different class of paleosol-related ore in which subsurface holes in limestone coincidentally or subsequently become filled with metal-precipitating solutions. Research subsequent to this book has shown that the dissolution of ore-bearing limestone probably shortly preceded precipitation of the lead-zinc minerals and that both the limestone-dissolving and metal-precipitating fluids rose from deep in the Earth.

Wright, V. P., and Alfred Fischer, eds. *Paleosols: Their Recognition and Interpretation.* Princeton, N.J.: Princeton University Press, 1986. This volume emphasizes paleosols that occur outside the United States, particularly those in the British Isles, Spain, and New Zealand. Chapter 6 reviews selected paleosols in the western United States.

PROTEROZOIC EON

The Proterozoic eon is the interval between 2.5 billion and 544 million years ago. During this period in the geologic record, processes presently active on Earth first appeared, notably the first clear evidence for plate tectonics. Rocks of the Proterozoic eon also document changes in conditions on Earth, particularly an apparent increase in atmospheric oxygen.

PRINCIPAL TERMS

ARCHEAN EON: the period of geologic time from about 4 billion to 2.5 billion years ago

OROGENIC BELT: a belt of crust that has been severely compressed, deformed, and heated, probably by convergence of crustal plates

PALEOMAGNETISM: the study of magnetism preserved in rocks, which provides evidence of the history of Earth's magnetic field and the movements of continents

PHANEROZOIC EON: the period of geologic time with an abundant fossil record, extending from about 544 million years ago to the present

PLATE TECTONICS: the theory that the outer surface of Earth consists of large moving plates that interact to produce seismic, volcanic, and orogenic activity

PRECAMBRIAN: the collective term for all geologic time before the Phanerozoic, that is, before about 544 million years ago

RADIOMETRIC DATING: the use of radioactive elements that decay at a known rate to determine the ages of the rocks in which they occur

TERRANE: a structurally distinct block of crust added to a continent by plate tectonic processes

EARTH'S THREE EONS

The largest subdivision of geologic time is the eon. Three eons comprise the known history of Earth: Phanerozoic, Proterozoic, and Archean. The Phanerozoic, whose abundant fossil record permits fine subdivisions of geologic time and intercontinental correlation of strata, is the interval about which the most is known. The Proterozoic, by contrast, is generally characterized by a sparse fossil record of simple life-forms that do not permit the sort of stratigraphic subdivision possible in the Phanerozoic. Ages of rocks in the Proterozoic can be established only by radiometric dating. For rocks as old as the Proterozoic, the inherent uncertainty in even the best dating is on the order of 10 million to 20 million years, comparable in length to some Phanerozoic periods.

The Proterozoic has been subdivided only into intervals comparable to eras (several hundred million years), and these subdivisions are not global but are defined in terms of major geologic events in separate regions. The principal Proterozoic subdivisions are the following: for the United States, Precambrian X (2.5 to 1.6 billion years ago), Precambrian Y (1.6 to 0.9 billion years ago),

and Precambrian Z (0.9 billion to 544 million years ago); for Canada, Aphebian (2.5 to 1.8 billion years ago), Helikian (1.8 to 1.0 billion years ago), and Hadrynian (1.0 billion to 544 million years ago); for Europe, Svecofennian (2.5 to 2.0 billion years ago), Gothian or Karelian (2.0 to 1.6 billion years ago), and Riphean (1.6 billion to 544 million years ago); and for Australia, Nullaginian (2.5 to 1.9 billion years ago), Carpentarian (1.9 to 1.4), and Adelaidean (1.4 billion to 544 million years ago). In addition to these regional terms, the term "Eocambrian" or "Ediacaran" is often used for the latest Proterozoic, from about 800 to 544 million years ago.

Much of the significance of the Proterozoic derives from the first appearance of processes that are still operating on Earth. The Archean, in contrast, may have experienced a quite different set of processes. It is therefore impossible to discuss the Proterozoic without also discussing the Archean to some extent. There are many unanswered questions about major Earth processes in the Archean. In particular, the Archean lacks structures that closely resemble Phanerozoic orogenic belts. Whether plate tectonics operated in its present form during the Archean is still unresolved. Dynamic processes within the Earth may have undergone a significant change about 2.5 billion years ago.

The contrast between Archean and Proterozoic geology is striking if still imperfectly understood. The Archean is dominated by two principal types of regional structure: greenstone belts and gneiss-migmatite terrains. Greenstone belts are troughs of volcanic rocks and deep-water sedimentary rocks intruded by elliptical granite bodies. Gneiss-migmatite terrains are bands of very highly metamorphosed and deformed rocks. Structures similar to both the greenstone belts and the gneiss-migmatite terrains have formed throughout the history of the Earth and can form through conventional plate tectonic processes, but the almost complete lack of other types of structure in Archean time is perplexing. It is widely (though not universally) believed that the Archean crust was more mobile than at later times and that large masses of continental crust did not exist then. The first extensive shallow-water rocks deposited in a continental-shelf or stable continent setting occur in the latest Archean rocks of South Africa and are abundant throughout the Proterozoic and Phanerozoic. The appearance of such rocks may mark the first appearance of stable continental crust.

CRUSTAL AND ATMOSPHERIC EVENTS

During the Proterozoic, there is the first widespread evidence for plate tectonics and orogeny (mountain building) comparable with that of the Phanerozoic. There seem to have been two major periods of plate collision during which large continents were assembled out of small continental plates. North America was assembled between about 1.9 and 1.6 billion years ago by the collision of several smaller Archean blocks plus the accretion of many smaller terranes. This episode is known as the Trans-Hudson Event. The overall process was very similar to the accretion that is known to have added large areas to western North America during the Phanerozoic. Between about 900 and 600 million years ago, South America and Africa also appear to have been assembled in much the same way to form part of the early Phanerozoic supercontinent of Gondwanaland. The sequence of events that assembled South America and Africa is called the Pan-African Event.

Some Proterozoic crustal events are still poorly understood. In particular, there was a widespread heating event that resulted in the intrusion and eruption of silica-rich igneous rocks (granite and rhyolite) between about 1.5 and 1.3 billion years ago across much of North America. This interval also resulted in the intrusion of large bodies of anorthosite, an igneous rock made mostly of feldspar minerals, on many continents. Anorthosite is otherwise uncommon, and the reason it was formed so extensively in the Proterozoic is not known.

A number of significant developments took place on the surface of the Proterozoic Earth. There is evidence for increasing oxygen content of the atmosphere, as indicated by the appearance of the first "red beds," red sandstone and conglomerate colored by iron oxide. Also, there were a number of major ice ages during the Proterozoic. One occurred about 2.3 billion years ago, and there is widespread evidence for a series of ice ages between about 900 and 600 million years ago.

The early Earth probably lacked free oxygen, except in minor amounts. Free oxygen is not a component of the raw materials that formed the planets, and oxygen is a highly reactive gas that would rapidly have combined with other substances. Oxygen originally was given off as a waste product by early organisms (and is actually toxic to some organisms even today). How and when the present oxygen level of the atmosphere was attained is controversial. Many scientists consider that increasing oxygen was related to the abrupt expansion of life at the start of the Phanerozoic. There is also evidence that a significant threshold in oxygen level was crossed about 2.0 to 1.8 billion years ago.

SEDIMENTARY DEPOSITS

Several types of sedimentary deposit, all closely related to the availability of free oxygen, either ceased or began to be deposited about 1.8 billion years ago. Detrital uranium deposits and banded iron formations, both believed to indicate low oxygen levels, become very uncommon in the geologic record after that time, while red beds appear at roughly the same time. Detrital uranium deposits are sandstone deposits in which dense minerals were concentrated by current action. This process is common today, but the Proterozoic deposits include uranium minerals, which are highly suscep-tible to weathering, and even pyrite (iron sulfide), which oxidizes extremely quickly and which is al-most never found in recent sedimentary deposits.

A detrital uranium deposit at Okolo, in the Af-rican nation of Gabon, is remarkable for another reason. The uranium ore from the deposit was found to be greatly depleted in uranium 235, the isotope used in nuclear reactors. Further investi-gation showed that the ore deposit had been a natural nuclear reactor. Present-day uranium is too poor in uranium 235 to sustain a nuclear chain reaction except under very controlled con-ditions. Yet 2.2 billion years ago, when the Okolo deposit formed, uranium 235 was far more abun-dant, abundant enough for a sustained chain reac-tion to begin when sedimentary processes col-lected enough uranium minerals in one place. So far the remarkable deposit at Okolo is unique; other detrital uranium deposits were studied for effects of natural chain reactions, and no other ex-amples were found.

Another type of ore deposit common in early Proterozoic rocks is banded iron formation. Banded iron formations consist of layers of iron oxides and silicates interbedded with fine-grained silica, or chert. The significance of these ore de-posits, apart from their great importance as sources of iron, lies in the degree of oxidation of the iron. Iron in nature has two oxidation states. Ferrous iron, the less oxidized state, consists of iron atoms that have lost two electrons and thus have a positive electric charge of +2. Ferric iron at-oms have lost three electrons and have a +3 charge. Common rust is mostly ferric iron oxide, as would be expected in the Earth's presently oxy-gen-rich atmosphere. The iron in banded iron for-mations, however, is mostly ferrous, even though it was deposited on the surface of the Earth. The ferrous iron in banded iron formations presents one of the strongest arguments for an oxygen-poor atmosphere on the early Earth. The iron was originally dissolved in seawater and was probably precipitated by microorganisms. Indeed, some of the best-preserved Proterozoic fossils are those of microorganisms preserved in the chert of banded iron formations. Ferrous iron is far more soluble in water than is ferric iron, so the Proterozoic seas may have been richer in dissolved iron than are the present oceans. Iron deposits precipitated by

microorganisms have also formed in the Phanero-zoic, but the iron is mostly ferric.

Both detrital uranium deposits and banded iron formations become rare in the geologic re-cord after 1.8 billion years ago. At about the same time, red sandstones appear in abundance. Red sandstones owe their color to ferric (highly oxi-dized) iron and are most commonly deposited on land or in shallow water. One of the earliest exten-sive red bed deposits is in northwestern Canada. These rocks were deposited shortly after the crust of the region had experienced a major orogeny about 1.8 billion years ago and represent debris eroded from the newly formed mountain range. A deposit of this sort is termed molasse. These red beds furnish not only evidence for the oxygen level of the atmosphere but also some of the first evidence for topographically high mountains.

GLACIATION

The Proterozoic experienced a number of ma-jor glacial events about 2.3 billion years ago and again from 900 to 600 million years ago. The in-tervening period appears to have been largely ice-free. The glaciation 2.3 billion years ago is best documented in the Gowganda formation of On-tario, where virtually every type of glacial deposit occurs. Other deposits in Wyoming and Quebec have been interpreted as glacial deposits of the same age, but the Gowganda formation remains the clearest evidence for a 2.3-billion-year-old gla-ciation.

The glacial deposits 900 to 600 million years old are far more problematical. Many of these rocks are diamictites, fine-grained sedimentary rocks with scattered large pebbles. Diamictites cannot have been simply deposited by running water; slow-moving water could not have trans-ported the pebbles, while sediment deposited by fast-moving water would have a much higher pro-portion of coarse material than do diamictites. Only a few plausible ways of forming diamictites exist. One way is for floating glacial icebergs to melt and drop trapped rocks into otherwise fine-grained sediment. When the sediments are finely layered, the occurrence of so-called dropstones is generally taken as clear evidence of glaciation. Diamictites can also form from nonglacial pro-cesses. For example, submarine landslides can mix a small amount of coarse debris into a large amount of fine sand and silt. The glacial origin of many diamictites has been controversial, and some of the late Proterozoic diamictites may not be glacial.

An additional problem with the late Proterozoic glaciation is the geographic distribution of the deposits. Glacial deposits by themselves indicate cold climates, but in many cases there are nearby carbonate rocks (dolomite and limestone) of similar age that typically form in warm climates. Also, paleomagnetic studies indicate that some areas with glacial deposits, notably Australia, were at low latitudes during the late Proterozoic. A number of possible explanations, none entirely satisfactory, have been proposed. A few geologists find a flaw in the evidence, either in the climatic indicators or the paleomagnetic evidence. Others propose that the late Proterozoic Earth was abnormally cold, still others that the temperature decrease with altitude was much greater than at present, allowing warm climates at sea level but glaciers at even moderate elevations.

Throughout most of the Proterozoic, simple life-forms dominated on Earth. The first life-forms, already established in the Archean, were prokaryotes, organisms without a cell nucleus, such as bacteria and blue-green algae. About 2 billion years ago, eukaryotes, or organisms with a cell nucleus, appeared. The only common Proterozoic fossils of large size are stromatolites, domelike masses of calcium carbonate or silica deposited by colonies of algae in shallow water. The first widespread evidence for large multicelled organisms appears in rocks about 800 million years old. These organisms are rare as fossils because they lacked hard body parts but seem to have resembled jellyfish and marine worms. Many are unrelated to present-day organisms and seem to represent extinct evolutionary lines.

Steven I. Dutch

CROSS-REFERENCES

BIBLIOGRAPHY

Dott, Robert H., Jr., and Donald R. Prothero. *Evolution of the Earth.* 5th ed. New York: McGraw-Hill, 1994. This basic textbook on historical geology is aimed at students of geology. However, it is very readable by anyone with a background in science. Presents an up-to-date account of the Earth's history from the viewpoint of plate tectonics. Includes a glossary.

Hoffman, Paul F. "United Plates of America, the Birth of a Craton: Early Proterozoic Assembly and Growth of Laurentia." *Annual Review of Earth and Planetary Sciences* 16 (1988): 543-603. Probably the most comprehensive general account of the Trans-Hudson Event that assembled North America between 1.9 and 1.6 billion years ago. Uses fairly technical language. Suitable for advanced college students.

Kasting, James B., O. B. Toon, and J. B. Pollack. "How Climate Evolved on the Terrestrial Planets." *Scientific American* 258 (February, 1988): 90-97. This article compares the evolution of Earth, Venus, and Mars. Its relevance to the Proterozoic is that the early Sun was fainter than is the present Sun, and the Earth's atmosphere must have been different to have trapped enough heat to allow liquid water to exist on the early Earth.

McMenamin, Mark. *Discovering the First Complex Life: The Garden of the Ediacara.* New York: Columbia University Press, 1998. This entertaining study of the earliest complex life-forms on the planet details the author's work on these organisms. Written for the interested student but understandable by the general reader.

Medaris, L. G., C. W. Byers, D. M. Michelson, and W. C. Shanks, eds. *Proterozoic Geology: Selected Papers from an International Proterozoic Symposium.* Memoir 161. Boulder, Colo.: Geological Society of America, 1983. One of the most general available overviews of the Proterozoic, with papers on tectonics, glaciation, and the geology of selected regions. Papers vary in reading level from relatively simple to advanced.

Meyer, Charles. "Ore Deposits as Guides to Geologic History of the Earth." *Annual Review of Earth and Planetary Sciences* 16 (1988): 147. A synthesis of a lifetime of experience as a mining geologist. Meyer documents the changing patterns of ore deposits through geologic time, with strong emphasis on the Precambrian. Uses fairly technical language. Suitable for college students.

Press, Frank, and Raymond Siever. *Understanding Earth.* 2d ed. New York: W. H. Freeman, 1998. This comprehensive physical geology text covers the formation and development of the Earth. Readable by high school students, as well as by general readers. Includes an index and a glossary of terms.

Schopf, J. William, ed. *Major Events in the History of Life.* Boston: Jones and Bartlett, 1992. An excellent overview of the origin of life, the oldest fossils, and the early development of plants and animals. Written by specialists in each field but at a level that is suitable for high school students and undergraduates. Although technical language is used, most terms are defined in the glossary.

Van Schmus, W. R., and W. J. Hinze. "The Midcontinent Rift System." *Annual Review of Earth and Planetary Sciences* 13 (1985): 345-383. A comprehensive description of a major rifting event in central North America about 1.1 billion years ago. Uses fairly technical language. Suitable for advanced college students.

STRATIGRAPHIC CORRELATION

Stratigraphic correlation is the process of determining the equivalence of age or stratigraphic position of layered rocks in different areas. It is critical to understanding Earth history because stratigraphic correlation is one of the principal methods by which the succession and synchrony of geological events are established.

PRINCIPAL TERMS

BIOSTRATIGRAPHY: the identification and organization of strata based on their fossil content and the use of fossils in stratigraphic correlation

CORRELATION: the determination of the equivalence of age or stratigraphic position of two strata in separate areas; more broadly, the determination of the geological contemporaneity of events in the geological histories of two areas

INDEX FOSSIL: a fossil that can be used to identify and determine the age of the stratum in which it is found

LITHOLOGY: loosely used by many geologists to refer to the composition and texture of a rock

SEDIMENTARY ROCKS: rocks formed by the accumulation of particles of other rocks, organic skeletons, chemical precipitates, or some combination of these

STRATIGRAPHY: the study of layered rocks, especially of their sequence and correlation

STRATUM (pl. STRATA): a single bed or layer of sedimentary rock

METHODS OF STRATIGRAPHIC CORRELATION

Stratigraphic correlation is the process of determining the equivalence of age or stratigraphic position (position in a vertical sequence of rock layers) of strata in different areas. Strata thus determined to be of the same age or in the same stratigraphic position are called "correlative" or "correlated."

The most obvious method of stratigraphic correlation is to demonstrate that strata are continuous over a given area. This is achieved by tracing the layers (often by walking along them) throughout the area to demonstrate their physical continuity. In areas where strata are continuously exposed (the Grand Canyon in Arizona is a good example), this direct method is simple and best. Over much of the Earth's surface, however, the strata that scientists wish to correlate are covered by younger rocks or vegetation, or erosion has removed them over some portion of their geographic extent, making this method impractical.

A commonly used method of stratigraphic correlation that is not subject to this drawback is correlation by lithologic similarity, or similarity in rock composition and texture. Thus, strata of the same lithology (rock type) in different areas may be considered correlative. In effect, this method attempts to identify the same layer in different regions without demonstrating the continuity of (that is, without tracing) the layer. The major problem with this method, however, is that there are many layers in the Earth's crust that are of similar lithology. In eastern Colorado, for example, it is not easy to tell a 90-million-year-old sandstone layer from a 70-million-year-old sandstone from lithology alone. Thus, mistakes in determining age equivalence are easy to make when correlating only by lithologic similarity.

A third method of stratigraphic correlation relies on the idea that two layers in different areas that occupy the same stratigraphic position are correlative. This method can be referred to as correlating by position in the stratigraphic sequence. For example, if a limestone layer overlies sandstone A in one region and a shale layer overlies sandstone A elsewhere, scientists may conclude that the limestone and shale are correlative. This method, however, is not without its pitfalls. In the example, it might be that the limestone layer at the first location was deposited right after sand-

1153

stone A was deposited, whereas at the second location, the shale layer was deposited many millions of years after sandstone A. To correlate the limestone and shale in this case is to recognize their equivalence of stratigraphic position but not to demonstrate their age equivalence.

Age equivalence in stratigraphic correlation is almost always based on fossils. Stratigraphic correlation by fossils depends on biostratigraphy, the recognition of strata by their fossil content. If a fossil represents an organism characteristic of a particular interval of geologic time, then it is referred to as an index fossil. Thus, if an index fossil is collected from a stratum, then that stratum is the same age as strata elsewhere that contain the index fossil.

Besides fossils, there are some other, less often used, methods of determining the age equivalence of strata. The most common of these methods is to obtain a numerical estimate (usually in millions of years) of the age of the strata. This can rarely be undertaken unless the strata contain volcanic material, which contains chemical elements needed to obtain a numerical age. These elements are not normally present in sedimentary rocks; volcanic ash beds are not nearly as common in sedimentary rocks as are fossils. That is why fossils have been and continue to be the primary means for determining the age equivalence of strata. To undertake stratigraphic correlation, it thus is necessary to define the lateral extent of strata in a given region, characterize their lithology, determine their stratigraphic sequence, and collect and establish the stratigraphic ranges of fossils. Once these data have been collected in one or more regions, a geologist can attempt to correlate the strata within an area or between disparate areas.

DEVELOPMENT OF STRATIGRAPHIC CORRELATION

The development of stratigraphic correlation began in 1669 when Nicolaus Steno, a Danish naturalist, recognized that the sequence of strata is directly related to their relative ages. Steno's principle of superposition thus identified the oldest layers as those at the bottom and the youngest strata as those at the top of a sequence of strata. Steno's principle and the recognition of fossils as the remains of past life allowed William Smith in England and Georges Cuvier and Alexandre Brongniart in France to undertake the first stratigraphic correlations based on fossils during the early nineteenth century. Smith's correlations resulted in the first geologic map of England (1815).

Cuvier and Brongniart were able to correlate stratigraphically the rock layers in the vicinity of Paris and thereby reconstruct the geological and biological history of this area. Previously, during the late eighteenth century, the German mining engineer Abraham Werner and his students had laid the basis for stratigraphic correlation by lateral continuity and lithologic similarity by arguing for the continuity of strata of a particular lithology over a broad area. Thus, by the early nineteenth century, stratigraphic correlation by lateral continuity, lithologic similarity, position in the stratigraphic sequence, and fossils was already being practiced by European geologists.

A folded mountain (Grassy Mountain, Alberta), with Jurassic and Lower Cretaceous clastic rocks exposed in the pit below. Layers of coal, sandstone, schist, and carbon have been deformed from their original horizontal positions due to deformations in the Earth's crust. (Geological Survey of Canada)

APPLYING STRATIGRAPHIC CORRELATION

Stratigraphic correlation plays a central role in understanding geological history. By allowing geologists to determine the synchrony or diachrony of strata in different areas, the geological and biological events recorded in these strata can be ordered in time. This ordering is the basis of the chronology of geological and biological events during the last 3.9 billion years of Earth history. Stratigraphic correlation is thus one of the methods by which the relative geological time scale of eons, eras, periods, epochs, and ages was constructed.

A good example of stratigraphic correlation helping to decipher geological history and leading to the discovery of a giant oil field comes from the Guadalupe Mountains of western Texas and eastern New Mexico. There, in the years before World War II, geologists and paleontologists used stratigraphic correlation (especially biostratigraphy) to unravel the complex geological history of the strata that form these mountains. This history revealed that huge barrier reefs had developed as the sea encroached from the south about 260 million years ago. When it was later learned that hydrocarbons often accumulate around reefs, interest in the petroleum potential of these rocks was aroused. Because drilling the rocks in the rugged Guadalupe Mountains was impractical, an effort was made to trace the reef strata into the nearby lowlands. There, drill cores brought up rocks and fossils from deep beneath the surface. Stratigraphic correlation, by lithologic similarity, position in stratigraphic sequence, and fossils, was used to identify the strata adjacent to the reef in what is now called the Delaware basin, south and east of the Guadalupe Mountains. Successful drilling for petroleum soon turned the Delaware basin into one of the world's giant oil fields, thanks to the identification of petroleum source rocks through stratigraphic correlation.

By ordering events in the history of the Earth, stratigraphic correlation has been critical to the development of a global geologic time scale. Such a time scale provides geologists with a shared temporal framework within which to view their observations. A good example of the use of stratigraphic correlation in the development of the global geological time scale comes from the concept of the Cambrian period. The British geologist Adam Sedgwick coined the term "Cambrian" (from "Cambria," the ancient Roman name for Wales) in 1835 to refer to rocks in northern England that he believed contained the oldest fossils. Geologists now recognize the Cambrian period (544 to 505 million years ago) worldwide because stratigraphic correlation proved that an interval of geologic time that corresponds to Sedgwick's original Cambrian rocks can be identified across the globe. One of the most distinctive aspects of Cambrian strata is the abundance and types of trilobites they contain. Correlation of the Cambrian strata of England with strata elsewhere has been mostly undertaken on the basis of these trilobite fossils. Indeed, Sedgwick himself first conducted such stratigraphic correlation in continental Western Europe.

By the 1880's, when American geologist Charles Doolittle Walcott identified Cambrian trilobites in North America, the Cambrian was well on its way to becoming a geologic time period recognized worldwide. Stratigraphic correlation, especially using fossils (biostratigraphy), thus played a significant role in the recognition of the Cambrian. All the geologic time periods now recognized worldwide are similarly rooted in stratigraphic correlation.

Spencer G. Lucas

CROSS-REFERENCES

BIBLIOGRAPHY

Ager, Derek V. *The Nature of the Stratigraphical Record*. 2d ed. New York: Halsted Press, 1981. A witty and unabashed look at stratigraphy. Much of the discussion focuses on problems of stratigraphic correlation. The bibliography is extensive, there is an index, and a few, well chosen illustrations illuminate the text.

Berry, William B. N. *Growth of a Prehistoric Time Scale*. Rev. ed. Palo Alto, Calif.: Blackwell Scientific Publications, 1986. Provides an excellent review of the principles of stratigraphic correlation but is mostly devoted to the history of how the global geological time scale was formulated. As such, it well relates many of the stratigraphic correlations upon which the time scale is based. Well illustrated, with a good bibliography and an index.

Brenner, Robert L., and Timothy R. McHargue. *Integrative Stratigraphy Concepts and Applications*. Englewood Cliffs, N.J.: Prentice-Hall, 1988. Chapters 11-13 provide a comprehensive look at all facets of stratigraphic correlation. Well illustrated, with extensive reference lists and an index.

Dott, Robert H., Jr., and Donald R. Prothero. *Evolution of the Earth*. 5th ed. New York: McGraw-Hill, 1994. This basic textbook on historical geology is aimed at students of geology. However, it is very readable by anyone with a background in science. Presents an up-to-date account of the Earth's history from the viewpoint of plate tectonics. Includes a glossary.

Hedberg, H. D., ed. *International Stratigraphic Guide*. New York: John Wiley & Sons, 1976. The international "rule book" for stratigraphy. It sets procedures and standards to be met when naming stratigraphic units. It also defines many terms used in stratigraphy and has an extensive bibliography. Chapter 6 is devoted to biostratigraphy.

Press, Frank, and Raymond Siever. *Understanding Earth*. 2d ed. New York: W. H. Freeman, 1998. This comprehensive physical geology text covers the formation and development of the Earth, as well as stratigraphic correlation. Readable by high school students, as well as by general readers. Includes an index and a glossary of terms.

Prothero, Donald R. *Bringing Fossils to Life*. Boston: McGraw-Hill, 1998. This well-illustrated and entertaining text covers a broad range of paleontological topics, including stratigraphy. Glossary, bibliography, and index.

TRANSGRESSION AND REGRESSION

Transgression and regression are two of the most common phenomena of the geologic record. The changes in sea level that they represent have a major impact on the interpretation of Earth history and in the reconstruction of ancient environments. Understanding these phenomena in the geologic past is essential if present sea-level changes are to be dealt with effectively.

PRINCIPAL TERMS

CONTINENTAL SHELF: the margin of the continents, usually covered by shallow seas

EUSTACY: any change in global sea level resulting from a change in the absolute volume of available sea water

OCEAN BASINS: the large worldwide depressions that form the ultimate reservoir for the Earth's water supply

REGRESSION: the retreat or withdrawal of the sea from land areas and the evidence of such a withdrawal

TRANSGRESSION: the extension of the sea over land areas and the evidence of such an advance

GLOBAL PROCESSES

Transgression and regression are two phenomena by which changes in sea level are recognized. The two are not processes as such but rather are the effects of changes in sea level. Transgression is any rise of the sea over a land area (including the continental shelf) or any change in the physical marine conditions involving a progression into a deeper water environment. Regression is a withdrawal of the sea from any land area or any change resulting in a progression from deep water to shallow water or nonmarine environments. The concepts of transgression and regression are themselves independent of a causal mechanism. They are the results of the processes that can cause sea-level change.

Transgression and regression are very diverse phenomena and are the result of many different processes. Some of these processes involve an actual change in global sea level, or eustacy. Because transgression and regression are effects, however, a transgression or regression can be produced by a local process rather than a global one. It is important to understand the various processes involved in transgression and regression by examining first the global processes, then the local ones.

The global, or eustatic, processes resulting in transgression and regression are those that cause a change in the volume of seawater on the conti-

nents. Such global changes are accomplished by two major processes: first, an actual change in the volume of the global seawater supply and, second, a change in the volume or holding capacity of the ocean basins. A change in the amount of seawater would not seem an easy task to accomplish. The Earth's environment is essentially a closed system and, although small amounts of water are added from volcanism and by human activity, that is not enough to account for wide fluctuations in global sea level. Even if enough new water could be added to the system to account for a sea-level rise, the problem of getting rid of large amounts of water to account for a sea-level fall remains. Rather than add or subtract new water, a redistribution of existing water at the Earth's surface seems a more likely way to change the volume of seawater. One way that may be accomplished is by the growth and melting of the Earth's continental glaciers.

Glacial ice is different from sea ice. Sea ice forms in high latitudes over open ocean. As such, there is a constant connection with the oceans and a seasonal interchange as the ice melts in the summer and freezes in the winter. Glacial ice, however, forms from accumulated snow on the continents. Once a glacier grows, the snow that forms glacial ice is no longer available to the oceans. It is locked on the continents for the duration of the glacial episode. As glaciers grow, more water is locked within them. As glaciers melt, the meltwater is available to the oceans. The effect on sea level can be dramatic. During a glacial advance (an "ice age"), sea level will fall. During a glacial melt, sea level will rise. This sea-level mechanism has characterized the geologic record of sea-level change over the past 20 million years and perhaps longer. It has also operated during various other times in Earth history as long as there was continental-scale glaciation.

Another process by which transgression and regression can occur is by a change in the geometry of the ocean basins. The ocean basins are the ultimate reservoir of standing water on Earth. The depths in these basins are great, and it would seem that a change in the volume would not have a great effect on sea level. Yet if the volume of the ocean basins decreases, for example, the excess water must go somewhere. When that happens, a flooding of the continents will occur. Once the volume of the ocean basins increases, the water will withdraw from the continents. The actual mechanism in the ocean basins is related to heat flow from the Earth's interior at the mid-ocean ridges. As the ridges become more volcanically active, an increase in heat flow occurs; the high heat flow causes an increase in the elevation of the mid-ocean ridge. This increase in elevation causes a decrease in the volume of the ocean basins, resulting in a transgression. As the volcanic activity decreases, heat flow decreases, and the elevation of the mid-ocean ridge drops so that the volume of the ocean basins is increased. This results in a regression. Sea-level changes brought about by this process produce large-scale fluctuations resulting in the flooding not only of the continental shelves and the coastal plains but also of the interiors of the continents. Much of the geologic record contains rocks deposited in seas in continental interiors. This process may result in transgressions where ancient sea level could be as much as 300 meters above present sea level. This process is not the dominant process at the present. The last sea-level transgression apparently related to this process was during the early Paleocene epoch (about 66 to 58 million years ago). The sea reached from the Gulf of Mexico up the Mississippi Valley into present-day southern Illinois and reached as far west as the Dakotas.

LOCAL PROCESSES

While the global changes in sea level would seemingly result in uniform transgression or regression throughout the world, that is not apparent when the sea-level record at any given place is studied. The reason for that is the effect of local processes on sea level. In fact, many transgressions and regressions are local phenomena and may not be recognizable in other regions.

Two local processes that control sea level are sedimentation and tectonics. By operating in coastal regions, these two processes can mask the global trend of sea level. Sedimentation is the process whereby sediment is deposited by water or wind. Sedimentation can affect local sea level if the accumulation rate is greater than is the global sea-level rise. As the sediment builds up in the coastal regions, the sediments displace the ocean, thus "filling in" the former seafloor. As an area that was once a part of the sea is now land, a regression has occurred. This process is presently operating in the modern environment. A very good example of this phenomenon is the modern Mississippi Delta in southern Louisiana. In this area, sedimentation rates are very high. Although global sea level is rising as the remaining continental glaciers melt, new land is being created around the Mississippi Delta. At this location, sediment is accumulating faster than sea level is rising. Thus, a regression at the delta is taking place. Away from the delta, sea level is still rising, and more land each year in southern Louisiana is being flooded by marine waters. Therefore, even though the Mississippi Delta is a site of high sedimentation rates and locally affects sea-level change, that is not enough to alter the global process.

Tectonism is another local process by which the pattern of sea-level change can be controlled. Although tectonics is a global process, the effects of this process can be very localized, and some include phenomena not related to global tectonics at all. For sea-level change, the processes of interest are uplift and subsidence. Uplift and subsidence in a coastal region can be related to many causes, but it is the result rather than the process that is important to transgression and regression. If the process of uplift is more rapid than is global sea-level rise, a regression will be apparent. Similarly, if subsidence is greater than is global sea-level fall, a transgression will be the interpretation. These two examples illustrate the extremes of the tectonic effects, but many variations exist. In the modern environment, global sea level is rising, but the apparent change in any specific area is controlled by the tectonic setting. An example is from the Baltic sea coast in Sweden, where uplift resulting from glacial rebound is not reversing the trend of global sea level but is slowing the apparent rate of sea-level transgression. Tectonic factors are one of the most difficult effects to take into consideration when examining transgression and regression because of the geographic variability and the nature of the tectonic processes themselves.

Transgression and regression are seemingly simple phenomena. Sea level is either rising or falling. Yet the contribution to the observed pattern by local processes is significant and, in some cases, is the dominant effect observed. In order to gain meaningful insight into the global pattern of sea-level change from local data, a thorough understanding of the geologic history of the geographic region being studied is essential.

Transgression and regression appear to be very dramatic and perhaps even catastrophic phenomena when their effects are considered. Neverthe-less, both must be kept in the proper time context. Transgression and regression are not unique phenomena in the geologic record. Sea level has been in constant fluctuation throughout most of geologic time, and there is very little evidence that there were times of global sea-level stasis. Transgressions and regressions generally occur over long periods of time and, while some sea-level changes such as glacially induced ones occur over a relatively short span of geologic time (several hundred thousand years), many occur over millions of years. Thus, the observed effect of sea-level change may appear minor unless viewed over long intervals of geologic time.

Richard H. Fluegeman, Jr.

CROSS-REFERENCES
Archean Eon, 1087; Biostratigraphy, 1091; Cenozoic Era, 1095; Cretaceous-Tertiary Boundary, 1100; Fossil Record, 1009; Geologic Time Scale, 1105; Ice Ages, 1111; Land Bridges, 1116; Mesozoic Era, 1120; Paleoclimatology, 1131; Paleogeography and Paleocurrents, 1135; Paleoseismology, 1139; Paleosols, 1144; Paleozoic Era, 1126; Proterozoic Eon, 1148; Stratigraphic Correlation, 1153; Unconformities, 1161; Uniformitarianism, 1167.

BIBLIOGRAPHY

Cooper, John D., Richard H. Miller, and Jacqueline Patterson. *A Trip Through Time: Principles of Historical Geology.* Columbus, Ohio: Merrill, 1986. This book is designed for first-year college students but is written in such a way that senior high school students should be able to read it with little difficulty. Contains a specific section on sea-level change with good illustrations of concepts. Summaries of sea-level change through time are also throughout the book. A glossary is included.

Dott, R. H., Jr., and R. L. Batten. *Evolution of the Earth.* 4th ed. New York: McGraw-Hill, 1988. A book written for first-year college students in historical geology. Clearly written and well illustrated. Of special interest is a section comparing local and global transgressions and regressions. Although written at a higher level than are many other books, it will be useful to anyone interested in sea-level change.

Hallam, A. *Phanerozoic Sea-Level Changes.* New York: Columbia University Press, 1992. Deals with the succession of alternating transgressions and regressions of the sea over the continents through time. Discusses the causes of these events and their relationship to other events, such as mass extinctions. Bibliography and index.

Lemon, Roy R. *Principles of Stratigraphy.* Columbus, Ohio: Merrill, 1990. This college-level textbook deals with all aspects of stratigraphy, including sea-level changes and transgressions and regressions. Glossary and index.

Levin, H. L. *The Earth Through Time.* 3d ed. Philadelphia: Saunders College Publishing, 1988. This book is an excellent reference for the general reader. Although designed as a first-year college text, it has a writing style such that readers from high school on will find it a valuable reference. A well-illustrated section on sea-level change as well as summaries of transgression and regression through time are presented.

Stanley, Steven M. *Earth System History*. New York: W. H. Freeman, 1999. A comprehensive overview of the development of the Earth. Explains the mechanism of transgression and regression and its effect on the depositional history of the Earth. Illustrations, glossary, bibliography, and CD-ROM. Readable by high school students.

UNCONFORMITIES

Unconformities are surfaces of erosion, nondeposition, or both that have, through renewed sedimentation, been covered by younger rock. Where they occur, they represent significant portions of geologic time not recorded by rock and, as such, serve to divide the geologic history of a region into chapters of more or less continuous deposition separated by depositional breaks.

PRINCIPAL TERMS

ANGULAR UNCONFORMITY: an unconformity in which the beds below are at an angle to the beds above

DIASTEM: a short depositional break of geologically unmeasurable duration caused by a limited period of nondeposition

DISCONFORMITY: an unconformity separating parallel strata and exhibiting physical evidence of erosion

INTRUSIVE CONTACT: contact resulting from injection of molten igneous rock into previously formed rock

NONCONFORMITY: an unconformity in which rock below is homogeneous igneous or metamorphic rock and rock above is layered

PARACONFORMITY: an unconformity for which the only evidence is a gap in the fossil record

REGRESSION: a shift of sedimentary facies away from land, commonly caused by a drop in sea level

SEDIMENTARY FACIES: different types of sediment that collected simultaneously in adjacent areas of the seafloor

TRANSGRESSION: a shift of sedimentary facies toward land, commonly caused by a rise in sea level

ROCK RECORD BOUNDARIES

The units that compose the rock record of a region are separated by two fundamental types of boundary: conformable and unconformable. Rock layers are said to be conformable if deposition of the layer above followed the one below without an appreciable lapse of time. For example, periodic delivery, as during storms, of similar sediment to a depositional site will produce a sequence of homogeneous layers separated from one another by simple bedding planes that represent the short, geologically unmeasurable periods between storms. Such layers are considered conformable. Rocks of contrasting types can also exhibit a conformable relationship. In present oceans, an appreciable area of the sea bottom exhibits different types of sediment collecting simultaneously at adjacent sites. The type of sediment that accumulates at a particular site depends on the type of grains being supplied and the wave and current energy operating there. Adjacent environments of relatively quiet water, such as a lagoon, and agitated water, such as a beach, will collect fine and coarse sediment, respectively. The fine and coarse sediments are said to represent sedimentary facies (*facies* is the Latin word for the general appearance or aspect of something). Should the agitated conditions spread so as to come to operate in the region formerly occupied by quiet-water conditions, then the coarser grains will be delivered to the site in which previously only finer grains accumulated. At the same time, because of the increase in energy, the finer grains would no longer be allowed to settle where they formerly did but would be moved on to another site. This change in energy within the area would result in deposition of a layer of coarse material over fine material with essentially no break in sediment accumulation between the two layers.

Most conformable contacts between unlike rock layers in the sedimentary record are interpreted to result from such shifts in sedimentary facies through time. These shifts often accompany a rise or fall in sea level. A shift of sedimentary facies toward land during a rise in sea level is termed transgression, and one in which facies are displaced away from land during a lowering of sea level is termed regression. A conformable se-

1161

quence of strata will generally exhibit an interlayering of grain types or a gradation of grain types between layers.

The rock record exhibits many examples of widespread marine regression that, for a period of time, exposed previously formed rock to subaerial weathering and erosion. In these examples, subsequent transgression of the sea initiates renewed sedimentation, covering the erosional surface with younger layers. Because the time involved allows for broad changes in the distribution of geographic elements and life-forms, these new sediments characteristically differ in grain type and fossil content from those below. The rocks of the younger series above the erosional surface are said to exhibit an unconformable relationship to those below, and the buried erosional surface separating the two series is termed an unconformity. The erosional surface, which is to say the unconformity, records an interval of geologic time not represented by the rock succession of the region.

Unconformities divide the history of a region into conformable groups of rock that are separated from one another by depositional breaks. They are the divisions between the major chapters of geologic history. The conformable units between a pair of unconformities compose a related group of sedimentary facies that have shifted through time and become intimately interbedded with one another. Because Earth history is understood and narrated in terms of the broad divisions outlined by unconformities, the identification and analysis of depositional breaks constitute one of the most important pursuits undertaken by geologists who specialize in the study of the stratigraphic record.

DEVELOPMENT OF UNCONFORMITY STUDIES

Although unconformable relationships were observed and sketched by sixteenth and early seventeenth century naturalists, the credit for understanding their significance to Earth history is generally given to James Hutton, the Scottish geologist whose basic observa-

tions are often considered as the beginning of modern geological investigations. As early as 1786, Hutton observed stratal relations in which the overlying series of beds rested upon the tilted, eroded edges of the underlying series. In the succeeding two years, Hutton found other examples of angular relations between strata in northern Scotland. He interpreted the sequence of events necessary to develop such a relationship as beginning with marine deposition of the underlying beds, followed by their lithification, deformation (tilting), uplift, and exposure to subaerial erosion, and concluding with submergence by the oceans and deposition of the overlying series of horizontal beds. Hutton was particularly impressed by the time he perceived necessary to produce the angular relationship. In fact, such observations strengthened his belief that the Earth was of great antiquity, a discovery generally accredited to Hutton by science historians.

Hutton did not apply the term "unconformity" to the stratal relationship he elucidated. S. I. Tomkeieff credits the origin of the term to a translation, in the first decade of the nineteenth century, of the German phrase *abweichende Lagerung* ("deviating bedding") into the English word "unconformity," which carried the same meaning as the ecclesiastical term "nonconformity" (dissent from the established church). Throughout the nineteenth century, the term "unconformity" was applied exclusively to the angular relation described by Hutton, although some, particularly Charles Darwin (in

A small mesa near San Lorenzo Arroyo in Arizona shows an unconformity overlying upthrust rock. (U.S. Geological Survey)

order to explain gaps in the record of the evolution of life), argued that the Earth's strata were replete with subtler, yet equally significant, depositional breaks that lacked discordance.

The first half of the twentieth century was a period of describing criteria for the recognition of unconformities, identifying the types of stratal relations represented by unconformities, and proposing terminology to be applied to the different types of relations. Building on a century of study, Eliot Blackwelder in 1909 clearly outlined the three basic types of relations found at erosional surfaces: angular discordance between beds above and below the erosional surface (the "unconformity" of Hutton), parallel strata above and below the erosional surface, and layered rocks resting on an erosional surface developed on homogeneous metamorphic or plutonic igneous rock, such as granite, which lacks bedding.

Following Blackwelder, a variety of terms was suggested to describe the different types of unconformable relations, with the term "unconformity" itself most commonly retained to designate Hutton's original angular relationship. For American geologists, the question of terminology essentially was settled by Carl Dunbar and John Rodgers in their popular advanced geology textbook *Principles of Stratigraphy* (1957). In a review of the subject, these authors accepted the term "unconformity" in a generic sense to refer to all types of significant depositional breaks in the geologic record. They identified four types of structural relations that previous geologists had associated with these breaks. For three of these, they recommended retaining previously proposed terms: angular unconformity, for the discordant relation of Hutton; nonconformity, for situations where layered rock rests on eroded, unlayered igneous or metamorphic rock; and disconformity, for an erosional surface of "appreciable relief" separating parallel strata. They introduced the term "paraconformity" for the fourth structural type, which had been considered by earlier geologists as a special kind of disconformity.

IDENTIFYING UNCONFORMITIES

Disconformities and paraconformities are similar inasmuch as both are interpreted as erosional surfaces separating parallel strata. They differ in that a disconformity exhibits physical evidence of

UNCONFORMITIES

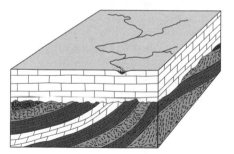

Angular Unconformity

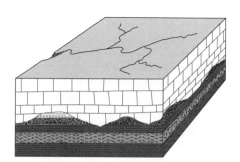

Disconformity

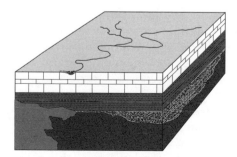

Nonconformity

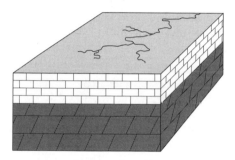

Paraconformity

An unconformity is visible where the horizontal layer meets the angled layers of volcanic ash beds below. (© William E. Ferguson)

erosion, preferably an undulatory surface that truncates beds and features in the rock below, whereas a paraconformity is manifested as a simple bedding plane essentially lacking relief. The paraconformity's interpretation as a surface that represents a significant portion of geologic time not recorded by rock is based on an abrupt change in fossil types, commonly accompanied by a sharp change in rock composition, across the boundary. Because abrupt changes in fossil content and rock composition can be explained by natural processes not involving loss of record, paraconformities are the most subjective of unconformities. It is for this reason that Dunbar and Rodgers recommended that these apparent breaks in sequence, which otherwise exhibit no evidence of erosion, be separately identified from disconformities.

Angular unconformities are the most vivid and unequivocal of depositional breaks. The special significance of angular unconformities is that they record crustal movement prior to the development of the unconformity. Care must be taken not to confuse angular beds resulting from crustal movement with beds naturally deposited at an angle, the so-termed cross beds characteristic of subaqueous and subaerial sand dunes.

Nonconformities are also reasonably simple to identify, but they need to be distinguished from intrusive contacts in which igneous rock has been emplaced as a molten body (magma) beneath layered rock. Although both nonconformable and intrusive contacts produce similar results, that is, layered rock above unlayered plutonic rock, the two differ fundamentally in their history. The former qualifies as an unconformity because the older igneous rock was exposed to weathering and erosion at the Earth's surface for some time prior to the deposition of the layered rock. In the intrusive relationship, which is not an unconformity, the younger igneous rock was intruded into previously formed layered rock. Geologists separate the two by examining the evidence at the contact. Indications of baking of the sedimentary rock would suggest an intrusive relation, whereas a weathered rind along the margin of the igneous rock would point toward a nonconformable contact.

Evidence for unconformities within parallel strata is more subjective than that for angular unconformities and nonconformities. The criteria developed fall into three broad categories: sedimentological, structural, and paleontologic. Sedimentological criteria revolve around phenomena that develop when a land surface that has undergone weathering and erosion for a period of time is transgressed by the seas. As an example, portions of the soil horizon developed on the layers below the unconformity can be preserved, generally significantly modified in appearance. If the underlying rock contains quartz sand or chert nodules, both relatively insoluble by the weak organic acids found in nature, these grains will, in the weathering process, become scattered through the soil. The energy of the advancing sea will remove the finer materials of the soil, and the coarser insoluble grains will become concentrated in the base of the unit overlying the erosional surface, generally in sharp contact with the underlying unit.

Structural criteria involve demonstration of

truncation (erosion) of features in the rock underlying the postulated erosional surface. Commonly, that involves, in the base of the overlying unit, channels cutting through whole beds in the underlying sequence, demonstrating the presence of topography—in some examples with relief of several hundred feet—on the erosional surface. On a smaller scale, fossil shells and other relatively coarse grains and structures can be cut off at the erosional surface.

Paleontologic evidence for an unconformity is constituted by a sharp change in fossil species, commonly associated with missing forms known to be present in the established sequence of fossils in nearby regions. The thought is that if sedimentation had been continuous, a complete paleontologic sequence would be present. Paleontologic evidence is almost always the only indication of the time value of an unconformity. If it is the only evidence for an unconformity, then the postulated erosional surface is most properly called a paraconformity.

EVALUATION OF DISCONFORMITIES AND PARACONFORMITIES

Because most evidence for unconformities in parallel strata can be explained also by natural processes not involving loss of record, some postulated disconformities and many paraconformities are the subject of debate among geologists. For example, the presence of a sandy layer with a sharp basal contact could be the result of a rapid shift in depositional facies rather than grains concentrated through weathering of the rocks below. A stream can break out of its channel and cut a new channel in the adjacent floodplain. Truncation of the bedding of the floodplain sediments would be pronounced. Yet the contact between the channel deposits and those of the floodplain into which the channel scoured would not be considered unconformable because the age difference between the two would at most be only a few years, a geologically insignificant value.

Paraconformable surfaces are the most difficult to evaluate. Those involving large gaps in the fossil record, expressible in terms of units of the geologic time scale, generally exhibit some sedimentological or structural evidence when examined closely over a large geographic extent. They are nevertheless enigmatic inasmuch as the indication

is that large areas lay exposed to subaerial erosion for considerable lengths of time without the development of distinct criteria for weathering and removal of rock. As the fossil gap represented at paraconformities decreases, interpretation becomes more contentious. The problem is that fossils were once life-forms, all of which are controlled, at least to some extent, by environmental conditions. Absence of a few fossil forms at a boundary could result from the absence of the environments to which the forms, when living, were adapted, rather than from loss of record through erosion. Norman Newell suggests that some paraconformities might develop during a prolonged standstill of sea level on a shallow shelf adjacent to a low-lying land area. Under such conditions, sedimentation on the shelf would eventually raise the sediment surface into water so shallow that the available wave energy would not allow more sediment to accumulate but instead would move it across the shelf to deeper water. If such conditions of nonaccumulation persisted long enough, ultimately surfaces that represented a significant extent of geologic time, but which exhibited no physical evidence of erosion, could develop. These surfaces would qualify as paraconformities.

It is apparent that no one criterion serves as unequivocal proof of an unconformity in parallel strata. In attempting to establish the position of an unconformity in undisturbed beds, geologists pursue several lines of sedimentological, structural, and paleontologic evidence. Geologists have become more and more aware that the stratigraphic record contains a spectrum of sedimentological breaks ranging from a few days of nondeposition to several hundred million years involving uplift, subaerial exposure, and erosion. In general, they agree that the term "unconformity" should apply only to those surfaces that involve a "significant" amount of time, although just how much time qualifies as significant is undefined. It is apparent that if an area is undergoing erosion, the sediment produced must be deposited somewhere. Therefore, all unconformities diminish laterally and eventually pass into conformable surfaces. In practice, geologists apply the term "unconformity" to those surfaces where the missing record can be expressed in terms of the regional stratigraphic column or in units of the international geologic time scale. Sedimentological breaks too small to be expressed in local or international terms are commonly referred to as diastems.

William W. Craig

CROSS-REFERENCES

BIBLIOGRAPHY

Blackwelder, Eliot. "The Valuation of Unconformities." *Journal of Geology* 17 (1909): 289. The earliest comprehensive discussion of unconformities, well written and still a valuable introduction to the subject. Suitable for college and advanced high school students. The journal should be available in most university and comprehensive technical libraries.

Lemon, Roy R. *Principles of Stratigraphy.* Columbus, Ohio: Merrill, 1990. This college-level textbook deals with all aspects of stratigraphy, including unconformities. Glossary and index.

Prothero, Donald R. *Bringing Fossils to Life.* Boston: McGraw-Hill, 1998. This well-illustrated and entertaining text covers a broad range of paleontological topics, including the development of sedimentary rocks and their use in stratigraphy. Glossary, bibliography, and index.

Stanley, Steven M. *Earth System History.* New York: W. H. Freeman, 1999. A comprehensive overview of the development of the Earth. Explains the development of unconformities and their use in stratigraphy. Illustrations, glossary, bibliography, and CD-ROM. Readable by high school students.

Tomkeieff, S. I. "Unconformity: An Historical Study." *Proceedings of the Geologists' Association* 73 (1962): 383. The most comprehensive examination available on the development of ideas about unconformities. Well written and illustrated. Easy to follow. Suitable for advanced high school students. Periodical available in most comprehensive university libraries.

Wicander, Reed, and J. S. Monroe. *Historical Geology.* 2d ed. St. Paul, Minn.: West, 1993. A beginning college-level textbook with a clear, succinct discussion of unconformities and facies (pages 47-51). Well illustrated. Suitable for advanced high school students.

UNIFORMITARIANISM

Uniformitarianism, in its most basic form, describes a methodology used in the study of Earth history. It assumes that the Earth can be interpreted in terms of natural processes that are operational today. In its broader context, uniformitarianism also embraces certain conclusions about Earth history: that the rates at which these processes operate are constant through time and that the Earth has experienced a cyclic history.

PRINCIPAL TERMS

CATASTROPHISTS: adherents to the belief that the present-day slow rates of natural processes are not sufficient to explain many features of the rock record

DILUVIAL THEORY: the belief that the Mosaic flood was responsible for shaping many of the Earth's surface features

DOCTRINE OF FINAL CAUSES: the belief that a purpose or design is revealed in the organization of nature

METHODOLOGICAL UNIFORMITARIANISM: that aspect of uniformitarianism that describes a procedural approach to the study of the Earth—that is, that ancient phenomena can be interpreted in terms of present-day natural processes

NEPTUNISTS: adherents to Abraham Werner's belief that granite and basalt formed by chemical precipitation from seawater

PLUTONISTS: adherents to James Hutton's belief that granite and basalt formed by crystallization from molten mineral matter

ROCK CYCLE: the path of Earth materials as erosional products of rock are deposited and reformed again into rock, which then can be eroded again

SUBSTANTIVE UNIFORMITARIANISM: that aspect of uniformitarianism that draws conclusions about the history of the Earth—that is, that natural processes have been taking place at constant rates through time and that the Earth has had a cyclic history

SURFICIAL GRAVELS: alluvium found at different places on the Earth's surface, originally interpreted as the product of the Mosaic flood

UNIFORMITY OF NATURE: Charles Lyell's formulation describing a nature in which processes are both consistent and constant through space and time; a mixture of methodological and substantive uniformitarianism

ORIGIN OF UNIFORMITARIANISM

Uniformitarianism owes its origin and development to three British geologists: James Hutton, who laid the groundwork for the concept; John Playfair, who elucidated and elaborated on Hutton's ideas; and Charles Lyell, who championed the uniformitarian cause in the mid-nineteenth century.

At the end of the eighteenth century, there was fairly close agreement between theories of the Earth based on direct, though limited, observations of Earth material and structure and the account of Earth history revealed in the biblical book of Genesis. The Earth was thought to be of no great antiquity, considered by most to be not much older than six thousand years. The surface features of the Earth, its mountains with their tilted rock, its river valleys, and its ubiquitous surficial gravels, were attributed to a great debacle, assumed to be the biblical flood of Noah. These features existed in equilibrium with the elements, possibly waiting for the next great debacle. Some erosion was happening, but not enough to alter the topography greatly in the short amount of time available. It was against this intellectual backdrop that Scottish scientist James Hutton presented to the Royal Society of Edinburgh a paper entitled *System of the Earth* (1785), which contained the seeds of a controversy that continued for fifty years. The paper was published in the first volume of the Society's Transactions under the title *Theory of the Earth: Or, An Investigation of the Composition, Dissolution, and Restoration of the Land Upon the Globe* (1788).

Hutton's work essentially outlined the rock cycle. He recognized, as others had before him, that the fossiliferous strata of the continents had formed on the bottom of the sea as unconsolidated sediment. In order to be added onto the continent, in places at angles of inclination too steep to represent the original repose of water-charged sediment, a mechanism was needed to fuse the sediment into rock and uplift the sea bottom. In Hutton's scheme, the energy required to do that was derived from the internal heat of the Earth. Once exposed to the elements, the rocks were eroded to form sediment again. In Hutton's view, the present land was derived from a former continent, and the disintegration and erosion of rock observed at its surface were providing material out of which a future continent would be constructed. Hutton emphasized that these processes that shaped the Earth were natural and operated very slowly, hardly causing noticeable change during the history of humankind. He eschewed calling upon preternatural causes in explaining natural phenomena.

Hutton's 1788 paper contains the major elements of what was later termed uniformitarianism. The Earth was to be interpreted in terms of present-day natural processes, excluding supernatural explanations. The processes operated very slowly and probably at fairly constant rates, although Hutton was not clear on this latter point. The major emphasis on constant rates was added by Charles Lyell almost fifty years later. Lastly, the Earth had gone through a cyclic development, or, as Hutton phrased it, a "succession of worlds," with the destruction of one world providing the material for the next.

Hutton's view of cyclic "worlds" destroyed and reconstructed by slow processes led him to conclude that the Earth is very old, in his phraseology, an Earth with "no vestige of a beginning" and, in reference to the cyclicity, "no prospect of an end." Hutton's "discovery" of an Earth of inscrutable age is considered his major contribution to geology. Hutton is also credited as the first to recognize the true origin of those rocks now known as igneous. His belief that these bodies were injected as molten masses into stratified rock molded his attitude toward the efficacy of heat in driving his cycles.

NEPTUNISM AND PLUTONISM

The prevailing view of these rocks at the time was that of the German mineralogist and natural philosopher Abraham Gottlob Werner, whose science of geognosy maintained that basalt and granite had been chemically precipitated from an ocean that from time to time had flooded the Earth. Hutton's adherents became known as Plutonists and Werner's as Neptunists. A few years after publication of Hutton's 1788 paper, his ideas were acrimoniously attacked by the Neptunists. Although the argument ostensibly was over the origin of basalt, a major undercurrent centered on Hutton's timeless Earth, a story that was not in line with that of the Scriptures. An ailing Hutton attempted to answer these attacks in 1795 with a wordy expansion of his ideas that is notorious for its unreadability. He died in 1797.

Fortunately, Hutton's good friend and sometime field companion John Playfair engaged the battle with the Neptunists. Playfair's prose was as facile as Hutton's was cumbersome, and it is from Playfair's *Illustrations of the Huttonian System of the Earth* (1802) that comes most of posterity's knowledge of Hutton's thoughts. In the period following Playfair's elucidation, most geologists came to accept the igneous origin of granite and basalt and to adjust their thinking in line with the unavoidable conclusion that the Earth is certainly much older than the six thousand years afforded it by a literal reading of Genesis.

The most important discovery in the first two decades of the nineteenth century was that the Earth's strata record a progression of life-forms, with primitive types in lower layers and more advanced types above. Foremost among researchers of the fossil record was the French comparative anatomist Georges Cuvier, who in 1812 published a monograph that was to serve as a major impetus for renewed controversy surrounding uniformitarianism. Cuvier, one of the first naturalists to reconstruct fossil vertebrates, was recognized as the premier scientist of his day. His early work had been on fossil vertebrates, the mammoth among many others, found in the surficial gravels. Because he felt that the extinction of this fauna could not be explained by processes presently acting on the surface of the Earth, he concluded that their demise resulted from a sudden but prolonged and localized inundation of the land by the sea brought about by unknown natural causes. He referred to the event as a revolution. Later research in the rocks around Paris caused him to expand his idea of Earth history to include several revolutions.

With his theory of periodic revolutions, Cuvier traditionally has been considered in English-speaking countries as a proponent of an Earth shaped through divinely instituted catastrophes, which in turn were followed by special creations to replace the extinguished life. (Martin J. S. Rudwick, in an excellent account entitled *The Meaning of Fossils*, points out the injustice of this characterization. Cuvier was a strict empiricist who believed that science and religion should be pursued separately.) Cuvier's research was immediately seized by William Buckland, England's most prestigious geologist, who took the latest of the revolutions and transformed it from its original form, a localized inundation of long duration, into a catastrophic, short-lived universal deluge—in short, the flood of Noah. Buckland's flood theory, the so-called diluvial theory, gained much popularity in England in the 1920's but was essentially abandoned toward the end of the decade as it became clear that the surficial gravels, the supposed deposits of the deluge, did not exhibit the distribution or character expected of a worldwide catastrophic event.

CATASTROPHISM

Still, a contingent of prominent British Diluvialists believed that the observations on which Cuvier's revolutions were based could not be explained by any of the processes in operation on the Earth. In response to the threat posed to the uniformity of nature theory by the position of these individuals, who became known as catastrophists, Charles Lyell published his landmark book *Principles of Geology* (1830-1833), a work built upon the basic tenets of Hutton's theory of the Earth. The ensuing debate between the catastrophists and Uniformitarians, as Lyell and his followers became known, was a conflict in viewpoint among scientists, not a struggle between science and the church, as is commonly assumed. In many important respects, the two sides were in agreement. The catastrophists willingly accepted the idea of an ancient Earth and were devoted in most situations to the methodology of uniformitarianism.

They were firmly opposed, however, to the interpretive aspects that the doctrine imposed on Earth history: namely, that the slow rates of present-day processes are representative of all time (steady rate) and that Earth history is cyclic, with one cycle looking much like another (steady state).

On the surface, the catastrophists' argument against steady-rate uniformitarianism seemed to be a simple request for accelerated rates to explain phenomena such as those on which Cuvier's revolutions were based. It is clear from their writings, however, that the catastrophists were anxious to interpret these episodes as times during which natural law was suspended by the regulatory hand of God. Their view of God was as both creator and referee, willing to intervene in the course of his creation if it veered off track. To them, any movement to remove the punitive hand of God from the course of things would lead to moral decay and the disintegration of British society. Lyell, on the other hand, was unwilling to concede any portion of Earth history, no matter how small, to a supernatural cause that fell outside the pale of scientific inquiry. As succinctly analyzed by Charles Coulston Gillispie in his revealing account of the times entitled *Genesis and Geology* (1951), the catastrophists were apprehensive that without any cataclysms there would be no God, whereas Lyell was concerned that without the uniformity of nature there would be no science. Lyell's opinion eventually prevailed, and supernatural explanations, for at least the physical world, were expunged from the methodology of historical interpretation. As a companion to this development, the efficacy of gradual processes in explaining all historical phenomena became firmly established in geologic dogma, but not without some unfortunate side effects.

To the catastrophists, Lyell's stubborn support of steady-state uniformitarianism not only seemed palpably false in the light of a fossil record that showed progression but also was contrary to physical laws that required the Earth to evolve as its energy dissipated. In the light of today's knowledge, Lyell's arguments against the evidence of the fossil record seem foolish. He went so far as to predict that ichthyosaurs, marine reptiles seemingly extinct for more than 65 million years, would return to populate the seas when their "world" once again rolled around. Stephen Jay Gould, in his short, provocative book, *Time's Arrow, Time's Cycle* (1987), argues that differing views of history occupied a central position in the conflict between the two sides. Lyell's repetitious Earth (time's cycle) was essentially ahistorical because it developed no unique configurations through time. Opposed to this was the catastrophists' vision of an ever-changing Earth (time's arrow), unique at each stage and thus amenable to sophisticated historical analysis. Lyell abandoned his stand on steady-state uniformitarianism shortly after Charles Darwin's convincing documentation that life on Earth had evolved.

William W. Craig

CROSS-REFERENCES

Archean Eon, 1087; Biostratigraphy, 1091; Cenozoic Era, 1095; Cretaceous-Tertiary Boundary, 1100; Evolution of Life, 999; Fossil Record, 1009; Geologic Time Scale, 1105; Ice Ages, 1111; Land Bridges, 1116; Mesozoic Era, 1120; Paleoclimatology, 1131; Paleogeography and Paleocurrents, 1135; Paleoseismology, 1139; Paleosols, 1144; Paleozoic Era, 1126; Proterozoic Eon, 1148; Sediment Transport and Deposition, 2374; Stratigraphic Correlation, 1153; Transgression and Regression, 1157; Unconformities, 1161; Weathering and Erosion, 2380.

BIBLIOGRAPHY

Ager, Derek V. *The Nature of the Stratigraphical Record.* New York: Halsted Press, 1981. A short, readable book emphasizing the incompleteness of the stratigraphic record (record of layered rocks) and the prominent role of natural cataclysms in Earth history. Well illustrated and suitable for college-level students.

Geikie, Archibald. *The Founders of Geology.* London: Macmillan, 1905. Reprint. Mineola, N.Y.: Dover, 1962. A long book, providing complete coverage on ideas about the Earth and its development extending back to the Greeks. Good discussions of Hutton, Playfair, and Lyell, and debates between Uniformitarians and opposing groups, are provided by a geologist not far removed from the times. Well written and understandable at the high school level.

Gould, Stephen J. *Time's Arrow, Time's Cycle: Myth and Metaphor in the Discovery of Geological Time.* Cambridge, Mass.: Harvard University Press, 1987. Treating what he calls the "Hutton myth," one of the country's leading geological scholars reassesses the contributions to geology of Hutton and Lyell and refocuses the debate between the catastrophists and Uniformitarians. This short, cleverly written, provocative book, which contrasts sharply with most previously written analyses, should be read only after an introduction to the subject from other sources. It is suitable for college-level students.

Hutton, James. "Theory of the Earth: Or, An Investigation of the Laws Observable in the Composition, Dissolution, and Resolution of the Land Upon the Globe." *Royal Society of Edinburgh Transactions.* Vol. 1, 1788. Reprint, with a foreword by George W. White. Darien, Conn.: Hafner, 1970. This article began the debate over what became known as uniformitarianism. This short, readable account of Hutton on his own ideas is understandable at the high school level.

Lemon, Roy R. *Principles of Stratigraphy.* Columbus, Ohio: Merrill, 1990. This college-level textbook deals with all aspects of stratigraphy. Chapter 2 discusses "the new uniformitarianism" and its relation to modern views on catastrophism. Glossary and index.

Lyell, Charles. *Principles of Geology, Being an Attempt to Explain the Former Changes of the Earth's Surface by Reference to Causes Now in Operation.* 3 vols. London: John Murray, 1830-1833. A well-written argument in support of uniformitarianism, directed toward the catastrophists. Understandable at all levels and an excellent example of nineteenth century scientific prose, the book went through several editions, one of which should be available in most university libraries.

Playfair, John. *Illustrations of the Huttonian The-*

ory. Edinburgh, Scotland: William Cheech, 1802. Reprint, with an introduction by George W. White. Urbana: University of Illinois Press, 1956. Beautiful prose describes Hutton's theory in detail. The book provides an excellent insight into the knowledge and ideas of geology of the time and is understandable at all levels.

Prothero, Donald R., and Fred Schwab. *Sedimentary Geology*. New York: W. H. Freeman, 1996. This college-level textbook discusses the development of sedimentary rocks and their use in stratigraphy. Includes a chapter on uniformity and its application in lithostratigraphy. Bibliography and index.

Rudwick, Martin J. *The Meaning of Fossils: Episodes in the History of Paleontology*. 2d ed. Chicago: University of Chicago Press, 1985. A fine, well-written account by a leading scholar of science history of the development of the study of paleontology, this book is of moderate length and is easy reading. Contains a well-balanced analysis of the meaning of uniformitarianism and is suitable for advanced high school and college levels.

Stanley, Steven M. *Earth System History*. New York: W. H. Freeman, 1999. A comprehensive overview of the development of the Earth. The first chapter deals with the principle of uniformity, and chapter 6 discusses its importance in stratigraphy. Illustrations, glossary, bibliography, and CD-ROM. Readable by high school students.

EARTH SCIENCE

ALPHABETICAL LIST OF CONTENTS

CATEGORIZED LIST OF CONTENTS

VOLCANOES AND VOLCANISM

WATER AND WATER PROCESSES

WEATHER AND METEOROLOGY

Categorized List of Contents